Conservation Biology of
Hawaiian Forest Birds

Conservation Biology of Hawaiian Forest Birds

EDITED BY
THANE K. PRATT,
CARTER T. ATKINSON,
PAUL C. BANKO,
JAMES D. JACOBI &
BETHANY L. WOODWORTH

Implications for Island Avifauna

YALE UNIVERSITY PRESS
NEW HAVEN & LONDON

Published with assistance from the United States
Geological Survey Headquarters, Reston, Virginia

Set in Joanna by Princeton Editorial Associates, Inc.,
Scottsdale, Arizona
Printed in the United State of America by Sheridan Books

Library of Congress Control Number: 2009928176
ISBN 978-0-300-14108-5 (hardcover : alk. paper)

A catalogue record for this book is available from the
British Library.

This paper meets the requirements of ANSI/NISO
Z39.48-1992 (Permanence of Paper).

10 9 8 7 6 5 4 3 2 1

Contents

Foreword

Mark Twain characterized the Hawaiian Islands as "the loveliest fleet of islands that lies anchored in any ocean." The isolation of the archipelago, 2,400 miles from the nearest continent, provided an evolutionary stage upon which Hawaii's diverse flora and fauna evolved. Hawai'i lacked many of the competitors, predators, diseases, and parasites found elsewhere. While this evolutionary scenario led to diverse ecological and morphological adaptations in the islands' plants and animals, isolation left Hawaii's forest bird species ill equipped to survive the challenges presented by the nonnative diseases, predators, parasites, browsers, grazers, and competitors that would come.

And come they did, 1,200 years ago when the isolation of the Hawaiian Archipelago was broken by Polynesian voyagers. With that first wave of humans came new ecological and evolutionary forces that would significantly shape the future of Hawaii's unique flora and fauna. The impacts on Hawaii's avifauna were direct, immediate, and long term. The impacts of man's fellow travelers, such as rats, pigs, and dogs, included habitat modification as well as predation on eggs, nestlings, and adults. Direct impacts included the habitat loss and fragmentation associated with agriculture and development. Hunting contributed to additional species extinctions and further reductions in bird populations. Nearly all the flightless birds vanished.

The arrival of Europeans 230 years ago brought a new wave of threats including cattle, sheep, goats, European pigs, two additional species of rats, mongooses, and cats, as well as new parasites, competitors, disease vectors, and diseases. In the 900 years between the first landfall by man and the arrival of Europeans, some 50 species of endemic Hawaiian birds became extinct.

Between the first landing of Captain Cook in 1778 and the passage of the Endangered Species Act nearly 200 years later, 26 species and subspecies of birds were driven to extinction and the last of the flightless birds vanished. Since passage of the Endangered Species Act in 1973, the previously mentioned threats have continued to weaken the viability of Hawaii's birds. These threats have led to further loss of species and reductions in the size and number of populations of many of the remaining species.

The plight of the Hawaiian avifauna did not go unnoticed. Of the species on the Department of the Interior's first endangered species list, 25% were Hawaiian birds. The response of the U.S. Fish and Wildlife Service and other groups to the status of Hawaii's threatened and endangered birds was immediate. Endangered species biologists were assigned to the islands, where they worked in cooperation with state biologists, academics, and conservation groups to take on the task of conserving Hawaii's endangered avifauna. Recovery plans identifying information gaps and providing guidelines for habitat restoration, predator control, ungulate removal, and public education have been written and rewritten.

Studies of the ecology, distribution, abundance, limiting factors, and other information needed to effectively mount a recovery effort for Hawaii's avifauna were conducted. Today the initial information gaps have largely been filled. Additionally, Hawaii's threatened and endangered species managers have not been idle. Information from researchers has been used to establish natural area reserves, national wildlife refuges, and other protected areas. Control programs for introduced pigs, goats, and sheep resulted in dramatic improvements in habitat quality, nowhere more evident than in Hawai'i Volcanoes and Haleakalā national parks and in Hakalau Forest National Wildlife Refuge. Predator control programs have demonstrated that removal of rats, cats, and mongooses results in

increased reproductive success and adult survival in endemic birds. Habitat restoration activities have resulted in healthy stands of koa trees that have attracted endemic honeycreepers, including the endangered 'Akiapōlā'au. These accomplishments provide clear testimony to what can be accomplished when researchers and managers work together.

While the conservation efforts are impressive and successes have been achieved, the scale of the endeavor, with few exceptions, has not matched either the scale of the threats or the vision of our conservation goals. We are losing, not gaining, ground in the race against extinction. This reality was made clear in the U.S. Fish and Wildlife Service's 2000 status report to Congress (U.S. Fish and Wildlife Service 2000). According to the report, nine forest bird species, ten with the putative loss of the Po'o-uli, had become extinct since passage of the Endangered Species Act. One species, the 'Alalā or Hawaiian Crow, was known only from captivity, and five other species were either declining or of unknown status. Not a single species was characterized as increasing. Thus, in spite of the previously mentioned management efforts, we are not getting ahead of the extinction curve.

Why is this? It is not due to a lack of knowledge; our knowledge of the ecology, breeding biology, distribution, and abundance of endemic forest birds meets that called for in recovery plans. We know enough to act. However, implementation of management actions has not kept pace. For 86% of Hawaiian birds, 50% or fewer of the recovery objectives were achieved, while for 63% of species, 25% or fewer of the recovery tasks were accomplished. We have a management implementation gap.

An emerging crisis in the status of endemic forest birds on the Alaka'i plateau of Kaua'i reveals just how large that gap is. Predator control and ungulate removal in the Alaka'i, management tasks essential to the survival and recovery of Kauai's endangered forest birds called for in the 1983 Kauai Forest Bird Recovery Plan, have yet to be implemented. Thus, for 25 years Kauai's endangered species have gone without needed management actions, actions that were known to be effective. As a result, the endemic forest birds and their habitats continued to decline, resulting in the extinction of several species and the need to list two others.

This story is not unique to Kaua'i but rather has been repeated on every island. The management actions identified in forest bird recovery plans written two decades ago have not been implemented at ecologically relevant scales. It is neither lack of information nor lack of planning that stands in the way of saving Hawaii's endangered avifauna; it is lack of action. When we have acted, we have done so at scales that have influenced nesting pairs or fractions of populations, not populations, metapopulations, subspecies, and species. Our chances to effectively act on behalf of Hawaii's endangered forest birds are rapidly decreasing as time passes. It is the Hawaiian people, collectively, in whose hands rest the future of Hawaii's forest birds. Successful efforts to restore the higher-elevation forests must be made across tens of thousands of acres, not hundreds, with predator control activities at scales consistent with recovering viable populations, subspecies, and species rather than maintaining populations on the brink of extinction.

The chapters in this book provide a synthesis of the information needed to implement the next era of forest bird conservation in Hawai'i, an era in which researchers and managers must join together to frame conservation management objectives. Designing actions to accomplish conservation objectives in an adaptive management mode, implemented at scales that are ecologically meaningful, may accomplish recovery goals while reducing the uncertainty about the impacts of threats and the efficacy of management actions. These things can be done; *they have to be done*. However, absent conservation actions at scales that are biologically relevant, this book will be one of several read on the 100th anniversary of the Endangered Species Act about an extinct avifauna that we failed to protect.

J. Michael Scott, Senior Scientist
U.S. Geological Survey, Idaho Cooperative
    Fish and Wildlife Research Unit
College of Natural Resources, University
    of Idaho, Moscow, Idaho

Preface

Birds are among the most mobile animals on earth, so it is not surprising that they have reached and colonized every island that can sustain them. Seabirds live over the open ocean but depend on islands as predator-free havens for nesting. Land birds vary in their ability to find islands, ranging from those species that will not cross open water at all to wandering, so-called tramp species in the South Pacific—certain pigeons, kingfishers, and flycatchers, to name a few—that have colonized the smallest and most remote islands. Some migratory continental birds even travel to islands annually, and they do so flying nonstop. Few sights are more awe inspiring than the late summer spectacle of Pacific Golden-Plovers arriving in the Hawaiian Islands from Alaska. Descending from a great height on set wings, the plovers convey nothing of the effort it took to steadily fly for 4,000 kilometers, a distance covered in about 50 hours. Although golden-plovers migrate widely to and from the tropical Pacific, those that winter in Hawai'i have reached the most remote islands on the planet.

Despite the regularity with which migratory shorebirds and waterfowl pass back and forth seasonally between Hawai'i and North America, the trip is far too great a challenge for the vast majority of land birds. Perhaps fewer than 30 colonizing species gave rise to the original land-based avifauna of somewhat more than 110 species in Hawai'i. Dispersal over such a great distance is a barrier for all terrestrial animals and plants, and it is impossible or highly unlikely for most continental species to reach the Hawaiian Islands, even over millions of years. Thus, from a bird's perspective, the islands were a paradise originally free of many of their worst enemies and competitors: rodents, snakes, ants, mosquitoes, most diseases, and even many kinds of birds. Birds that could cross the vast oceanic expanse entered a very different world from their homeland, one that invariably led to evolutionary change for the settlers. As a result, the Hawaiian land-based avifauna consists of the highest percentage of endemic forms (98%) anywhere in the world.

Indeed, biological isolation is what best characterizes the evolutionary environment for Hawaiian birds. Evolution in isolation primarily accounts for the unique avifauna of endemic species. It has also led to bizarre adaptations such as unique bill shapes and flightlessness, to the explosive adaptive radiation of the Hawaiian honeycreepers, to unusually "slow" demographic and life history patterns, and ultimately to the vulnerability of these insular birds.

Sadly, Hawaiian birds have suffered terrible losses since the arrival of people at least 1,200 years ago. Entire bird groups have disappeared: flightless waterfowl, ibises, and rails with stunted wings; a sea-eagle; several long-legged, bird-hunting owls; and the renowned 'ō'ō with their coveted golden plumes and piping song. Of the more than 50 species of honeycreeper present when the first person set foot on a Hawaiian shore, a meager 17 remain. In the past 50 years, 10 bird species have disappeared in the wild, and the 'Alalā exists only in captivity. The last known Po'o-uli died in 2004.

Extinction of Hawaiian forest birds has resulted mainly from habitat loss and the ecological impacts of introduced species. People have cleared so much native vegetation that less than 10% of the land area remains as undisturbed forest. Introduced species with ecological effects on birds include a diverse host of plants, animals, and diseases. Some examples are trees that invade and replace native forest, feral livestock that devour vegetation, rats and cats that prey on birds, alien birds that compete for food, tiny parasitoid wasps that deplete the insect prey base of native birds, and avian pox and malaria that decimate susceptible native hosts. An outcome of this calamity is that the Hawaiian avifauna has gained an unfortunate reputation for extinction.

This picture is discouraging, but keep in mind that *the surviving species are still here*. It is incorrect to assume that *all* species are beyond saving. To understand this point, consider this: some Hawaiian forest birds are still abundant. The 'Apapane has populations on all the forested islands, and these exceed one million birds. As I write, one of these crimson honeycreepers is probing flowers in the azalea bush blooming just outside my office window. Bird communities at high elevations on Hawai'i Island harbor astonishing numbers of birds. The dawn chorus at places like Hakalau Forest National Wildlife Refuge can be breathtaking! It is a wondrous thing that in such times of wrenching change to their environment, an abundance of birds still arises with each generation, for these small birds have a short life cycle by human standards, with only a few years to reproduce. Even an endangered species can seem puzzlingly resistant. How do Maui Parrotbills sustain potential losses to predators and disease when their annual reproduction is at best only one offspring per pair? To make that low rate of reproduction work, adults must survive and reproduce for many years, and such an apparently high level of survival begs the questions what are they doing right, and what environmental conditions allow them to persist?

Although many Hawaiian forest birds are diminishing in total number, remarkably, a few are adapting and expanding into man-made habitats. To see an O'ahu 'Amakihi one need go no further than the suburbs of Honolulu. Slowly, over the past several decades, 'amakihi have advanced out of the wooded hills behind the city. They now reside and reproduce, albeit marginally, in urban habitats miles from the nearest stick of native vegetation. Here they face seemingly insurmountable threats: rats, cats, diseases, and aggressive nonnative birds, not to mention more mundane hazards such as picture windows and speeding automobiles.

Despite an optimistic outlook for the 'amakihi, most native Hawaiian forest birds are in decline, and the threats to their survival are severe. Although the persistence of forest birds owes much to the characteristics and adaptability of the remaining species, it also depends on progress in research and conservation. Before the 1980s, we knew very little about Hawaiian forest birds. Early research had focused on determining the status of populations and documenting their habitat preferences. During that period, Hawaii's now famous case of avian malaria came to light. Over the past two decades, we have expanded our research efforts to cover bird species' comparative life histories; their population ecology, particularly in relation to limiting factors; many aspects of disease epidemiology; experimentation with removal of predators; the development of techniques for reintroductions; and first-time captive propagation for the majority of Hawaiian songbirds.

We have also made enormous progress in conservation: we are restoring many thousands of hectares of forest, allowing former pastureland to convert to forests, controlling feral ungulates and predators, and captive-rearing forest birds in new facilities. Never before has Hawai'i seen so much conservation activity for native species and ecosystems. Yet compared with endangered bird species on the mainland United States, Hawaiian bird species lag far behind in the recovery effort each receives. Most habitats are still subject to damage from feral ungulates, alien weeds, and plant pests; control of predators is not sustainable even in small areas; and nowhere have we routed two big killers, avian malaria and pox. Clearly much more needs to be done.

This book is concerned with what are known in Hawai'i as "forest birds," specifically, all the native passerines (songbirds or perching birds): the four 'ō'ō species and the Kioea (traditionally classified as honeyeaters but now recognized as Hawaii's only endemic bird family), the 'Alalā (Hawaiian Crow), the 'Elepaio (a monarch flycatcher), the Millerbird (a sylviid warbler), the 'Ōma'o and its relatives (thrushes), and the hugely diverse Hawaiian honeycreepers (cardueline finches) (Plates 1–9, 12–13). For purpose of comparison we include all native songbirds, even those nonforest species that dwell in coastal shrubland on the Northwestern Hawaiian Islands. The native songbirds, every one an endemic species, are bound by long evolutionary histories in the islands. Many have similar habitat requirements, and all face the same threats.

Also included as forest birds are the Hawaiian Hawk ('Io) and our native population of the cosmopolitan Short-eared Owl (Pueo), because these two raptors are coevolved predators of native songbirds (Plate 1). We exclude the shorebirds, waterfowl, and seabirds, whether migratory or resident, and also the mostly terrestrial Hawaiian Goose (Nēnē), because, for the most part, their evolutionary histories in the islands have been shorter than those of the forest birds, they mainly occupy habitats other than forest, and they live very differently. As for the diverse and growing fauna of introduced birds, this book covers species that have penetrated the forest and compete with native songbirds (Plates 10–11).

This book reviews the biology and conservation of Hawaiian forest birds. The first part introduces the Hawaiian avifauna. Here we describe the original evolutionary environment of the birds, outline patterns of avian evolution in the islands, recount and analyze patterns of extinction since the arrival of people, then summarize the cultural uses of and beliefs about forest birds among the indigenous Hawaiian people. Part 2 consists of chapters that review the biology of Hawaiian forest birds and factors limiting their populations. Part 3 considers the conservation of forest birds and their supporting ecosystems. Part 4 includes a series of case studies that illustrate how the biology and conservation setting of an endangered species influence the course of action necessary to recover the population. We conclude with two chapters looking at trends and future directions for research and conservation. In these chapters we attempt to answer these questions: Can Hawaiian forest birds survive? What will it take on our part to ensure their survival?

Hawaiian forest birds are at a crucial turning point. The sad, uninterrupted history of bird declines continues. Yet for the first time, local declines have been halted and in some cases reversed thanks to field management and research. Can we act in time to save the remaining Hawaiian forest birds? It is the editors' fervent hope that the information contained in this book will inspire an accelerated effort to pull these birds back from the brink of extinction.

Thane K. Pratt

Acknowledgments

The first idea for this book, and the impetus for it, came in 1998, prompted by John D. "Doug" Buffington and a review panel at the U.S. Geological Survey (USGS). Our research center, the Pacific Island Ecosystems Research Center (PIERC), a local branch of the USGS, had a long history researching Hawaiian forest birds, and the time had come to write a comprehensive reference work synthesizing the findings of our avian research group.

By ornithological analogy, it could be said that, like a fledging island bird, this book has had a relatively long incubation and nestling development. The period from inception to delivery was filled with other writing projects to move the results of research into print. All the while, many of us were engaged in the year-round field schedule typical of tropical ornithology. In view of all these distractions, we thank former PIERC directors William W. M. Steiner and David A. Helweg, and present director Loyal A. Mehrhoff, for their support and patience. David Helweg reviewed early drafts of all chapters and gave helpful suggestions for improving them. We also thank Frank S. Shipley (deputy regional biologist) and Mark K. Sogge (ecologist and acting deputy regional biologist) at the USGS Western Regional Office for their encouragement and for seeking book funds. We thank the USGS headquarters in Reston, Virginia, for funding publication of the book. We especially thank Ronald E. Kirby, senior advisory biologist and bureau approving official—biology at the Western Regional Office, for reviewing all sections of the book draft and offering editorial advice.

The editors were all research biologists at PIERC when the book project began. Thane Pratt coordinated the book project.

He thanks the other editors for their guidance and long hours in bringing this book to print. The coeditors are listed in alphabetical order. Although it is tempting to single each out for his or her individual contribution, it is best said that we functioned as a team. However, Thane particularly thanks Bethany Woodworth for her special interest in the book, for taking on responsibilities when Thane was away, and for her assistance in seeking a publisher.

We are especially grateful to our authors and coauthors for their commitment and hard work. This book is their contribution of knowledge gained from many years' experience in conservation biology of Hawaiian birds and ecosystems. The following people reviewed chapters for the book: Robert A. Askins, Donna L. Ball, Jonathan Bart, Gregory S. Butcher, Karl Campbell, Phillip Cassey, Sheila Conant, Susan Cordell, Gretchen C. Daily, Peter Daszak, Reginald E. David, Scott R. Derrickson, Duane R. Diefenbach, David C. Duffy, Guy Dutson, Fern P. Duvall II, Steven G. Fancy, Andrew Fenton, Jeffrey T. Foster, Samuel M. O. Gon III, B. Rosemary Grant, Curtice R. Griffin, Jeffery S. Hatfield, David A. Helweg, Richard N. Holdaway, Brenden S. Holland, Gregg Howald, John Innes, Lloyd F. Kiff, Carolyn M. King, Tony Leukering, Alan A. Lieberman, Julie L. Lockwood, Timothy Male, Kepa Maly, Thomas E. Martin, John M. Marzluff, Jannet McAllister, Hamish McCallum, Brian K. McNab, Arthur C. Medeiros, Donald V. Merton, Marie P. Morin, Eugene S. Morton, Michael P. Moulton, Jay T. Nelson, James D. Nichols, Storrs L. Olson, A. Townsend Peterson, David N. Phalen, Stuart Pimm, H. Douglas Pratt, Jonathan P. Price, C. John Ralph, Fred L. Ramsay, Niel Reimer, William K. Reisen, Robert E. Ricklefs, Gordon H. Rodda, Roger G. Rose, Craig Rowland, Oliver A. Ryder, Michael D. Samuel, Joseph J. Schall, J. Michael Scott, Robert J. Shallenberger, Carol Silva, Daniel Simberloff, David G. Smith, Thomas B. Smith, Thomas J. Snetsinger, Eric B. Spurr, Jeffrey A. Stratford, Kirsty J. Swinnerton, P. Quentin Tomich, Peter M. Vitousek, Patricia Warren, David S. Wilcove, Gary A. Wobeser, Thierry M. Work, Eric A. VanderWerf, and Robert M. Zink. John M. Marzluff and two anonymous reviewers read the book manuscript for Yale University Press and offered suggestions for improving it.

We thank Jack Jeffrey for his many fine photographs and for help in arranging the color illustrations. Herb Kāne, H. Douglas Pratt, and Rob Shallenberger kindly permitted use of their artwork or photographs. Keola Awong and Bobby Camara (Hawai'i Volcanoes National Park) contributed the pronunciation of Hawaiian bird names for the figure captions. We thank Mike Scott for his foreword and for his encouragement and inspiration throughout the course of this book project. Keith Leber, acquiring editor at the University of Hawai'i Press, gave initial guidance, instruction, and encouragement on preparing an edited volume. Amy Miller assisted in the colossal job of compiling and formatting the manuscript. Marcos Gorresen assembled the illustrations and redrafted the graphics where necessary. Jean E. Thomson Black (executive editor for science and medicine at Yale University Press), Jack Borrebach (the production editor for the Press), and their colleagues did much to promote our book and guide it through the many steps of publication. Peter Strupp and Princeton Editorial Associates performed a superb job copy editing the manuscript and transforming it into a book.

PART 1

Introduction

Origins,
Historic Decline,
and Culture

CHAPTER ONE

Origins and Evolution

THANE K. PRATT

In the extremely isolated Hawaiian Archipelago, a unique environment has promoted rapid evolutionary divergence among native birds and, in the case of the Hawaiian honeycreepers, spectacular diversification leading to new trophic morphologies and life histories. These morphologies and life histories have often repeated patterns common elsewhere; for example, nectar-feeding Hawaiian honeycreepers converge with honeyeaters of Australasia and the sunbirds of Africa and tropical Asia. But sometimes unique solutions have emerged, such as the double-tooled bill of the bark-excavating 'Akiapōlā'au or the pincering beak of the mysterious Lāna'i Hookbill, which have no counterpart among the world's 10,000 bird species (see Table 1.1 for the scientific names of native birds and Plates 1–9, 12–14 for illustrations of them).

This chapter introduces the Hawaiian land bird avifauna: the evidence for its origins and the evolutionary setting against which later human-caused changes and stresses can be compared. It also explores the peculiarities of Hawaiian birds in the context of the evolution of birds on islands. These topics include reduced dispersal, high population densities, increased competition within species, adaptive radiation, and loss or reduction of adaptive defenses against specific predators and diseases. A major theme here and throughout the book is that avian evolution in the absence of continental factors has led to adaptations that may still be advantageous in modern insular environments but are more often maladaptive as these conditions change.

Table 1.1. Native land-based bird species from the Hawaiian Islands

	Ly	N	K	O	Mo	Ln	Ma	H
Family Anatidae: Ducks, Geese, and Swans								
True Geese (Nēnē and Relatives)								
Nēnē (Hawaiian Goose, *Branta sandvicensis*)			F		F	F	F	HF
†*Branta hylobadistes*							F	
Moa-nalo								
†*Chelychelynechen quassus*			F					
†*Thambetochen xanion*				F				
†*Thambetochen chauliodous*					F		F	
†*Ptaiochen pau*							F	
Ducks								
Koloa (Hawaiian Duck, *Anas wyvilliana*)			HF?	HF?	HF?		HF?	HF?
Laysan Duck (*Anas laysanensis*)	H		F	F?	F?		F	F
Family Ardeidae: Herons, Bitterns, and Allies								
ʻAukuʻu (Black-crowned Night-Heron, *Nycticorax nycticorax*)			HF	H	H	H	H	H
Family Threskiornithidae: Ibises and Spoonbills								
†*Apteribis glenos*					F			
†*Apteribis brevis*							F	
Family Accipitridae: Hawks, Kites, Eagles, and Allies								
†White-tailed Eagle (*Haliaeetus albicilla*)				F	F		F	
†*Circus dossenus*				F	F			
ʻIo (Hawaiian Hawk, *Buteo solitarius*)			F		F			H
Family Rallidae: Rails, Gallinules, and Coots								
†Laysan Rail (*Porzana palmeri*)	H							
†Moho (Hawaiian Rail, *Porzana sandwichensis*)								HF
†*Porzana ziegleri*				F				
†*Porzana menehune*					F			
†*Porzana keplerorum*							F	
†*Porzana ralphorum*				F				
†*Porzana severnsi*							F	
ʻAlae ʻula (Common Moorhen, *Gallinula chloropus sandvicensis*)			H	H	H		H	H
ʻAlae keʻokeʻo (Hawaiian Coot, *Fulica alai*)			HF	HF	H		H	H
Family Recurvirostridae: Stilts and Avocets								
Aeʻo (Black-necked Stilt, *Himantopus mexicanus knudseni*)			HF	HF	H		H	H
Family Strigidae: Typical Owls								
Pueo (Short-eared Owl, *Asio flammeus sandwichensis*)		H	HF	HF	H	H	H	H
†*Grallistrix auceps*			F					
†*Grallistrix orion*				F				
†*Grallistrix geleches*					F			
†*Grallistrix erdmani*							F	
Family Meliphagidae: Honeyeaters (traditional) or Mohoidae: a new family with no common name as yet (Fleischer et al. 2008)								
†ʻŌʻōʻāʻā (Kauaʻi ʻŌʻō, *Moho braccatus*)			HF					
†Oʻahu ʻŌʻō (*Moho apicalis*)				HF				
†Bishop's ʻŌʻō (*Moho bishopi*)					HF		H?F?	
†Hawaiʻi ʻŌʻō (*Moho nobilis*)								HF
†Kioea (*Chaetoptila angustipluma*)								HF

Table 1.1. Continued

	Ly	N	K	O	Mo	Ln	Ma	H
Family Corvidae: Crows and Jays								
'Alalā (Hawaiian Crow, *Corvus hawaiensis*)								H
†*Corvus impluviatus*				F				
†*Corvus viriosus*				F	F			
Family Monarchidae: Monarchs								
Kaua'i 'Elepaio (*Chasiempis sandwichensis sclateri*)			HF?					
O'ahu 'Elepaio (*Chasiempis s. ibidis*)				HF?				
Hawai'i 'Elepaio (*Chasiempis s. sandwichensis*)								HF?
Family Sylviidae: Old World Warblers								
Millerbird (*Acrocephalus familiaris*)	H	H						
Family Turdidae: Thrushes								
†Kāma'o (*Myadestes myadestinus*)			HF					
†'Āmaui (*Myadestes woahensis*)				HF				
†'Oloma'o (*Myadestes lanaiensis*)					HF?	H	F?	
'Ōma'o (*Myadestes obscurus*)								HF
Puaiohi (*Myadestes palmeri*)			HF					
Family Fringillidae (Finches), Tribe Drepanidini: Hawaiian Honeycreepers								
Laysan Finch (*Telespiza cantans*)	H			F	F			
Nihoa Finch (*Telespiza ultima*)		H			F			
†*Telespiza persecutrix*			F	F				
†*Telespiza ypsilon*					F		F	
†'Ō'ū (*Psittirostra psittacea*)			HF	HF	H	H	H	H
†Lāna'i Hookbill (*Dysmorodrepanis munroi*)						H		
Palila (*Loxioides bailleui*)			F	F				H
†*Loxioides kikuchi*			F					
†*Rhodacanthis litotes*				F			F	
†Lesser Koa-Finch (*Rhodacanthis flaviceps*)								H
†*Rhodacanthis forfex*			F				F	
†Greater Koa-Finch (*Rhodacanthis palmeri*)								H
†*Chloridops wahi*			F	F			F	
†Kona Grosbeak (*Chloridops kona*)								H
†*Chloridops regiskongi*				F				
†*Orthiospiza howarthi*							F	
†*Xestospiza conica*			F					
†*Xestospiza fastigialis*				F	F		F	
Maui Parrotbill (*Pseudonestor xanthophrys*)					F		HF	
Hawai'i 'Amakihi (*Hemignathus virens*)					H	H	H	HF
O'ahu 'Amakihi (*Hemignathus flavus*)				HF?				
Kaua'i 'Amakihi (*Hemignathus kauaiensis*)			HF					
'Anianiau (*Hemignathus parvus*)			HF					
†Greater 'Amakihi (*Hemignathus sagittirostris*)								H
†Lesser 'Akialoa (*Hemignathus obscurus*)								HF
†Kaua'i Greater 'Akialoa (*Hemignathus ellisianus stejnegeri*)			HF					
†O'ahu Greater 'Akialoa (*Hemignathus e. ellisianus*)				H				
†Maui-nui Greater 'Akialoa (*Hemignathus el. lanaiensis*)					F?	H	F?	
†*Hemignathus upupirostris*			F	F				
†Kaua'i Nukupu'u (*Hemignathus lucidus hanapepe*)			HF?					

continued

Table 1.1. Continued

	Ly	N	K	O	Mo	Ln	Ma	H
†Oʻahu Nukupuʻu (*Hemignathus l. lucidus*)				HF?				
†Maui Nukupuʻu (*Hemignathus l. affinis*)					F?		HF?	
ʻAkiapōlāʻau (*Hemignathus munroi*)								H
†*Hemignathus vorpalis*								F
ʻAkikiki (*Oreomystis bairdi*)			HF					
Hawaiʻi Creeper (*Oreomystis mana*)								H
†Oʻahu ʻAlauahio (*Paroreomyza maculata*)				HF				
†Kākāwahie (*Paroreomyza flammea*)					H			
Maui ʻAlauahio (*Paroreomyza montana*)						H	HF	
†*Vangulifer mirandus*							F	
†*Vangulifer neophasis*							F	
†*Aidemedia chascax*				F				
†*Aidemedia zanclops*				F				
†*Aidemedia lutetiae*					F		F	
ʻAkekeʻe (*Loxops caeruleirostris*)			H					
ʻĀkepa (*Loxops coccineus*)				H			H	H
†*Ciridops tenax*			F					
†ʻUla-ʻai-hāwane (*Ciridops anna*)					H?F?			H
ʻIʻiwi (*Vestiaria coccinea*)			H	HF	H	H	HF	HF
†Hawaiʻi Mamo (*Drepanis pacifica*)								H
†Black Mamo (*Drepanis funerea*)					H		F	
ʻĀkohekohe (*Palmeria dolei*)					H		HF	
ʻApapane (*Himatione sanguinea sanguinea*)			HF	HF	H	H	HF	HF
†Laysan Honeycreeper (*H. s. freethii*)		H						
†Poʻo-uli (*Melamprosops phaeosoma*)							HF	

Sources: This list is adapted from James and Olson (1991: Table 14) and Olson and James (1991: Table 5) and updated from Cooper et al. (1996), Olson and James (1997), Fleischer et al. (2000), Burney et al. (2001), and James and Olson (2003, 2005, 2006).

Notes: Taxonomy and listing are fitted to the Checklist of North American Birds (American Ornithologists' Union 1983 and updates through 2007). Unfortunately, this arrangement is outdated with respect to newer sources (James 2004, Pratt 2005) and results in an awkward placement for some species. Subspecies likely to be elevated to species rank are listed (Pratt 2005, VanderWerf 2007). The name *Geochen rhuax* is treated here as a nomen dubium (Paxinos, James, Olson, Sorenson, et al. 2002). Species presence on each island is indicated as historic, H, or subfossil, F. Some subfossil records require more study when two or more similar species are endemic to neighboring islands; these are indicated as F? Island name abbreviations are Ly, Laysan; N, Nihoa; K, Kauaʻi; O, Oʻahu; Mo, Molokaʻi; Ln, Lānaʻi; Ma, Maui; H, Hawaiʻi.

†Indicates an extinct endemic species or an extinct Hawaiian population of an indigenous species.

ORIGINS

South Pacific Sources

The Hawaiian avifauna is most unusual, and, significantly, it is just as much defined by what kinds of birds are absent as by those present. What sorts of birds do we find in Hawaiʻi? Where did they come from?

To the south, a vast realm of oceanic islands extends from the Caroline and Mariana islands in the west, to New Caledonia and the Fiji Islands in the south, and to Pitcairn and Henderson islands to the east. Although this tropical South Pacific region spans 10,000 km—about two and a half times the distance from Hawaiʻi to North America—its scattered islands share a land bird fauna that is remarkably similar throughout. Related species of rails (in the genera *Porzana, Gallirallus, Porphyrio*), pigeons (*Ducula, Ptilinopus, Gallicolumba*), swiftlets (*Aerodramus*), kingfishers (*Todirhamphus*), honeyeaters (*Myzomela*), cuckoo-shrikes (*Coracina, Lalage*), monarchs (*Rhipidura, Monarcha, Pomarea*), reed-warblers (*Acrocephalus*), starlings (*Aplonis*), white-eyes (*Zosterops*), and parrot-finches (*Erythrura*) span most, or all, of this region. So similar are the birds of these island groups that a person from Palau traveling to Samoa 5,500 km away would recognize most land birds encountered there. However, the Marquesans who colonized Hawai'i long ago would have recognized only 4 types of land birds in their new home: ducks (*Anas* spp.), crakes (*Porzana* spp.), a monarch flycatcher ('Elepaio), and a reed-warbler (Millerbird). Of the more than 23 land bird species that reached the Hawaiian Islands and gave rise to the Recent avifauna, only 4 are known with some certainty to have arrived from the South Pacific islands, although perhaps more lineages did, because they are present in both the South Pacific and the northern hemisphere continents (Table 1.2). However, the ancestors of most Hawaiian birds came not from the tropical South Pacific but from the temperate Holarctic region to the north: North America and eastern Asia (Mayr 1943, Fleischer and McIntosh 2001). At least 12 ancestral lineages, and perhaps many more, originated from these two continents, although it has often been impossible to determine which was the source (Fig. 1.1). Hawaiian forest birds with Holarctic origins include the I'o, Pueo, crows, thrushes, very likely the 'o'ō and kioea, and the Hawaiian honeycreepers.

These hypotheses of origin are made by identifying the bird species or genus most closely related to each Hawaiian lineage and inferring that the geographic range of the ancestral species was the same. Traditionally, phylogenetic relationships have been based on morphologically unique characteristics (shared derived characters). Recently, molecular techniques, particularly sequencing of mitochondrial DNA, have yielded additional independent evidence for sorting out the origins of Hawaiian birds (Fleischer and McIntosh 2001). For example, the skeletal remains of a subfossil Hawaiian eagle could not be distinguished from its two continental relatives. However, DNA studies matched this bird with the White-tailed Eagle from temperate Eurasia rather than the Bald Eagle (*H. leucocephalus*) from North America (Fleischer et al. 2000). Similarly, DNA comparisons discovered that the Hawaiian Rail and Laysan Duck are more closely related not to northern species but instead to South Pacific ones, the Spotless Crake (*P. tabuensis*) and Pacific Black Duck (*A. supercilliosus*), respectively (Fleischer and McIntosh 2001).

Circumstantial evidence indicates that birds in the South Pacific, including ancestors of the Laysan Duck, Hawaiian Rail, 'Elepaio, and Millerbird, disperse between islands, presumably to search for new habitat. None of these species is migratory, but occasionally a bird (or birds) will be seen striking out to sea or alighting on a ship with no land in sight. Such reports are not infrequent from places such as the Melanesian Islands (Mayr and Diamond 2001).

Why didn't more pan-Pacific land birds settle in Hawai'i? Species from oceanic islands in the South Pacific closest to Hawai'i would not be expected to have reached the Hawaiian Islands regularly for several reasons. First, Hawai'i is simply too far away. The high islands of the South Pacific are essentially no closer to Hawai'i than is North America, 4,000 km distant (see Fig. 1.1). This distance seems to be insurmountable for nearly all nonmigratory land birds. The nearest major source of southern land birds would be the Marshall and Gilbert archipelagos, 3,000 km away, whose tiny islands have relatively few bird species. Only minuscule Johnston Atoll lies between them and Hawai'i. Second, the size of source populations on South Pacific islands is very much smaller than those on continents,

Table 1.2. Lineages of the resident Hawaiian land bird avifauna, including waterbirds

Arriving Taxon	Source Area, Time of Arrival (millions of years ago)	Derived Hawaiian Taxa
Goose (*Branta* aff. *canadensis*)	NA, <1 mya	Nēnē (*B. sandvicensis*) and 2+ †*Branta* spp.
Duck (*Anas* or related genus)	Unknown, 3.2–3.7 mya	†Moa-nalo, 3 genera, 4 spp.
Duck (*Anas* aff. *platyrhynchos*)	NA or AS, <1.5 mya	Koloa (*A. wyvilliana*)
Duck (*Anas* aff. *superciliosa*)	SP, ~5 mya	Laysan Duck (*A. laysanensis*)
Black-crowned Night-Heron	NA or SA, time unknown	Black-crowned Night-Heron (*N. n. hoactli*)
Ibis (*Eudocimus*)	NA or SA, 1.1–2.1 mya	†*Apteribis*, 2 spp.
Crake (*Porzana* aff. *pusilla*)	AS or SP, 0.5–1.5 mya	†Laysan Rail (*P. palmeri*)
Crake (*Porzana* aff. *tabuensis*)	SP, 1–3 mya	†Moho (*P. sandwichensis*)
Crake (*Porzana*), one or more	Unknown	5+ †undescribed spp.
Common Moorhen (*Gallinula chloropus*)	NA, AS, or SA	'Alae 'ula (*G. c. sandvicensis*)
Coot (*Fulica*)	Unknown	'Alae Ke'oke'o (*F. alai*)
Black-necked Stilt (*Himantopus mexicanus*)	NA, 0.35–1.0 mya	Ae'o (*H. m. knudseni*)
White-tailed Eagle (*Haliaeetus albicilla*)	AS, 0.75 mya	†White-tailed Eagle (*H. abicilla*)
Hawk (*Buteo* aff. *brachyurus* or *swainsoni*)	NA, 0.3–1.1 mya	'Io (*B. solitarius*)
Harrier (*Circus*, Acciptridae)	Unknown	†*C. dossenus*
Owl (Strigidae)	Unknown	†*Grallistrix*, 4 spp.
Short-eared Owl (*Asio flammeus flammeus*)	NA or AS, perhaps recent	Pueo (*A. f. sandwichensis*)
Monarch (*Monarcha, Pomarea*)	SP, >1.5 mya	'Elepaio (*Chasiempsis sandwichensis*)
Crow (*Corvus*), one or more	NA or AS, 3.7–4.7 mya	'Alalā (*C. hawaiiensis*) and 2+ †*Corvus* spp.
Reed-Warbler (*Acrocephalus*)	SP, age unknown	Millerbird (*A. familiaris*)
Thrush (*Myadestes*)	NA, 3.4 mya	*Myadestes* 5 spp.
Honeyeater (Meliphagidae)	SP, age unknown	†'Ō'ō (*Moho* 4 spp.), †Kioea (*Chaetoptila* 1 sp.)
Or alternatively:		
Silky-flycatcher (Ptilogonatitidae) or relative	NA, 14–17 mya	New family, Mohoidae, including the 'ō'ō and Kioea
Finch (Fringillidae)	NA, AS, or SA, <5–6 mya	Hawaiian Honeycreepers (Drepanidinae), 21 genera and more than 50 species

Source: Adapted from Fleischer and McIntosh (2001).

Notes: Evidence for biogeographic origins comes from two sources. First, a search outside the Hawaiian Islands for the closest relatives of Hawaiian birds can indicate their geographic origins (Fleischer and McIntosh 2001, Lovette et al. 2002, Filardi and Moyle 2005). These relationships are revealed by studies of comparative morphology or molecular genetics. Second, the historic record of migration to the islands gives some measure of the frequency with which potentially colonizing lineages arrive. Although a few Hawaiian lineages originated in the South Pacific (SP), most came from North America (NA) and Asia (AS). Some colonist lineages from North America also occur in South America (SA), and the possibility exists that South America was instead the source area. Times of arrival are based on measures of genetic divergence and follow methods of Price and Clague (2002). These estimates are tentative and approximate at best, but they indicate that the avifauna is no older than the high islands, with the possible exception of the 'ō'ō and Kioea. Because the relationships of the 'ō'ō and Kioea have recently been disputed (Fleischer et al. 2008.), two alterative origins for them are listed in this table.
†Indicates extinct lineages.

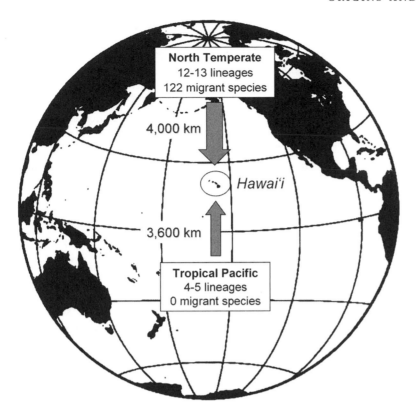

Figure 1.1. Map of the Pacific Ocean showing routes of origin for Hawaiian land birds. The majority of founding lineages originated as migrants from northern temperate North America and Asia (Holarctic region), as suggested by phylogenetic relationships. Present-day migrants are entirely from the Holarctic region. No species of land bird from the tropical South Pacific (Australasian region) has ever been reported from Hawai'i. Data on lineages are from Table 1.2.

decreasing the chance that wayward birds from those islands might reach Hawai'i. Source populations in the South Pacific are even smaller now, reduced by the same causes that have decimated Hawaiian birds, and many species there are likewise extinct. Third, insular species are usually endemic to one or a few islands, a limited range that suggests they no longer have a tendency to disperse. Finally, even with regard to "tramp" species that apparently do disperse regularly, little is known about how far they are able to disperse, but it seems likely that flights are limited to a few hundred kilometers at most, the greatest dis-

tance between neighboring islands in the South Pacific. Whatever the explanation, the fact remains that historically not one South Pacific vagrant has ever turned up in Hawai'i (Pyle 2002).

Northern Hemisphere Sources

The northern origin of much of the Hawaiian avifauna and the present-day flow of migrants show that seasonal migration rather than chance dispersal brought colonizers to the islands. Migratory birds are well adapted for long-distance flight. Upon departure they are physiologically and behaviorally prepared for a long journey, and they lay on fat reserves before departure. They fly at high altitudes to increase their speed, follow a relatively direct course, and all head in the same direction. The precise timing of migration strengthens the probability that more than one bird will arrive at a time in Hawai'i, a necessary condition for colonization.

Thousands of northern migrants successfully complete the 4,000 km oceanic crossing to the Hawaiian Islands every year, many of them on repeat visits. Once the final continental shore recedes over the horizon behind them, the legion of migrants is faced with nothing but sky and ocean until their Hawaiian landfall. There is not a speck of land that could offer relief midvoyage. Although most are shorebirds and waterfowl, among the migrants are rare land and freshwater vagrants that demonstrate how often potential colonists arrive. Herons, ibises, crakes, coots, seaeagles, hawks, harriers, the Short-eared Owl, and even a finch, the Common Redpoll (*Carduelis flammea*)—all relatives of living or extinct native Hawaiian birds—have turned up historically (Pyle 2002). Many groups of migrants to Hawai'i have never produced resident populations: grebes (two genera), cormorants, falcons, cranes, plovers, sandpipers, cuckoos, nighthawks, swifts, kingfishers, swallows, pipits, and sparrows. So far, an astounding 122 species of north temperate migrants have been recorded in Hawai'i, and the list is still growing.

The reason that more species have not established Hawaiian populations must be either that the same migratory instinct that brings them south to the islands in the fall takes them back north in the spring, or that they are unable to breed because the islands do not offer suitable breeding habitat. The only two species known to stay and reproduce in the past 100 years are the Blue-winged Teal (*Anas discors*) and the Pied-billed Grebe (*Podilymbus podiceps*), but these pioneers failed to establish themselves. Colonization by potential residents is far rarer than their migratory arrival.

The Youthfulness of the Hawaiian Avifauna

However rare the addition of new birds to the Hawaiian avifauna may be, it has been a relatively recent and ongoing process, at least as measured by the age of these bird lineages versus the age of the archipelago. Based "on the relatively low level of molecular divergence between Hawaiian taxa and their closest non-Hawaiian relatives," Fleischer and McIntosh (2001: 59) conclude that "none of these Hawaiian lineages split from mainland ancestors earlier than about 6.4 Ma (million years ago)," and their estimates of divergence fall "well within the period of formation of the current set of main islands." Interestingly, the same can be said of nearly all other Hawaiian animal and plant lineages so far studied (Price and Clague 2002). The Hawaiian honeycreepers, crows, and perhaps the moa-nalo and thrushes, are the oldest known lineages, although one enigmatic group, the 'o'o and Kioea, could also be the most ancient (R. Fleischer et al. 2008).

The Hawaiian Archipelago predates the current main islands, as shown by the remnants of much older islands, now atolls or seamounts submerged beneath the ocean surface. The archipelago is formed by a stationary, volcanically active hotspot in the earth's crust that at present is under the largest island, Hawai'i (Wilson 1963, Carson and Clague 1995). Progressively older, volcanically inactive islands in the chain stretch away from the hotspot toward the northwest. As islands are formed, they are carried off by the northwestward movement of the Pacific Plate in conveyor belt fashion. Thus, there have been islands situated in approximately the same location for perhaps as long as 32 million years. This geological history suggests the possibility that birds first reached much older islands and then simply moved over to the new islands as they emerged to the southeast. Conceivably, resident bird populations could be maintained indefinitely by island hopping among these large and numerous islands. The presence of some species across the Hawaiian chain suggests that bird dispersal among the Hawaiian Islands has been an active and continuous process. So it seems odd that the land bird lineages should be relatively young, especially considering that much older lineages have been identified for other Hawaiian organisms such as fruit flies and lobelias (Price and Clague 2002).

Why, then, are Hawaiian forest bird lineages so young? Several explanations are possible. First of all, geologic history shows that "the current landscape of large, closely spaced islands [was] preceded by a period with smaller, more distantly spaced islands," and this evidence has led to a well-supported theory that "much of the present species pool is probably the result of recent colonization from outside the archipelago" (Price and Clague 2002: 2429). In other words, the earlier archipelago of smaller, more distantly spaced islands received fewer colonists than the present larger and closer islands.

Second, lineages of island birds frequently are more prone to extinction in a natural state because their populations are small, confined to a limited geographic area (MacArthur and Wilson 1967, Ricklefs and Bermingham 2002). Nevertheless, the mechanisms of natural (versus human-caused) extinction are poorly understood for island species (Ricklefs and Bermingham 2002).

Another explanation for the youth of the Hawaiian forest bird lineages is that the fauna is essentially a byproduct of the development of the great bird migrations in the northern hemisphere. The oldest founders could have arrived when birds first began regularly migrating to Hawai'i. Perhaps these migrations began in the Pliocene, with increasing seasonality at northern latitudes and the onset of the ice ages. This timing coincides roughly with the formation of the present Hawaiian Islands and the age of the oldest Hawaiian bird lineages. Such a hypothesis could also have implications for the timing of the arrival of plant lineages, the majority of which were transported to the islands attached to or inside birds (Fosberg 1948, Green et al. 2002, Price and Wagner 2004).

The Attenuation and Disharmony of the Hawaiian Avifauna

Colonization by temperate migrants has had important evolutionary outcomes. The colonists must have been very mismatched to their new environment. Hawaiian tropical forests and shrublands bear little resemblance floristically, structurally, or in seasonality to the breeding habitat of these birds in coniferous and deciduous temperate forest, tundra, and steppe. (However, the wintering habitat of ancestral species was mainly tropical.) Arrivals from the South Pacific, such as the ancestors of the 'Elepaio, would have been more at home. The Hawaiian forests resemble their Pacific equatorial habitat both structurally and floristically. Indeed, much of the tree flora originated in the Indo-Pacific (Fosberg 1948; Wagner, Herbst, et al. 1999). (It is telling that birds ubiquitous in the South Pacific, such as fruit-pigeons, lories, honeyeaters, and starlings, are prohibited from importation into Hawai'i because of concerns that they would easily become established.) The northern migratory ancestors of Hawaiian forest birds (a raptor, a crow, and probably a thrush and a finch) all had to adapt to an unfamiliar tropical Hawaiian environment, which led to substantial evolutionary change with time.

A concept illustrated by the formation of the Hawaiian avifauna is faunal attenuation, the filtering effect of long-distance colonization which eliminates certain groups of species less suited to oversea colonization. The successful species are particularly adept travelers, not just a random selection from a pool of potential species at the source. Thus, insular fauna are not necessarily representative subsets of continental fauna but are instead "disharmonic," with many continental groups predictably absent. Consider, for example, the bird fauna of Fiji. That archipelago is made up of oceanic islands similar in size to those of Hawai'i but closer to the continental sources of Australia and New Guinea and presumably subject to more frequent colonization by birds. The native, land-based avifauna of Fiji includes members of about 24 families, versus 13 for Hawai'i, but still lacks many of the same bird families that are poor island colonizers worldwide. This greater phylogenetic diversity more closely approximates that of a typical continental bird community, its component species exploiting various feeding niches much as they would on the continents where they originally evolved.

In Hawai'i, where the establishment of bird species has been so infrequent, gaps in the bird community have been filled not by later immigrating species but by the adaptation and speciation of birds already present, in a process known as adaptive radiation. The more than 23 lineages established in Hawai'i gave rise to more than 110 species (93 described species from Table 1.1 plus approximately 20 species yet to be described from already collected fossil remains). Surprisingly, nearly half arose from a single colonization and radiation: the Hawaiian honeycreepers descended from a cardueline finch (Pratt 2005).

No doubt the colonists' ecological characteristics to some degree predetermined their evolutionary direction. The first Hawaiian honeycreeper was likely an opportunist that speciated into an impressive array of feeding niches, including nectarfeeders, seed-crackers, a frugivore, and an intriguing host of insectivores. By contrast, nearly all other bird lineages maintained their characteristic mode of life. For example, the five species of Hawaiian thrushes are still ecologically typical thrushes that feed on fruit and insects. Among the honeycreepers, the extreme range of variation in bill morphology and feeding behavior reflects the family's exaggerated ability to evolve novel morphologies throughout their continental range when compared with thrushes (Lovette et al. 2002).

In summary, then, the Hawaiian avifauna is derived from a few colonizing species,

mostly migrants arriving from very different temperate environments in the northern hemisphere. Through rapid speciation, these lineages produced a diverse tropical avifauna. How did the new island environment influence the evolution of forest birds in Hawai'i?

HAWAI'I, THE ULTIMATE PARADISE FOR BIRDS

Limiting Factors

The factors that limit bird populations are familiar to anyone who has watched and thought about birds. Dominant biotic factors are habitat, food, and competition with other birds, particularly members of the same species. Predators—typically cats and other carnivores, rodents, and snakes—raid bird nests and capture adults when they can, and they may depress populations below the carrying capacity of the food supply, either directly or by influencing their prey's behavior. Parasites and diseases take a toll but perhaps contribute less to mortality rates in normal, coevolved situations than do predation and food shortage (Newton 1998).

As a result of the filtering effect of dispersal, limiting factors in the original, prehuman Hawaiian environment differed dramatically from those in continental environments, and they differed to a fatal degree from those operating in Hawai'i today. Many biotic limiting factors never made it across the 4,000 km oceanic barrier. The list of no-shows includes many animals we take for granted: all mammals (except two species of bat), reptiles, and amphibians; most birds, as mentioned earlier; and many key groups of insects and parasites.

Release from Competition, Except from Other Birds

Most competitors for the insect food of birds, particularly the predatory ants and social wasps and bees (Wilson 1996), were absent from Hawai'i until brought by people. Parasitoids—the insect parasites of caterpillars and other arthropods, a major control of their host populations— were few in prehuman Hawai'i (Zimmerman 1948; this volume, Chapters 2 and 7). Lizards and frogs can be the most abundant vertebrates in the tropics (Rodda et al. 2001), but these competitors, too, were missing. The only mammals were two species of bat that hawked the night skies for moths and beetles. Thus Hawaiian birds were largely released from competition for invertebrate food, and they became the dominant predators of all animal life, large and small, in the islands' forests.

As for flower nectar, Hawaiian birds faced competition from no other vertebrate nectar feeders, such as bats (the Hawaiian bats are insectivorous only). Similarly, birds were the only fruit eaters and seed dispersers; in tropical forests elsewhere, birds compete for fruit mainly with mammals such as bats and monkeys. At the community level in Hawai'i, use of flowers by nectar-feeding birds can be nearly universal across the woody flora. For example, virtually all species of flowers in a Maui cloud forest are visited by Hawaiian honeycreepers, even when the flowers show characteristics for insect pollination (Berlin et al. 2001). The high incidence of bird pollination has influenced the evolution of flowers in the islands. Of the 14 lineages of Hawaiian plants that have experienced an evolutionary shift in pollination method, 11 evolved flower morphologies to enhance bird pollination as an adaptive shift away from dependence on insects for this service (Price and Wagner 2004).

The majority of woody plants in Hawai'i, at both the community and regional levels, are bird dispersed, which is not surprising as this characteristic follows a worldwide pattern for tropical flora (Ricklefs and Renner 1994, Berlin et al. 2000, Price and Wagner 2004). Remarkably, in Hawai'i there has been a de novo evolution of fleshy fruits ingested by birds that disperse seeds in their feces (Price and Wagner 2004). These fleshy fruits evolved in four lineages (lobelias, mints, pinks, and a coffee relative) that more often produce dry capsules releasing abiotically dispersed seeds.

Thus, in Hawaiian ecosystems a food web existed whereby birds were the sole vertebrate consumers. Their populations were abundant as a result, and they were exposed to novel evolutionary options, about which more will be said later.

Release from Vertebrate Predators, Except Other Birds

Another form of ecological release was freedom from nonavian predators of adult birds and their nests and a shift to predation solely by birds. Continental predators of forest birds include a diversity of small mammals, snakes, lizards, and birds. Originally, Hawai'i was without predators of birds apart from hawks, owls, crows, and possibly ibises and rails.

The suite of surviving native raptors includes the 'Io or Hawaiian Hawk, a generalist predator nevertheless deft at catching birds; the Short-eared Owl, an occasional hunter of birds and robber of nests; and several rare species of overwintering migratory hawks and falcons (Chapter 12). Extinct endemic raptors notable for their bird-catching morphology are a harrier and a genus of owls (Olson and James 1991; see also this chapter, Table 1.1). Subfossil remains of owl pellets bear evidence to a diet of honeycreepers. Besides the raptors, other bird predators were nest-robbing corvids, the Hawaiian Crow or 'Alalā, and several extinct raven species (James and Olson 1991; Banko, Ball, et al. 2002; see also this chapter, Table 1.1).

Birds face predation while foraging, resting in the daytime, sleeping at night, and sitting on the nest. Nest contents may also end up as a predator's meal. Each of these situations creates a different vulnerability for the bird and responses from it. Depredation of active or sleeping birds is rarely witnessed and therefore most difficult to study. However, indirect evidence such as a high annual survivorship rate indicates that tropical birds are adept at thwarting predators by avoiding encounters in the first place and by escaping attacks (Stutchbury and Morton 2001).

The same cannot be said of nests. Nests are hidden, but they cannot flee when discovered. Nests are a center of parental and chick activity, and this activity is noticed by predators using visual, auditory, and olfactory cues (Skutch 1949; Conway and Martin 2000; Martin, Martin, et al. 2000). In continental forests, rates of nest loss typically account for more than half of nesting attempts, and most of that loss is due to predation (Martin 1995, Newton 1998).

In a variety of nest studies, diurnal predators, mainly birds and particularly crows and their relatives, were identified as the most frequent predators (Martin, Martin, et al. 2000). However, with the advent of video cameras at nests, snakes are now being appreciated as perhaps more important in some ecosystems (Weatherhead and Blouin-Demers 2004). Certainly the efficacy of a nocturnal, tree-climbing, bird-eating snake has been demonstrated by the introduction of the brown treesnake (*Boiga irregularis*) to Guam and the subsequent extirpation of most of the island's avifauna (Rodda et al. 1997). Snakes are poor island colonizers and have not yet been successfully introduced to Hawai'i (apart from a tiny, burrowing species). Arboreal mammals are also prevalent nest predators on continents, especially rodents such as squirrels (Siepielski 2006), because of their high population numbers. Rats, cats, and mongooses introduced to many tropical islands, including Hawai'i, have decimated or extirpated bird populations and altered the behavior of their prey (Atkinson 1977; Burger and Gochfeld 1994; this volume, Chapter 11).

Release from Brood Parasites

Avian brood parasites are birds, such as cuckoos, that lay their eggs in the nests of other birds to be hatched and reared by the host parents. Brood parasitism often leads to dramatically lower reproductive output by the host parents (Davies 2000). The most familiar brood parasites from a conservation perspective are the New World cowbirds, which endanger novel host species, particularly on Caribbean islands (Davies 2000). There are no brood parasites in Hawai'i.

Release from Arthropod Vectors of Disease

As important as all other forms of ecological release was the original absence of nearly all biting, parasitic arthropods. By themselves, biting insects can affect the survival and breeding of birds, but they are also infamous vectors of disease for birds and people (Newton 1998). Hawai'i was fortunate to be largely free of these ectoparasites originally, and to a diminishing extent it still is (van Riper and van Riper 1985).

The main vectors of blood-borne diseases are biting flies, of which mosquitoes are the most familiar. But the world has a surprising diversity of other biting flies: black flies, sand flies, biting midges, biting muscid flies, and deer flies. All of these flies spread specific diseases, and all were absent from Hawai'i. The failure of mosquitoes, black flies, sand flies, and biting midges to reach Hawai'i can probably be attributed to the small size of adults and their short lifespan, which does not give them enough time to make a lengthy saltwater crossing. Some of these groups, mosquitoes and biting midges for example, do extend naturally by island-hopping far into the tropical South Pacific, potentially resulting in a more typical evolutionary environment of disease transmission and host resistance in these islands (Belkin 1962, Jarvi et al. 2003). The Hawaiian Islands and the Galápagos Islands may have been the only sizeable tropical islands free of mosquitoes.

Some disease-transmitting, blood-feeding arthropods live on birds and, because they are carried about by their hosts, could more easily be brought to islands. In Hawai'i, louse flies, an endemic flea, and several tick species were originally associated with seabirds and their rookeries (van Riper and van Riper 1985). The few possibly native species of louse flies occur on seabirds and the Short-eared Owl, and one transmits Haemaproteus to frigatebirds (van Riper and van Riper 1985, Work and Rameyer 1996). Regardless, louse flies are very rarely present on forest birds, and those that are, are probably not native. Blood-sucking lice and kissing bugs were absent.

It is worth mentioning here other arthropod parasites that are not necessarily associated with disease transmission but nevertheless can be problems themselves. Absent from Hawai'i are parasitic blowflies and botflies, whose maggots feed on nestling birds. Botflies recently made the news as a conservation issue when they were discovered to be widely prevalent in Galápagos finches (Fessel and Tebbich 2002). Chewing lice and mites are the main external parasites of native Hawaiian forest birds, but they have been little studied (van Riper and van Riper 1985).

Release from Most Diseases

Birds have existed and coevolved with diseases and their vectors in the continental tropics for millions of years. These disease systems are prevalent, diverse, and complex. For instance, a survey of birds from tropical Australia and New Guinea found infections of two vector-borne diseases, Plasmodium (malaria) and Haemoproteus (another blood protozoan), in 69 of the 105 forest bird species sampled and in 44% of 428 individuals surveyed (Beadell et al. 2004). Thus most bird species in this continental system had blood parasites, and their prevalence among individuals was high. Furthermore, numerous lineages of the parasites were detected. Yet despite a high disease prevalence in these forests, the consequences of infections in the wild birds were difficult to detect. Birds with

long evolutionary histories of parasite exposure seem to rarely become sick and die, although chronic disease can reduce fitness (Møller 1994).

Without insect vectors to spread them, many bird diseases never became established in Hawai'i (van Riper and van Riper 1985). Hawaiian forest birds originally lacked avian malaria, Haemoproteus, Leucocytozoon, filarial worms, and other vector-transmitted parasites. In fact, the only native parasitic protozoa are a single genus of coccidia, Isospora spp., which inhabit the intestines of native 'Elepaio, thrushes, and honeycreepers.

Very few other native internal parasites have been discovered: two genera of nematodes and one species of acanthocephalan. Trematodes have been reported from native birds, but it is not known if they are indigenous or introduced. Tapeworms in honeycreepers now seem to be nonnative. Only a few cosmopolitan, but possibly native, bacterial and fungal diseases have been reported. Viral diseases, including arboviruses, seem to be lacking. More parasites of native birds perhaps await discovery, but at present the list is very short indeed.

Certainly the original absence of insect vectors inhibited the spread of disease in Hawai'i, but other factors may also be operating. Intuitively, we assume that diseased birds cannot fly the 4,000 km needed to bring pathogens to Hawai'i. Yet some birds can carry infections with no apparent effect on their health, and migratory birds elsewhere have even been reported to have higher haematozoan parasite loads than do nonmigratory ones (Figuerola and Green 2000). In some cases, the reason for the failure of parasites to become established in prehuman Hawai'i may be that they require specific hosts and are unable to shift from the hosts that brought them (e.g., waterfowl or shorebirds) to novel passerine (songbirds or perching birds) hosts. Furthermore, some parasites have complicated life histories, requiring specific alternate hosts missing in the islands.

One review of bird parasites and the diseases they cause states, "Existing information would suggest that, in most bird species, parasites cause little mortality among adults or young and are much less prevalent as a major limiting factor for breeding numbers than are predation or food-shortage" (Newton 1998: 284). However, the review concludes that major impacts on a population "can occur when bird species, mainly through human action, have come into contact with generalist parasites to which they have no evolved resistance" (Newton 1998: 286). The premicre example of this process has been the introduction of mosquitoes, avian malaria (*Plasmodium relictum*), and avian pox virus (*Avipoxvirus* spp.) to the Hawaiian bird fauna. Without a recent history of exposure, Hawaiian birds succumbed to these diseases (Chapter 9).

Evolutionary Change Brought on by an Insular Suite of Limiting Factors

The absence of so many limiting factors prevalent on continents had major consequences for the evolution of Hawaiian birds. Most bird competitors could not fly the distance or failed to colonize, and thus the colonists that did succeed faced little competition with other bird species for resources. Competitive release was enhanced by the absence of mammals, reptiles, frogs, and certain insects that would otherwise have exploited the food sought by birds. All predators of birds were missing except for other birds, and disease and parasites were also largely absent, setting the stage for loss of defensive adaptations in an evolutionary environment free, or nearly free, of these selective pressures. In the section that follows we explore some of the evolutionary changes brought about by the extremely insular and highly peculiar ecological setting of the Hawaiian Islands, particularly changes that make birds vulnerable in the not-so-insular environment of modern Hawai'i.

EVOLUTION IN THE HAWAIIAN INSULAR ENVIRONMENT

An Archipelago of Geographic Diversity

An important component of the Hawaiian evolutionary setting is the unique geographic environment of the islands. The Hawaiian Islands are mostly large, 115–10,433 km^2. The eight main islands are aligned from the youngest and most volcanically active, Hawai'i, in the southeast, to the oldest and most weathered, Kaua'i, to the northwest (Fig. 1.2). Importantly, four islands—Maui, Moloka'i, Lāna'i, and Kaho'olawe—formed a single island, "Maui Nui," as recently as 10,000–20,000 years ago, when a lower sea level exposed the shallow channels between them (Price and Elliott-Fisk 2004). There are a few small islands (<1 km^2) adjacent to the large ones, but none now supports native forest birds. Continuing up the chain, the nine islands or groups of islands of the Northwestern Hawaiian Islands are either atolls or fragmentary remnants of now submerged volcanic islands. Two of the Northwestern Hawaiian Islands, Nihoa and Laysan, were inhabited by five taxa of forest birds, of which the Laysan Millerbird and Laysan Honeycreeper are now extinct. Two species, the Laysan and Nihoa finches, once inhabited the main islands also, as indicated by fossil evidence (James and Olson 1991). With the exception of the Millerbird, the passerine avifauna of the Northwestern Hawaiian Islands is closely related to and derived from that of the main islands, which hold the large majority of Hawaiian forest birds (James and Olson 1991). For the archipelago as a whole, the historical, land-based avifauna was 55 species, and the total fauna, including prehistoric species known from subfossil remains, was at least twice that number (see Table 1.1).

The floristics and geography of the vegetation on the main islands are tropical and only moderately seasonal (Chapter 6). The islands are more biogeographically diverse

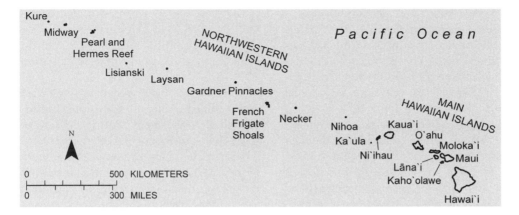

Figure 1.2. Map of the Hawaiian Islands.

than other islands in Oceania. Mountainous topography creates an extreme gradient in rainfall that in turn yields an equally dramatic gradient in vegetation. Although rainforest grows on the windward mountain slopes, dry forest and shrublands cover the leeward side (Plates 15–16). These plant communities are quite varied and at their extremes are completely different from each other floristically and structurally. The gradient can be so steep in places that rainforest and desert lie only a few kilometers apart. Vegetation patterns are repeated from island to island, differing with the amount of topography and rainfall. The highest islands, Maui and Hawai'i, support nearly all life zones the tropics can offer, and in this respect they can be viewed as continents in miniature. As mentioned earlier, Hawaiian habitats bear no resemblance to habitats in the higher northern latitudes, the source of most of the ancestral Hawaiian birds. Thus, Hawaiian birds have had to adapt to two fundamental conditions of their environment—first, that it is tropical and diverse, and second, that it is insular.

Adapting to a Tropical Environment

The special traits of life history and ecology that characterize tropical birds and distinguish them from birds of temperate zones have been brought to light in recent years. Key attributes of tropical birds generally, including Hawaiian forest birds, are sedentary (nonmigratory) behavior, reduced activity (energy conservation), longer lifespan, delayed maturation, smaller clutch size, longer developmental time, and extended breeding season, among others (Stutchbury and Morton 2001, Wiersma et al. 2007; this volume, Chapter 8). One factor driving the evolution of these traits is decreased seasonality. In an environment with a less pronounced growing season and, conversely, a season of milder hard times, food is more constantly available year-round, enhancing survival and promoting extended breeding. But decreased seasonality also allows populations of competitors and predators to build and enhances disease transmission, requiring birds to evolve better defenses against competition, predation, and disease.

Adapting to an Insular Environment

Special life history traits and ecological characteristics of birds on islands versus those on continents have also been the subject of recent study (Grant 1998b, Murray 2001). Oddly, some of the evolutionary outcomes are similar to those seen in tropical versus temperate comparisons —for example, sedentary behavior, energy conservation, greater longevity, delayed maturation, and decreased clutch size— but these characteristics arise for reasons in addition to those explaining patterns in tropical birds. These topics will be explored further in Chapter 8. Other evolutionary trends are prevalent on islands, such as reduction of large body size, increase in small body size, increased energy conservation, tameness and other reduced antipredator behavior, and even flightlessness in some species (Clegg and Owens 2002, McNab 2002). The main factors driving these changes for island birds seem to be competitive release and reduced mortality to predation and disease, factors that were discussed earlier. On islands, these factors lead to an increase in life expectancy, population density, and competition among species members, resulting in demographic, life history, and morphological responses to crowding (MacArthur et al. 1972, Murray 2001, Clegg and Owens 2002).

Restricted-Range Species

In tropical environments with a relatively constant food supply, individuals of many bird species are able to find all they need to eat year-round in their own patch of habitat. Birds develop territorial behavior to defend that patch. Sedentary and territorial behavior could be particularly acute for insular birds living at relatively high population densities. Most Hawaiian forest birds do occupy small home ranges year-round (with some notable exceptions, as discussed later), although surprisingly few of the surviving species are actually territorial (Pratt, Simon, et al. 2001). As a result of permanent residency and sedentary lifestyles, tropical and insular birds typically lose their inclination and ability to travel long distances. (This generalization applies best to adults, because juvenile birds do disperse in search of a place to settle.) Accompanying morphological changes in sedentary rainforest species on continents include shorter wings and reduced power

of flight (Keast 1996). For insular birds, reduction of traits to help them disperse is a common condition, as it is generally for both plants and animals on islands (Grant 1998b).

With increasingly sedentary behavior and reduced ability to make long flights, Hawaiian forest birds are more likely to find the distance between the main islands a barrier. The channels between the main islands, at 17–72 km in width, amount to far greater distances than most Hawaiian forest birds travel in the course of day-to-day or seasonal activity. There may be exceptions among the nectar-feeding Hawaiian honeycreepers, specifically ʻŌʻū, ʻIʻiwi, and ʻApapane. These species wander over their home islands in search of seasonal and patchy food sources (Perkins 1903, Ralph and Fancy 1995) and are, or were, the most likely candidates to fly between islands. However, for most Hawaiian forest birds, the channels are probably formidable barriers. It is difficult to demonstrate movement of these birds, or the lack of it, between islands in modern Hawaiʻi, particularly now that the lowlands are generally devoid of native forest birds poised to make the crossing. Nevertheless, it is still significant that no species of Hawaiian passerine endemic to only one island has ever been recorded as a vagrant away from its home island.

An important outcome of minimal interisland dispersal has been the evolution of species and subspecies of birds endemic to single islands. Genetic differentiation of geographically separated populations is believed to be the usual form of speciation among birds and seems to be practically the universal outcome of long-term residency of birds on islands (Mayr 1942, Newton 2003). A species may disperse from its parent island to colonize one or more neighboring islands and eventually produce new species there. Evidence for this pattern of colonization is demonstrated by the distribution of closely related species, one to an island, along the Hawaiian chain

(Amadon 1950, Fleischer et al. 1998). The majority of genera of Hawaiian forest birds are distributed in this way, and the outcome is a core community shared among the islands, with representatives displaying the same ecological morphology (Freed 1999). Thus almost every island had its 'ō'ō, thrush, 'amakihi (Hemignathus sp.), 'akialoa (Hemignathus sp.), 'Apapane, and so on. In addition, some genera did not disperse and are restricted to one island, for example, the Maui Parrotbill, 'Ākohekohe, and Po'o-uli, all on Maui Nui.

Frequent speciation within the Hawaiian Islands has resulted in rather similar avifaunas among the islands, each composed mainly of endemic forms. The genetic separation of island populations of Hawaiian forest birds has recently been quantified by molecular techniques, supporting the recent trend in systematics of Hawaiian birds in recognizing numerous former subspecies as distinct species (Pratt and Pratt 2001). Fifty-three historic taxa (species and subspecies) of historically known Hawaiian forest birds are, or were, endemic to single islands, whereas seven occur, or occurred, on more than one island (taking into account the known subfossil distributions of these species; James and Olson 1991). Only two species show obvious plumage differences on an island, the Maui 'Alauahio on Maui and Lāna'i (which until recently were both part of the single island Maui Nui), and the 'Elepaio, with three intergrading subspecies on Hawai'i. Single-island endemics such as these present special problems to conservation.

A disproportionate percentage of restricted-range species worldwide are threatened (31% versus only 4% for birds that do not have restricted ranges; Stattersfield et al. 1998: 27–28). Restricted-range species are defined by the International Union for Conservation of Nature (IUCN) as those species "judged to have a breeding range of 50,000 km^2 or less throughout historical times" (Stattersfield et al.

1998: 27). By this definition, all Hawaiian forest birds have restricted ranges; however, for our purposes the comparison is just as meaningful for species with ranges restricted to a single island.

Not only is a restricted-range species more vulnerable than a widespread one because it has less habitat to lose to human activity, but it also possesses intrinsic vulnerabilities that result from a small range and population size. For example, a species endemic to Moloka'i might become extinct because mosquitoes and avian disease can reach the island's summit at 1,514 m. However, if the species occurred throughout Maui Nui, it might still survive at 1,500–3,000 m on Mt. Haleakalā, above the reach of avian disease. Year-to-year fluctuations in weather, food availability, and the population cycles of predators and disease theoretically are more likely to drive small, restricted populations to extinction. Most historic extinctions of Hawaiian birds have involved species or subspecies restricted to one island (32 taxa), with the exception of the formerly widespread and now extirpated 'Ō'ū. Of present concern, all 17 species or subspecies of endangered Hawaiian forest birds are now restricted to one island (IUCN designations in Bird-Life International 2000; this volume, Chapter 5). On the other hand, of the taxa listed as not threatened or vulnerable, 7 are restricted to one island and 4 are more widespread—the Short-eared Owl, Maui 'Amakihi (Hemignathus virens wilsoni), 'I'iwi, and 'Apapane. Given this preponderance of single-island species, forest bird conservation in Hawai'i focuses on single-island issues, small populations, and recovery in situ rather than on reintroductions to other islands (U.S. Fish and Wildlife Service 2006).

Adaptive Radiation of the Hawaiian Honeycreepers

Often a publication or lecture on Hawaiian birds opens with a preamble extolling the adaptive radiation of the Hawaiian honeycreepers. Over time this device would wear thin if not for the superlative diversity of the honeycreepers, a complex bird world of its own, built in such a short time, in such a small place (Plate 19). The present count is 50 species (see Table 1.1), and that number is still growing with additions from the recent fossil record (James and Olson 2003, 2005, 2006; James 2004; Pratt 2005). Information forthcoming every year deepens appreciation for these favorite Hawaiian birds (Pratt 2005).

Adaptive radiation is "the evolution of ecological diversity within a rapidly multiplying lineage" (Schluter 2000a: 1). Hawai'i is particularly rich in radiations of plants and animals, including the silversword alliance (composite family), lobelias, mints, and assemblages in the coffee, citrus, and African violet families, and, among insects, *Drosophila* pomace flies, bees, and crickets, among many other groups (Wagner and Funk 1995). The Hawaiian Islands are isolated, large, and numerous enough so that evolutionary processes drive the construction of diverse flora and fauna through the creation of new species. By contrast, on small or less remote islands elsewhere, colonization and extinction predominate in shaping the biota (Losos 1998). As for the Hawaiian honeycreepers, they are the largest radiation of birds on an oceanic archipelago anywhere, and they surpass, by a considerable margin, radiations on much larger islands, such as the Antilles, New Zealand, Madagascar, or the Solomon Islands. Darwin's finches of the Galápagos Islands, with 14 species, are a small radiation compared with the Hawaiian honeycreepers.

Adaptive radiation typically results in a lineage's becoming a characteristic or even dominant component of the ecological community in the region where it evolves. In the Hawaiian Islands, honeycreepers are really the only diverse radiation of birds. Other forest bird lineages—'ō'ō, crows, monarchs, and thrushes—have produced just a few, mainly allopatric, species. The diversity of honeycreepers is many times larger, and they outnumber other native birds in individuals and species everywhere in Hawai'i today.

Adaptive radiation is believed to develop in response to intense intraspecific competition and relaxed interspecific competition (Schluter 2000a). As evolving species adapt to exploit novel food sources, they partially escape competition from more conservative feeders. Shifts in feeding behavior lead to selection for changes in bill morphology to take specific foods. Life history changes accompany specialization in foraging.

The radiation of the honeycreepers has been first about food and bills, as is discussed in Chapter 7. In many cases, these shifts have resulted in morphologies and life histories that have replicated solutions shown by continental birds, for instance the thin, curved bill of a nectar feeder or the tweezers bill of an insect feeder. Extreme or completely novel solutions have also evolved, illustrated by the massive seed-cracking bill of the Kona Grosbeak or the twisted bill of an 'Ākepa, designed for opening leaf buds. Specialization seems to have prevented successful hybridization and genetic introgression among species of Hawaiian honeycreepers (Grant 1994). By comparison, the more recently radiated and generalist Galápagos finches have hybridized frequently. Specialization is one of the themes of adaptive radiation, and with specialization comes inherent vulnerability to change (Chapter 2).

Dependency on a narrow range of foods and competitive superiority at taking those foods are advantageous as long as conditions do not change rapidly. But if the food source is depleted by natural or human-caused perturbations to the environment, the birds will ultimately starve. For example, Palila are such specialized foragers on māmane seeds (*Sophora chrysophylla*) that the species' fortune declined when sheep devoured the māmane forests (Chapter 23).

In other cases, specialization may be accompanied by vulnerable life history characteristics (Chapters 2 and 8). For example, when a species with inherently low reproductive potential (e.g., one young bird per season) is exposed to higher rates of mortality, say from a novel disease, it cannot adequately compensate for such losses through recruitment. Generally speaking, risk of extinction increases with lower fecundity (Bennett and Owens 1997). Specialization and finely tuned adaptation to the Hawaiian environment are no longer strengths when components of the ecosystem themselves succumb to pressures from invasive species, whether browsing by livestock, plant diseases, or new predators of the insect prey base. Because of the birds' long-coevolved and special linkage to other native biota, conservation of Hawaiian birds strongly depends on a good defense of native ecosystems and their many constituent species.

Evolution in the Absence of Mammalian and Reptilian Predators

Predators do more than take eggs, young, and adults and reduce the size of bird populations. Predation operates as natural selection on the heritable biology and behavior of prey, and it also influences, through experience, the learned behavior of prey. This selection pressure ultimately makes prey more resistant to predation if it does not first drive the prey to extinction. Predation is now understood to be as important as food in influencing avian life histories (Martin 1995).

What have the original predation pressures in the islands meant for Hawaiian bird ecology today? A suite of native raptors has hunted birds on the wing and no doubt kept its prey's avoidance skill attuned. Throughout their evolutionary history, Hawaiian birds have been exposed to fairly normal selective pressures of predation by diurnal and nocturnal birds of prey and corvids. Predatory birds find their prey by

sight and hearing. Their prey avoid capture in risky situations by being visually cryptic or silent or by escaping on timely and appropriate recognition of a predator. In addition to sight and hearing, mammalian and reptilian predators employ the sense of smell to find their prey, and mammals also use touch. Hawaiian birds with long evolutionary histories in the islands may have lost some of their ability to evade capture when these novel predators were later introduced.

TAMENESS

The reputed tameness of island birds is generally not now shown by Hawaiian species, although Laysan and Nihoa finches are ridiculously bold on the mammal-free Northwestern Hawaiian Islands (Morin and Conant 2002). Other Hawaiian forest birds may naïvely ignore creeping dangers in the night, with fatal outcomes, thereby accounting for the loss or endangerment of some species. Surviving species may have incrementally regained appropriate fear responses with each introduction of a predatory mammal. Fear response in Hawaiian birds is one of the many facets of predation ecology awaiting study.

ODOR

Odor is a peculiar trait of Hawaiian honeycreepers. Likened to the smell of moldy canvas or steam rising from an asphalt road, this heavy, clinging odor is characteristic of nearly all honeycreepers (Perkins 1903, Pratt 1992). Not only are the birds aromatic, but so are their nests and droppings. The substance producing this odor has not been identified, nor is it understood why the birds possess it. A worldwide survey of birds odorous to humans found that few species are bad smelling and that most passerines have no odor, with Hawaiian honeycreepers the most notable exception (Weldon and Rappole 1997). That study did not determine whether this odor functions as a deterrent to predators or is simply a metabolic by-product. But surely if people can smell honeycreepers, the mammalian enemies of birds can detect them even better. Where mammalian predators are present, scent may defeat the attempt by a roosting or nesting birds to hide, and the best defense would be to keep out of reach.

FLIGHTLESSNESS

Flightlessness is frequently an adaptation of birds on mammal-free islands and was common in Hawai'i, where perhaps more than 20 species of flightless waterfowl, ibises, and rails evolved (Olson and James 1991; this volume, Plate 14). Reduction of the wings and other elements of the flight apparatus saves the bird energy and resources (McNab 1994a, 1994b). In many cases, but not all, evolution of flightlessness may be driven by selection for larger body size, which confers the independent advantages of increased social dominance and, for herbivorous species, increased gut. A larger gastrointestinal track allows for a diet high in bulk and low in nutrition, which necessitates longer digestion. Of course, flightlessness commits a bird to a terrestrial life in situations where food and other resources can be found year-round within walking distance.

All Hawaiian flightless birds are now extinct, easy victims of predation by people, rats, cats, and mongooses. The last surviving species, the Laysan Rail, disappeared from Midway Island in 1945, only two years after the accidental introduction of black rats (*Rattus rattus*; Olson 1999b).

ROOSTING BEHAVIOR

Roosting behavior (where and how birds sleep or rest) is one of the most neglected aspects of the biology of tropical birds, particularly for species that roost solitarily, and there are no studies with which to make insular versus continental comparisons (Skutch 1989). Night-roosting behavior of Hawaiian forest birds is virtually unknown. All species are believed to roost in tree crowns, except in the Northwestern Hawaiian Islands, where in the absence of trees birds must roost on or near the ground. A radio-tracking study of the Palila, a tree-roosting honeycreeper, showed that the behavior of birds varies individually, with some birds shifting roosts every few nights while others faithfully return to the same site night after night for weeks (Banko, Johnson, et al. 2002). Repeated use of a roost site could catch the attention of a visually alert predator, either a raptor or a mammal, whereas the considerable accumulation of droppings beneath a frequently reused roost, noted in the Palila study, could attract an olfactory oriented mammalian predator.

NESTING BEHAVIOR

Predation by mammals threatens nesting more than any other stage in the life cycle of a forest bird. Predation is thought to influence the evolution of nesting behavior by (1) strongly increasing the diversity of nest types and nesting sites among species to make nests harder for predators to find, (2) reducing the number of parental visits to nests to minimize the risk of discovery, (3) shortening the nestling period to reduce the time of vulnerability to nest predators, and (4) possibly decreasing clutch size, not only to reduce activity at the nest but also to save the parents' effort for later nesting attempts (Skutch 1949; Martin 1988, 1993, 1995; Martin, Martin, et al. 2000). More will be said in Chapters 8 and 11 about the ecology of Hawaiian forest birds with respect to the present community of predators, but it is worth highlighting two issues. First, the nest sites of Hawaiian forest birds are not distributed throughout the available range of locations. All Hawaiian forest birds nest in the tree canopy or in tree cavities, essentially none in the shrub layer or on the ground where they would be more vulnerable to small mammals. Nesting success in these canopy sites is higher than for tropical birds on continents. Second, Hawaiian forest birds on average have longer nestling periods than their continental relatives. Prolonged nestling periods may have promoted survival with low rates of food delivery in an environment of higher within-species competition. Now, however, these unusually long nestling periods may expose the chicks to greater risk of predation by mammals. Other evolutionary questions with respect to the nesting of Hawaiian forest birds and predation by mammals remain to be studied.

Evolution in the Absence of Pathogens and Their Vectors

The loss of host resistance to pathogens on islands—in our case, the vulnerability of Hawaiian forest birds to avian malaria and avian pox virus—is still poorly understood. The depletion of island populations of birds caused by diseases has been demonstrated only for Hawai'i, where malaria did not previously exist (although a similar situation is suspected for the Society and Marquesas islands). Chapter 9 explores this topic in detail. Parallels can be sought with human examples, including the history of the native Hawaiian people, whose population declined rapidly from disease-caused mortality on exposure to novel pathogens (Stannard 1989).

Falling from Abundance: The Shift from an Insular to a Continental Environment

The last general point to make about the reduced number of limiting factors in the original Hawaiian environment is that because of the ecological release afforded by these factors, birds were likely very abundant, much more so than on continents. Early naturalists commented on the abundance of Hawaiian forest birds (Chapter 2). Surprisingly, some Hawaiian birds are still locally superabundant in mountain forests above the lowland zone of disease transmission and on the Northwestern Hawaiian Islands (Chapter 5). This abundance follows the pattern of high densities of organisms often observed on islands— termed *excess density compensation*—whereby a reduced suite of competitors, predators, and disease take a lesser toll, a situation that promotes crowding (MacArthur et al. 1972, Schluter and Repasky 1991, Terborgh et al. 2001, Rodda and Dean-Bradley 2002).

The introduction of novel species to Hawai'i causes more than a decline in the abundance of native species and replacement of native biota. Although bird numbers would be expected to fall even without the handicap of insular adaptations that are now maladaptive, the net effect of new limiting factors forever alters the evolutionary environment from an insular one to one that is continentlike. Theoretically, adaptation by birds to new limiting factors would take precedence over the old, now diminished challenge of coping with intense intraspecific competition at high population densities. Should Hawaiian birds be fortunate enough to adapt to the new threats (e.g., through better predator avoidance and disease resistance), they will still become less and less like island birds.

SUMMARY: WHY ARE HAWAIIAN FOREST BIRDS SO ENDANGERED?

In this chapter we have explored how the extremely remote geography of the Hawaiian Islands determined the makeup of the avifauna and how evolution on islands has shaped the life history and ecology of birds. We can now summarize the ways that biogeography and insular evolution influence the vulnerability and conservation of Hawaiian forest birds.

The Hawaiian Islands are the most remote landmass on earth and have been difficult to settle even by strong-flying birds.

A small number of founding species gave rise to a large avifauna through speciation and adaptive radiation. The ancestors of Hawaii's birds were primarily derived from migrants southbound from temperate North America and Asia, and, once settled, these colonists changed over a few million years into tropical, insular species. By contrast, only four species found their way to Hawai'i from tropical islands in the South Pacific, arriving already well adapted to Hawaiian environments.

Because the founding species were few and because other vertebrate competitors were absent, the stage was set for adaptive radiation, producing in Hawaiian honeycreepers the most diverse community of species ever to evolve from a single insular bird species. The large size of the Hawaiian Islands and the open water barriers between the islands have led to the evolution of numerous species restricted to only one or a few islands. Restricted-range species are especially vulnerable to extinction. Furthermore, specialization in exploiting native sources of food has led to vulnerability when the food sources, themselves island biota, have been reduced or eliminated.

The vast oceanic barrier between Hawai'i and the continents kept out most competitors, predators, and diseases of birds. Evolutionary changes in the absence of mammals—for example, loss of their fear response and evasive roosting and nesting behavior—have predisposed Hawaiian forest birds to predation by these newcomers. Hawaiian birds are also much less resistant than continental birds to novel diseases. In the chapters that follow, we will examine what has happened to Hawaiian forest birds as a result of ecological changes brought about by people and the biota accompanying them.

ACKNOWLEDGMENTS

I am grateful to colleagues and numerous visitors to Hawai'i Volcanoes National Park for many discussions and ideas on the evolutionary biology and conservation of Hawaiian birds. Individuals who helped by supplying hard-to-find information or directing me to novel sources are C. T. Atkinson, P. C. Banko, R. C. Fleischer, H. F. James, D. A. LaPointe, L. W. Pratt, J. P. Price, G. Rodda, B. L. Woodworth, and T. M. Work. M. P. Gorresen and J. P. Price assisted me in drafting figures. T. E. Martin, B. K. McNab, H. D. Pratt, L. W. Pratt, and T. B. Smith reviewed the manuscript and suggested improvements.

CHAPTER TWO

Historic Decline and Extinction

WINSTON E. BANKO AND PAUL C. BANKO

Extinction of plant and animal species has become one of the leading and most urgent environmental issues of our time. Understanding and reversing the causes of population decline are vital to preserving global biodiversity. Since the year 1500, 129 bird species worldwide have vanished; 4 more exist only in captivity, and this deterioration continues (Butchart et al. 2004). Today, 1 in 8 species of birds (a total of 1,186 species) is at risk of becoming extinct by the year 2100 (BirdLife International 2000). The great majority of these threatened species are tropical forest-dwelling birds, mainly on islands. Since 1800, more than 90% of bird species extinctions have involved island birds, and their special vulnerability continues.

The study of island extinctions reveals the ecological and evolutionary vulnerabilities of species to changing conditions, and change has never been more dramatic than on remote islands following human colonization. Hawaiian ecosystems were profoundly transformed first by Polynesians beginning a millennium ago and then by Westerners in the past two centuries. Old bird skins, pinned insects, herbarium sheets, pollen deposits, and nearly forgotten manuscripts document what we thought was the richness of Hawaiian ecosystems. However, recent discoveries of subfossil remains of previously unknown Hawaiian and other island species have revealed the greater diversity of insular biota and its loss from human impacts (Perkins 1903, 1913; Zimmerman 1948; James and Olson 1991; Olson and James 1991; Athens et al. 1992; Steadman 1995; Wagner, Herbst, et al. 1999; Burney et al. 2001). As a result, the Hawaiian Islands, the world's most isolated archipelago, are foremost in species extinctions. Before humans arrived, at least twice as many species of birds inhabited these islands as were found at the time of Western contact, and by one estimate, less than 10% of the original avifauna remains today (Pimm et al. 1995). At least 71 species or subspecies of Hawaiian birds disappeared before the arrival of Captain James Cook in 1778, and since then 24 more have disappeared (Banko et al. 2001). Among the Hawaiian passerines (songbirds or perching birds), 24 (69%) of the 35 remaining species or subspecies are federally listed as Endangered Species, although 10 of these may already be extinct (Chapter 5).

Hawaiian birds have obviously been unable to cope with the enormous ecological upheaval that was triggered by humans over 1,000 years ago (Banko et al. 2001, van Riper and Scott 2001). Understanding the causes and circumstances of native bird decline is a critical first step in their conservation (Temple 1986, Johnson and Stattersfield 1990). Our purpose here is to explain how and why Hawaiian forest bird populations declined during the past two centuries. We summarize and correlate the sometimes sketchy and disjointed historical information to reveal new insights into patterns and processes of extinction. Although we primarily discuss the population histories of passerine species in this chapter, we also briefly consider the two native birds of prey, a hawk and an owl, that used forest habitat (see Chapter 12 for more on birds of prey). We focus our analyses on the later stages of the thousand-year bird decline because observations of the behavior and habitats of birds have been recorded only within the past two centuries. Although naturalists of the early historical era left us with crucial firsthand accounts about the ecology of Hawaiian forest birds, we also draw upon the expanding body of knowledge of the ancient avifauna as it is known from subfossilized bones ("fossils") to provide context for understanding what happened in the historical period.

Our description of the history of decline and extinction differs from other accounts in several ways. First, we use information from written accounts and specimen labels to describe the historical trends of bird populations, not just species. This approach allows a more detailed evaluation of the geographical and temporal patterns of decline. Second, population trends are correlated with the fundamental feeding strategies of species because feeding ecology largely shaped the morphology, behavior, energetics, and demography of Hawaiian forest birds (Pratt 2005; this volume, Chapter 7). Intimate relationships between species and particular kinds of nectar, fruits, seeds, and invertebrate prey are at the crux of the adaptive radiation of the Hawaiian honeycreepers (Fringillidae: Drepanidinae), the most prevalent forest birds in the islands. Using this ecological perspective rather than the more traditional phylogenetic (taxonomic) approach better enables us to consider overlooked or underappreciated limiting factors, such as agents that disrupted food webs and feeding ecology. Framing our analyses in terms of the feeding ecology of species also provides new insights into the impacts and interactions of all of the factors threatening Hawaiian forest birds.

The decline of Hawaiian forest birds is often explained as the nonadaptive response of ecologically constrained or behaviorally naïve species to the arrival of new diseases and predators and to habitat disturbance (Warner 1968, Berger 1972, Atkinson 1977, Ralph and van Riper 1985, Pratt 1994, van Riper and Scott 2001). However, the reduced availability of important foods and disruption of food webs are also considered important factors contributing to this decline (Banko and Banko 1976; this volume, Chapter 7). McNab (2002) has stressed the importance of energetic constraints on the ecology of island vertebrates, and the potential effects of energetics and feeding ecology on the distributions of two Hawaiian bird species has recently been modeled (Porter et al. 2006). More fully understanding the combined and interrelated impacts of limiting factors in the context of Hawaiian bird ecology can guide the protection and restoration of remaining populations.

PATTERNS OF POPULATION DECLINE

To identify population trends for the historical period (defined here as 1778–1983), we analyzed about 10,000 records of the status and distribution of forest bird species compiled from specimen labels and published or unpublished reports and field notes (see the Avian History report series of Banko [1980–1987] and W. Banko and P. Banko [1980]). Because field effort has varied greatly over the past two centuries, we compare the distribution and relative abundance of species during two 16-year periods of intensive survey activity: 1887–1902 and 1968–1983 (Table 2.1). The geographic area covered included the 12 main forest districts in the islands (Table 2.2, Fig. 2.1), as identified and surveyed by the Hawai'i Forest Bird Survey (HFBS; Scott et al. 1986). Not surprisingly, there are many caveats to consider when comparing the results found and interpretations formed during thin slices of time separated by more than six decades. Although the goals and methods employed by these different generations of biologists were not the same, relevant insights into population decline emerged because the changes in populations and habitats were so dramatic. Additional perspectives about the status of the avifauna and the condition of habitats before and during Polynesian colonization of the islands have been derived from studies of ancient bird bones and other biological material (James and Olson 1991, Olson and James 1991, Burney et al. 2001). Population status and trends since 1983 are discussed in Chapter 5.

Geographical Influences on Population Decline

SPECIES-AREA RELATIONSHIPS

The fossil record reveals the distribution of land bird species on different islands before human colonization, and though not yet completely documented, it indicates that the number of species on each island seems to have been unrelated to island size (Olson and James 1982a, 1982b). Nevertheless, the number of species surviving into historic times is correlated with island size (Juvik and Austring 1979), and most species today occupy large, high-elevation habitats (Scott et al. 1986: 334, Fig. 312). Bird populations have disappeared primarily from small habitat islands and low elevations since the arrival of humans. For example, historical bird populations have declined precipitously on Moloka'i and Lāna'i, the two smallest forested islands, and in habitat islands containing the smallest amount of high-elevation (>1,000 m) forest (e.g., the two mountain ranges on O'ahu and Puna and Kohala on Hawai'i; Table 2.3). Conversely, populations have persisted longer on the relatively small island of Kaua'i, where there is substantial montane forest.

ELEVATION AND HABITAT TYPE

The fossil record shows that Hawaiian forest bird species and populations disappeared or declined drastically in dry, lowland habitats prior to Western contact. Evidence of a high level of species richness in lowland areas is particularly evident on Maui, where fossil collection sites are relatively widespread. There, new fossil honeycreeper species were represented more frequently in lava tubes at five lowland sites (<1,000 m) than at six upland sites (data from James and Olson 1991).

The decline and disappearance of native passerine populations continued in

Text continues on page 37

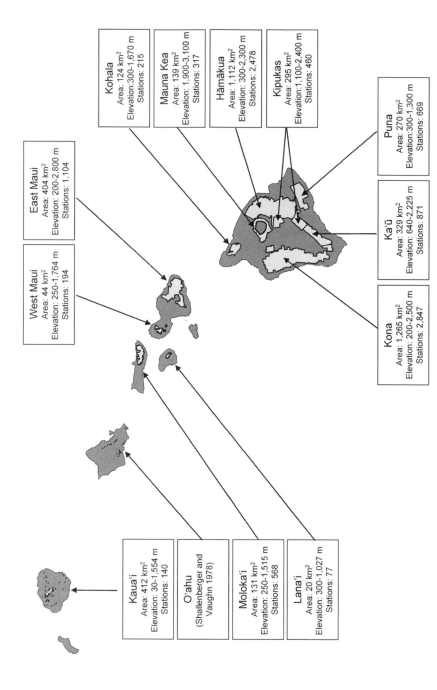

Kohala
Area: 124 km²
Elevation:300–1,670 m
Stations: 215

Mauna Kea
Area: 139 km²
Elevation: 1,900–3,100 m
Stations: 317

Hāmākua
Area: 1,112 km²
Elevation: 300–2,300 m
Stations: 2,478

Kipukas
Area: 295 km²
Elevation:1,100–2,400 m
Stations: 460

Puna
Area: 270 km²
Elevation:300–1,300 m
Stations: 669

East Maui
Area: 404 km²
Elevation: 200–2,800 m
Stations: 1,104

West Maui
Area: 44 km²
Elevation: 250–1,764 m
Stations: 194

Ka'ū
Area: 329 km²
Elevation: 640–2,225 m
Stations: 871

Kona
Area: 1,265 km²
Elevation: 200–2,500 m
Stations: 2,847

Kaua'i
Area: 412 km²
Elevation: 30–1,554 m
Stations: 140

O'ahu
(Shallenberger and
Vaughn 1978)

Moloka'i
Area: 131 km²
Elevation: 250–1,515 m
Stations: 568

Lana'i
Area: 20 km²
Elevation: 300–1,027 m
Stations: 77

Figure 2.1. Hawaiian forest bird survey districts. Polygons depict areas surveyed by Scott et al. (1986) from 1976 to 1983. Dots depict additional areas surveyed on Kaua'i by J. Sincock (in Scott et al. 1986) and on O'ahu by Shallenberger and Vaughn (1978). These districts delimit the populations listed in Table 2.1 and throughout the chapter.

Table 2.1. Historic (1887–1902) and recent (1968–1983) status and distribution of Hawaiian forest birds

Family	Species	Food Guild—Degree of Specialization	Habitats—Prehistoric and Historic	Population Distribution by District	Historic Population Status 1887–1902	Recent Population Status 1968–1983	Year Last Observed	U.S. Fish and Wildlife Service Listing
Accipitridae	'Io (Buteo solitarius)	Vertebrate and Arthropod—Generalist	LD, LM, LW, MD, MM, MW, SD	Kona—H.I. Ka'ū—H.I. Puna—H.I. Hāmākua—H.I. Mauna Kea—H.I. Kohala—H.I. Kaua'i (fossil) Moloka'i (fossil)	Common Common Common Common Common Common	Common Common Common Common Common Common		Endangered
Strigidae	Pueo (Short-eared Owl, Asio flammeus sandwichensis)	Vertebrate and Arthropod—Generalist	LD, LM, LW, MD, MM, MW, SD	All	Present	Present		
Corvidae	'Alalā (Corvus hawaiiensis)	Fruit and Seed—Intermediate	LD, LM, MD, MM, MW	Kona—H.I. Ka'ū—H.I. East Maui (fossil?)	Common Common	Rare Extinct	2002 1924	Endangered (exists in captivity)
Meliphagidae (or Mohoidae)	'Ō'ō'ā'ā (Moho braccatus)	Nectar—Intermediate	LD, LM, LW, MM, MW	Kaua'i	Common	Rare	1987	Endangered
	O'ahu 'Ō'ō (Moho apicalis)	Nectar—Intermediate	LD, LM, LW, MM, MW	O'ahu	Extinct (rare earlier)		1837	
	Bishop's 'Ō'ō (Moho bishopi)	Nectar—Intermediate	LD, LM, LW, MM, MW	Moloka'i East Maui	Rare Rare?	Extinct Rare?	1904 1981?	
	Hawai'i 'Ō'ō (Moho nobilis)	Nectar—Intermediate	LD, LM, LW, MM, MW	Kona—H.I. Ka'ū—H.I. Puna—H.I. Hāmākua—H.I.	Common Rare Rare Common	Extinct Extinct Extinct Extinct	1892 1902 1899 1902	

continued

Table 2.1. Continued

Family	Species	Food Guild—Degree of Specialization	Habitats—Prehistoric and Historic	Population Distribution by District	Historic Population Status 1887–1902	Recent Population Status 1968–1983	Year Last Observed	U.S. Fish and Wildlife Service Listing
Meliphagidae (or Mohoidae) (continued)	Kioea (Chaetoptila angustipluma)	Nectar—Intermediate	LM, MM	Puna?—H.I. Kipukas?—H.I. O'ahu (fossil) East Maui (fossil)	Extinct Extinct		1859 1841	
Monarchidae	'Elepaio (Chasiempis sandwichensis)	Arthropod—Generalist	LD, LM, LW, MD, MM, MW, SD	Kaua'i O'ahu Kona—H.I. Ka'ū—H.I. Puna—H.I. Hāmākua—H.I. Mauna Kea—H.I. Kohala—H.I.	Abundant Abundant? Abundant Abundant Abundant Abundant Unspecified Abundant?	Common Rare– Common? Common Rare Rare Common Rare Common		Endangered
Sylviidae	Millerbird (Acrocephalus familiaris)	Arthropod—Generalist	CD	Laysan Nihoa	Abundant Abundant	Extinct Abundant	1913	Endangered
Turdidae	Kāma'o (Myadestes myadestinus)	Fruit and Seed—Generalist	LD, LM, LW, MM, MW	Kaua'i	Abundant	Rare	1985	Endangered
	'Āmaui (Myadestes woahensis)	Fruit and Seed—Generalist	LD, LM, LW, MM, MW	O'ahu	Extinct (rare earlier)		1825	
	Oloma'o (Myadestes lanaiensis)	Fruit and Seed—Generalist	LD, LM, LW, MM, MW	Lāna'i Moloka'i East Maui (fossil)	Abundant Common	Extinct Rare	1934 1980	Endangered
	'Ōma'o (Myadestes obscurus)	Fruit and Seed—Generalist	LD, LM, LW, MD, MM, MW, SD, AD	Kona—H.I. Ka'ū—H.I. Puna—H.I. Hāmākua—H.I.	Abundant Abundant Abundant Abundant	Extinct Common Common Common	1902	

Species	Diet—Niche	Code	Location	Status	Status	Date	Status
Puaiohi (*Myadestes palmeri*)	Fruit and Seed—Intermediate	LD, LM, LW, MM, MW	Mauna Kea—H.I.	Common	Extinct	1892	Endangered
			Kohala—H.I.	Common	Extinct	1892	
			Kaua'i	Rare	Rare		
Fringillidae: Drepanidinae							
Laysan Finch (*Telespiza cantans*)	Fruit and Seed—Generalist	CD	Laysan	Abundant	Abundant		Endangered
			O'ahu (fossil)				
			Moloka'i (fossil)				
Nihoa Finch (*Telespiza ultima*)	Fruit and Seed—Generalist	CD	Nihoa	Abundant	Abundant		Endangered
			Moloka'i (fossil)				
'Ō'ū (*Psittirostra psittacea*)	Fruit and Seed—Specialist	LD, LM, LW, MD, MM, MW	Kaua'i	Abundant?	Rare	1989	Endangered
			O'ahu	Rare	Extinct	1899	
			Moloka'i	Abundant	Extinct	1907	
			Lāna'i	Abundant?	Extinct	1931	
			West Maui	Abundant?	Extinct	1894	
			East Maui	Abundant?	Extinct	1901	
			Kona—H.I.	Abundant	Extinct	1933	
			Ka'ū—H.I.	Abundant	Extinct	1896	
			Puna—H.I.	Abundant	Rare	1979	
			Hāmākua—H.I.	Abundant?	Rare	1987	
			Kohala—H.I.	Abundant?	Extinct	1892	
Lāna'i Hookbill (*Dysmorodrepanis munroi*)	Unknown—Specialist	LM	Lāna'i	Rare	Extinct	1918	
Palila (*Loxioides bailleui*)	Fruit and Seed—Specialist	LD, MD, MM, SD	Kona—H.I.	Common	Extinct	1950	Endangered
			Mauna Kea—H.I.	Common?	Rare		
			Kaua'i (fossil)				
			O'ahu (fossil)				
Lesser Koa-Finch (*Rhodacanthis flaviceps*)	Fruit and Seed—Specialist	MM	Kona—H.I.	Rare	Extinct	1891	

continued

Table 2.1. Continued

Family	Species	Food Guild—Degree of Specialization	Habitats—Prehistoric and Historic	Population Distribution by District	Historic Population Status 1887–1902	Recent Population Status 1968–1983	Year Last Observed	U.S. Fish and Wildlife Service Listing
Fringillidae: Drepanidinae (continued)	Greater Koa-Finch (*Rhodacanthis palmeri*)	Fruit and Seed—Specialist	MM	Kona—H.I.	Common	Extinct	1896	
				Ka'ū—H.I.	Rare	Extinct	1937	
				Hāmākua—H.I.	Rare	Extinct	1892	
	Kona Grosbeak (*Chloridops kona*)	Fruit and Seed—Specialist	MM	Kona—H.I.	Rare	Extinct	1892	
	Maui Parrotbill (*Pseudonestor xanthophrys*)	Arthropod—Specialist	LD, LM, LW, MM, MW	East Maui	Rare	Rare		Endangered
				Moloka'i	Fossil			
	Hawai'i 'Amakihi (*Hemignathus virens*)	Arthropod—Generalist	LD, LM, LW, MD, MM, MW, SD, AD	Moloka'i	Abundant	Rare		
				Lāna'i	Abundant	Extinct	1976	
				West Maui	Abundant?	Common		
				East Maui	Abundant	Common		
				Kona—H.I.	Abundant	Abundant		
				Ka'ū—H.I.	Abundant	Abundant		
				Puna—H.I.	Abundant	Common		
				Hāmākua—H.I.	Abundant	Common		
				Mauna Kea—H.I.	Abundant?	Abundant		
				Kohala—H.I.	Abundant?	Abundant		
	O'ahu 'Amakihi (*Hemignathus flavus*)	Arthropod—Generalist	LD, LM, LW, MM, MW	O'ahu	Common	Common?		
	Kaua'i 'Amakihi (*Hemignathus kauaiensis*)	Arthropod—Intermediate	LD, LM, LW, MM, MW	Kaua'i	Common	Rare		

Species	Foraging guild	Islands	Location			Date	Status
'Amianiau (Hemignathus parvus)	Arthropod—Generalist	LD, LM, LW, MM, MW		Abundant	Common		
Greater 'Amakihi (Hemignathus sagittirostris)	Arthropod—Intermediate	LW, MW	Hāmākua—H.I.	Rare	Extinct	1901	
Lesser 'Akialoa (Hemignathus obscurus)	Arthropod—Specialist	LD, LM, LW, MD, MM, MW	Kona—H.I.	Common	Extinct	1896	Extinct
			Ka'ū—H.I.	Abundant	Extinct	1902	
			Puna—H.I.	Rare	Extinct	1900	
			Hāmākua—H.I.	Abundant	Extinct	1940	
			Kohala—H.I.	Rare?		1892	
Greater 'Akialoa (Hemignathus ellisianus)	Arthropod—Specialist	LD, LM, LW, MM, MW	Kaua'i	Common	Extinct	1969	Endangered
			O'ahu	Rare	Extinct	1939	
			Lāna'i	Rare	Extinct	1894	
			Moloka'i (fossil)				
			East Maui (fossil)				
Nukupu'u (Hemignathus lucidus)	Arthropod—Specialist	LD, LM, LW, MM, MW	Kaua'i	Rare	Rare or Extinct	1990s?	Endangered
			O'ahu	Extinct		1860?	
			East Maui	Rare	Rare	1996	Endangered
			Moloka'i (fossil)				
'Akiapōlā'au (Hemignathus munroi)	Arthropod—Specialist	LD, LM, LW, MD, MM, MW, SD	Kona—H.I.	Abundant	Rare		Endangered
			Ka'ū—H.I.	Common	Rare		
			Puna—H.I.	Abundant	Extinct	1977	
			Hāmākua—H.I.	Common	Rare		
			Mauna Kea—H.I.	Common	Common		
'Akikiki (Oreomystis bairdi)	Arthropod—Intermediate	LD, LM, LW, MM, MW	Kaua'i	Common	Common	2004	Unspecified Rare 2004 Petitioned for Listing

continued

Table 2.1. Continued

Family	Species	Food Guild—Degree of Specialization	Habitats—Prehistoric and Historic	Population Distribution by District	Historic Population Status 1887–1902	Recent Population Status 1968–1983	Year Last Observed	U.S. Fish and Wildlife Service Listing
Fringillidae: Drepanidinae (continued)	Hawai'i Creeper (*Oreomystis mana*)	Arthropod—Intermediate	LM, LW, MM, MW, SD	Kona—H.I.	Abundant	Rare		Endangered
				Ka'ū—H.I.	Abundant	Rare	1973	
				Puna—H.I.	Rare	Extinct		
				Hāmākua—H.I.	Common	Rare	1975	
				Mauna Kea—H.I.	Unspecified	Extinct	1972	
				Kohala—H.I.	Unspecified	Extinct		
	O'ahu 'Alauahio (*Paroreomyza maculata*)	Arthropod—Intermediate	LD, LM, LW, MM, MW	O'ahu	Common	Rare	1978	Endangered
	Kākāwahie (*Paroreomyza flammea*)	Arthropod—Intermediate	LM, LW, MM, MW	Moloka'i	Common	Extinct	1963	Endangered
	Maui 'Alauahio (*Paroreomyza montana*)	Arthropod—Intermediate	LD, LM, LW, MM, MW	Lāna'i	Common	Extinct	1937	
				West Maui	Abundant	Extinct	1896	
				East Maui	Abundant	Common		
	'Akeke'e (*Loxops caeruleirostris*)	Arthropod—Specialist	LM, LW, MM, MW	Kaua'i	Common	Rare		Petitioned for Listing
	'Ākepa (*Loxops coccineus*)	Arthropod—Specialist	LM, LW, MM, MW	O'ahu	Rare	Extinct	1893	Endangered
				East Maui	Common	Rare	1980	Endangered
				Kona—H.I.	Abundant	Rare		
				Ka'ū—H.I.	Common	Rare		
				Puna—H.I.	Rare	Extinct	1943	
				Hāmākua—H.I.	Common	Rare		
				Kohala—H.I.	Common	Extinct	1892	

Species	Feeding	Code	Location			Year/Status
Ula-ʻai-hāwane (Ciridops anna)	Fruit and Seed—Intermediate	LW, MW	Kona—H.I. Hāmākua—H.I. Kohala—H.I. Molokaʻi (fossil)	Extinct Rare Rare	Extinct Extinct	1893 1892
ʻIʻiwi (Vestiaria coccinea)	Nectar—Intermediate	LD, LM, LW, MD, MM, MW, SD	Kauaʻi Oʻahu Molokaʻi Lānaʻi West Maui East Maui Kona—H.I. Kaʻū—H.I. Puna—H.I. Hāmākua—H.I. Mauna Kea—H.I. Kohala—H.I.	Abundant Common Abundant Abundant Abundant Abundant Abundant Abundant Abundant Abundant Unspecified Abundant	Common Rare Rare Extinct Rare Rare Rare Common Rare Abundant Rare Rare	1928
Hawaiʻi Mamo (Drepanis pacifica)	Nectar—Specialist	LW, MW	Kona—H.I. Hāmākua—H.I. Kohala—H.I.	Rare Rare Rare	Extinct Extinct Extinct	1890 1899 1892
Black Mamo (Drepanis funerea)	Nectar—Specialist	LW, MW	Molokaʻi East Maui (fossil)	Rare	Extinct	1907
ʻĀkohekohe (Palmeria dolei)	Nectar—Intermediate	LW, MW	Molokaʻi East Maui	Abundant Abundant	Extinct Rare	1907 Endangered

continued

Table 2.1. Continued

Family	Species	Food Guild—Degree of Specialization	Habitats—Prehistoric and Historic	Population Distribution by District	Historic Population Status 1887–1902	Recent Population Status 1968–1983	Year Last Observed	U.S. Fish and Wildlife Service Listing
Fringillidae: Drepanidinae (continued)	'Apapane (Himatione sanguinea)	Nectar—Generalist	LD, LM, LW, MD, MM, MW, SD, AD	Laysan	Abundant	Extinct	1923	
				Kaua'i	Abundant	Abundant		
				O'ahu	Common	Common		
				Moloka'i	Abundant	Abundant		
				Lāna'i	Abundant	Rare		
				West Maui	Abundant	Abundant		
				East Maui	Abundant	Abundant		
				Kona—H.I.	Abundant	Common		
				Ka'ū—H.I.	Abundant	Abundant		
				Puna—H.I.	Abundant	Abundant		
				Hāmākua—H.I.	Abundant	Abundant		
				Mauna Kea—H.I.	Unspecified	Rare		
				Kohala—H.I.	Abundant	Common		
	Po'o-uli (Melamprosops phaeosoma)	Arthropod—Intermediate	LM, MW	East Maui	Rare?	Rare	2004	Endangered

Sources: Fossil records are reported in James and Olson (1991), Olson and James (1997), and Burney et al. (2001).

Notes: Assignments of food guilds, degree of specialization, and habitat types are further described in Chapter 7. Food guilds represent the most important types of foods: arthropods (including snails), fruits and seeds, nectar, and small vertebrates (for birds of prey). Degree of specialization includes generalists, intermediates, and specialists. Habitat types (Chapter 6) are as follows: AD, alpine dry shrubland; CD, coastal dry shrubland (Laysan and Nihoa); LD, lowland dry forest; LM, lowland mesic forest; LW, lowland wet forest; MD, montane dry forest; MM, montane mesic forest; MW, montane wet forest; SD, subalpine dry woodland. Survey districts are mapped in Fig. 2.1, although "Kipukas" is not included in this table, with the exception of Kioea, because this region was unknown to early naturalists. H.I., Hawai'i Island. Categories of population status (rare, common, abundant) in the historic and recent periods indicate the relative abundance of the hawk and passerine species; these categories are not satisfactorily applied to the owl, for which the category "present" is used instead. Since 1983, some species have become extinct in the wild or extinct globally, as reported in Chapter 5 and as suggested in the "Year Last Observed" column of this table.

Table 2.2. Relative survey effort by early and recent observers

Survey District	1887–1902 (person-weeks)	1968–1983 (team-weeks)
Kaua'i	49	24
O'ahu	42	12
Moloka'i	27	2
Lāna'i	18	1
West Maui	12	4
East Maui	22	13
Kona	43	9
Ka'ū	7	5
Puna	13	4
Hāmākua	25	11
Mauna Kea	<1	1
Kohala	22	2

Sources: The number of weeks spent by observers in each survey district was estimated from published accounts, field diaries, and specimen labels for the early naturalists (1887–1902; Wilson and Evans 1890–1899; Rothschild 1893–1900; Henshaw 1902; Perkins 1903, 1913; Munro 1944; time measured in person-weeks) and from Scott et al. (1986) for recent observers (1968–1983; time measured in team-weeks).

lowland habitats throughout the historical period. Species richness and, with very few exceptions, population densities of native birds were greater at higher elevations, as reported by the HFBS (Scott et al. 1986) and by Shallenberger and Vaugn (1978). Only 8 of 26 native passerine species were encountered below 500 m elevation, whereas 7 species were found exclusively above 1,000 m elevation (Scott et al. 1986). The ecological characteristics of these two groups of species are discussed in later sections.

Relevant to the loss of ancient bird populations in lowland habitats is the fact that most fossil sites are located in relatively dry or seasonally dry habitats. Hawaiian dry forests have been largely destroyed historically (Chapter 6), but dryland plant and invertebrate communities were famously diverse, and they supported rich communities of birds (Rock 1913, Zimmerman 1948, Christensen and Kirch 1986, Athens et al. 1992, Burney et al. 2001, Sakai et al. 2002). The high level of bird diversity in ancient dry forests is revealed by the fossil remains of many species (e.g., 'Ō'ō'ā'ā, Kāma'o, Puaiohi, 'Ō'ū, Maui Parrotbill, Nukupu'u, Greater 'Akialoa, 'Akikiki, O'ahu 'Alauahio, and Po'o-uli; the scientific names of native birds are given in Table 2.1) that historically were associated only with mesic (moderately wet) or wet forests (Olson and James 1982a; Pratt, Kepler, et al. 1997; Burney et al. 2001).

Although the gross features of Hawaiian geography and habitat have influenced the persistence of bird populations, population distribution and abundance were probably affected by a variety of more subtle, interacting factors. In a landmark study, Scott et al. (1986) reported that the distributions of species found on their extensive surveys were most strongly and consistently influenced by elevation, moisture, and tree biomass. Although they analyzed bird distributions with respect to habitat characteristics and some other ecological variables (e.g., mosquito distribution), it was beyond the scope of their surveys to include many other relevant ecological variables, such as the presence of predators, ungulates, and arthropod prey.

Evolutionary and Ecological Influences on Population Decline

The evolution of Hawaiian forest birds has been greatly influenced by the extreme isolation of the archipelago, the diverse array of environmental conditions, a relatively depauperate resource base, and the infrequency of colonization from outside sources (Chapter 1). The Hawaiian avifauna is therefore characterized by a high level of endemism (the geographical uniqueness of its species) resulting from rapid speciation within the archipelago (Fleischer et al. 1998, Pratt 2005). Additionally, the adaptive radiation of Hawaiian honeycreepers is among the most famous examples of bird evolution on oceanic islands. This group of

Table 2.3. Numbers of historically known passerine species surviving to the recent period (1976–1983) in each survey district

Island—District	Elevation Range (m)	Recent Forest Bird Habitat[a] (km²)	Original Forest Bird Habitat <1,000 m Elevation[b] (km²)	Original Forest Bird Habitat >1,000 m Elevation[b] (km²)	Historical Populations	Recent Populations	Percent Surviving
Kaua'i	30–1,554	412	1,274	133	13	12	92
O'ahu			1,460	2	11	6	55
Moloka'i	250–1,515	131	640	28	10	4	40
Lāna'i	300–1,027	20	362	0	8	1	13
West Maui	250–1,764	44			5	3	60
East Maui	200–2,800	404			10	9	90
Maui—Total	200–2,800	448	1,419	421	15	12	80
Hawai'i—Kona	200–2,500	1,265			17	8	47
Hawai'i—Ka'ū	640–2,225	329			13	8	62
Hawai'i—Puna	300–1,300	270			12	6	50
Hawai'i—Hāmākua	300–2,300	1,112			15	9	60
Hawai'i—Mauna Kea	1,900–3,100	139			9	6	67
Hawai'i—Kohala	300–1,670	124			12	4	33
Hawai'i—Total	200–3,100	3,239	5,136	4,226	78	41	53
Total	200–3,100				135	76	56

Notes: Some values are missing for lack of data.
[a]The totals for areas surveyed by the Hawai'i Forest Bird Survey (Scott et al. 1986) or on Kaua'i by J. Sincock (unpubl. data in U.S. Fish and Wildlife Service 1983b) were estimated by map planimetry, which does not account for topographic relief. O'ahu is not included because it has not been systematically and comprehensively surveyed.
[b]Jonathan P. Price, University of Hawai'i, Hilo, unpubl. data.

closely related birds, whose common ancestor colonized Hawai'i about 5 million years ago (Fleischer et al. 1998), exhibits a variety of bill sizes and shapes as well as ecological adaptations that span nearly the entire range of such variation described in passerines worldwide (Pratt 2005) (Plate 19). The great variety of bills and feeding ecology of Hawaiian honeycreepers are what distinguish them from all other groups of birds, although mating system, social behavior, vocalization, and plumage are intrinsically interesting and important aspects of honeycreeper biology (Freed et al. 1987). Variation in feeding ecology, in turn, affects the demography, behavior, distribution, abundance, and conservation of Hawaiian birds in critical ways (Pratt 2005; this volume, Chapter 7).

Hawaiian honeycreeper feeding ecology is characterized by specialization on particular sources of nectar, fruit, seed, and invertebrate prey (Perkins 1903). Bird species that feed mainly on any one of these types of resource are not necessarily more prone to extinction, but food specialization does present some ecological challenges (Chapter 7). Specialists may be more constrained energetically and less efficient in provisioning offspring than are generalists. For example, specialized honeycreeper species tend to be larger than generalists (with some notable exceptions), yet they produce relatively small eggs and clutches, lay fewer clutches per season, and incubate their eggs and raise their offspring over a longer period. Low reproductive potential compromises the ability of species to withstand and recover from environmental setbacks, and it may limit their genetic diversity, making species less able to cope with habitat degradation and novel diseases, predators, and competitors (Freed 1999; this volume, Chapter 10). Additionally, specialist species face greater risks when resources decline, because they are less able to switch to alternate foods (Chapter 7). Compounding these difficulties is that the resource requirements of larger-bodied specialists are greater, which results in smaller, more vulnerable populations (Brown and Lomolino 1995, McNab 2002).

Feeding ecology affects other life history traits in ways that may make species more vulnerable to threats and environmental change. For example, some species exploit resources across large areas and gradients of elevation, whereas others are highly sedentary and territorial. Also, because specialists use relatively fewer habitat types, their ranges may be more restricted in size and elevation than those of generalists. On small islands, therefore, patterns of movement, dispersal, and home range have potentially serious consequences for the persistence of bird populations.

As predicted by their presumed ecological advantages, more populations of generalists persisted during the historic period, regardless of whether they ate mostly arthropods, fruits and seeds, or nectar. In contrast, relatively few populations of specialist nectar eaters or fruit and seed eaters survived into recent times; and though about 40% of the populations of arthropod specialists persisted until recently, nearly all are considered endangered, and some may be extinct. A more detailed analysis of historic changes in populations of generalists and specialists follows. We compare generalists and specialists without considering phylogenetic contrasts because a widely accepted phylogeny of the Hawaiian honeycreepers is not available, and assignment even to a genus is controversial for many species (Pratt 2005).

Changes in Population Distribution and Abundance

HISTORICAL CHANGES IN THE DISTRIBUTIONS OF SPECIES

If specialists were more vulnerable to changes in their environment, the ecological turmoil of the historical era should have affected their distributions dramatically. In fact, comparisons of the collection records and observations of early naturalists (1887–1902) and the results of recent population surveys (1968–1983) show that specialists were highly vulnerable. By the late 1800s, naturalists often recorded the locality or at least the general area or district where they collected or observed birds. However, they often assumed that species commonly found in well-known locales would have occupied nearby regions that were less frequently visited, such as Kohala, and we make the same assumption. Problematic species are Kioea, 'Ula-'ai-hāwane, Hawai'i Mamo, and other species that were very rare and close to extinction during the early survey period. Early naturalists assumed that these species occurred in a district if they deemed secondhand reports credible.

On the main Hawaiian islands, the degree of feeding specialization among songbirds significantly affected the probability of survival of populations through the historic period: 36 of 42 (86%) generalist populations persisted compared to 23 of 43 (53%) of intermediate populations and only 17 of 50 (34%) specialist populations (χ^2 test, P = 0.03). Although population persistence was not significantly affected by foraging guild, only 32% of fruit- and seed-eating populations survived while 63% of arthropod-eating populations and 66% of nectar-eating populations survived into recent times. Many seed eaters disappeared during the period of Polynesian settlement (James 1995), reducing the number that could be subjected to modern threats.

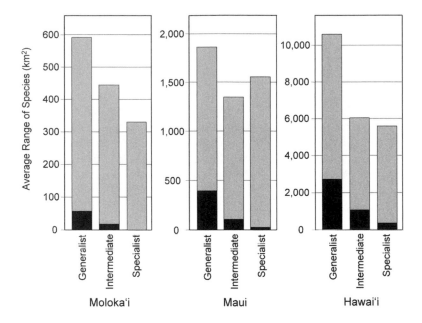

Figure 2.2. Average ranges (km²) of passerine species with generalist, intermediate, or specialist feeding ecology on three islands during ancient (prehuman) and recent (1968–1983) times, on gray and black bars, respectively. Kaua'i and O'ahu were omitted because estimates of islandwide ranges were not available for the recent period; on Lāna'i extinctions were too numerous to allow comparison.

Distributions of species typically contract to the periphery of their ranges as threats spread (Channell and Lomolino 2000), and accordingly, many Hawaiian birds disappeared first in lowland habitats where the impacts of settlement and invasive species became concentrated. Scott et al. (1986: Table 9) estimated the sizes of species' original ranges from general associations of bird distributions (using historical and fossil evidence) and broad categories of vegetation communities. Although only approximations can be made, the ancient ranges of generalists and specialists seem to have been similar in size on each island (Fig. 2.2). In other words, specialists may have been about as widespread as generalists in ancient times, even if they may have been less abundant. To estimate the ranges of contemporary populations,

Scott et al. (1986: Table 10) related the distributions of birds and vegetation using finer-scale data and methods. Ranges have shrunk about 76% overall (average ancient range = 2,848 km²; average recent range = 671 km²). Even more troubling is that the ranges of extant species (those encountered by the HFBS) have contracted on average to merely 16% of their original (prehuman) size.

Not surprisingly, the ranges of species were smaller on smaller islands, and populations there have been reduced more sharply by invasive threats (see Fig. 2.2). On small islands or in small habitat patches, populations are closer to the threshold of viability, as indicated by their more drastic contractions of range and higher rates of extinction, especially in the case of specialists. On Moloka'i and Lāna'i, for example, the HFBS did not encounter specialists or some intermediate species, and the ranges even of generalists were dangerously small, two orders of magnitude smaller than the ranges of generalists on Hawai'i, the largest island. Additionally, small-island generalists occupied only a fifth to a third as much of their ancient ranges as did generalists on Hawai'i.

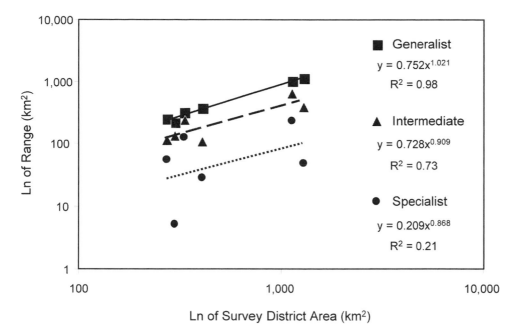

Figure 2.3. Effect of native forest area on range sizes of passerine species with generalist, intermediate, or specialist feeding ecology in the recent period (1968–1983). The effect of forest area on bird range size is strong for generalist species, significant for intermediate species, but negligible for specialist species, suggesting that generalists occupy larger portions of forest tracts than do specialists because they are less constrained by resource availability. Range sizes of bird species were compared among survey districts larger than 250 km^2 (Scott et al. 1986: Table 11), including East Maui and, on Hawai'i Island, Kona, Puna, Ka'ū, Kipukas, and Hāmākua. Generalist: 'Elepaio, 'Oma'o, Hawai'i 'Amakihi, and 'Apapane. Intermediate: 'Alalā, Hawai'i Creeper, Maui 'Alauahio, 'I'iwi, 'Ākohekohe, and Po'o-uli. Specialist: 'Ō'ū, Palila, Maui Parrotbill, 'Akiapōlā'au, Nukupu'u, and 'Ākepa.

Populations on larger islands are not immune from the ecological disasters that have wiped out populations on smaller islands. On Maui and Hawai'i, for example, the range sizes of specialist populations have become desperately small, and some generalist populations seem to be shrinking in range or numbers (Chapter 5). In the larger survey districts (>250 km^2) of Scott et al. (1986), the fractions of districts occupied by generalist, intermediate, and specialist passerines were significantly different (Kruskal-Wallis $\chi^2 = 14.44$, P = 0.0007), with generalists occupying much larger ranges than specialists. Comparing relative range sizes, therefore, provides a general gauge of the greater vulnerability of specialist species.

The positive relationship between island size (or gross availability of habitat) and range size of bird species yields another insight into forest bird ecology. Because generalists can exploit a wider assortment of resources for themselves and their young, they occupy multiple habitat types and range widely across native forests (Fig. 2.3). In contrast, specialists occupy relatively few types of habitats, but these tend to be high in quality because the birds' narrow niche requirements exclude them from low-quality habitats. This may partially explain why Scott et al. (1986) found specialist species mainly in larger, more intact forests at higher elevations (>1,500 m) above the influence of mosquito-borne diseases and many other threats. However, even where diseases are not prevalent (e.g., Kipukas district), specialists occupy much smaller ranges than generalists, suggesting their greater sensitivity to habitat quality.

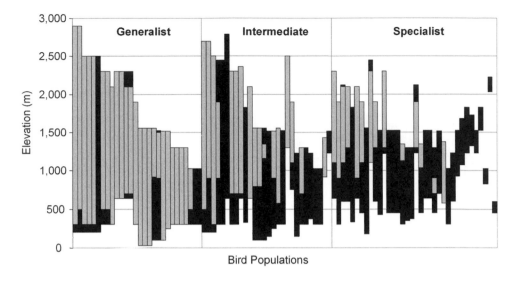

Figure 2.4. Ranges of elevation occupied by passerine populations historically (1887–1902) and recently (1968–1983). Gray bars represent recent elevation distribution, and black bars represent the additional amount of range occupied historically. Extinct populations are represented by all-black bars. Populations within each category of food specialization (see Chapter 7) are arranged first by descending rank order of historical elevation range, then by descending rank order of recent elevation range. Survey districts included in the analysis were Kaua'i, Moloka'i, Lāna'i, East Maui, and, on Hawai'i, Kona, Ka'ū, Puna, and Hāmākua. Other districts were excluded because the historical information is inadequate.

Therefore, specialist populations should respond more to improvements in habitat quality, whereas generalist populations may spread as habitats simply expand in area.

Differences in the altitudinal limits of populations also demonstrate that specialists have become more restricted than have generalists. However, the upper and lower limits of bird distributions in the early historical period are difficult to establish and to compare with recent information. To depict coarse changes in altitudinal distribution, we estimated from Banko (1980–1987) and W. Banko and P. Banko (1980) the upper and lower limits of the range of passerine populations from specimen records or reports referring to an elevation or locality. HFBS coverage ex-

tended approximately from the tree line to the lowest limits of continuous native vegetation in each district except Kaua'i and O'ahu (Scott et al. 1986). We assumed that all species historically occurred at least within the lower and upper limits of their recent ranges, as determined by the HFBS or, on Kaua'i, by Sincock (in Scott et al. 1986). However, the 'Ō'ū challenges our assumption about the lower limits of distribution because historical populations may not have extended to the lowest edge of the forest in some districts (Banko 1986).

Populations of generalist feeders tended to be distributed over a greater range of elevation than were specialists both early and late in the historical period (Fig. 2.4). The narrower altitudinal distribution of specialists during the 1887–1902 early period suggests that their ranges began contracting earlier or faster than those of generalists. Populations that were distributed narrowly over elevation in historical times were prone to extinction, regardless of whether they occurred primarily in the lowlands or the highlands. The comparatively narrower ranges of specialists today indicate that these populations continue to be more vulnerable. During 1968–1983 surveys, the elevation ranges of specialist feeders were only about half those of generalists, having shrunk since the early period by an average of 464 m compared to 138 m for generalists. Only 29% (4/14) of specialist populations were found below 1,000 m elevation, whereas all 26 generalist populations and 82% (14/17) of intermediate populations extended below 1,000 m elevation.

Analysis of altitudinal distributions also revealed that range contractions were mainly the result of birds' vacating the lowlands. Retreat from the lowlands was more pronounced for specialist populations, although generalists also lost ground (see Fig. 2.4). In recent times, for example, specialist populations were found about 285 m higher in elevation than were generalists. Therefore, as specialists retreat farther up the slopes of their volcanic homes, their distributions shrink in size as well as altitudinal range, and their populations become further isolated.

HISTORICAL CHANGES IN
THE ABUNDANCE OF POPULATIONS

The vulnerability of specialists to ecological change should be demonstrated by declining numbers as well as shrinking ranges. Early naturalists did not systematically survey or quantify forest bird populations, but they did characterize their relative abundance in generally meaningful terms, and their accounts usually agree among themselves, lending credibility to these observations. Even as they hunted for species new to science, the naturalists were keenly aware that some populations were disappearing rapidly (Henshaw 1902, Perkins 1903, Bryan 1908). Because of their comprehensive, if not quantitative, studies, we can compare population abundance during the early (1887–1902) and late (1968–1983) historical periods. To standardize the terminology for analysis of early accounts, we characterized passerine populations as "abundant," "common," "rare," or "extinct," based on Banko (1980–1987) and W. Banko and P. Banko (1980). We

then applied these categories to densities of birds reported by Scott et al. (1986) or, in the case of Kaua'i, by Sincock (in U.S. Fish and Wildlife Service 1983b). To demonstrate the ecological significance of population trends, we grouped species according to their degree of feeding specialization.

Despite methodological differences between early and recent surveys, the drastic decline of most forest bird species is clear. During 1887–1902, naturalists considered abundant or common all 28 populations of generalists, 23 of 31 (74%) populations of intermediates, and 20 of 35 (57%) populations of specialists. Recent surveys (1968–1983) indicated that only 19 of 28 (68%) generalist populations, 4 of 31 (13%) intermediate populations, and no specialist populations were abundant or common. The results of a test comparing the number of populations of generalists ($n = 78$), intermediates ($n = 66$), and specialists ($n = 67$) that remained unchanged or decreased in abundance by one, two, or three categories were significant ($\chi^2 = 6.81$, $P = 0.033$). Fewer than the expected number of generalist populations decreased in status during the historical period, whereas more populations of specialists declined in abundance and by more categories than expected by chance. Moreover, we estimated from Scott et al. (1986) that the mean population densities of generalists (185 birds per km^2) were almost five times higher than the densities of intermediates (40 birds per km^2) and more than 60 times higher than the densities of specialists (3 birds per km^2).

Temporal Patterns of Population Decline

The temporal patterns of forest bird decline during the historical period are difficult to evaluate because populations were not systematically or frequently inventoried prior to the HFBS study (1976–1983) and because O'ahu and portions of Kaua'i were not surveyed by the HFBS. Following the exploratory surveys of 1887–1902, Hawaiian forest birds were largely ignored for decades. The unevenness of the historical record, therefore, may give the impression that populations declined in discrete episodes, suggesting the impact of a new predator, disease, or other threat. Although patterns of decline might have appeared rather more continuous had populations been surveyed more frequently and systematically, the historical record does not indicate that populations necessarily declined steadily.

As modern surveys (Chapter 5) and the experience of contemporary biologists demonstrate, estimates or impressions of population abundance can be highly variable over time, space, and species. Contributing to the difficulty of detecting trends is that populations may seem to have declined rapidly when instead birds may have shifted to another portion of their range in search of food. For example, Perkins (1903) suspected that Palila had declined during the several years between his visits to North Kona. Although Palila numbers possibly did drop abruptly, birds also may have moved in response to the availability of their favorite food, seeds of the māmane tree (*Sophora chrysophylla*). One could easily form the same impression as did Perkins by making scattered visits to Mauna Kea, where Palila move about regionally (Hess et al. 2001; Banko, Johnson, et al. 2002; Banko, Oboyski, et al. 2002; this volume, Chapter 23).

Establishing the time of extinction, except for that of closely monitored populations such as ʻAlalā and Poʻo-uli, becomes increasingly difficult as birds become rarer (for criteria see International Union for Conservation of Nature 2001). The last specimen collected, captured, or photographed is obvious proof of population persistence to that time, but last observations of birds are often open to doubt and depend on the experience (and luck) of the observer, the distinctiveness of the species, and the circumstances under which observations are made (Pratt and Pyle 2000). Nevertheless, last observations of species are the only way to approximate when populations have declined beyond hope of recovery, even though they may persist long after the last recorded observation (Banko 1968, Fitzpatrick et al. 2005). Population viability may decline to the point of no return, if not to actual extinction, from years to decades after the last observation (Grant et al. 2005). Historically, many Hawaiian forest birds have been relatively inaccessible, and relict populations may have persisted for a long time without notice. Given these considerations, we lump the last reported observations of species by 50-year periods to coarsely examine temporal patterns of population decline.

For more than a century after Western contact, the scientific awareness of Hawaiian birds was so fragmentary and incomplete that many extinctions could have been overlooked (Medway 1981, Olson and James 1994, Pimm et al. 1995). Some species collected in the early historic period are known from just a few specimens, suggesting that others may have escaped detection by naturalists. Because only two populations are reported to have vanished before 1850, it may be tempting, but is probably incorrect, to believe that the major die-off of species during ancient Hawaiian times had slowed to a trickle (Pimm et al. 1995). During 1887–1902, many bird populations were quite rare and restricted in range, suggesting that they may have been declining for some time, possibly since before Western contact. In fact, 27 passerine populations apparently vanished or declined severely before 1901. Although the decline subsequently seemed to have slowed somewhat (22 populations lost during 1901–1950 and 17 more during 1951–2000), 8 populations have recently disappeared or may soon (Fig. 2.5). Nevertheless, the most vulnerable populations are already gone, and the remaining populations may persist for a long time if protected.

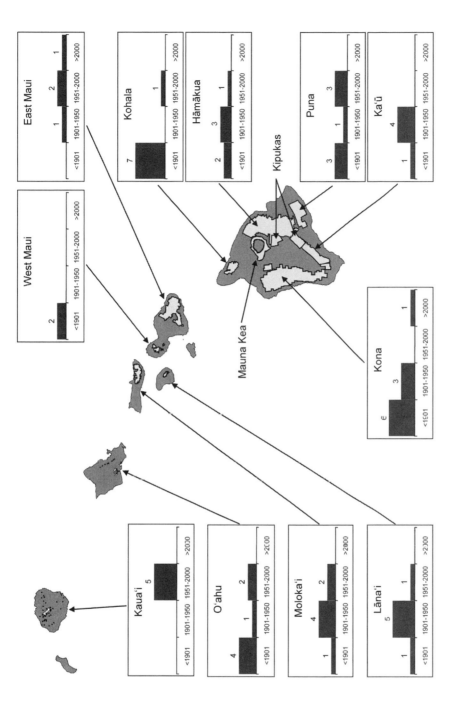

Figure 2.5. Attrition of passerine populations by district during the historical period. Populations (shown by bars in each graph) presumably vanished during the 50-year period in which the last observation was reported (Banko 1980–1987; also see Table 2.1). Survey districts as in Fig. 2.1.

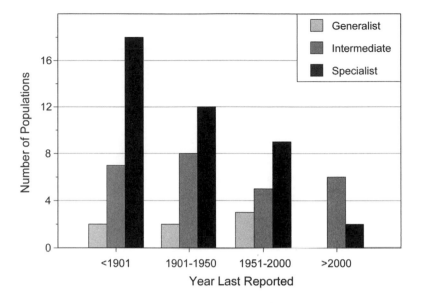

Figure 2.6. Attrition of specialist, intermediate, and generalist passerine populations during the historical period. Data were derived from Table 2.1. Most populations disappearing before 1901 were of specialist species, and their rate of attrition has declined as the most vulnerable populations have vanished. Attrition of intermediate and generalist populations has been relatively constant, but a larger proportion of extinctions expected before 2050 will involve intermediate and generalist species as the number of specialist populations dwindles. Two species have disappeared so far in the 2000–2050 period: the 'Alalā and the Po'o-uli, both intermediate species. Extinctions of certain populations of two intermediate species (Hawai'i Creeper and 'I'iwi) and two specialist species ('Akiapōlā'au and 'Ākepa) are anticipated soon.

All survey districts lost passerine populations historically, but 27 populations were last reported before 1901 in Kohala (7), Kona (6), O'ahu (4), Puna (3), Hāmākua (2), West Maui (2), Ka'ū (1), Lāna'i (1), and Moloka'i (1) (see Fig. 2.5). Between 1901 and 1950, 22 populations were reported for the last time on Lāna'i (5), Moloka'i (4), Ka'ū (4), Kona (3), Hāmākua (3), Puna (1), East Maui (1), and O'ahu (1). Only on Kaua'i did all historically known populations persist past 1950, but 5 populations there have recently disappeared. Extinctions of 7 populations of five species are impending or may have oc-

curred in five districts (O'ahu, Moloka'i, West Maui, East Maui, and Kona); however, more districts may soon lose their small populations of 'I'iwi and other intermediate species. The early losses of populations in Kona are noteworthy because of the district's large size, narrow distribution of wet forest, and broad distribution of dry and mesic forest. The leeward position of Kona suggests the continuation into historical times of processes that were responsible for numerous prehistoric extinctions in dry, lowland habitats.

Specialists suffered the most extinctions overall, and it is not surprising that they account for two-thirds of all extinctions by 1901. Although more than half of the extinctions recorded during the twentieth century were of specialists, their relative rate of attrition slowed because most were already extinct (Fig. 2.6). On the other hand, the attrition of intermediate and generalist populations has been relatively steady. Now that most districts have lost many of their specialists, however, proportionately more extinctions will likely involve intermediate and generalist populations. Two intermediate species, 'Alalā and Po'o-uli, disappeared from the wild after 2000 (Chapter 5), and other populations

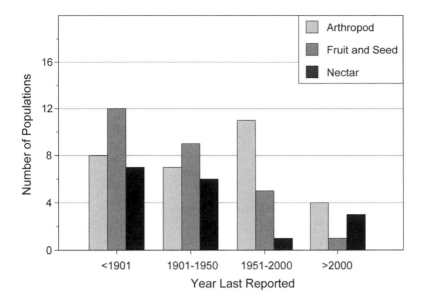

Figure 2.7. Attrition of passerine populations according to foraging guild. Data were derived from Table 2.1. Most fruit and seed eaters and nectar eaters disappeared before 1901 or 1950, whereas arthropod eaters have disappeared somewhat more evenly throughout the historical period. The only extinctions so far in the 2000–2050 period have been the fruit-eating 'Alalā and an arthropod eater, the Poʻo-uli. However, extinctions of three populations of arthropod eaters (the 'Akiapōlā'au, Hawai'i Creeper, and 'Ākepa) and three populations of a nectar-eater, the 'I'iwi, are pending.

of intermediates (Hawai'i Creeper and 'I'iwi) are likely to disappear before 2050. Concern also has been raised for the 'Akikiki and 'Akeke'e, two Kaua'i endemics (Chapter 5). Impending extinctions of specialist populations involve 'Akiapōlā'au and 'Ākepa (Chapter 5).

Rates of attrition have also varied among foraging guilds (Fig. 2.7). Extinction of fruit and seed eaters and nectar eaters occurred primarily before 1901 or 1950, whereas arthropod eaters disappeared somewhat more evenly throughout the historical period. The extinctions so far since 2000 have been of the fruit-eating 'Alalā (now surviving only in captivity) and apparently the arthropod- and snail-eating Poʻo-uli. However, local population extinctions of four arthropod eaters

('Akiapōlā'au, Hawai'i Creeper, 'Ākepa, and possibly 'Akikiki) and of a nectar eater ('I'iwi) are impending. There are no indications that populations of the two birds of prey are in jeopardy (Chapter 12).

MECHANISMS AND AGENTS OF POPULATION DECLINE

Small ranges and numbers of individuals predispose populations to extinction, so bird communities on small islands are highly vulnerable to anthropogenic changes in ecosystems (Manne et al. 1999, Hughes 2004, Sodhi et al. 2004). Even on relatively recently settled islands, such as the Galápagos, rare populations can disappear with little warning (Grant et al. 2005). On remote islands, bird species that evolved with limited numbers and kinds of resources, predators, diseases, and competitors can easily fall victim to the many invasive species that eventually arrive with people (Simberloff 1995a; this volume, Chapter 1).

Although the impacts of causal factors are sometimes difficult to isolate (Grant 1995), vertebrate predators stand out as early and universally recognized exterminators of island birds (Blackburn et al. 2004). Rats, cats, and other mammals have been implicated in the destruction of many bird populations on islands everywhere, but their impacts in New Zealand have perhaps been documented most thoroughly (Atkinson 1989, Lovegrove 1996a, Holdaway 1999). Island birds can also be threatened by nonmammalian vertebrates, such as the brown treesnake (*Boiga irregularis*), which has wiped out nearly all native land birds on Guam (Savidge 1987). Habitat destruction and degradation are also generally accepted as major threats to island bird populations, because the level of endemism of the avifauna is high, habitat availability is dramatically limited, and habitat quality can be affected by many invasive species (Temple 1986, Johnson and Stattersfield 1990, Pimm and Askins 1995, BirdLife International 2000, Brooks et al. 2002). A striking feature of the decline of Hawaiian birds, however, is the importance not only of introduced predators and habitat loss beginning soon after human colonization but also of disease and food web disruption during historical times (Chapters 7 and 9). Although disease, competition, or food availability have not been explicitly linked to the decline of whole communities of birds on other islands, their impacts probably are felt by some insular species (Diamond 1984, BirdLife International 2000).

The roles of disease and predation in the decline of Hawaiian birds have been considered in detail, but these factors remain uncontrolled (Chapters 17 and 18). Habitat destruction is often acknowledged in the decline of native birds, and the impacts of vertebrate browsers and rooters are well understood; however, additional studies would help demonstrate the roles of nonvertebrate agents, interactions between agents, and the mechanisms by which habitat changes have affected bird populations (Chapters 6 and 7). Despite the evolution of many extraordinary feeding specialists, disruptions to food webs are usually dis-

missed as not having played a significant role in the historical decline of Hawaiian and other island birds for want of evidence or because other factors seemed more obvious, more powerful, or better correlated with the perceived chronology and geography of bird extinctions (Warner 1968, Berger 1972, Atkinson 1977, Ralph and van Riper 1985, Pratt 1994, van Riper and Scott 2001). However, a more comprehensive ecological context for understanding population decline is emerging as the primary agents, mechanisms, and ecological consequences of food web disruption are investigated (Banko and Banko 1976; this volume, Chapter 7). In the following paragraphs we summarize the four major factors limiting Hawaiian passerine populations as they relate to historical patterns of decline (Table 2.4).

Predation

Originally threatened only by native raptors, Hawaiian forest birds became prey for small mammals when Polynesian rats (*Rattus exulans*) and later feral cats (*Felis catus*), Norway rats (*R. norvegicus*), black rats (*R. rattus*), and small Indian mongooses (*Herpestes auropunctatus*) arrived from outside the archipelago (Atkinson 1977; this volume, Chapter 11). The locations and substrates used by birds for nesting, roosting, and foraging may expose some bird species to higher risks of predation (Thibault et al. 2002). After humans and introduced mammals arrived in Hawai'i, nearly all flightless species disappeared (Olson and James 1991), and the remaining forest bird species nested as high in trees as possible (Chapter 8). Specialists may be particularly vulnerable to predators, because their incubation and chick-rearing periods tend to be longer (Chapter 7). Additionally, specialists tend to have less capacity for renesting after predator attacks, and the mortality of adults may have graver demographic consequences (Chapter 8). For example, 'Elepaio, a generalist species, can

Table 2.4. Agents and activities that threaten or potentially threaten Hawaiian forest birds

Type of Threat	Agent or Activity	Period of Activity or Impact	Geographical Scope of Impacts	Impacts on Forest Birds and Their Habitats
Agriculture, settlement, commerce	Polynesian agriculture and settlement	Ancient era–1900	All islands and regions; coastal–500 m elevation	Degradation, fragmentation, and destruction of lowland forests.
	Bird harvest for food and feathers	Ancient era–1900	All islands, regions, elevations (seabird harvest); various localities (feather harvest)	Overharvesting and extinction of edible species, especially flightless birds. Disruption of sea-to-land nutrient cycle through loss of guano-producing seabirds. Depletion of certain feather-producing species, a process that increased with introduction of firearms.
	'Iliahi or Sandalwood (Santalum spp.) harvest	1800–1830	All islands; dry–mesic regions; low to middle elevations	Degradation and destruction (mostly by burning) of low to midelevation forests containing sandalwood.
	Logging (mostly koa, Acacia koa)	1830–present	All islands; all regions; all elevations	Degradation, fragmentation, and destruction of forests.
	Tree fern (Cibotium spp.) harvest	1830–1880	Regional impacts mainly on Hawai'i (Puna, Ka'ū, Kohala); lowland wet forests	Reduction of understory cover, causing local alteration of vegetation when tree ferns were toppled for fine fiber (pulu).
	Firewood, historically	1820–1900	All islands; primarily lowlands near ports and settlements	Degradation, fragmentation, and destruction of lowland forests; ōhi'a primarily harvested.
	Sugar and other Western agriculture and settlement	1870–present	All islands and regions; low elevations	Most Western agriculture and settlement overlaid areas already cleared or degraded by Hawaiians, but new areas of habitat also cleared and fragmented. Introduction of biocontrol agents.
	Ranching	1830–present	All islands, regions, elevations	Habitat degradation through removal of forest understory, spread of grasses and weeds, reduced tree regeneration, disruption of invertebrate communities, erosion; fragmentation and destruction of forest cover; dry and mesic forests especially vulnerable.
	Weeds	1840–present	All islands, regions, elevations	Habitat degradation and replacement through competition with native plants, changes in the structure and composition of native vegetation, increased fire threat, and changes in arthropod communities; some weeds provide additional resources for native birds.

continued

Table 2.4. Continued

Type of Threat	Agent or Activity	Period of Activity or Impact	Geographical Scope of Impacts	Impacts on Forest Birds and Their Habitats
Agriculture, settlement, commerce (continued)	Fire	Ancient settlement to present	All islands; mainly dry–mesic regions; all elevations	Habitat degradation and loss; recent increases in fire frequency and intensity following disturbance of vegetation by feral ungulates and consequent invasion by alien grasses.
	Plant pests and pathogens	1900–present	All islands, regions, elevations, but impacts are relatively local at present	Reduction of vigor, reproduction, and survival of various native plant species at various spatial scales but usually locally; potentially devastating if new pests or pathogens were to arrive.
Ungulates	Feral cattle (Bos taurus)	1800–present	All islands, most regions and elevations; impacts declining to local levels since 1950	Degradation, fragmentation, and destruction of native vegetation by browsing except where protected by rough lava; severe impacts on nearly all species of native trees and other plants.
	Feral sheep (Ovis aries), mouflon sheep (O. gmelini musimon), and hybrids	Feral sheep: 1800–present; mouflon: 1960–present	Hawai'i Island and Lāna'i (mouflon only); dry and mesic regions	Habitat degradation, fragmentation, and destruction on all substrates; severe impacts on many native plant species.
	Feral goats (Capra hircus)	1800–present	All islands (extirpated on Lāna'i); all elevations; impacts declining to local levels since 1970	Habitat degradation, fragmentation, and destruction on all substrates; severe impacts on nearly all native plant species.
	Axis deer (Axis axis), Black-tailed Deer (Odocoileus hemionus)	Axis: since 1870; Black-tailed: since 1960	Axis: O'ahu (extirpated), Moloka'i, Lāna'i, and Maui; Black-tailed: Kaua'i; both species usually in dry and mesic regions, all elevations	Habitat degradation, fragmentation, and destruction on all substrates; severe impacts on many native plant species.
	Feral pig (Sus scrofa, European domestic)	1800–present	All islands (extirpated on Lāna'i); range expanded in some regions since 1950; all habitats and elevations	Degradation of native vegetation by rooting and eating of understory plants, facilitating spread of weeds and creating breeding sites for mosquitoes that transmit malaria and pox to birds; Polynesian variety of pig not widespread or greatly destructive in native forests.

Category	Species	Date	Distribution	Impact
Competitors	Ants (numerous species)	1820–present	All islands; most dry–mesic habitats; lowland wet habitats	Disruption of native arthropod communities; reduction of native arthropod availability; potential reduction of nectar availability.
	Predaceous wasps (e.g., *Vespula pensylvanica*)	1980–present (variable or different islands)	All islands; mesic and dry habitats, local; *Vespula* mostly at high and middle elevations	Disruption of native arthropod communities through predation on a wide variety of prey species, especially caterpillars; disruption of predator-prey and plant-herbivore interactions; reduced availability of caterpillar prey.
	Parasitoid wasps and flies (several families)	1890–present	All islands, regions, elevations	Disruption of native arthropod communities through mortality on a wide variety of host species; disruption of predator-prey and plant-herbivore interactions; reduced availability of caterpillar prey.
	Honey bees (*Apis mellifera*)	1890–present	All islands, regions, elevations	Alteration of pollination ecology by pollination of some native plants no longer visited by native insect or bird pollinators; reduction of nectar for birds.
	Birds (various species)	1890–present	All islands, regions, elevations	Competition with native birds for various foods; reduced availability of arthropod prey, nectar, fruits; depredation of nest contents and adults.
Predators	Polynesian rats (*Rattus exulans*), black rats (*R. rattus*)	Polynesian rat: ancient era–present; black rat: since 1880	All islands, regions, elevations	Predation on eggs, nestlings, adults; degradation of native vegetation through herbivory on a wide variety of plants; consumption of flowers, fruits, seeds, seedlings, bark; changes in arthropod and snail communities through herbivory on host plants and through predation; increases in cat densities.
	Feral cats (*Felis catus*)	1850–present	All islands, all native bird habitats	Predation on nestlings and adult birds. Although introduced rodents are prey, cats probably do not appreciably reduce rodent impacts.
Disease and vectors	Mosquitoes	1830–present	All islands, <1,500 m in wet forest, <1,300 m in dry forest	Infection of birds of all species and ages with pox or malaria.
	Avian pox virus	1850–present	All islands, primarily with mosquito distribution	Lesions may debilitate birds; secondary infection may lead to sickness and death, particularly when malaria is also involved; all bird species affected to some degree.
	Avian malaria	1920–present	All islands, with mosquito distribution	All birds variously suffer sickness and mortality.

nest up to eight times in a season as nests are depredated or otherwise fail (Sarr et al. 1998, VanderWerf 1998a). In contrast, the specialist Palila may renest only once or twice after their first nest failure of the season (Banko, Johnson, et al. 2002). Female Palila are not usually killed at the nest, but females of other species may be at higher risk (Chapter 23).

The rates of bird mortality and nest destruction are likely to be higher in habitats that support large numbers of predators, even if other resources are the main attraction for predators (Chapter 11). Strong associations between birds and habitats that were also favored by predators may have exposed populations of 'Ula-'ai-hāwane and 'Ō'ū to unusually high population levels of rats for long periods. 'Ula-'ai-hāwane apparently preferred to forage in native hāwane or loulu palms (Pritchardia spp.), and 'Ō'ū were strongly associated with 'ie'ie (Freycinetia arborea; Perkins 1903). Polynesian rats may have contributed to the extirpation of loulu stands that once dominated portions of lowland O'ahu and presumably other islands, and black rats were strongly attracted to 'ie'ie fruit (Perkins 1903, Athens et al. 1992). Forests with abundant trees and shrubs in the understory probably support large populations of rats (Chapter 11), and the level of predation on birds may be high in such resource-rich habitats. Feral cats are abundant in relatively dry habitats, where they threaten Palila and other native birds (Banko, Johnson, et al. 2002; this volume, Chapter 23). Elsewhere, birds nesting near the edge of habitat patches sometimes suffer predation more often (Chalfoun et al. 2002), and that may be the case in the Hawaiian Islands as well.

Hawaiians captured forest birds for ornamental feather work, and though they may not have eaten their most prized quarry frequently, they routinely ate less valuable specimens in addition to other species that were incidentally caught (Emerson 1895; Athens et al. 1991; this volume, Chapter 3). After firearms became avail-

able, large numbers of Hawai'i 'Ō'ō were shot for their feathers, possibly hastening the decline of local populations (Henshaw 1902). However, there is no indication that other forest bird species were shot frequently early in the historical period.

Disease

Avian pox virus (Avipoxvirus spp.) and avian malaria (a protozoan parasite, Plasmodium relictum) are widely suspected to have had major impacts on historical populations of forest birds (Chapter 9). A deadly strain of pox not related to fowl pox arrived in Hawai'i at least 100 years ago (Jarvi, Triglia, et al. 2008), and suggestive lesions historically observed (mostly during 1887–1902) on native forest birds have been implicated in the decline of populations in lowland habitats (van Riper et al. 2002). The blood parasite that causes avian malaria is thought to have arrived in the islands around 1920 (but see Pratt [2005] for an argument for an earlier arrival), and its transmission (and to a lesser extent that of pox) depends entirely on birds being bitten by infected mosquitoes (van Riper et al. 1986). Research into the impacts of malaria on bird populations has taken a range of approaches, including identifying pathogens and mosquito vectors, evaluating the pathogenicity of malaria to various bird species in challenge trials or by observing infected wild birds held in captivity, investigating the ecology of the mosquito vector, assessing geographical and phenological patterns of disease prevalence across a range of native and introduced bird species, and associating the prevalence of mosquito vectors with low densities of native birds (Warner 1968; Scott et al. 1986; van Riper et al. 1986, 2002; Vander-Werf et al. 2001; this volume, Chapter 9). Interactions between bird hosts, mosquito vectors, and disease organisms have recently been explored in detail. The results reveal a widespread susceptibility to malaria in native songbirds, with resistance possibly evolving in only a few lowland populations (Woodworth et al. 2005; this volume, Chapters 9 and 14).

How do the results of disease research fit with contrasting patterns of historical decline of generalist and specialist populations? Could the greater loss of specialists, which occurred as smaller populations, have been due to reduced survival, growth, maintenance, and reproduction resulting from a more precarious nutritional and energetic status (Chapter 7)? Challenge experiments indicate that Hawaiian forest bird species may differ innately in their susceptibility to malaria, although less can be said about variation in susceptibility to pox (van Riper et al. 2002; this volume, Chapter 9). Specialists, which are endangered, cannot be systematically compared with generalists, but their rates of mortality from malaria would be expected to match or exceed those of generalists and intermediate species, whose mortality rates may approach 100% (van Riper et al. 1982: 11; this volume, Chapter 9). Of the native species challenged by malaria in laboratory trials, intermediate feeders ('I'iwi and Maui 'Alauahio) suffered more serious effects and a higher rate of mortality than generalists ('Ōma'o, Hawai'i 'Amakihi, and 'Apapane) (Atkinson, Lease, et al. 2001; this volume, Chapter 9). Whether this difference could have resulted from variation in the nutritional states of different species, especially during the early developmental period, is not known. Likewise, pathways and mechanisms by which pathogens may be excluded (resistance) or accommodated (resilience) in birds under natural conditions in Hawaiian forests are poorly understood (but for genetic perspectives see Chapter 10).

Nutritional and energetic factors may work synergistically with diseases to reduce survivability (Klasing 1998b), especially during the embryonic and early nestling periods. Differences in nutrition during the breeding season, therefore, may be pivotal to understanding variation in the survival and productivity of generalists and specialists where the prevalence of diseases is high. For example, birds may reproduce at lower capacity because of the stress of fighting infection (Bonneaud et al. 2003), and nestlings may fare poorly when they receive fewer maternal antibodies or are poorly provisioned by parents that are physically compromised by disease or food inavailability (Boag 1987, Hoi-Leitner et al. 2001). Well-nourished birds, therefore, likely survive and reproduce at higher rates than those less well nourished when challenged by disease (Klasing 1998b). Although young 'Apapane become lethargic and feed at reduced rates when experimentally infected with malaria (Yorinks and Atkinson 2000), some chronically infected populations of generalists persist (Vander-Werf, Burt, et al. 2006) and reproduce successfully (Kilpatrick et al. 2006), perhaps in part because they can forage for a wider range of foods. The rate of mortality is highest among young birds during malaria and pox epidemics (C. Atkinson, unpubl. data), suggesting that at least some adults had survived previous challenges and that susceptibility and foraging ability may be related.

Interactions between feeding ecology and disease susceptibility may help explain why some generalists and some intermediate species continue to occupy or are recolonizing mosquito-infested lowland habitats. Resistance or tolerance to malaria may be evolving in generalists, such as Hawai'i 'Amakihi, partly because a broader diet and greater capacity to provision offspring help to sustain their populations in lowland habitats (Woodworth et al. 2005; this volume, Chapter 7). Moreover, recent surveys indicate that important bird prey, such as caterpillars, are less abundant in lowland forests (Peck et al. 2008), suggesting that only some generalist species are able to occupy these habitats.

Habitat Destruction and Degradation

Forest loss and degradation lead inexorably to extinction as preferred habitats, foraging substrates, and foods of birds become increasingly scarce, and though species that naturally tend to be area-sensitive, isolated, and rare are most affected, nearly all birds are impacted by massive deforestation (Davies et al. 2000, Zanette et al. 2000, Sodhi et al. 2004). Rats, fire, and agriculture were the forces most likely to have changed forest size, structure, and composition during pre-Western times (Cuddihy and Stone 1990; this volume, Chapter 6). Since then, introduced cattle, goats, sheep, pigs, and deer have been the major destructive force in Hawaiian forests (see Table 2.4). Where ungulates (hoofed mammals) have not yet destroyed forests entirely, they have removed understory vegetation, thereby facilitating the spread of landscape-dominating and fire-promoting weeds and changing habitat conditions for invertebrate communities (Chapter 6). Also, human settlement, logging, burning, road building, and a variety of agricultural and silvicultural activities have fragmented and marginalized bird habitats.

Bird populations in low elevations or relatively dry habitats presumably declined rapidly as native vegetation was altered by ancient Hawaiians and later by Westerners. For example, the Po'o-uli disappeared long ago from the dry, low- and mid-elevation forests of leeward Maui and historically became confined to cloud forest high on the windward slope of Maui at the margin of its range (Pratt, Kepler, et al. 1997). Birds in areas with little suitable vegetation above 1,000 m elevation (e.g., O'ahu, Moloka'i, Lāna'i, West Maui, Puna, and Kohala) were probably the ones most affected by lowland forest reduction and fragmentation. Specialists seem especially sensitive to habitat degradation, whereas generalists respond directly to the overall availability of habitat. The longer persistence of specialist birds on Kaua'i, for example, may be partly due to fewer historical extinctions of plant species and relatively low levels of overall disturbance to plant communities (Sakai et al. 2002).

Forest fragmentation and reduction pose major problems for species that follow seasonally available resources along environmental gradients such as elevation and rainfall (Chapter 7). Of the species known for resource tracking, generalist species ('Apapane) seem least affected by forest fragmentation and loss, whereas intermediate and specialized species ('Alalā, 'Ō'ū, Palila, and 'I'iwi) have not fared as well. Sedentary species also suffer when habitats are fragmented, because they may not readily cross gaps and are forced to remain in smaller and often degraded patches. The destruction of small trees, shrubs, and herbaceous plants by ungulates and rats particularly threatens species that forage in the understory. Those most affected would have been the 'Alalā, 'ō'ō, thrushes, 'Ō'ū, 'I'iwi, Hawai'i Mamo, Black Mamo, and Po'o-uli (Chapter 7). Potentially, habitat fragmentation provides avenues for the dispersal of predators and other invasive species. Although Chalfoun et al. (2002) provide mixed support for increased rates of nest predation near habitat edges, mosquitoes likely penetrate deeper into bird ranges (Chapter 17), and weeds spread readily in forest openings (Chapter 6). Additionally, birds near habitat edges may respond in various ways to changes in plant composition, structure, and demography resulting from variation in microclimate and the distribution of light, moisture, nutrients, and organisms (Ries et al. 2004).

Evidence that forest destruction and degradation harmed native bird communities is most compelling for O'ahu, where the earliest historic bird extinctions were documented, suggesting that the removal and thinning of forests, mainly by cattle, was one of the early problems on that island (Perkins 1903, 1913).

Food Web Disruption and Food Depletion

Food webs have been significantly disrupted by alien parasitic wasps and flies (known as parasitoids because they eventually kill their hosts) and by predatory ants and wasps (Perkins 1913; Zimmerman 1958a, 1958b; this volume, Chapter 7).

Significantly, predatory and parasitic Hymenoptera (the insect order that includes ants and wasps) attack caterpillars, which are important to nestlings of all species and to many adults (Perkins 1903). Fewer caterpillars available to feed nestlings would have impacted specialist birds more than generalists (Chapter 7).

Ants have not been observed attacking Hawaiian forest birds, but early in the historic period some, including the big-headed ant (*Pheidole megacephala*), invaded native arthropod communities in low- and midelevation habitats, and later other species, especially the Argentine ant (*Linepithema humile*), spread to dry and mesic habitats at high elevations (Cole et al. 1992; this volume, Chapter 7). Predacious wasps invaded native forests rather late historically, and the Western yellowjacket (*Vespula pensylvanica*), which poses the greatest threat to bird foods, became locally abundant at middle and high elevations (Howarth 1985, Gambino et al. 1987). Parasitoids capable of attacking a wide range of hosts were particularly devastating to many caterpillar populations, and they invaded forests soon after being released to control agricultural pests or following their accidental arrival in the islands even before 1900 (Perkins 1913; Zimmerman 1948, 1958a, 1958b; Bianchi 1959). Parasitoid impacts probably were initially greater near lowland agricultural areas, but eventually environmentally diverse and remote areas were invaded (Henneman and Memmott 2001, Brenner et al. 2002, Oboyski et al. 2004, Peck et al. 2008). The host preferences of native parasitoid wasps are not well known, but alien species far outnumber them and are more often reared from native caterpillars (Zimmerman 1958b, Henneman and Memmott 2001, Nishida 2002, Peck et al. 2008).

Imported birds with generalist foraging habits potentially compete with Hawaiian birds for important arthropod prey when they invade native forests. The Common Myna (*Acridotheres tristis*), which was imported to Hawai'i to control outbreaks of defoliating caterpillars, irrupted briefly but spectacularly in native forests shortly before 1900 but then retreated to disturbed habitats (Chapter 13). Populations of other alien birds, such as the Japanese White-eye (*Zosterops japonicus*), have potentially competed with native birds longer and with greater impact (Chapter 13). Based on correlations of population distributions, Mountainspring and Scott (1985) concluded that competition for food by alien birds probably contributed to the decline of native birds. However, alien birds seem less invasive and less competitive in relatively intact habitats that are dominated by native birds (Diamond and Veitch 1981, Moulton and Pimm 1983, Scott et al. 1986).

Other introduced vertebrate animals causing major disruption to forest bird food webs are introduced rats. Rats consume much of the fruit and seed crops of plants that are important to native birds, and rats eat many species of invertebrate prey (Cuddihy and Stone 1990; this volume, Chapters 7 and 11).

Interactions among Agents of Decline and Cumulative Impacts on Bird Populations

Predation, disease, habitat degradation, and competition alone could have reduced bird populations given their impacts elsewhere (Diamond 1984, Savidge 1987, Mack et al. 2000), but together they were devastating in Hawai'i. Agriculture and settlement destroy habitat outright, resulting in unambiguous impacts on bird populations, but some human activities degrade habitats and resources in less obvious ways. Many invasive species enter the islands through commercial transportation and shipping (Chapter 15), and early biocontrol agents introduced to protect crops from pests attack important prey of birds. Cattle and other ungulates change the microhabitats of plants and arthropods that are important to birds, and they eventually destroy forests by suppressing tree regeneration (Chapter 6). Some native plant species are especially sensitive to browsing or rooting, resulting in changes to the structure and composition of the vegetation and the reduction of nectar-producing flowers, fruits, and seeds that are major resources for native birds (Chapter 7). Similarly, native invertebrate prey that depend on particular host plants and microhabitats can also be affected by changes in vegetation (Chapter 7). Following disturbance to soil and vegetation by ungulates, many weeds and pests invade bird habitats. When alien grasses invade, for example, fire threats increase and the regeneration of many native shrubs and small trees declines (Chapter 6). By excavating the starchy core of fallen tree ferns, feral pigs not only transform native vegetation but create breeding sites for introduced mosquitoes that transmit pox and malaria to birds (Chapters 9 and 17).

In more direct ways, a daunting variety of invasive species prey upon native birds, cause disease, and compete for critical resources. Rats both prey on birds and change the structure and composition of forests by removing some native plant species and reducing the availability of some arthropod prey, either directly through predation or indirectly through vegetation disturbance (Chapters 7 and 11). In turn, as forest bird populations decline, their interactions with plant host species and arthropod and snail prey change, potentially affecting pollination, seed dispersal and predation, and predator-prey relationships, all of which accelerate habitat degradation (Carlquist 1974, Diamond 1984, Meehan et al. 2002, Price and Wagner 2004). The overall result of these interactions has been to make birds even more vulnerable to new threats by reducing their populations and genetic diversity (Chapter 10) and thereby diminishing their capacity to adapt to changing conditions.

Early on, some Hawaiian forest birds may have succumbed to relatively few factors, but modern populations are increasingly affected by a complex suite of threats that encompass nearly all the major factors stressing forest birds globally (Owens and Bennett 2000). Indeed, the Hawaiian Islands are a microcosm of negative human impacts on island and even continental ecosystems. The synergistic effects of invasive predators, diseases and vectors, habitat modifiers, and food competitors make it difficult to evaluate the impacts of individual threats in Hawai'i (Grant 1995) and elsewhere (Thibault et al. 2002, Grant et al. 2005). Further obscuring their individual impacts on forest bird populations, the effects of various invasive species overlap and vary over time and space (Diamond 1984), but evaluating multiple factors at different geographical scales can help identify management priorities (Chapter 16). Notwithstanding studies that correlate native bird distributions and abundances with environmental factors and the prevalence of invasive species, there is little detailed or quantitative knowledge about how dynamic interactions among invasive threats impact native bird populations (Scott et al. 1986). Research aimed at understanding multiscale interactions among native birds, predators, disease organisms and vectors, habitat factors, and food resources has begun (Chapter 14).

POPULATION TRAJECTORIES AND CONSERVATION STRATEGIES

The bird species most sensitive to ecological change disappeared early during human colonization of the Pacific islands, and more recently settled island groups retained a greater proportion of their avifauna (Pimm et al. 1995). Of the ancient Hawaiian birds that disappeared during Polynesian colonization of the islands, most were flightless, ground-nesting, large-bodied, or specialized in their feeding habits (James and Olson 1991, Olson and James 1991, James 1995, Boyer 2008). Hawaii's agricultural and commercial development during the past two centuries resulted in many additional extinctions, although the small native hawk and owl flourished due to their generalized foraging habits. Most endangered Hawaiian forest bird species are specialists whose populations are declining. Preventing the extinction of specialist populations is challenging because specialists have suffered greater impacts from the many historical threats and environmental changes, and they tend not to rebound quickly, if at all, from ecological disturbances, which are themselves difficult to mitigate or repair. Because so many specialists have already disappeared or are nearly extinct, conservation efforts will increasingly focus on populations of intermediate and generalist species, as in some other tropical environments (Castelletta et al. 2000).

Efforts to conserve Hawaiian forest birds have been concentrated mostly in high-elevation areas, because most endangered species have been extirpated from lower elevations (Scott et al. 1986, Banko et al. 2001). Nevertheless, opportunities to protect and enhance lowland habitats for generalist and intermediate species should not be overlooked. To this end, contiguous expanses of forest extending from near the coast to the tree line could be managed with varying degrees of intensity and with different objectives and strategies. At low and middle elevations, initial management might focus on increasing the forest cover overall and reducing predation and disease transmission to increase generalist populations. It has been suggested that reducing the impacts of predators and habitat stressors could facilitate disease resistance in midelevation bird populations (Kilpatrick 2006). At high elevations, improving habitat quality, expanding habitat through reforestation, and reducing predation should aid the recovery of specialists. The survival of nectar eaters and fruit and seed eaters might improve as host plant abundance increases, but increasing the availability of caterpillars and other arthropod prey for nestlings would likely make a difference in the reproduction and survival of all species.

SUMMARY

Following the millennium of Polynesian-influenced extinctions, at least one population of nearly every Hawaiian forest bird species succumbed to the myriad forces unleashed by later cultures. Nevertheless, some species and populations fared worse than others. Suffering most of all were the ecological specialists, whose capacity to rebound from ecological adversity is low due to reduced reproductive potential and diminished flexibility in foraging behavior and habitat requirements. Specialists more often than generalists declined severely in numbers and range during the historical period, and only a third of specialist populations persisted into recent times. Remnant populations of specialists are seriously threatened by extinction, and most endangered forest bird species are specialists.

Survival was most difficult for specialists and generalists alike in relatively small forest districts and lowland habitats. Historical populations may also have suffered greater losses in drier, leeward areas, where the fossil record indicates that many species disappeared prior to Western contact. Unfortunately, we know little about the distribution and abundance of Hawaiian birds before 1887. In the first century following Western contact, some species may even have disappeared before they could be discovered. An active period of ornithological exploration from 1887 to 1902 yielded the first contemporary and comprehensive description of the islands' avifauna and its distribution. Then another large gap in our knowledge occurred during the first half of the twentieth century, obscuring the timing and geographical patterns of bird declines. Any perceived slowing of extinction may be largely the result of the earlier disappearance of the most vulnerable populations, which left mainly the smaller and more generalized and intermediate species to represent the once dazzling and diverse avifauna.

Although the native hawk and owl benefited from much of the disturbance to native ecosystems historically, the many factors that have battered Hawaiian passerine populations are individually powerful and difficult to mitigate, and their interactions and combined impacts pose enormous challenges to restoration. Predation by Polynesian rats, which continued into historical times following their introduction by ancient Hawaiians, was later supplanted by predation by black rats and cats. Habitat degradation, which began with forest clearing and change by the Polynesian rat and the ancient Hawaiians themselves, was also greatly intensified and expanded by overwhelming populations of the feral ungulates that accompanied Westerners. Avian pox was first noticed in Hawaiian forest birds during the 1880s. Although mosquitoes arrived early in the historical period, malaria may not have afflicted native passerines until about 1920. Feral pigs facilitated the spread and persistence of mosquitoes by creating breeding sites as a consequence of their feeding habits. Direct threats to caterpillars, key foods in the diets of all passerine nestlings, came first from predaceous ants, which arrived early in the historic period and had profound effects on native arthropod communities in lowland forests. Parasitic wasps and flies were the next major threats to caterpillars, potentially reducing their availability to all forest birds.

Recovering relict populations of specialists will be difficult, but protecting and restoring populations of generalist and many intermediate species should be much more feasible. Although some populations of generalists have declined and even disappeared historically, others tend to withstand and rebound from all but the most severe and sustained ecological disturbances. Increasing the availability of habitat, eliminating feral pigs and other habitat stressors, reducing populations of disease vectors and predators, and even modestly mitigating other threats from invasive species should go far in restoring populations of generalists in areas outside the upland retreats of the specialist species.

ACKNOWLEDGMENTS

This research was made possible in part by support from the U.S. Geological Survey Invasive Species Program and Wildlife and Terrestrial Resources Program; the National Science Foundation (Grant DEB00-83944: Biocomplexity of Introduced Avian Disease in Hawai'i); and the U.S. Fish and Wildlife Service and National Park Service (support to W. E. Banko during 1965–1987). We thank K. Brinck for statistical assistance.

CHAPTER THREE

Hawaiian Culture and Forest Birds

VERNA L. U. AMANTE-HELWEG
AND SHEILA CONANT

Hānau ka Hualua ka makua
Puka kana keiki he Manu, lele . . .
Hānau ka 'Alae ka makua
Puka kana keiki ka 'Apapane, lele
Hānau ka 'Alalā ka makua
Puka kana keiki he Alawi, lele
Hānau ka 'E'ea ka makua
Puka kana keiki he 'Alaiaha, lele
Hānau ka Mamo ka makua
Puka kana keiki he 'O'o, lele. . . .

Born was the egg, the parent
Out came its child, a bird, and flew . . .
Born was the Mudhen, the parent
Out came its child, an 'Apapane bird, and
flew
Born was the Crow, the parent
Out came its child, an Alawi bird, and
flew
Born was the 'E'ea bird, the parent
Out came its child, an 'Alaaiaha bird, and
flew
Born was the Mamo honey-sucker, the
parent
Out came its child, an 'O'o bird, and
flew. . . .

Thus is the ancestry of Hawaiian birds depicted in the Kumulipo, the Hawaiian creation chant (Beckwith 1951: lines 295–296 and 303–310).

The kia manu felt the stirrings in the forest, warm mist rising along the rays of the rising sun. He roused his son, smiling at the boy but gesturing to him to be quiet. Working together, they skillfully applied kēpau (sticky glue) and rich red lehua blossoms to the kia (pole trap) his son had made out of 'ulei wood. They hoisted the kia up among the 'ōhi'a branches. As they sat together hidden in the 'ōhi'a foliage, the man cupped his hands and called to the birds, imitating the bell-like whistle of 'I'iwi. There was a flash of red, and an

'I'iwi joined them, filled with boundless energy. After a few moments, it perched on the kia and was ensnared by the kēpau. With quiet excitement, they lowered the kia and removed the bird with bright red feathers. More birds followed—Hawai'i 'Ō'ō, 'I'iwi, and 'Apapane. Because mankind comes last in the Kumulipo (creation chant), they took only what was needed. Later, on the walk back to the village, they thanked the akua and their 'aumakua (gods and family spirits) for their good fortune. The colors of the hulu (feathers) were especially ornate and would bring pride to their family as the kia manu presented containers filled with hulu for the ali'i (chiefs).

In this chapter we hope to provide an understanding of the extent to which forest birds were intertwined with the lives of Hawaiians in traditional culture. Forest birds played many roles; they were important for both utilitarian purposes, such as food and implements, and for ritual and metaphysical purposes. Exploring the cultural dimensions of Hawaiian society and their relationship to forest birds may enable us to comprehend the philosophical, psychological, and emotional elements of Hawaiian customary practices and lore. In addition, exploration into each of these dimensions may shed some light on values placed on native forest birds and provide an even stronger motivation to conserve these rare species. Please refer to the Hawaiian Pronunciation Guide and Glossary for Hawaiian words and all scientific names.

CULTURAL ANALYSIS FRAMEWORK

Every culture maintains that there is a purpose for the existence of living things, and individuals perceive their relationship with animals and interpret the behavior of animals in the light of their culture. In Hawaiian culture, nature and animals have direct spiritual, moral, and social signifi-

cance and are an integral part of human existence (e.g., see Beckwith 1951). Our understanding of the cultural relationships between traditional Hawaiians and forest birds may be enhanced through a social scientific framework that identifies psychological dimensions of culture (Box 3.1).

The material in this chapter was constrained by certain conditions. First, to keep the chapter consistent with the theme of this book, we have confined our discussion to forest birds. In fact, artifacts and legends also included many species other than forest birds. For example, feathers from domesticated fowl (moa), native waterfowl, and seabirds were incorporated into items such as kāhili (feathered standards symbolic of royalty; e.g., Rose et al. 1993). Second, we have attempted to confine our discussion to Hawaiian cultural traditions reported before the extensive contact with continental cultures that altered both the components and the uses of feathered objects. For example, feathers from introduced species were incorporated into various material goods produced for ceremony and everyday usage. Moreover, clothing such as 'ahu'ula (feathered cloaks and capes), formerly worn only by male warriors or ali'i, came into use by both men and women of high rank and by officials (Brigham 1899–1903). Third, instead of relying on publications in which cultural material was paraphrased or partially quoted, we sought information from original English-language sources wherever possible. Finally, because cultural views and practices began to change following contact with the outside world, it has been challenging for us to sort out the original relationships between the Hawaiian people and forest birds based on the historic record. While this record no doubt is colored by writers, generally foreign writers, witnessing a changing culture, it nevertheless brings us close to the unique cultural conditions developed in the islands.

The historical, social, and psychological elements of Hawaiian relationships with

Box 3.1.   **Worldviews: A Framework for Psychology of Culture**

Cultural worldviews are belief systems that affect the beliefs, attitudes, and behavior of each person in a given group (Kluckhohn and Stodtbeck 1961, Geertz 1973, Altman and Chemers 1980, Smart 1983, Rohner 1984, Shweder 1991). Cultural beliefs affect how individuals perceive their relationship with the natural world and interpret the behavior of animals and natural phenomena (Beidler 1979, Smart 1983, Noske 1989). In many cultures, nature and animals have a direct spiritual, moral, and social significance and are an integral part of human existence (Fraser 1922, Neihardt 1961, Ross 1964, White 1964, Ortiz 1972, Campbell 1973, Rensberger 1977, Heelas and Lock 1981).

Cultural ties to the natural world are expressed in customs, lore, and artifacts, and it is practical to interpret and classify these using social scientific dimensions that identify and describe important aspects of culture (Fraser 1922, Campbell 1973, Smart 1983, Harris 1985). In the *mythical* dimension of culture, gods or powers are assigned control over various aspects of nature and human life. This aspect of culture is closely related to spirituality or religiosity. In this dimension, deities or legendary figures are often symbolized in animal forms. In the *social* dimension of culture, interactions between individuals and groups influence codependency. Social interaction and individual actions not only ensure that the bond between the individual and the group is kept alive and strong but also ensure enculturation of the practices and customs acquired by each member of the group. In the *experiential* dimension of culture, events intensify and increase awareness of individuals' relationships with others, their natural surroundings, and wildlife. Many cultures believe that certain events or experiences between humans and nature or animals are predetermined and that the bond between them connects the welfare of one with the other. In the *doctrinal* dimension of culture, principles or laws established by a group are related in stories and myths, and they are made meaningful and relevant through the interpretive art of the skilled storyteller. In some doctrinal teachings, allegories related to animal behavior are used to teach cultural mores (social norms and ethics). In the *ritual* dimension of culture, ceremonial or social events serve to channel and heighten awareness of the importance of personal development. The ritual dimension highlights the passage from one stage of life to another. Many cultures utilize animals during ceremonial proceedings or traditional events.

forest birds are far more extensive than those covered in this chapter. Forest birds are referred to, directly and metaphorically, in some Hawaiian proverbs or folk sayings ('ōlelo no'eau), which are worthy of extended research. Archival material in Hawaiian language sources remains to be explored. Study of this material would be an excellent and valuable undertaking. Records of practices, myths, and legends vary from island to island, and in the pages allocated to this chapter we have chosen to provide examples that transcend specific islands rather than incompletely describing the diverse and rich island- and family-specific stories. The sophistication and depth of traditional Hawaiian culture are readily apparent in the literature and in practice. We encourage the reader to spend an afternoon in the reading room of a library exploring the full text of the myths and legends that we have summarized.

Box 3.2.   Feathers and Hawaiian Implements

When we think of Hawaiian feathered objects, the colors red and yellow immediately come to mind. What birds did these beautiful feathers come from? Were other colors of feathers used? We will never know the full spectrum of the bird feathers that were used in the construction of Hawaiian feather artifacts because so few of these objects were collected and brought home by the first Europeans to arrive in Hawai'i. In her catalog of precontact material culture objects from the Pacific, Kaeppler (1978) described 87 feather objects that were collected during Cook's three voyages to Hawai'i. These objects included 30 'ahu'ula (cloaks and capes), 16 mahiole (helmets), 7 kāhili (feathered standards), 19 head ornaments and lei hulu (lei of feathers), 8 feathered images, and 1 feathered temple. These objects represent for us the nature of Hawaiian feather artifacts before European contact influenced their construction. The awe and admiration of feathered objects expressed by the Europeans prompted Hawaiians to immediately and dramatically change their construction of these objects, particularly the 'ahu'ula. After European contact, few capes were constructed, but cloaks were much more numerous, even larger in size, and featured almost exclusively the feathers of forest birds, whereas previously feathers from chickens, 'Iwa, and Koa'e'ula had been used.

The red feathers in the vast majority of both pre- and postcontact feather artifacts were from 'I'iwi. A small number of the crimson red feathers of 'Apapane appear in one of the images and, still attached to the birds' skin (see Fig. 67 in Kaeppler 1978), on one cape collected by Cook (Conant, unpubl. notes). Why Hawaiians preferred 'I'iwi feathers to those of 'Apapane is not known. It may

have been because 'Apapane were more common and a less brilliant shade of red, and thus less valuable, or because of the symbolic importance of these different birds in Hawaiian culture. Strangely, the striking scarlet feathers of adult male Kākāwahie and the yellow or orange-red feathers of adult male 'Ākepa have not been observed in any pre- or postcontact objects examined by Conant (unpubl. notes).

Five species of honeyeaters were collected and named by Western naturalists. Of these, four—'Ō'ō'ā'ā, O'ahu 'Ō'ō, Bishop's 'Ō'ō, and Hawai'i 'Ō'ō—could have provided both yellow and black feathers. Examination of 75 kāhili in the Bishop Museum (Rose et al. 1993) and of most of the feathered objects described by Kaeppler (1978) reveal that both the black and the black and white (pīlali) tail feathers and wing feathers of the Hawai'i 'Ō'ō were used in the construction of kāhili (Rose et al. 1993, Conant unpubl. notes). The axillary plumes ('ē'ē) of Hawai'i 'Ō'ō were the most commonly used yellow feathers in all of the objects examined by Conant (unpubl. notes), including both pre- and postcontact objects.

A Hawaiian honeycreeper, the mamo (Hawai'i Mamo is the ornithological name), had yellow upper tail coverts (ko'omamo) and undertail coverts. Mamo feathers were used in many objects. For example, Conant (2005: 283) estimated that it would have taken feathers from 80,000 mamo to manufacture the famous Kamehameha cloak (Plate 21). In the words of Malo (1951: 77), "An ahu-ula made only of mamo feathers was called an alaneo and was reserved exclusively for the king of a whole island, ali'i ai moku; it was his kapa wai-kaua or battle cloak." Some precontact 'ahu'ula had

---

Box 3.2.   *Continued*

both 'ē'ē (Hawai'i 'Ō'ō feathers) and pue (mamo feathers); however, the feathers were grouped separately. For example, Fig. 66 in Kaeppler (1978) shows a small 'ahu 'ula with both mamo and Hawai'i 'Ō'ō feathers. The 'ē'ē appear in diamond-shaped patches at the top and sides of the cape, while pue form both a lining on the outer edge of the cape and a border between the red and yellow on the top and sides of the cape and on its back, which has both Koa'e'ula and 'Iwa feathers (Conant, unpubl. notes).

Although green feathers appear in lei hulu and one 'ahu'ula that was lost during World War II (Kaeppler 1970), it is difficult to know which of the many Hawaiian honeycreepers the green feath-ers may have come from. Malo (1951) mentions 'amakihi and 'Ō'ū, but there were numerous other honeycreepers with green feathers. Using DNA from the feath-ers in artifacts to identify bird species could answer the question of which green birds' feathers were used.

There has been much speculation about how the collection of feathers by Hawai-ians may have contributed to the decline and extinction of Hawaiian forest birds. We will never be able to quantify this loss. Rose et al. (1993) and Conant (2005) discuss this issue in detail and assert that the postcontact introduction of alien diseases and the use of firearms to col-lect birds were likely to have had a much greater effect than precontact collection.

FOREST BIRDS IN HAWAIIAN MATERIAL CULTURE

The biological and ethnographic data col-lected on Hawaiian forest birds provide information about the roles of forest birds in Hawaiian customs and practice (Box 3.2) (Davis 1978, Kaeppler 1978, Sinoto 1978, Olson and James 1982b, Clark and Kirch 1983, Kirch 1985, James et al. 1987, James and Olson 1991, Olsen and James 1991, Nakamura 1999). Artifacts that were collected during the first foreign explo-ration of the islands by Captain James Cook confirm that forest birds were prominently featured in the cultural practices and ma-terial culture of the Hawaiian community. Examination of feather artifacts allowed identification of the forest birds that were used in the making of 'aumākua hulu manu (icons of royal battle gods), mahiole (feath-ered helmets), 'ahu'ula, kāhili, lei (neck-laces), malo (loincloths), and ornaments in pre- and post-contact periods (Kaeppler 1970, 1978; Rose et al. 1993; Conant 2005; this volume, see Plates 20 and 21). Ethnographic records report that many forest birds were considered a source of food (Keauokalani 1865, Emerson 1895, Malo 1951, Nalimu no date, Te Rangi Hiroa 1957).

Various authors—Keauokalani (1865), Alexander (1891), Emerson (1895), Brig-ham (1899–1903), Malo (1951), Beck-with (1970), Davis (1978), Clark and Kirch (1983), Nakamura (1999)—verify that the following species of Hawaiian forest birds were used in Hawaiian prac-tice: 'Alalā, 'akialoa (several species), 'Aki-apōlā'au, 'amakihi (four species), 'Apapane, 'Elepaio, 'I'iwi, 'Io, Kākāwahie, Hawai'i Mamo, Nēnē, 'Ōma'o, 'ō'ō (four species), 'Ō'ū, and Pueo. Table 3.1 lists forest birds known by ornithologists that were used in traditional Hawaiian feather work or for food. Findings were derived from ethno-graphic, historical, and biological refer-ences (Keauokalani 1865; Teauotalani 1865; Kalakaua 1888, 1972; Emerson 1895; Munro 1944; Malo 1951; Kamakau 1964;

Table 3.1. Selected forest birds in traditional Hawaiian culture

Species	Akua 'Aumākua[a]	Food	Feather Work
Nēnē—Goose		✓	✓
'Io—Hawaiian Hawk	✓	✓	✓
Pueo—Short-eared Owl	✓	✓	
'Ō'ō—Honeyeater	✓	✓	✓
'Alalā—Hawaiian Crow	✓	✓	✓
'Elepaio—Monarch Flycatcher	✓	✓	
'I'iwi—Honeycreeper	✓	✓	✓
Mamo—Honeycreeper	✓	✓	✓
'Apapane—Honeycreeper	✓	✓	✓

[a]Akua 'aumākua were extensions of spiritual power.

Handy and Handy 1972; Kaeppler 1978; Rose et al. 1993; Nakamura 1999; Conant 2005).

Although physical and ethnographic research provides details on the use of Hawaiian forest birds for food and feather work, it offers only limited descriptions of the cultural relationship of the community with these birds. As noted earlier, appreciation of the relationships of native Hawaiians with native forest birds can be gained by investigation of the role of Hawaiian forest birds in the mythical, doctrinal, experiential, social, and ritual dimensions of the Hawaiian culture (see Box 3.1). We will use the framework of these five dimensions to explore Hawaiian cultural relationships with forest birds.

FOREST BIRDS IN HAWAIIAN COSMOGENY

Hawaiian Gods and Spirits

At the time of continental contact, Hawaiian society was built around a caste system, with the ali'i (royalty) tracing their lineage to the gods, and maka'āinana (commoners) working to serve the ali'i. Spirituality, social class, and social custom were entwined, and there were likely substantial differences in the way the ali'i and maka'āinana perceived and treated the animals and plants in their environment. To grasp the broad relationship of traditional Ha-

waiian society with animals and nature, we begin with the area of cosmogeny and mythology.

The mythical dimension of traditional Hawaiian culture emphasized devotion to the gods and ancestors, because the cosmogonic genealogy links people's origins to the lineage of their gods and ancestors. Hawaiian mythology describes recognized forms of spiritual forces or gods called akua and 'aumākua (Malo 1951; Kamakau 1964; Pukui et al. 1972, Vol. 1). Akua are forces that are highest in rank and power, while 'aumākua are lesser gods or family spirits. Both could appear in physical form (bird, shark, plant, rock, rainbow, and so on).

Although akua were high-ranking gods, they could at times appear as animal entities that acted as guardians (akua 'aumākua) of ali'i or maka'āinana (Malo 1951); the akua also could call upon lesser gods to appear in these roles. Hence, akua 'aumākua were extensions or manifestations of the akua and personified in two ways, as akua pili kino (ancestral gods of chiefs and maka'āinana) or akua 'ohana (family spirits or personal gods) (Kamakau 1964). Kū, Kāne, Kanaloa, and Lono were akua 'aumākua who had the ability to transform themselves into akua pili kino or akua 'ohana. When summoned or appealed to, they could alter their appearance and become the akua 'ohana who aided the 'ohana (family) who worshiped them. In many instances, they would appear in the form of a manu (bird).

It is important to note that although a family's 'aumakua normally appeared as a particular object or animal, it still had kino lau (the ability to take many forms or bodies) (Pukui et al. 1972, Vol. 1). An 'aumakua, therefore, could change its appearance from a rock to a plant to an animal at will. Some of the 'aumākua chose the form of animals, including birds (i.e., 'Alalā, Pueo, 'Elepaio, etc.) and made their presence known by calling or flying nearby.

Because 'aumakua played a vital role in Hawaiian society, often in the form of animals, it follows that animal themes were interwoven throughout Hawaiian society. As Hawaiian forest birds could be 'aumākua, their whole being was considered sacred. Because the hulu (feathers) are part of the bird's body, they and feathered objects derived from them were infused with the mana (sacred spiritual essence) of the akua and 'aumākua. Hawaiian forest birds were acknowledged as 'aumākua, ancestors of the family from generation to generation.

Birds in the Spiritual Life of Royalty

For the ali'i, birds were both the symbol and living personification of their 'aumākua. As recounted in the Kumulipo, birds were akua 'aumākua of the gods, whose powers could influence the outcome of future ventures (Malo 1951, Kamakau 1992). Beckwith (1951: 70) noted that the Kumulipo, "contains allusion to the birds Halulu and Kīwa'a, whose feathers, attached to images of the gods, are supposed to rise or fall to predict success or failure of a war party." Objects such as the kāhili (feathered standards), used to demarcate sacred space, herald royalty, and fan ali'i, may have been thought to provide spiritual protection (Hawaiian Ethnographic Notes 1922: 2751; Rose et al. 1993).

Throughout the course of war and conquest, Kamehameha I believed that the mana of Kūkā'ilimoku ("Kū the land-grabber," whose war god represented in a feathered image was displayed during expeditions) provided him protection and victory (Alexander 1891, Malo 1951, Kamakau 1992). Moreover, his success in battle substantiated his belief that feathered images representing the god(s) were sacred and powerful. If the 'ahu'ula (feathered capes and cloaks) worn in battle prevented or mitigated injuries from weapons, this protection supported Kamehameha I's belief that objects and garments fashioned with feathers embodied spiritual support and strength of akua (Brigham 1899–1903; Kaeppler 1978, 1985).

The Symbolic Significance of Feathered Objects

Because the significance of the feathered objects was based on honoring the gods and entreating divine intervention, the feathered objects presented to Captain James Cook and other members of his crew were very important offerings (Kaeppler 1978). It seems clear that Captain Cook was first perceived to be "Lono akua maoli" (the god Lono in human form) (see "Hawaiian Gods and Spirits," this chapter), and accordingly was given gifts of feathered objects from the ali'i. For the ali'i, the offerings of feathered objects were not just a friendly gesture: these objects had considerable material value and mana (spiritual power). The ali'i gave Captain Cook some of the most valuable possessions they owned. The gifts were, in fact, a show of respect—a display of humility, honor, and reverence. Written accounts on this topic support that this was the case. Beaglehole (1967: Part 1: 512, footnote) reported that on January 26, 1779, at Kealakekua Bay, Captain Cook was given at least seven cloaks and capes: "King [Kalaniopuu] got up & threw in a graceful manner over the Captns Shoulder the Cloak he himself wore, & put a feather Cap upon his head, & a very handsome fly flap in his hand: besides which he laid down at the Captains feet 5 or 6 Cloaks more, all very beautiful, & to them of the greatest Value." Imagine, then, the potential psychological impact the ali'i and kāhuna (priests) may have felt when, after giving so much of their mana in the feathered objects, they realized that Cook was not Lono akua maoli (Pukui et al. 1972, Vol. 2: 209–210). Nevertheless, stories contain accounts that Cook's body was buried with great customary respect (e.g., Pukui et al. 1972, Vol. 2: 211).

Birds in the Spiritual Life of Common Folk

For the maka'āinana, forest birds had a notable status as 'aumākua or akua 'ohana (see "Hawaiian Gods and Spirits," this chapter). Pukui et al. (1972, Vol. 2: 123) noted, "The closest man-with-god relationship came in the 'aumakua (spiritual beings, guardian, protector)." Each family had its own personal deities or 'aumākua who were compassionate and approachable spirits and not associated with the gods of the ruling class (Malo 1951; Kamakau 1964; Beckwith 1970; Pukui et al. 1972, Vol. 1).

Naturally, having evidence of the spiritual life force of deceased relatives was comforting to the family: "For 'ohana (family), loyalty continued into eternity" (Pukui et al. 1972, Vol. 2: 123), and the death of the body did not necessitate the death of the 'uhane (spirit) (Penukahi 1871). In traditional Hawaiian society, the belief was that an 'aumakua came into being because the spirit of a deceased relative elected to become 'aumakua and/or because the 'ohana decided that a family member should be deified as one (Pukui et al. 1972). If a family decided that a deceased relative should be worshiped as an 'aumakua, a special ceremony called kākū'ai was performed for the deceased relative (Malo 1951; Beckwith 1970; Pukui et al. 1972, Vol. 1). In this ceremony, the family could choose the 'aumakua form their deceased relative would embody. An example of kākū'ai can be found in the legend

of Pumaia (Fornander 1916–1920, Beckwith 1970), in which Kū-ali'i killed Pumaia, a pig farmer, because Pumaia refused to give up his favorite pig. Pumaia's spiritual voice told his wife where his bones were located. In a ceremony, his wife and daughters wrapped and honored Pumaia's bones through offerings and prayer. Upon completion of this ritual, Pumaia's spirit was transformed into a very powerful 'aumakua (in the form of a Pueo) that wreaked havoc on Kū-ali'i and his men until retributions were made to Pumaia and his family.

Birds as Mediums to the Spirit World

The ability to communicate with 'aumākua was very important in traditional Hawaiian society because 'aumākua were believed to have a direct line of communication to the akua (Malo 1951; Pukui et al. 1972, Vol. 1; Kamakau 1991). Pukui et al. (1972, Vol. 1: 35) called this role of an 'aumākua that of a "spiritual go between," as the 'aumākua could pass on the prayers of the family to the akua. 'Aumākua could be beckoned with prayers and rituals, and, if an 'aumakua chose, it could respond to this calling while the individual was awake or asleep. Birds were viewed as instruments, representatives, and manifestations of gods through which people were able to communicate with their akua or 'aumakua.

An animal form of 'aumakua worshiped by a number of families was the Pueo (Alexander 1891, Malo 1951, Beckwith 1970, Pukui et al. 1972). The families of this 'aumakua say that Pueo delivered messages of warning to their human relatives via signs, portents, or agitated behavior. For example, Pueo 'aumākua were known to signal warnings of imminent misfortune (e.g., a house was near collapse), danger (e.g., a person was about to be in an accident), and death (e.g., a person was near death or had just died). These messages were given in the form of the Pueo's cry or hovering behavior. Pukui et al. (1972, Vol. 1: 55) noted that "hovering birds may convey a protective warning . . . but birds that single out a person or house may mean disaster." Pukui related a personal experience, the memory of the strange way an owl cried while she was visiting an aunt. She immediately decided that Pueo, her ʻaumakua, was telling her to go home. When she returned home, she found that her hānai (adopted child) was very sick with fever. Similarly, feelings of foreboding were expressed by several of Pukui's client families. In one account, birds were observed to habitually perch on a client's home. The family interpreted the birds' conduct as a hōʻike (exhibition or warning) of imminent disaster and promptly moved out. In the experience of another client, birds hovering around a house were associated with the death of the man who lived there.

The ʻaumākua of traditional Hawaiian culture were believed to possess the powers of ʻānai, a severe curse that could cripple, sicken, or kill those who violated kapu (forbidden and sacred privilege) or who were engaged in unlawful behavior (Malo 1951; Beckwith 1970; Pukui et al. 1972, Vol. 1). It was believed that ʻānai could be delivered through visitation by the ʻaumakua in the form of a forest bird. Moreover, those who continued to behave badly in life knew that their ʻaumakua would ensure that they were justly punished in the realm of Milu (dark eternity) (Malo 1951; Beckwith 1970; Pukui et al. 1972, Vol. 1).

Birds in Hawaiian Social Class and Spirituality

It is apparent that there were spiritual and mythological differences in the way in which the aliʻi and the makaʻāinana interacted with forest birds. There is no doubt, however, that both classes perceived forest birds and feathered objects as more than elegant artisanship. For the aliʻi, whose main concern was power and order within their society, feathered objects were the embodiment of the gods and provided the aliʻi a sense of personal strength that could not be gained by mere physical existence. Their loyalty to and faith in their akua and ʻaumakua were reflected in the high value of their feathered objects. They also recognized that feathered objects not only symbolized the grand power of the gods but also aroused warriors' confidence in and loyalty to the aliʻi on the battleground.

For the makaʻāinana, who were mainly concerned with protection and with life after death, the belief that their deceased relatives existed as ʻaumākua helped families cope with the sadness associated with death. They were comforted in the belief that the loyalty, love, and concern of departed relatives continued to exist and appear in other forms. Needless to say, there was an understanding that if one were good and moral while alive, ʻaumakua would be waiting to welcome them into the realm of Pō (darkness). As each family's well-being was closely linked to the well-being and safekeeping of its ʻaumakua, the identity and dwelling place of ʻaumākua were closely guarded and known only to that family. "Hawaiians didn't go around talking about their ʻaumakua" (Pukui et al. 1972, Vol. 1: 41). The belief that a departed loved one could exist in any one of the creatures of nature was reassuring. So, as one would expect, families' rapport and interactions with birds and their habitats increased every person's awareness of, sense of belonging with, and affection for plants and creatures in their natural world.

FOREST BIRDS AND TRADITIONAL HAWAIIAN SOCIETY

The social dimension in traditional Hawaiian culture included the rapport and bonds individuals formed with their akua, aliʻi, ʻohana, nature, and animals. In this aspect of Hawaiian culture, social gatherings were opportunities for reverence and merriment.

The social element of Hawaiian culture helped to strengthen individuals' affinity for one another and solidified the members' allegiance to their ali'i and gods (Brigham 1899–1903; Malo 1951; Kamakau 1964; Pukui et al. 1972, Vol. 1; Kirch 1985).

The Makahiki Festival

The annual Makahiki Festival brought all Hawaiians together and was celebrated as a time free from strife and for gathering strength (Kamakau 1992). The Makahiki included an 'aha'aina ho'omana'o (great feast of commemoration) and was sanctified as the time for rest and worship outside the ordinary religious observances (Keauokalani 1865, Malo 1951, Te Rangi Hiroa 1957, Kamakau 1964). It was an opportune time for people to come together and give thanks to the gods who provided them with lush and fertile land. During the Makahiki Festival, feathered objects were prominently displayed and honored. Glorification of these feathered objects was important because they were representations of the Hawaiian gods, ancestors, and the divine lineage of the ali'i. To demonstrate the magnificence of his ancestral line, the ali'i nui (king) built feathered structures (see Kaeppler 1978: front cover and Fig. 60) that paid tribute to his ancestral gods (Kamakau 1992). By doing so, he made the akua and the ali'i the central focus of the festivities (see also "Hawaiian Forest Birds in Ceremonial Rituals," this chapter).

Feathered images representing the god of sports (akua pā'ani) were used to pay homage to the Makahiki gods and were presented at the opening of each of the sporting events that was part of the festival. At one of the proceedings, an image of the god Lono (akua poko) was displayed on a staff in the form of a man wearing a mahiole (feathered helmet) (Malo 1951, Kamakau 1964). The structure signaled the god's presence and sanction of the competition. Offerings and prayers to the gods were performed to bless each athletic event or formal observance.

Feathers as Symbols of Royalty

Although the Makahiki Festival was an annual event in which feathers were showcased, feathered objects were also featured in the pageantry at other royal affairs and celebrations. During these occasions, the feathered garments and accessories of the royal families denoted their lineage to the akua and akua 'aumākua and accentuated the degree of wealth and power (material and spiritual) enjoyed by the monarchy (Alexander 1891, Kaeppler 1985).

Tribute to the ali'i lineage to the gods was reflected in the color and design of each feathered item and indicated the class of the person wearing it (Kalakaua 1888, 1972). Specific colors were reserved for the ali'i, special warriors, or kāhuna (priests and specialist practitioners) (Kaeppler 1985). Yellow was the kapu color of the king, red the color for the priesthood, and mixtures of these colors were used for lesser members of the nobility (Kalakaua 1888, 1972). Feathered ornaments and ceremonial clothing were made mainly with the feathers of 'Apapane, 'amakihi, 'I'iwi, Hawai'i Mamo, and 'ō'ō as well as the feathers of moa (chickens) and various seabirds (Alexander 1891; Malo 1951; Kaeppler 1978, 1985; Kamakau 1992; Conant 2005). However, certain species (particularly 'I'iwi, 'ō'ō, and mamo) and their feathers were reserved for the manufacture and ornamentation of clothing and implements for the ali'i, perhaps as were gems in European cultures (Kalakaua 1888, 1972; Emerson 1895; Malo 1951). For example, the mamo and 'ō'ō were valued for their yellow feathers, which could be used only by royalty (Kalakaua 1888, 1972; Malo 1951). The mamo's anthropomorphic traits (smartness, cunning, and pride) caused it to be regarded as the "chief of small, mountain birds" (Keauokalani 1865: 1144), and its personality may have been another reason that its feathers were used in a number of feather works made for the royal families.

Moreover, status was denoted by types of clothing. Among the ali'i, only the ali'i nui was allowed to use full-length 'ahu'ula (feathered cloaks). Lesser chiefs, kāhuna, and chosen warriors were permitted to use short feathered 'ahu'ula (capes) and mahiole. The high chiefesses decorated their hair with hulu-kua (feathered combs); hakupapa—headbands of feathers of 'Ō'ū, 'amakihi, 'I'iwi, 'ō'ō, and mamo; or a mixture of flowers and feathers (Kalakaua 1888, 1972; Malo 1951).

Other feathered items highly visible during royal gatherings were the kāhili that surrounded the royal court (Kalakaua 1888, 1972; Malo 1951; Te Rangi Hiroa 1957; Rose et al. 1993). Young ali'i or kahu (honored attendants) carried feathered kāhili when accompanying their ruling chief and chiefess (Kalakaua 1888, 1972; Malo 1951). The kāhili is a staff that has a feathered cylinder made up of au (branches), a kumu (pole), pā'ū (skirt), and pāpale (cap) (Rose et al. 1993). Kāhili were significant at formal events in that they marked the presence of the ali'i and spiritually protected the ali'i. Kāhili included a variety of feathers from forest birds such as 'ō'ō, 'I'iwi, and 'Alalā, as well as native seabirds, and, after European contact, exotic and domesticated birds.

Feathers as Currency and Tribute

Because of their spiritual connection to the akua and 'aumākua, Hawaiian forest birds and their feathers were highly prized and valuable to the ali'i. Moreover, because the ali'i could substantiate their lineage to the gods, feathers from forest birds were sanctioned as property of the ali'i only and protected by kapu (sacred reservation) (Malo 1951, Kamakau 1992). Forest birds' feathers were also highly valued because it takes tens of thousands of bird feathers to create a single large feathered object (Kalakaua

1888, 1972; Kaeppler 1970; Rose et al. 1993). The hulu of 'I'iwi, Hawai'i Mamo, and 'ō'ō were used in payment of tribute for land tenure (levied on each ahupua'a, a traditional land management unit) (Keauokalani 1865, Alexander 1891, Emerson 1895, Malo 1951, Te Rangi Hiroa 1957, Kamakau 1964). Payment of tribute was also an expression of respect and appreciation for the king and chiefs, who safeguarded the people's way of life.

In the social dimension of traditional Hawaiian culture, we find that forest birds and feathers were used in honoring the gods, supporting the nobility, blessing activities, and displaying pageantry. The status given to forest birds and feathered artifacts was an acknowledged and important social aspect of Hawaiian culture.

INDIVIDUAL EXPERIENCE WITH HAWAIIAN FOREST BIRDS

In the experiential dimension, personal encounters with Hawaiian forest birds were special and may have been interpreted as having divine importance. Extended exposure to the birds intensified and raised awareness of the bond of the ali'i or certain individuals with these animals. Because of their birthright, the ali'i were committed to creating images, artifacts, and costumes befitting the gods, their ancestors, and their status. The production of feathered items for the ali'i was supported by artisans and craftspeople with a variety of specialties. These people likely had a very different relationship with the birds than did the ali'i. The most experienced were po'e kia manu, the bird catchers.

Po'e Kia Manu and Their Understanding of Forest Birds

The kia manu was in the service of the ali'i nui with the specific task of bringing back plumage from forest birds for feathered objects. It is not surprising that the kia manu's relationship with the forest birds differed from that of other artisans or craftspeople, not only because of his extended exposure to the birds in their habitat but also because he was answerable to the king. For example, although particular forest birds may have been 'aumakua for the kia manu, he probably did not consider the task of capturing or killing these animals as an act of defiance or blasphemy because the ali'i nui's direct lineage to the akua superseded the kia manu's 'aumakua, and any directive not carried out had serious repercussions. In any case, the kia manu knew his task was to collect feathers that were suitable for the ali'i. He also understood that feathers needed to be harvested carefully in order to maintain future supplies. In an account related to snaring, Keauokalani (1865: 1137) noted, "All of the birds are not taken . . . only about two or three. . . . Bird catchers say that they always leave some behind to grow up."

The kia manu's knowledge of and experience with forest birds and their habitats was extraordinary. He recognized the birds not only by their appearance (e.g., Malo 1951) but also by their behavior or demeanor (Keauokalani 1865, Teauotalani 1865, Emerson 1895). Here we give some brief examples of the kia manu's bird lore.

The 'Io was described as clever and had a confident yet poised demeanor. Because of Io's commanding disposition and great skill, it was perceived to be the "king of the forest birds" and was named after the king 'Io-lani-ka-io-nui-maka-lanau-moku (Keauokalani 1865: 1128, Teauo-talani 1865: 192–193). The 'Alalā was known to have an excitable nature and was perceived to be a bold animal because it was unafraid to attack man (Keauokalani 1865: 1134; Teauotalani 1865: 196).

The Hawai'i Mamo was cited as the prince or king of Hawaiian plumage birds (Emerson 1895: 108). It was reputed to have a proud, audacious, shrewd, watchful, and suspicious personality (Keauokalani 1865: 1144). (It is interesting to note that Keauokalani [1865: 1144] was in error when he described the Mamo as an "all-yellow bird." Mamo were predominantly black, with some yellow feathers on their rump and vent areas.)

The 'I'iwi was described as a small, handsome bird with a beautiful singing voice. The kia manu knew of three types of 'I'iwi: 'i'iwi polena, 'i'iwi alokele, and 'i'iwi popolo. The 'i'iwi polena was observed as a bird with yellowish juvenal feathers and yellow irises, legs, and claws, hence its name. The 'i'iwi alokele was observed as having the same feather combination as the first; however, it had reddish-brown eyes. The 'i'iwi popolo was observed as having purely bright red feathers. The 'I'iwi was noted for its happy personality and endless energy (Keauokalani 1865).

The 'ō'ō was described as a beautiful bird with deep-black feathers and bits of yellow on the wings and tail. Because it was "beautiful to look at," it was called the "royal bird" (Keauokalani 1865: 1146). The kia manu noticed that the 'ō'ō had a crying type of call ("cko"), and because this bird dwelt in the mountains, it was labeled "the bird that cries in the mountain" (Keauokalani 1865: 1146). The male was called an 'ā'ā, while the female was called a pīpī. The 'ō'ō was said to have a jealous and domineering personality and a "most musical and melancholy" voice.

The Kia Manu's Work

The po'e kia manu began their work with a prayer to Kū-huluhulu-manu (the god of kia manu and feather workers), their ancestors, and their 'aumakua and a pledge to offer their first catch to these beings (Emerson 1895; Malo 1951; Pukui et al. 1972, Vol. 2). Prayers may have helped to allay any fear of reprisal from their 'aumakua.

The po'e kia manu were aware of the patience required in mastery of their bird and forest craft. Although many practiced the art of snaring in the 'ōhi'a lehua forests, each kia manu's technique was tailored to local climatic and vegetative features (Keauokalani 1865, Emerson 1895, Malo 1951). In addition, each was cognizant of the proper handling of the feathers. To preserve the collected plumage of his birds from wet weather, the kia manu wore a long, thatched rain cloak made out of a mesh of olonā, onto which ti-leaves were draped (Emerson 1895: 111).

The equipment and other items that were used in bird catching included the kia (bird-catching stick or pole made out of kauila or 'ūlei), hāpapa (two sticks made out of 'ōlapa branches and joined together to form a cross), a net or snare (made out of olonā or 'ie'ie fiber), stones, and kēpau (birdlime: a sticky, gluelike substance) (Keauokalani 1865, Emerson 1895, Malo 1951, Handy and Handy 1972, Nalimu no date). The po'e kia manu would select one or more of these items, depending on what and where they were going to hunt.

The term birdlime (an adhesive, viscous substance used to catch small birds) has origins in Britain, where bird catchers traditionally used extracts from British trees (e.g., mistletoe, elder) to catch birds (Webster's Revised Unabridged Dictionary 1913). The term has been adapted to describe an analogous substance used by native Hawaiians. The kēpau (birdlime) used by the kia manu may have been taken from the pāpala kēpau, 'oha, 'ulu, or kukui (Emerson 1895, Degener 1930, Stone and Pratt 2002, Nalimu no date).

Ethnographic records describe Hawaiian techniques used to capture forest birds. Snaring, netting, clubbing, and pelting with stones were methods mostly used on bigger birds (e.g., 'Alalā, Nēnē, and Pueo) (Keauokalani 1865, Malo 1951, Nalimu no date). Smaller birds such as "the akakane, iiwi, and oo were caught . . . with fine line made of olonā" (Nalimu no date: 824). In this method, a snare was smeared with kēpau.

In the netting method, a large mesh net was laid either over a bird's nesting area or in a place where live maunu (bait, i.e., a rat or small bird) could be observed by the larger forest birds such as the 'Io or Pueo (Keauokalani 1865, Emerson 1895, Malo 1951). The birds of prey were caught when they landed on or in the net in pursuit of the live bait or upon returning to their nesting site.

Large and small forest birds were also caught by the kia or the hāpapa method (Keauokalani 1865, Emerson 1895, Malo 1951, Handy and Handy 1972). If the kia (a wooden pole approximately 6 m in length [Namilu no date]) was used, the tip was smeared with kēpau. The kia manu would call to the bird, and when the bird was close enough, the kia manu sneaked up on it and raised his kia to the bird's breast. As the kia touched the bird, the bird was held fast by the sticky substance and was then carefully retrieved.

In the hāpapa method, two sticks were formed into a cross with a hook fashioned at the intersection for hanging purposes. The upper portion of the hāpapa was smeared with kēpau and covered with nectar-laden flowers or a colorful banner made from strips of kapa (bark cloth). For a larger bird (i.e., 'Io or Pueo), maunu (live bait) was tied to the prong at the tip of the hāpapa with some plant material to lure the bird to the hāpapa (Keauokalani 1865, Emerson 1895, Handy and Handy 1972). If a small bird was used as the maunu to lure other small birds, kēpau was smeared on the perch at such a distance as to not ensnare the live decoy if it struggled or attempted to fly (Emerson 1895). Once the hāpapa was set in place, the kia manu moved away from the apparatus and called the desired bird by imitating its cry. Birds were caught and at times killed as soon as they landed and stuck to the hāpapa (Keauokalani 1865, Emerson 1895, Degener 1930).

The literature presents conflicting reports as to whether the birds were killed and consumed or released after feathers had been extracted. Degener (1930) reported that kukui nut oil was used to wash away the kēpau or sticky substance before birds were liberated if they were not killed. Emerson (1895: 111) noted that King Kamehameha I reproached his kia manu for taking the life of the birds they caught, stating, "The feathers belong to me, but the birds themselves belong to my heirs."

Observations made by Captain Cook and the surgeon (William Anderson) and artist (John Webber) on his ship, HMS *Resolution*, during Cook's first landing at Waimea Bay, Kaua'i, on January 18, 1778, indicated that birds were killed and their feathered skins sold at the marketplace. Anderson, who made note of Hawaiian feather 'ahu'ula, recorded his observations of 'I'iwi brought for sale by the Hawaiians (Cook and King 1784, Vol. 2: 207–208): "We were at a loss to guess from whence they could get such a quantity of these beautiful feathers: but were soon informed, as to one sort; for they afterward brought great numbers of skins of small red birds for sale, which were often tied up in bunches of twenty or more, or had a small wooden skewer run through their nostrils. At the first, those that were bought, consisted only of the skin from behind the wings forward; but we, afterward, got many with the hind part, including the tail and feet.... The red bird of our island was judged by Mr. Anderson to be . . . about the size of a sparrow; of a beautiful scarlet colour, with a black tail and wings; and an arched bill, twice the length of the head, which, with the feet, was also of a reddish colour."

It is doubtful that a bird could have survived if a large proportion of feathers were removed, and Emerson (1895: 110) believed it would have been "an act of cruelty . . . to set it loose in such a condition." Malo (1951: 38–39) stated that the birds "were delicious when eaten," and it is likely that the forest birds were used as food by the kia manu (and his family) during extended hunting trips (Keauokalani 1865, Emerson 1895, Henshaw 1902, Perkins 1903, Munro 1944, Nalimu no date).

Women and Forest Birds

We could find very little information in the literature describing women's experience with Hawaiian forest birds. Feathered artifacts for women's use or adornment indicate that some forest bird feathers were used for the chiefess's hulu-kua, lei, and haku-lei (Kalakaua 1888, 1972; Malo 1951). According to Emerson (1895: 106), although the po'e kia manu typically worked alone, their wives would sometimes accompany them, making "kapa (bark cloth) from the delicate fibres of the māmaki (*Pipturus albidus*) bark, perhaps to aid in plucking and sorting of feathers." Emerson's implication that kapa was made during hunting expeditions is interesting, because the making of kapa requires a heavy kua (anvil) and pounders. Moreover, beating bark to make kapa is noisy and would likely have disturbed forest birds. Thus if Emerson was accurate, the women may have made kapa at some distance from where the kia manu was working.

Interestingly, however, legend tells us that the first 'ahu'ula (feather cloak) was made by a young woman named Kanikaniaula on the island of Maui (Nakuina 1907). In this legend, Kanikaniaula had died, but her spirit appeared to the kahuna (priest) Eleio. During this appearance, her spirit offered the kahuna the feathered cloak she had been making, telling him that it could be found at her family's home. The kahuna traveled to her home, whereupon he happened upon her fallen body and restored her to life. As promised, after the final touches were put on the cloak, Kanikaniaula and the feathered cloak were given to him by her parents. Eleio presented both the feather cloak and Kanikaniaula to King Kaka'alaneo. The king, who was taken with both the woman and the cloak, married Kanikaniaula. This legend notes that the "highest chiefs of the land" trace their descent to this royal couple.

HAWAIIAN FOREST BIRDS IN MYTHS AND LEGENDS

In the doctrinal dimension of traditional Hawaiian culture, animals were portrayed in mythology and legends. In this aspect of life, the principles or laws established by the group were related in parables made meaningful and relevant through the interpretive skill of the storyteller. The animals' behavior and reactions not only depicted subjective forms of reality but also highlighted the beliefs, values, and practices of the people. Thus, attributions of the animals' behavior, thoughts, and feelings were likely based on human experiences and conduct. The philosophy and ideals emphasized in the doctrinal dimension of Hawaiian culture may have been vital for individual and group safety and benefit and for their continued way of life.

As in other cultures, there may have been underlying historical realities that formed the origins of Hawaiian mythology and legends. The experiences, struggles, and sentiments of the animals portrayed in Hawaiian legends corresponded to the feelings, concerns, and hardships people experienced in real life. In the myths and legends, the trials and tribulations confronted by the humans and animals were limitless. Forest birds were often portrayed as powerful, knowledgeable, and experienced beings, and family dynamics and the relationship between humans and animals were important features in many stories. The intense emotional exchanges among the people and between humans and animals represented dire, heart-wrenching, and puzzling situations faced by humans.

Pueo in Allegories

Actual incidents and human-animal interactions were reflected and embellished throughout Hawaiian mythology and legends. Personal accounts, for example, described a number of extraordinary escapes with the help of Pueo (Alexander 1891, Beckwith 1970). In one account, a warrior of Kamehameha claimed that he would have plummeted over a cliff had a Pueo not appeared and thrust his spear into the earth to keep him from falling. In another, a man related that he would have been captured had a Pueo not chased him out of the cave where the enemy was searching for him. In a different account, there was a report that Napaepae of Lahaina, who had capsized his boat in the Pailolo Channel, would have swum all night had a Pueo not flapped its wings and pointed him to land. These personal experiences are analogous to legends in which the akua 'aumakua (i.e., Kū, Kāne, Lono, or Kanaloa) gave warning of the enemy's approach (see the legend of Kai'iwi, the sacred Pueo, this section), in which Pueo 'aumakua were summoned by families who hoped to gain protection or help during a crisis (see the legend of Kaili and Nailima, this section), and in which Pueo 'aumakua guided Maui to safety (see the legend of Maui and the owl, this section).

A manifestation of the god Kāne (Kāne i ka pahua) was known to appear as a Pueo. In another legend, Pueo nui akea (owl god of Maui) was recognized as the deity that brought souls back to life and as a protector in battle and danger—ka Pueo kani kaua ("the owl who sings of war") (Beckwith 1970). In addition, Pueo, a child of the goddess Hina, was instrumental in saving the life of his brother Maui (Fornander 1916–1920; Pukui et al. 1972, Vol. 2).

Kihapa and Lakapu and the Battle of the Owls

Two legends in particular advocate altruism and self-sacrifice. In the legend of husband and wife Kihapa and Laukapu (Irwin 1936), the couple was very kind to the birds of the forest (mamo, 'ō'ō, and 'I'iwi). Thousands of beautiful feathers were showered on the couple as a wedding gift from the birds (we should not forget that feathers were a valuable commodity for paying taxes).

In another legend, "Kapoi's Story" or the "Battle of the Owls," Kapoi was rewarded for his unselfishness (Henshaw 1902, Thrum 1907, Metzger 1929, Armitage and Judd 1944, Knudsen 1946, Pukui 1951, Kamakau 1991). Kapoi was very poor but happened upon Pueo eggs and took them. Upon discovering that her eggs were missing, the mother Pueo begged to have her eggs back. Despite being hungry, Kapoi returned them to her, and she became his 'aumakua. Kapoi built a heiau ki'i (sacred temple, altar), which the king considered an act of rebellion and ordered Kapoi's execution. His Pueo 'aumakua, named Kūkauakahi, enlisted the help of Pueo from all the islands. They defeated the king and his warriors, and Kūkauakahi was declared the most powerful 'aumakua.

These legends appear to address ideals related to loyalty and goodwill and imply that rewards and recognition will come to those who are unselfish and humble.

A Maui Legend

One of the Maui legends tells of the pride the demigod Maui took in the forest birds (Colum 1924, 1937; Pukui 1960). A visitor attempted to outclass the worldly riches of Maui with his own stories. The visitor, however, became awestruck by the music produced by invisible singers and conceded that Maui was supreme. Upon his concession, Maui allowed his magnificent singers, the forest birds ('I'iwi, 'Apapane, 'ō'ō, and mamo) to appear to the visitor. This legend describes the foolishness of self-importance and arrogance and suggests that in the end one may find that there is someone greater than oneself.

Alakolea, Pumaia, and Mokulehua

The legends of Alakolea (Green 1928), Pumaia (Fornander 1916–1920, Beckwith 1970), and Mokulehua (Kamakau 1964) condemn greed, power, and suicide. Alakolea, a kia manu, was warned by a neighbor that he was being watched by the bird god because he was catching more birds than he needed. Alakolea ignored his neighbor's warning and continued his bird-catching activities. The birds killed him, and he fell into an imu (ground oven) and was destroyed.

In the legend of Pumaia, there was a power struggle between a prophet and Pueo nui o kona, 'aumakua of Pumaia and Wakaina. In this story, the souls of Pumaia and Wakaina were being chased by the prophet. Pueo nui o kona questioned the prophet about the chase. Pueo nui o kona explained that he could return Pumaia and Wakaina to human existence if he chose. The prophet, unhappy about this possibility, challenged Pueo nui o kona and was killed.

In the legend of Mokulehua, his wife Pueo had committed suicide and as a result was confined in Milu, a dark and everlasting realm. Mokulehua, heartsick about his wife's situation, sought help from his god, Kanikaniaula. Through the mana of Kanikaniaula, he located his wife and escaped with her from the cruel depths of Milu.

These stories reinforce the Hawaiians' intolerance of offensive behavior and immoral acts. However, the legend of Mokulehua and his wife also includes the theme "love will overcome."

Lā'ie-i-ka-wai, Kai'iwi, Kaili and Nailima, and Maui and the Owl

The legends of Lā'ie-i-ka-wai (Kalakaua 1888, 1972), Kai'iwi (Apo 1946), Kaili and Nailima (Armitage and Judd 1944, Pukui 1951), and Maui and the Owl (Fornander 1916–1920; Pukui et al. 1972, Vol. 2) focus on escape from imminent danger, rescue, and guidance to safety.

The legend of the princess Lā'ie-i-ka-wai begins with her special relationship with the forest birds (viz., 'Alalā, 'I'iwi, 'Apapane, 'Elepaio). The forest birds were her guardians, and their presence was made known by their singing. When an unworthy prince, with the help of his five sisters, was unable to win the heart of Lāʻie, he abandoned his sisters and returned home. His sisters befriended the princess, and they, along with the forest birds, protected her from further advances of the prince.

In another legend, the bird Kaiʻiwi watched over the community and signaled impending danger. A kahuna observed Kaiʻiwi circle the moon, screeching in distress and scattering feathers about, which alerted the people of an enemy's approach. The warriors were readied for battle, and Kaiʻiwi was honored.

In the legend of Kaili and Nailima, a young boy named Kaili was captured by warriors. His sister Nailima witnessed the capture. Nailima prayed to Pueo, their ʻaumakua. Pueo located the boy and, by flapping his wings, advised him how to escape without being detected. Kaili and Nailima arrived home safely because of Pueo's guidance.

In the legend of Maui and the Owl, Hina (mother of Maui) had given birth to another child in the form of a Pueo. Once when Maui was taken prisoner, Pueo searched for Maui, rescued him, and led him to safety.

These legends address the power of ʻaumākua and show that families should trust in their ʻaumakua to provide guidance and protection.

Tragedies in Legend

There are legends in which individuals are destined to live tragic lives. The legend of the princess of Mānoa is one such legend (Kalakaua 1888, 1972; Thrum 1907; Westervelt 1963; Skinner 1971; Pukui et al. 1972, Vol. 2). The princess of Mānoa, Kahalaopuna, was slain and buried by her jealous fiancé, Kauhi, because he believed the lies of two lesser kings that they had been intimate with her. Pueo, Kahalaopuna's ʻaumakua, witnessed the incident,

unearthed her body, and restored her life. This cycle of murder, burial, and restoration to life occurred several more times. On his last attempt, Kauhi buried Kahalaopuna under a hala or koa tree (depending on the version), where she became entangled in its roots. Pueo, unable to disentangle her from this grave, regretfully departed. ʻElepaio, another ʻaumakua of Princess Kahalaopuna, recounted the incident to her parents. Before her parents could free her from the tree roots, a young chief named Makana was able to remove her body. The princess recovered and eventually married Makana. Kauhi and the liars were put to death.

All the legends recounted thus far address a number of themes: punishment of bad behavior, control or lack of control over one's destiny, and the importance of faith and love in distressful situations.

Kahuoi and the Sacred Spear Point

The legend of Kahuoi (Fornander 1916–1920) and that of the sacred spear point or magic spear (Kalakaua 1888, 1972; Skinner 1971) touch upon the importance of individual self-worth, hard work, and productivity. In the first, a farmer's son, Kahuoi, refused to help his parents cultivate their land. Unable to get his son to work, the farmer decided to send him away. In his new place, Kahuoi could not get enough to eat and soon learned that he needed to tend the land in order to survive. An ʻElepaio prophesied that Kahuoi's plantation would become famous if Kahuoi cultivated the land. Kahuoi became industrious, and his plantation became well known.

The second legend is about Kaululaʻau, a prince of Maui, who was recklessly wild and mischievous. Because of his unproductive ways, he was exiled to a land filled with evil spirits. In that land, a kahuna gave Kaululaʻau a sacred spear point with which he performed many good deeds and destroyed many monsters. In another story, the relatives of a malicious chief performed a ceremony that produced what was thought to be a Pueolike monster, Pueo-ali'i. The monster was slain by Kaulula'au, and he was hailed a hero.

Clearly, the ideals of Hawaiian culture were highlighted in myths and legends. The connection of birds to heroes and hero-ines, even tragic ones, accentuates their central, sacred status. The birds serve as the symbols supporting the idea that the people in the stories are not trivial or common. Because some of the myths and legends paralleled or were consistent with true-life incidents, it seems plausible that they may have influenced the psycho-logical and ethical principles of traditional Hawaiian culture. It stands to reason that the legends provided an avenue through which people could freely discuss feelings related to intimacy, anger, fear, and joy. Did the attribution of superhuman abili-ties to animals reflect the high regard of Ha-waiian people for courage and strength of character? Were the animals used as tools to advocate guardianship and loyalty to family? The legends and myths described in this chapter address consequences of positive and negative behavior and use the forest birds as mechanisms to convey these messages.

HAWAIIAN FOREST BIRDS IN CEREMONIAL RITUALS

In the ritual dimension of traditional Ha-waiian culture, ceremonial or social events highlighted the passage from one stage of life to another and legitimized and sanc-tioned formal traditional practices. This aspect of life gave rise to the spiritual and ethical importance of group harmony and achievements. Participation in the rituals was demanded by the kāhuna and signi-fied the people's trust and confidence in their religious convictions. In Hawaiian cul-ture, it appears that ritualistic acts and ora-tions were mechanisms used to call upon the spiritual power of nature or animals

and to connect with and pay homage to the gods and ancestors.

Hawaiian rituals, like those of many other cultures, involved animals, including forest birds. Forest birds likely inspired feelings of safety and were thought to be helpful for predicting good or bad for-tunes. The birds were the medium used to avert evil, appease deities, and influence future events. Because the birds were linked to the powers of the 'aumākua, their pres-ence was significant in some formal rituals. Rituals that were known to include forest birds included the lifting of kapu, canoe making, 'aumakua worship, and the con-struction of heiau ki'i (temples).

Canoe Making

A well-known example of how forest birds were used in the ceremonial process was canoe making (Alexander 1891; Malo 1951; Beckwith 1970; Pukui et al. 1972, Vol. 2). The canoe-making ritual began with showing reverence to Kū-ka-'ōhi'a-Laka (a forest god of 'ōhi'a trees), whose image was adorned with feathers. After giv-ing honor to this god, the kahuna prayed and called upon the assistance of the god-dess Lea, who appeared in the form of an 'Elepaio. When she appeared, the canoe builders paid very close attention to her behavior. If the 'Elepaio ran up and down a tree, this indicated that the wood was of good quality. If it stopped and pecked at the tree, its action was a sign that the wood was not suited for making a canoe.

The Bond between 'Ohana and 'Aumākua

In previous sections we explained that people were able to communicate with their akua or 'aumakua through birds. In 'aumakua rituals, a special ceremony called kākū'ai was performed by living family members for their deceased relatives (Beck-with 1951; Malo 1951; Pukui et al. 1972, Vol. 2). This ritual could transform a de-ceased member of the family into a special class of 'aumākua, one with the form of a Pueo. In this ritual, the living members of the family would wrap the bones of their deceased relative in kapa and offer them to the god, who appeared in Pueo form. From that point on, the Pueo was considered the family 'aumakua.

Disruption of Rituals by Birds

Pule 'aha (prayer or direct appeal to the gods) at the ceremony for the lifting of kapu had to be performed with utmost perfection. It required absolute attention and silence from the worshipers and animals. During the construction of a heiau, the behavior of the forest birds could have an enormous impact on the ritual (Alexander 1891, Malo 1951, Kamakau 1964). For instance, Kamakau (1991) and Malo (1951) noted that because silence was such a crucial part of this ritual, any sound made by a wild Pueo had the power to ruin the success of any appeal.

Hawaiian rituals, like those of many other cultures, were solemn and earnest. These rituals increased the value of the activities or occasions being celebrated. Ritual performances heightened participants' spiritual awareness and sense of purpose and brought the community or families closer together. In the ritual dimension of Hawaiian culture, birds represented and validated the belief that all things, physical and spiritual, were connected. While participation in rituals was expected, they also provided a sense of security to those concerned about potentially dangerous activities (e.g., ocean travel) and about issues related to life after death. Given their spiritual core, rituals and ceremonies were an essential part of the traditional Hawaiian lifestyle.

CONSERVATION: CAN NATIVE FOREST BIRDS BE HEARD ABOVE THE DIN?

Eventually, colonization by outside cultures contributed to loss of the spiritual ideol-ogy associated with native manu and their hulu. The value and utility of the forest birds and feather work were replaced by gems and other material possessions. The ali'i and maka'āinana became fascinated by and interested in acquiring material goods from foreign countries, and the spiritual significance of feather work faded with time.

Nineteenth-Century Pleas for Conservation

The subfossil record (e.g., Olson and James 1982b, James et al. 1987, James and Olson 1991) tells us that many forest bird species went extinct between the time of the first Polynesian settlement and that of Hawaiians' contact with Europeans. Extinction rates increased dramatically postcontact. Both pre- and postcontact habitat alteration (e.g., forest clearing, introduction of alien species and diseases) were the most likely causes of extinction, although we do know that predation by both Hawaiians and later settlers may have dramatically affected ground-nesting birds (Chapter 2). Rose et al. (1993) speculated that the capture of birds for collection of feathers did not have a significant effect on bird populations until guns were introduced (see Box 3.2).

In the mid-nineteenth century, private clubs (e.g., Hui Manu) and local agencies began introducing other bird species to the islands (Castle 1935, Dillingham 1938). In 1871, an observant and concerned Hawaiian, Penukahi (1871), noticing the impact of these foreign bird species on native bird populations, made a plea to the legislature:

> My friends must be wondering who these lost natives are and think perhaps that they are old men. No, not they. Some have gone on the usual way of all earthly beings and we knew of their going. These natives that I am talking about, we know not where they are gone. It is this, the native of our upland,

the iiwi, the o-u, the akakane, the amak-ihi, the oolomao, the elepaio, these are the natives that are lost. Some of you may ask, "What is the reason for their being lost?" I will tell you, it is because of the increase of the bad birds from foreign lands on our plains, mountains, mountain tops, vallies, cliffs, forest, taro patch borders, shores and streams. That is why these natives are lost because the bad birds are increasing among us and they harm the plants, and the wild food plants. . . . They were small birds with beautiful voices and feathers. We enjoyed watching them when we were small children. When a gale blew here, the birds of the mountains came out and gathered before the doors of the houses. It was fun to see the leaves of the ilima move when we were little and these were our playmates when we were small children. Before, when the other birds had not come, there were many iiwi, amak-ihi, akakane, o-u, oolokela, and elepaio right around here, on the cannas, on the hau trees, on the small noni trees and farther up there were flocks of them among the blossoms of the mountain apples, on the low lehua trees and on the tall ohia trees. They were the interesting things of the upland but now, they are lost. . . . When the legislature is in session again, the law relating on these birds from foreign lands can be repealed for they are harmful.

Subsequently, an article that appeared in the newspaper *Paradise of the Pacific* echoed Penukahi's worst fears. In this article, titled "Hawaiian Birds," Munro (1895: 50) stated that Hawaiian forest birds, unused to a ground enemy, easily fell prey to cats, rats, or mongooses. He noted that the Common Mynah were destroying the nests of native birds and that the destruction of the forests by cattle and deer was playing an important part in their extinction.

Although challenges to the existence of Hawaiian birds were recognized for nearly

a century, very little action was taken to rectify the plight of the forest birds until the conservation movement arose in the middle of the twentieth century. A description of this sad history is found in Chapter 2. Every day native Hawaiian birds face the challenge of surviving alongside these and other alien species amid the ecological changes associated with their presence. However, the resurgence in Hawaiian cultural interest in native birds has stimulated a broader awareness of and concern about their status. We hope this will also result in a recapturing of the value of ancient Hawaiian relationships with forest birds.

SUMMARY

In this chapter we have described many roles that forest birds played in traditional Hawaiian practices and how they were perceived by Hawaiian people. It is apparent that the birds were important both for utilitarian purposes, such as food and implements, and for ritual and metaphysical purposes. The significance of Hawaiian forest birds has been highlighted using a framework of cultural dimensions that emphasize multiple sociopsychological aspects of individual life in traditional Hawaiian society. In the social dimension of culture, forest birds and their feathers were used to denote and separate the ruling and common classes. The craft of skilled bird catchers and the use of birds as food were central to the experiential dimension. The mythical dimension identified the role of forest birds as deities and manifestations of gods, and in the doctrinal dimension, forest birds were woven into legends and provided allegories for good and bad behavior. The experiences of traditional Hawaiians with nature and forest birds were mirrored in mythology and legends and reflected social values of cooperation, kindness, unselfishness, pride, safety, worthiness, responsibility, and loyalty. Inherently, forest birds were essential players in the success or failure of traditional customs and practices. Moreover, their beauty and behavior were recognized and appreciated by the ali'i and maka'āinana alike.

We have attempted to provide an understanding of the extent to which forest birds were interwoven with the lives of Hawaiians in traditional culture. A sense of the Hawaiian worldview can inspire and deepen our enjoyment and appreciation of the cultural importance of Hawaiian birds every time we hear the call of an 'I'iwi or 'Apapane while walking through the wao akua (upland, often uninhabited areas) of a Hawaiian forest. Today the last of the forest birds are in peril. Their fate will be decided in part through individual choice of action or inaction, driven by the beliefs, attitudes, and knowledge of every person in Hawai'i, be they kama'āina (native-born), malihini (newcomers), or visitors.

ACKNOWLEDGMENTS

Working together on this chapter was good fun. We are most grateful to Pat Bacon (Bernice Pauahi Bishop Museum), Kepā Maly (Kumu Pono Associates), Samuel M. Gon III (The Nature Conservancy–Hawai'i), and two anonymous reviewers for their comments and suggestions.

PART 2

Status, Biology, and Limiting Factors

CHAPTER FOUR

Monitoring Hawaiian Forest Birds

RICHARD J. CAMP,
MICHELLE H. REYNOLDS,
BETHANY L. WOODWORTH,
P. MARCOS GORRESEN,
AND THANE K. PRATT

With the threat of extinction looming over many island species, the urgency for accurate and reliable species assessments has pushed Hawai'i toward the forefront of estimating forest bird populations. Estimating species distribution and density is especially important for habitat protection in the islands because most endemic species have limited ranges, specific habitat requirements, and disease vulnerabilities and occur as single populations (Scott et al. 1986; this volume, Chapters 1 and 2). Moreover, the phenomenal pace of ecosystem change in Hawai'i makes effective monitoring an integral component of effective and efficient conservation. To meet the special challenges posed by rare birds inhabiting dense forests over rugged terrain, Hawaiian researchers have contributed to the development of sampling theory and methodology for determining forest bird distribution and density (Ramsey and Scott 1979; Reynolds et al. 1980; Scott, Jacobi, et al. 1981; Scott, Ramsey, et al. 1981). Over the past several decades, the results of forest bird surveys have guided the placement and design of preserves (Chapter 16), provided the information for the process to list or delist endangered species (U.S. Fish and Wildlife Service 2006), yielded insight into the dynamics of an evolving avian malaria and forest bird transmission and resistance system (Woodworth et al. 2005), and allowed for assessment of the effectiveness of management actions such as predator control and reforestation (VanderWerf and Smith 2002; Camp, Pratt, et al. 2009; U.S. Geological Survey Pacific Island Ecosystems Research Center [USGS-PIERC], unpubl. data).

Monitoring measures changes in the distribution, size (relative or actual abundance), or structure of a population. Most large-scale monitoring projects worldwide use methods that provide indexes of population size, such as relative abundance of detections (where the total number of birds detected is related to their actual abundance). One of the most commonly used methods is point counts, wherein the observer tallies the number of birds seen or heard at each station or point but does not record the distance to each bird (not to be confused with the point transects of distance sampling, discussed later). Examples include the U.S. Breeding Bird Survey and programs in Australia, Norway, and France and adjacent countries (Blondel et al. 1970, Bystrak 1981, Barrett et al. 2003, Husby 2003). Although point counts are useful tools for monitoring bird populations (Rosenstock et al. 2002), they do not provide estimates of actual population sizes.

Increasingly, large-scale monitoring programs are adopting distance sampling to generate estimates of population size. Distance sampling, whereby the counter records the distance to the bird from a station or transect, allows for calculating detection probabilities and density estimates from which population sizes can be generated. Line transect methods, a form of distance sampling, are used to monitor breeding birds throughout the United Kingdom on the Breeding Bird Survey (Marchant 1994), but the U.S. Breeding Bird Survey uses different methodology (see Peterjohn et al. 1995). This program was initiated in 1992 and replaced the point-count-based Common Birds Census in 1994. Line transect sampling is also applied for monitoring migratory birds in British Columbia, Canada (Martin and Ogle 2000). The Rocky Mountain Bird Observatory (RMBO) uses point transect (also called variable circular plot [VCP]) and line transect methods to survey breeding birds in Colorado, Wyoming, and other areas in the United States (Leukering et al. 2000; T. Leukering, RMBO, pers. comm.). Distance sampling methods are primarily used for monitoring bird densities and distribution, although they are also used for monitoring ratios of relative abundance over time and space.

The baseline for forest bird monitoring in Hawai'i has been the Hawai'i Forest Bird Survey (HFBS). In 1976–1983, the U.S. Fish and Wildlife Service (USFWS) conducted a systematic and comprehensive survey of forest birds and habitat on most of the main Hawaiian islands (Scott et al. 1986). Observers were deployed to tally birds at stations (points) along transects using VCP counts. The HFBS provided the first quantitative population estimates for all Hawaiian forest birds. Since then, VCP counts have become the standard for subsequent forest bird surveys in Hawai'i. The HFBS revealed that the largest populations of forest birds and prime habitat fell outside the best-protected areas (U.S. Fish and Wildlife Service 1982b). Because of these results, new key reserves, such as Hakalau Forest National Wildlife Refuge and Hanawī Natural Area Reserve, were created on lands identified as hotspots by the HFBS. Unfortunately, the utility of the HFBS results faded with time as population estimates and range maps became outdated. Up-to-date information on the status and trends of forest bird populations in Hawai'i was needed, particularly as feedback on management programs for restoring habitat.

The Hawai'i Forest Bird Interagency Database Project (HFBIDP) was created in 1999 by the USGS and its cooperators to fill this need. The HFBIDP brought together data from almost 600 surveys of forest birds conducted over the past quarter century by numerous governmental and private organizations throughout Hawai'i. The HFBIDP has now compiled and analyzed these data for bird distribution, density, and trends and has made them available to managers and decision makers (Gorresen et al. 2005; Camp, Gorresen, et al. 2009). Highlights of the results of this synthesis, including information on distribution, abundance, and trends in native bird populations in key areas, are presented in Chapter 5. In this chapter we present an overview of the history and development of forest bird surveys in Hawai'i, describe the survey and analysis techniques most commonly used, and provide an assessment of current monitoring throughout the state. We intend for this review to contribute to the conservation of Hawaiian forest birds by highlighting the importance of forest bird monitoring and presenting guidelines for improvement of monitoring in the future.

BIRD SURVEYS IN HAWAI'I

There is an extensive legacy of surveying forest birds in Hawai'i. Attempts to identify and describe the populations began with the historical explorations and discoveries of early naturalists (Rothschild 1892, Henshaw 1902, Perkins 1903) and have culminated today in systematic searches for rare birds (Reynolds and Snetsinger 2001) and in large-scale monitoring programs (Scott et al. 1986; this volume, Chapter 5). Although these and other surveys have had various objectives, they have documented species distributions and abundances and, when conducted over sufficiently large areas and time periods, have provided estimates of population sizes and trends (e.g., Scott et al. 1984, 1986; Fancy et al. 1996; Jacobi et al. 1996; Leonard et al. 2008; Camp, Pratt, et al. 2009).

Early Surveys

The first systematic survey to ascertain the status of Hawaiian birds across the archipelago was conducted from 1935 to 1937 by the pioneer naturalist G. C. Munro (1944) after he became aware of the devastating population crashes at the beginning of the twentieth century. Munro's work provided an abundance index (extremely numerous, abundant, uncommon, etc.) for native species and documented range contractions and population declines. In the years that followed, bird sightings

(or lack of them) helped chronicle further declines and range contractions of native birds and the expansion of alien species. However, it was not until a survey of the Alaka'i Plateau of Kaua'i in 1968–1973 by John Sincock that estimates of population sizes and associated standard errors were produced for Hawaiian forest birds (U.S. Fish and Wildlife Service 1983b, Scott et al. 1986). This work initiated a new paradigm for monitoring forest birds in Hawai'i and ushered in the use of standardized sampling techniques to obtain statistically robust density and population estimates. In addition, with Sincock's surveys the paradigm shifted from indexes of abundance to quantitative estimates based on detection probabilities. Other early examples of Hawaiian bird surveys using quantitative approaches include those of Conant (1975), Ramsey and Scott (1979), and Reynolds et al. (1980).

The Hawai'i Forest Bird Survey

The basis for long-term population monitoring in Hawai'i was established by the landmark HFBS (Scott et al. 1986). The principal objective of the HFBS was to determine for each species its distribution, population size, and density pattern by habitat using VCP sampling (Scott et al. 1986). This monumental eight-year survey spanned 4,114 km² in 12 study areas on five forested main islands, excluding only O'ahu (Figs. 4.1–4.3). Nearly 10,000 stations were sampled along transects that traversed the most rugged and inaccessible terrain in the state.

The HFBS was the first to apply VCP sampling on a large scale, and the methodology was developed, applied, and statistically supported simultaneously by much of the early research conducted in Hawai'i (Emlen 1971; Ramsey and Scott 1978, 1979; Reynolds et al. 1980; Scott, Ramsey, et al. 1981). Because few native forests remained below about 600 m, the HFBS concentrated on surveying all of the mid- and high-

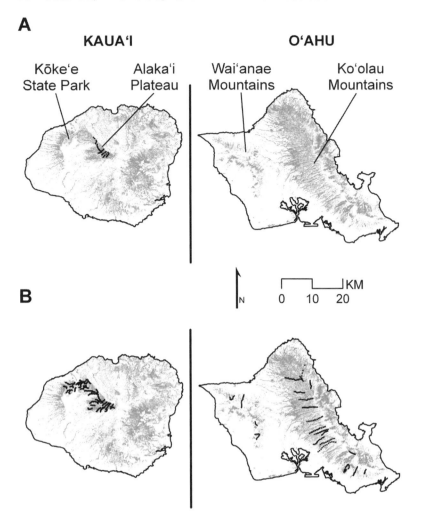

Figure 4.1. Locations of survey transects (black points) on (A) Kaua'i (1981) and O'ahu (not surveyed) for the Hawai'i Forest Bird Survey and (B) subsequent surveys (1989–2008). On Kaua'i, the 2000 survey greatly expanded coverage. O'ahu had been surveyed only once, in 1991. Shading indicates areas designated as forest habitat by the National Oceanic and Atmospheric Administration, Coastal Change Analysis Program (1995).

elevation forests. Transects were spaced 3–5 km apart across the upper reaches of the mountainsides (see Figs. 4.1–4.3), connecting a series of stations spaced evenly and far enough apart that the same individual birds would typically not be recounted, generally at 134 m intervals (later surveys have generally used 150 m intervals). Surveys were conducted in May–August using 8-minute counts. The observer recorded the distances from the center of the station to all birds seen or heard, the time, and the weather conditions. In analysis, the number of birds counted was adjusted for detectability, where detectability varied as a function of distance from the observer (Fig. 4.4). Estimates of abundance for each species were then used to calculate density by study area and habitat type (Box 4.1). The HFBS generated the first, and for many species the only, rangewide data in Hawai'i and established the baseline from which to monitor population trends.

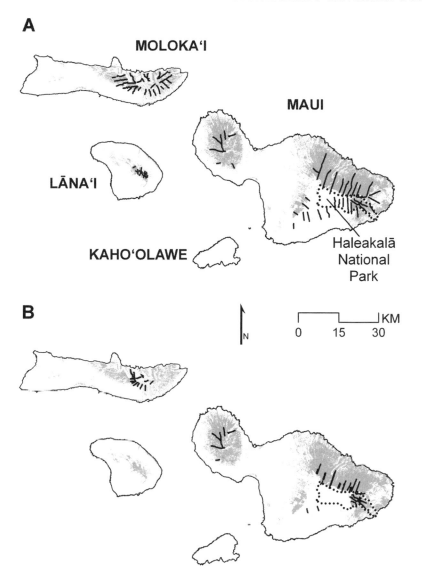

Figure 4.2. Locations of survey transects on Maui, Lāna'i, and Moloka'i for (A) the Hawai'i Forest Bird Survey (1979–1980) and (B) subsequent surveys (1988–2004). Haleakalā National Park is delineated with dotted lines.

Monitoring since the Hawai'i Forest Bird Survey

Since the HFBS, some of the original transects have been resampled and many new transects have been established (see Figs. 4.1–4.3). In 1988, the Hawai'i State Division of Forestry and Wildlife (DOFAW) initiated a program to monitor long-term population trends by resampling HFBS transects with the goal of resurveying each of the major regions once every five years. Surveys have focused on areas that support core populations of forest birds, and to date they have covered approximately 17% of the original HFBS study area.

The Palila (*Loxioides bailleui*), an endangered finch-billed Hawaiian honeycreeper restricted to the subalpine dry forests of Mauna Kea (see Fig. 4.3), Hawai'i Island, is the subject of the longest-lasting continuous

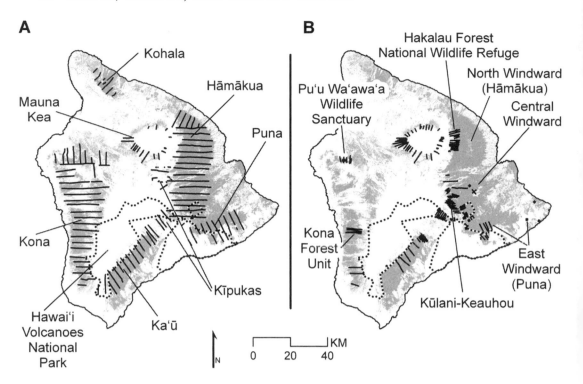

Figure 4.3. Locations of survey transects (solid lines), region names, and native and exotic forests and woodlands (shaded areas) on Hawai'i Island for (A) the Hawai'i Forest Bird Survey (HFBS) and (B) subsequent surveys. The HFBS was conducted between 1976 and 1983 with transect coverage closely matching the extent of mid- and high-elevation forest. Subsequent surveys were conducted between 1978 and 2008 and focused on management areas and populations of special concern. Hawai'i Volcanoes National Park is delineated with dotted lines.

forest bird–monitoring program in the islands (Jacobi et al. 1996; Johnson et al. 2006; Leonard et al. 2008; this volume, Chapter 23). Indeed, the Palila is one of the longest-monitored passerines (perching birds) anywhere. The Palila monitoring program, initiated in 1980, spans a 29-year period and now incorporates 32 transects covering 105 km². Surveys are conducted during the nonbreeding season.

Another long-term monitoring program has been established and maintained at Hakalau Forest National Wildlife Ref-

uge (Hakalau), on the windward slope of Mauna Kea, Hawai'i Island (see Fig. 4.3). The 133 km² refuge is the largest actively protected area of mid- to high-elevation rain forest on the island of Hawai'i, and it encompasses core populations of three endangered species, the 'Akiapōlā'au (*Hemignathus wilsoni*), Hawai'i 'Ākepa (*Loxops coccineus*), and Hawai'i Creeper (*Oreomystis mana*). Annual surveys have been conducted at Hakalau since 1987 in forests between 1,000 and 2,000 m in elevation, sampling 350 stations on 15 transects and covering an area of 55 km². In 1999, the surveys were extended by 11 km² into the World Union Parcel to examine trends at lower elevations on the refuge. Survey results indicate that populations of all eight native forest birds are stable or have increased in the central core of the refuge (Camp, Pratt, et al. 2009). Further research will explore the effects on bird populations of management actions such as ungulate removal, rodent control, and forest regeneration.

A

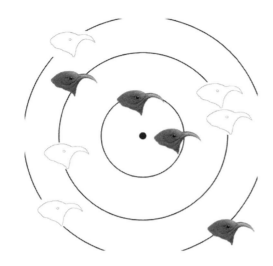

B

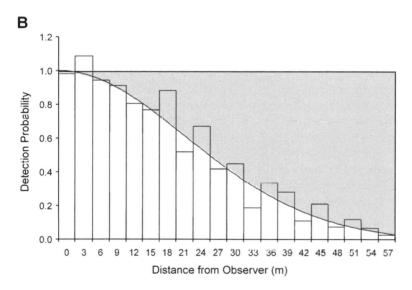

Figure 4.4. Data acquired by point transect distance sampling. This method adjusts the estimated abundance by determining the relationship of observed birds with a probability of detection. (A) Hypothetical bird distribution at a sampling station. The bull's-eye represents the station center point, and the circles are distance intervals from the center point. Solid 'I'iwi represent birds detected, while open 'I'iwi are undetected birds. (B) Histogram of distance data and best-fitting detection function for 'I'iwi at Hakalau Forest National Wildlife Refuge, Hawai'i Island, from surveys conducted between 1986 and 2000. The curve represents the corresponding estimated detection function. Fewer 'I'iwi are detected at greater distances from the center point. The estimate of the proportion of 'I'iwi missed during sampling is shown in the shaded area. Density is then calculated as the number of birds detected divided by the expected proportion of 'I'iwi in the effective area of detection. *Source:* Illustration of 'I'iwi © H. D. Pratt.

Box 4.1. Variable Circular Plot Analysis

VCP and other distance sampling methods are unique because they can provide estimates of density, and thus population size, in contrast to point counts and other index methods that yield only relative abundance. To estimate densities, distance measurements are modeled to calculate a species-specific detection function and a probability of detection, which is subsequently used to estimate bird density and abundance (Buckland et al. 1993, 2001). Distance sampling assumes that all birds are detected with certainty at the station center point, that birds are detected prior to any responsive movement, that distances are measured without error, and that species are correctly identified. Distance sampling theory takes into account that some birds away from the station center point go undetected and that the farther a bird is from the station the greater the probability it will be missed. Therefore, density is the number of birds detected adjusted by the detection probability and divided by the area surveyed (see Fig. 4.4).

Distance sampling is not without difficulties. In particular, meeting the estimator's assumptions and ascertaining detection probabilities are both potential problem areas that are subject to current debate. Although a full discussion of the advantages and limitations of distance sampling is beyond the scope of this chapter, the interested reader may consult Cassey and McArdle (1999); Barry and Welsh (2001); Rosenstock et al. (2002); Bart, Droege, et al. (2004); Fewster and Buckland (2004); Fewster et al. (2005); and references therein. To summarize, when distance sampling methods and analyses are applied correctly and the assumptions are met, the estimator has been shown to perform to expectation and produce reliable population estimates

that are unbiased and precise (Fewster et al. 2005). Current debate centers on the extent to which it is possible to meet the model assumptions in practice and how best to handle deviations or violations for the most reliable estimator.

Many factors in addition to distance may influence the probability of detecting birds. Some of these factors can be standardized in the analysis, such as habitat, sampling conditions (cloud cover, rain, wind, gusts, and time of day), and observer variability (Buckland et al. 2004, Thomas et al. 2006). Other factors that affect detection of individuals, such as sex, age, and breeding status, cannot be accounted for. Thus biologists must be aware of which segment of the population they are monitoring.

Precision in population estimates can be greatly improved by stratifying on habitat variables, a statistical procedure whereby stations within similar habitats are pooled to produce estimates. For example, Scott et al. (1986) stratified the 139 km^2 Mauna Kea study area into two habitat types (māmane [*Sophora chrysophlla*] and māmane-naio [*Myoporum sandwicense*]), two canopy cover classes (<25% and ≥25% tree cover), and six elevation strata (200 m intervals between 1,900 and 3,100 m). Both the detection function and the habitat are assumed to be constant within a stratum but to vary across strata. Habitats are, however, rarely homogeneous. Furthermore, large numbers of stations are required for each stratum to produce reliable density estimates.

An alternative to a stratum-based approach is to predict species occurrence and population size across an entire study area using geographic information systems (GIS) and regression-based species-habitat models, accounting for covariables (Thomas et al. 2006, Gorresen

---

Box 4.1. *Continued*

et al. 2009). This approach uses habitat variables derived from satellite imagery and extracted for each survey station. Using regression, species-habitat models are developed relating density estimates and habitat characteristics for each survey station. A density surface is then predicted for the study area from the species-habitat models and satellite imagery in a GIS (Plate 22). This predictive modeling holds great promise for improving population estimates, and it has been utilized by the Hawai'i Forest Bird Interagency Database Project, USGS-PIERC.

In 1990, another set of surveys in the Kūlani-Keauhou region was established to monitor bird trends in mid- to high-elevation rain forest on windward Mauna Loa, Hawai'i Island, including public and private lands adjoining Hawai'i Volcanoes National Park (see Fig. 4.3). The avifauna is similar to that of Hakalau and harbors core populations of 'Akiapōlā'au, Hawai'i 'Ākepa, and Hawai'i Creeper. Taken together, these monitoring programs, established independently by Kamehameha Schools, the National Park Service (NPS), and the USGS, have conducted annual surveys of native-dominated forests between 750 and 2,100 m elevation up to the present, although spatial and temporal coverage have not always been complete. In 1994, the 'Ōla'a-Kīlauea Partnership was established to coordinate management, research, and monitoring within this complex of private and public conservation lands.

There have been many smaller-scale (<50 km²) or short-term (<10 years) surveys throughout Hawai'i, including at Pu'u Wa'awa'a Wildlife Sanctuary (USGS-PIERC), Haleakalā National Park (NPS), and the Kona Forest Unit of Hakalau Forest National Wildlife Refuge (see Figs. 4.1–4.3). Despite the limited scale of these surveys, taken together they have represented an opportunity to combine scattered geographic coverage and examine wider patterns in species status and trends (this volume, Chapter 5).

Other Surveys and Monitoring Studies

A variety of other methods have been employed to estimate the population size, relative abundance, or demographic parameters of Hawaiian forest birds. These methods include point counts, spot mapping, censuses (direct counts of all individuals in the population), line transects, and mark-recapture studies. One of the best-known nationwide monitoring programs, the Christmas Bird Count (CBC) (Butcher 1990, Ralph et al. 1995) has been conducted annually throughout the state since 1944 (R. L. Pyle, pers. comm.), but its coverage takes in mostly accessible urban and agricultural areas and little native forest bird habitat. Although the CBC is not used for monitoring native Hawaiian forest birds, it is the best source of data in Hawai'i for birds in habitats other than forest environments and has documented the spread of alien birds that have invaded forests. (Hawai'i is not a participant in the other nationwide monitoring program, the U.S. Breeding Bird Survey.)

Fixed radius point counts (FRPCs) are similar to VCP counts except that distances are recorded only as within or outside of a fixed-radius circle of the observer (Hutto et al. 1986). The radius is determined through preliminary surveys in each habitat, and ideally it encompasses an area that can be completely surveyed, minimizing missed birds (Hutto et al. 1986). FRPCs provide an index of population size; therefore, they cannot be used to estimate densities. They have been used in only a few studies in Hawai'i (Shallenberger and Vaughn 1978, Reynolds et al. 2003).

Spot mapping, or territory mapping, is based on the exclusive spacing of territorial birds and is a fairly intensive method requiring repeated visits to the study area within a breeding season. Population size is derived from the number of individual territories in a study area (Bibby et al. 2000). In Hawai'i, spot mapping has rarely been used to calculate population size because most native species are not territorial (Pratt, Simon, et al. 2001). The few species for which the method has been applied include Puaiohi (*Myadestes palmeri*) (Snetsinger et al. 1999) and O'ahu 'Elepaio (*Chasiempis sandwichensis ibidis*) (VanderWerf et al. 1997). Because Puaiohi inhabit isolated stream drainages and are seasonally vocal, VCP methods have proven ineffective for this species, whereas O'ahu 'Elepaio are too rare and patchily distributed to be well surveyed by VCP. In addition, Nelson and Fancy (1999) used spot-mapping methods to validate VCP techniques for a translocated 'Ōma'o (*Myadestes obscurus*) population.

Although VCP has been the primary method used to survey birds on a large geographic scale, it is not appropriate for all species, and it provides poor estimates for extremely rare birds (Buckland et al. 2001). In recognition of this problem, the Rare Bird Search (RBS) was undertaken in 1994–1996 to update the status and distribution of 13 "missing" Hawaiian forest birds (Reynolds and Snetsinger 2001). RBS surveyors made 23 expeditions to forests on Kaua'i, Maui, Moloka'i, and Hawai'i. The RBS located four critically endangered species or populations but failed to find other species (Reynolds and Snetsinger 2001). The RBS methods applied in Hawai'i are a modified form of the area-search technique (Ralph et al. 1993, Bibby et al. 2000), whereby transects or search routes are surveyed instead of plots. Perhaps most important, rare bird searches can be used to quantify the probability of detection within a species' range to evaluate the likelihood of extinction and the search effort needed to be confident that a species present in the area was not missed (Reynolds and Snetsinger 2001, Scott et al. 2008).

Where a population is extremely small, such as that of 'Alalā (*Corvus hawaiiensis*) or Po'ouli (*Melamprosops phaeosoma*), researchers have conducted targeted searches for colorbanded individuals to obtain a total census of the population (this volume, Chapters 20 and 21).

Distance sampling may also be used with line transects, instead of VCP. Line transect methods, whereby the observer walks along a transect and records the perpendicular distance to each bird seen or heard, have been used to determine the population density of the 'Io (Hawaiian Hawk, *Buteo solitarius*), Nihoa Millerbird (*Acrocephalus familiaris*), Nihoa Finch (*Telespiza ultima*), and Laysan Finch (*Telespiza cantans*) (Conant et al. 1981; Morin and Conant 1994; this volume, Chapter 12). Distance sampling with line transects is most useful for open and contiguous habitats such as grass and shrubland habitats that are easily traversed and for surveys of conspicuous species.

Finally, mark-recapture methods, where birds are captured in mist nets and banded and are later identified upon recapture or resighting, have been used by a number of researchers in Hawai'i to measure and monitor relative abundance (birds per 100 mist net–hours) and demographic characteristics of populations such as movements, survival, and fecundity. Studies of variable length and intensity have been conducted in high-elevation native forests at Hakalau (e.g., Hart 2001, Woodworth et al. 2001, Freed et al. 2008), Haleakalā (Simon et al. 2000, 2001), Keauhou Ranch (Ralph and Fancy 1994a, 1994b, 1994c, 1995, 1996), windward Mauna Loa (van Riper et al. 1986, Woodworth et al. 2005), subalpine Mauna Kea (Lindsey et al. 1995), low-elevation Oʻahu and Hawaiʻi Islands (Shehata et al. 2001, Woodworth et al. 2005), and other locations. Although mark-recapture methods are available for estimating population size (e.g., Otis et al. 1978, Pollock et al. 1990) and population growth rates (Pradel 1996), these methods are more labor intensive than distance sampling methods; more effort is required to capture birds in mist nets than to detect them by sight and sound. Although monitoring birds with distance sampling may be more efficient over large areas, there are situations, such as those involving rare and endangered birds, in which tracking demographic characteristics will yield better insight into population trends (Steidl 2001).

RECENT ADVANCES IN DATA ANALYSIS AND THEIR APPLICATIONS IN HAWAIʻI

Hawaiian monitoring programs use a distance-sampling framework that allows for calculating the detection probabilities of individual species in order to estimate population size and variance (Reynolds et al. 1980, Buckland et al. 2001). Three recent developments in distance sampling have improved parameter estimation: (1) using geographic information systems to model population estimates, (2) incorporating covariates, and using (3) an information-theoretic approach in model selection.

Previous Hawaiian studies have estimated population sizes based on a combination of habitat and elevation strata (e.g., Scott et al. 1986, Jacobi et al. 1996, Banko et al. 1998). Stratification groups members of a population into relatively homogeneous subgroups, in this case geographically. This approach relies on the homogeneity of sampling unit estimates within a stratum to reduce variability. Although this assumption is plausible, it is rarely met even for narrowly defined strata, and in the past, computing technology severely limited the number of factors that could be used to define a stratum. Present computing capabilities permit the use of multiple biotic factors (e.g., habitat type, canopy height, and closure) and abiotic factors (e.g., rainfall, elevation, and slope) in a geographic information system to produce population estimates (Gorresen et al. 2007, 2009). The HFBIDP is applying these techniques to estimate the population distributions and densities of native and alien birds on Hawaiʻi Island (Gorresen et al. 2007; this volume, Plate 22), and intends to continue this process for forest birds on other islands.

The second development is the use of maximum likelihood methods to account for the effects of covariates on detection functions (Marques and Buckland 2003, Thomas et al. 2006). Methods previously used to account for covariates calibrated the distance measurements to a set of reference conditions (Fancy 1997). The current procedures adjust the scale of the detection function by the covariates instead of modifying the raw data (Marques and Buckland 2003). Covariates adjust the scale parameter in detection probability estimation and are assumed to affect the rate at which detectability decreases as a function of distance. The covariates do not influence the shape parameter, which is conditional on the key function and adjustment terms. If the proportion of birds available—those singing or visible so that they can be counted—varies in a temporal manner, a separately fitted detection function is needed (inclusion of a season covariate is not sufficient). If a seasonal effect results in birds not being detected with certainty at the station center point, the model will underestimate the population size. Thus, seasonal detection probabilities need to be evaluated, and sampling should be conducted during the season when birds are most detectable.

Previously used model selection procedures relied on methods that required subjective model comparisons, such as binning data into intervals, assessing fit with goodness-of-fit tests, and selecting the model that did not reject the null hypothesis. Instead, an information-theoretic approach relies on a formulation that produces classes with the smallest loss of information about the data and thus does not require the subjective assessment of similarity. Conceptually, one generates a suite of relative plausibility candidate models and then, using an information criterion, selects the best-fit model from a suite of candidate models that describes the data as efficiently as possible. The best-fit model is the model that best approximates the data under the principle of parsimony (Burnham and Anderson 2002). Population size and variance are then calculated from this best approximating model, or, in a situation in which multiple reasonable models exist, model averaging may be used to generate estimates.

ASSESSMENT OF PAST AND CURRENT MONITORING PROGRAMS IN HAWAI'I

Monitoring is not simply surveillance. It is a program of repeated surveying with clearly defined measurable objectives, assessment to baseline parameters, standardized field and analytical methods, and a mechanism to implement actions (Greenwood et al. 1994). Carefully designed studies will ensure that monitoring programs attain their objectives and detect biologically meaningful population trends. Forest bird monitoring programs in Hawai'i strive to provide land managers, conservationists, and scientists with information on the status and trends of bird populations and the outcomes of management actions, but this information can be obtained only from well-designed monitoring programs. Evaluation of existing programs will elucidate limitations and provide insight for the improvement and design of future programs (Cochran 1977). How have past and current monitoring programs met these ob-jectives, and how can they be modified in the future to be more effective?

What Have We Accomplished?

The development of centralized data repositories is one of the most difficult, yet most important, tasks undertaken by monitoring projects (Bart 2005). Without an institutionalized means of processing and reporting the data, evaluation of monitoring programs cannot be conducted. The establishment of the HFBIDP in 1999 marked an important chapter in forest bird monitoring in Hawai'i. The HFBIDP represents the combined efforts of 11 organizations and agencies, and now contains data on almost 600 surveys (over one million detections of individual birds) conducted over 30 years. The HFBIDP will ensure data archiving, permit synthesis of data from multiple surveys, and provide a basis from which to assess species status and monitoring effectiveness.

Below, we take a two-fold approach to address what has been accomplished in monitoring Hawaiian forest birds. First, we review the spatial and taxonomic extent of past surveys, and second, we assess whether monitoring objectives are being met.

THE SPATIAL AND TAXONOMIC
EXTENT OF SURVEYS

Overall, there are approximately 4,500 km² of forest on the main Hawaiian islands (National Oceanic and Atmospheric Administration, Coastal Change Analysis Program 1995). Of these, approximately 1,900 km² (42%) have been surveyed for forest birds using quantitative methods. In order to examine population trends in areas of interest, survey transects and stations must be resampled over time. Approximately 600 km² (13% of forests) have been sampled two or more times, allowing for the analysis of population trends (Fig. 4.5).

Overwhelmingly, the HFBS sampled more habitat on each island than any other

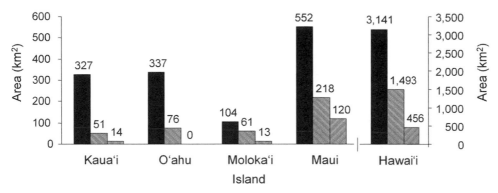

Figure 4.5. The amount of forest bird habitat surveyed on the main Hawaiian Islands. A small fraction of forest habitat has been sampled using variable circular plot methods to assess species distributions and trends. The dark solid bar indicates forest cover derived from the National Oceanic and Atmospheric Administration, Coastal Change Analysis Program (1995), and includes forest and woodland cover types. The left-leaning hatch marks indicate forest cover sampled at least once. The right-leaning hatch marks indicate the total extent of forest sampled two or more times, allowing for analysis of trends. The area was determined by arbitrarily delineating a 1 km buffer around survey stations and summing the amount of forest within the buffer. The area surveyed on Lāna'i was almost one-half the forest cover (13 of 27 km²); however, Lāna'i has not been resurveyed.

survey, with the exception of Kaua'i, where the HFBS coverage of 32 km² was expanded to 51 km² by a survey in 2000. Proportional to its area, Moloka'i has received the most extensive coverage, with 61 km² (59%) of its forests surveyed, whereas Kaua'i has had the least coverage, with just 51 km² (16%) of its forests surveyed (see Fig. 4.5). Of the original HFBS transects and stations, about 30% have been resampled. The amount of resampled area has varied by island, from O'ahu and Lāna'i, which have not been resampled at all, to Maui, where 55% of the survey area (120 of 218 km²) has been resampled (see Fig. 4.5).

Hawaiian forests harbor at least 22 species of native forest birds that are currently extant; 10 of these are endangered or otherwise of concern; a further 15 species

are endangered but probably no longer extant (Chapter 5, Table 5.1). Are we adequately monitoring each of these species? Bart, Burnham, et al. (2004) proposed a standard requiring that at least two-thirds of a species' range be repeatedly sampled for trends. All of the USFWS-designated endangered bird species in Hawai'i are currently subject to monitoring programs using VCP, rare bird searches, surveillance of banded birds, or spot mapping. However, these programs frequently cover only a portion of the species' ranges. Species with ranges <100 km² have been well surveyed for trends on all islands, except O'ahu (Fig. 4.6). On Maui, more than 66% of the ranges of Maui Parrotbill (*Pseudonestor xanthophrys*), Maui 'Alauahio (*Paroreomyza montana*), and 'Ākohekohe (*Palmeria dolei*) have been repeatedly sampled for trends. On Kaua'i, 55% and 54%, respectively, of Puaiohi and 'Akikiki (*Oreomystis bairdi*) ranges have been sampled for trends. In addition to being surveyed by VCP, roughly two-thirds of the Puaiohi population is being actively monitored using spot-mapping methods (P. Roberts and T. Savre, pers. comm.). Although the O'ahu 'Elepaio has not been monitored in repeated VCP surveys, monitoring of <10% of the species' range using spot-mapping techniques is conducted regularly (E. VanderWerf, pers. comm.). Two of five extant endangered species on Hawai'i, Palila and Hawai'i 'Ākepa, with ranges of 100–350 km², have received more than 50% repeated sampling coverage, with the Palila receiving more

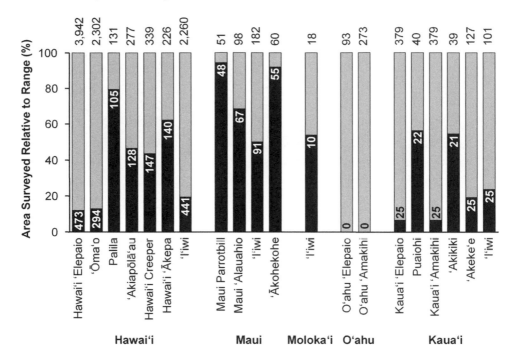

Figure 4.6. Proportion of species' ranges repeatedly surveyed (dark gray) and not sampled (light gray) for select species. A modest fraction of most species' ranges has been repeatedly sampled to permit assessment of change in distribution and abundance. Size of species' ranges (values above bars, km²) were determined by manually delineating records of species occurrence (Camp, Gorresen, et al. 2009), including and subsequent to the Hawai'i Forest Bird Survey. Inclusion of unoccupied and unsuitable habitats may inflate range estimates. Repeatedly surveyed areas (values within bars, km²) were delineated by a minimum convex polygon around coincident surveys (for details, see Camp, Gorresen, et al. 2009). Roughly two-thirds of the Puaiohi population (P. Roberts and T. Savre, pers. comm.) and less than one-tenth of the O'ahu 'Elepaio range (E. Vander-Werf, pers. comm.) are actively monitored using spot mapping techniques.

than 66% coverage. In contrast, ranges of 'Akiapōlā'au, Hawai'i Creeper, and 'Akeke'e (*Loxops caeruleirostris*) have not been repeatedly sampled as thoroughly (proportion of coverage 46%, 43%, and 19%, respectively).

Species with ranges >350 km² have been less well sampled, as was expected for broadly distributed nonendangered species. Only 6–13% of the ranges of Hawai'i and Kaua'i subspecies of 'Elepaio (*Chasiempis sandwichensis*) and 'Ōma'o and

of Kaua'i 'Amakihi (*Hemignathus kauaiensis*) have been sampled in multiple years (see Fig. 4.6). 'I'iwi (*Vestiaria coccinea*), a species recently showing declines throughout much of its range (Chapter 5), occurs on all Hawaiian islands except Lāna'i, and the proportion of its range repeatedly sampled has varied between 0 and 54% per island. The 'I'iwi range on Maui and Moloka'i has received more than 50% coverage. However, like other species on Hawai'i and Kaua'i islands, 'I'iwi has received relatively poor coverage (<25% of its range repeatedly sampled; see Fig. 4.6). The birds of O'ahu have been sampled only once using VCP methods; repeating the VCP survey or sampling using other monitoring methods is needed before we can understand changes in distribution and trends for 'I'iwi and the other extant native species, O'ahu 'Elepaio, O'ahu 'Amakihi (*Hemignathus flavus*), and 'Apapane (*Himatione s. sanguinea*). In summary, only Palila on Mauna Kea and the rare Maui forest birds meet the two-thirds standard for sampling coverage, and statistically all other species are inadequately sampled for trends by the criteria of Bart, Burnham, et al. (2004).

Aside from an accounting of spatial and taxonomic extents surveyed, an assessment of what has been accomplished should include an evaluation of how well surveys have attained monitoring objectives, such as tracking population changes over time. In Hawai'i, monitoring objectives typically fall into two broad types: (1) surveillance monitoring and (2) monitoring response to management actions. In general, most monitoring in Hawai'i is surveillance monitoring; management action monitoring is more recent and much less developed. From each of the organizations responsible for surveying Hawaiian forest birds, we requested copies of their monitoring objectives. However, most programs had not formulated explicit, quantifiable monitoring objectives. Instead, most of the objectives focused on the frequency of sampling (U.S. Fish and Wildlife Service 2006) or assessed the success of forest management at increasing populations of native Hawaiian birds, such as at Hakalau. Without quantifiable limits, it is difficult to measure the success and progress of monitoring programs. This problem has been recognized, and several organizations responsible for surveying Hawaiian forest birds are currently developing monitoring-specific objectives, for example, monitoring by the NPS Inventory and Monitoring Program and comprehensive conservation planning by the USFWS National Wildlife Refuge System.

By the time the HFBS began, most native birds were concentrated in mid- and high-elevation rain forests; therefore, surveys have focused on these forests. As a result, most alien-dominated and low-elevation forests have never been surveyed. Recent research reveals that low-elevation forests are locally occupied by one or more species of common native birds and may be important habitats for coevolutionary processes with respect to diseases (Shehata et al. 2001, Reynolds et al. 2003, Woodworth et al. 2005, Eggert et al. 2008).

Frequency of sampling, a critical component of our ability to detect trends, varies widely among monitoring programs and ranges from annual surveys, such as at Hakalau, to those conducted once every five years, including DOFAW surveys in Ka'ū, West Maui, and Kaua'i. Only three Hawaiian monitoring programs had sampled more than 10 times as of 2009: Palila (29 times at annual intervals), Hakalau Forest National Wildlife Refuge (22 times), and Kūlani-Keauhou (16 times).

Where Can We Improve?

Monitoring efforts in Hawai'i have made the most of limited resources by focusing on surveying the core ranges of rare native species and areas managed for forest birds, and the establishment of a centralized database to house all forest bird monitoring data ensures that the data will be used and stored. In this section we address aspects of the Hawaiian monitoring program that, if improved, would greatly increase our ability to detect and respond to changes in bird populations and address management concerns. Of particular interest are (1) defining monitoring objectives, (2) increasing spatial coverage and consistency, (3) increasing temporal duration, frequency, and consistency, and (4) increasing analytical power to detect trends. We go on to discuss a framework for (5) replacing the current piecemeal approach with a coordinated scheme aimed at monitoring important habitats at a variety of nested scales and expanding the current emphasis on surveillance to include explicit management effects (see the section headed "The Next Era in Forest Bird Monitoring in Hawai'i").

Monitoring without realistic, specific, and measurable objectives is simply surveillance and is not advisable. Ideally, objectives should be developed early in the monitoring process in consultation with statisticians to determine quantifiable limits, sampling duration and allocation, and statistical or spatial limits (e.g., target population, inference, and study area). Thus, objectives should guide and control how data are sampled and interpreted and should provide a measure of accountability and track progress toward attaining performance goals.

The ecological monitoring literature commonly presents three types of objectives: monitoring, sampling, and management. Monitoring objectives typically lack quantifiable limits; thus, they are ambiguous statements of resource surveillance. Sampling objectives, in contrast, are unambiguous and include a set of specific measures (e.g., specified change, target levels of precision, power, and acceptable Type I and II error rates). For example, the NPS Inventory and Monitoring Program objective for monitoring long-term trends in land bird abundances is to detect, with an 80% probability, a 50% change in species' densities over a 25-year period, with a Type I error rate of 20%, in forest and woodland habitats of Hawai'i Volcanoes National Park (Camp et al. in press). Sampling objectives are usually developed as companion objectives to management objectives and provide quantifiable measures of the influence of management actions on population change. Threshold or change limits regarding the desired state or condition of a resource are the focus of management objectives. For example, in order to restore native vegetation and ultimately increase native bird densities, Hakalau initiated an ungulate management plan with objectives to control cattle and pigs (the Feral Ungulate Management Plan; U.S. Fish and Wildlife Service 1996b). These management objectives are specific and measurable, for example, terminate all grazing in the refuge by 1997, "eradicate wild cattle from units within six months of fence completion," eliminate pigs by 1999 in select units, and "eliminate pigs four years after completion of fencing for remaining

feral ungulate management units" (U.S. Fish and Wildlife Service 1996b: 40–41). These objectives are supported by monitoring objectives to sample vegetation every five years and forest birds every year (U.S. Fish and Wildlife Service 1996b). Assessing the impacts of feral ungulate eradication could be better served if the vegetation and bird monitoring objectives were developed into sampling objectives; the existing surveillance data could be used to set realistic and specific quantitative limits.

INCREASING SPATIAL COVERAGE
AND CONSISTENCY

One recurring problem in monitoring Hawaiian forest birds over the long term is that limited resources often constrain monitoring to only small areas of interest. It is of little value to compare population densities or estimates from different time periods unless the same areas are surveyed each time. Otherwise, differences may be attributable to change in the area surveyed rather than to real population changes. Thus, although a large area may have been originally surveyed, such as in the HFBS, we can draw conclusions only about the population that falls within the smaller, resampled area (Chapter 5). These subsequently surveyed areas often are quite small (see Figs. 4.1–4.3, 4.5), limiting our ability to monitor populations at meaningful scales (see the section headed "The Next Era in Forest Bird Monitoring in Hawai'i").

Second, although the focus on high-quality habitats at higher elevations is understandable, the flip side is that we are currently unable to adequately assess patterns that may be occurring in other habitats, such as low-elevation forests, or at the edges of species' ranges. With few exceptions, such as the Palila monitoring program, current monitoring programs in Hawai'i cannot determine rangewide population patterns without combining multiple surveys across time and space—and often not even then (see Fig. 4.6). Species often begin to decline at the edges of their ranges, where they are ecologically stressed (Wilcove and Terborgh 1984), and these are precisely the areas least frequently surveyed. Thus it is difficult to detect range contractions until a species disappears from entire portions of its range, as did the Hawai'i Creeper from Hawai'i Volcanoes National Park (Conant 1975) or the 'Akikiki from Kōke'e State Park (Foster et al. 2004).

Third, with the advent of a centralized database in Hawai'i, it has become clear that lack of spatial consistency within a study area is a problem in many cases, particularly for detecting changes over time. Groups undertaking repeat surveys have often, either accidentally or out of necessity, conducted subsequent counts at stations different from those used in the original survey. This precludes the use of some of the more powerful statistical methods, such as repeated measures analysis (Sokal and Rohlf 1995), and introduces additional variance into the results (see the section headed "Increasing Analytical Power").

Finally, assessing birds' abundance over large spatial extents may prove difficult. In addition to density, another measure of abundance is the frequency of occurrence, or the proportion of stations at which a particular species has been observed. Trends in frequency of occurrence are often closely matched to trends in density (e.g., Gorresen et al. 2005). However, using raw frequency data in such analyses does not account for the fact that species may have been present at stations where they have not been detected, and it implicitly assumes that detection probability equals 1. Recently, strides have been made to account for variable detection probabilities among species and habitats.

Proportion of area occupied (PAO) analysis is an analytical method that uses occurrence data—birds' presence or absence—while taking into account variable detection probabilities among species and habitats to calculate occupancy rates. In general, species occurrence at a specific loca-

tion is recorded from multiple visits within a season. Absence may be the result of the species not occupying the location or, alternatively, being present but missed. Multiple observations allow for estimation of the detection function based on a closed-population, mark-recapture framework (MacKenzie et al. 2002). The accuracy and precision of the detection function, in general, increases with increasing sampling effort; visiting a minimum of 20 locations 3–10 times within a season is recommended (MacKenzie et al. 2002). PAO analysis has not been utilized in Hawai'i because survey stations are usually visited only once per season and pooling across years is problematic (violating the closed-population assumption and confounding detection probabilities). However, PAO analyses can provide insight into changes in a species' range and may be more sensitive to trend detection than are density estimates (MacKenzie et al. 2002, 2003).

INCREASING TEMPORAL COVERAGE AND CONSISTENCY

The effectiveness of a monitoring program will be influenced by its duration, sampling frequency, and consistency. In order to adequately track population changes, monitoring programs need to be of sufficient duration to capture and follow both population trajectories and cycles, which typically require decades of surveying (Gunderson and Folke 2003, Redman and Kinzig 2003). Therefore, detecting trends from data sets with fewer than 10 surveys can prove difficult, and highly variable species may require substantially more surveys before population patterns are confidently understood (Hatfield et al. 1996). Many of Hawaii's managers and researchers have made long-term commitments to forest bird monitoring; the value and strength of these programs will become clear in the coming decades.

In addition to program duration, sampling frequency and consistency are important. In most cases, programs that sample frequently and consistently—ideally sampling annually without missing years—are likely to reveal trends more quickly than programs that survey infrequently (Gerrodette 1987; see the section headed "Increasing Analytical Power"). Serial surveys can also detect smaller changes in population sizes and facilitate the tracking of population responses to management actions. Similarly, if the same transects and stations are sampled consistently, analyses can include all transects and increase the power of tests.

Finally, it is important that monitoring programs select the best time of year to meet their objectives and consistently survey during that time of year. Hawaiian forest birds are generally more vocal and therefore more detectable during their courtship and breeding season, generally December–May for most species (Ralph and Fancy 1994b, Simon et al. 2002). Peak vocalization varies among species and may differ seasonally. If the timing of surveys varies among years, the variance among surveys will increase, making it more difficult to detect trends. Perhaps more important, variation of timing may lead to erroneous trend results. Estimates of forest birds taken from the HFBS, which was conducted during the summer months (May–August), are sometimes significantly lower than those from subsequent surveys conducted 10 years later during the spring (January–May), and it is difficult to discern whether populations have increased or we are simply observing a seasonal effect (Chapter 5). One approach for dealing with seasonal variation in detection probabilities is to include season as a covariate in density estimation (see the section headed "Recent Advances . . .").

INCREASING ANALYTICAL POWER

The analytical power of a monitoring program is its probability of detecting a change in the population over time when one truly exists (β). The power of monitoring programs has been increasingly appreciated (Cohen 1988, James et al. 1996, Gibbs 2000), although the appropriate methodology for quantifying it is controversial (Thomas 1997, Steidl 2001). We use bird density to illustrate power.

The power of a monitoring program is determined by the following parameters: variance, sample size, monitoring duration, magnitude, and statistical significance (α). Variance is a measure of variability in bird density over time. The coefficient of variation (CV) of density is a measure of the variability in estimates of density from year to year, and it varies among species due to differences in population dynamics, seasonality, vocalization behavior, and factors that affect detectability. Note that the CV is spatial variance in relation to density. Therefore, the more uniform the detections of a particular species, the smaller the coefficient of variation. The major implication is that common species and those that tend to vocalize frequently and regularly will have lower CVs than rare species and those that vocalize less. Second, the larger the sample size, or number of stations surveyed, the smaller the CV. Thus, in general, sampling more stations is better but with diminishing returns (discussed later).

Magnitude (effect size) may refer to either the rate of change in a population over time (e.g., 1% decline per year) or an absolute change over a time period (e.g., 30% decline in 10 years). As might be expected, the larger the magnitude of change one is trying to detect, the greater the power of the test. For example, it is easier to detect a 50% decline than a 5% decline over the same period. Likewise, the longer the monitoring duration, the greater the power of the test as the cumulative magnitude increases. The magnitude of change that can be detected is thus conditional on the sampling rate and monitoring duration. For example, when all factors remain constant, it is easier to detect a small rate

Table 4.1. The power to detect a negative trend in density for native bird species of the Hakalau Forest National Wildlife Refuge

Species	Coefficient of Variation	Power by Rate of Annual Decline		
		3%	5%	10%
'I'iwi	0.35	52	79	100
Hawai'i 'Amakihi	0.47	41	60	96
'Ōma'o	0.55	39	57	96
Hawai'i 'Elepaio	0.65	34	47	84
'Apapane	0.69	33	45	81
Hawai'i 'Ākepa	0.72	33	44	80
Hawai'i Creeper	0.73	33	44	79
'Akiapōlā'au	1.35	27	32	52

Notes: Power (percentage) is the probability of detecting a trend when one actually occurs. Declines of 3%, 5%, and 10% in the mean annual density were calculated for a prospective 10-year monitoring period, given linear trends, one-tailed significance levels (alpha) of 0.20, and observed coefficient of variation. The alpha and power ratio was evaluated at a 1:1 ratio (20-80 convention) because making a Type II error in this situation is more costly than making a Type I error (Cohen 1988, Taylor and Gerrodette 1993, Di Stefano 2003). Power ≥80% is adequate to detect a negative trend.

of change over a long period of time (a 5% decline with 20 years of data) than it is to detect a large rate over a short period (a 50% decline with 2 years of data). Alpha (α, the Type I error) is the significance of a statistical test and corresponds to the probability of incorrectly stating that density has changed when it has not. Alpha and analytical power are related such that relaxing α (e.g., 0.05 to 0.10) increases analytical power, although this relationship is not linear (see Cohen 1988).

What sort of analytical power can we expect from the current monitoring programs used in Hawai'i? To answer this, we evaluated the power for a subset of surveys from Hakalau Forest National Wildlife Refuge, which has one of the most extensive monitoring programs in the state. Hakalau has conducted surveys of 196–391 stations annually in February–May for 14 years (1987–2000). We used a 10-year prospective power analysis that estimated the probability that the existing monitoring program would be able to detect a de-

cline in a population of a given magnitude. Variance was estimated by regression of annual density estimates for each species. The regression standard error was converted to a CV and input into program Trends (Gerrodette 1987).

The CV of density for species at Hakalau ranged from 0.35 for 'I'iwi to 1.35 for 'Akiapōlā'au (Table 4.1). The results of the analysis were sobering. For relatively common and vocal species, 'I'iwi and Hawai'i 'Elepaio, for example, the existing monitoring program in Hakalau could detect only catastrophic declines (>60%) of the population over 10 years (or a 10% annual decline; power >80%; $\alpha = 0.20$). Moreover, even this magnitude of change could not be detected for 'Akiapōlā'au and Hawai'i Creeper trends (power <80%). Thus, the current monitoring program in Hakalau possesses sufficient power to detect devastating declines only for species with low levels of variability, at CV less than approximately 0.7, and lacks sufficient power to detect less drastic declines. Given the historical data and current efforts, most Hawaiian monitoring programs do not provide adequate power to detect statistical trends for species possessing moderate or greater levels of variability, typically those species of most interest to conservation biologists.

Improving programs' power to detect trends may be accomplished through both study design and analytical methods. The most effective approaches for meeting monitoring objectives are to conduct species- and site-specific power analyses that ensure sample sizes are appropriate and to commit to frequent and consistent sampling over an extended period. Furthermore, maximizing trends detection can be achieved by repeating surveys at consistent stations over time. Equally important to improving power are training observers, matching survey times to peak vocalizations, and choosing realistic magnitude effects and significance levels (see Gibbs et al. 1998, Steidl 2001, Di Stefano 2003). Unfortunately, for certain highly variable or rare species, implementing one or more of these procedures may not be sufficient. For these species, other survey methods, such as spot mapping or demographic studies, may need to be employed, as these methods may be more sensitive and better reflect population changes (Steidl 2001).

DEVELOPING ALERT LIMITS

Hawaiian monitoring programs currently lack criteria that alert managers to departures from normal population fluctuations. We recommend that alert limits and threshold levels be developed for each monitoring objective and included in the monitoring program (U.S. Environmental Protection Agency 2000). Alert limits will provide criteria by which to judge when a species or population requires increased monitoring, management actions, or additional research to determine likely causes of the problem (Marchant et al. 1997, Dunn 2002).

Alert limits have been put in place by some programs. The United Kingdom Breeding Bird Survey established two alert limits, a high limit to indicate when bird abundance declines by at least 50% over 25 years and a medium alert limit to indicate when a trend declines between 25 and 49% over 25 years (these limits are part of the British Trust for Ornithology's Integrated Population Monitoring program; Crick et al. 1997). The U.S. Breeding Bird Survey uses the same high alert limit threshold for detecting downward trends but also monitors for increasing abundance, a doubling over 25 years (Peterjohn et al. 1995). Given the observed annual fluctuations in Hawaiian forest birds, an alert limit based on a 50% change in density or abundance over 25 years appears to be a realistic threshold for long-term trends.

Determining statistically significant long-term population declines is difficult and inefficient because it requires frequent surveys over long periods. For Hawaiian forest bird populations, there is little statistical power in measuring moderate changes before trends are detected and conservation actions are implemented. Alert limits could also be established for factors relevant to the species' status other than long-term declines in population size. These factors might include short-term declines, range contractions, and minimum abundance levels, which we explain next. The advantage of including these factors in a monitoring program would be that the future trajectory of a population might signal even more serious concerns than long-term declines, and immediate management actions and further investigation could be implemented (e.g., Leonard et al. 2008).

The Palila population has fluctuated from year to year around an average 27 birds per km^2 (Chapter 5). Since 2003, however, the Palila population has steadily declined from 6,600 to 3,800 birds (Leonard et al. 2008). This short-term downward trajectory had not been observed in the previous 24 years of surveying and was not sufficient to change the long-term trend. However, it indicates a noticeable change in the stability of the Palila population. An alert limit based on declines for four or more years in a row (regardless of variability) would recognize such a trajectory; appropriate responses might include continued monitoring and identifying the cause(s) of the decline before implementing remedial actions.

In instances in which monitoring completely or mostly encompasses surveys of a species' range—for example, that of 'Akikiki —it is possible to identify range contraction and set alert limits. The 'Akikiki range has declined almost 60%, from 88 km² in 1970 to 36 km² in 2000, and this decline may be continuing (Foster et al. 2004; this volume, Chapter 5). Identifying contraction in a species' range is challenging—it is difficult to identify temporary versus permanent absence and easy to confound a change in abundance with range contraction (Rodriguez 2002)—but could work well for sedentary species. For example, if the range has contracted >25% between two consecutive surveys, additional monitoring focused on sampling the periphery of the range could be implemented. This action would confirm the contraction, and management designed to halt the contraction could be employed. Furthermore, if the range has contracted >50% over three surveys or over a 10-year period, additional management actions could be activated.

A third factor for which alert limits could be established might be a species reaching a minimum abundance level, regardless of variability. Criteria for defining thresholds for minimum abundance levels could follow the conservation status limits defined by the International Union for Conservation of Nature (2001) for a population size based on mature individuals. A species that has declined to fewer than 1,000 individuals would receive one set of remedial actions, whereas a species' further decline to <250 individuals would trigger more aggressive actions. By combining

distribution and abundance, alert limits could be defined based on changes in both factors. This would capture the interaction between these factors and could be crucial for determining the most appropriate remedial actions.

THE NEXT ERA IN FOREST BIRD MONITORING IN HAWAI'I

Over the past three decades, researchers and managers have been primarily concerned with obtaining baseline information on Hawaiian forest bird populations and determining whether those populations were stable, increasing, or declining. At the beginning of this period, forest bird habitats across the islands were passively managed and mostly in a deteriorating state (Chapter 16). Surveying and monitoring in these years yielded invaluable data and insights for conservation work, the highlights of which are presented in the next chapter (Chapter 5). However, this picture is changing.

The tide may be turning for the fortunes of Hawaiian forest birds now that large-scale active restoration has commenced for native forests, particularly at the high elevations above the zone of active transmission of mosquito-borne disease. The creation of a broad-scale, unified monitoring scheme for forest birds would provide the opportunity to move beyond surveillance monitoring and develop science useful for tracking the response of birds to restoration activity. The goal would be to structure forest bird monitoring in Hawai'i so that it can provide broad-reaching predictions for the effects of specific management actions on bird populations.

A Unified Approach to the Problem of Spatial Scale

According to Briker and Ruggiero (1998: 326–327), "A single approach to monitoring, at one scale, is not sufficient for all questions." The 30 years of surveillance monitoring in Hawai'i has addressed many applied research questions, such as species status, trends, distribution, and habitat requirements. Although the current coverage of forest bird surveys in Hawai'i is impressive, the myriad efforts of multiple agencies working with different objectives have led to challenges in understanding patterns at the islandwide and regional scales (Chapter 5). We believe that forest bird conservation in Hawai'i would be well served by implementing a coordinated scheme aimed at monitoring important habitats at a variety of nested scales.

Briker and Ruggiero (1998) have proposed sampling at three general levels of spatial scale. Level 1 surveys involve sampling across a species' entire regional distribution (landscape scale), which measures patterns across the entire range or region. This level is essential for understanding species' range contractions and expansions and for determining trends in species populations overall. Level 2 studies census a subset of a region. Certain locations or habitats within a region may influence the region overall. For example, bird population fluctuations in Hakalau influence population patterns in the north windward region of Hawai'i Island. Measurements at this level are essential for understanding processes that occur at regional scales. Level 3 research, such as demography studies or monitoring response to management actions, intensively samples specific sites, providing information at a local scale. Measurements at this level are more likely to reveal the causes of changes detected at Levels 1 and 2.

LEVEL 1 SAMPLING

In Hawai'i, Level 1 surveys include the DOFAW forest bird monitoring program with respect to endangered but not more widespread species. This program aims to survey the most important habitat, taking in the complete ranges of endangered species and the core range of common species, on all islands except Hawai'i Island, which has been only partially monitored due to its extensive size. A second example is the monitoring program for Palila, which completely surveys all of that species' range annually.

Spatial coverage by the current programs is inadequate at the landscape scale. Our analyses showed that only 13% of Hawaiian forests are resampled over time (see Fig. 4.5). Additionally, repeated surveying has met the criteria of Bart, Burnham, et al. (2004), two-thirds standard sampling coverage, for only 4 of 22 Hawaiian birds (18%; see Fig. 4.6). Increased coverage is required on all of the Hawaiian islands to sample across species' ranges.

Monitoring at the landscape scale is usually logistically constrained by limited funding and personnel; therefore, minimizing the sampling effort while maximizing the survey extent must be carefully evaluated. We suggest that surveys at the landscape scale follow a multiyear rotation scheme similar to that used by the DOFAW forest bird monitoring program. It has also been suggested that sampling for occupancy (presence or absence) (e.g., PAO analysis) be considered if sampling for density at the landscape scale is cost-prohibitive (MacKenzie et al. 2002, 2003).

LEVEL 2 SAMPLING

Most surveys since the HFBS have covered a subset of a region with respect to endangered birds on Hawai'i Island (except the Palila) and all common, widespread forest birds, and thus have been conducted at a Level 2 spatial scale. For monitoring at this scale, surveys should be conducted annually, especially if regional patterns fluctuate widely or are different. A notable Level 2 program is the one responsible for the Hakalau survey discussed earlier. Annual surveys encompass the core populations (but not the range edges) of eight native forest birds. Existing and proposed monitoring programs at this level should be carefully evaluated for their contributions to understanding Level 1 processes and avian biology. A cautionary note is necessary for Level 2 monitoring: the core range needs to be delineated, and it should be unlikely to change over time.

Priorities for Level 2 monitoring would include continuation of the current programs at Hakalau and the Kūlani-Keauhou areas (Hawai'i) and continuation with increased frequency at Pu'u Wa'awa'a Wildlife Sanctuary, Ka'ū and Kapāpala regions (Hawai'i), and Haleakalā National Park (Maui). We also recommend that this survey level be established to sample the remaining populations of all threatened and endangered species and species of concern (specific recommendations for survey areas appear in U.S. Fish and Wildlife Service 2006). Again, sampling across the entire range of a species, even for endangered Hawaiian birds, can be prohibitive. Urquhart and Kincaid (1999) and McDonald (2003) suggest integrating site visitation with a temporal sampling design (referred to as a panel design) to optimize trend detection, spatial coverage, and sampling efficiency. This approach has merit for monitoring Hawaiian birds at the regional scale and for integrating population properties and patterns at the landscape scale.

LEVEL 3 SAMPLING

Level 3 surveys illuminate specific population processes at a local scale and provide information on the causes of changes detected at Levels 1 and 2. Frequency of sampling is variable; however, most studies require annual or more frequent surveys. For example, Woodworth et al. (2001) studied the survival rate and other parameters of Hawai'i Creeper at three study sites within Hakalau during 1994–1999. This research provided information at Level 3 that was used to corroborate increases in population density within Hakalau (Level 2; Woodworth et al. 2001) and adjoining regions (Level 1; Camp, Pratt, et al. 2009). Level 3 studies are most informative for rare species that cannot be effectively monitored by VCP studies for want of statistical power. Thus Level 3 studies would be very useful for monitoring all extant species listed in the *Revised Recovery Plan for Hawaiian Forest Birds* (U.S. Fish and Wildlife Service 2006).

Beyond Surveillance Monitoring

Management and monitoring have not been integrated in the past, and without substantial restructuring of Hawaiian forest bird monitoring, the prospects for an integrated program are not promising. Integrating monitoring with management actions would set Hawaiian forest bird monitoring within the framework of active adaptive management, wherein population patterns improve our understanding of ecological systems (Shea et al. 2002). For this to happen, we recommend that a unified Hawaiian forest bird monitoring program be developed that includes objectives with specific and quantifiable limits. The monitoring program would coordinate survey design. Consistent sampling would be conducted over three nested spatial scales (Levels 1–3). In addition, the monitoring program would track populations over a long duration. Surveillance monitoring can provide the information necessary to identify study areas within each spatial scale and how population properties and patterns relate through the nested scales. Results from surveillance monitoring can also be used to parameterize prospective power analyses to determine quantifiable measures for objectives and realistic sampling efforts.

SUMMARY

There has been an extensive legacy of bird inventories and surveys in Hawai'i, starting with historical explorations in the 1800s and culminating with statistically robust long-term population monitoring. We describe and provide examples of the numerous sampling methods applied in Hawai'i, with an emphasis on distance sampling methods, particularly variable circular plot counts. Population monitoring in Hawai'i commenced with the Hawai'i Forest Bird Survey (Scott et al. 1986), which was the first to apply VCP counts on a large scale. Subsequently, almost 600 surveys in Hawai'i have used VCP methodology.

The field of population estimation is advancing rapidly, and we discuss the most relevant improvements as they pertain to Hawaiian monitoring. Notable developments have included modeling species-habitat associations in a geographic information system, methods to account for the effects of sampling covariates in density estimation, and strategies for objective model selection.

It is evident that detecting meaningful population distribution, density, and trends is difficult (Chapter 5), and our understanding of the relationships between management actions and bird patterns is limited. The development of a centralized data repository, the Hawai'i Forest Bird Interagency Database, has allowed us to evaluate Hawaiian forest bird monitoring. Less than half of forest bird habitat has been surveyed, and only a small fraction (13%) of forest bird habitat has been repeatedly surveyed, allowing for trends assessment. Only 4 of 22 Hawaiian birds have been repeatedly surveyed in more than two-thirds of their species' ranges, and broadly distributed species have been less well surveyed (<25% of their ranges). In addition, we have said that most existing programs can be characterized as surveillance monitoring not associated with management actions. Furthermore, most programs are not based on explicit, quantifiable monitoring objectives; however, several organizations recognize this limitation and are revamping their programs.

By restructuring Hawaiian forest bird monitoring while preserving the monitoring already in place, we may address management questions and speed the recovery of Hawaiian forest birds. The foremost change needed is to transition from surveillance monitoring to programs based on quantifiable sampling and management objectives. Monitoring populations at meaningful spatial and temporal scales is needed and can be achieved through a unified approach to monitoring at a variety of nested scales. Furthermore, sampling consistency would improve monitoring by permitting the use of more powerful statistical methods and reducing parameter variance. The statistical power to detect trends in most Hawaiian forest birds is low in the current monitoring programs. We identify both study design and analytical methods that could improve trends detection. Establishing alert limits and threshold levels is necessary to identify departures from normal population fluctuations and to integrate sampling objectives with management actions. Therefore, we encourage the adoption of a unified, long-term monitoring program at three levels of spatial scale (landscape, regional, and population) based on objectives with specific and quantifiable limits.

ACKNOWLEDGMENTS

This chapter was produced under the auspices of the Hawai'i Forest Bird Interagency Database Project, a project of the USGS Pacific Island Ecosystems Research Center, and the following cooperating agencies: the National Park Service, USGS Pacific Basin Information Node (NBII), Hawai'i Cooperative Studies Unit at the University of Hawai'i at Hilo, State of Hawai'i Division of Forestry and Wildlife, Hawai'i Gap Analysis Program, Kamehameha Schools, Hawai'i Natural Heritage Program, The Nature Conservancy–Hawai'i, U.S. Fish and

Wildlife Service, and U.S. Forest Service. The concept of an interagency database to synthesize all VCP data in Hawai'i originated with S. Fancy, and he has continued to contribute in various ways to its development. We especially thank the managers and field biologists who collected the data and worked so hard to maintain a core group of trained counters, as well as the numerous interns who assisted with the preparation of data described herein. We also thank H. D. Pratt for the 'I'iwi image used in Fig. 4.4 and J. Bart, J. Citta, D. R. Diefenbach, J. J. Duda, J. S. Hatfield, T. Leukering, J. D. Nichols, and F. L. Ramsey for comments on earlier drafts of the chapter.

CHAPTER FIVE

Status and Trends of Native Hawaiian Songbirds

P. MARCOS GORRESEN, RICHARD J. CAMP,
MICHELLE H. REYNOLDS,
BETHANY L. WOODWORTH,
AND THANE K. PRATT

The conservation crisis facing much of the Hawaiian avifauna has been met by concerted efforts to survey and monitor the birds' distribution, abundance, and trends throughout the islands. Between 1976 and 1983, the U.S. Fish and Wildlife Service (USFWS) conducted systematic surveys of forest birds and plant communities on the main Hawaiian islands as part of the Hawai'i Forest Bird Survey (HFBS) (Scott et al. 1986). The results of this historic effort have provided the basis for many plant and bird recovery plans and land acquisition and management decisions in Hawai'i. Unfortunately, the HFBS population estimates and range maps eventually became outdated, and the lack of current data hindered bird recovery and conservation. Moreover, a considerable amount of data from subsequent surveys remained to be compiled, analyzed, and published (Chapter 4). Much of this is now being processed and studied by the Hawai'i Forest Bird Interagency Database Project (HFBIDP), a cooperative effort of the U.S. Geological Survey Pacific Island Ecosystems Research Center (USGS-PIERC) and numerous partners (see the acknowledgments).

This chapter is a review and synthesis of the data acquired by these surveys and an appraisal of the current status of 32 native passerines (songbirds or perching birds) in the main Hawaiian islands and the Northwest Islands chain. We focus on population changes that occurred since the beginning of quantitative forest bird surveys in Hawai'i in the late 1960s. Included are all native forest birds extant at the time of these surveys. Not treated are more than 30 species of forest birds that became extinct prior to this period (Chapter 2).

The assessment of population status is a continuing endeavor undertaken in hopes of answering several questions crucial to managing and conserving these unique birds: How have native Hawaiian passerines fared over the past 35 years? Which species are increasing or decreasing in number or distribution? Are patterns in status and trends evident among and between geographic areas? Is land protection or other management leading to recovery of endangered bird populations? The assessment of status is made particularly urgent by the rapid decline in distribution and population size of many species during the past several decades.

METHODS

To assess the current status and trends of Hawaiian forest birds, we have drawn on material from the literature, unpublished reports, the original HFBS data, and additional data from recent and ongoing statewide surveys. In particular, we have made use of the count data from almost 600 surveys conducted between 1976 and 2008. Although comprehensive statistical analyses and detailed density estimates are beyond the scope of this chapter, the supporting data, methods, and results have been reported by Camp, Gorresen, et al. (2009). Population trends from long-term survey data ($\geq$5 years) were derived from regression analyses, and infrequent or short-term surveys were evaluated using two-sample tests of initial and most recent density estimates. Where multiple surveys were partially coincident in their sampling coverage, only the subset of stations shared by all years was included in trend analyses. Population size was quantified from measures of abundance or density extrapolated to suitable habitat area available (e.g., Scott et al. 1986) or from densities statistically modeled using habitat correlates (Gorresen et al. 2007, Gorresen et al. 2009). The variances associated with population estimates are reported either as 95% confidence intervals (indicated by a range

between two values) or as standard errors (indicated with the symbol $\pm$). The number of populations, contiguity, and range extent were qualitatively assessed using maps of occurrence and distribution. For place names mentioned in this chapter, refer to Figs. 5.1–5.3. Unless otherwise indicated, use of the term *endangered* corresponds to the federal listing designation (U.S. Fish and Wildlife Service 2006), and *nonendangered* is applied to those species not federally listed. The term *extinct* is derived from the designation of the International Union for Conservation of Nature (IUCN) (BirdLife International 2004) or from less formal assessments of a bird's status in other literature sources. These and other status terms are presented for each species in Appendix 5.1.

The challenges inherent in assessing population trends are many, including limited spatial and temporal coverage, high levels of variability, small sample sizes, low statistical power to detect trends, and so on (reviewed in Chapter 4). Assessing how abundance changes over time is also complicated by differences in the seasons during which surveys were conducted. Note that most of the HFBS surveys were conducted in the summer months (May–July), somewhat past the peak breeding period for most forest birds, whereas subsequent surveys have usually been conducted in spring (January–May). Because forest birds are generally more vocal and therefore more detectable in spring, comparisons of HFBS data with data collected later may show apparent changes in population size that must be interpreted with caution. Further, in order to make comparisons across years, we had to use identical methods for analyzing all surveys, and in some cases this use of the same methods made it necessary to reexamine older data sets, particularly those from the HFBS. As a result, some of the population estimates reported here are slightly different from those reported in the original sources. Despite such challenges and limitations, these data are a

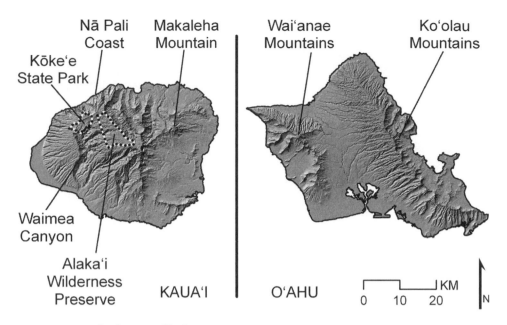

Figure 5.1. Geographical names of bird survey regions
on the islands of Kaua'i and O'ahu.

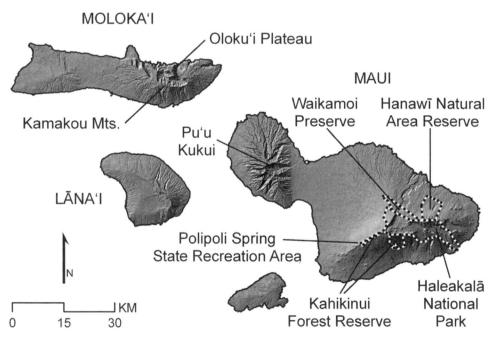

Figure 5.2. Geographical names of bird survey regions
on the islands of Moloka'i, Lāna'i, and Maui.

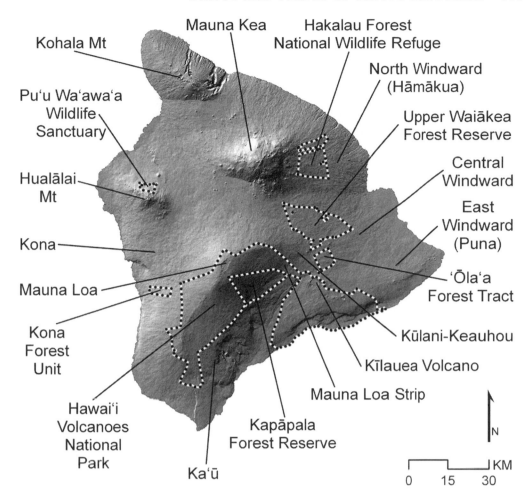

Mauna Kea

Hakalau Forest
National Wildlife Refuge

Kohala Mt

North Windward
(Hāmākua)

Pu'u Wa'awa'a
Wildlife
Sanctuary

Upper Waiākea
Forest Reserve

Hualālai
Mt

Central
Windward

East
Windward
(Puna)

Kona

Mauna Loa

'Ōla'a
Forest Tract

Kona
Forest
Unit

Kūlani-Keauhou

Kīlauea Volcano

Mauna Loa Strip

Hawai'i
Volcanoes
National
Park

Kapāpala
Forest Reserve

N

Ka'ū

KM
0   15   30

Figure 5.3. Geographical names of bird survey regions on Hawai'i Island.

phenomenal resource that have not previously been fully analyzed or synthesized. The analytical methods and results for much of the information presented here are detailed by Camp, Gorresen, et al. (2009). See Chapter 4 for a discussion of survey methods and recommendations for their improvement in the monitoring of bird populations.

SPECIES ACCOUNTS

Family Meliphagidae: Honeyeaters

This family originally consisted of five species in two endemic genera, *Chaetoptila*

and *Moho*, all now extinct. Traditionally classified as honeyeaters (Meliphagidae), the Hawaiian species have recently been proposed on the evidence of molecular analyses to be an unrelated endemic family, the Mohoidae (Fleischer et al. 2008).

KAUA'I 'Ō'Ō

The 'Ō'ō'ā'ā (Kaua'i 'Ō'ō, *Moho braccatus*) was the last of five species of the endemic honeyeaters to persist in any numbers and the only one endemic to Kaua'i (Scott et al. 1986, Sykes et al. 2000). This acrobatic bird, the most insectivorous of the honeyeaters, had sooty black plumage, yellow thighs, and a haunting, flutelike song. The Kaua'i 'Ō'ō was once common in lowland and montane native forests but disappeared from the lowlands in the early 1900s and was rare and restricted to the interior of the Alaka'i Plateau by the 1930s (Munro 1960). Surveys between 1968 and 1973 resulted in a population estimate of 36 ± 22 (U.S. Fish and Wildlife Service 1983b), and only one breeding pair was located during the 1981 HFBS survey (Scott et al. 1986). The species was last seen in 1985 and last heard in 1987 (Pyle 1985b, 1987). Intensive surveys during 1995–1996 and ongoing fieldwork have failed to detect the species (Reynolds and Snetsinger 2001, Foster et al. 2004). It is presumed extinct.

BISHOP'S 'Ō'Ō

Endemic to Moloka'i, Bishop's 'Ō'ō (*Moho bishopi*) was last recorded there in 1904 (Scott et al. 1986, Sykes et al. 2000). A few observations of 'ō'ō-like birds in the upland forests of windward Maui were made in the 1970s and 1980s and may have been of this species (Sabo 1982, Sykes et al. 2000). However, the 1979–1980 survey of the HFBS and numerous subsequent surveys did not confirm these reports. The species is presumed extinct (Reynolds and Snetsinger 2001).

Family Corvidae: Crows and Jays

Three endemic species of this family have been described in the genus *Corvus*, of which only the endangered 'Alalā survives.

'ALALĀ

The 'Alalā (Hawaiian Crow, *Corvus hawaiiensis*) is a large, omnivorous crow that once ranged widely in old-growth 'ōhi'a (*Metrosideros polymorpha*) and koa (*Acacia koa*) forests of western and southeastern Hawai'i Island (Chapter 20). It has undergone a dramatic decline in numbers and is now extinct in the wild. The 1976–1979 survey of the HFBS yielded a population estimate of 76 ± 9 birds (Scott et al. 1986), and demographic studies at that time by W. Banko

and P. Banko (1980) indicated that there were at least 53 birds in the core breeding population in central Kona. By 1992, the species existed as a single population of 11 birds (Banko and Jacobi 1992), and intensive searches and surveys between 1992 and 2003 failed to detect additional 'Alalā (Reynolds and Snetsinger 2001). Twenty-seven captive-raised birds released between 1993 and 1999 bolstered the population temporarily (Kuehler et al. 1995; this volume, Chapter 20). However, because of a high rate of mortality, the remaining captive-raised birds were removed from the wild in 1999. The last wild birds were seen in 2002, and 60 birds were managed in captivity as of 2009 (A. Lieberman, pers. comm.).

Family Monarchidae: Monarchs

This family has just one species in Hawai'i, in the endemic genus *Chasiempis*, differentiated into five subspecies, one of which is endangered.

'ELEPAIO

The 'Elepaio (*Chasiempis sandwichensis*) is an insectivorous monarch flycatcher locally common on the islands of Hawai'i and Kaua'i and uncommon to rare on O'ahu (VanderWerf 1998a). Five subspecies are currently recognized, three of which occur on Hawai'i Island (*C. s. sandwichensis, C. s. ridgwayi,* and *C. s. bryani*) and one each on Kaua'i (*C. s. sclateri*) and O'ahu (*C. s. ibidis*). The 'Elepaio is fairly adaptable and able to use a wide variety of native and nonnative forest habitats. It has sizeable populations on Kaua'i and Hawai'i islands (152,000 and <200,000 birds, respectively), and its densities are increasing on Kaua'i and appear stable in some regions on windward Hawai'i Island (Camp, Gorresen, et al. 2009). However, its densities have decreased on leeward and midelevation windward Hawai'i since the HFBS. Moreover, the small fragmented population (<2,000 birds; range <55 km^2) on O'ahu is rapidly declining and has been listed as an endangered subspecies (VanderWerf et al. 2001, U.S. Fish and Wildlife Service 2006).

Family Sylviidae: Old World Warblers

This subfamily in Hawai'i originally consisted of one species of *Acrocephalus* reed-warbler, including two subspecies, one of which is extinct and the other endangered.

NIHOA MILLERBIRD

The endangered Nihoa Millerbird (*Acrocephalus familiaris kingi*) is endemic to Nihoa Island in the Northwestern Hawaiian Islands (Morin et al. 1997). The subspecies *A. f. familiaris* previously was found on Laysan Island but became extinct in the early 1900s as a consequence of habitat destruction by feral rabbits (*Oryctolagus cuniculus*) (Sincock and Kridler 1977). This "secretive denizen of brushy hillsides" (Pratt et al. 1987) feeds on insects and larvae, particularly lepidopterans, in the low, shrubby vegetation that comprises about 40 of the 63 ha of Nihoa. Annual population estimates ranged from 31 to 731 birds (mean = 380 birds) for surveys conducted between 1967 and 1996 (Conant and Morin 2001). No long-term trends were evident, but the population appears to fluctuate in correspondence with rainfall events. There are 100 to 200 territories on the island, plus an unknown number of nonterritorial birds.

Family Turdidae: Thrushes

In Hawai'i, this family was originally made up of five endemic species of *Myadestes* solitaires: three now extinct, one endangered, one nonendangered.

KĀMA'O

A frugivorous solitaire, the Kāma'o (*Myadestes myadestinus*) was considered the most common forest bird on Kaua'i during the late 1800s but declined drastically in range and numbers in the early 1900s (Richardson and Bowles 1964). Surveys between 1968 and 1973 yielded a population estimate of 337 ± 243 birds (U.S. Fish and Wildlife Service 1983b). By the time of the 1981 HFBS survey, the Kāma'o population had declined to 24 ± 20 (Scott et al. 1986). Kāma'o were reliably sighted until 1985, and unconfirmed sightings were reported until 1991 (Pyle 1985a, 1985b, 1993). None has since been detected during intensive searches or surveys, and the species is most likely extinct (Reynolds et al. 1997, Foster et al. 2004).

OLOMA'O

The Oloma'o (*Myadestes lanaiensis*) was once ubiquitous throughout the mesic and wet forests of Moloka'i, Lāna'i, and possibly Maui (Wakelee and Fancy 1999). It was likely extirpated from Maui by the late 1800s and from Lāna'i by the early 1900s and was presumed extinct on Moloka'i shortly thereafter. Following its rediscovery on Moloka'i in 1963 (Pekelo 1963), there were two or three sightings in 1975 (Scott et al. 1977), three detections during the 1980 HFBS survey (Scott et al. 1986), and an unconfirmed report in 1988 (Reynolds and Snetsinger 2001). These records have all been from the same small area of dense rain forest above 1,000 m. Surveys in 1988, 1995, and 2004 did not encounter Oloma'o, and although the remote Oloku'i Plateau has remained unsurveyed since the HFBS, the species is likely extinct (Reynolds and Snetsinger 2001).

'ŌMA'O

Endemic to the island of Hawai'i, the locally common 'Ōma'o (*Myadestes obscurus*) has fared best of all the native solitaires (Scott et al. 1986, Wakelee and Fancy 1999). This predominantly frugivorous species formerly occurred throughout the wet and mesic forests of Hawai'i Island, but it was extirpated from the Kohala and Kona regions in the early 1900s and presently occurs only in the windward regions of the island on about 30% of its former range (van Riper and Scott 1979). The species has a contiguous and sizeable population (170,000 birds; Scott et al. 1986), and its densities remain stable in the larger upper-elevation tracts of forest habitat in Ka'ū and within the Hakalau Forest National Wildlife Refuge (NWR) on eastern Mauna Kea (Camp, Gorresen, et al. 2009). It is worth noting, however, that its densities have decreased in the central and east windward regions (eastern Mauna Loa and Kīlauea Volcano) since the 1977 and 1979 HFBS surveys. Nonetheless, the 'Ōma'o is one of the few native species that persist at middle elevations and has been observed as low as 300–500 m in elevation (Reynolds et al. 2003). The species' presence at middle elevations and its limited susceptibility to experimental infection provide encouraging evidence that the 'Ōma'o may be developing resistance to avian malaria (Atkinson, Lease, et al. 2001).

PUAIOHI

The Puaiohi (Myadestes palmeri) is an endangered solitaire endemic to the Alaka'i Plateau on the island of Kaua'i. Predominantly frugivorous, this secretive, cliff-nesting species occupies high-elevation (1,000–1,280 m) riparian habitats of native rain forests dominated by 'ōhi'a and a dense understory of mostly native shrubs and ferns. About 75% of the breeding population is concentrated within an area of about 20 km² (Snetsinger et al. 1999). The early naturalists considered Puaiohi very rare (Munro 1960), and by the 1970s the species was thought to be nearly extinct (Banko 1980). The population was estimated at 177 ± 96 from surveys conducted between 1968 and 1973 (U.S. Fish and Wildlife Service 1983b). Scott et al. (1986) estimated that there were 97 ± 129 individuals within their 25 km² study area centered on the central and southern

plateau. Intensive searches in 1995–1996 identified 145 ± 19 individuals from six of eight river drainages surveyed (Reynolds et al. 1997). As a result of more extensive targeted searches from 1999 to 2004, the entire population is currently thought to number 300–500 individuals (Snetsinger et al. 1999; this volume, Chapter 22), but the population trend is unknown.

Family Fringillidae, Subfamily Drepanidinae: Hawaiian Honeycreepers

This endemic subfamily was originally made up of at least 50 species in 21 genera: 18 species described from subfossil remains, another 9 extinct in early historic period, 14 endangered (of these, 5 are probably now extinct), and 9 nonendangered.

LAYSAN FINCH

The endangered Laysan Finch (Telespiza cantans) is endemic to Laysan Island in the Northwestern Hawaiian Islands (Morin and Conant 2002). A population translocated in 1967 also persists on the Pearl and Hermes Atoll, but an earlier translocation to Midway Atoll was extirpated shortly after the accidental introduction of black rats (Rattus rattus) in the 1940s. In part due to its omnivorous feeding habits, the species survived the catastrophic degradation of native vegetation following the introduction of rabbits to Laysan in the early 1900s. Estimates for the Laysan population ranged from 5,201 to 20,802 between 1966 and 1998 (Morin and Conant 2002), and the fluctuations reflect the effects of weather on vegetation and invertebrate productivity (Morin 1992a). The 212 ha of vegetated habitat on Laysan Island has an estimated carrying capacity of 10,000 individuals and in 1998 supported a population of 9,911 ± 895 birds (Morin and Conant 2002), comparable to the 1968–1990 mean of 11,044 ± 3,999 birds (Morin and Conant 1994). The 32 ha of habitat on Pearl and Hermes Atoll is able to support about 500 birds, and population estimates between 1968 and 1998 ranged from 108 to 876 (mean = 411 birds). In 1998, however, the birds on two islets (North and Seal-Kittery) were found to have disappeared, and the remaining populations on the islets of Southeast and Grass stood at 383 birds.

NIHOA FINCH

The endangered Nihoa Finch (*Telespiza ultima*) is another Nihoa Island endemic (Morin and Conant 2002). This omnivorous species uses the low shrub- and grasslands that comprise the island's vegetation, and it nests in rock crevices. A population translocated in 1967 to French Frigate Shoals was extirpated by 1984, and the species presently consists of a single population on Nihoa Island. Surveys conducted from 1967 to 1996 yielded annual population estimates ranging from 946 to 6,686 (mean = 3,177 birds). The population in 1996 was estimated at 2,362 ± 455 individuals, and the island's 40 ha of habitat is capable of supporting about 3,000 birds. The wide range of estimates may, in part, reflect sampling error. However, fluctuations may also be caused by weather-related changes in food availability, as seen in the closely related Laysan Finch. Although no trend was detected during the last 10-year survey period (1985–1996) reported by Morin and Conant (2002), densities from 1967 to 1996 revealed a negative trend (−3.3% per year; P = 0.004; reanalysis by USGS-PIERC).

'Ō'Ū

The 'Ō'ū (*Psittirostra psittacea*) is a finch-billed honeycreeper once common and widespread in the main Hawaiian islands (Snetsinger et al. 1998). Primarily frugivorous, the species used a wide range of habitats but was most abundant in mid-elevation 'ōhi'a forests with 'ie'ie vines (*Freycinetia arborea*), from which it sought much of its food. The 'Ō'ū was extirpated from O'ahu, Moloka'i, and Maui by the early 1900s and from Lāna'i by the 1930s (Banko 1986). With only 33 detections, the 'Ō'ū was the rarest species detected on Hawai'i Island during the 1977–1979 HFBS survey (Scott et al. 1986). At that time, the population was down to an estimated 394 ± 166 birds, mostly restricted to the forested slopes of northeastern Mauna Loa. Despite occasional unconfirmed reports, subsequent surveys and intensive rare bird searches failed to detect 'Ō'ū, and the last confirmed sighting was made in the 'Ōla'a Forest Tract of Hawai'i Volcanoes National Park (NP) in 1987 (Snetsinger et al. 1998, Reynolds and Snetsinger 2001).

Already imperiled on Kaua'i by the 1960s (Richardson and Bowles 1964), the 'Ō'ū was found by an islandwide survey between 1968 and 1973 to be restricted to the Alaka'i Plateau and to number 62 ± 41 birds (U.S. Fish and Wildlife Service 1983b). The 1981 HFBS survey detected only three birds and confirmed the species' catastrophic decline (Scott et al. 1986). Two 'Ō'ū were seen on Kaua'i in 1989 prior to the extensive habitat loss caused by Hurricane 'Iniki in 1992 (Pyle 1989). No confirmed sightings have been made since, and the species is probably extinct (Reynolds and Snetsinger 2001).

PALILA

The Palila (*Loxioides bailleui*) is an endangered, seed-eating, finch-billed honeycreeper dependent on māmane trees (*Sophora chrysophylla*) for all aspects of its biology (Lindsey et al. 1995; Banko, Johnson, et al. 2002; this volume, Chapter 23). Palila were historically distributed on Hawai'i Island at elevations from 1,200 to 3,000 m on the volcanoes Mauna Kea, Hualālai, and Mauna Loa. However, by 1975 the species occupied only 10% of its former historic range and was estimated at 1,595 ± 231 birds (van Riper et al. 1978). Palila are now found only above 2,000 m in about 136 km² of subalpine and dry forest fringing Mauna Kea (Banko, Johnson, et al. 2002). Of this area, a mere 30 km^2 on the western and southwestern slope harbors 96% of the total population. Surveys during 1980–1996 and 1997–2004 yielded population estimates of 1,913 ± 196 and 2,542 ± 173 birds, respectively. Although the species' densities appear to have been stable or increasing since 1980 in the population core (Camp, Gorresen, et al. 2009), Jacobi et al. (1996) detected a decreasing number of birds in areas of low densities at the eastern margin of the species' range, and this result suggests that the species' range may be contracting (Chapter 23). Moreover, Leonard et al. (2008) report the recent, precipitous decline in Palila numbers—from a mean of 6,633 in 2003 to 3,862 in 2007.

MAUI PARROTBILL

The Maui Parrotbill (*Pseudonestor xanthophrys*) is an endangered Hawaiian honeycreeper notable for the massive parrotlike beak with which it bites open bark and wood in pursuit of insect prey (Simon et al. 1997). Prehistorically found on Moloka'i and in low-elevation habitats in East Maui, the species is now restricted to a single population occupying 50 km^2 of mesic and wet forest above 1,200 m on northern Haleakalā Volcano. The current range may encompass suboptimal habitat because of the relative scarcity of koa, a favored foraging substrate (U.S. Fish and Wildlife Service 2006). The population was estimated at 502 ± 116 individuals based on the 1980 HFBS survey (Scott et al. 1986). Surveys from 1995 to 1997 at Hanawī, a study site located in the core of the species' range, showed that the Maui Parrotbill occurred there at approximately the same density (40 birds per km^2) as in 1980 (Simon et al. 2002). However, subsequent surveys across the species' range have not conclusively shown that its densities are stable (Camp, Gorresen, et al. 2009).

HAWAI'I 'AMAKIHI

Endemic to the islands of Hawai'i (*Hemignathus virens virens*), Lāna'i, Maui, and Moloka'i (*H. v. wilsoni*), the Hawai'i 'Amakihi is an omnivore found in a wide range of native and nonnative habitat types (Lindsey et al. 1998). 'Amakihi were last seen on Lāna'i in 1976 and were not detected during the 1979 HFBS survey (Scott et al. 1986). On Hawai'i, the species has a large population (>800,000 birds) with a fairly contiguous and extensive range (~3,000 km^2). On leeward Hawai'i Island, 'amakihi densities in north Hualālai have markedly increased since the 1978 HFBS survey and are now among the highest throughout its range (in 2003, mean = 2,189 ± 77 birds per km^2). Densities in central Kona at the Kona Forest Unit of the Hakalau Forest NWR and in south Kona seem stable.

On windward Hawai'i Island, 'amakihi densities have remained largely unchanged with two important exceptions. First, surveys through the mid-1990s showed densities to have decreased to near zero in the midelevation region east of Kīlauea Volcano (Camp, Gorresen, et al. 2009). Second, by contrast, although 'amakihi were once found to be generally rare below elevations of about 1,300 m (Scott et al. 1986, Reynolds et al. 2003), recent work has found resident and breeding individuals at numerous locations in native forests below 350 m elevation in lower Puna District in the east windward region (Woodworth et al. 2005, Spiegel et al. 2006). These birds exhibit serological evidence of having survived prior malaria infections, and surveys show a marked increase in 'amakihi numbers since the the mid-1990s. These observations and genetic research suggest that the species may be evolving a resistance to avian malaria at lower elevations (Foster et al. 2007, Eggert et al. 2008).

Scott et al. (1986) estimated that the eastern Maui population of Hawai'i 'Amakihi numbered about 44,000 birds within a 340 km² range on the wet windward and dry southern slopes of Haleakalā Volcano. However, density estimates of 'amakihi have increased there almost three-fold between 1980 (358 birds per km²) and 1997–2001 (1,007 birds per km²), and the population may be considerably larger than previously estimated (Camp, Gorresen, et al. 2009).

The West Maui population is restricted to an area of 36 km² and was estimated at only about ~2,800 birds in 1980 (Scott et al. 1986). 'Amakihi densities on West Maui also appear to have increased since the 1980 HFBS survey (from 89 birds per km² to 130 birds per km² in 1997) (Camp, Gorresen, et al. 2009).

Based on the 1979 HFBS survey, Molokaʻi was estimated to harbor a population of nearly 2,000 'amakihi within an area of 37 km² (Scott et al. 1986). Although Lindsey et al. (1998) stated that the population may be declining, surveys of upland forest in 1988–1989 and 1995 showed no apparent downward trend (Camp, Gorresen, et al. 2009). Recent low-elevation (<250 m) detections may indicate a larger range than previously realized and the development of resistance by the population to avian malaria (Chapter 9).

Oʻahu ʻAmakihi

Oʻahu ʻAmakihi (Hemignathus flavus) are distributed as two disjunct populations on the Waiʻanae and Koʻolau mountain ranges (Lindsey et al. 1998). Honolulu Christmas Bird Counts between 1958 and 1985 showed a decline in their numbers (Williams 1987). However, recent surveys have detected ʻamakihi at elevations lower than previously noted (Conry 1991, VanderWerf 1997) and may indicate their development of a resistance to avian malaria (Shehata et al. 2001), an increasing population trend, and reoccupation and use of low-elevation nonnative habitat throughout the outskirts of Honolulu (Lindsey et al. 1998). Although a 1991 survey found

'amakihi restricted to habitat above 500 m in the southern Waiʻanae range, the species was recorded as low as 100 m in the Koʻolau Mountains (Camp, Gorresen, et al. 2009). Extrapolation of the observed densities to occupied habitat in the Koʻolau range and the south Waiʻanae region yields estimated populations of 49,500 ± 4,400 and 2,300 ± 900, respectively.

Kauaʻi ʻAmakihi

The Kauaʻi ʻAmakihi (Hemignathus kauaiensis) occurs in native forest above about 600 m on Kauaʻi, with a range comprising the Alakaʻi Plateau and the adjacent Kōkeʻe State Park, the upper reaches of surrounding drainages such as Waimea Canyon, and the Makaleha Mountains (Scott et al. 1986). The islandwide population was estimated at about 11,000 in 1968–1973 (U.S. Fish and Wildlife Service 1983b) and 15,000–20,000 by the late 1980s (Ellis et al. 1992). Although each of these surveys was conducted using different methods, more rigorous evidence comes from an area of 25 km² in the interior Alakaʻi Wilderness Preserve. Six surveys were conducted on consistent routes between 1981 and 2008 and indicated that, although highly variable, Kauaʻi ʻAmakihi numbers have increased relative to the 1981 HFBS survey (Camp, Gorresen, et al. 2009). The islandwide population in 2008 was estimated at about 51,000 (39,800–62,700) birds.

ʻAnianiau

The ʻAnianiau (Hemignathus parvus) is a common forest bird on Kauaʻi that feeds on nectar and forages for arthropods on flowers and the foliage of trees and shrubs. ʻAnianiau occur in the greatest numbers in native forest above 450 m (Richardson and Bowles 1964), but they also have been found in native and nonnative forests in drainages along the northwest coast down to 100 m (Scott et al. 1986, Lepson 1997). The main population occurs on the Alakaʻi

Plateau, in Nā Pali Coast valleys, and at Kōke'e State Park, with a small isolated population on the Makaleha Mountains. The U.S. Fish and Wildlife Service (1983b) estimated an islandwide population of 24,230 ± 1,514 birds. However, densities of 'Anianiau in the interior Alaka'i Wilderness Preserve have markedly increased since the 1981 HFBS survey. Surveys across a more extensive area comprising the entire Alaka'i Plateau in 2000, 2005, and 2007 yielded an islandwide population estimate of about 37,500 (30,300–44,600) birds for 2008 (Camp, Gorresen, et al. 2009).

KAUA'I GREATER 'AKIALOA

The Kaua'i Greater 'Akialoa (*Hemignathus ellisianus stejnegeri*) is one of three subspecies of the Greater 'Akialoa, which also includes the extinct O'ahu Greater 'Akialoa (*H. e. ellisianus*) and the Maui-nui Greater 'Akialoa (*H. e. lanaiensis*). This 'akialoa is a large-bodied Hawaiian honeycreeper with a dramatically long and decurved bill it used to probe for arthropods and take nectar from 'ōhi'a and lobelia flowers (Lepson and Johnston 2000). Once common and widespread on Kaua'i, the subspecies occupied all forest types above 200 m. Following population declines in the 1800s, the Kaua'i 'Akialoa was rare by the 1920s, although accounts indicate that it persisted in the interior of the Alaka'i Plateau as late as the 1960s (Munro 1960, Richardson and Bowles 1964, Conant et al. 1998). Intensive surveys in the region since then have not resulted in any additional detections. The Kaua'i Greater 'Akialoa is presumed extinct (Reynolds and Snetsinger 2001).

NUKUPU'U

Equipped with long, decurved bills, the three subspecies of Nukupu'u (*Hemignathus lucidus*) primarily fed on insects and spiders and historically occupied montane forests (Pratt, Fancy, et al. 2001). The O'ahu subspecies (*H. l. lucidus*) has been extinct since

at least the late 1800s. Known historically only from leeward mesic and wet forests above 600 m, Kaua'i Nukupu'u (*H. l. hanapepe*) have been extremely rare since 1900. Unconfirmed sightings were made from 1960 to 1996; however, intensive rare bird searches and surveys since then have failed to detect the subspecies (Pratt and Pyle 2000, Reynolds and Snetsinger 2001), and it is very likely extinct. Only one Maui Nukupu'u (*H. l. affinis*) was detected during the 1980 HFBS survey (Scott et al. 1986), and one bird was found in the Hanawī Natural Area Reserve on the northeastern slope of Haleakalā during the 1994–1996 Hawai'i Rare Bird Search (Reynolds and Snetsinger 2001). Despite considerable ongoing survey effort in the region, the last sighting was made in 1996, and this subspecies also is likely extinct (Pratt and Pyle 2000).

'AKIAPŌLĀ'AU

The 'Akiapōlā'au (*Hemignathus munroi*) uses its elaborate two-part beak to extract insect larvae from the trunks and branches of large trees. It forages preferentially on koa and nests almost exclusively in 'ōhi'a (Scott et al. 1986; Pratt, Fancy, et al. 2001). This endangered endemic of Hawai'i Island is distributed as four disjunct populations in the windward Hawai'i, Ka'ū, Mauna Kea, and Kona regions, and it is considered to be declining overall in range and abundance. Scott et al. (1986) estimated the entire population of the species at 1,496 ± 318 birds in 1977, and Fancy et al. (1996) estimated it at 1,148 ± 33 individuals.

'Akiapōlā'au counts are characterized by a low number of detections and high variability, making trend assessment problematic. Only the north windward population, centered on the Hakalau Forest NWR, an area of active koa forest restoration, shows convincing evidence of improving long-term trends (Camp, Pratt, et al. 2009). The 'Akiapōlā'au numbers in the refuge were projected at about 410 birds as of

2007, and additional habitat immediately to the south may harbor a comparable number of birds. In contrast, a 2002 survey of the Upper Waiākea Forest Reserve in the central windward region did not record 'Akiapōlā'au in areas in which they had been detected during the 1977 HFBS survey, indicating that the species' range may have contracted in that region (Camp, Gorresen, et al. 2009). However, regenerating koa in degraded or deforested areas on Kamehameha Schools' Keauhou Ranch has recently been observed to support fairly high densities of 'Akiapōlā'au (Pratt, Fancy, et al. 2001; Pejchar et al. 2005). Thus, the overall trend in and size of the central windward population are unclear.

The 'Akiapōlā'au in Ka'ū were estimated at 533 ± 163 birds in 1976 (Scott et al. 1986) and only 44 individuals in 1993–1994 (95% confidence interval = 0–94 birds; Fancy et al. 1996). Although a 2005 survey yielded a more encouraging population estimate of 1,073 (616–1,869) birds, Tweed et al. (2007) caution that the differences among estimates are likely a reflection of sampling error related to year-to-year variability in vocalization and detectability rather than to actual changes in population size.

In contrast to the mixed status of the windward populations, those in leeward Hawai'i are clearly in trouble. A small, relict population in central Kona (centered on the Kona Forest Unit of the Hakalau Forest NWR) has been nearly extirpated based on surveys from 1995–2001 (Camp, Gorresen, et al. 2009). Likewise, surveys on Hualālai between 1990 and 2003 have not detected 'Akiapōlā'au in areas for which there are historical records of its presence (van Riper 1973). Taken together, these results indicate that the species' range may soon no longer include leeward Hawai'i Island.

Until recently, the disjunct Mauna Kea subalpine population was concentrated in two clusters on the western (Pu'u Lā'au) and eastern (Kanakaleonui) slopes of the mountain (Scott et al. 1986). However, 'Akiapōlā'au have not been observed in these areas since 2004 and are likely extirpated from the region (Pratt, Fancy, et al. 2001; this volume, Chapter 2).

'AKIKIKI

The 'Akikiki (*Oreomystis bairdi*), or Kaua'i Creeper, is a warblerlike Hawaiian honeycreeper that gleans insects mainly from tree trunks and branches, and it appears to depend on tall trees upon which to forage (Foster et al. 2000). Once common and widely distributed, this imperiled Kaua'i endemic is now limited to native montane forests above 800 m (Scott et al. 1986). The 'Akikiki population was estimated at 6,832 ± 966 birds in 1973, when the species' 88 km² range encompassed Kōke'e State Park and the Alaka'i Plateau and included a small isolated population on Lā'au Ridge. 'Akikiki disappeared from the Kōke'e region by 1981, and by 2000 the population numbered only 1,472 ± 680 birds and was limited to an area of 36 km² in the interior Alaka'i Wilderness Preserve (Foster et al. 2004). Mean densities varied from 29 to 99 birds per km² between 2000 and 2008, and as of 2008, the population was estimated at 3,568 (2,369–5,011) birds (Camp, Gorresen, et al. 2009). However, the high variability and low number of 'Akikiki detections make assessments of trend and population size difficult. The rapid range contraction and small population size indicate that the 'Akikiki may be in danger of extinction.

HAWAI'I CREEPER

The Hawai'i Creeper (*Oreomystis mana*) is an arboreal, bark-gleaning insectivore that is most abundant in mature 'ōhi'a and koa-'ōhi'a forests above 1,500 m (Scott et al. 1986, Lepson and Woodworth 2002). This endangered species is endemic to Hawai'i Island, where it is distributed as four disjunct populations in the windward

Hawai'i, Ka'ū, Kona, and Hualālai regions. Based on the results of the 1977 HFBS survey, Scott et al. (1986) estimated that the entire population of the species was 12,501 ± 1,440 birds. Of this total, 10,102 ± 827 individuals were thought to occur on the windward slopes of Mauna Kea and Mauna Loa. However, Hawai'i Creeper densities are stable or increasing in the Hakalau Forest NWR, averaging 120 birds per km^2 since 1987, with a population projected at 5,956 (3,621–9,818) birds in 2008 (Camp, Pratt, et al. 2009). Additional habitat immediately south of the refuge may support a comparable number of birds.

The species may not be faring as well elsewhere in windward habitats. Surveys in the Keauhou-Kīlauea region from 1972 to 1975 demonstrated an average density of 31 birds per km^2 (Conant 1975). Subsequent surveys showed similar densities; however, trend assessments were inconclusive (Camp, Gorresen, et al. 2009). A 2002 survey of the Upper Waiākea Forest Reserve did not record Hawai'i Creepers in areas in which they had been detected during the 1977 HFBS survey. The species has also been absent from Hawai'i Volcanoes NP since at least the early 1970s (Conant 1975, P. Banko and W. Banko 1980). These observations suggest that the species' range is contracting in the central windward region.

The second largest Hawai'i Creeper population is located in Ka'ū and was estimated at 2,102 ± 540 birds based on the 1976 HFBS survey (Scott et al. 1986). Although densities ranged as high as 139 ± 33 birds per km^2 in 2002, the low number of detections and high level of variability complicate trend assessment (Camp, Gorresen, et al. 2009). The species' 64 km^2 range in Ka'ū is currently limited to a narrow band of forest above 1,500 m, and given the density observed in 2005, this range was estimated by Tweed et al. (2007) to support about 2,268 (1,159–4,438) individuals.

The populations on Hualālai and central Kona were estimated to number about 220 and 75 individuals, respectively, based on the 1978 HFBS survey (Scott et al. 1986). However, Hawai'i Creeper detections have rapidly declined in leeward Hawai'i Island in the past several decades, and these relict populations may be nearly extirpated (Camp, Gorresen, et al. 2009). A 2003 survey of Hualālai detected a single individual, and surveys within the Kona Forest Unit of the Hakalau Forest NWR show that Hawai'i Creepers persist there only at elevations above 1,500 m and at low densities (18 ± 11 birds per km^2 in 2001). A high degree of philopatry (VanderWerf 1998b) may limit the connectivity between the isolated leeward and Ka'ū populations and further exacerbate downward trends.

O'AHU 'ALAUAHIO

The O'ahu 'Alauahio (*Paroreomyza maculata*), or O'ahu Creeper, is a warblerlike insectivorous Hawaiian honeycreeper (P. Baker and H. Baker 2000). Common in the late 1800s, the O'ahu endemic was rare by the 1930s (Munro 1960). Only nine credible sightings were reported from 1941 to 1975 (Shallenberger and Pratt 1978), and all were from mixed introduced and koa-'ōhi'a forests in the middle to upper elevations of the Ko'olau Mountains (P. Baker and H. Baker 2000). Intensive surveys from 1976 to 1978 detected only three birds (Shallenberger and Pratt 1978). Several unconfirmed sightings were made between 1985 and 1990 (P. Baker and H. Baker 2000), but a 1991 survey did not detect O'ahu 'Alauahio (USGS-PIERC, unpub. data), and the species may now be extinct.

KĀKĀWAHIE

The Kākāwahie (*Paroreomyza flammea*), or Moloka'i Creeper, was a brilliant scarlet (males) or rusty brown (females) honeycreeper endemic to Moloka'i (P. Baker and

H. Baker 2000). This curious and active bird picked over trunks, branches, and leaves in search of insects. Once widely distributed at both low and high elevations, it was still common as late as 1907 but declined rapidly thereafter and became rare by the 1930s (Perkins 1903, Munro 1960). The last sightings of Kākāwahie were from 1961 to 1963 (Pekelo 1963). The 1979–1980 HFBS survey and subsequent surveys yielded no further records, and the species is presumed extinct (Reynolds and Snetsinger 2001, U.S. Fish and Wildlife Service 2006).

MAUI ʻALAUAHIO

The Maui ʻAlauahio (*Paroreomyza montana*), or Maui Creeper, is the only surviving species of its genus. It occupies both native and alien forests and ranges into subalpine woodland and scrubland (H. Baker and P. Baker 2000). Historically widespread on Maui and Lānaʻi islands, the species is now restricted to three populations on East Maui. The largest contiguous population extends from Waikamoi Preserve eastward to Kipahulu Valley on the windward slopes of Haleakalā Volcano. Two small, isolated populations are located on the dry leeward slopes at Polipoli Spring State Recreation Area and Kahikinui Forest Reserve. The total population of Maui ʻAlauahio during the 1980 HFBS survey was estimated at 34,839 ± 2,723 birds (Scott et al. 1986). However, subsequent surveys of windward Haleakalā have recorded ʻAlauahio at significantly higher densities (731 ± 64 birds per km^2 during the 1980 HFBS survey versus 1,167 ± 74 birds per km^2 between 1997 and 2001) and may indicate a larger population than previously estimated (Camp, Gorresen, et al. 2009). On the other hand, susceptibility to avian malaria may be causing the species' range to contract upslope, with few individuals found below 1,600 m (H. Baker and P. Baker 2000).

ʻAKEKEʻE

The ʻAkekeʻe (*Loxops caeruleirostris*), or Kauaʻi ʻĀkepa, is an insectivorous honeycreeper that forages in ʻōhiʻa canopy foliage (Lepson and Pratt 1997). The Kauaʻi endemic is common and widely distributed in native forest above 1,000 m. When first comprehensively surveyed from 1968 to 1973, the ʻAkekeʻe existed in two populations totaling 5,066 ± 840 birds: a main population extending from Kōkeʻe State Park to the Alakaʻi Plateau and a small isolated population on the Makaleha Mountains (U.S. Fish and Wildlife Service 1983b). The Makaleha population has not been resurveyed, and its current status is unknown. The 1981 HFBS survey of a 25 km^2 area in the eastern Alakaʻi Wilderness Preserve revealed a density of 48 birds per km^2. Estimates thereafter were substantially greater but have trended down from a peak of 204 birds per km^2 in 1989 to 96 birds per km^2 in 2008, and the absence of detections since 2000 in the western part of the Alakaʻi Plateau indicates that the species' range has contracted (Camp, Gorresen, et al. 2009). Forest regrowth following two destructive hurricanes in 1982 and 1992 may have contributed to an initial increase in ʻAkekeʻe numbers (Pratt 1994). Extrapolation of the density recorded across the Alakaʻi Plateau in 2008 (62 per km^7) to the species' 127 km^2 range produces a population estimate of about 7,887 (5,220–10,833) ʻAkekeʻe (Camp, Gorresen, et al. 2009).

ʻĀKEPA

The ʻĀkepa (*Loxops coccineus*) is an insectivore that forages almost exclusively on the terminal leaf clusters of ʻōhiʻa and among koa leaves and pods. It is most abundant in mature ʻōhiʻa and koa-ʻōhiʻa forests above 1,500 m (Scott et al. 1986, Lepson and Freed 1997). One of three recognized subspecies, the Oʻahu ʻĀkepa (*L. c. rufus*) probably became extinct in the 1930s (Lepson and Freed 1997). The Maui 'Ākepa (L. c. ochraceus) has been rare since the early 1900s (Perkins 1903), and only eight individuals were detected in the high-elevation windward forests of Hanawī Natural Area Reserve during the 1980 HFBS survey (Scott et al. 1986). The last unconfirmed sighting occurred in 1988, and the subspecies is likely extinct (Lepson and Freed 1997, Reynolds and Snetsinger 2001).

The endangered Hawai'i 'Ākepa (L. c. coccineus) occurs as disjunct populations in north and central windward Hawai'i, Ka'ū, Kona, and Hualālai (Scott et al. 1986, Lepson and Woodworth 2002). Based on the HFBS survey, the subspecies' entire population was estimated at 13,892 ± 1,825 birds (Scott et al. 1986). About half this number was projected to occur in the north windward region (eastern Mauna Kea). However, like Hawai'i Creeper densities, those of 'Ākepa have increased about 1% per year in the Hakalau Forest NWR since the refuge's establishment in 1987, although densities appear to have been decreasing since 1999 (Camp, Pratt, et al. 2009). As of 2008, the refuge was projected to harbor about 6,839 (5,184–9,044) birds. Habitat immediately south of the refuge may support a smaller number of additional birds.

The range of Hawai'i 'Ākepa in the central windward region has excluded the nearby Hawai'i Volcanoes NP since at least the 1970s (P. Banko and W. Banko 1980). A 2002 survey of the Upper Waiākea Forest Reserve in the region also did not record the subspecies in areas in which it had been detected during the 1977 HFBS survey. Its presence in the region may now be limited to the Kūlani-Keauhou area (Camp, Gorresen, et al. 2009). In addition, 'Ākepa density in this area may be decreasing, although a high level of variability currently prevents detection of a trend.

The Ka'ū region supports the second largest population of Hawai'i 'Ākepa, estimated to number 5,293 ± 780 birds and

to occupy a range of 180 km² in 1976 (Scott et al. 1986). However, range contraction and highly variable estimates of densities make an assessment of population size and trend difficult. For example, density estimates increased from the 34 ± 11 birds per km² observed during the 1976 HFBS survey to 156 ± 30 birds per km² in 1993 and 107 ± 29 birds per km² in 2002 but then decreased to 35 ± 9 birds per km² in 2005 (Camp, Gorresen, et al. 2009). In 2005, the species' range in Ka'ū was thought to be only 80 km² and to no longer extend below 1,500 m, and the population was estimated at 2,556 (1,340–4,876) birds (Tweed et al. 2007). However, as in the cases of other rare species, the differences among density estimates may be the result of sampling error rather than changes in population size, and the regional population may be greater than currently estimated.

'Ākepa occur as disjunct and relict populations on Hualālai and central Kona. Based on the 1978 HFBS survey, Scott et al. (1986) estimated a combined population of 661 ± 126 birds. However, densities have declined in leeward Hawai'i Island in the past several decades (Camp, Gorresen, et al. 2009). Recent surveys have found a few 'Ākepa still present on north Hualālai (Pu'u Wa'awa'a Wildlife Sanctuary). A 1988 survey encountered at least six birds adjacent to the Kona Forest Unit of the Hakalau Forest NWR (Pratt et al. 1989). Subsequent surveys have detected few birds (Camp, Pratt, et al. 2009).

'I'IWI

The 'I'iwi (Vestiaria coccinea) is a nectarivorous honeycreeper that occurs on the five largest islands of Hawai'i and is most abundant in closed-canopied, high-stature 'ōhi'a and koa-'ōhi'a forests above 1,500 m (Fancy and Ralph 1998). The largest population (>340,000 birds) and range (~2,000 km²) occurs on Hawai'i Island (Scott et al. 1986). Windward 'I'iwi populations there have generally shown downward trends except in high-elevation forests, specifically the main unit of Hakalau Forest NWR, Kūlani-Keauhou, and possibly Ka'ū, where densities appear to be stable (Camp, Gorresen, et al. 2009; Camp, Pratt, et al. 2009). 'I'iwi numbers in the central windward region have been declining, and the species' range is contracting upslope, with few occurrences below 1,100 m during the breeding season (Gorresen et al. 2005). On leeward Hawai'i Island, 'I'iwi densities appear to have declined in the Hualālai and Kona regions (Camp, Gorresen, et al. 2009). 'I'iwi densities have declined at lower elevations (below 1,500 m) in the Kona Forest Unit of the Hakalau Forest NWR, but in the upper elevations they appear stable.

The species is distributed in two disjunct populations on Maui (Scott et al. 1986). Based on the 1980 HFBS survey, the East Maui population on the windward slopes of Haleakalā was estimated at 18,812 ± 1,006 individuals. Later surveys have revealed higher 'I'iwi densities, and the population is probably greater than previously thought (Camp, Gorresen, et al. 2009). The West Maui population was estimated to number 176 ± 74 birds and to be restricted to 16 km² of habitat about 30 km distant from the eastern population (Scott et al. 1986). Surveys subsequent to the 1980 HFBS survey show that the population persists at very low densities (Camp, Gorresen, et al. 2009).

Twelve 'I'iwi were detected during the 1979 HFBS survey of Moloka'i, and, based on these results, Scott et al. (1986) estimated that 80 ± 33 birds were distributed on the Kamakou Range and Oloku'i Plateau. However, surveys in 1988, 1995, and 2004 detected only 2, 1, and 3 birds, respectively, and indicate that the Moloka'i population is at high risk of disappearing altogether (Reynolds and Snetsinger 2001; Camp, Gorresen, et al. 2009).

The species' precipitous decline on O'ahu was evident by the early 1900s (Fancy and Ralph 1998). A 1991 survey failed to detect 'I'iwi (Conry 1991; Camp, Gorresen, et al. 2009), and surveys from 1994 to 1996 recorded only 8 individuals dispersed in three isolated populations in the Wai'anae and Ko'olau ranges (VanderWerf and Rohrer 1996). Estimated to number <50 birds in 1991 (Ellis et al. 1992), the O'ahu population likely faces extirpation.

The 'I'iwi population on Kaua'i appears to be in decline. In the early 1970s, 'I'iwi occurred down to about 900 m, and the entire population numbered about 26,000 ± 3,000 birds and spanned a 140 km² range (U.S. Fish and Wildlife Service 1983b). By 2000, the population had decreased to 9,985 ± 960 birds (Foster et al. 2004), and the species' range was 100 km², contracting especially at the western margins and with occurrence mostly restricted to above 1,100 m (Camp, Gorresen, et al. 2009). Based on the 1968–1973 surveys, the core population in the interior Alaka'i Plateau (above 1,200 m) was estimated at 7,800 ± 2,300 birds (Scott et al. 1986). Subsequent surveys in this region have yielded highly variable densities but indicate that this portion of the population is stable at present (Camp, Gorresen, et al. 2009).

'ĀKOHEKOHE

The 'Ākohekohe (*Palmeria dolei*), or Crested Honeycreeper, is an endangered, nectarivorous Hawaiian honeycreeper restricted to wet and mesic native forest above 1,100 m (Berlin and VanGelder 1999). Extirpated from Moloka'i in the early 1900s, 'Ākohekohe now occur only on the northeastern slope of Haleakalā Volcano on Maui. The population is restricted to 58 km² (5% of its original range on Maui) and was estimated at 3,753 ± 373 individuals based on the 1980 HFBS survey (Scott et al. 1986). The HFBS and subsequent surveys of the 'Ākohekohe range yielded densities of 81 ± 10 birds per km² in 1980, 98 ± 11 birds per km^2 from 1992 to 1996, and 116 ± 14 birds per km^2 between 1997–2001 (Camp, Gorresen, et al. 2009). Densities in the core of the species' range within the Hanawī Natural Area Reserve were 183 ± 59 birds per km^2 in 1988 and 290 ± 10 birds per km^2 from 1995 to 1997 (Berlin and VanGelder 1999). These results indicate that the species' rangewide and core densities have both increased, and the current population may be larger than previously estimated.

'APAPANE

The 'Apapane (*Himatione s. sanguinea*) is a common and widespread nectarivore found in native forests on all five of the main Hawaiian islands (Scott et al. 1986). On Hawai'i Island, the species has a large population (>1,000,000 birds) with a fairly contiguous and extensive range (~3,000 km^2). 'Apapane densities have markedly increased or remained stable throughout much of the island since the late 1970s (Camp, Gorresen, et al. 2009). The sole exception is east of Kīlauea Volcano, where densities have decreased at middle elevations.

Based on the 1980 HFBS survey, the 'Apapane on East and West Maui were estimated at about 94,000 and 16,000 birds, respectively (Scott et al. 1986). The eastern population is distributed in an area of 370 km^2 spanning the wet windward and dry southern slopes of the Haleakalā range. Surveys have recorded higher densities since the 1980 HFBS survey (Camp, Gorresen, et al. 2009), and the population is likely larger than that estimated by Scott et al. (1986). The West Maui population occurs in 41 km^2 of forest habitat on northwest Pu'u Kukui. Surveys of West Maui in 1980 and 1997 detected similar densities, indicating that the population is stable (Camp, Gorresen, et al. 2009).

'Apapane are widely distributed above 1,000 m on Kaua'i (Fancy and Ralph 1997) and were estimated at about 163,000 in-

dividuals from surveys conducted from 1968 to 1973 (U.S. Fish and Wildlife Service 1983b). Surveys since the 1981 HFBS survey have shown that the birds' densities are highly variable, and Foster et al. (2004) speculated that 'Apapane were adversely affected by Hurricane 'Iniki in 1992 but have since recovered.

Although the proportions of extinct native species on Lāna'i, O'ahu, and Moloka'i are the highest in Hawai'i (Pratt 1994), 'Apapane persist on these islands. The 'Apapane is now the only honeycreeper remaining on Lāna'i (Fancy and Ralph 1997). Estimated at only 540 ± 213 birds during the 1979 HFBS survey, the Lāna'i population has not been resurveyed, although the species is still present in low numbers (Walther 2006). On O'ahu, 'Apapane were fairly widespread, particularly at middle elevations, during a 1977–1978 survey of the leeward Ko'olau range (Shallenberger and Vaughn 1978). Although absent from the north Wai'anae Mountains, 'Apapane were detected at low densities in the southern part of the range (65 ± 35 birds per km^2) and in the leeward Ko'olau range (121 ± 13 birds per km^2) during a 1991 survey (Camp, Gorresen, et al. 2009). Extrapolation of the observed densities to occupied habitat in the Ko'olau range (~200 km^2) and the south Wai'anae region (~11 km^2) yielded estimated populations of about 24,000 ± 2,600 and 715 ± 385 birds, respectively. Based on the 1979 HFBS survey, east Moloka'i was estimated to harbor about 39,000 individuals (Scott et al. 1986). However, the bird's densities have greatly increased in upland forest since 1979, and recent low-elevation (<250 m) detections may indicate that some individuals have survived and acquired immunity to further exposure from avian malaria (Chapter 9).

PO'O-ULI

The Po'o-uli (*Melamprosops phaeosoma*) is a critically endangered honeycreeper discovered a mere 30 years ago, at which time the species was rare and confined to a single area of wet 'ōhi'a forest above 1,400 m on windward Haleakalā Volcano, Maui (Casey and Jacobi 1974). Po'o-uli forage on tree branches of the subcanopy and understory and feed primarily on small snails, insects, and spiders (Pratt, Kepler, et al. 1997). Based on three birds detected during the 1980 HFBS survey, a population of 141 ± 141 individuals was estimated to occur within a range of 13 km^2 (Scott et al. 1986). However, the species' density has undergone a dramatic decline, from an estimated 76 ± 8 birds per km^2 in 1975 to 15 ± 7 birds per km^2 in 1981 and 8 ± 4 birds per km^2 in 1985 (Mountainspring et al. 1990). Six birds were detected during intensive searches in 1994–1995, and only 3 birds were located between 1997 and 2000 (Pratt, Kepler, et al. 1997; Reynolds and Snetsinger 2001). Attempts to bring these birds into captivity and maintain them were unsuccessful. The species was last seen in 2004 and is feared extinct (Chapter 21).

SUMMARY

Status, Threats, and Recovery Options

When quantitative bird surveys were first begun in Hawai'i in the late 1960s, 14 forest bird species were considered at high risk of extinction, and probably most existed in populations of fewer than 500 individuals each. Their present status is, with a few exceptions, very disheartening. Eleven of these species—Kaua'i 'Ō'ō, Bishop's 'Ō'ō, Kāma'o, Oloma'o, 'Ō'ū, Kaua'i Greater 'Akialoa, Nukupu'u (both Kaua'i and Maui forms), O'ahu 'Alauahio, Kākāwahie, Maui 'Ākepa, and Po'o-uli—are almost certainly extinct. The hope for these birds is that they will be rediscovered in a remote corner of the Hawaiian wilderness through perseverance, targeted searches, and luck (Chapter 21). In preparation for such an event, a "Rare Bird Dis-

covery Protocol" has been established that calls for intensive field data collection, multiple agency coordination, and immediate intervention (U.S. Fish and Wildlife Service 2006).

Three of the highest-risk bird species are still extant today, although their futures are by no means secure: the 'Alalā, Millerbird, and Puaiohi. Captive propagation has stabilized the 'Alalā population following rapid decline and extirpation in the wild. 'Alalā recovery exemplifies the need for captive propagation and reintroduction to be undertaken in the context of high-quality habitat conservation (Chapter 20). Finding and securing this habitat is currently a significant challenge. Analysis of the population viability of the Millerbird indicates that the probability of extinction is high for the single small extant population (Conant and Morin 2001). Rigorous quarantine to prevent new alien introductions to Nihoa Island is an essential component of Millerbird management and conservation, and translocation and establishment of a second population has been proposed (Morin et al. 1997). On a brighter note, recent surveys for Puaiohi in their remote streamside habitat have shown that members of the species are more numerous and widespread than previously thought, though still fewer than 500 birds. Captive breeding and release are also improving the prospects for Puaiohi recovery (Chapter 22).

The status of seven other imperiled species or subspecies numbering between 500 and 5,000 individuals is mixed (see Fig. 5.4; Appendix 5.1). The O'ahu 'Elepaio, Nihoa Finch, Palila, Maui Parrotbill, 'Akiapōlā'au, 'Akikiki, and 'Ākohekohe demonstrate some or all of the hallmarks of endangered species: small population size, declining densities and population size, restricted distribution, contracting range, and isolated subpopulations. These species are listed as endangered by the USFWS or the IUCN and are the focus of ongoing efforts at ameliorating threats and the risk of

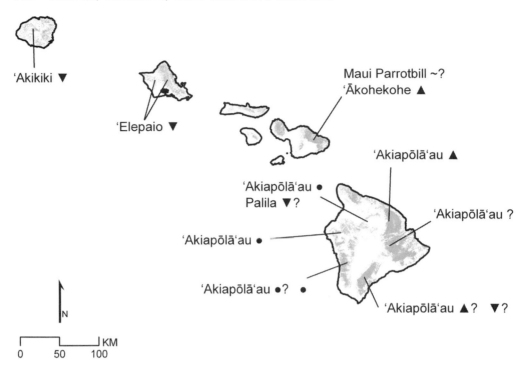

Figure 5.4. Population trends for forest bird species listed as endangered by either the USFWS or the IUCN (Bird-Life International 2004) and with populations numbering between 500 and 5,000 individuals. Trends are based on current changes in estimated density, population size, and species' range. The symbols used indicate the following: ▲, increasing trend; ▼, decreasing trend; ●, absence; ~, apparently stable population; and ?, uncertainty in the trend assessment resulting from the high variability of observed densities. The pair of symbols for central Kona and Ka'ū, Hawai'i Island, refer to trends above and below 1,500 m. Shading indicates areas designated as forest habitat by the National Oceanic and Atmospheric Administration, Coastal Change Analysis Program (1995). See Figs. 5.1–5.3 for regional and local names.

extinction. Experimental rat eradication to reduce nest predation rates appears effective at reducing demographic decline and stabilizing local populations of O'ahu 'Elepaio (VanderWerf and Smith 2002). Small population size and large fluctuations in numbers predispose the Nihoa Finch to a high probability of extinction (Morin and Conant 2002). Introduction of a second population to Midway Atoll has been considered but currently is not being pur-

sued. Preventing the establishment of alien species and population monitoring are the management options presently available for the Nihoa Finch. The Palila population on west Mauna Kea is likely to benefit from ungulate control, habitat restoration, and predator reduction (Banko et al. 2001; this volume, Chapter 23), although poorly understood recent declines in abundance are cause for concern (Leonard, Banko, et al. 2008). Captive propagation, bird translocation, and the establishment of a resident group separate from the core population are being pursued as means of reducing the vulnerability of the species to catastrophic events such as fire (Chapter 23). Maui Parrotbill and 'Ākohekohe recovery centers on the protection of native high-elevation forests from the destructive effects of feral pigs, the reforestation of montane pastures on Maui, and the proposed establishment of additional populations by means of captive propagation and translocation (U.S. Fish and Wildlife Service 2006). Recent observations of 'Akiapōlā'au using young koa at the Hakalau

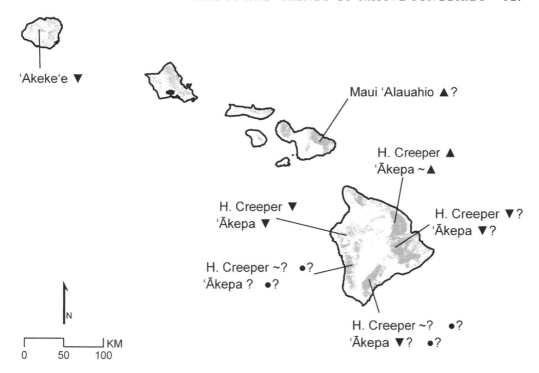

Figure 5.5. Population trends for forest birds listed as endangered by either the USFWS or the IUCN and with populations greater than 5,000 individuals. See Fig. 5.4 for an explanation of the symbols used.

Forest NWR and in koa plantations at the Kamehameha Schools' Keauhou Ranch indicate that forest restoration of pastures above 1,500 m and near existing 'Aki-apōlā'au populations may significantly contribute to the recovery of this species (Pratt, Fancy, et al. 2001; Pejchar et al. 2005). 'Akikiki recovery is complicated by the fact that although the causes for its decline have not been identified, two hurricanes in the past 25 years have toppled much of the species' foraging substrate (Foster et al. 2004). However, the development of captive propagation and reintroduction techniques for the Hawai'i Creeper may serve as a model for use with the 'Akikiki.

The Laysan Finch, Hawai'i Creeper, Maui 'Alauahio, 'Akeke'e, and Hawai'i 'Ākepa have populations greater than 5,000 individuals (see Fig. 5.5; Appendix 5.1) but remain vulnerable to a variety of threats

and are also listed as endangered by the USFWS or the IUCN. Like the Millerbird and Nihoa Finch, the Laysan Finch runs a high risk of extinction due to the fragility of its insular habitat and the potential for alien introductions and catastrophic weather events (Morin and Conant 2002). Restricted human access, weed eradication, and population monitoring currently are the most effective measures for management of the Laysan Finch. Hawai'i Creeper and Hawai'i 'Ākepa populations are stable or increasing in the larger tracts of high-elevation forest habitat in north windward Hawai'i but are diminishing in smaller, more fragmented, disturbed habitats in central Kona and Hualālai. Reducing disease transmission and restoring high-elevation forests would benefit these endangered species (U.S. Fish and Wildlife Service 2006). Densities of Maui 'Alauahio and 'Akeke'e appear stable, but the range of the 'Akeke'e has contracted. Both continue to be threatened by the encroachment of exotic plants, ungulates, and the upward spread of disease driven by global

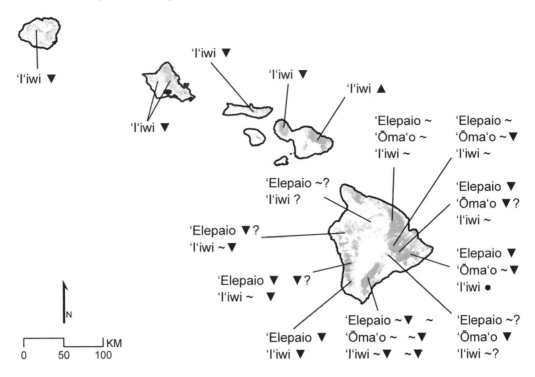

Figure 5.6. Population trends for forest bird species not listed as endangered by either the USFWS or the IUCN but that are nevertheless generally considered species of concern. See Fig. 5.4 for an explanation of the symbols used.

warming (Lepson and Pratt 1997; H. Baker and P. Baker 2000; Benning et al. 2002). These species are expected to respond well to ungulate removal and habitat restoration above elevations harboring mosquitoes.

The Hawai'i 'Elepaio, 'Ōma'o, and 'I'iwi are not listed as endangered by the USFWS but are nevertheless considered species of concern. These species have large populations but are experiencing range contraction and negative trends in many parts of their ranges (see Fig. 5.6; Appendix 5.1). The 'I'iwi in particular, with its bright scarlet plumage and long, curved, orange beak, is the "poster child" for Hawaiian forest birds susceptible to malaria. Fully 90% of 'I'iwi bitten by a single infected mosquito perish from the disease (Atkinson et al. 1995), and this susceptibility is widely considered the cause of the limited

distribution and gradual decline in 'I'iwi numbers (Chapter 9). The creation of high-elevation refugia may not be sufficient to safeguard the bird. 'I'iwi, like the closely related 'Apapane, make seasonal foraging flights over the landscape in search of nectar, and these flights often bring them into contact with mosquitoes at lower elevations. Furthermore, expansion of avian malaria into higher-elevation habitats through introduction of cold-tolerant mosquitoes, land-use changes, and global warming may well spell disaster for this familiar bird.

Six Hawaiian forest bird species with large populations show stable or improving trends (see Fig. 5.7; Appendix 5.1). These include the Kaua'i 'Elepaio, three species of 'amakihi, the 'Anianiau, and the 'Apapane. In contrast to the status of 'Elepaio subspecies on O'ahu and parts of Hawai'i Island, the Kaua'i 'Elepaio population appears to be increasing. The long-term prospects for this adaptable subspecies much depend on the degree to which it can withstand habitat degrada-

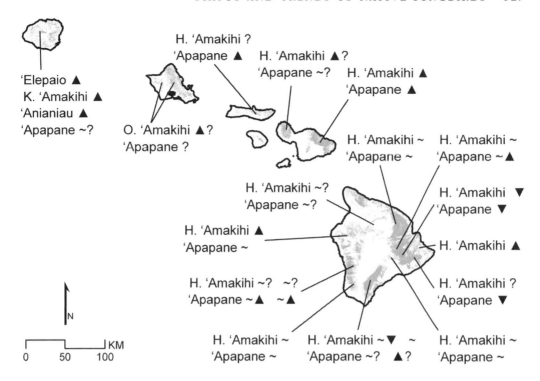

'Elepaio ▲
K. 'Amakihi ▲
'Anianiau ▲
'Apapane ~?

H. 'Amakihi ?
'Apapane ▲   H. 'Amakihi ▲?
'Apapane ~?   H. 'Amakihi ▲
'Apapane ▲

O. 'Amakihi ▲?
'Apapane ?

H. 'Amakihi ~   H. 'Amakihi ~
'Apapane ~   'Apapane ~▲

H. 'Amakihi ~?
'Apapane ~?   H. 'Amakihi ▼
'Apapane ▼

H. 'Amakihi ▲
'Apapane ~   H. 'Amakihi ▲

H. 'Amakihi ~? ~?
'Apapane ~▲   ~▲   H. 'Amakihi ?
'Apapane ▼

N

0   50   100   KM

H. 'Amakihi ~
'Apapane ~   H. 'Amakihi ~▼ ~
'Apapane ~? ▲?   H. 'Amakihi ~
'Apapane ~

Figure 5.7. Population trends for forest bird species not listed as endangered by either the USFWS or the IUCN and that show stable or positive trends overall. See Fig. 5.4 for an explanation of the symbols used.

tion and the threats associated with alien introductions. Likewise, 'Anianiau density has increased considerably on Kaua'i in the past several decades. The three 'amakihi species share several traits that bode well for the long-term survival of this group, including generalized habitat requirements, flexible foraging behavior, and potential for further expansion into lowland areas. The 'Apapane remains widespread and common in native forests, and the species exhibits increasing trends throughout much of its range.

Prospects for the Future

A number of important generalities are evident from the multiplicity of species-specific trends. Notably, many native passerines, particularly endangered or special-concern species, appear to be stable or increasing in areas with large tracts of high-elevation native forest even while decreasing in more fragmented or disturbed areas at middle to low elevations. The overall result is that native birds are increasingly restricted to high-elevation forest and woodland refugia (Box 5.1, Figs. 5.8–5.10). It is these upland habitats that require intense, sustained efforts at conservation and restoration and from which long-term recovery strategies may be based. For example, the eight native species resident within the Hakalau Forest NWR—the Hawai'i 'Elepaio, 'Ōma'o, Hawai'i 'Amakihi, 'Akiapōlā'au, Hawai'i Creeper, 'Ākepa, 'I'iwi, and 'Apapane—have shown significant increases in density or stable trends since 1987 (Camp, Pratt, et al. 2009). The 13,252 ha refuge on windward Hawai'i was established specifically for the protection of native forest birds. Many of its management actions, particularly habitat protection, ungulate removal, and koa forest restoration, appear to be paying off.

A second notable development is the apparent persistence or recolonization of

---

Box 5.1.   **Bird Extinction on Smaller Main Islands**

Within the time frame of this study (1968–2008), five endemic bird species on Kaua'i are believed to have gone extinct: Kaua'i 'Ō'ō, Kāma'o, 'Ō'ū, Kaua'i 'Akialoa, and Kaua'i Nukupu'u. The disproportionate loss of this island's avifauna is probably due to the fact that only about 150 km² of forest habitat in the Alaka'i region is situated above the elevation (~1,000 m) at which the level of avian malaria transmission is high (Chapter 9). Although the elevation of this area was high enough to have previously protected montane bird populations from avian disease, global warming may now be eroding this safety zone (Benning et al. 2002). In 1982 and 1992, Kaua'i also had the misfortune of being in the direct path of two major hurricanes that demolished old-growth forest and, in turn, facilitated the encroachment of nonnative vegetation and alien birds (Foster et al. 2004). Moreover, the lowlands, in which montane forest birds shelter during severe storms, now expose them to avian disease (Pratt 1994).

Are the remaining Kaua'i forest birds following the same path to extinction as those already lost? Of the eight remaining native species, four (Kaua'i 'Elepaio, Kaua'i 'Amakihi, 'Anianiau, and 'Apapane) have fared well, maintaining sizeable populations (>37,000 birds) and stable or increasing numbers (Foster et al. 2004; Camp, Gorresen, et al. 2009). In general, these species or related taxa are also doing well elsewhere in the Hawaiian Islands. The 'Akeke'e population, although

of moderate size (7,900 birds) appears to be contracting in distribution. The endemic 'Akikiki has varied four-fold in number since 1981, and its range has diminished to half its prehurricane extent. The 'I'iwi also experienced a decline and contraction, although not as severe. Finally, the Puaiohi, perhaps the most endangered bird on Kaua'i, remains rare but apparently stable at present.

The avifauna on Moloka'i has also recently experienced a similar knockout punch, presumably from the upslope advancement of avian disease. With only about 30 km² of forest above 1,000 m, Moloka'i has lost the last populations of two species, Oloma'o and Kākāwahie, and the 'I'iwi is now less and less common. Notably, Hawai'i 'Amakihi are more widespread in windward valleys at low elevations than in the uplands. 'Apapane still occupy most rain forest from summit to sea level.

O'ahu, the most populous and developed of the main Hawaiian islands, has almost no forest habitat above 1,000 m and consequently had lost most of its native forest birds by the mid-1900s (Chapter 2). The island's population of 'I'iwi is also in imminent danger of extinction, and without additional conservation action, the O'ahu 'Elepaio may be close behind. Encouragingly, the O'ahu 'Amakihi and 'Apapane persist in most forests, and the former has expanded its range into nonnative vegetation in the suburbs of Honolulu.

---

lowland forests by O'ahu 'Amakihi and Hawai'i 'Amakihi (Lindsey et al. 1998). The presence of O'ahu 'Amakihi at low elevations, where avian malaria is presumably common, suggests that they may be evolving a resistance to the disease (Shehata et al. 2001; this volume, Chapter 9). The Hawai'i 'Amakihi is breeding and even increasing in low-elevation Hawai'i despite the highest prevalence of malaria found anywhere in the islands (Woodworth et al. 2005), and individuals from low eleva-

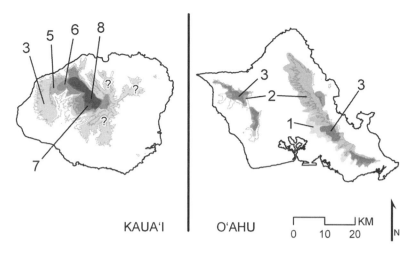

KAUA'I | O'AHU

Figure 5.8. Distribution of native passerines on Kaua'i and O'ahu. The shading corresponds to overlapping ranges, and the numbers refer to the number of species. Elevation is shown in 500 m contours.

tions survive the acute malaria challenge better than their high-elevation conspecifics (Chapter 9). This remarkable adaptation, however, may be eclipsed by the continued loss of much of the remaining areas of lowland native habitats to development and invasive plants. Efforts to protect high-elevation habitat must be coupled to the conservation of native habitat at lower elevations to ensure that the existing disease-tolerant genotypes evolve and retain the potential to serve as founders for recovering bird populations.

ACKNOWLEDGMENTS

This chapter was produced by the Hawai'i Forest Bird Interagency Database Project, a cooperative effort of the Pacific Island Ecosystems Research Center of the U.S. Geological Service; U.S. National Park Service; U.S. Geological Survey Pacific Basin

Figure 5.9. Distribution of native passerines on Moloka'i, Lāna'i, and Maui. Shading, numbers, and contours as in Fig. 5.8.

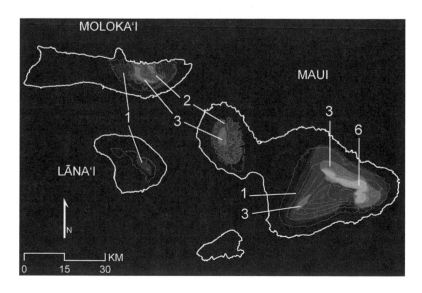

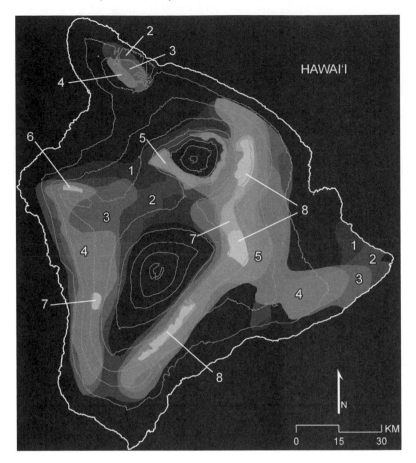

Figure 5.10. Distribution of native passerines on Hawaiʻi Island. Shading, numbers, and contours as in Fig. 5.8.

Information Node; University of Hawaiʻi–Pacific Cooperative Studies Unit; State of Hawaiʻi–Division of Forestry and Wildlife; Gap Analysis Program–Hawaiʻi; Kamehameha Schools; Hawaiʻi Natural Heritage Program; The Nature Conservancy of Hawaiʻi; U.S. Fish and Wildlife Service; and U.S. Forest Service. We especially thank the numerous interns who assisted with the preparation of data for the various studies described herein. Three anonymous reviewers provided useful comments on an early draft.

APPENDIX 5.1. STATUS SUMMARY OF EXTANT AND RECENTLY EXTINCT HAWAIIAN PASSERINE BIRDS

Species	Island Distribution	USFWS	IUCN	Population Size	Number of Wild Populations	Comments
Kaua'i 'Ō'ō *Moho braccatus*	(K)	E	EX	—	?	Last sighting in 1985; last audio detection in 1987
Bishop's 'Ō'ō *Moho bishopi*	(Mo, Ma?)		EX	—	0	Last detected on Moloka'i in 1904; unconfirmed reports in the 1980s from Maui
'Alalā *Corvus hawaiiensis*	H	E	EW	60	0	Entire population in captivity; extinct in the wild
Kaua'i 'Elepaio *Chasiempis sandwichensis sclateri*	K	—	(E)	152,000	1	Common above 600 m in native and exotic forest; density increasing
O'ahu 'Elepaio *Chasiempis s. ibidis*	O	E	(E)	<2,000	2	Range 55 km²; 6+ subpopulations on two mountain ranges; numbers rapidly decreasing
Hawai'i 'Elepaio *Chasiempis s. sandwichensis, C. s. ridgwayi, C. s. bryani*	H	—	(E)	<200,000	~5	Densities decreasing in Hualālai, Kona, and east windward Hawai'i Island; stable in lower elevation Ka'ū and at Hakalau Forest NWR
Nihoa Millerbird *Acrocephalus familiaris kingi*	NW	E	CR	<400	1	Very small range (<0.5 km²); marked variation in population, but appears stable over the long term; high risk of extinction
Kāma'o *Myadestes myadestinus*	(K)	E	EX	—	?	Last detections (unconfirmed) in 1985
Oloma'o *Myadestes lanaiensis*	(Mo)	E	(CR-PE)	—	?	Last detection during 1980 HFBS; unconfirmed report in 1980
'Ōma'o *Myadestes obscurus*	H	—	VU	170,000	1	Extirpated from Kona and Kohala; possibly declining in central and east windward Hawai'i Island; stable in Ka'ū and Hakalau Forest NWR

continued

Species	Island Distribution	USFWS	IUCN	Population Size	Number of Wild Populations	Comments
Puaiohi *Myadestes palmeri*	K	E	CR	300–500	1	Small range (<20 km^2); narrow habitat requirements; captive propagation ongoing
Laysan Finch *Telespiza cantans*	NW	E	VU	10,000	2	Small range (240 ha total); marked variation in population, but appears stable over the long term
Nihoa Finch *Telespiza ultima*	NW	E	CR	3,000	1	Very small range (40 ha); marked variation in population, possible long-term decline; high risk of extinction
'Ō'ū *Psittirostra psittacea*	(All)	E	CR-PE	—	?	Rapid decline on Kaua'i and Hawai'i; last confirmed sighting on Hawai'i in 1987 and on Kaua'i in 1989
Palila *Loxioides bailleui*	H	E	E	3,900	1	Population within 30 km^2; range contracting and density decreasing from 2003 to 2007; west Mauna Kea habitat recovering but vulnerable to fire
Maui Parrotbill *Pseudonestor xanthophrys*	Ma	E	CR	500	1	Single, small range (50 km^2); density appears stable
Hawai'i 'Amakihi *Hemignathus virens virens*	H	—	(LC)	>800,000	1	Density variable in central and south Kona, declining in midelevation windward Hawai'i Island but stable or increasing elsewhere; expanding range locally at low elevations
Hawai'i 'Amakihi *Hemignathus v. wilsoni*	Ma, Mo, (L)	—	(LC)	50,000	3	Small disjunct population on West Maui; increasing densities on East Maui; stable on Moloka'i; extirpated from Lāna'i in the 1970s; expanding range locally at low elevations
O'ahu 'Amakihi *Hemignathus flavus*	O	—	VU	52,000	2	Density possibly increasing; expanding range locally into lower elevation and non-native habitats

Species	Island Distribution	USFWS	IUCN	Population Size	Number of Wild Populations	Comments
Kaua'i 'Amakihi *Hemignathus kauaiensis*	K	—	VU	51,000	1	Densities increasing
'Anianiau *Viridonia parva*	K	—	VU	37,500	1	Densities increasing
Kaua'i Greater 'Akialoa *Hemignathus ellisianus stejnegeri*	(K)	E	EX	—	?	Last reported in 1969
Kaua'i Nukupu'u *Hemignathus lucidus hanapepe*	(K)	E	(CR-PE)	—	?	Unconfirmed reports up to mid-1990s
Maui Nukupu'u *Hemignathus lucidus affinis*	(Ma)	E	(CR-PE)	—	?	Unconfirmed detection in 1996
'Akiapōlā'au *Hemignathus munroi*	H	E	E	1,900	2	Density increasing in Hakalau Forest NWR and stable in upper Ka'ū; likely decreasing in central windward Hawai'i Island; extirpated from subalpine Mauna Kea and probably Kona districts; isolated populations; range contracting
'Akikiki *Oreomystis bairdi*	K	C	CR	3,600	1	Small population and range (36 km²); range contracting
Hawai'i Creeper *Oreomystis mana*	H	E	E	14,000	4	Density increasing in Hakalau Forest NWR and possibly stable in upper Ka'ū; likely decreasing in central windward Hawai'i Island; nearly extirpated from Hualālai and central Kona
O'ahu 'Alauahio *Paroreomyza maculata*	(O)	E	CR-PE	—	?	Last confirmed sighting in 1978
Kākāwahie *Paroreomyza flammea*	(Mo)	E	EX	—	?	Last confirmed sighting in 1963
Maui 'Alauahio *Paroreomyza montana newtoni*	Ma	—	(E)	>35,000	2	North population density increasing, but range may be contracting; southwest population small and trends unknown

continued

Species	Island Distribution	USFWS	IUCN	Population Size	Number of Wild Populations	Comments
'Akeke'e *Loxops caeruleirostris*	K	—	E	7,900	1	Densities fluctuate widely and range contracting; Makaleha Mountain population status unknown
Maui 'Ākepa *Loxops coccineus ochraceus*	(Ma)	E	(E)	—	?	Last sightings in 1980
Hawai'i 'Ākepa *Loxops coccineus coccineus*	H	E	(E)	12,000	4	Density possibly increasing in Hakalau Forest NWR and possibly stable in upper Ka'ū; likely decreasing in central windward Hawai'i Island; nearly extirpated from Hualālai and central Kona
'I'iwi *Vestiaria coccinea*	All, (L)	—	NT	360,000	5	Density decreasing throughout Hawai'i but stable in Hakalau Forest NWR and increasing on East Maui; range contracting at lower elevations
'Ākohekohe *Palmeria dolei*	Ma	E	CR	3,800	1	Small population and range (~60 km^2); density increasing
'Apapane *Himatione s. sanguinea*	All	—	LC	1,300,000	6	Densities increasing or stable in much of range but decreasing in midelevation east windward Hawai'i Island; expanding range locally at low elevations
Po'o-uli *Melamprosops phaeosoma*	Ma	E	CR	—	?	Rapid population decline and range contraction; last seen in the wild in 2004

Source: Table modified from Jacobi and Atkinson (1995), but data originate from this chapter.

Notes: Species distributions include all major Hawaiian islands (All), Kaua'i (K), O'ahu (O), Moloka'i (Mo), Lāna'i (L), Maui (Ma), Hawai'i Island (H), and the Northwest Hawaiian Islands (NW). Species are presumed extinct where the island is indicated in parentheses. Listing designations by the U.S. Fish and Wildlife Service (2006) and the International Union for Conservation of Nature (IUCN) (BirdLife International 2004) include extinct (EX), extinct in the wild (EW), critically endangered (CR), critically endangered–potentially extinct (CR-PE), endangered (E), vulnerable (VU), near threatened (NT), candidate for federal listing (C), of least concern (LC), or not listed as endangered or threatened by the U.S. Fish and Wildlife Service (—). Acronyms in parentheses indicate a listing designation at the species level. Population size is the most recent estimate or long-term survey average, and number of populations refers to the number of geographically distinct groups, regardless of genetic connectivity.

CHAPTER SIX

Loss, Degradation, and Persistence of Habitats

LINDA W. PRATT AND JAMES D. JACOBI

Forest bird species in Hawai'i have evolved with the islands' endemic tropical flora over several million years and have adapted in myriad ways to exploit these forest environments. The fate of the forest birds of the Hawaiian Islands is intimately linked with that of the forests. What happened to transform the original vegetation of the Hawaiian Islands? Why is so much of the modern landscape vegetated with alien plant species? How has history shaped the ecology and geography of the forest bird habitat we have today? The goals of this chapter are to describe the natural vegetation of Hawai'i and to illustrate the story of its decline and persistence.

ORIGINS OF THE HAWAIIAN FLORA

The flora of the Hawaiian Islands is relatively small but is the most highly endemic of that belonging to any floristic province (Wagner, Herbst, et al. 1999), and the natural forest types are unique in the world. Like the flora of other isolated oceanic islands, that of Hawai'i is disharmonic (Eliasson 1995) or taxonomically uneven, which means it lacks many plant groups common on the nearest continents (Mueller-Dombois and Fosberg 1998). Consequently, few tree and shrub species make up the canopy and secondary tree layer of Hawaiian forests and woodlands, and most of the natural plant diversity is contained in the understory (Mueller-Dombois et al. 1981). With the exception of coastal vegetation, which is typically composed of indigenous species widespread in the Pacific, the combination of woody species in extant forest communities is not duplicated in other Pacific island forests (Mueller-Dombois and Fosberg 1998). The native flowering plants and ferns of Hawai'i are derived primarily

137

from the Indo-Pacific and Austral regions, and fewer than 300 angiosperms and about 125 ferns are thought to have been the progenitors of the modern flora of the Hawaiian Islands (Fosberg 1948, Wagner 1995).

North America is the continent closest to the Hawaiian Islands, but America and the Boreal region are thought to have provided only 18% and 3% of the original plant immigrants, respectively (Fosberg 1948). Even though the flora of the Hawaiian Islands has this slight affinity with those regions, a passerine bird (songbird or perching bird) flying west from North America or east from northern Asia would find virtually no familiar trees in Hawaiian rain forests. Thus the bird ancestral to Hawaii's endemic honeycreepers, thought to have been a cardueline finch that arrived 5–6 million years ago (Fleischer and McIntosh 2001), had to adapt to completely novel habitats with unfamiliar foods, nesting sites, and plant cover (Chapter 1).

GEOGRAPHY AND CLIMATE PATTERNS THAT INFLUENCE PLANT DISTRIBUTION

The main Hawaiian islands fall entirely within the tropics, just south of the Tropic of Cancer. The climate of the high-elevation Hawaiian Islands is distinctly tropical, and seasonality in temperature or rainfall is not pronounced. Even though a few months are drier than most others (Giambelluca et al. 1986), there is no regularly recurring cycle of wet and dry seasons as is experienced in many other tropical lands. However, the El Niño–Southern Oscillation pattern, which occurs every four to seven years, causes trade winds to subside, suppresses storms, and reduces winter rainfall (Giambelluca and Schroeder 1998). Although dry periods occur yearly in most low leeward regions and throughout the subalpine zone of the Hawaiian Islands, most wet forest types do not experience

prolonged drought, except during severe El Niño events.

For the purpose of explaining the distribution of natural vegetation in Hawai'i (Table 6.1), the most important feature of the climate is the prevalence of northeast trade winds, which interact with the mountain slopes of the high islands to create an orographic rainfall pattern. Orographic rainfall is precipitation on windward mountain slopes generated from moist air blown upslope by trade winds to a level where air is cooled and clouds form; such rainfall supplies most of Hawaii's water (Giambelluca and Schroeder 1998). Correspondingly, the windward sides and summits of mountains support wet vegetation types on the islands of Kaua'i, O'ahu, Moloka'i, and Maui (Plates 23–25). The lower island of Lāna'i has wet forest only near the summit. East Maui and Hawai'i exhibit wet vegetation below a temperature inversion zone that fluctuates near 1,800 m elevation (1,500–3,000 m) and prevents moisture-laden warm air from progressing up the mountain slope. The trade wind temperature inversion is caused by rising (warm) air and sinking (cold) air meeting in a layer that acts as a ceiling. Above this ceiling, air becomes warmer as altitude increases, and dry conditions prevail on the high mountain slopes (Giambelluca and Schroeder 1998).

The small islands of Ni'ihau and Kaho'olawe, as well as the western half of Moloka'i, are in the rain shadows of adjacent larger islands or volcanoes and have dry climates that do not currently sustain wet forest vegetation. However, there is botanical evidence from early collections implying that a moist forest formerly existed on Ni'ihau (Wichman and St. John 1990) and in west Moloka'i (Rock 1913).

Leeward slopes are seasonally dry during summer months, except on Hawai'i Island, where a particular "Kona" pattern of high summer rainfall operates. This convectional rainfall pattern is caused by diurnal surface heating and the resulting

Table 6.1. Vegetation classification system for Hawai'i

Zone	Elevation Range (m)	Moisture Regime	Vegetation Community Example	Island Distribution	Forest Bird Habitat?
Coastal	3–30	Dry	'Āweoweo (*Chenopodium oahuense*) Shrubland	NW	Yes
		Mesic	Hala (*Pandanus tectorius*) Forest	K, O, Mo, Ma, H	No
		Wet	Hau (*Hibiscus tiliaceus*) Shrubland	All except Ni, L, Ka	No
Lowland	0–1,000	Dry	Pili Grassland	All, leeward	No
		Mesic	Lama/'Ōhi'a Forest	K, O, Mo, Ma, H	Yes
		Wet	'Ōhi'a/Uluhe Forest	All except Ni, Ka	Yes
Montane	Above 1,000–2,000	Dry	'Ōhi'a/'A'ali'i Montane Shrubland	K, Ma, H	Yes
		Mesic	Koa/'Ōhi'a Montane Mesic Forest	K, Ma, H	Yes
		Wet	'Ōhi'a/Hāpu'u Forest	Windward H	Yes
		Wet	'Ōhi'a Montane Wet Forest	K, Mo, Ma, H	Yes
Subalpine	Above 2,000–3,000	Dry	Māmane Forest	Ma, H	Yes
		Mesic	*Deschampsia* Grassland	Ma, H	No
		Wet	*Deschampsia/Carex/Oreobolus* Mixed Bog	Ma	No
Alpine	Above 3,000	Dry	'Āhinahina (*Argyroxiphium* spp.)/ Na'ena'e (*Dubautia* spp.) Shrubland	Ma, H	No

Source: Adapted from Pratt and Gon (1998).

Notes: Binomials are given in the chapter text except for those listed here. H, Hawai'i Island; K, Kaua'i; Ka, Kaho'olawe; L, Lāna'i; Ma, Maui; Mo, Moloka'i; Ni, Ni'hau; NW, Northwest Hawaiian Islands; O, O'ahu.

rising warm air, which creates upslope winds and draws in moist air from the ocean (Giambelluca and Schroeder 1998). Transition areas of moist (mesic) vegetation occur at the thermal inversion zone, the interface of windward and leeward slopes, and the edges of mountain rain shadows. The subalpine and alpine zones of the high islands also have low precipitation and support dry vegetation (Gagné and Cuddihy 1999).

NATURAL CHANGES AND SUCCESSION

The Hawaiian Islands are volcanic in origin, and the youngest island, Hawai'i, has on its southeastern flank the most active volcano in the world. Kīlauea Volcano erupts vast quantities of lava, and more than 90% of its surface is covered by fresh lava every 1,100 years (Holcomb 1987). The adjacent massive Mauna Loa Volcano has had 40% of its surface buried by lava in the past 1,000 years (Lockwood and Lipman 1987). Primary succession is therefore an important phenomenon explaining vegetation patterns on active volcanoes (Plate 17).

Historic flows (<200 years old) are prominent in the geology of Hawai'i Island (Wolfe and Morris 1996). The rift zones of Kīlauea and Mauna Loa have produced immense expanses of lava flows over the past 400 years (Plate 26), resulting in the separation of forested areas by new lava surfaces without tree cover. Flows may be as wide as 10–15 km, a distance that might daunt some forest bird species. Kīpuka, which are islands of older vegetation surrounded by more recent flows, are important refugia of mature forests with a high degree of plant diversity (Plate 17). They are also valuable bird habitats and serve as links for birds, invertebrates, and plants between once-contiguous forests. The slopes of the active volcanoes are highly hetero-geneous in substrate ages and forest devel-opment stages. This juxtaposition of dif-ferent forest types within flying distance makes available to forest birds a greater diversity of resources than could be found in a single forest stage.

The Natural Fire Regime

Fire did not play a large role in the long-term history of vegetation development in Hawai'i (Mueller-Dombois 1981). Except for glacial periods with dry and cool cli-matic conditions (Hotchkiss 1998), when fires may have been relatively frequent, the natural fire regime of Hawai'i was charac-terized by infrequent, relatively small fires (Vogl 1977). Except on Hawai'i and in East Maui, the Hawaiian volcanoes are extinct, and volcanic activity no longer plays a role in vegetation succession. Before the inva-sion of alien, fire-adapted grass species, fine fuels were not continuous in most natural vegetation, and fire did not penetrate far into native forests. Few endemic species respond positively to wildfire, and none seems to be fire-dependent, although some trees, like koa (*Acacia koa*) and 'a'ali'i (*Do-donaea viscosa*), are fire-tolerant (Smith and Tunison 1992). Past forest fires must have resulted in temporary loss of bird feeding and breeding habitat, but the small sizes of prehistoric fires would have limited the severity of the impact on forest birds.

Succession

The principal tree species of the Hawaiian Islands, 'ōhi'a lehua (*Metrosideros polymorpha*) is all-important to the endemic nectivo-rous birds because its showy red, orange, or yellow flowers are the most abundant source of high-quality nectar (Carpenter and MacMillen 1973; this volume, Chap-ter 7). Leaf buds of the tree are also im-portant feeding substrate for native birds such as the 'Ākepa (*Loxops coccineus*). With its dense canopy foliage and small, twiggy branches, 'ōhi'a provides ideal nest sites for forest birds (Chapter 8). 'Ōhi'a lehua is a remarkably polymorphic species includ-ing eight distinct varieties that differ in leaf size, shape, and degree of pubescence (hairs on leaf surface) (Dawson and Stem-mermann 1999). Morphologically dif-ferent varieties replace each other along elevational, moisture, and substrate age gra-dients. Recent studies comparing the phys-iology and morphology of 'ōhi'a lehua varieties grown in a common garden and in natural stands suggested a potential ge-netic basis for morphological traits of leaf size and an environmental cause for traits of leaf thickness and pubescence (Cordell et al. 1998). Varieties of 'ōhi'a lehua with hairy (pubescent) leaves are adapted to the colonization of fresh lava and tephritic sur-faces, where seedlings may be seen within four years posteruption (Smathers and Mueller-Dombois 1974). Succession pro-ceeds with the invasion of additional tree, shrub, and fern species until a mature for-est stand is achieved, typically within 400 years in the wet lowland zone of Hawai'i Island (Atkinson 1970). This closed forest is not a "climax" community, for species continue to change even after 9,000 years of succession. A study of a chronosequence of eight lava flows on Mauna Loa, Hawai'i (Kitayama et al. 1995), revealed that smooth-leaved (glabrous) varieties of *Metro-sideros* replace pubescent types after 3,000 years. The matted fern uluhe (*Dicranopteris linearis*) is an early invader that is succeeded within 300 years by hāpu'u tree ferns (*Ci-botium* spp.), which achieve their greatest dominance on deep ash substrates 1,000 years old (Becker 1976). Tree fern cover declines after 3,000 years, to be replaced by woody plants.

Long-term succession on older sub-strates (3.5–5.6 million years old) of the northernmost high islands of Kaua'i and O'ahu (Clague 1998) has resulted in a great diversity of forest understory species. This diversity is displayed by several common forest tree and shrub genera, such as Myrsine, Melicope, Dubautia, and Cyrtandra, which have a greater number of species on old versus young islands (Carr 1985; Wagner, Herbst, et al. 1999). Nonetheless, the principal canopy trees of all wet and most mesic forests remain 'ōhi'a and koa, except in diverse forest communities that cover little area today (but may have been more widespread in prehuman times). Dry forest succession involves replacement of the ubiquitous 'ōhi'a by tree species better adapted to compete for water on older flows, such as lama (Diospyros sandwicensis) (Stemmermann and Ihsle 1993). Older forests may have correspondingly diverse communities of forest birds. Even on the youngest island of Hawai'i, montane forests on old substrates are characterized by rare birds, such as the 'Ākepa and Hawai'i Creeper (Oreomystis mana), species that are restricted to forests with a dense, closed canopy of large-diameter trees or woodlands near such sites. These large trees serve as important nesting sites (Chapter 8).

'Ōhi'a Dieback

'Ōhi'a dieback is a large-scale perturbation characterized by thinning of the tree crown, defoliation, and eventual death of most mature trees in even-aged stands or cohorts. Forest decline was first reported on Maui 100 years ago (Lyon 1909), and by the mid-twentieth century it was considered a serious problem on more than 50,000 ha of Hawai'i Island forests (Jacobi et al. 1983). Dieback was particularly conspicuous on young lava flows where even-aged stands of trees died synchronously, but the phenomenon had also been observed on Kaua'i substrates older than a million years (Mueller-Dombois 1986). After years of research on insects or plant pathogens as possible causative agents (Papp et al. 1979, Hodges et al. 1986), the disease hypothesis was abandoned, and abiotic stress triggered by drought, wet soils, or wind was implicated as the cause of dieback (Mueller-Dombois 1985).

Concurrent research on the composition and structure of dieback forests supported a hypothesis of 'ōhi'a dieback as a natural successional pattern, similar to canopy dieback processes elsewhere in the Pacific (Mueller-Dombois 1980, Jacobi 1983). Changes in permanent vegetation plots within dieback areas revealed a high level of 'ōhi'a regeneration, indicating that the dominant forest tree was replacing itself (Jacobi et al. 1983). Subsequent evaluation of permanent plots confirmed continued 'ōhi'a regeneration at many study sites but revealed that displacement by alien plants or endemic tree ferns was occurring in some forests (Jacobi et al. 1988). Nutrient deficiency was investigated and rejected as the primary cause of 'ōhi'a dieback (Gerrish et al. 1988).

If 'ōhi'a dieback is a natural successional phenomenon, endemic forest birds have certainly coped with such periodic changes in their habitat. However, the presence of novel alien plant species in dieback areas may lead to the conversion of large tracts of formerly native forest over a relatively short time. Even in predominantly native forest, there may be a persistent seed bank of alien species that can quickly respond to disturbance (Drake 1998). 'Ōhi'a dieback should be considered during the design and planning of natural areas, which should be large enough to allow for forest succession, including dieback, without excessive fragmentation or alien plant invasion (Jacobi et al. 1988).

Forest Diseases

Although 'ōhi'a dieback is not a disease phenomenon, there are plant diseases and insects that affect both 'ōhi'a and other important native forest matrix species. Recently, a Central American guava rust (Puccinia psidii) has invaded O'ahu and begun to affect 'ōhi'a seedlings (Hauff 2005); the long-term impacts on native forests are unknown. The disease, also known as Eucalyptus rust, is considered a serious worldwide threat to Eucalyptus plantations (Coutinho et al. 1998). Koa has been shown to be the host of several native rust fungi (Hodges and Gardner 1984, Gardner 1997a), and a vascular wilt fungus (Fusarium oxysporum f. sp. koae) has been implicated in koa dieback. This phenomenon appears to be increasing in severity on Hawai'i Island (Anderson et al. 2002). Most other recognized pathogens of native Hawaiian tree and shrub species appear to be native (Borth et al. 1990; Gardner 1997a, 1997b).

HABITAT LOSS DURING THE PRECONTACT HAWAIIAN PERIOD

Background

Polynesian voyagers arrived in the Hawaiian Islands around A.D. 800 (Athens 1997, Kirch 2007). Polynesian culture had developed in the southwest Pacific by the first millennium B.C., and Polynesians probably came to Hawai'i from the Marquesas >4,000 km to the southeast. There may also have been later arrivals from Tahiti (Kirch 1985).

During their first few centuries in Hawai'i, Hawaiians lived primarily at the coast and exploited marine resources. Analysis of midden piles on O'ahu and Moloka'i indicated that early Hawaiians were subsisting mainly on marine mollusks, fish, and seabirds rather than cultivated plants and animals (Pearson et al. 1971, Kirch and Kelly 1975). The initial settlement phase lasted perhaps 300–500 years, during which human population growth was gradual, most settlements were on the seashore, and resource use focused on the sea and coastal lowlands (Kirch 1985). During this phase, there was little disturbance of natural vegetation other than exploitation of wood for structures and cooking fires.

The early impact of humans on the Hawaiian avifauna was the use of birds for food, resulting in the reduction and extirpation of seabird populations, loss of the

Nēnē (Hawaiian Goose, Branta sandvicensis) and Koloa (Hawaiian Duck, Anas wyvilliana) from several islands, and extinction of flightless geese, flightless moa-nalo derived from duck ancestors (Chelychelynechen quassus, Thambetochen spp., Ptaiochen pau), many species of flightless rails (Porzana spp.), and other ground-dwelling birds (Olson and James 1982a, Sorenson et al. 1999). There can be little doubt that humans utilized the easily available food source of flightless, large-bodied birds just as they did on other Pacific island groups like New Zealand (Worthy and Holdaway 2002). The frequency of bird bones at archaeological sites increases with depth and age, indicating that this food resource was locally exhausted during the early settlement phase (Kirch 1985). Almost 40 species of land birds went extinct during the prehistoric Hawaiian period and were discovered as fossils only recently. The losses of small passerine species are thought to be primarily the result of habitat alteration rather than direct human predation (Olson and James 1982a, 1982b; Kirch 1983). Secondary predation from rats, pigs, and dogs that arrived with Hawaiian settlers may have also contributed to the decline or loss of many lowland birds (Olson and James 1982b), particularly those that nested on the ground or in understory vegetation. For details of the decline of forest birds in Hawai'i, see Chapter 2.

Shifting Agriculture

After the colonization period, Hawaiians moved toward the establishment of permanent settlements and agriculture. During this 500-year interval, the human population continued to grow, and agriculture expanded upland from the original settlement sites (Kirch 1985). Shifting agriculture was practiced using the slash-and-burn technique with long fallow periods. As the human population grew, fallow periods decreased, erosion of upland slopes became more pronounced (Allen 1997), and agriculture expanded beyond the large valleys (Kirch 1982).

Crops varied from wet to dry areas, but starchy staples such as kalo (taro, *Calocasia esculenta*) and 'uala or sweet potato (*Ipomoea batatas*) were the most important (Abbott 1992). Approximately 30 plant species were introduced to Hawai'i by Polynesian settlers, among them food and fiber crops, dye and medicinal plants, and several wetland weeds (Nagata 1985, Athens 1997). None of these plants can be considered a serious forest pest today. Only kukui (*Aleurites moluccana*) has established significant cover in windward lowlands, where the tree likely displaced native koa forests and riparian vegetation more than a millennium ago (Fosberg 1972).

Permanent Agriculture

Between A.D. 1100 and 1650, there was a period of expansion and agricultural intensification in Hawai'i that accompanied an increase in the human population (Kirch 1985). During that period the lowland vegetation of the Hawaiian Islands below 500 m elevation was almost entirely replaced by cultivated fields, dispersed settlements, and anthropogenic grasslands caused by repeated fires, while upland sites remained little disturbed (Burney and Burney 2003).

Between 1500 and 1800, agricultural fields coalesced into permanent systems within extensive belts of fertile, relatively well-watered land on Hawai'i Island at Kohala (Rosendahl 1972, Vitousek et al. 2004), Kona (Newman 1972), Waimea (Clark and Kirch 1983), and Ka'ū (Ellis 1827) and on Maui at Kahikinui (Kirch 1985). Cultivated fields were associated with drier grasslands downslope and with partially forested resource extraction areas at higher elevations (Kelly 1983, McEldowney 1983).

Intensification of agriculture beyond windward valleys resulted in deforestation, erosion, infilling of water basins, and con-

version of forests to grasslands (Kirch and Kelly 1975, Allen 1997). The low islands of Ni'ihau and Kaho'olawe, as well as western Moloka'i and the lower slopes of Lāna'i, had lost much of their original forest cover by the time European explorers arrived (Ellis 1827, Beaglehole 1967). Permanent agriculture thus transformed the lowlands of the Hawaiian Islands by the end of the eighteenth century.

The Use of Fire by Hawaiians

Fire was the primary tool used by Hawaiians to clear lands for cultivation (Kirch 1982). Several of the large field systems have pronounced burn layers that represent either the original removal of native tree cover (Murakami 1983) or the use of fire for clearing fallow fields. Fire was also used in marginal cultivations to burn off vegetation (Handy and Handy 1972) and to increase the cover of 'ama'u ferns (*Sadleria cyatheoides*) used as pig feed (Kirch 1982). Large expanses of the lowlands were regularly burned to clear woody vegetation and stimulate indigenous pili grass (*Heteropogon contortus*), which was used as thatching material in dry areas (Menzies 1920). Favored for its pleasant odor and red color, pili thatching required replacement about every 5–10 years (Abbott 1992), so there was a continuous demand for large quantities of grass.

A Summary of the Impacts of Hawaiians

By the time that European explorers saw the Hawaiian Islands, there was a ring of cleared vegetation, grass, and cultivated fields surrounding the larger islands, and well-developed forests were inconspicuous on or absent from the smaller islands. Captain James Cook commented on the lack of forest vegetation on Kaua'i in his journal of 1778: "We saw no wood but what was up in the interior part of the island and a few trees about the villages" (Beaglehole 1967: 264). Others on this first visit of Europeans to the Hawaiian Islands recorded similarly bleak views of Kaho'olawe, western Moloka'i, and Ni'ihau, which was described as entirely bare of trees. In 1792–1794, botanist Archibald Menzies, who accompanied Captain George Vancouver, estimated a distance of 11 km from Kealakekua to the forest edge on Hawai'i Island (Menzies 1920). Several decades later, Ellis (1827) observed that the forest began 8–10 km from the windward shore near Hilo, Hawai'i Island.

There can be little doubt that Hawaiians transformed the lowlands (below 500 m elevation) of the islands and replaced natural forests with anthropogenic vegetation (Athens 1997), converting "a natural ecosystem into an actively manipulated cultural landscape" prior to European arrival (Kirch 1982: 4). According to Kirch (1982: 5), "There is scarcely an area in the lowlands (if it receives greater than 500 mm rainfall and is not a steep cliff) that upon archaeological reconnaissance does not yield evidence of indigenous Polynesian agricultural use" (Plate 27).

What were the ultimate impacts on native birds of a relatively large human population subsisting on the resources of land and sea in the confined space of the Hawaiian Islands? Replacement of the lowland forests with cultivated fields removed habitat necessary to forest birds restricted to the lowlands. The loss of perhaps two-thirds of the original avifauna of the islands (Olson and James 1982b) is evidence that the lowlands, particularly in the dry leeward regions, were important habitat for many bird species. The fossil record also indicates that many bird species now restricted to high-elevation wet sites had formerly ranged into the lowlands.

We can only guess at the losses in the Hawaiian flora, but a recent list of endangered plant species (Wagner, Bruegmann, et al. 1999) includes many plants of dry and mesic ecosystems. The estimated extinction rate of 8% for native plant taxa (Mehrhoff 1998) may be low, for it is based on losses sustained in modern times. The pollen record indicates that several species extremely rare today played a large role in prehistoric lowland forests and shrublands. For example, *Kanaloa kahoolawensis*, a nearly extinct shrubby legume now restricted to a sea stack off Kaho'olawe (Lorence and Wood 1994), was an important element in the pollen assemblages of dry lowland sites on O'ahu (Athens 1997). Analysis of pollen from other low-elevation sites on O'ahu revealed high percentages of loulu (*Pritchardia* spp.) pollen, indicating that the palms were major components of the community (Athens 1997, Hotchkiss and Juvik 1999). Today such palms are found only in small groves scattered on mountain ridges and cliffs (Wagner, Herbst, et al. 1999). Evidence from a fossil site on the south coast of Kaua'i indicated that the original lowland vegetation included many tree species now restricted to high-elevation forests (Burney et al. 2001). Clearly, the vegetation of O'ahu and Kaua'i has been much altered in the past millennium.

Despite the environmental changes caused by Hawaiians, no highly invasive plant species were present to take advantage of disturbance. The uplands of the main Hawaiian islands remained largely intact throughout more than 1,000 years of Hawaiian inhabitation, and it was the introductions and practices of the modern post–European contact era that created the landscape patterns of forest vegetation today (Plate 25).

HABITAT LOSS AFTER WESTERN CONTACT

Early Trade and Crop Production

Modern agriculture in Hawai'i began with the production of foodstuffs for trade with whaling vessels arriving in the Hawaiian Islands starting in 1819 (Abbott 1992). This early trade provisioned hundreds of visiting ships with food and the firewood essential for rendering whale oil (Kuykendall 1938). Because early trade crops were probably cultivated in traditional agricultural areas, this early postcontact agriculture did not result in significant expansion of cultivation and had little impact on the remaining native vegetation.

SANDALWOOD

The same cannot be said of the trade in sandalwood, or 'iliahi (*Santalum* spp.), which represented a brief but destructive phase in Hawaiian history that resulted in the loss of lowland dry forest that had survived the precontact expansion of agriculture on Hawai'i and several of the larger islands. Sandalwood, valued for its fragrance, was intensively harvested and exported to China from 1815 (St. John 1947) until the 1870s (Schmitt 1977). When sandalwood became scarce, fires were set in dry forests so that trees or fallen wood could be located by their scented smoke (St. John 1947). Such wholesale destruction of forests to obtain one valuable commodity was unsustainable, and the sandalwood trade collapsed within 50 years of its inception. The losses of lowland dry and mesic forest due to sandalwood harvesting are difficult to quantify, but this activity clearly contributed to the conversion of the islands' dry forests to alien grassland by the twentieth century. Such leeward lowland forests formerly supported diverse communities of native birds, as documented in the fossil record (Olson and James 1982a). Some now extinct birds that favored the leeward side of Hawai'i Island, such as the Greater and Lesser koa-finches (*Rhodacanthis* spp.), may also have ranged into forests destroyed during the sandalwood trade era.

SUGAR CANE

Although sugar cane, or kō (*Saccharum officinarum*), was introduced to Hawai'i by Polynesians, it was not a major food (Abbott 1992). Sugar was recognized as a potential commercial crop in Hawai'i as early as 1825 (Kuykendall 1938). By the 1840s,

more than 20 sugar plantations were operating in the islands (Culliney 1988), and the land reform laws following the Great Māhele of 1848 gave impetus to the industry by clarifying land ownership and allowing the sale of agricultural land to non-Hawaiians (Kuykendall 1938).

Nineteenth-century commercial sugar plantations were largely in the wet lowlands on the windward sides of the islands. In the 1870s, Victorian traveler Isabella Bird (1906) described one such plantation at Onomea on Hawai'i Island, where fields extended to 460 m elevation, above which hauling tracks continued into the forest. By 1880, the number of sugar plantations had increased to more than 60 and included more than 50,000 ha (Kuykendall 1967). Plantations continued to be developed into the twentieth century, particularly on marginal lands of Hawai'i Island. By 1980, sugar was cultivated on more than 100,000 ha on four of the Hawaiian islands (Morgan 1983), after which cropland declined to less than 25,000 ha (Kelly 1998).

Today, sugar cane is no longer cultivated on Hawai'i Island, where former fields have been converted to diversified agriculture, cattle pastures, and tree plantations. Less productive cane lands have been abandoned to a cover of alien trees. Cultivation of other modern crops has had little impact on native forests, other than the clearing of dry and mesic forests for macadamia nut (*Macadamia integrifolia*) plantations on leeward Hawai'i and the small-scale development of lowland wet forests for cultivation of papaya (*Carica papaya*) in Puna District.

Although many sugar plantations incorporated lands previously disturbed by Hawaiian agriculture and burning (Newman 1972), low-elevation forests were almost certainly cleared above the limits of Hawaiian cultivation, and the impacts extended beyond the fields cleared. The need for firewood to fuel the sugar mills (Judd 1927a) and the diversion of water for irrigation led to losses of integrity in forests adjacent to plantations (Culliney 1988).

One positive aspect of the sugar industry was the creation of forest reserves to provide watersheds for the plantations (Giffard 1918). The establishment of reserves, which was strongly supported by the sugar industry and the U.S. Forest Service (Hall 1904, Hosmer 1959), was the single most important governmental action ever for the protection of native Hawaiian forests (Cuddihy and Stone 1990). The forest reserves are the nucleus of the remaining forest bird habitat, and their creation in the early twentieth century was the beginning of the modern conservation movement in Hawai'i. Without this effort, Hawai'i would have far less forest today.

Forestry and Logging

Most forests of the Hawaiian Islands are dominated by 'ōhi'a and koa, and these were the species exploited by early logging practices. Koa has always been the preferred wood for timber. Commercial koa logging began in the 1820s and expanded in the 1830s, particularly in the Kona and Hāmākua districts of Hawai'i (Jenkins 1983). Much of the koa extraction was concurrent with the development of cattle ranches. Such logging may not have been considered a sustainable industry but rather a necessary precursor to pasture improvement. Today, almost treeless pastures of alien grasses cover the upland slopes of windward Mauna Kea and Waimea, Hawai'i Island, as well as the west slope of Haleakalā, Maui, replacing the nineteenth-century koa forests. East Maui forests provided koa lumber and firewood for lowland sugar mills until the accessible timber was depleted by the late 1800s (Jenkins 1983). Koa logging continues in forested ranches of the South Kona District, Hawai'i Island, where lack of water for milling delayed full exploitation.

The consequence of the destruction of large tracts of forest was the loss of habitat for forest birds, plants, and native invertebrates. Several birds known mostly from the koa forests of Kona disappeared or became rare in the late 1800s (Banko 1986). The loss and degradation of native forests continued into the twentieth century. Based on analysis of fossil sites, dry and mesic forests, particularly those of the lowlands, were especially rich in native forest bird species (Olson and James 1982a).

INTRODUCTION OF TIMBER SPECIES

Because of the need to maintain forests as watersheds and reforest areas damaged by grazing and feral animals (Giffard 1918), tree planting was an integral part of early territorial forestry practice, particularly during the Great Depression years, when Civilian Conservation Corps labor was available (Rodwell 2001). Between 1910 and 1960, more than 1,000 alien tree and shrub species were planted in Hawaiian forest reserves (Skolmen 1979). More than a million native koa trees were used, but the bulk of the plantings were fast-growing nonnative tropical species, including 2 million each of Australian swamp mahogany (*Eucalyptus robusta*), silk oak (*Grevillea robusta*), and paperbark (*Melaleuca quinquenervia*) (Nelson 1965). Early plantings were intended to be of no commercial value to ensure that watershed forests were not cut (Lyon 1918), but trees were later selected for their potential commercial value. Most of these introductions have not naturalized, but some of the worst invaders of native forests became established in many regions of Hawai'i as intentional plantings in forest reserves, including faya (*Morella faya*), strawberry guava (*Psidium cattleianum*), and Christmas berry (*Schinus terebinthifolius*) (Skolmen 1979). The role of such plantings in the dynamics of weed invasion has not been thoroughly investigated.

MODERN COMMERCIAL LOGGING AND SILVICULTURE

Despite past plans for commercial timber extraction from 35,000–166,000 ha of forest (Hawaii Department of Land and Natural Resources and Department of Planning and Economic Development 1976), a major forest products industry based on existing timber plantations has not developed in Hawai'i (Cannarella 1998). These alien forests, though temporary, provide habitat for common native forest birds, benefiting the remaining nectarivores and insectivores.

Modern commercial logging consists largely of koa extraction from private lands, which in the past resulted in a decrease in forest cover and loss of habitat for endangered plant and bird species (Warshauer and Jacobi 1982). One study documented a decrease in native bird densities and recolonization by alien birds after native forest clearing for a koa silviculture project (Sakai 1988). In recent years, logging and ranching practices have evolved toward more sustainable methods. Today, there is interest in koa silviculture on deforested lands to supply future demand for this valuable mahoganylike hardwood (Skolmen and Fujii 1980), despite the uncertain economics of such ventures (Skolmen 1986). If successful, koa silviculture practiced on old pasturelands may result in increased forest bird habitat that will be beneficial to both endangered and common forest birds during the 40- to 50-year tree maturation period. Research on the use of young koa plantations by the endangered 'Akiapōlā'au (Hemignathus munroi) has shown a promising upward trend for native forest birds in 25-year-old koa stands (Pejchar 2004). Issues of timber harvest in forests supporting endangered birds must be resolved for silviculture to result in both economic returns and effective bird conservation.

Ranching

The use of native forests for cattle pasture has disturbed far more forest bird habitat in Hawai'i than has farming or forestry (Plate 18). Land currently or formerly used as pasture covers large expanses of Hawai'i Island, particularly on Mauna Kea and in leeward Kohala and Kona districts (Plate 28), as well as significant portions of East Maui. Lowland pastures were developed on former agricultural land, but upland pastures were forests prior to the introduction of cattle and feral animals little more than 200 years ago. In some areas, logging was a precursor to formal ranching (Jenkins 1983). A number of modern ranches still support open forest whose canopy tree species are important for forest birds (Scott et al. 1986).

Cattle (Bos taurus) were introduced to Hawai'i by Captain Vancouver in 1793–1794. Allowed to multiply, feral herds quickly spread into the uplands (Tomich 1986). An industry based on hides and tallow developed around wild cattle, and by 1830 Parker Ranch was operating on formerly forested lands near Waimea (Henke 1929). In the mid-nineteenth century, land reforms led to the development of ranches on all the main islands (Kuykendall 1938). On Kaua'i, O'ahu, and Moloka'i, early ranches were associated with sugar plantations (Henke 1929). The cattle ranches on Hawai'i and Maui were large, produced most of the territory's cattle, and were likely to include montane forests (Philipp 1953). At the peak of ranching activity, ranches covered 25% of the total state land area (Hugh et al. 1986).

The long-term impacts of cattle in Hawai'i are ultimately the conversion of forest to grassland; the effects are the same whether the cattle are domestic or feral (although feral animals are more difficult to remove from forested land). One recent study reported the decline of forest and the doubling of open grassland throughout Pu'uwa'awa'a Ranch on leeward Hawai'i over 40 years of grazing (Blackmore and Vitousek 2000). Losses of tree cover may be assumed for other ranches due to both logging and grazing (Warshauer and Jacobi 1982). Even without active clearing, tree cover decreases in grazed areas, for cattle interfere with the reproduction of canopy tree species such as koa in montane forests (Baldwin and Fagerlund 1943) and māmane (Sophora chrysophylla) in dry subalpine forests (Scowcroft 1983). Compaction of soils is also a negative effect of cattle (Scowcroft 1971); the shallow roots of koa may be easily damaged by grazing (Judd 1927b). Recovery of forests after grazing pressure is removed varies with the degree of degradation; heavily disturbed dry forests invaded by alien grasses may exhibit no native tree reproduction even after decades of protection (Cabin et al. 2000). By contrast, montane koa forests have been observed to expand and recover quickly following the removal of cattle (Tunison, McKinney, et al. 1995).

Fire

Direct physical damage to native plants is not the only negative impact of cattle grazing. More than 200 alien grass species have been introduced to Hawai'i in support of the ranching industry (Whitney et al. 1939). Other invasive grasses and forbs have entered Hawai'i as contaminants of imported grass seed. Some introduced pasture grasses, such as molasses grass (Melinis minutiflora) and jaragua (Hyparrhenia rufa), are highly invasive and flammable and greatly increase the likelihood, frequency, and size of wildfires in invaded forests and woodlands. These and other fire-adapted tropical grasses have altered the natural fire regime in Hawai'i and expanded grasslands by carrying fire into native forests whose woody species may not survive burning (D'Antonio and Vitousek 1992, Smith and Tunison 1992), particularly by repeated fires. The cover of native trees declines greatly in burned woodlands, and native woody plants virtually disappear after successive fires (Hughes et al. 1991; Tunison, Loh, et al. 1995).

The intensification of grass cover was due not only to the introduction of forage and weed species. Ornamental fountain grass (Pennisetum setaceum) was introduced around 1914 and now dominates many dry communities of Hawai'i Island (Smith 1985), where it has the greatest altitudinal range of any grass (Williams et al. 1995). Fountain grass grows faster and produces more biomass than do native lowland grasses (Williams and Black 1994), produces more viable seeds than native pili (Goergen and Daehler 2001), and alters ecosystem properties involving light, water, and nutrients (Cordell et al. 2002). The potential distribution of fountain grass has been predicted to include more than a third of the forested area of Hawai'i Island (Jacobi and Warshauer 1992). Fountain grass is less widespread on other islands, and there is an ongoing effort to prevent its establishment on Maui by a systematic program of early detection and eradication of small populations (L. Loope, pers. comm.).

Because of fountain grass, as well as a suite of African and European forage species and accidentally introduced American grasses such as bush beardgrass (Schizachyrium condensatum) and broomsedge (Andropogon virginicus), the risk of fire is greatest in dry habitats of Hawai'i Island (D'Antonio and Vitousek 1992). Forest bird habitats of upper Mauna Kea, Hualālai, and leeward Mauna Loa are particularly vulnerable to fire. Although the wet forests of windward Hawai'i are not at great risk, woodlands dominated by the matted fern uluhe may be vulnerable to fire during drought periods (Tunison et al. 2001). Rain forests with tree fern understory are the least susceptible to fire of all natural vegetation types. Nonetheless, there is evidence that fires occurred, albeit rarely, in wet upland vegetation prior to any disturbance by humans (Burney et al. 1995).

Removal of invasive grass cover is practicable only in small areas (Cabin et al. 2000), although research continues to improve the cost-effectiveness of grass control (Cordell et al. 2002). Attempts to arrest the spread of particularly invasive grasses have been locally successful (Tunison et al. 1994), and rehabilitation of burned areas by introducing fire-tolerant natives is being researched (Tunison et al. 2001). However, fire suppression remains the most effective way to prevent further loss of bird habitats in dry areas, and large-scale restoration of grass-invaded forests is unlikely without effective grass removal (Chapter 16).

Feral Ungulates

The Hawaiian Islands have no native land mammals other than an endangered bat (Hawaiian Hoary Bat, *Lasiurus cinereus semotus*) (Tomich 1986), and the isolated Hawaiian flora has developed without the evolutionary influences of mammalian grazers and browsers. There are numerous examples of Hawaiian plant groups whose endemic members lack the thorns, prickles, poisons, and strong-smelling foliage of their continental relatives (Carlquist 1980, Ziegler 2002). When prickles do exist, as

on foliage of the lobelioid *Cyanea*, they have been interpreted as a deterrent to browsing by extinct flightless birds (Givnish et al. 1994).

Polynesian voyagers brought to Hawai'i three domestic animals, including the pig (*Sus scrofa*) (Nagata 1985, Kirch 2007). Domestic ungulates were introduced by the first European explorers of Hawai'i, starting with Captain Cook, who left pigs and goats (*Capra hircus*) on Ni'ihau in 1778 (Tomich 1986). Captain Vancouver introduced cattle, sheep (*Ovis aries*), and more goats to Hawai'i in 1792–1794. Four domestic species achieved feral populations on multiple islands by the early 1800s: cattle, pigs, sheep, and goats (Table 6.2). Later twentieth-century introductions of ungulates were designed to increase game-hunting opportunities, and the species released were wild animals rather than domestic stock (Fig. 6.1).

The development of wild populations of domestic animals (feralization) has been a continuing process on ranches and farms

Table 6.2. Year of ungulate introduction and distribution of feral animals in the Hawaiian Islands

Ungulate Species	Island							
	Ni	K	O	Mo	L	Ka	Ma	H
Axis Deer (*Axis axis*)	—	—	1868 ext	1868*	1920*	—	1959*	—
Cattle (*Bos taurus*)	—	*	—	*	—	ext	*	1793*
Donkey (*Equus asinus*)	1825 ext	1900 ext	—	—	—	—	—	<1847*
Goat (*Capra hircus*)	1778 ext?	1792*	*	*	ext	ext	1793*	1778*
Mouflon (*Ovis gmelini musimon*)	—	1958 ext	—	—	1954*	—	—	1962*
Mule Deer (*Odocoileus hemionus*)	—	1961*	—	—	—	—	—	—
Pig (*Sus scrofa*)	1778 ext?	*	*	*	ext	<1840 ext	*	1778?*
Sheep (*Ovis aries*)	<1893	1791 ext	—	—	<1893 ext	ext	—	1793–1794*

Sources: Dates and presence of feral populations are taken from MacCaughey (1918), Tomich (1986), and Anderson (1999b).
Notes: The status of most feral animal populations on Ni'ihau is uncertain. Ni, Ni'ihau; K, Kaua'i; O, O'ahu; Mo, Moloka'i; L, Lāna'i; Ka, Kaho'olawe; Ma, Maui; H, Hawai'i Island; *, wild or feral populations present in forest bird habitat; ext, extirpated from island; —, presumed not present.

Figure 6.1. Feral ungulates in Hawai'i. Top left: feral pig; top right: mouflon; bottom left: feral goat; bottom right: axis deer. Sources: Photos © Jack Jeffrey except for axis deer, which is © Richard A. Cooke III.

proximate to native vegetation. Complete feralization of the pig may have been delayed in upland forests by an early absence of earthworms (an important source of protein) and alien fruit-bearing plants, particularly strawberry guava (Loope and Mueller-Dombois 1989).

FERAL PIGS

The feral pig may be the most important agent of damage currently found in Hawaiian forest bird habitat. Pigs are highly adaptable to different conditions and are distributed from dry coastal lowlands to the sparsely vegetated alpine zone (Stone 1985). On Hawai'i Island, feral pigs reach their greatest densities in montane rain forests with dense tree fern cover (Giffin 1973). Montane wet and mesic forests are also the principal habitat for remaining

forest bird populations on Hawai'i, Maui, and Kaua'i (Scott et al. 1986; this volume, Chapter 5).

In forest habitats, feral pigs consume many native plants (Giffin 1973, Cooray and Mueller-Dombois 1981, Diong 1982), and some favored species, such as lobelioids, are important nectar sources for birds. Tree ferns, with a core of high-energy, digestible starch, are also preferred forage. Pigs extract the starch by chewing open the tree fern trunk, leaving water-holding troughs that develop into mosquito breeding sites (Diong 1982, Atkinson et al. 1995). The importance of pigs in the mosquito-bird disease cycle cannot be overemphasized (Chapter 17).

Perhaps more damaging to native plants than actual consumption is the disturbance of the forest floor by pigs in search of earthworms, roots, and other food. Cooray and Mueller-Dombois (1981) measured the amount of digging as comprising 38% of the forest floor at one site on Hawai'i Island. Other researchers have estimated rates of ground cover disturbance from pig rooting as high as 60–75% (Katahira 1980, Anderson and Stone 1994). Continued rooting of the forest floor by pigs invariably leads to the establishment of alien plants where weed populations occur near impacted forests. Studies of intensively pig-dug sites in native forest on Hawai'i Island documented short-term increases in alien plant cover, particularly grasses, ferns, and aggressive shrubs (Loh and Tunison 1999, Pratt et al. 1999). Weed invasions may be exacerbated by the loss of the shade-producing tree fern understory, which suppresses weedy plant cover (Burton 1980). Feral pigs have been implicated as dispersers of some of the worst rain forest weeds, such as strawberry guava (Diong 1982) and banana poka (*Passiflora tarminiana*) (La Rosa 1984).

Exclosure studies have demonstrated recovery of common native forest plants following the removal of pigs (Loope and Scowcroft 1985). The level of prior disturbance and condition of surrounding forest contribute to the degree of recovery in protected areas (Stone et al. 1992). Because 'ōhi'a trees have survived in many animal-damaged and grazed former forests, birds dependent on this species may be less severely impacted by pigs, at least in the short term.

FERAL SHEEP AND MOUFLON

Domestic sheep have been raised on four islands (see Table 6.2), but feral sheep populations have developed largely on Kaho'olawe and Hawai'i, where they persist today on Mauna Kea and Hualālai (Tomich 1986). Warner (1960) eloquently described the destruction of the Mauna Kea māmane forest brought about by a dense population of feral sheep used as game animals.

Browsing pressure from sheep prevents reproduction of the dominant māmane and alters forest composition, leading to an increase in less palatable naio (*Myoporum sandwicense*) and restriction of favored species to inaccessible rocky sites (van Riper 1980a). Feral sheep graze on both native and alien grass species, but māmane trees are preferred browse near the tree line (Giffin 1976). The endangered Palila (*Loxioides bailleui*) is completely dependent upon the māmane forests of Mauna Kea (Chapter 23). In 1980, after a protracted court battle, feral sheep reductions were initiated to protect critical habitat for Palila (Juvik and Juvik 1984). Even after the removal of most feral sheep, tree bark stripping continued, māmane reproduction did not recover evenly, and many sites had low native sapling density and dense alien grass cover (Hess et al. 1999).

Mouflon sheep (*Ovis gmelini musimon*), an ancient form of domestic sheep from Corsica and Sardinia (Clutton-Brock 1989), were introduced to Mauna Kea in the 1960s because they were thought to be less destructive than feral sheep, with which it was hoped they would hybridize (Tomich 1986). However, a postrelease study of food habitats revealed that mouflon also favored māmane and stripped the bark off native trees (Giffin 1980). Mouflon are now being removed from Mauna Kea, along with feral sheep. Mouflon were also introduced to Kahuku Ranch on Mauna Loa in the 1960s and have expanded their range around the mountain (Hess et al. 2006). Mouflon threaten forest bird habitat throughout their range in forests and subalpine woodlands; they also foray into rain forests.

The nimble mouflon are much harder to exclude from conservation areas than feral sheep. The expansion of mouflon populations on the island of Hawai'i now threatens areas previously fenced to keep out goats. The existing fences are too short to exclude mouflon, and barriers must be built higher, at great expense, to be effective against these leaping animals.

FERAL GOATS

After the introduction of goats more than 200 years ago, feral herds quickly became common on most of the Hawaiian islands. Recently, goats were removed from Niʻihau, Lānaʻi, and Kahoʻolawe, but they remain on the five larger islands (Tomich 1986), where they are locally distributed from coastal lowlands to the mountains. Invaders of dry and mesic communities, they also impact wet forests, particularly during dry periods (Stone 1985). Once large herds of feral goats become established in natural areas, they are difficult to eradicate (Stone and Loope 1987).

Feral goats suppress native plants in lowland vegetation (Mueller-Dombois and Spatz 1975), but their primary impact on forest bird habitat is interference with the reproduction of important tree species, such as koa (Spatz and Mueller-Dombois 1973) and māmane (Loope and Scowcroft 1985). Food habit studies have demonstrated that goats are largely browsers of trees and shrubs; bark-stripping and seedling destruction may severely damage montane mesic forests (Baker and Reeser 1972) and subalpine woodlands (Yocom 1967). The long-term impacts of goats include accelerated erosion, increased alien plant cover, and lower diversity of native plants (Loope and Scowcroft 1985, Scowcroft and Hobdy 1987, Stone et al. 1992). Although feral goats have been removed from much native forest bird habitat on Maui and Hawaiʻi, their continued presence in unprotected forests reduces habitat stability and long-term value.

INTRODUCED DEER

Few deer species have been introduced to Hawaiʻi, and only axis deer (*Axis axis*) are well established on more than one island. Currently axis deer inhabit Lānaʻi, Molokaʻi, and Maui; their introduction to Hawaiʻi Island was proposed but met with strong opposition and was rejected (Tomich 1986). On eastern Molokaʻi, axis deer are implicated in the destruction of a large expanse of wet forest that would

otherwise be forest bird habitat (Scott et al. 1986). The role of deer in destroying Molokai's native forest was recognized as early as 1897, when expert hunters were recruited from California to reduce the deer population (Black 1906). Despite the removal of thousands of deer, a large population persists on windward Molokaʻi. Axis deer were released on Maui in 1960, where few animals were reported until recently when the deer population was estimated as >4,000 animals (Anderson 1999a). The reproductive potential of axis deer, their extreme habitat flexibility, and their ability to use diverse forage (Anderson 1999b) make them a serious threat to Maui forest bird habitat.

Mule deer (*Odocoileus hemionus*) are restricted to Kauaʻi, where they have persisted in low numbers for >40 years (Tomich 1986). Mule deer browse native tree species, such as koa, ʻōhiʻa, and pilo (*Coprosma* spp.). They inhabit dry and mesic forest but range to 1,220 m elevation and in dry periods penetrate the native bird habitat of the Alakaʻi Swamp (Telfer 1988).

Rats

Three species of alien rats and the house mouse (*Mus musculus*) are present in Hawaiʻi (Stone 1985). The Pacific or Polynesian rat (*Rattus exulans*) arrived more than 1,200 years ago (Steadman 1995). The European black rat (*R. rattus*) was a relatively late arrival in Hawaiʻi in ca. 1870 (Atkinson 1977) but is now the most abundant rat in montane forests that are essential bird habitat. Predation of forest birds by rodents is discussed in Chapter 11. Indirect effects of rodents on forest bird habitat include seed predation of native trees and shrubs (Sugihara 1997) and bark stripping of favored species (Russell 1980, Scowcroft and Sakai 1984). Observers of rat damage to rare trees of wet and dry forests have reported seed predation, flower feeding, and bark stripping (Baker and

Allen 1979, Medeiros et al. 1986, Male and Loeffler 1997). Rats have negatively influenced the reproduction of a host of native plants and thus have changed the natural composition of the forests they inhabit. Their impacts will not be fully appreciated until long-term studies of ratfree areas are accomplished.

Alien Insects

Thousands of alien insect species have arrived in Hawai'i in modern times (Eldredge and Evenhuis 2003), and more continue to reach the islands each year. Although most new insect immigrants may not harm native species or invade native forests, recent arrivals include a number of potential pest species (Beardsley 1979), and a few have already had serious negative impacts on endemic tree species. A recent example of an alien insect that is damaging to vegetation is the two-spotted leafhopper (Sophonia rufofascia). This polyphagous phloem feeder was first detected on O'ahu in 1987 and quickly spread to all the main islands. In Hawai'i, it occurs to an elevation of 1,500 m and has been documented on more than 300 plant species, including 97 natives. 'Ōhi'a, particularly smooth-leaf varieties, are susceptible to damage from two-spotted leafhoppers. Such damage is pronounced in areas heavily infested with faya trees, a favored host (Lenz 2000). Large-scale 'ōhi'a dieback has not been attributed to this insect, but its impacts on 'ōhi'a forests are a subject worthy of study. Two-spotted leafhoppers have been implicated in the death of large patches of uluhe ferns, particularly on O'ahu (Jones et al. 2000). Although uluhe is capable of eventually replacing itself in dieback areas, the temporary openings are often invaded by fast-growing weed species, such as Koster's curse (Clidemia hirta), a shrubby melastome that is difficult to control (Follett et al. 2003). The leafhopper does not have the same deleterious effect on forms of uluhe

with hairy fronds, so large expanses of uluhe-dominated woodlands on Hawai'i Island have been spared disturbance and replacement.

Introduced insect herbivores that feast on native plants or are predators of native seeds are too numerous to discuss here. There are also many alien insects—primarily social yellowjackets (Vespula pensylvanica), other wasps, bees, and ants—that impact native forests indirectly by predating native pollinators or by competing with them for nectar and pollen, as does, for example, the nonnative honeybee (Apis mellifera). The problem of escalating numbers of invertebrate invaders (including slugs) and the lack of adequate quarantine measures to intercept them have been discussed in depth by others (Beardsley 1979; Howarth 1985; this volume, Chapter 15). Invasive insects that attack large suites of native tree species or forest dominants such as 'ōhi'a and koa have the potential to greatly reduce the quality and extent of remaining forest bird habitat. The introduction of invasive insects or diseases that attack 'ōhi'a and prevent its flowering might be catastrophic for many Hawaiian forest birds.

Invasive Alien Plant Species

Invasive nonnative plant and animal species are collectively the greatest threat to Hawaii's remaining natural forests (Medeiros et al. 1995) (Plate 25) and forest bird habitat (Scott et al. 1986). The Hawaiian Islands provide an enormous number of examples of introduced plant species that have invaded native forests. The rate of establishment of alien plant species has increased since the arrival of Hawaiians (Loope and Mueller-Dombois 1989), who brought approximately 30 food and fiber plants with them along with a few associated wetland weeds. Wagner, Herbst, et al. (1999) recognized 869 naturalized species of flowering plants in the islands, a number that grows every year with the publication of new records (Eldredge and Miller 1995, Eldredge and Evenhuis 2003).

In support of agriculture, forestry, and urban gardens, thousands of cultivated and ornamental species have been introduced. The most recent treatment of vascular plants cultivated in Hawaiian gardens lists >8,000 species, exclusive of native plants and weeds (Staples and Herbst 2005). Although many cultivated plants are unlikely to naturalize and become pests, Staples et al. (2000) recognized 469 that may be invasive based on their dispersal mechanisms. Smith (1985) identified 86 invasive plants that have already become pests in natural areas of Hawai'i. No definitive prioritized list of plants invasive in Hawai'i exists, but much recent work has been accomplished on a Hawai'i weed risk assessment project (Daehler et al. 2004). The list developed and a rating of evaluated alien species are available online at www .HEAR.org.

Characteristics that contribute to the success of invasive plants are small seed size, early reproduction (Rejmanek and Richardson 1996), adaptability to different habitats, tolerance of variable conditions and disturbance, fast growth, and ease of dispersal (Staples et al. 2000). Several plant groups have shown a propensity to establish in Hawaiian forests, in part conforming to the life forms of the most threatening invaders recognized by Daehler (1998). In wet forests of Hawai'i, tropical trees and shrubs have been successful invaders, among them melastomes (*Miconia calvescens, Tibouchina* spp.), guavas (*Psidium cattleianum, P. guajava*), and raspberries (*Rubus ellipticus* var. *obcordatus, R. glaucus*). Kāhili ginger (*Hedychium gardnerianum*) and vines such as banana poka have been among the most disruptive nonwoody groups (Table 6.3). Dry forests have been invaded by African and American grasses, a suite of tropical woody shrubs, and nitrogen-fixing trees, such as faya. Upper-elevation woodlands are susceptible to invasion by temperate grasses, vines (cape ivy or *Delairea odorata*), and biennials,

such as common mullein (*Verbascum thapsus*). Although many weeds are threats to forest bird habitat in Hawai'i, one in particular has the capacity to completely replace native forest. The recent history of miconia is presented as a case study of an invasive tree in forests of Hawai'i.

MICONIA: HAWAII'S MOST INVASIVE TREE?

Native to tropical America, velvet tree (*Miconia calvescens*), known locally as miconia, is a recent invader of the Hawaiian Islands, where it was introduced as an ornamental in 1961 and planted in several botanical gardens (Medeiros et al. 1997). Although not listed as naturalized in 1990 (Wagner, Herbst, et al. 1999), miconia was spreading from plantings on O'ahu and Kaua'i and was well established at multiple lowland sites on windward Maui and Hawai'i islands by 1996. The potential distribution of the species, based on known moisture requirements, encompasses most of windward Maui and Hawai'i below 1,830 m elevation (Plate 29), almost all of Kaua'i, and the entire Ko'olau and northern Wai'anae ranges of O'ahu (Medeiros et al. 1997).

The impacts of miconia on tropical forests are well known from the example of Tahiti (Meyer 1998). After its introduction to Tahiti in 1937, the species expanded to dominate forests on 65% of the island (Meyer 1996) and displaced native vegetation through its rapid growth and production of deep shade (Meyer and Malet 1997). Miconia trees have tremendous reproductive potential; they may flower more than half the year and produce more than 40,000 berries per tree (Meyer 1998). In Tahiti, a large seed bank may persist in soil for four years (Meyer and Malet 1997), and extremely high densities of seedlings (>17,000 per m^2) may result (Medeiros et al. 1997). In Hawai'i, as in Tahiti, several alien bird species are available to disperse the abundant seeds (Medeiros et al. 1997). Most remaining native Hawaiian forest bird

Table 6.3. Highly invasive alien plant species in forest bird habitat of the Hawaiian Islands

Scientific Name	Common Name	Life-Form	Distribution
Andropogon virginicus	Broomsedge	Grass	K, O, Mo L, Ma, H
Anemone hupehensis, var. japonica	Japanese anemone	Herb	H
Bocconia frutescens [now *Macleaya cordata*]	Plume poppy	Shrub	Ma, H
Cestrum nocturnum	Night cestrum	Shrub	K, O, Ma, H
Clidemia hirta	Koster's curse	Shrub	K, O, Mo, L, Ma, H
Dactylis glomerata	Orchard grass, Cocksfoot	Grass	K, O, Mo, Ma, H
Delairea odorata	Cape ivy, German ivy	Vine	Ma, H
Ehrharta stipoides	Meadow ricegrass	Grass	K, O, Mo,Ka, Ma, H
Falcataria moluccana	Albizia	Tree	K, O, Mo, L, Ma, H
Grevillea robusta	Silk oak	Tree	All main HI except Ni
Hedychium gardnerianum	Kāhili ginger	Herb (large)	K, O, L, Ma, H
Ilex aquifolium	English holly	Tree	Ma, H
Juncus effusus	Japanese mat rush	Rush	O, Mo, Ma, H
Melinis minutiflora	Molasses grass	Grass	All main HI except Ni
Miconia calvescens	Velvet tree, Purple plague	Tree	K, O, Ma, H
Morella faya	Faya, Firetree	Tree	K, O, L, Ma, H
Olea europaea ssp. cuspidata	African olive	Tree	K, M, H
Paspalum conjugatum	Hilo grass	Grass	K, O, Mo, L, Ma, H
Passiflora ligularis	Sweet granadilla	Vine	K, O, L, Ma, H
Passiflora tarminiana	Banana poka	Vine	K, Ma, H
Pennisetum clandestinum	Kikuyu grass	Grass	K, O, Mo, L, Ma, H
Pennisetum setaceum	Fountain grass	Grass	K, O, L, Ka, Ma, H
Psidium cattleianum	Strawberry guava, Waiawi	Tree	K, O, Mo, L, Ma, H
Psidium guajava	Common guava	Tree	K, O, Mo, L, Ma, H
Rubus argutus	Prickly blackberry	Shrub	K, O, Mo, Ma, H
Rubus ellipticus var. obcordatus	Yellow Himalayan Raspberry	Shrub	O, H
Schefflera actinophylla	Octopus tree	Tree	K, O, Mo, L, Ma, H
Schinus terebinthifolius	Christmas berry	Shrub	K, O, Mo, L, Ma, H
Schizachyrium condensatum	Bush beardgrass	Grass	O, H
Setaria palmifolia	Palmgrass	Grass	K, O, Mo, L, Ma, H
Spathodea campanulata	African tuliptree	Tree	K, O, Mo, L, Ma, H
Sphaeropteris cooperi	Australian tree fern	Fern	K, O, L, Ma, H
Tibouchina herbacea	Cane tibouchina	Shrub	O, Mo, L, Ma, H
Ulex europaeus	Gorse	Shrub	Mo, Ma, H
Verbascum thapsus	Common mullein	Herb	Ma, H

Sources: Adapted from Smith (1985); Wagner, Herbst, et al. (1999); Staples et al. (2000); and Big Island Invasive Species Committee (2005).
Notes: HI, Hawaiian Islands; H, Hawai'i; K, Kaua'i; Ka, Kaho'olawe; L, Lāna'i; Ma, Maui; Mo, Moloka'i; Ni, Ni'ihau; O, O'ahu.

populations will be at risk from miconia if efforts to control and eliminate the pest fail (Chimera et al. 2000).

The efforts of the decade-old Melastome Action Committee and the Maui Invasive Species Committee in planning, obtaining funding for, and organizing control efforts have led to early success in containing and reducing miconia infestations on Maui. Efforts to control the less widespread infestations on O'ahu and Kaua'i have also been successful (Medeiros et al. 1997). The larger infestation on Hawai'i Island is proving more difficult to manage, but a program to control miconia and other invasive plant species has been developed by the Big Island Invasive Species Committee (2005).

OTHER ALIEN PLANTS

More than 80 other alien plant species threaten the integrity of forest bird habitat on all four of the larger Hawaiian islands (Smith 1985). The continued expansion of yellow Himalayan raspberry (*R. ellipticus*) and kāhili ginger threatens the habitats of all forest birds other than those of the subalpine zone. The ability of invaded forests to maintain native tree composition and reproduction may be seriously impaired. Every possible attempt should be made to prevent the establishment of yellow raspberry on other Hawaiian islands (Loope 1992) and to treat localized infestations of kāhili ginger before they spread. Although strawberry guava may not be rapidly expanding its range, managers would be prudent to prevent incursions into important forest bird habitat. Significant and widespread control of many alien plant invaders will likely require the successful establishment of biocontrol agents (Smith 1985; this volume, Chapter 16).

INTERACTIONS AMONG ALIEN SPECIES
AND ECOSYSTEM ALTERATION

Although any one of the habitat-altering alien plant species or feral animals is a threat to the stability of intact forests, alien species are rarely unaccompanied in invaded habitats. Suites of alien plants with similar life forms are often capable of taking advantage of disturbance and become established in the same areas. Alien organisms often facilitate the invasions of other aliens in complex ways. The simplest example of this is the disturbance of rain forest floors by the feeding and rooting of feral pigs, which allows easily dispersed, fast-growing, light-loving weeds to establish themselves in otherwise densely vegetated and shady areas. More subtle are the interactions of inconspicuous earthworms with feral pigs. Earthworms provide a source

of protein that allows the development of large populations of pigs in native forests that would otherwise supply poor-quality forage. Many interactions involving predation of insect pollinators, seed predation by introduced small mammals, and competition among alien and native plant species for various resources have been only superficially investigated. The dispersal of weed seeds by native and alien birds has been intensively studied only recently (Medeiros 2004; Foster and Robinson 2007; this volume, Chapter 13). Much remains to be learned in this realm of Hawaiian conservation biology.

The alteration of ecosystem processes by invasive alien species in Hawai'i has been well documented over the past two decades (Vitousek 1990). Although the effects of invaders is context-dependent (D'Antonio and Hobbie 2005), in Hawai'i nitrogen-fixing alien plants (particularly faya and albizia or *Falcataria moluccana*) alter nitrogen cycling by greatly increasing the inputs of nitrogen and phosphorus into naturally nutrient-poor systems, such as young lava flows and early successional forests (Vitousek and Walker 1989, Hughes and Denslow 2005). Native plant species may then be outcompeted and replaced by alien plants more capable of using the increased nutrients, such as strawberry guava. Even forest dominants may be lost in this way. Alien grasses increase fire frequencies in forests and woodlands and lead to losses of nitrogen from volatilization (D'Antonio and Vitousek 1992); grasses may also change the abiotic environment in invaded Hawaiian woodlands (Mack and D'Antonio 2003).

SUMMARY: WHICH HABITATS ARE LEFT, AND WHY?

The original vegetation of Hawai'i was unique, and what remains is precious for its constituent plant species as well as for its value as the home of native forest birds and invertebrates. The losses and alterations of vegetation have not affected all types of Hawaiian forests equally. As is true in many other tropical lands (Janzen 1988), in Hawai'i dry forests have been heavily impacted. Dry and mesic forests contain 75% of the listed endangered plant taxa of Hawai'i and have been hardest hit by the combined perturbations of fire, ranching, feral goats and sheep, and invasion of alien grasses (Stemmermann and Ihsle 1993). Despite the former wide distribution of xeric vegetation in prehuman Hawai'i (Price 2004), very little dry forest remains today, and none may be considered "pristine." Dry forests persist at high elevations or on young substrates unsuitable for either farming or cattle grazing; these remnants are useful to forest birds, because malaria incidence is low in native dry woodlands (van Riper et al. 1986).

By contrast, large expanses of wet native forest exist, particularly on the islands of Hawai'i and Maui. The wet forests that remained after the imposition of Hawaiian agriculture and the practice of burning are primarily above 600 m elevation. Some low-elevation wet forests are still found at rocky sites that were marginal for either Hawaiian or modern agriculture. Large blocks of extant native wet forest persist because they were sources of upland plants and birds in the ancient Hawaiian ahupua'a (wedge-shaped land units) and because they had great value as watersheds for modern agriculture, particularly for growing sugar. Territorial foresters who had the foresight to establish large forest reserves and protect them from continued forest clearing and alien ungulates deserve much credit for what remains.

Even though less than 10% of the land area of the Hawaiian Islands remains as undisturbed native forest (Gagné 1988) and far less might be considered pristine, the largest island of Hawai'i retains much vegetation predominantly or entirely native in tree composition. Data from the Hawai'i Forest Bird Survey conducted more than 25 years ago indicated that almost 200,000 ha of Hawai'i Island were covered with forest and shrubland in which both the canopy and the understory were dominated by native plants; more than half of this predominantly native vegetation was categorized as wet forest (Jacobi and Scott 1985). Today, many native forests are integrated into large wilderness blocks because of recent public-private partnerships that have linked together forests under different ownerships into larger, more defensible conservation units (Chapter 16).

Fragmentation of Hawaii's remaining forests has certainly occurred, but its effects on Hawaiian forest birds have been fewer than we might expect from the example of temperate North America. Fragmentation of eastern American deciduous forests has led to increased edge compared with interior forest, and this enhanced edge has resulted in an increase in brood parasites and nest predators, dry conditions, and a reduced invertebrate prey base for forest birds (Askins 2000). Fragmentation and forest bird declines have also occurred in mainland tropical forests, although the threats faced by tropical birds may differ from those living in temperate forests (Stratford and Robinson 2005). Edge effects on forest birds have not been specifically studied in Hawai'i. However, we might predict that any impacts are less negative than those reported for mainland habitats because Hawai'i has no brood parasites or natural vertebrate predators other than birds (Chapter 1). Predators of Hawaiian forest birds are found throughout their forest habitats and are not restricted to edges, although the relative densities of predators have rarely been the object of study in Hawai'i. An exception is a recent study of the endangered 'Io (Hawaiian Hawk, *Buteo solitarius*) that found the greatest hawk densities in pastures with scattered trees (Klavitter et al. 2003; this volume, Chapter 12). Most mammalian predator studies in Hawai'i have focused on black rats (Chapter 11), and for this species, high densities have been reported from closed native forests (Stone 1985, Sugihara 1997). The topics of edge effects, the relationship of bird densities to variations in habitat, and predators' effectiveness in different vegetation types are fertile fields for future study.

In Hawai'i, all extant forest bird species are associated with old-growth, mature native forest. This situation contrasts with that in North America and the continental tropics, where there are suites of species adapted to early successional forest. Some of the processes responsible for creating patches of natural second-growth vegetation are missing from Hawai'i; examples are the lack of large grazing and browsing mammals and the low natural fire frequency in the original vegetation. Natural processes that do create early stages of vegetation succession in Hawaiian landscapes are landslides that are invaded by the matted fern uluhe or lava flows that are reinvaded by 'ōhi'a trees. However, the early successional vegetation types created by these two processes are not characterized by bird communities that differ from those of surrounding forests.

The primary threats to Hawaiian forests today are alien species, both plants and animals, rather than forest clearing for human uses. Continued alterations of the native plant communities have severe consequences for native Hawaiian forest birds. Most rare forest bird populations are found only at upper elevations (>1,500 m), and even the common native species are now generally confined to habitats above 900 m, except in lowland areas that retain a native forest canopy (Chapters 5 and 16). From an ecosystem perspective, the loss or reduction of forest bird populations in these habitats may also seriously jeopardize important functions such as seed dispersal and pollination.

Despite extensive modifications to native Hawaiian ecosystems, the habitats that remain have been identified as essential areas for conservation, and many are being actively managed to maintain and enhance their integrity. The most important steps are to prevent the introduction of new invasive species into Hawai'i (Chapter 15), eliminate or control the impacts of the most serious introduced plant and animal species, and assist with the restoration of both matrix and uncommon elements of the native ecosystems. Conservation efforts are compromised as species like axis deer and mouflon sheep become more widely established on the islands of Maui and Hawai'i, respectively. Renewed efforts to eliminate these problems before they become intractable are essential to maintain habitats suitable for viable populations of native forest birds and all other components of these unique ecosystems.

In recent years more emphasis has also been placed on securing lowland conservation areas that can support forest bird habitats if the impacts of avian disease can be reduced (Chapter 17). An extremely important aspect of land protection has been to link together habitat parcels through ownership or jurisdictional partnerships to form large conservation landscapes that can be managed at the ecosystem level (Chapter 16). Although we have lost many species and some ecosystems have shown signs of coming unraveled, the prospects for conservation of Hawaiian forest birds within the context of their natural ecosystems are good if we can continue to manage forest bird habitats at the landscape level.

ACKNOWLEDGMENTS

We thank all the individuals who provided us with suggestions and topical information for this chapter. Particular thanks are due to S. Hess, D. LaPointe, and T. Pratt of the U.S. Geological Survey Pacific Island Ecosystems Research Center. S. Gon III of The Nature Conservancy and S. Juvik and J. Juvik of the University of Hawai'i at Hilo allowed us to use and adapt their maps and figures. Review comments were provided by R. Askins, S. Cordell, S. Gon III, A. Medeiros, and J. Price.

CHAPTER SEVEN

Evolution and Ecology of Food Exploitation

PAUL C. BANKO AND WINSTON E. BANKO

Food is critical to the survival and reproduction of animals, and how individuals secure food strongly influences the ecology of populations, evolution of species, and radiation of lineages (Schluter 2000a, Grant and Grant 2002). Nowhere have food resources shaped the physical features and lifestyles of birds more strikingly than in the remote Hawaiian Islands. There, forest bird species evolved a dazzling array of bill shapes and sizes to exploit the relatively limited assortment of animal and plant foods.

Our purpose here is to explore relationships between foods and the bird species that evolved to eat them. We mainly discuss the food ecology of passerine bird species, the so-called songbirds or perching birds, but we also briefly consider the birds of prey, or raptors, which include a small endemic species of hawk and an indigenous species of owl (see Chapter 11 for more on the feeding ecology of birds of prey). The many novel or extreme adaptations of the Hawaiian honeycreepers to particular foods are renowned manifestations of adaptive radiation among birds (Pratt 2005), and we suggest how some key resources may have facilitated or subsidized the honeycreeper radiation. The demographic and ecological consequences for species that became food specialists are then discussed relative to the agents that have altered the habitats and reduced the availability of important resources, especially insect foods that were critical to young, growing birds. Understanding how food webs became disrupted may be helpful in developing strategies to restore native forest bird populations. Although a variety of factors threaten forest bird populations, the role of food web disruption has not been assessed. It is our hope to begin filling this gap in knowledge and thereby stimulate new conservation solutions.

HABITATS, FOODS, AND FORAGING GUILDS OF HAWAIIAN FOREST BIRDS

Richness of Food Resources

In ancient Hawai'i, passerine birds found food among 956 known species of flowering plants (Wagner, Herbst, et al. 1999), at least 5,449 insect and 313 spider species (Howarth 1990, Nishida 2002), and about 750 land snails (Cowie et al. 1995), although a very few single taxa became abundant throughout the archipelago. For example, canopies of Hawaiian forests are dominated by fewer than 15 tree genera, and only some of these were noted to have been prominently used by honeycreepers (Perkins 1903). Although birds forage on many shrubs and lianas, detailed trophic relationships are difficult to reconstruct because native ecosystems, especially at lower elevations, have changed markedly since Polynesian colonization (Burney et al. 2001). Among the changes to ecosystems have been the disappearance or decline of native food species and the appearance of new resources. However, there are only minor examples of birds incorporating new food items into their diets except in the case of the native birds of prey, whose diets expanded from birds and arthropods to include rodents when rats and mice were introduced by humans.

Habitat Associations

Birds usually spend their entire lives in association with particular plant and invertebrate communities, developing intimate relationships with unique suites of resources. In Hawai'i, substantial vegetation complexity and variability exist across the archipelago due to sequential island building and senescence, dramatically variable topography, steep rainfall gradients, and variable substrate age and structure (Chapters 1 and 6). Despite this varied geography, a relatively small number of plant species dominate large areas of Hawaiian forests, and relatively few forest bird species became restricted to particular habitats. Therefore, birds have access to important habitat elements over large areas.

The two most common forest trees, 'ōhi'a lehua and koa, supported most bird species known historically (for scientific names of native plants and birds, see Appendix 7.1). However, fossil evidence indicates that many bird species that became extinct during the early period of Hawaiian settlement thrived particularly in dry to mesic habitats (James and Olson 1991), where high levels of tree and shrub diversity (Rock 1913, Price 2004) provided the greatest variety of flowers, fruits, and seeds and supported rich assemblages of arthropods (Zimmerman 1948, Swezey 1954) and probably land snails (Cowie 1995). The details of how and where overlaps occurred in species richness of native plants, invertebrates, and birds are not yet known, but studies are beginning (Chapters 6 and 16).

Interactions between Birds and Their Foods

Birds derive more benefit from some foods than from others, a fact that is crucial to understanding how intimate associations developed between birds and their plant and invertebrate foods. Birds can extract energy and nutrients from nectar, flower parts, seeds, and leaves, and they also eat insects, spiders, snails, and other invertebrates that are sustained directly or indirectly by plants. When birds eat leaves, buds, and seeds, for example, they potentially reduce the ability of the host plant to photosynthesize, grow, and reproduce. As a result, the plant may evolve protective measures (Rhoades 1979). For example, plants that sequester chemicals making their seeds less palatable or difficult to digest may be less favored by seed eaters. Plants also may produce far more seeds than can be eaten by birds in the area. On the other hand, some birds promote plant reproduction by pollinating flowers and dispersing seeds across the landscape. Additionally, plants may benefit when birds eat insects that feed on their leaves, seeds, or other vital parts (Marquis and Whelan 1994, Van Bael et al. 2003).

Because birds were the dominant vertebrate colonists in the Hawaiian Islands and, by all accounts, existed in very dense populations, their high energetic demands must have exerted especially strong selective pressures on the plants and invertebrate animals that they ate. Gram for gram, birds eat more food each day than any other kind of vertebrate; for example, they eat about 31% more food than mammals (Nagy 2001). The impact of bird predation on invertebrates is suggested by the cryptic appearance of most Hawaiian insects, spiders, and snails and by changes in the composition of the arthropod community when birds are prevented from foraging in trees (Perkins 1913, Gruner 2004).

High rates of food consumption by birds also influenced the evolution of Hawaiian plant species and lineages. Some plants shifted from insect pollination to bird pollination, for example, and some became adapted for bird dispersal of their seeds through the evolution of fleshy fruits (Carlquist 1974, Price and Wagner 2004). The flowers of most native woody plants are often pollinated by birds seeking nectar, and their seeds are dispersed (or destroyed) by birds (Chapter 1). The most abundant, widespread tree species, 'ōhi'a, is adapted for both bird and insect pollination, and there is competition for pollinators between 'ōhi'a individuals as well as with other plant species (Carpenter 1976).

Bird Energetics and Phenology

To satisfy their high metabolic demands, birds typically require foods rich in carbohydrates and lipids. However, to meet the demands of growth, reproduction, molt, and health maintenance, birds also need foods rich in nutrients such as amino acids, vitamins, and minerals (Klasing 1998a). Moreover, events in their annual cycle should overlap relatively little (at least for individual birds) or occur slowly to reduce conflicting physiological demands. Similar to most birds, Hawaiian species typically renew their flight feathers after nesting, when their energy demands are likely greatest (Ralph and Fancy 1994b). In Hawai'i, breeding seasons are typically long because arthropod prey are available throughout the year; however, the majority of nesting occurs when day lengths are increasing and parents and fledglings have more time to forage (Peck 1993; this volume, Chapter 8). Even without the pronounced seasonality in weather, balancing energy is important to Hawaiian birds. Studies have suggested that the lower-than-expected metabolic rates observed in some Hawaiian species may reduce their energy costs under stressful thermal conditions and when foraging opportunities are limited (MacMillen 1981, Weathers and van Riper 1982).

To better understand patterns of habitat and resource use and the timing of reproduction and molt, more research is needed into the energy and activity budgets of Hawaiian birds and the energetic and nutritional quality of their foods, although some research has proceeded along these lines (Carpenter and MacMillen 1976b; MacMillen 1981; Sakai and Carpenter 1990; Banko, Cipollini, et al. 2002).

Feeding Guilds

Some foods are gifts to consumers and are meant to be eaten. For example, birds are enticed by plants to eat nectar and fleshy fruits and, as a result, flowers may be pollinated and seeds dispersed. Other foods, however, are not so easily obtained. Seeds may be toxic or protected by coverings, and arthropods and snails make themselves hard to find, capture, or extract. When birds rely mostly on invertebrate prey and seeds, feeding adaptations for hunting and overcoming defenses are expected to evolve. Birds that eat mainly nectar or fruits, however, tend to develop adaptations for tracking or defending these resources against competitors (Carpenter and MacMillen 1976a). Foraging primarily for one sort of food—nectar, fruits, seeds, or invertebrate prey—identifies the guild to which a bird species belongs, but most species tend to sample across at least two guilds, and all become arthropod hunters when they have young to feed. In the following pages we summarize adaptations of species to the guilds in which they spend most of their time (Box 7.1).

NECTAR EATERS

Many Hawaiian forest bird species frequented flowers for nectar historically, but only 10 depended heavily on this high-energy but low-protein resource. All 4 species of 'ō'ō relied on nectar, whereas only 5 of the 32 historically known Hawaiian honeycreepers made nectar their primary diet choice. Species with relatively short bills, such as 'Apapane and 'Ākohekohe, obtain nectar primarily from 'ōhi'a, but they also sample the flowers of a variety of other plants (Perkins 1903, Berlin et al. 2001). The close match between the longer, curved bills of the extant 'I'iwi and the extinct Hawai'i Mamo and Black Mamo and the flowers of some endemic lobelias and mints and the characteristics of many Hawaiian flowers indicates that birds and plants influenced the ecology and evolution of one another (Perkins 1903, Givnish 1998). Nevertheless, even the extraordinary extinct sickle-billed 'akialoa species occasionally took 'ōhi'a nectar (Perkins 1903). The intermediate length and curvature of the 'I'iwi bill indicate adaptation to the tubular flowers of lobelias and mints, but these nectar sources have become rare. The 'I'iwi apparently has switched to 'ōhi'a nectar as its main source of energy, and this dietary change may account for the apparent slight reduction in the bird's upper bill length during the historic period (Smith et al. 1995). The influence of 'ōhi'a on the bill length and foraging ecology of the other sickle-billed honeycreepers is an open question, but 'ōhi'a nectar seems, at a minimum, to have been an important supplemental energy source for many species.

FRUIT AND SEED EATERS

Birds that frequently eat fruits or seeds interact with plants as seed dispersers and seed predators. Fruits are lower in energy than nectar, but they sometimes contain significant amounts of protein (Sakai and Carpenter 1990). On the other hand, seeds are generally higher in protein and lipids, but they are often difficult to extract (Banko, Cipollini, et al. 2002). All five species of thrushes (Turdidae), the crow, and two honeycreepers were primarily fruit eaters, and six honeycreepers specialized in eating seeds. Heavy, conical bills were necessary to extract seeds from hard shells or tough, fibrous coverings that required great force to crush or tear open; thus, seeds were seldom, if ever, dispersed (Perkins 1903). For example, immature seeds of māmane, the primary food of Palila, are contained inside fibrous pods. Seeds of koa also are protected inside tough pods and were preferred by the extinct koa-finches. However, the extinct Kona Grosbeak cracked the hard shells of naio to extract the embryos within. The Laysan Finch and Nihoa Finch, once distributed in the main Hawaiian islands, probably ate (usually destructively) the seeds of a variety of coastal and lowland plants (Morin and Conant 2002). On the other hand, species tending to eat fruits were more likely to have dispersed seeds. The 'Ō'ū, for example, ate a variety of fruits in addition to its main food, the fruit of 'ie'ie. Likewise, the native thrushes have been observed to eat a variety of fruits

Box 7.1.   **Characterizing Feeding Specialization in Hawaiian Forest Birds**

The degree of foraging specialization of Hawaiian forest birds can be characterized as generalized, intermediate, and specialized. Criteria for deciding in which category a species best fits are based primarily on bill shape and size but also on range of habitats, host plants, foraging substrates, foraging maneuvers, and apparent diet breadth (Appendix 7.1). These criteria are more readily applied within foraging guilds: nectar eaters, fruit and seed eaters, arthropod eaters (and snail eaters), and birds of prey (or raptors). Within foraging guilds, species are grouped with closely related taxa or with species that have bills of similar shape and function. A particular foraging guild is not considered more or less indicative of specialization than another, but species that significantly occupy more than one guild are considered less specialized than species with similar bills but stronger associations with only one guild. Our analysis differs from that of Ralph and Noon (1988), who categorized the degree of specialization by the range of foraging behaviors. The bill characteristics of species within each foraging guild can be evaluated whether or not anything else is known about the species' feeding ecology, allowing inclusion of historically known species.

The figures of bird heads and bills are copied from Frohawk (in Rothschild 1893–1900; figures for a few species were not available) to illustrate gross variation in bill form and size within and between foraging guilds. Although this approach falls short of representing the full range of variation in bill shape and size within species, the often dramatic differences between species demonstrates the relative degree of divergence in the evolution of bill forms. The symbols in square brackets associated with figure labels designate the species' conservation status (U. S. Fish and Wildlife Service): [E], endangered; [X], extinct; no symbol, not listed as threatened or endangered.

		Nectar-Eating Guild	
	Generalist	Intermediate	Specialist
Meliphagidae (or Mohoidae)		ʻŌʻōʻāʻā [E]	
		Oʻahu ʻŌʻō [X]	
		Bishop's ʻŌʻō [X]	
		Hawaiʻi ʻŌʻō [X]	
		Kioea [X]	

continued

	Generalist	Intermediate	Specialist
Fringillidae: Drepanidinae	'Apapane	'Ākohekohe [E]	Hawai'i Mamo [X]
		'I'iwi	Black Mamo [X]

Fruit- and Seed-Eating Guild

	Generalist	Intermediate	Specialist
Corvidae		'Alalā [E] (not shown)	
Turdidae	'Ōma'o	Puaiohi [E]	
	'Oloma'o [E]		
	Kāma'o [E]		
Drepanidinae		'Ula-'ai-hāwane [X]	'Ō'ū [E]
	Laysan Finch [E]		Palila [E]
	Nihoa Finch [E] (not shown)		Lesser Koa-Finch [X]
			Greater Koa-Finch [X]
			Kona Grosbeak [X]

Arthropod- (and Snail)-Eating Guild

	Generalist	Intermediate	Specialist
Monarchidae	'Elepaio		
Sylviidae	Millerbird (not shown)		

Box 7.1.   Continued

Box 7.1.   Continued

	Generalist	Intermediate	Specialist
Drepanidinae	'Anianiau	Kaua'i 'Amakihi	
	Hawai'i 'Amakihi	Greater 'Amakihi [X]	
	O'ahu 'Amakihi		
		Maui 'Alauahio	Hawai'i 'Ākepa [E]
		O'ahu 'Alauahio [E]	'Akeke'e
		Kākāwahie [X]	
		'Akikiki	
		Hawai'i Creeper [E] (not shown)	
			Maui Parrotbill [E]
			Nukupu'u (Kaua'i [E], O'ahu [X], Maui [E])
			'Akiapōlā'au [E]
			Lesser 'Akialoa [X]
			Greater 'Akialoa (Kaua'i, O'ahu, and Lāna'i) [X]
		Po'o-uli [E] (not shown)	Lāna'i Hookbill [X]

(Perkins 1903, Snetsinger et al. 1999, Wakelee and Fancy 1999). The 'Alalā, largest of all the historically known fruit eaters, now extinct in the wild, consumed the greatest variety and size range of fruits, dispersing many species (Sakai et al. 1986; Banko, Ball, et al. 2002).

ARTHROPOD EATERS

All Hawaiian forest birds historically ate caterpillars and other arthropods or fed them to their young, but many species foraged primarily or exclusively on insects and spiders, rarely or never taking nectar or fruit. The 'Elepaio (Monarchidae), Millerbird (Sylviidae), and a majority of honeycreepers (19 of 32 species) mainly ate arthropods. Some arthropod eaters, particularly the 'Elepaio and the 'amakihi species, have numerous but relatively ordinary methods of feeding, such as gleaning from the surface of leaves and bark (Perkins 1903). Others, however, evolved extraordinary bills and behavior for capturing prey, including probing and flaking (species of 'akialoa and Nukupu'u), prying ('Akeke'e and 'Ākepa), and gouging, breaking, and hewing wood (Maui Parrotbill and 'Akiapōlā'au) (Perkins 1903). Included in the arthropod-eating guild is the Po'o-uli, which also consumed land snails as well as arthropods (Pratt, Kepler, et al. 1997). Many Hawaiian forest birds eat land snails on occasion, and snails may have been prey more frequently in former times when they were abundant.

BIRDS OF PREY

Small birds and arthropods were the main prey available to the native hawks and owls until the arrival of rats, first with the Polynesians and later when Westerners brought in other rat species and mice (Chapter 11). The 'Io evolved no remarkably novel features for capturing prey, but this small hawk may be somewhat unusual by virtue of its heavy consumption of spiders and larvae of moths and beetles earlier in the historic period (Henshaw 1902, Perkins 1903). The Pueo, which in physical characteristics and behavior has diverged little from continental forms of the same species, the Short-eared Owl, may have become established or more abundant in Hawai'i only after rats had become available as prey and forests had been opened up by the Polynesians (Olson and James 1982a).

EVOLUTION OF FOOD SPECIALIZATION

Adaptive Radiation of Hawaiian Honeycreepers and Other Forest Birds

Adaptive radiation is the proliferation of divergent species from a single ancestral species (Schluter 2000a, 2000b). Species produced in adaptive radiations are distinguished from one another by variation in physical and behavioral traits used for exploiting the resources and environments available to them. Traits diverge further in response to competition for resources, which promotes the exploitation of new resource types and environments that may be underutilized by other populations or species.

Species radiation has been more dramatic in the Hawaiian honeycreepers than in any other group of birds in the Hawaiian Archipelago (Plate 19). Originating from a single cardueline finch ancestor, 50 species of honeycreepers are recognized, and more await description based on subfossil remains (James and Olson 1991, James 2004). Fleischer and McIntosh (2001) propose that honeycreepers began diverging from their common ancestor soon after the emergence of Kaua'i (5.1 million years ago), the oldest of the large islands.

Seed eaters appeared prolifically and early in the honeycreeper radiation (Fleischer et al. 2001), presumably because of their finch ancestry. Over a third of the honeycreeper species described so far had "finch-like" bills suited to a diet of seeds or fruits

(James and Olson 1991). Nevertheless, adaptations for eating arthropods may also have evolved early from the finch-billed lineage, as perhaps also happened in Darwin's finches (Burns et al. 2002, James 2004). Additionally, land snails became important prey of at least one basal finch-billed species, the Po'o-uli. The relatively small number of nectar eaters, on the other hand, likely appeared later in the radiation, perhaps only 2–3 million years ago (Fleischer and McIntosh 2001). The greater diversification of arthropod-eating honeycreeper species, although it may have preceded the radiation of nectar eaters, is consistent with the finding that character displacement occurs at higher frequencies within the animal-eating guild, where interspecific competition may be strongest (Schluter 2000b).

Competition for a limited assortment of suitable foods has long been considered a major force driving the rapid and highly divergent radiation of Hawaiian honeycreepers (Perkins 1903, Amadon 1950). Additionally, plant and arthropod species that colonized the archipelago during the honeycreeper radiation could have presented new opportunities for exploitation and specialization by species experiencing strong competition for existing resources. The impetus behind the evolution of new lineages of honeycreepers may become clearer when the time of arrival of more important resources is known. However, even if the radiation occurred after all major resources were already in place, interactions between plant and animal communities would have been dynamic as they responded to global climate and sea level fluctuations and as islands emerged, grew, coalesced, separated, eroded, and subsided (Price and Clague 2002).

The radiation of Hawaiian honeycreepers may also have been influenced by native raptors. Various combinations of an eagle, a harrier, hawks, and owls on each island (Olson and James 1991) may have generated selective pressures on the foraging behavior, habitat use, and other characteristics of the bird species they hunted. Evolutionary responses of prey species to shared predators have been modeled (Abrams 2000); however, the competitive interactions of prey in more complex ancient Hawaiian communities would seem especially difficult to resolve.

In contrast to the explosive radiation of honeycreepers, the Hawaiian thrushes produced only a few similar species. The greater diversity of honeycreeper bill shapes might reflect greater capacity for bill variation within the cardueline finch family (Lovette et al. 2002). If, however, the thrush progenitor arrived after the honeycreeper radiation was well under way, as suggested by Fleischer and McIntosh (2001), competition may also have constrained the thrush radiation. Competition for fruit and arthropods would likely have been sharpened somewhat by the slight overlap in size between the two groups of birds.

The one or more ancestors of the Hawaiian crows were larger than any of the honeycreepers, and they arrived after the honeycreeper ancestor (Fleischer and McIntosh 2001). The crow radiation produced about five quite different species, suggesting that their larger size reduced competition with honeycreepers and other forest birds for foods. The 'Alalā and four fossil species varied most from one another in the size and shape of their bills, indicating strong competition for food among them (James and Olson 1991; Banko, Ball, et al. 2002). The ancestors of the 'Elepaio and Millerbird also arrived after the honeycreeper ancestors, and speciation among them was limited.

Just as our book was in final stages of completion, R. Fleischer et al. (2008) reported DNA evidence that the birds of the 'ō'ō species together with the Kioea are the oldest lineage of Hawaiian bird, dating back to 14–17 million years ago. It seems remarkable that this lineage of five historic species was not more diverse. The paucity of 'ō'ō species and relatives causes one to consider whether they may have been supplanted to some degree by the later radiation of the honeycreepers.

Evolutionary Aspects of Specialization

Specialization and generalization bracket the spectrum of strategies used by consumers to find and secure food, and these terms are conveniently used to compare and classify different types of consumers in terms of their ecological niche (Futuyma and Moreno 1988). Specialization infers niche restriction in terms of habitat use, diet, or foraging behavior. Modifications in species' bill, palate, or tongue for feeding on particular food types indicate that they have attained adaptive peaks of fitness (Benkman 1993). Feeding specializations can be expected to arise through natural selection when the availability of different food types fluctuates such that consumers must rely on alternative foods when preferred foods become scarce (Schluter and Grant 1984, Grant and Grant 1989). Under these conditions, competition for limiting resources promotes divergence in feeding behavior and morphology among consumers (Boag and Grant 1981). Specialization is predicted to evolve when premier resources are easily exploited and alternative resources are difficult to use without special morphological, behavioral, or physiological adaptations (Robinson and Wilson 1998). As mentioned previously, opportunities to exploit new resources or habitats also promote specialization. The degree of specialization increases if species are able to occupy successively higher adaptive peaks over time, hybridization is rare or nonexistent, and environments are geographically and ecologically diverse as well as climatically stable (Grant 1994, 1998c). However, because specialization involves trade-offs in foraging efficiencies (Smith 1987, Benkman 1993, Konuma and Chiba 2007), the risk of extinction increases for specialists that are slow or unable to adapt to changes in habitats and foods (Kuris and Norton 1985).

Applying this body of theory to Hawaiian honeycreepers, we suggest that abundant, conspicuous species of caterpillars should be among the most preferred foods of birds because they are nutritious and easy for birds to capture, regardless of bill shape or size. When abundant, some caterpillars are easily found and captured even by birds with highly modified bills and specialized foraging behaviors. However, when caterpillars become scarce due to either seasonal environmental variation or competition for them among birds, nectar, fruit, seeds, and other invertebrate prey become more valuable even though they may ordinarily be more difficult or less profitable to exploit than caterpillars. Therefore, birds should forage heavily on caterpillars when they are abundant, but they should switch to alternative foods when caterpillars become scarce. Diet switching may have costs, however, for digestive capabilities may adapt to particular diets, making diet switching at least temporarily inefficient or difficult (Levey and Karasov 1989).

This theory explains why very specialized birds, such as the ʻakialoa species, took both hard-to-get resources (e.g., *Oodemas* beetles, *Prognathogryllus* crickets, lobelia nectar), which required a long, curved bill, and more easily obtained food (e.g., foliage caterpillars, ʻōhiʻa nectar), which did not require such a modified bill (Perkins 1903). Similarly, fruit and seed specialists, such as ʻŌʻū, Palila, and the Greater Koa-Finch, were also historically observed foraging on abundant caterpillar species.

Although these examples illustrate that feeding versatility was not entirely compromised as bird species evolved specialized bills and feeding behaviors, specialists should not be as efficient at exploiting many foods taken by more generalized foragers. Given different foraging efficiencies and other costs of switching diet, specialists should usually rely on foods for which they are particularly adapted, and generalists should exploit foods that are easy to capture. Diet overlap for preferred foods, therefore, does not preclude the coexis-

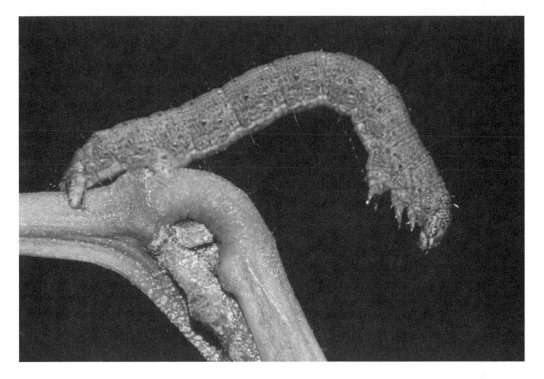

Figure 7.1. A *Scotorythra* caterpillar—an important food source for Hawaiian forest birds. This *S. rara* from Volcano, Hawai'i, is about 3 cm long. *Source:* Photo by R. Peck for the U.S. Geological Survey.

tence of generalists and specialists, and premier resources need not be even periodically abundant for specialists to coexist with or replace generalists (Robinson and Wilson 1998). Nevertheless, specialists are more likely to suffer the consequences of disruptions to food webs from environmental changes. Not surprisingly, the honeycreeper radiation gave rise to all of the specialist Hawaiian bird species, most of which fed primarily on arthropods or fruit and seeds (see Box 7.1).

Foods That Subsidized Specialization

Subsidizing the radiation and specialization of Hawaiian honeycreepers were certain caterpillar species and other arthropods that were widespread, abundant, and easily captured by birds with bills of any shape and size. Providing essential nutrients for nestling growth and supplementing the diets

of older birds, caterpillars were premier foods of all forest species. Leaf-eating *Scotorythra* spp. caterpillars (Geometridae; Fig. 7.1) were especially important foods of Hawaiian forest birds because they were essentially universal in nestling diets (Henshaw 1902; Perkins 1903, 1913).

Based on the degree of their morphological divergence from closely related moth species, *Scotorythra* probably originated before Kaua'i rose above the sea (M. Heddle, pers. comm.); if so, they may have preceded the arrival of the honeycreeper ancestor. Many of the 38 *Scotorythra* species described historically (Nishida 2002) were abundant and distributed on many host plants used by foraging birds (Perkins 1913). Just as the availability of caterpillars influences the timing and amount of breeding of Darwin's finches in the Galápagos (Grant and Grant 1989), *Scotorythra* could have strongly affected the reproduction and demography of Hawaiian forest birds. Although a few cardueline finch species feed their young largely or entirely on seeds, a rare behavior among birds generally, insects are usually provided because they are more easily digested than seeds and contain more protein (Newton 1972). It is doubtful that so many honeycreeper species would have become specialized on nectar and fruit, which are low in protein, unless caterpillars were readily obtainable. Spiders were another common protein-rich food of forest birds (Perkins 1903, 1913; Baldwin 1953). As previously mentioned, the nectar of 'ōhi'a is an example of a high-energy food that is abundant and accessible to many bird species.

ECOLOGY OF FOOD SPECIALIZATION AND GENERALIZATION

Resource Tracking and Movements

When preferred foods become scarce, birds may either move or switch to alternative resources. Two mobile, nectar-eating species, 'I'iwi and 'Apapane, track the 'ōhi'a and māmane bloom along elevation gradients and across landscapes (Perkins 1903, Baldwin 1953, Carpenter and MacMillen 1980, Ralph and Fancy 1995, Berlin et al. 2001). Both species also move far around the volcano Mauna Kea, from wetter, montane 'ōhi'a forests on the eastern slope to dry, subalpine forest on western and other slopes, to take advantage of the seasonal māmane bloom (Hess et al. 2001). 'Ākohekohe and Hawai'i 'Amakihi, in contrast, usually shift to other resources when nectar becomes less available (Baldwin 1953, van Riper 1984, Berlin et al. 2001, Hess et al. 2001).

Some specialized fruit and seed eaters also moved in response to resource availability. 'Ō'ū and 'Alalā, for example, moved in and out of leeward, midelevation forests according to the seasonal availability of 'ie'ie fruit (Perkins 1903). Among the seed specialists, Palila move up and down the slopes of Mauna Kea over the course of months to track the availability of māmane pods (Hess et al. 2001; Banko, Oboyski, et al. 2002).

Although arthropod eaters do not move far in search of prey, some forest birds responded to the once common irruptions of koa-defoliating caterpillars (i.e., *Scotorythra*) (Perkins 1903). Among the species of birds observed feasting on caterpillar outbreaks were the fruit-eating 'Ō'ū and 'Ōma'o and the seed-eating Greater Koa Finch. The fruit eaters each traveled relatively far from their usual habitats to caterpillar outbreaks. The opportunistic exploitation of caterpillars, however, underscores the value of relatively accessible prey to birds that were better adapted for other foods.

Resource Defense

Some species of Hawaiian birds vigorously defend their access to flowering trees and shrubs, an obvious indication of competition for nectar. Dominating the foraging hierarchy are larger species and older individuals (Carpenter and MacMillen 1980; Pimm and Pimm 1982; Carothers 1986a, 1986b). Species of 'ō'ō frequently chased the smaller honeycreepers from flowering trees, just as the larger honeycreepers defended their nectar resources from smaller species (Perkins 1903). 'Ōhi'a nectar, which was accessible to birds with any sort of bill shape and size, was defended even by Black Mamo and possibly Hawai'i Mamo, both of which had bills that were particularly well suited for feeding at lobelia flowers (Perkins 1903). Now that the larger, more specialized nectar-eating species have all disappeared and many nectar resources have become scarcer, surviving nectar eaters are likely to be observed competing mainly for 'ōhi'a and māmane nectar, which are still comparatively common and widespread. In contrast, Hawaiian forest birds do not typically defend rich patches of fruits, seeds, or arthropods against encroachment by other individuals, although some species protect their access to food by maintaining territories for purposes other than breeding (Pratt, Simon, et al. 2001).

Table 7.1. Egg and clutch production of Hawaiian honeycreepers

	Estimated Fresh Egg Weight (mean ± SE g)	Ln Fresh Egg Weight / Ln Female Weight (mean ± SE g)	Clutch Size (mean ± SE)	Ln Estimated Clutch Weight / Ln Female Weight (mean ± SE)
Generalist[a]	2.00 ± 0.128	0.27 ± 0.015	2.8 ± 0.17	0.68 ± 0.031
Intermediate[b]	2.39 ± 0.161	0.31 ± 0.012	2.1 ± 0.18	0.58 ± 0.027
Specialist[c]	2.81 ± 0.431	0.33 ± 0.030	1.7 ± 0.20	0.51 ± 0.032

Notes: When adjusted for body weight, clutches of specialist species are lightest and contain one fewer egg than do those of generalists. The values for clutch size and female body weight were obtained from Birds of North America accounts (Poole and Gill 1992–2002); estimated egg weights were calculated from Hoyt (1979) using the egg dimensions in the Birds of North America accounts. Egg and female weights were log (ln) transformed to facilitate comparisons of the ratios of egg to female weight among species of different sizes. The resulting log linear relationship between egg weight and female weight was strong except that the ratios for four species differed substantially from the overall trend. The eggs of these four outlier species were measured by one person, who was also the sole source for two other species that were consistent with the relationship. To remove the bias that might be associated with this observer, we adjusted the ratios of all six species by first recalculating the log linear regression without them and then applying the new regression equation to revise the egg weights. Thus, the ratios of the four outlier species changed substantially to conform to the overall relationship, but the ratios of the two other species were only slightly changed. This procedure was recently validated when two new sets of eggs of one outlier species, the 'Apapane, were measured and found to conform closely to the overall relationship between egg weight and female body weight (P. Banko, unpubl. data). This adjustment of the six species had the overall effect of reducing the estimated mean egg weights of generalist and intermediate species but not those of specialist species. The significance (P ≤ 0.1) of differences between categories of specialization for each variable was determined by Kruskal-Wallis rank sum test: egg weight (P = 0.33), egg weight / female weight (P = 0.405), clutch size (P = 0.0242), clutch weight / female weight (P = 0.0257).
[a]Hawai'i 'Amakihi, 'Anianiau, 'Apapane ('Anianiau and 'Apapane egg weights were adjusted, as described above).
[b]Kaua'i 'Amakihi, Hawai'i Creeper, Maui 'Alauwahio, 'Akikiki, 'I'iwi, 'Ākohekohe (Kaua'i 'Amakihi, 'Akikiki, and 'I'iwi egg weights were adjusted, as described above).
[c]Palila, Maui Parrotbill, 'Akiapōlā'au, 'Ākepa, 'Akeke'e ('Akeke'e egg weights were adjusted, as described above).

Life History, Demography, and Resource Use

REPRODUCTION

With fewer options than generalists have for feeding themselves and their offspring, specialist honeycreepers tend to produce lighter clutches of eggs for their body weight, and they require more time to incubate their eggs and rear their young (Tables 7.1 and 7.2). When reproduction is limited by food availability or feeding behavior, evolutionary theory predicts that parents should produce small clutches of large eggs and that they should take more time to incubate the eggs and rear the chicks (Martin 2004). This theory points to a major factor underlying differences in the reproductive behavior of honeycreeper specialists and generalists, although specialists produce eggs that are not significantly heavier than the eggs of generalists. It follows, therefore, that the annual productivity of specialists is relatively low because they are less able than generalists to acquire energy and convert it into eggs and offspring.

When adjusted for body weight (and ignoring phylogenetic contrasts), generalist females produce clutches that are 33% heavier than those of specialists (see

Table 7.2. Incubation and parental care in Hawaiian honeycreepers

	Incubation Period (days)	Incubation Efficiency (mean days of incubation per gram of clutch ± SE)	Mean Fledging Period (days)	Mean Postfledging Period (weeks)
Generalist[a]	13.7	2.5 ± 0.26	16.9	13
Intermediate[b]	15.4	3.1 ± 0.39	19.7	9
Specialist[c]	15.9	4.0 ± 0.77	21.1	17.5

Notes: Incubation (eggs in nest) and fledging (chicks in nest) periods tend to increase with the degree of feeding specialization, especially when adjusted for clutch weight. Values were derived from Birds of North America accounts (Poole and Gill 1992–2002); clutch weights were calculated as in Table 7.1. The significance ($P \leq 0.1$) of differences between categories of specialization for each variable was determined by Kruskal-Wallis rank sum test: incubation period ($P = 0.158$), incubation efficiency ($P = 0.239$), fledging period ($P = 0.062$), postfledging period ($P = 0.565$). Although differences between foraging guilds are significant only for the fledging period, the trends of other variables support the hypothesis of slower, less efficient development.

[a]Hawai'i 'Amakihi, 'Anianiau, 'Apapane ('Anianiau excluded in analysis of postfledging period).
[b]Kaua'i 'Amakihi, Hawai'i Creeper, Maui Alauwahio, 'I'iwi, 'Ākohekohe (Kaua'i 'Amakihi, Hawai'i Creeper, Maui Alauwahio excluded from analysis of postfledging period).
[c]Palila, Maui Parrotbill, 'Akiapōlā'au, 'Ākepa ('Akiapōlā'au excluded from analysis of incubation period and incubation efficiency).

Table 7.1; note that here and elsewhere we compare generalists and specialists without considering phylogenetic contrasts because a widely accepted phylogeny of the Hawaiian honeycreepers is not available [Pratt 2005]). Illustrating the dramatic contrast in energetic commitment to egg production are the two smallest honeycreepers, 'Ākepa (female 10.4 g) and 'Anianiau (female 10.3 g). 'Ākepa, which are arthropod specialists, produce eggs weighing 17% and clutches weighing 34% of their body weight. 'Anianiau, which are generalists that exploit small arthropods and nectar, produce eggs weighing 38% and clutches weighing 116% of their body weight.

The evolutionary trajectories of generalists and specialists should also be affected by the energetic costs of incubation and chick rearing. An indication of the greater parental efficiency of generalists is that they commit only 2.5 days of incubation to each gram of clutch weight, whereas specialists require 4.0 days (see Table 7.2). Overall, generalists spend 2.2 fewer days incubating their clutches than do specialists. The

idea that generalists should be more efficient at chick rearing than specialists is supported by their shorter fledging time. Offspring of generalists leave the nest an average of 4.2 days earlier than do chicks of specialists.

Fledglings tend to forage inefficiently and are more sensitive to food limitation than are adults (Weathers and Sullivan 1989). Therefore, fledglings of specialists should require extended parental care, in terms of the provisioning of food subsidies or tutoring to develop their foraging skills. Parental dependency tends to be longer for specialists (17.5 weeks versus 13 weeks for generalists; see Table 7.2), potentially reducing opportunities for nesting again in the same season.

Therefore, the overall advantage that generalists have over specialists in terms of their reproductive capacity includes larger clutch size, faster development of embryos and chicks, and earlier independence of fledglings. This advantage largely explains why generalist populations are better able to withstand and recover from environmental and biological threats (Chapter 2).

BODY SIZE AND ENERGETICS

The body size of Hawaiian honeycreeper species, both within and between foraging guilds, is related to their degree of food specialization. Specialists' wings average 20 mm longer than the wings of generalists (Kruskal-Wallis rank sum test, Bonferroni-corrected $P = 0.014$). The tendency toward larger size in specialist species may be related to their having longer or more powerful bills and the associated musculature for extracting foods that are otherwise difficult to exploit. Some seed specialists obviously require stout bills and strong musculature. Among the nectar eaters, however, size is probably more related to position in the dominance hierarchy and its consequences for access to rich concentrations of nectar (Pimm and Pimm 1982). Exceptions to the trend of larger specialists, however, are the tiny 'Akeke'e and 'Ākepa, which spend much of their time feeding at the delicate tips of leafy branches, a foraging niche requiring agility, finesse, and light weight.

SEX RATIO

Comparing the sex ratios of honeycreeper specimens collected early in the historic period reveals that the proportion of females is lower in specialist species (Table 7.3). Although biases in collecting males and females might occur in species whose males are more conspicuous, the tendency for lower female representation among specialists holds even for monochromatic species in which males and females are indistinguishable to humans. There is little indication that sex ratios vary substantially among foraging guilds, although honeycreeper species that eat mostly fruits or seeds are all specialists, which may account for lower female representation. The female survival rate may be lower in specialist species because they expend more energy and time while nesting.

Greater Vulnerability of Specialized Birds to Changes in Habitat and Foods

The vulnerability of specialized feeders to changes in habitat and food availability has

Table 7.3. Proportion of females in museum collections of Hawaiian honeycreeper species, by degree of feeding specialization and foraging guild

	Arthropods	Fruits and Seeds	Nectar	All
Generalist[a]	0.45		0.48	0.46 (n = 4)
Intermediate[b]	0.50		0.32	0.45 (n = 9)
Specialist[c]	0.32	0.32	0.25	0.32 (n = 14)
All	0.42 (n = 17)	0.32 (n =5)	0.37 (n = 5)	0.40 (n = 27)

Notes: Specialist species were represented by few females compared to species with generalist or intermediate feeding ecology. Differences between foraging guilds were not apparent. Values were calculated by separately summing numbers of females and males for all age classes combined for all species considered to be generalist, intermediate, or specialized in their feeding ecology (see Appendix 7.1 for details). Most specimens were collected before 1910, and data from specimens collected after 1950 were excluded from analysis. Specimen label data were obtained from Banko (1979) and the B. P. Bishop Museum (C. Kishinami, pers. comm.). The results of a χ^2 test indicate that sex ratio differences by feeding specialization are highly significant (P < 0.001).

[a]Hawai'i 'Amakihi, O'ahu 'Amakihi, 'Anianiau (arthropod eaters); 'Apapane (nectar eater).

[b]Kaua'i 'Amakihi, Greater 'Amakihi, 'Akikiki, Hawai'i Creeper, O'ahu Alauwahio, Kākāwahie, Maui Alauwahio (arthropod eaters); 'I'iwi, 'Ākohekohe (nectar eaters).

[c]Maui Parrotbill, Lesser 'Akialoa, Greater 'Akialoa, Nukupu'u, 'Akiapōlā'au, 'Akeke'e, 'Ākepa (arthropod eaters); 'Ō'ū, Palila, Lesser Koa-Finch, Greater Koa-Finch, Kona Grosbeak (fruit and seed eaters); Hawai'i Mamo, Black Mamo (nectar eaters).

long been recognized (Perkins 1903). Even when new resources in the form of alien species arrived, they were not heavily exploited by specialists, although native birds of prey flourished after the introduction of rodents, and some native passerine species exploited introduced arthropods, nectars, and fruits (Perkins 1903; Baldwin 1953; R. Peck and P. Banko, unpubl. data). Of greater significance demographically, however, is the quality of nestling and fledging diets and the capacity of parents to obtain adequate food. Not surprisingly, therefore, and consistent with theory (Cody 1974), generalists are more common and widespread, particularly in unpredictable environments, regardless of whether the adults exploit mainly nectar, fruit and seeds, or arthropods in their diet (Chapter 2).

Food specialists are not put at risk solely because particular resources disappear altogether; instead, food availability need only decline below the threshold at which exploitation is energetically profitable. Studies of competition among nectar eaters clearly demonstrate this principle, because it is relatively easy to measure the number of flowers available in a patch of habitat and because, at certain levels of resource availability, dominant species exclude or reduce access to the richest sites by chasing smaller species away (Carpenter and MacMillen 1976a, 1976b). Studies indicate that nectar scarcity leads to more aggressiveness and territoriality, less daily movement, partitioning of resources, and increased foraging efficiency (Carpenter and MacMillen 1980). Of evolutionary importance is the effect that resource limitation has on the poorer competitors, such as the ʻamakihi and other small species, driving them toward marginal habitats, less productive foraging patches, and more generalized diets (Pimm and Pimm 1982). When resources decline, therefore, specialist nectar eaters, which also tend to be the larger, dominant competitors, are at greater risk because they cannot exploit their preferred resources efficiently and

because they are less able to switch to alternative resources (Pimm and Pimm 1982). Because larger species have greater resource requirements, their populations tend to be smaller, which also increases their vulnerability to changing conditions and other threats (Brown and Lomolino 1995, McNab 2002). Supporting this view is the fact that the large, highly specialized birds, such as the various species of ʻōʻō and the Hawaiʻi Mamo, Black Mamo, and ʻĀkohekohe, suffered greater rates of extinction and population decline than the smaller, more generalized species, such as the ʻIʻiwi, ʻApapane, and various ʻamakihi (Carpenter and MacMillen 1980, Pimm and Pimm 1982).

MECHANISMS, PATTERNS, AND IMPACTS OF FOOD DECLINE

Disruption of Forest Communities and Threats to Fruits, Seeds, and Nectar Resources

Lava flows, natural fires, hurricanes, and dieback and replacement of ʻōhiʻa stands are some natural events that alter Hawaiian landscapes, but human activities and alien species change forest ecosystems more substantially and permanently (Cuddihy and Stone 1990; this volume, Chapter 6). Lowland forests are largely gone, and many tracts of montane forests have been reduced in size by agricultural use. The cumulative effect of agriculture and timber harvest has been to shrink and fragment native forest areas and potentially reduce the availability of habitat and food to birds. Many forests surviving agricultural conversion have been seriously degraded by browsing and other disturbance by feral ungulates: cattle, pigs, goats, feral sheep, mouflon sheep, and deer. The worst impacts occurred in dry forests, ʻieʻie habitat, and the koa forests at middle and high elevations and elsewhere in the mesic belt of forest on all islands. Feral and domesticated browsers and grazers affected the availability of bird foods by removing or diminishing understory vegetation, halting or slowing regeneration of canopy trees, and modifying soil and light conditions to promote the invasion of alien weeds. Weeds have replaced native plants, usually without substituting their value as food resources, while at the same time changing habitat conditions once favorable to native invertebrate species. Changing of fire regimes is another consequence of alien plant invasion that poses threats to bird resources (Cuddihy and Stone 1990; D'Antonio and Vitousek 1992; this volume, Chapter 6). As forests changed or disappeared, so did many of the food resources that were critical to the survival and reproduction of native birds.

Because of the notoriety of alien rodents as predators of island forest birds, their ability to transform bird habitats and devour many of their foods is sometimes overlooked. Rats change the structure and composition of Hawaiian forests by reducing plant populations and reproduction (Cuddihy and Stone 1990; this volume, Chapters 6 and 11). Polynesian rats (*Rattus exulans*) and black rats (*R. rattus*) eat the flowers, fruits, seeds, bark, stems, and whole seedlings of a variety of species, in some cases reducing or eliminating their populations. Important examples of plant species whose fruits are eaten by both birds and rats are 'ie'ie, loulu, hō'awa, olopua, and pilo. 'Ie'ie, in particular, was hit especially hard by black rats, leading R. C. L. Perkins (1903) to speculate that the destruction of 'ie'ie may have contributed to the early extinction of 'Ō'ū on O'ahu. Rats also reduce the recruitment of koa, one of the most important resources of birds, when they strip the bark of saplings (Scowcroft and Sakai 1984).

Alien insects also threaten foods of Hawaiian birds by reducing the productivity and vigor of native trees and shrubs. Already established in Hawaiian forests are twig borers, seed predators, leaf eaters, and sap suckers (Cuddihy and Stone 1990).

Furthermore, many alien insects that are established in native forests could serve as vectors for a variety of plant diseases (Howarth 1985).

Diseases of forest trees and shrubs, though poorly documented, threaten bird resources and habitats in obvious ways. Because forests are dominated by only a few tree species, a highly pathogenic plant disease could bring an end to many bird species. Prolonged drought or changes in environmental conditions that would stress trees could set the stage for forest die-off at the local or landscape level. For example, a forest disease that is diminishing resources critical to Palila is *Armillaria mellea*, an introduced fungus that attacks the roots of many trees and is involved in, if not responsible for, substantial mortality of māmane trees and saplings on Mauna Kea (Gardner and Trujillo 2001).

Disruption of Forest Invertebrate Communities and Threats to Important Prey

CHANGES IN POPULATION DYNAMICS AND AVAILABILITY OF *SCOTORYTHRA* CATERPILLARS

Although entomologists still commonly collect *Scotorythra* in native forests on the larger Hawaiian islands (Giffin 2007), their efforts provide little systematic measure of population abundance. However, recent surveys for *Scotorythra* revealed that some species believed to have disappeared since Perkins's time are still extant (Heddle 2003). Additionally, *Scotorythra* caterpillars were the most important constituent, along with spiders, of the arthropods collected from koa (but not 'ōhi'a) canopies in a montane habitat that supported one of the most intact native bird communities on Hawai'i (Gagné and Howarth 1981). Nevertheless, some notable changes are evident in the population dynamics of this group since the time of Perkins. A century ago, for example, *Scotorythra* were highly conspicuous in native forests, and some species periodically irrupted over large areas, severely defoliating koa trees in some localities (Henshaw 1902; Perkins 1913; Swezey 1926, 1954). Since 1902, however, *Scotorythra* irruptions have been noted infrequently on Maui and Hawai'i islands. Trying to interpret changes in the frequency of koa defoliation events is pointless without detailed field investigations, because the many factors potentially contributing to insect outbreaks may interact in complex ways (Barbosa and Schultz 1987).

ALIEN PARASITIC WASPS AND FLIES

Many species of small wasps (Hymenoptera) and flies (Diptera), known as parasitoids, whose larvae typically consume the eggs, larvae, or pupae of other insects, arrived accidentally or were introduced long ago to Hawai'i to control agricultural pests. Alien parasitic wasps greatly outnumber native parasitoid species (Nishida 2002). Many alien parasitoid species had profound impacts on populations of moths and other native species (Zimmerman 1948, Swezey 1954). Parasitoid species were sometimes observed in abundance during outbreaks of *Scotorythra*, but their roles in regulating defoliator populations has not been studied (Swezey 1926, 1927; Fullaway 1947). Nevertheless, 40 years ago foresters generally assumed that populations of *Scotorythra* seldom built up to damaging levels because they were being regulated by biocontrol agents (e.g., Whitesell 1964).

The host preferences of native parasitoids, distributed mainly in high-elevation native forests, are poorly known but do include late-instar *Scotorythra* larvae (Peck and Banko, unpubl. data). The broad host range of many of the early accidentally or intentionally introduced parasitoids made them a major threat of nearly the entire spectrum of native caterpillars (Zimmerman 1958b). Moreover, many alien wasps and flies exerted persistent pressure on Hawaiian caterpillars because their generalist habits enabled them to switch hosts when a particular species became scarce (Zimmerman 1948, 1958a).

Although the most damaging generalist parasitoids apparently were introduced before the 1940s, some biocontrol agents that were imported later also spread into native forests (Bianchi 1959). Research on the Alaka'i Plateau of Kaua'i indicates that some of the 122 or more alien parasitoids released on the Hawaiian Islands during the past 100 years still strongly impact native caterpillar populations, including *Scotorythra* and *Thyrocopa* (Henneman and Memmott 2001). On the windward slopes of Mauna Loa, Hawai'i, alien parasitoid wasps were found throughout native forest but predominately at low- and mid-elevation sites (Peck et al. 2008). The results of studies in the dry, subalpine woodland of Mauna Kea, Hawai'i, indicate that native *Cydia* spp. (Tortricidae), *Scotorythra*, and other caterpillar prey of Palila are parasitized at high frequencies (Brenner et al. 2002, Oboyski et al. 2004). Habitats surveyed on the Alaka'i Plateau, windward Mauna Loa, and subalpine Mauna Kea are quite different ecologically, geographically, and climatically from one another and especially from the agricultural areas where biocontrol agents have been released or have escaped since the 1890s. The high abundance of parasitoids and rates of parasitism of native caterpillars in these disjunct and disparate areas indicate that alien wasps and flies continue to impact the insect prey of native birds.

Other arthropod prey of native birds have been affected historically by alien parasitoids. For example, the larvae of *Plagithmysus* spp. beetles (Cerambicidae), the most frequently taken food of Maui Parrotbill, were heavily parasitized by an alien wasp in the 1890s (Koebele 1901, Perkins 1913).

ALIEN ARTHROPOD PREDATORS

The sometimes staggering impacts of predatory ants on arthropod communities derive from the sheer size of ant colonies and the thoroughness with which ant workers forage, their efficiency at maximizing energy, and their cooperative retrieval of prey (Hölldobler and Wilson 1990). Because the Hawaiian Archipelago is one of the few places on earth that was not naturally colonized by ants, it is understandable that some of the nearly 100 species that historically invaded the islands have had enormous impacts on many native arthropods (Wilson and Taylor 1967, Nishida 2002). More than a century ago, Perkins (1913) had much to say about the difficulty of collecting native insects of all sorts within the range of a notoriously predaceous ant (*Pheidole megacephala*: Formicidae).

Pheidole was thought to be the single most important cause of the decline and extinction of native insect populations in the lowlands, wreaking more damage even than the destruction of forest habitats, so it is not surprising that this ant devastated lowland populations of *Scotorythra* caterpillars (Perkins 1913). The time of the arrival of *Pheidole* and other ants on the islands is unknown, but ants ravaged bird specimens being prepared by Andrew Bloxam as early as 1825, and ant predation on the caterpillars and pupae of moths was noted in the mid-1800s (Newcomb 1852, Jones 1925). Zimmerman (1948) later corroborated Perkins's observations and conclusions regarding the destructiveness of *Pheidole* to native insects, and Gagné (1979) suggested that *Pheidole* and other ants were limiting endemic arthropods below about 800 m elevation. The severe impact of ants on native spider communities has also been noted (Gillespie 1999).

Predaceous wasps are other social insects with a great capacity for disrupting native arthropod communities. Notoriously effective at hunting arthropod prey is the invasive Western yellowjacket, *Vespula pensylvanica* (Vespidae). Established first on Kaua'i in 1919 and as late as 1978 on Hawai'i, this voracious predator has become locally abundant at middle and high elevations and may have far-reaching consequences for native arthropod communities in habitats that are especially important to native birds (Howarth 1985; Gambino et al. 1987; D. Foote, pers. comm.). Entomologists noticed a marked decline of native moths as Western yellowjackets became abundant in Volcano, Hawai'i Island (W. Mull and F. Howarth, pers. comm.).

DISEASES OF NATIVE ARTHROPODS

Although little is known about the diseases of arthropod prey of native Hawaiian birds, a species of fungus has been observed destroying *Scotorythra* caterpillars and pupae, suggesting the possibility that disease may sometimes regulate populations (Koebele 1901, Perkins 1913).

RODENTS

Polynesian rats apparently consume arthropods more frequently than do black rats in Hawaiian rain forests, but both species can noticeably impact invertebrate populations and communities on the ground (Lindsey et al. 1999; this volume, Chapter 11). Although limited studies suggest that black rats mainly eat plant material, their stomachs frequently contain arthropods (Cuddihy and Stone 1990). Indirect evidence that rats impact arthropod communities is the conspicuousness of large, flightless arthropods on Nihoa, where rats are absent and few other alien species have invaded until recently (Conant et al. 1984, Latchininsky 2008). Rats also consume land snails, which were important prey of one honeycreeper species (Perkins 1903, 1913).

ALIEN SNAILS

The once rich and abundant native terrestrial snail fauna of 760 species has declined dramatically, and many species have disappeared from Hawaiian forests, suffering not only from many of the same invasive species that threaten all other major groups of native plants and animals but also from predatory snails introduced for biocontrol, especially the rosy cannibal snail (*Euglandina rosea:* Spiraxidae) (Cowie 1998). The garlic snail (*Oxychilus alliarius:* Zonitidae), another predator of native snail species, has become common in the habitat of Po'o-uli. However, Po'o-uli, which heavily exploit native snails, do not seem to eat garlic snails (Chapter 21).

ALIEN BIRDS

Alien birds are the most obvious direct competitors of native birds for arthropod prey and other foods, because they forage in the same places and in many of the same ways as do native birds. However, very few species have become established or abundant in native forests (Chapter 13). The Common Myna (*Acridotheres tristis*), Japanese White-eye (*Zosterops japonicus*), Red-billed Leiothrix (*Leiothrix lutea*), and Japanese Bush-Warbler (*Cettia diphone*) are the species whose generalist diets and opportunistic habits make them most likely to affect food availability in native forests (Chapter 13). The cumulative effects on the resources of native forest birds by successive waves of generalist invaders may be substantial, even if diffuse and difficult to evaluate. For example, Perkins wrote in a letter that defoliation of koa by *Scotorythra* caterpillars became less frequent when the numbers of Common Myna reached peak levels (Swezey 1954).

Changes in Geographical and Temporal Patterns of Food Availability

The diverse array of ecosystem modifiers discussed so far strongly suggests that bird populations today do not likely encounter the same suite of foods in generally the same abundance as did their predecessors a century ago or earlier. Nevertheless, all evidence indicates that ecosystem changes occurred early and most profoundly in low-elevation and dryland forests (Kirch 1982, Cuddihy and Stone 1990). Therefore, bird resources would have been af-

fected sooner and more severely on islands and volcanoes with relatively little forested area above 500 m elevation: Moloka'i, Lāna'i, West Maui, and O'ahu (Pratt and Gon 1998). Not only geographical patterns but also seasonal patterns of flower and fruit availability were truncated with the disappearance of lowland forests. Additionally, populations of arthropods would have changed as lowland forests declined. For example, recent research along an elevation gradient on windward Hawai'i indicates that arthropod prey are most available at high elevations, where native bird communities are most intact (Peck et al. 2008). Thus, unlike some tropical areas where arthropod numbers and species richness are reduced at high elevations (Janzen 1973), the Hawaiian lowlands may not have arthropod resources adequate to support populations of many native birds.

The scale on which changes in bird resources has occurred is massive, but more changes are inevitable as new alien threats, such as frogs and reptiles, arrive and begin to compete with established pests (Chapter 15). An additional concern is the impact of global warming trends on bird resources. If droughts increase in frequency and severity, for example, important bird foods would likely become scarcer.

Impacts of Changes in Food Availability on Native Bird Populations

RESPONSES OF SPECIALISTS, GENERALISTS, AND FORAGING GUILDS

All bird species would have suffered demographically from reductions in the availability of caterpillars and other abundant, easily obtained prey to feed their young; however, specialized birds were most vulnerable, because they would have had few, if any, suitable alternatives. Similarly, specialized adult birds would have felt the greatest impacts from reduced caterpillar subsidies when the availability of their specialty foods also declined, either temporarily or for the long term. Generalists could have compensated for an overall decline in the availability of *Scotorythra* by switching to alternate prey, but specialists would have been less successful in doing this. It is significant that about 50 years after Perkins noted their importance, *Scotorythra* were thought to be infrequently eaten by several nonspecialist forest birds, the Hawai'i 'Amakihi, 'Apapane, and 'I'iwi (Baldwin 1953).

Baldwin (1953) acknowledged that many native moths were less common in the 1940s than in the 1890s, but he suggested that Hawai'i 'Amakihi, 'Apapane, and 'I'iwi could eat more abundant species of native and introduced moths as well as Homoptera and other arthropods to compensate for the loss. The three relatively versatile foragers that Baldwin studied evidently had adjusted their diets, but the diets of specialist species have been little studied since Perkins's time, and many are now extinct or are too rare to study. The essential question is this: Are *Scotorythra* still the widespread and abundant resource that Perkins observed?

The many disruptions to the availability of important foods did, in fact, impact populations of specialized species more severely and frequently than those of generalists during the historic period (Chapter 2). Although nearly all generalist species persisted into recent times, fewer than half of specialist species and fewer than a third of specialist populations survived through the historic period. Additionally, all historically known populations of the Hawaiian Hawk, a generalist, have survived on the island of Hawai'i. It should not be surprising that generalists, with their greater foraging versatility and higher reproductive capacity, fared better as ecosystems changed and resources became less available. The persistence of populations across foraging guilds was not uniform; many populations of arthropod eaters (63%) and nectar eaters (66%) survived, whereas relatively few populations of fruit eaters and seed eaters

(32%) persisted into recent times. At the population level, generalists that ate mostly arthropods, fruit, or nectar survived relatively frequently (86%). However, no specialist nectar eaters and few fruit and seed eaters (22%) survived, and although about 46% of all populations of arthropod specialists survived, all are listed as endangered or are being considered for listing.

What ecological factors could have favored the overall success of nectar eaters and arthropod eaters? One reason for the arthropod guild's success may have been their morphological and behavioral adaptations for obtaining animal protein, both for their nestlings and for themselves. Individuals foraging for arthropods on a daily basis would be expected to have developed search images whereby they more easily recognize particular types of prey and know where to find them by using visual cues learned through experience. Competence in capturing a variety of arthropod prey, including alien species, would have been especially advantageous when important native prey species began to decline. The level of persistence of nectar eaters was relatively high in part because all populations of 'Apapane, the only generalist, survived into recent times. The widespread and abundant distribution of 'ōhi'a may also have helped to sustain populations of some nectar eaters. On the other hand, populations of specialized species, such as Kioea, species of 'ō'ō, Hawai'i Mamo, and Black Mamo were apparently already contracting rapidly early in the historic period.

RESPONSES OF PERCHING BIRDS IN THE
NORTHWESTERN HAWAIIAN ISLANDS

Land bird communities on the tiny, remote Northwestern Hawaiian Islands dramatically demonstrate the persistence of generalist foragers in the face of massive ecological upheaval. Despite the scarcity or lack of trees, perching birds eke out a living on Laysan and Nihoa Islands. The vegetation of both islands has been greatly modified due to human disturbance, especially on Laysan, where introduced rabbits (*Orycolagus cuniculus*) denuded the island of vegetation from 1903 to 1923 (Chapter 6). The loss of vegetation precipitated the decline or extinction of moth species that were important prey of all the birds there (Munro 1944, Butler and Usinger 1963). Ants and parasitoid wasps also became threats to moth populations (Morin and Conant 1998).

Apparently because of its omnivorous foraging habits, only the Laysan Finch survived the period when rabbits decimated plant life on Laysan. Animal foods eaten by the birds there today, even after the vegetation has recovered to a large extent, include a variety of invertebrates, carrion and associated maggots, and bird eggs (Morin and Conant 2002). In addition, the Laysan Finch eats a wide range of seeds, fruits, flowers, leaves, seedlings, stems, and roots. The diet of nestlings likely includes seeds in addition to arthropods, when they are available. Although the Laysan Honeycreeper (*Himatione sanguinea freethii*) and the Millerbird were generalist foragers on Laysan, they probably were unable to find suitable food alternatives and died out when populations of caterpillars and moths declined with the loss of their host plants (Ely and Clapp 1973).

On Nihoa, the vegetation has been altered by early human occupation and fire, but it was never so completely destroyed as on Laysan, and no perching birds have disappeared during historic times. The Nihoa Millerbird forages for a wide variety of native and alien arthropod prey (Morin et al. 1997). The Nihoa Finch, like its counterpart on Laysan, is an omnivorous generalist (Morin and Conant 2002).

NEW AND DIFFERENT
COMPETITIVE INTERACTIONS

Although we have already discussed the ramifications for generalists and specialists of the decline of important foods, the question remains: What happens when a species suddenly must share its niche space with many new food competitors? Limited insight into this question comes from examining the competition between alien and introduced birds. Mountainspring and Scott (1985) determined that alien bird species competed sufficiently with native birds to occasionally affect their distribution and density, taking into account the effect of habitat structure. They found that the widespread and abundant Japanese White-eye played the greatest role in competitively displacing native species. Compared to a single species, however, the latent impacts of the total spectrum of other organisms invading native bird habitats and usurping resources would seem to have been far more insidious over time.

INTERACTIONS BETWEEN
FOOD AND DISEASE

Variable susceptibility to introduced bird diseases suggests that strong selection pressures are being exerted on Hawaiian host species (van Riper et al. 1986; this volume, Chapter 9). Even if birds do not die from disease, they may suffer reproductively because of the stress of fighting infection (Bonneaud et al. 2003). Additionally, the body condition of nestlings can be improved by the amount and quality of food they receive, and the immune capacity of nestlings may depend partly on how the nutritional status of the female affects her ability to transfer maternal antibodies to her eggs (Boag 1987, Hoi-Leitner et al. 2001). To the extent that their immune systems and overall physiological function are affected by food availability and diet quality, birds that are well nourished are likely to survive and reproduce at higher rates when they are challenged by disease (Klasing 1998b). Nevertheless, when sick, generalist species should be expected to breed more successfully than specialist species, because they can potentially forage for a wider range of foods. However, the interaction of nutrition and immune competence are most likely to impinge on the fitness of young birds that are recently independent of parental care but still learning how to make their own living, especially the young of specialist species, whose foraging abilities are probably least developed compared with those of experienced adults.

Suspected variation in the susceptibility of bird populations and species to avian malaria under natural conditions suggests that, among many other factors, there may be a relationship between foraging behavior, food availability, nutritional status, and survival in areas where the level of disease transmission is high (Chapter 14). Of the honeycreeper species challenged by avian malaria in laboratory trials, 'I'iwi and Maui 'Alauahio suffered the most serious effects and the highest levels of mortality (Chapter 9). However, despite clinical evidence of their greater susceptibility to malaria, food availability and quality may also be important in limiting their distribution and abundance. Both species are relatively intermediate in terms of foraging behavior (see Box 7.1), and it is not surprising that their populations are constrained to a greater degree than are those of generalist species, such as Hawai'i 'Amakihi and 'Apapane. Is it mere coincidence that lowland habitats are occupied almost exclusively by generalist species but not by any specialist species (Chapter 2)? Mounting evidence of resistance or tolerance to malaria by Hawai'i 'Amakihi suggests that broader diets and greater capacity to provision offspring are critical advantages to surviving in lowland habitats and adapting to disease (Chapter 9).

SUMMARY AND LESSONS
FOR CONSERVATION

The food resources available to native Hawaiian birds were limited to those few plant and animal species that flourished in or dominated landscapes following the infrequent colonization of their distant-traveling founders. This small base of resources resulted in strong competition for food, which gave rise to a prolific radiation of species in one lineage, the honeycreepers. Supporting this radiation were caterpillars and other abundant and easily captured arthropod prey that provided the protein needed for growth by young birds. Virtually all historically known forest passerine species exploited the widespread and abundant caterpillars of the endemic moth genus, *Scotorythra*, to feed their nestlings and themselves. Therefore, *Scotorythra* probably were crucial to the evolution of radical, and in some cases unprecedented, feeding adaptations for nectar, fruit, seeds, and invertebrate prey that were not accessible to species with ordinary bills and foraging habits. The limits to which species evolved extreme bill forms were likely set by the availability of such arthropod prey and the ability of parent birds to capture them. No other groups of moths or other arthropods seem to have replaced *Scotorythra* as preeminent foods of Hawaiian forest birds.

Intricate relationships evolved between consumers and their resources, but these were unraveled by a long succession of invasive species and by forest destruction following human colonization by Polynesians and later immigrants. In this ecological tumult, birds with the most extreme bill forms and foraging behaviors were disadvantaged relative to generalists, which were able to exploit a broader base of resources to sustain themselves and their young. Compared to specialists, generalists are smaller, more fecund, faster growing, and more abundant and widespread, presumably because their energetic and nutritional requirements are more easily met by their feeding strategies. Differences in feeding ecology, reproduction, and life history adaptations largely explain why generalist species tend to better survive habitat changes and challenges from diseases, predators, and other threats while specialists are primarily reduced to relict populations in habitats where threats are less pervasive overall.

To the extent that food availability is related to bird survival and reproduction in habitats where disease or other threats are prevalent, conservation strategies should aim to improve habitat and foraging conditions. Eliminating feral ungulates will go far in restoring forest canopy and under-story plants, but by itself, this action is likely to fall short of replenishing the food resources needed by birds. For example, some important plant and animal resources may require active restoration at the same time that other harmful factors, such as rodents and predatory snails, are reduced. Controlling threats to important arthropod prey from alien predators or parasitoids presents enormous challenges, and a great deal of innovative research will be needed

APPENDIX 7.1. HABITAT AND FORAGING ECOLOGY OF HAWAIIAN FOREST BIRDS

Family	Species	Food Guild[a]	Degree of Specialization[b]	Historic and Prehistoric Habitats[c]
Accipitridae	ʻIo (Buteo solitarius)	R&A	G	LD, LM, LW, MD, MM, MW, SD
Strigidae	Pueo (Short-eared Owl, Asio flammeus sandwichensis)	R&A	G	LD, LM, LW, MD, MM, MW, SD
Corvidae	ʻAlalā (Corvus hawaiiensis)	F&S	I	LD, LM, MD, MM, MW
Meliphagidae (or Mohoidae)	ʻŌʻōʻāʻā (Moho braccatus)	N	I	LD, LM, LW, MM, MW
	Oʻahu ʻŌʻō (Moho apicalis)	N	I	LD, LM, LW, MM, MW
	Bishop's ʻŌʻō (Moho bishopi)	N	I	LD, LM, LW, MM, MW

ACKNOWLEDGMENTS

This research was made possible in part thanks to support from the U.S. Geological Survey Invasive Species Program and Wildlife and Terrestrial Resources Program; the National Science Foundation (Grant DEB00-83944: Biocomplexity of Introduced Avian Disease in Hawai'i); and the U.S. Fish and Wildlife Service (support to W. E. Banko during 1965–1978). We thank C. Kishinami, Bishop Museum, Honolulu, for data on the sex and age of bird specimens; K. Brinck for statistical assistance; and R. Peck and D. Pollock for entomological assistance and the use of unpublished data.

to develop management strategies and techniques. A step forward, however, is to minimize the potential menace from future invasive species.

Historically Known Major Habitat Type or Host Plants[d]	Foraging Substrate and Microhabitat[d]	Foraging Behavior[d]	Diet[d]
'Ōhi'a,[e] a variety of other native and non-native forest and agricultural lands	Tree perches (high or low) and soars	Stoop on larger prey from perch or aerially; may follow quarry on foot; take arthropods while perched	Spiders, beetle larvae, caterpillars, mice, rats, birds
Grasslands, shrublands, woodlands, forests	Hunt on the wing, close to the ground, occasionally a perch	Hunt aurally and visually; forage while quartering, hovering, or sometimes perching	Mice, rats, birds, arthropods
'Ōhi'a, koa forests; 'ie'ie	Tree canopy, subcanopy, vines, ground; forage over twigs, foliage, large branches, trunks	Pluck, swallow whole or piecemeal, or hammer open fruits; glean, probe, or flake bark for arthropods; rob nectar	'Ie'ie, a variety of small to large fruits; nectar; caterpillars, other insects; spiders; isopods; small bird eggs, nestlings
'Ōhi'a, koa forests	Trees, shrubs for nectar; tree trunks, branches for arthropods	Probe flowers; probe, flake bark for arthropods; defend nectar resources	'Ōhi'a, other nectar; spiders; crickets, cockroaches, beetles, caterpillars (*Scotorythra*), true bugs; millipedes; small land snails; fruit
Forest	No information	No information	Nectar, probably arthropods
'Ōhi'a, koa forests	Trees, shrubs	Probe flowers; defend nectar resources	Lobelia, 'ōhi'a, banana nectar; arthropods; banana fruit

continued

Family	Species	Food Guild[a]	Degree of Specialization[b]	Historic and Prehistoric Habitats[c]
Meliphagidae (or Mohoidae)	Hawai'i 'Ō'ō (*Moho nobilis*)	N	I	LD, LM, LW, MD, MM, MW
	Kioea (*Chaetoptila angustipluma*)	N	I	LM, MM
Monarchidae	'Elepaio (*Chasiempis sandwichensis*)	A	G	LD, LM, LW, MD, MM, MW, SD
Sylviidae	Millerbird (*Acrocephalus familiaris*)	A	G	CD
Turdidae	Kāma'o (*Myadestes myadestinus*)	F&S	G	LD, LM, LW, MM, MW
	'Āmaui (*Myadestes woahensis*)	F&S	G	LD, LM, LW, MM, MW
	'Oloma'o (*Myadestes lanaiensis*)	F&S	G	LD, LM, LW, MM, MW
	'Ōma'o (*Myadestes obscurus*)	F&S	G	LD, LM, LW, MD, MM, MW, SD, AD
	Puaiohi (*Myadestes palmeri*)	F&S	I	LD, LM, LW, MM, MW
Fringillidae: Drepanidinae	Laysan Finch (*Telespiza cantans*)	F&S	G	CD

Historically Known Major Habitat Type or Host Plants[d]	Foraging Substrate and Microhabitat[d]	Foraging Behavior[d]	Diet[d]
ʻŌhiʻa, koa forests; māmane forest	Trees, shrubs	Probe flowers; defend nectar resources; moderate movements to exploit nectar	ʻŌhiʻa, lobelia, māmane nectar; caterpillars, beetle larvae, other insects; ʻieʻie flower bracts, fruit; fruit
Forest	No information	Probe flowers	Nectar; probably arthropods
ʻŌhiʻa forest; a variety of vegetation types, including alien-dominated	Branches, foliage, rotten wood (standing or lying), around trees on ground	Branch glean, flight-glean, aerial fly-catch	Spiders; insects (especially adult and larval beetles, moths); millipedes; centipedes; slugs
ʻĀweoweo, pōpolo, ʻilima shrubs, bunchgrass	Leaves, twigs of shrubs; ground litter	Glean	*Agrotis* caterpillars, moths ("millers"); *Nysius* bugs, flies, beetles
Closed or open forest dominated by koa, ʻōhiʻa	Entire vertical range of forest canopy; shrubs, epiphytes, subcanopy trees	Pluck fruits; swallow fruits whole	Many small–medium fruits; caterpillars (*Scotorythra*); probably other invertebrates
No information	No information	No information	Probably small–medium fruits; probably caterpillars, other invertebrates
Closed, open forest; shrubs, trees	Entire vertical range of forest canopy; shrubs, epiphytes, subcanopy trees	Pluck fruits; swallow fruits whole	Small–medium fruits; caterpillars (*Scotorythra*), other invertebrates
Closed, open forest; shrubs, small and large trees	Entire vertical range of forest canopy; shrubs, epiphytes, subcanopy trees	Pluck fruits; swallow small fruits whole, eat larger fruits piecemeal; glean, fly-catch arthropods; scratch, probe ground litter	Many small–medium fruits; many insects (*Scotorythra* caterpillar outbreaks, true bugs, beetles); spiders; other invertebrates
Closed, open forest; steep, densely vegetated ravines, stream beds	Subcanopy trees, shrubs	Pluck, hover-glean fruits; swallow fruits whole, sometimes piecemeal; glean arthropods from vegetation; scratch, probe ground litter	Small–medium fruits; *Rhyncogonus* beetles, caterpillars (*Scotorythra*), *Megalagrion* damselflies; spiders; land snails
Coastal shrub-grassland	All vegetation (especially shrubs, vines, grasses)	Pick, crack, extract seeds; pluck fruits; bite leaves, buds, flowers; glean arthropods from foliage, twigs; crack open bird eggs; dig with bill in sand; scavenge carcasses	A variety of seeds, fruits, leaves, flowers, stems, seedlings, roots; carrion; invertebrates; bird eggs

continued

Family	Species	Food Guild[a]	Degree of Specialization[b]	Historic and Prehistoric Habitats[c]
Fringillidae: Drepanidinae	Nihoa Finch (*Telespiza ultima*)	F&S	G	CD
	'Ō'ū (*Psittirostra psittacea*)	F&S	S	LD, LM, LW, MD, MM, MW
	Lāna'i Hookbill (*Dysmorodrepanis munroi*)	A?	S	LM
	Palila (*Loxioides bailleui*)	F&S	S	LD, MD, MM, SD
	Lesser Koa-Finch (*Rhodacanthis flaviceps*)	F&S	S	MM
	Greater Koa-Finch (*Rhodacanthis palmeri*)	F&S	S	MM
	Kona Grosbeak (*Chloridops kona*)	F&S	S	MM
	Maui Parrotbill (*Pseudonestor xanthophrys*)	A	S	LD, LM, LW, MM, MW

Historically Known Major Habitat Type or Host Plants[d]	Foraging Substrate and Microhabitat[d]	Foraging Behavior[d]	Diet[d]
Coastal, lowland shrub-grassland	All vegetation (especially 'āweoweo, pōpolo, 'ilima shrubs)	Pick, crack, extract seeds; pluck fruits; bite leaves, buds, flowers; glean arthropods from foliage, twigs; crack open bird eggs	A variety of seeds, fruits, leaves, flowers; insects; bird eggs
'Ie'ie-'ōhi'a forest	Terminal portions of vines, subcanopy trees, shrubs; canopies of koa, māmane trees for caterpillars	Piecemeal extraction, consumption of 'ie'ie, other larger fruits; glean caterpillars	Fruit of 'ie'ie, lobelias, many other small fruits; a variety of agricultural fruits; caterpillars (*Scotorythra* outbreaks on koa; caterpillars on māmane); 'ōhi'a nectar, young leaves, and flower buds
Dry forest with ōpuhe, 'akoko	No information	No information	Ōpuhe fruits; likely other fruits and invertebrates
Māmane, naio	Terminal branches in canopies of large, medium trees	Pick, shred pods; pick buds; rob nectar; glean branches	Mainly immature māmane seeds, also flower parts, nectar, young leaves; other seeds and fruit; caterpillars (*Uresiphita polygonalis, Cydia, Scotorythra*)
Tall koa forest with naio, māmane, 'a'ali'i, 'iliahi	Terminal branches in canopies of large trees	Pick open pods to extract seeds	Immature koa seeds; caterpillars
Tall koa forest with naio, māmane, 'a'ali'i, 'iliahi	Terminal branches in canopies of large trees	Pick open pods to extract seeds; glean caterpillars	Immature koa seeds (adult, nestling diet); caterpillars (*Uresiphita polygonalis, Scotorythra*)
Recent lava flows with naio	Terminal branches in canopies of medium-sized trees	Crack hard shells to extract embryos	Naio embryos; probably caterpillars
'Ōhi'a, koa forests	Branches of trees and shrubs	Bite open wood; split twigs; pry up bark; bite open fruit for caterpillars; glean branches, foliage	Beetle larvae, pupae (*Plagithmysus, Proterhinus, Nesotocus*); caterpillars (*Hyposmocoma, Carposina*)

continued

Family	Species	Food Guild[a]	Degree of Specialization[b]	Historic and Prehistoric Habitats[c]
Fringillidae: Drepanidinae	Hawai'i 'Amakihi (*Hemignathus virens*)	A	G	LD, LM, LW, MD, MM, MW, SD, AD
	O'ahu 'Amakihi (*Hemignathus flavus*)	A	G	LD, LM, LW, MM, MW
	Kaua'i 'Amakihi (*Hemignathus kauaiensis*)	A	I	LD, LM, LW, MM, MW
	'Anianiau (*Hemignathus parvus*)	A	G	LD, LM, LW, MM, MW
	Greater 'Amakihi (*Hemignathus sagittirostris*)	A	I	LW, MW
	Lesser 'Akialoa (*Hemignathus obscurus*)	A	S	LD, LM, LW, MD, MM, MW
	Greater 'Akialoa (*Hemignathus ellisianus*)	A	S	LD, LM, LW, MM, MW
	Nukupu'u (*Hemignathus lucidus*)	A	S	LM, LW, MD, MM, MW

Historically Known Major Habitat Type or Host Plants[d]	Foraging Substrate and Microhabitat[d]	Foraging Behavior[d]	Diet[d]
A wide variety of forest, shrub communities	Terminal branches of trees, shrubs; vines	Glean arthropods; probe, nectar-rob flowers	Insects (especially true bugs, caterpillars); spiders; ʻōhiʻa, māmane, a variety of other nectar; small fruits, juices; occasionally sap flux
A wide variety of forest, shrub communities (native and nonnative)	Terminal branches of trees, shrubs	Glean arthropods; probe flowers	Arthropods; nectar; small fruits, juices; occasionally sap flux
ʻŌhiʻa, mixed ʻōhiʻa-koa forests	Trunks, branches of trees; terminal branches of trees, shrubs	Glean arthropods; probe flowers	Arthropods; small fruits, juices; nectar
ʻŌhiʻa, koa forests	Terminal branches of trees, shrubs; vines; fern fronds	Glean arthropods; probe, nectar-rob flowers	A variety of arthropods; ʻōhiʻa, other nectar; small fruits, juices
Wet ʻōhiʻa forest with ʻieʻie vines	ʻIeʻie leaf axils; under side of bark of ʻōhiʻa	Probe plant parts; pry up bark	Crickets (*Paratrigonidium*), caterpillars, beetles; spiders, other arthropods; ʻōhiʻa nectar
ʻŌhiʻa, koa forests	Bark; cavities in trunks, branches, fronds, ʻieʻie vines; decaying wood; litter at base of ʻieʻie, hala pepe leaves; hāpuʻu tree fern trunks, epiphytes	Probe tree substrates, flowers; peck; hop along branches, creep up tree trunks	Beetle larvae, adults; caterpillars (Geometridae [probably *Scotorythra*], Gelechiidae); crickets, spiders, other arthropods; lobelia, ʻōhiʻa nectar
ʻŌhiʻa, ʻōhiʻa-ʻōlapa, koa forests	Bark; cavities in trunks, branches, fronds, ʻieʻie vines; decaying wood; litter at base of ʻieʻie, hala pepe leaves; tree fern trunks, epiphytes	Probe tree substrates, flowers; peck; hop along branches, creep up tree trunks	Beetle larvae, adults; caterpillars (Geometridae [probably *Scotorythra*], Gelechiidae); crickets; spiders, other arthropods; lobelia, ʻōhiʻa nectar
Closed ʻōhiʻa, koa-ʻōhiʻa forests	On Kauaʻi: koa, ʻōhiʻa, understory trees; on Maui: ʻōhiʻa	Creep along branches, trunks; hammer on live, dead wood with lower bill, then probe and hook prey with flexible upper bill; pry bark loose to expose prey; glean arthropods from leaves; seldom probe flowers	Beetles (*Oodemas*), caterpillars (*Scotorythra*, *Thyrocopa*, *Hyposmocoma*); spiders

continued

Family	Species	Food Guild[a]	Degree of Specialization[b]	Historic and Prehistoric Habitats[c]
Fringillidae: Drepanidinae	'Akiapōlā'au (*Hemignathus munroi*)	A	S	LD, LM, LW, MD, MM, MW, SD
	'Akikiki (*Oreomystis bairdi*)	A	I	LD, LM, LW, MM, MW
	Hawai'i Creeper (*Oreomystis mana*)	A	I	LM, LW, MM, MW, SD
	O'ahu 'Alauahio (*Paroreomyza maculata*)	A	I	LD, LM, LW, MM, MW
	Kākāwahie (*Paroreomyza flammea*)	A	I	LM, LW, MM, MW
	Maui 'Alauahio (*Paroreomyza montana*)	A	I	LD, LM, LW, MM, MW
	'Akeke'e (*Loxops caeruleirostris*)	A	S	LM, LW, MM, MW
	'Ākepa (*Loxops coccineus*)	A	S	LM, LW, MM, MW
	'Ula-'ai-hāwane (*Ciridops anna*)	F&S	I	LW, MW

Historically Known Major Habitat Type or Host Plants[d]	Foraging Substrate and Microhabitat[d]	Foraging Behavior[d]	Diet[d]
Closed koa, koa-'ōhi'a, 'ōhi'a, māmane forests	Branches, trunks of koa, māmane, naio, kōlea, 'ōhi'a, 'ie'ie vines	Creep along branches, trunks; hammer on live, dead wood with lower bill, then probe and hook prey with flexible upper bill; flake loose bark, epiphytes; hammer and pry apart dried māmane and koa pods to expose prey; hammer 'ōhi'a bark to create sap wells	Beetles (*Oodemas*, *Plagithmysus*), caterpillars (*Scotorythra*, *Thyrocopa*, *Hyposmocoma*, *Cydia*), other insects; spiders; 'ōhi'a sap
'Ōhi'a, koa forests	Bark, dead wood, and foliage along trunks and branches of trees and shrubs	Glean, probe for arthropods while creeping, clinging	Caterpillars (*Scotorythra*), moths, beetles, other insects; spiders; other invertebrates
'Ōhi'a, koa forests	Bark, dead wood, and foliage along trunks and branches of trees and shrubs	Glean, probe for arthropods while creeping, clinging	Caterpillars (*Scotorythra*), moths, beetles (especially larvae), other insects, insect eggs; spiders, spider eggs
'Ōhi'a, koa forests	Bark, dead wood, and foliage along trunks and branches of trees and shrubs	Glean, probe for arthropods	Caterpillars (*Scotorythra*), moths, beetles, other insects; spiders; other invertebrates
'Ōhi'a forest	Bark, dead wood, and foliage along trunks and branches of trees and shrubs	Glean, probe for arthropods	Caterpillars (*Scotorythra*), moths, beetles (especially larvae), other insects; spiders; other invertebrates
'Ōhi'a forest; māmane savannah; shrubland; nonnative forest	Bark, dead wood, and foliage along trunks and branches of trees and shrubs	Glean, probe for arthropods	Caterpillars (*Scotorythra*), moths, beetles (especially larvae), other insects, insect eggs; spiders, spider eggs; rarely nectar
'Ōhi'a forest; 'ōhi'a-koa forest	Outer branches, terminal leaf clusters	Pry open leaf buds ('ōhi'a), silk-bound leaves (koa, other) for arthropods	Caterpillars (*Scotorythra*), other insects, insect eggs; spiders
Koa forest; koa-'ōhi'a forest; 'ōhi'a forest	Terminal leaf clusters and small branches, especially 'ōhi'a	Pry open leaf buds ('ōhi'a), silk-bound leaves (koa, other) for arthropods	Caterpillars (*Scotorythra*), other insects, insect eggs; spiders
'Ōhi'a forest with loulu (fan-palm)	Loulu flowers, fruit clusters, leaves	No information	Fruit, nectar, insects

continued

Family	Species	Food Guild[a]	Degree of Specialization[b]	Historic and Prehistoric Habitats[c]
Fringillidae: Drepanidinae	'I'iwi (*Vestiaria coccinea*)	N	I	LD, LM, LW, MD, MM, MW, SD
	Hawai'i Mamo (*Drepanis pacifica*)	N	S	LW, MW
	Black Mamo (*Drepanis funerea*)	N	S	LW, MW
	'Ākohekohe (*Palmeria dolei*)	N	I	LM, MW
	'Apapane (*Himatione s. sanguinea*)	N	G	LD, LM, LW, MD, MM, MW, SD, AD
	Po'o-uli (*Melamprosops phaeosoma*)	A	I	LM, MW

Notes: Knowledge of the range of habitats, host plants, foraging substrates, foraging maneuvers, and apparent diet breadth of bird species can sharpen our understanding of adaptations for feeding on arthropods, nectar, and fruits and seeds. Most of what is known about Hawaiian bird diets and foraging behavior, particularly for species now extinct, was determined by R. C. L. Perkins, a brilliant ornithologist and entomologist at the turn of the nineteenth century. Owing to historic ecological changes, more recent studies may somewhat distort our perspectives of the diet and foraging behavior of some species. We must nevertheless be cautious about stretching inferences made by Perkins and other early naturalists. Therefore, our summaries of the feeding ecology are based on the Birds of North America (Poole and Gill 1992–2002) species accounts, which, although they incorporate most of the information provided by Perkins and other early naturalists, also include more recent information. Illustrations have been copied from Frohawk (in Rothschild 1893–1900).

Historically Known Major Habitat Type or Host Plants[d]	Foraging Substrate and Microhabitat[d]	Foraging Behavior[d]	Diet[d]
'Ōhi'a, koa forests	Middle/upper canopy; terminal branches, foliage; understory; vines	Probe, nectar-rob flowers; glean over foliage, branches; track, defend nectar resources	'Ōhi'a, māmane, lobelia, other nectar; insects (especially *Scotorythra* caterpillars, true bugs); spiders
'Ōhi'a forest	Terminal branches of trees and shrubs	Probe flowers; probably defend nectar resources	Lobelia, 'ōhi'a, loulu nectar; probably arthropods
Very wet 'ōhi'a forest	Terminal branches of trees and shrubs	Probe flowers; defend nectar resources	Lobelia, 'ōhi'a nectar; probably arthropods
'Ōhi'a forest	Middle/upper canopy, seasonally in understory; terminal branches of trees, shrubs	Probe, nectar-rob flowers; glean over foliage, branches; defend nectar resources	'Ōhi'a, other nectar; caterpillars, flies; spiders
'Ōhi'a, koa forests	Middle/upper canopy; terminal branches, foliage	Probe flower; glean over foliage, branches; track nectar resources ('ōhi'a, māmane)	'Ōhi'a, māmane, other nectar; insects (especially true bugs, caterpillars); spiders
'Ōhi'a forest	Subcanopy, understory trees, shrubs; twigs, branches, foliage	Extract/crush snails; glean, probe; flake, pry, pull apart bark, lichens for arthropods	Native snails (*Succinea, Nesopupa, Tornatellaria, Tornatellides, Zonitoides*); beetles, other insects; spiders; fruit

[a]Foraging guilds represent the most frequently obtained or typical kinds of foods consumed by birds: A, arthropods (including snails); F&S, fruits and seeds; N, nectar; R, raptorial (for birds of prey).

[b]Degree of specialization: G, generalist; I, intermediate; S, specialist.

[c]Habitat types from Scott et al. (1986) following the classification found in Chapter 6: CD, coastal dry shrubland (Laysan and Nihoa island birds); LD, lowland dry forest; LM, lowland mesic forest; LW, lowland wet forest; MD, montane dry forest; MM, montane mesic forest; MW, montane wet forest; SD, subalpine dry woodland; AD, alpine dry shrubland.

[d]Information derived from Birds of North America accounts (Poole and Gill 1992–2002) and Perkins (1903).

[e]Scientific names of plants: 'a'ali'i (*Dodonaea viscosa*), 'akoko (*Chamaesyce* spp.), 'āweoweo (*Chenopodium oahuense*), banana (*Musa* spp.), hala pepe (*Pleomele* spp.), hāpu'u (*Cibotium* spp.), 'ie'ie (*Freycinetia arborea*), 'iliahi (*Santalum* spp.), 'ilima (*Sida fallax*), koa (*Acacia koa*), kōlea (*Myrsine* spp.), lobelia (Campanulaceae), loulu (also hāwane, *Pritchardia* spp.), māmane (*Sophora chrysophylla*), naio (*Myoporum sandwicense*), 'ōhi'a lehua (*Metrosideros polymorpha*), ōpuhe (*Urera glabra*), pōpolo (*Solanum americanum*).

CHAPTER EIGHT

Life History
and Demography

BETHANY L. WOODWORTH
AND THANE K. PRATT

The striking diversity in life history strategies among the world's organisms has inspired naturalists for centuries. Why do some species produce many offspring in a short time, adopting a "live-fast, die-young" strategy, while others assume a more slowly paced lifestyle wherein less energy is invested in current reproduction and more is saved for survival and production of future offspring? Birds have been the subjects of countless comparative and experimental investigations aimed at answering this question (reviewed in Ricklefs 2000, Martin 2004). A central problem has been to understand how ecological conditions such as resource limitation or predation pressure influence a species' life history (Moreau 1944, Lack 1947, Skutch 1949). Moreover, how do life history strategies translate into differences among species in population dynamics (Saether et al. 2002) and vulnerability to human-caused extinction (Bennett and Owens 1997, Freed 1999, Owens and Bennett 2000)? Island birds, including Hawaiian birds, have adapted to life with a plethora of available ecological niches, few predators or parasites, and high densities of conspecifics, so they present uncommon opportunities to address questions of life history evolution (Grant 1998a; this volume, Chapter 1).

The adaptive radiation of the Hawaiian honeycreepers into various foraging niches led to a spectacular array of bill shapes and functions (Chapter 7). Less appreciated is the diversity of the honeycreepers' life history traits such as survival rate, clutch size, postfledging dependency period, and fecundity. Sadly, many species of Hawaiian forest birds went extinct before we could discover where they built their nests or how they raised their young. But over the past 20 years, studies of avian population ecology have been undertaken for most of the remaining birds, and demographic information is now available for many endangered and nonendangered species. The results show intriguing differences among species, from the bark-probing 'Akiapōlā'au that raises a single young and tends it for over 10 months to the Puaiohi, a thrush that may raise up to four broods in a single season. (Scientific names of birds are given in Appendixes 8.1 and 8.2; those species not listed therein are given in the text.)

In the present chapter, we review the available data on Hawaiian forest bird breeding ecology, survival, nesting success, and annual productivity. We include data for 26 species in seven families, including 1 species of hawk, 1 honeyeater, 2 subspecies of monarch flycatcher, 1 species of crow, an Old World warbler, 2 thrushes, and 19 species of Hawaiian honeycreepers. Our first goal is a comparative analysis of life history traits that may uncover patterns in the population ecology of Hawaiian passerines (songbirds or perching birds). We examine the ways in which the tropical environment led to predictable adaptations among the forest birds. What changes have occurred in the course of their evolution from temperate finch to tropical finch or from temperate thrush to tropical thrush? We also examine whether the environmental conditions on an island have led to predictable island-associated adaptations. In other words, what demographic traits associated with living on islands do we observe in Hawaiian birds as a result of their evolution from continental forms to insular forms? Although we cannot disentangle the separate influences of tropical and insular environments within the scope of this chapter, we hope to shed light on how these selective pressures have shaped the population ecology of Hawaiian forest birds.

The environment in which we study Hawaiian forest birds today is drastically altered from that in which they originally diversified and evolved. Whereas predators once came from the skies, using their avian eyes and ears to seek prey, they now are more likely to climb from below and to have a sharp mammalian sense of smell. Lowland forests that once were productive habitats have become uninhabitable for many species of forest birds due to habitat destruction and avian disease. Countless introductions of insects, plants, and competing birds have forever altered the food resources available. To fully understand Hawaiian forest bird demography, we must consider both historic and current conditions. Our second goal, therefore, is to examine how new population drivers, particularly habitat loss and modification, mammalian predation, and disease, act upon bird life history and impact current demography. How has forest bird demography changed with changing ecological pressures, and how can such knowledge help us identify potential conservation and recovery actions?

BACKGROUND

Phylogeny and Population Status of Hawaiian Forest Birds

Most of the variation in life history patterns among living species of birds occurs at the family level and above (Bennett and Owens 2002); thus it is important to understand the phylogenetic relationships of the Hawaiian forest birds. The closest relatives of the 'Io (Hawaiian Hawk) are New World buteos, probably the Short-tailed Hawk or Swainson's Hawk (Riesing et al. 2003). The Hawaiian honeyeaters, long thought to be derived from Australasian honeyeaters (Meliphagidae), have recently been shown to be a remarkable example of convergent evolution. Their closest relatives are members of a songbird group probably from the Americas, and they have been proposed as a new family, Mohoidae (Fleischer et al. 2008). The Common Raven of North America is probably the closest relative of the sole remaining native corvid, the 'Alalā or Hawaiian Crow (Fleischer and McIntosh 2001). Three subspecies of 'Elepaio represent a single colonization of Hawai'i by Pacific flycatchers in the genus *Monarcha* (Filardi and Moyle 2005). The Millerbird is a differentiated form of a widely distributed genus of Old World warblers, *Acrocephalus*, that probably colonized Hawai'i from the Pacific (Fleischer and McIntosh 2001). A *Myadestes* thrush closely related to the North American solitaires underwent a limited speciation on the islands, resulting in a complex of five species (Fleischer and McIntosh 2001). Both molecular and osteological phylogenies support the view that the Hawaiian honeycreepers (Fringillidae: Drepanidinae) arose from the cardueline finches (Carduelinae) of North America or Asia (Fleischer and McIntosh 2001, James 2004).

Unfortunately, none of the Hawaiian forest bird families has escaped tragic losses (U.S. Fish and Wildlife Service 2006; this volume, Chapter 2). Five species of Hawaiian honeyeater survived into the historic period but are now extinct, and there are virtually no data on their life histories. The 'Alalā is extinct in the wild, as are two related species known from subfossils. Of the two subspecies of Millerbird, only the Nihoa Millerbird remains. Of the original five Hawaiian thrushes, only two are still extant, one of which is endangered. And finally, of the original 50-plus Hawaiian honeycreeper species, more than half are now extinct, and most of the remaining species are endangered.

Overview of the Breeding Ecology and Behavior of Hawaiian Forest Birds

Breeding ecology and behavior are intertwined with life histories. Most Hawaiian honeycreepers, although sedentary, do not maintain all-purpose breeding and foraging territories but rather defend only the immediate area around their nest, much as

do other members of the finch family. In contrast, birds in the other families, including the 'Io, 'Alalā, 'Elepaio, 'Ōma'o, and Puaiohi, hold traditional territories, as is typical of the continental relatives of these birds (see Pratt, Simon, et al. 2001 for a review). Thus, spatial use among Hawaiian birds appears to parallel that of family lines, suggesting that this trait is phylogenetically constrained.

Most Hawaiian forest birds are socially monogamous, stay together through successive breeding attempts within a season, and have the high rates of home range and mate fidelity expected of sedentary, nonmigratory birds (Stutchbury and Morton 2001, Pratt 2005). Only one species, the Maui 'Alauahio, has evolved cooperative breeding, with the young remaining in their natal territory through the following breeding season and helping to raise their siblings (H. Baker and P. Baker 2000). Puaiohi and Palila also have low frequencies of helpers at the nest, but they do not appear to be true cooperative breeders (Snetsinger et al. 1999; Banko, Johnson, et al. 2002). The infrequency of cooperative breeding in Hawaiian birds may be a conservative feature, because in an island situation where conspecific densities are often high and birds are long lived, we might expect opportunities to obtain a territory to be limited and juvenile dispersal to be delayed (Arnold and Owens 1998).

In general, the behavior of each species at the nest is similar to the behavior of its closest mainland relatives. Females of all but one species build the nest and incubate alone; females of the monarch flycatcher, the 'Elepaio, share incubation duties with the male (van Riper 1995; Wakelee 1996; Snetsinger et al. 1999; Clarkson and Laniawe 2000; Banko, Ball, et al. 2002; Pratt 2005). Males of all species contribute by feeding the female on the nest, defending the nest, and feeding the young, and such help can be crucial to nesting success (Kepler et al. 1996, Tweed et al. 2006). After the breeding season, native species may join in large, loosely structured, mixed-species flocks (Hart 2003).

Influences on Life History Traits

Central to life history theory is a generally observed trade-off between survival and fecundity across a wide range of taxa (e.g., Martin 1995). Most hypotheses suppose that variation in life history is due to variation among organisms in the degree of food limitation, nest predation, or adult mortality (e.g., Moreau 1944, Lack 1947, Skutch 1949). Energy investment is thought to be allocated toward points in the life history that contribute most to long-term reproductive success, namely those stages with the highest survival potential. For example, if the level of predation on nests is high, allocation should favor adaptations that enhance adult survival at the expense of offspring production. In this way, ecological conditions are thought to influence the evolution of clutch size, developmental periods, adult and juvenile survival rates, nest site selection, parental behavior, and other life history traits (Stearns 1992, Roff 2002).

Many life history traits are known to be related to body size and phylogenetic history. Survival rates are positively correlated with body size across a variety of vertebrates, including birds, whereas larger species of animals tend to have lower fecundity than smaller ones (Stearns 1992, Roff 2002). Phylogeny has been shown to be responsible for a large proportion of the variation in life history traits among birds (Stearns 1992, Roff 2002). Our ability to examine the influence of allometry and phylogeny on life history traits of Hawaiian forest birds was hindered because the range of variation within our sample was relatively small: with the exception of the hawk and crow, all are small Passeriform birds weighing between 10.3 and 49.1 g. Likewise, there was little variation

in clutch size (a standard measure of fecundity in birds; see Appendix 8.1). Moreover, the majority (73%) belong to a single subfamily, the Drepanidinae or Hawaiian honeycreepers. Data exist for only one or two representatives of each of the other six families, so our sample is not phylogenetically balanced. These constraints have led us to defer the task of a full comparative analysis that includes controlling for allometry and phylogenetic contrasts. In the present chapter we focus on summarizing available demographic data and comparing Hawaiian forest birds with their nearest continental relatives, which with few exceptions are temperate continental forms.

Life History Evolution of Birds in Tropical and Insular Habitats

Tropical birds generally share a suite of life history traits that includes relatively small clutch size, long development time (incubation, nestling, and postfledging dependent periods), high rates of nest loss to predators, and high annual rates of adult survival relative to temperate birds (Stutchbury and Morton 2001). The suite of tropical adaptations is thought to be the result of life history trade-offs in response to the more constant environmental conditions that prevail in the tropics, principally a benign tropical climate, a relatively low level of seasonality of resources, and short day length.

In addition to tropical adaptations, the Hawaiian avifauna has simultaneously evolved in an insular environment (Chapter 1). Because of their relative isolation and small size, islands have a low level of species richness, resulting in a greater availability of ecological niches. They also have a scarcity of predators, parasites, and interspecific ecological competitors. As a result, island species often exist at high population densities and experience heightened intraspecific competition (Grant 1998a). The most common evolutionary responses of both plants and animals to the altered pressures from insular environments are a reduction in dispersal ability (Grant 1998a), reduced size in large mammals and birds and increased size in small ones (Lomolino 1985, Clegg and Owens 2002), niche shifts in foraging and nesting strata (Whittaker 1998), greater vulnerability to introduced predators and parasites (King 1985), and reduced metabolic rates (McNab 2002). Differences between mainland and island forms may arise due to founder effects, genetic drift, bottlenecks, phenotypic response to habitat, and natural selection and adaptation (Whittaker 1998; this volume, Chapter 10).

Methods Used in This Review

To examine patterns of life history traits and demographic rates among Hawaiian forest birds, we compiled all available data on Hawaiian forest bird demography and life history traits found in peer-reviewed journals, books, technical reports, and unpublished field notes of ornithologists. For Hawaiian birds, sources were located by searching electronic databases (Cambridge Scientific Abstracts, BioSIS, PubMed, and the ProQuest Dissertations and Theses Database) using the search terms "Hawaii," "birds," and scientific names. We also consulted the Birds of North America accounts (Poole and Gill 1992–2002) and made direct contact with researchers in Hawai'i and elsewhere. For non-Hawaiian species, we consulted the Birds of North America species accounts, which are designed to be a comprehensive review of the life history of each species. Sources were excluded if they did not meet minimum standards for sample size or if they referred only to captive or captive-bred birds.

One difficulty in combining data from multiple studies is that data collection or analysis methods differ among studies, and each has its own biases and limitations. Although we cannot present the details of field methods for each of the more than 100 studies referred to here, we have dealt with this problem in two ways. First, we have included in tables information to indicate how data were gathered and discuss in the text how varying methodology may affect the results. For example, in Table 8.1 we indicate which survival rate estimates were obtained using mark-recapture, radiotelemetry, or resight (enumeration) methods. Second, where raw data were available, we consulted the original source, recalculated parameter values as necessary to obtain consistency across studies, and presented details on calculations in the relevant section. If values differed from those reported in the original source, we explained the difference in a footnote to the relevant table. When calculating average parameter values for groups of species, we first calculated a weighted average for species that were represented by multiple studies and then calculated separate averages for data gathered by distinctly different methods, for example, mark-recapture versus enumeration estimates of survival rates or fractional versus Mayfield estimates of nest success.

AVIAN LIFE HISTORIES AND DEMOGRAPHY IN HAWAI'I

Adult and Juvenile Survival

To examine patterns of survival rates among Hawaiian forest birds, we compiled data from 26 studies of 16 species and subspecies, most from high-elevation populations outside the range of introduced avian diseases. The adult survival rate refers to the annual survival of birds ≥1 year of age, with the exception of 'Alalā, Laysan Finches, and 'Ākepa. Separate estimates are presented for second-year (SY) 'Alalā and SY and third-year (TY) Laysan Finches and 'Ākepa. These species were treated differently because they have distinct subadult plumages and their survival rates differed significantly among age classes (Duckworth et al. 1992, Lepson and Freed 1995, McClung

2005). The studies used two general approaches, a fractional survival rate (enumeration) and capture-recapture model-based estimators. The fractional survival rate is the proportion of birds marked in one year or season that are recaptured or resighted in a subsequent year. The method assumes that any bird not resighted has died. In contrast, capture-recapture model-based estimators explicitly account for the possibility that a bird might be alive and in the study area but might not be resighted in a particular sample period (Pollock et al. 1990). Neither method can distinguish between departure from the study area and mortality, so they are said to measure "apparent survival."

The juvenile survival rate is the proportion of recently fledged birds surviving to 1 year of age, and studies of juvenile survival also used enumeration and capture-recapture models. Model-based estimates of juvenile survival rate may be higher than enumerative estimates because model-based estimates are based on a sample of hatch-year birds that are old enough to be mobile and captured in mist nets, whereas some enumeration estimates tracked birds banded as nestlings and thus included birds in the vulnerable postfledging dependent period.

Annual adult survival rates of Hawaiian forest passerines averaged 0.78 (± 0.03 SE, n = 16 species and subspecies; Table 8.1). A few Hawaiian forest birds have remarkably high annual survival rates, such as Hawai'i Creepers (0.88) and 'Ākohekohe (0.95). The apparently low annual adult survival rate of 'I'iwi (0.55 and 0.60 in two studies; see Table 8.1) may be an artifact of relatively high levels of dispersal and mobility in this species, which would bias survival estimates low. Alternatively, it might reflect their extreme vulnerability to avian malaria combined with their high level of mobility (Atkinson et al. 1995).

Survival rates of birds have long been believed to be higher in tropical habitats than in temperate latitudes. Although a study by Karr et al. (1990) called this observation into question, the general pattern has held up in a wider review (Sandercock et al. 2000). Hawaiian forest bird survival rates appear at least as high as those of other tropical island passerines studied to date: annual adult survival rates of seven Puerto Rican species averaged 0.68 (0.51–0.79; Faaborg and Arendt 1995), and those of 17 Trinidadian species averaged 0.65 (0.45–0.85; Johnston et al. 1997). By comparison, estimates from similar capture-recapture studies in the temperate zone were markedly lower (reviewed in Sandercock et al. 2000). Estimates of survival for mainland cardueline finches were about 38–47% per annum (Badyaev 1997), much lower than for the drepanidines.

Traditional explanations for high survival rates on islands point to lower predation rates (e.g., Karr et al. 1990), but this is not certain in present-day Hawai'i, where introduced cats, rats, and mongooses are all known to prey on adult birds (Chapter 11 and references therein). Hawaiian birds have a small clutch size relative to temperate birds (discussed later), which, coupled with the generally observed trade-off between fecundity and survival, might allow birds to survive over more breeding seasons. In addition, nonmigratory tropical or insular species might outlive their temperate counterparts because they do not pay the considerable price of either annual migration or residency in a harsh winter climate (e.g., Sillett and Holmes 2002). Interestingly, tropical and island birds have reduced metabolic rates, and some authors speculate that these lower rates would result in higher survival rates (McNab 2002, Wikelski et al. 2003).

The juvenile survival rates of Hawaiian birds averaged 0.32 (± 0.03 SE, n = 13; see Table 8.1). On average, the survival rates of juveniles were 42% those of adult birds (range 18.1–57.1%, n = 13 species, based on weighted averages of enumeration and model-based estimates combined for each species; see Table 8.1). Although there are

Table 8.1. Annual adult and juvenile survival rates of 16 species of Hawaiian forest birds estimated from mark-recapture, radiotelemetry, and resighting studies

Species	Adult Survival Rate[a]			Juvenile Survival Rate[b]			Method[d]	Source[e]	Comments
	Est.	SE	n[c]	Est.	SE	n[c]			
'Io (Hawaiian Hawk)	0.88	0.08	17	0.27	0.11	15	ER	1	Native-dominated habitat
	0.94	0.04	17	0.82	0.12	11	ER	1	Exotic-dominated habitat
'Alalā (Hawaiian Crow)	0.92	0.08[f]	12	—	—	—	E	2	
	0.84	0.10[f]	13	—	—	—	E	3	
	0.90[g]	0.06[f]	5 (27)	0.50	0.20[f]	6	E	4	Central Kona
	0.70[g]	0.09[f]	11 (29)	0.43	0.13[f]	14	E	4	Honaunau and Hualālai
	1.00	0.0	6	—	—	—	E	4	SY (age 1 to age 2)
O'ahu 'Elepaio	0.83[h]	0.09[f]	18	—	—	—	E	5	Females with predator control
	0.50[h]	0.13[f]	14	—	—	—	E	5	Females without predator control
	0.78[h]	0.04[f]	128	—	—	—	E	5	Males
Hawai'i 'Elepaio	0.82	0.04	129	0.25	0.10	30	M	6	
	0.90	0.09[f]	11	0.29	0.11[f]	17	E	7	
	0.79	0.05	—[i]	—	—	—	E	8	Females
	0.86	0.04	—[i]	—	—	—	E	8	Males
	—	—		0.33	0.08[f]	33	E	8	
	0.82	0.08[f]	27 (27)	—	—	—	E	9	
'Ōma'o	0.77	0.07	430	—	—	—	M	6	
	0.70[j]	0.08	28	—	—	—	E	10	Males
	0.38[j]	0.24	10	—	—	—	E	10	Females
Puaiohi	0.66	0.11[j]	153	0.40[j]	0.09	137	M	11	
	0.73	0.13[f]	11	0.25	0.07[f]	36	E	12	
Laysan Finch	—	—		0.04	0.01[f]	284	E	13	
	—	—		0.60	—	214	M	14	
	0.70	0.07[f]	48	—	—	—	M	14	SY males
	0.85	0.04[f]	65	—	—	—	M	14	ASY males
	0.74	0.08[f]	33	—	—	—	M	14	SY + ASY females
Palila	0.63	0.05	665	0.36	0.08	319	M	15	
Hawai'i 'Amakihi	0.71	0.04	148	0.59	0.20	21	M	6	
	0.85[k]	0.05[f]	70 (84)	—	—	—	E	16	
	0.75	0.03	110	—	—	—	M	17	
	—	—		0.16	0.04[f]	71	E	17	

	Adult survival rate[a]	SE	n	Juvenile survival rate[b]	SE	n			Comments
'Akiapōlā'au	0.71	0.03	20	—	—	—	M	18	
Hawai'i Creeper	0.88	0.03	43	—	—	—	M	19	
	—	—	—	0.32	0.09[f]	28	E	19	
	0.73	0.12	49	0.40	0.11[f]	20	M	20	
	0.74	—	—	0.22	—	—	E	20	
Maui 'Alauahio	0.95[l]	0.03[f]	15 (41)	0.45	—	—	M	21	
'Ākepa	0.70	0.27	61	—	—	—	E	22	
	—	—	—	0.23	0.08[f]	30	M	20	
	—	—	—	0.43	0.10	57	E	20	
	0.82[m]	0.04	36	—	—	—	M	23	ATY males
	0.57[m]	0.12	23	—	—	—	M	23	TY males
	0.85[m]	0.13	18	—	—	—	M	23	SY males
	0.80[m]	0.04	46	—	—	—	M	23	AHY females
	0.73	0.06[f]	64	—	—	—	E	24	
'I'iwi	0.55	0.12	123	0.09	0.05	212	M	25	
	0.60	0.05	134	0.24	0.06	44	M	6	
'Ākohekohe	0.95	0.10	36	—	—	—	M	26	
'Apapane	0.72	0.11	201	0.13	0.07	228	M	25	
Enumeration mean[n]	0.80	0.03	13	0.31	0.05	9			
Model-based mean[n]	0.75	0.03	11	0.36	0.07	8			
Overall mean[n]	0.78	0.03	16	0.32	0.03	13			

Notes: Scientific names are provided in Appendix 8.1.

[a]Adult survival rate is the annual survival rate of birds ≥1 year of age unless otherwise noted in the "Comments" column. Male and female survival rates, survival rates in different habitats, and survival rates with and without predator control are presented separately only if they were significantly different. "—" indicates that no comparable data were provided in the original source for that value. HY, hatch year (<1 year); SY, second year (age 1 to age 2); TY, third year (age 2 to age 3); AHY, after hatch year (>1 year); ASY, after second year (>2 years); ATY, after third year (>3 years).

[b]Juvenile survival rate is the annual survival rate of birds from fledging (in the case of enumeration studies) or first capture as hatch-year birds (in the case of mist netting studies) to 1 year of age. Where estimates of survival for other subadult classes are available and significantly different from that for adults (e.g., yearling 'Alalā, subadult 'Ākepa), they are listed in the "Adult Survival Rate" column with identifying notes under "Comments."

continued

c Sample size (n) is the number of birds included in the study. Studies that used enumeration methods sometimes presented n as the number of individual birds followed and sometimes presented it as the number of total bird-years (one bird observed over three sample periods [two capture intervals] would equal two bird-years). We present n as the number of individuals followed and give the number of bird-years in parentheses when this information was available. Sample size for averages across species (bottom three rows) is the number of species.

d Method: M, model-based capture-recapture estimates (Pollock et al. 1990). E, enumerative (fractional) estimates, the proportion of birds marked in one year that are recaptured or resighted the following year. Birds not resighted the following year but resighted two years following their marking are assumed to have survived at a constant rate over the two years (e.g., annual survival rate = the square root of the proportion that survived for two years). R, radiotelemetry. See text for an explanation of how the methods used may affect survival rate estimates, especially for juveniles.

e Sources: (1) Klavitter et al. 2003; (2) Griffin 1985; (3) Banko, Ball, et al. 2002; (4) W. Banko and P. Banko 1980, Duckworth et al. 1992; (5) VanderWerf and Smith 2002; (6) USGS, unpubl. data; (7) van Riper 1995; (8) VanderWerf 1998b, 2004; (9) Sarr et al. 1998: Appendix I; (10) Wakelee 1996; (11) Ralph and Fancy 1994c; (12) Snetsinger et al. 2005; (13) Morin 1992a; (14) McClung 2005: Tables 3.2 and 3.3; (15) Lindsey et al. 1995; (16) van Riper 1987; (17) Kilpatrick 2003, Kilpatrick et al. 2006; (18) Ralph and Fancy 1996; (19) Woodworth et al. 2001; (20) Ralph and Fancy 1994a; (21) J. Lepson in Lepson and Woodworth 2002; (22) H. Baker and P. Baker 2000; (23) Lepson and Freed 1995; (24) Hart 2001; (25) Ralph and Fancy 1995; (26) Simon et al. 2001.

f For enumeration estimates, if variance was not given by the original source, we calculated the standard error according to binomial probabilities (Zar 1996).

g Duckworth et al. (1992) reported both observer-weighted and time-weighted estimates for 'Alalā; we present the time-weighted estimates here for consistency with other studies. Adult survival rates of 'Alalā were significantly different in two areas (the Central Koha versus the Honaunau and Hualāli areas), and both are presented here.

h The survival rate of female O'ahu 'Elepaio in areas with predator control was significantly greater than that without predator control. Male survival rates were not significantly different with or without predator control and were combined as a weighted average.

i Survival rates of Hawai'i 'Elepaio from VanderWerf (2004) were based on a mean of 63 birds per year over six years, although it was not clear whether this refers to capture periods or intervals, and separate sample sizes for males and females were not reported.

j Wakelee's (1996) survival estimates for 'Ōma'o were generated by calculating survival probabilities over bimonthly intervals for each cohort and generating a survivorship curve. Ralph and Fancy's (1994c) estimate of the juvenile survival rate of 'Ōma'o includes second-year birds retaining some juvenile plumage. The standard error of the adult survival rate was reported as 0.08 in the abstract and 0.11 in the text; we used the value reported in the text.

k Van Riper (1987: 97) reported that 10/14 birds survived to two years after banding, and 48/56 survived one year; 0.85 is a weighted average of these two values, assuming constant survival of the original 14 birds over the two-year interval as described in footnote d.

l H. Baker and P. Baker (2000) reported that 13/15 birds (87%) survived year one, 13/13 birds survived year two, and 13/13 birds survived year three, for a survival rate of 95% over 41 bird-years.

m Lepson and Freed (1995: Table 2) presented multiple survival rate estimates for several age class and sex combinations; we present the estimates that the authors preferred (Lepson and Freed 1995: 409). Although adult male and female survival rates did not differ significantly, we present them separately in this table for several reasons: (1) adult male survival rate differed from that of third-year (TY) males; (2) adult male and female survival rates were estimated using different models; and (3) no estimates for the two sexes combined were presented.

n For those species for which multiple estimates were available, we calculated a weighted average for each species before calculating overall means so that each species is included only once in the analysis. When calculating weighted averages, we used the number of individual birds (which was available for almost all studies) rather than the number of bird-years. n = number of species.

good reasons to expect juvenile mortality to outstrip that of adults, it is also true that the dispersal of young is greater than that of adults (Greenwood and Harvey 1982), contributing to the difference in apparent survival. The juvenile survival rates of passerines are poorly known, especially for tropical birds, but are believed to range from 0.25 to 0.40 (Anders et al. 1997). Because Hawaiian birds are nonmigratory, young may have the opportunity to remain in or near their natal territory for extended periods, which might enhance their probability of survival (Karr et al. 1990, Badyaev 1997).

Age at First Breeding

The age at which a species initiates breeding is an important component of its life history strategy, with implications for both individual fitness and population growth rates. Of 16 species of Hawaiian forest birds studied, all but the hawk and crow are apparently capable of breeding at 1 year of age (as SY birds; Table 8.2). However, only a handful regularly do so; most do not breed routinely until their third year, as after-second-year (ASY) or older birds. Species that delay breeding until their third year are typically strongly sexually dichromatic (Fisher's exact test, P = 0.256; see Table 8.2; exceptions are the monochromatic 'Io, 'Alalā, and 'Ākohekohe), and all but one, the 'Alalā, exhibit delayed plumage maturation (in maturing 'Alalā, the colored mouth and iris darken with age). VanderWerf (2001c) provides an excellent review of delayed plumage maturation and its potential demographic effects in Hawaiian forest birds.

In general, proportionately fewer SY birds breed than do older birds, and this has been found in the 'Elepaio, Laysan Finch, Palila, 'Ākepa, and 'Ākohekohe (Morin 1992a; Lepson and Freed 1997; Berlin and VanGelder 1999; Banko, Johnson, et al. 2002; VanderWerf 2004). Moreover, females of some species are more likely to

breed as SY birds than are males (Vander-Werf 2004), suggesting that breeding opportunities for males are more limited than for females, perhaps as a result of male-biased sex ratios (discussed next). Finally, in at least one species, Maui 'Alauahio, delayed dispersal and breeding is associated with cooperative breeding (H. Baker and P. Baker 2000).

Sex Ratios

Sex ratios influence breeding opportunities and the evolution of mating systems (Emlen and Oring 1977). As in passerine birds in general (Breitwisch 1989), sex ratios in native Hawaiian forest birds are typically male-biased, ranging from 1.1:1 in a population of mist-netted 'Ākepa to 2.5:1 for mist-netted 'Ōma'o (see Table 8.2). Only the 'Akikiki has been reported to have a female-biased sex ratio (Foster et al. 2000). Based on museum specimens, Banko and Banko (Chapter 7) concluded that specialist honeycreeper species have a lower proportion of females than do generalist honeycreepers, and they propose that this may be due to the greater reproductive costs (egg weight as a proportion of body weight and incubation time per gram of clutch weight) to specialists compared with generalists.

One potential mechanism for male-biased sex ratios in passerine birds is the differential survival of males and females. Male-biased sex ratios might occur if females were more vulnerable to predation, disease, or starvation for physiological or behavioral reasons, for example, while carrying out incubation and brooding duties. Sex differences in the survival rates of males and females have been found in 'Elepaio and 'Ōma'o (Wakelee 1996, VanderWerf and Smith 2002, VanderWerf 2004) but not in 'Ākepa, Palila, or Hawai'i 'Amakihi (Lepson and Freed 1995, Lindsey et al. 1995, Kilpatrick et al. 2006). Although the sex ratios of eggs and nestlings of passerine birds are generally near unity (Breitwisch

Table 8.2. Sexual dichromatism, delayed plumage maturation, typical age at first breeding, sex ratios, nest sanitation behavior, and drepanidine odor of 28 species of Hawaiian forest birds

Species	Sexual Dichromatism[a]	Delayed Plumage Maturation[b]	Typical Age at First Breeding (earliest)	Sex Ratio M:F (n, method)[c]	Nest Sanitation	Odor[d]	Source
'Io (Hawaiian Hawk)	M	2-yr delay in both sexes	3–4 yrs	—	Clean	No	Clarkson and Laniawe 2000
'Alalā (Hawaiian Crow)	M	None	2–4 yr	—	Clean	No	Banko, Ball, et al. 2002
Hawai'i 'Elepaio	D	2-yr delay in both sexes	2–3 yr (1 yr)	Male biased[e]	Clean	No	van Riper 1995; Vander-Werf 1998a, 2001c, 2004
Millerbird	M	Unknown	—	—	—	No	Conant and Morin 2001
'Ōma'o	M	1-yr delay in both sexes	1 yr	2.5:1 (117, M) 3:1 (24, E)	Clean	No	Wakelee 1996, Wakelee and Fancy 1999
Puaiohi	M	1-yr delay in both sexes	1 yr	1.5:1 (15, E)	Clean	No	Snetsinger et al. 1999
Laysan Finch	D	1-yr delay in males	2 yr (1 yr)	—	Feces accumulate	No	Morin 1992a, Morin and Conant 2002
Palila	D	1-yr delay in males	Females: 1 yr Males: 2 yr	1.8:1 (594, E)	Feces accumulate	Yes	Lindsey et al. 1995; Banko, Johnson, et al. 2002
Greater Koa-Finch	D	Unknown	—	6.6:1	—	Yes	Olson 1999a
Kona Grosbeak	M	Unknown	—	1.6:1 (49, M)	—	Yes	Olson 1999a
Maui Parrotbill	D	1-yr delay in males	—	2:1[f]	Feces accumulate	Yes	Simon et al. 2000
Hawai'i 'Amakihi	D	1-yr delay in males	2 yr (1 yr)	1.36:1 (123, C)	Clean	Yes	van Riper 1987, Lindsey et al. 1998
Kaua'i 'Amakihi	D	Unknown	—	—	Clean	Yes	Eddinger 1970, Lindsey et al. 1998
'Anianiau	D	Unknown	—	1.3:1 (48, C)	Clean	Yes	Eddinger 1970
Lesser 'Akialoa	D	Unknown	—	1.55:1 (79, C)	—	Yes	Lepson and Johnston 2000
Kaua'i Greater 'Akialoa	D	Unknown	—	1.6:1 (73, C)	—	Yes	Lepson and Johnston 2000
'Akiapōlā'au	D	1-yr (possibly 2-yr) delay in males	2–3 yr (1 yr)	—	—	Yes	Pratt, Fancy, et al. 2001

'Akikiki	M	Unknown	—	1:1.35 (61, M)	—	Yes	Foster et al. 2000
Hawai'i Creeper	M	None	1 yr	1.41:1 (64, M)	Clean	Yes	Lepson and Woodworth 2002
O'ahu 'Alauahio	D	2-yr delay in both sexes	2 yr	1:1 (101, C)	—	Yes	P. Baker and H. Baker 2000
Maui 'Alauahio	D	2-yr delay in both sexes	2 yr	1:1[g]	Clean	No	H. Baker and P. Baker 2000
'Akeke'e	D	Unknown	—	1.5:1 (32, C)	—	Yes	Lepson and Pratt 1997
'Ākepa	D	2-yr delay in males only	2–3 yr (1 yr)	1.2:1 (106, M) / 1.14:1 (—, M) / 1.16:1 (149, M)	Feces accumulate	Yes	Lepson and Freed 1995, 1997; Hart 2001
'I'iwi	M	1-yr delay in both sexes	1 yr	1.44:1 (264, M) / 2.3:1 (—, C)[h] / 1.4:1 (39, C)	—	Yes	Fancy and Ralph 1998
'Ākohekohe	M	1-yr delay in both sexes	2 yr (1 yr)	1.25:1 (78, M)	Clean	Yes	Berlin and VanGelder 1999, VanGelder and Smith 2001
'Apapane	M	1-yr delay in both sexes	1 yr	1.7:1 (504, M) / 1.5:1 (182, M) / 1.3:1 (119, C)	Clean	Yes	Fancy and Ralph 1997, Nielsen 2000
Po'o-uli	M	1-yr delay in both sexes	—	—	Clean	No	Pratt, Kepler, et al. 1997

Notes: "—" indicates that no data were provided for that value.

[a]M, sexually monochromatic plumage; D, sexually dichromatic plumage.

[b]From VanderWerf 2001b.

[c]Sources of data for determining sex ratios include specimens in collections, either museum specimens or shooting records (C), birds captured by mist netting (M), and birds hatched from eggs collected in the wild (E). Museum samples may be male-biased due to the more conspicuous coloration and/or calling behavior of males (e.g., the Greater Koa-Finch; Olson 1999a). Mist-netting capture data may be male-biased due to the increased activity of males relative to females, especially during the breeding season. Sex ratios are of adult birds unless otherwise noted.

[d]Presence or absence of drepanine odor follows Pratt (1992).

[e]The population is slightly male-biased (VanderWerf 1998a).

[f]The sex ratio is estimated to be 1:2 [females to males] (Simon et al. 1997); no other information is available.

[g]The sex ratio is estimated to be 1:1 (H. Baker and P. Baker 2000); no other information is available.

[h]The sex ratio of adult and juvenile birds combined (Fancy and Ralph 1998).

1989), the sex ratio of 24 'Ōma'o eggs collected for captive propagation was 3:1 males (Wakelee 1996), a striking observation for a sexually monomorphic species. Finally, for behavioral reasons, the capture rates of males may be higher than those of females. Sex ratios can be adjusted for biases in capture rates, but this has yet to be done for Hawaiian birds.

Productivity

LENGTH AND TIMING OF BREEDING AND MOLT

Breeding season length is key to productivity, for species with longer breeding seasons have a greater opportunity to replace failed clutches and double-brood. To examine the length and timing of breeding and molting for Hawaiian birds, we consulted the Birds of North America species accounts (Poole and Gill 1992–2002), which compiled data from multiple sources (egg dates, mist-netting data, and fledging dates). Because these accounts combine data from multiple populations, study areas, and years, they may overestimate the total length of breeding and molt and the degree of molt-breeding overlap that occurs for a particular individual or population. In addition, the breeding seasons of poorly studied species may appear to be shorter due to small sample sizes. Despite these caveats, we believe these data provide insight into the timing and variability of breeding seasons of Hawaiian forest birds.

For most Hawaiian birds, peak breeding occurs in February–May, with variation among taxa related to diet (Fig. 8.1). Nectarivores typically begin breeding as early as November, whereas arthropod eaters and fruit and seed eaters typically begin nesting later in the season and continue into August. Hawaiian forest birds, like their mainland tropical counterparts, enjoy prolonged breeding seasons of about

5.5 months (mean 165 days, range 75–365, n = 25 spp.; see Fig. 8.1). This period is considerably longer than temperate passerines' breeding season of 3–4 months (Ricklefs 1966, 1969). However, there is substantial variation among taxa. The 'Ākepa, a hole-nesting, specialist insectivore, nests for only about three months relatively late in the season (Lepson and Freed 1995). At the other end of the spectrum is another specialist insectivore, the 'Akiapōlā'au, which fledges its single young at any time of year (Pratt, Fancy, et al. 2001). In wet montane forests on Hawai'i Island, insectivorous birds breed over a more prolonged period than do nectarivorous birds (Ralph and Fancy 1994c). We found that breeding season lengths were more variable among the insectivores, ranging from 92 days to year-round, as compared to those of nectarivores, all of which had breeding seasons of roughly 180 days (see Fig. 8.1).

Breeding activity in Hawaiian forest birds appears to coincide with food availability, but the exact nature of that timing varies among species. Peak Palila breeding activity generally coincides with periods of greatest availability of their primary food source, māmane (Sophora chrysophylla) pods (van Riper 1980b; Pratt, Banko, et al. 1997; this volume, Chapter 7). Similarly, in subalpine māmane forest, peak Hawai'i 'Amakihi breeding coincides with māmane flowering, with fine-tuned adjustments based on the year (van Riper 1987). In the nectarivorous 'Ākohekohe, nesting peaks when 'ōhi'a-lehua (Metrosideros polymorpha) are in flower (Berlin et al. 2001). 'Ākepa, a specialist insectivore, apparently times its breeding so that fledgling independence coincides with peak arthropod availability in late summer (Fretz 2000). Alternative hypotheses for the timing of breeding, for instance, that breeding is timed to avoid periods of peak predation or to allow young to complete molt before fall migration, have not been examined in Hawaiian

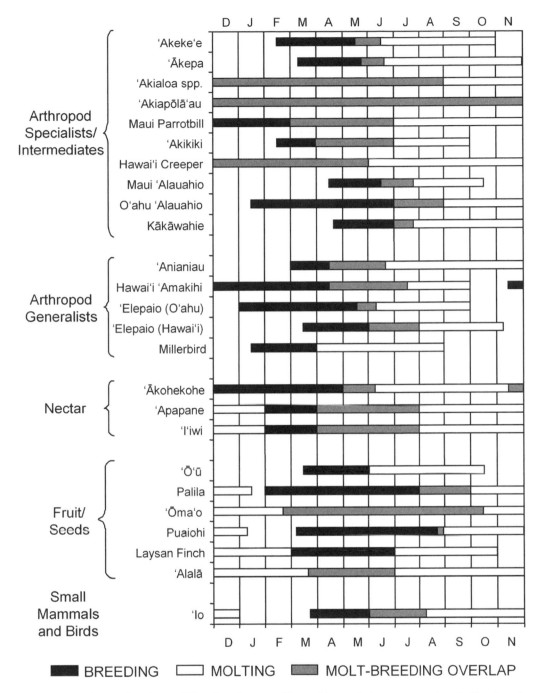

Figure 8.1. Breeding and molting of Hawaiian forest birds by month. The breeding season represents peak and off-peak breeding as determined from egg dates, mist-netting data, and fledge dates compiled in Birds of North America species accounts (Poole and Gill 1992–2002), and original sources can be found therein. The molting period represents peak and off-peak molt in both adults and juveniles, because only a few accounts distinguished between them. The molt-breeding overlap is limited to molt of flight feathers. Categories of foraging guild and specialization follow Banko and Banko (Chapter 7).

forest birds, although molt constraints are not likely to be a factor in this non-migratory avifauna.

Molt occurs over an extended period, generally July–December, during the nonbreeding season (see Fig. 8.1). Both breeding and molting birds may be present simultaneously over a period of months. However, individual Hawaiian forest birds generally partition these activities, so that only a small percentage of individuals (3.2–4.9%) is in breeding condition while simultaneously molting body or flight feathers (Ralph and Fancy 1994c, Wood-worth et al. 2001), which is consistent with the minimal molt-breeding overlap found in other tropical birds (e.g., 3.1–8.5% of individuals; Foster 1975).

CLUTCH SIZE

The fact that the clutch sizes of tropical birds are generally smaller than those of north temperate birds is well known and has formed the centerpiece of many theories about evolution of their life histories (Moreau 1944, Lack 1947, Skutch 1949). Clutch sizes also tend to be smaller on oceanic islands than on the nearest mainland in the temperate zone (Grant 1998a). However, this pattern does not hold in the tropics, where clutch size is about the same on islands as on the adjacent mainland (Lack 1976). Selective forces may be acting similarly on tropical islands and continents, or different pressures may be acting in both environments to reach the same outcome.

Consistent with these observations, the clutch sizes of Hawaiian forest birds are uniformly small across taxonomic groups, with mean clutch sizes of 1.0–3.2 (Table 8.3; Appendix 8.1). On average, Hawaiian species have clutch sizes that are one-third to two-thirds the size of north temperate mainland species in the same family (see Table 8.3) and similar to those of tropical passerines (Skutch 1985). The differences in clutch size in thrushes and finches

were significant (Mann-Whitney U-test, $P < 0.01$ and $P = 0.05$ for thrushes and finches, respectively), while for other families with a single Hawaiian representative, the clutch size of the island species was smaller than that of any representative of the north temperate group (see Table 8.3). Among the Hawaiian honeycreepers, generalists have larger clutch sizes for their body weight than do specialists (Chapter 7), and clutch size appears to vary with foraging guild: the mean clutch size of the arthropod-eating guild is 1.85 ± 0.2 SE compared with 2.15 ± 0.3 SE for nectarivores and 2.67 ± 0.3 SE for fruit and seed eaters (the small sample sizes of nectarivores and fruit and seed eaters precluded statistical analysis). Despite predictions that cavity nesters might have larger clutch sizes than open-nesting birds (Martin and Li 1992), the Hawai'i 'Ākepa, the only extant obligate cavity nester among the Hawaiian forest birds, also has a small clutch size (1.7). The clutch sizes of some Hawaiian forest birds vary among years depending on prevailing environmental conditions, showing a degree of phenotypic plasticity, such as in the Hawai'i 'Amakihi (van Riper 1987) and Laysan Finch (Morin 1992a).

DEVELOPMENT PERIOD

Hawaiian forest birds have, on average, longer development times than their north temperate relatives (see Table 8.3; Appendixes 8.1 and 8.2). Hawaiian honeycreepers incubate their eggs for 13–17 days (mean = 15.1 ± 1.2, n = 14), compared with 11.7–14 days (12.9 ± 0.7, n = 15) for north temperate cardueline finches (Mann-Whitney U-test $P < 0.01$; see Table 8.3). Similarly, nestling honeycreepers remain in the nest 16.1–25.3 days (20.1 ± 2.8, n = 15), longer than the usual 11.5–20 days (15.2 ± 3.3, n = 13) for north temperate finches ($P < 0.05$; see Table 8.3). Incubation and nestling periods are also significantly longer for Hawaiian *Myadestes* than

Table 8.3. Clutch sizes, incubation periods, and nestling periods for Hawaiian forest birds and their north temperate mainland relatives

Family	Mean Clutch Size (n, range)		Incubation Period in Days (n, range)		Nestling Period in Days (n, range)	
	Hawaiian	North Temperate Mainland	Hawaiian	North Temperate Mainland	Hawaiian	North Temperate Mainland
Accipitridae	1.0 (1, 1.0)	2.8 (8, 2.0–3.5)	38.0 (1, 38.0)	32.8 (8, 29.5–36.5)	61.0 (1, 61.0)	41.5 (7, 32.5–50.2)
Corvidae	2.8 (1, 2.8)	4.6 (7, 4.0–5.3)	20.0 (1, 20.0)	19.1 (5, 17.5–21.7)	42.0 (1, 42.0)	35.5 (5, 31.7–38.5)
Turdidae	2.0 (2, 2.0)	3.6 (9, 3.2–4.0)	14.8 (2, 13.5–16.0)	12.3 (9, 12.0–13.0)	18.7 (2, 18.3–19.0)	12.3 (9, 11.0–14.0)
Fringillidae	2.2 (19, 1.0–3.2)	4.1 (16, 3.0–5.2)	15.1 (14, 13.0–16.8)	12.9 (15, 11.7–14.0)	20.1 (15, 16.1–25.3)	15.2 (13, 11.5–20.0)

Notes: The mean, number of species (n), and range of values are shown. Data for individual species are in Appendixes 8.1 and 8.2.

for north temperate continental thrushes (Mann-Whitney U-test P = 0.05 for both; see Table 8.3). Overall development times (incubation + nestling periods) are on average approximately 7 days longer for the Hawaiian than the temperate mainland finches, 7 days longer for the corvid, 9 days longer for the thrushes, and an impressive 25 days longer for the hawk.

These findings are consistent with observations on tropical birds in general, which have long development times relative to temperate birds (Ricklefs 1969, Stutchbury and Morton 2001). Longer development times might have evolved in prehistoric Hawai'i, where the level of predation was relatively low (discussed later) and species would have been under less intense selection to hasten development. However, this would not explain why tropical mainland birds, with their higher rates of nest predation, also have long development periods. On islands, we also expect greater intraspecific competition for food, especially in a tropical environment. Reduced food availability in the face of stiff intraspecific competition might have led to delayed development.

Are longer development periods associated with slower growth rates in Hawaiian birds? Van Riper (1978) found that the growth rates of Palila and Hawai'i 'Amakihi were lower than the average growth rate for 21 Fringillids and less than for all passerines except the Formicariidae, Tyrannidae, and Corvidae (reviewed by Ricklefs 1968). It is not clear, however, whether slowed growth rates are typical of Hawaiian birds or what the proximate and ultimate causes are. For example, there is growing appreciation that feeding rates are important components of an animal's life history strategy because they influence growth, survival, and predation rates (Martin, Scott, et al. 2000; Martin 2004). Feeding rates have been studied for only a few Hawaiian species, such as Hawai'i 'Elepaio (Vander-Werf 1998b) and Palila (Laut et al. 2003). How do egg size, metabolic rates, diet,

predation rates, and parental attendance (incubation rhythms and feeding rates) interact? The answers to these questions await further research.

The duration of parental care following fledging varies widely among passerines, and general trends have shown that post-fledging parental care periods are longer for tropical birds, those that nest at high elevations, and those that have specialized foraging maneuvers (Ricklefs 1969, Badyaev and Ghalambor 2001). This prolonged period of parental care is thought to be an important component of a bird's life history strategy, influencing the survival of young birds but reducing the probability that parent birds will be able to raise multiple broods in a season (Badyaev 1997). Overall, information on postfledging behavior and dependency is lacking for most Hawaiian birds, and it is poorly known even for relatively well-studied temperate taxa. However, among the Hawaiian forest birds, the timing of natal dispersal seems to vary depending on foraging ecology and breeding systems. Dispersal after a few weeks is seen among nectarivores and frugivores, for example, 'Ōma'o (Wakelee and Fancy 1999), Puaiohi (Snetsinger et al. 1999), 'I'iwi (Fancy and Ralph 1998), and 'Apapane (Fancy and Ralph 1997). By contrast, species that require complicated foraging maneuvers are tolerated in their natal territories for up to 10 months or more, as are 'Elepaio, an aerial forager (VanderWerf 1994, 2004), and two specialist insectivores, the Maui Parrotbill (Simon et al. 1997) and 'Akiapōlā'au (Pratt, Fancy, et al. 2001). Helpers among the cooperatively breeding Maui 'Alauahio delay dispersal up to 20 months (H. Baker and P. Baker 2000).

NESTING SUCCESS

To examine patterns of nesting success among the Hawaiian forest birds, we compiled data from 36 studies of 19 species of Hawaiian forest birds (Table 8.4). Nesting success was most frequently presented as the proportion of nests that fledged young; however, studies differed in whether they included all nests, only active nests, or only active known-fate nests. We examined the original sources for each study and calculated the fractional nest success estimate as the proportion of *active, known-fate* nests that fledged at least one young. Only studies reporting on the fates of at least five active, known-fate nests were included. We excluded nests removed for captive propagation and those that failed due to human disturbance. For this reason, some fractional rates presented here differ from those in the original sources (indicated by footnotes in Table 8.4).

Fractional methods for estimating nesting success can bias estimates upward, because nests that fail early in the nesting cycle are less likely to be included in the sample of nests found. For this reason, we also included estimates of daily and overall survival rates using the Mayfield method (Mayfield 1961, 1975). Daily survival rates also have the advantage of measuring survival over a constant unit of time, allowing us to compare relative daily predation risks across species with different nesting cycle lengths. With one exception, the estimate of nesting success calculated using the Mayfield method was less than that using the fractional method (mean difference 10.2%, range 0.017–0.269; see Table 8.4). Mayfield estimates were available from 17 studies of 11 species. When testing for the influence of various factors on nesting success, we calculated a weighted average for each species so that each was included in the analysis only once.

Under the best of circumstances, it is difficult to determine the exact cause of nest failure, and most Hawaiian species build inaccessible nests high in the tree canopy. We found that studies varied in their criteria for classifying nest failures as "unknown cause." For example, some researchers assumed that nests that failed concurrent with an unusual weather event

had failed due to weather, while others more cautiously called the cause in these cases unknown. Because studies varied in the amount of detail provided, it proved infeasible to reclassify nest failures from raw data consistently across studies. Thus we have left failure classifications as reported in the original sources.

Nesting success averaged 53.4% (13–83%) based on 44 fractional estimates for 19 species and 46.3% (19–80%) based on 17 Mayfield estimates for 11 species. Nesting success thus varied widely among species and studies, including species with remarkably low rates of nesting success ('I'iwi, 19%) and high rates of success ('Ākohekohe, 75%) (see Table 8.4).

The numbers of present-day nest losses in Hawai'i, although likely increased since the introduction of small mammals to the forests, are still not as high as they are in continental neotropical forests and are more on a par with those observed in north temperate systems. Tropical species suffer losses of about two-thirds of nests (66%, range 27–88%, n = 46 species; Skutch 1985) compared with temperate species, which lose only about half of attempted nests (25–55%; Skutch 1985, Martin 1992b, Stutchbury and Morton 2001). We interpret this to mean that the small clutch sizes and long development times of Hawaiian forest birds are not the evolutionary result of high nest predation rates. Badyaev (1997) also found that variation in the life histories of cardueline finches across an elevational gradient could not be explained by variation in nest predation rates.

CAUSES OF NEST FAILURE

For some species, a large percentage of nests apparently never receive eggs. In 28 studies of 12 species, pairs abandoned an average of 16.2% of nests before laying eggs (range 0–39%) due to severe weather, encounters with potential predators, pilfering of nest-building material by the same

Table 8.4. Nesting success of 19 species and subspecies of Hawaiian forest birds

Species	Fractional Nesting Success (n)[a]	Daily Survival Rate of Nests (n)[b]	Overall Survival Rate of Nests[c]	Proportion of Nests or Eggs Failing Due to:						Island[j]	Source[k]
				Nest Predation[d]	Failure of Eggs to Hatch[e]	Starvation of Chicks or Death in Nest[f]	Weather[g]	Abandonment of Eggs after Laying[h]	Unknown or Other Causes[i]		
'Io (Hawaiian Hawk)	0.667 (30)	0.991 (—)[l]	0.409	—	0.033	—	—	0.100	0.200[l]	HI	1
'Alalā (Hawaiian Crow)	0.534 (88)*	—	—	—	—	—	—	—	—	HI	2
	0.167 (6)	—	—	—	—	—	—	—	—	HI	3
	0.100 (10)**	—	—	—	—	—	—	—	—	HI	3
	0.56 (57)[m]	—	—	—	—	—	—	—	—	HI	4
	0.44 (16)	—	—	—	—	—	—	—	—	HI	5
	0.64 (22)	—	—	—	—	—	—	—	—	HI	6
O'ahu 'Elepaio	0.130 (53)[n]**	—	—	X[o]	X	0.019	X[o]	0.075	X	OA	7
(with predator control)	0.58 (64)[p]	—	—	X	—	—	X	—	—	OA	8
(without predator control)	0.33 (30)[p]	—	—	X	—	—	X	—	—	OA	8
Hawai'i 'Elepaio	0.800 (25)	0.995 (31)	0.801	0.080	—	—	—	—	0.120	HI	9
	0.670 (15)*	—	—	—	—	—	—	—	—	HI	9
	0.395 (119)[q]	—	—	0.395	0.143	—	0.025	—	0.042	HI	10
	0.789 (19)	—	—	0.0	0.0	—	0.210	—	—	HI	11
	0.658 (38)**	—	—	0.0	0.25[r]	—	0.079[r]	—	—	HI	11
	0.650 (123)	—	—	—	—	—	—	—	—	HI	12
'Ōma'o	0.481 (28)	0.974 (33)	0.398	0.148	—	—	—	0.074	0.296	HI	13
	0.833 (12)	0.993 (12)	0.763	—	—	—	—	0.083	0.083	HI	9
	0.421 (19)	—	—	0.474	—	—	—	—	0.105	HI	14
Puaiohi (wild)	0.702 (94)[s]	0.974 (—)	0.433	0.104	0.010	—	0.010	0.031	0.063	KA	15
Puaiohi (captive-bred, reintroduced)	0.417 (24)	0.970 (27)	0.400	0.375	0.0	0.0	0.042	0.083	0.083	KA	16

Laysan Finch	0.285 (516)**	—	—	0.064	0.120	0.080	0.008	0.041	0.376	LA	17
Palila	0.591 (22)	—	—	—	0.045	0.090	0.045	0.136	0.091	HI	18
	0.500 (12)*	—	—	—	0.125	0.040	—	0.250	—	HI	18
	0.522 (23)**	—	—	—	0.130	—	0.087	0.217	0.043	HI	18
	0.435 (85)[t]	0.968[q](81)	0.251	0.165	≥0.12	0.0	0.0	≤0.141	0.0	HI	19
	0.350 (100)**	—	—	0.256	≥0.18	0.008	0.0	≤0.382	0.0	HI	19
	0.493 (280)	—	—	—	—	—	—	—	—	HI	20
	0.377 (536)**	—	—	—	—	—	—	—	—	HI	20
	0.558 (163)	—	—	—	—	—	—	—	—	HI	21
	0.57 (7)	—	—	—	—	—	—	—	—	HI	22
Maui Parrotbill	0.500 (8)	0.976 (8)	0.420	≤0.25	0.125	—	0.125–0.375	—	—	MA	23
Hawai'i 'Amakihi	0.472 (128)	—	—	0.076	0.016	0.04	0.056	0.280	0.024	HI	24
	0.347 (282)**	—	—	0.082	0.113	0.008	0.060	0.280	0.035	HI	24
	0.575 (26)	0.976 (28)	0.454	0.115	—	—	0.115	—	0.190	HI	9
	0.385 (13)*	—	—	—	—	—	—	—	—	HI	9
	—	0.981 (119)	0.542	—	—	—	—	—	—	HI	25
Kaua'i 'Amakihi	0.810 (21)	—	—	0.048	0.032	—	0.048	0.095	0.032	KA	26
	0.810 (63)**	—	—	0.032	—	—	0.048	0.048	0.037	KA	26
'Anianiau	0.630 (27)	—	—	0.111	0.0	—	0.222	0.0	0.143	HI	26
'Akiapōlā'au	0.571 (7)[u]	—	—	0.143	0.143	0.0	—	—	—	HI	27
Hawai'i Creeper	0.300 (33)*	0.960 (34)	0.246	—	—	—	0.121	—	0.576	HI	28
Maui 'Alauahio	0.255 (47)	—	—	0.404	[v]	0.020	0.043	0.128	0.170	MA	29
'Ākepa	0.792 (53)	—	—	—	—	—	—	—	—	HI	30
	0.500 (22)	—	—	—	—	—	—	—	—	HI	31
	0.560 (9)	—	—	—	—	—	—	—	—	HI	32
'I'iwi	0.533 (15)	—	—	0.053	—	—	0.067	0.067	0.267	KA	26
	0.533 (30)**	—	—	0.100	—	—	0.067	0.067	0.233	KA	26
	0.300 (110)	0.956 (111)	0.193	0.018	—	—	0.155	—	0.527	HI	9
	0.169 (65)*	—	—	—	—	—	—	—	—	HI	9
	0.480 (152)	0.973 (173)	0.431	—	—	—	0.191	—	0.329	HI	33
	0.373 (75)*	—	—	—	—	—	—	—	—	HI	33
'Ākohekohe	0.571 (21)	—	—	0.150	0.050	—	0.150	—	0.100	MA	34
	0.750 (40)	0.989 (42)	0.680	0.025	—	—	0.125	0.025	0.125	MA	35

continued

Table 8.4. Continued

Species	Fractional Nesting Success (n)[a]	Daily Survival Rate of Nests (n)[b]	Overall Survival Rate of Nests[c]	Proportion of Nests or Eggs Failing Due to:						Island[j]	Source[k]
				Nest Predation[d]	Failure of Eggs to Hatch[e]	Starvation of Chicks or Death in Nest[f]	Weather[g]	Abandonment of Eggs after Laying[h]	Unknown or Other Causes[i]		
'Apapane	0.702 (38)	—	—	0.273	0.0	0.0	0.091	0.273	0.364	KA	26
	0.578 (109)	0.976 (95)[w]	0.449[w]	0.097[v]	0.010[w]	0.0[w]	0.029[w]	0.0[w]	0.330[w]	HI	36
	0.485 (97)	0.973 (106)[w]	0.405[w]	—	—	—	—	—	—	HI	36
	0.649 (37)	0.983 (37)	0.590	0.081	—	—	0.135	—	0.135	HI	9
	0.500 (16)*	—	—	—	—	—	—	—	—	HI	9
Mean	0.534[x]	0.977	0.463	0.159	0.054	0.025	0.097	0.096	0.191		
SE	0.025	0.003	0.041	0.028	0.015	0.011	0.016	0.019	0.031		
N	44	17	17	23	15	10	22	17	25		

Notes: Data were recalculated from the original source where necessary for consistency (indicated in footnotes). "—" indicates that a particular source of mortality was reported but the data were insufficient to calculate proportions. "X" indicates that no data were available; "—" indicates that the data were insufficient to calculate proportions. Scientific names are listed in Appendix 8.1.

[a] The proportion of active, known-fate nests that fledged at least one young. Sample size (n) is the number of known-fate active nests found at all stages of the nesting cycle, excluding nests that failed due to human error or removal for captive propagation and nests that never received eggs. Estimates followed by an asterisk (*) include only nests found before incubation or in early incubation and are therefore unbiased. Estimates followed by a double asterisk (**) are based on the proportion of eggs, rather than nests, that were successful. If the original source calculated fractional nest success using nests that had been abandoned before egg laying, those disturbed by human activities, or those that had unknown fates, we recalculated the fractional nest success including only active, known-fate nests; thus, the fractional nesting success reported here may differ from that in the original source.

[b] The daily survival rate (DSR) of nests as calculated using the Mayfield (1961, 1975) estimate of nest success. Sample size (n) is the number of active nests rather than exposure days because the latter were frequently not reported. With one exception (Pletschet and Kelly 1990), DSRs during incubation and nestling periods were not significantly different, so the combined DSR is reported here.

[c] The survival rate of nests over the entire nesting cycle as calculated using the Mayfield (1961, 1975) estimate of nest success (the daily rate of nesting success raised to the power of the length of the nesting cycle in days).

[d] The proportion of known-fate, active nests that failed due to predation (proportion of eggs or nestlings if followed by **). Where sources reported the causes of nest mortality as a proportion of failed nests, we recalculated the figure as the proportion of known-fate, active nests.

[e] The proportion of known-fate, active nests that did not hatch due to infertility, inviability, or addling (proportion of eggs or nestlings if followed by **).

[f] The proportion of known-fate, active nests that failed because the chicks died in the nest due to starvation or other causes (proportion of eggs or nestlings, if followed by **).

[g] The proportion of known-fate, active nests lost due to severe weather (wind, rain) or, in the case of Laysan Finches, flooding (proportion of eggs or nestlings if followed by **).

[h] The proportion of known-fate, active nests lost due to abandonment during the incubation or nestling period (proportion of eggs or nestlings if followed by **).

[i] The proportion of known-fate, active nests lost due to unknown or other causes (proportion of eggs or nestlings if followed by **). Other causes include poor nest construction (Palila and Hawai'i 'Amakihi), intraspecific interference (Puaiohi), and chicks falling from the nest ('Io).

[k]Sources: (1) Griffin et al. 1998; (2) Klavitter et al. 2003; (3) Chapter 20; (4) W. Banko and P. Banko 1980; (5) Temple and Jenkins 1981; (6) Giffin 1983, as cited in Banko, Ball, et al. 2002; (7) Conant 1977; (8) VanderWerf and Smith 2002; (9) USGS, unpubl. data; (10) Sarr et al. 1998; (11) van Riper 1995; (12) VanderWerf 1998a, 2004; (13) Wakelee 1996; (14) Wakelee and Fancy 1999; (15) Snetsinger et al. 2005; (16) Tweed et al. 2006; (17) Morin 1992a, 1992b; (18) van Riper 1978, 1980b; (19) Pletschet and Kelly 1990; (20) Pratt, Banko, et al. 1997; (21) Bankc, Johnson, et al. 2002; (22) Laut et al. 2003; (23) Simon et al. 2000; (24) van Riper 1978, 1987; (25) Kilpatrick et al. 2006; (26) Eddinger 1970; (27) Pejchar et al. 2005; L. Pejchar, unpubl. data; (28) Woodworth et al. 2001; (29) H. Baker and P. Baker 2000; (30) Lepson and Freed 1995; (31) A. Medeiros and L. Freed, unpubl. ms.; (32) F. Hart, unpubl. data; (33) W. Kuntz, unpubl. data; (34) VanGelder and Smith 2001; (35) Simon et al. 2001; (36) Nielsen 2000.

[l]Griffin et al. (1998) provided Mayfield estimates for 1930 and 1981 and for incubation and nestling periods separately. The Mayfield estimate reported here is the un-weighted average of the two years' data (no exposure days reported) and the weighted average of daily nest success rates for the incubation and nestling stages. Unknown fates include four nests where chicks fell from the nest.

[m]Duckworth et al. (1992:Table 2.3) report a nesting success rate of 66% based on 50 nests for this data set. We follow the reanalysis of Banko, Ball, et al. (2002:Table 2) in this chapter.

[n]Values as reported by Conant (1977), although we were unable to replicate the percentages or numbers reported in that paper.

[o]Predation and weather were the primary causes of nest failure in both studies of O'ahu 'Elepaio (Conant 1977: 207; VanderWerf and Smith 2002).

[p]The percentage of nest success with predator control was significantly greater than without predator control (VanderWerf and Smith 2002).

[q]VanderWerf (1998a) reported a nesting success rate of 41% based on a sample size of 116 nests for this data set, but the original source reveals a sample size of 119 active, known-fate nests.

[r]The proportions do not sum to 1.0 because van Riper (1995) calculated proportions based on different samples, for instance, the percentage of eggs failing to hatch as the proportion of 44 eggs incubated to term, the percentage fledging as the proportion of 38 eggs, and the proportion of nestlings lost to weather as the proportion of 28 young followed through the nestling period (van Riper 1995: 522). The proportion failing due to weather reported here differs from that reported by van Riper because we re-calculated the figure based on the number of eggs laid and followed (3/38 eggs = 0.079).

[s]Fractional nesting success and the proportion of nest fates were based only on pairs for which seasonal fecundity was known (Snetsinger et al. 2005:Table 2).

[t]Fractional nesting success was calculated from the data of Pletschet and Kelly (1990:Table 2), which reports 37 nests successful over the nestling period (fledged young) and 85 total active nests (i.e., we assumed all active nests were of known fate). The mean DSR was back-calculated from the reported overall survival rate and reported ex-ponent.

[u]Fractional nesting success based on seven monitored nests is reported here. Pair follows of 25 territorial pairs over two years (n = 50 pair-years) showed that 20 pairs (48%) were successful in fledging a single young and raising it to 10 months of age (L. Pejchar, unpubl. data).

[v]No Maui 'Alauahio nests were known to have failed due to failure to hatch; a separate hatchability estimate for 26 eggs in 13 two-egg clutches was 96% (Baker and Baker 2000).

[w]The DSRs were back-calculated from overall survival rates using the interval lengths (exponents) given in Nielsen (2000: 33). The DSRs are weighted averages over two nestling periods (incubation and nestling) over three years (Hawai'i Volcanoes National Park, HAVO) and two years (Kona). Overall survival rates were obtained by raising the average DSR to the average interval length (33 days). Nielsen (2000) combined data from two study areas when describing the causes of failure of 28 nests where the causes were known; therefore, the proportions reported here are of 206 total active known-fate nests for both study areas combined.

[x]We treated each of the studies as an independent sample (n = 44 studies of 19 species). Treating the species as the independent unit (i.e., calculating a weighted average of nesting success for each species before calculating the overall average) changed the results very little (e.g., fractional nesting success estimate = 0.522 ± 0.034 SE, n = 19 species). Where more than one estimate of nest success was available from a single study (e.g, nest and egg success were both reported from one data set), we included only one of the estimates (the first one) in the calculations.

or different species, or other factors. The species most likely to have false starts were Hawai'i 'Amakihi (which abandoned 22–35% of nest starts; van Riper 1978; U.S. Geological Survey [USGS], unpubl. data), Hawai'i Creepers (25–33%; Vander-Werf 1998b, Woodworth et al. 2001), 'I'iwi (24–32%; Eddinger 1970; USGS, unpubl. data), and 'Apapane (0–39%; Nielsen 2000; USGS, unpubl. data). Once nests have received eggs, they are rarely abandoned (see Table 8.4); but note that with nests high in the canopy, it may be difficult to discern abandonment from other types of failure.

Once active, eggs or nestlings may befall a number of fates: nest predation, hatching failure, severe weather, and starvation are frequent causes of nest failure in Hawaiian forests. Overall, nest predation was identified as the reason for the failure of 16% of active, known-fate nests (see Table 8.4). In addition, some proportion of the nests that failed for unknown reasons (19% overall) undoubtedly failed due to predation. Thus, nest predation is the leading known cause of nest failure among Hawaiian birds, as it is in north temperate regions (Martin 1992b) and the humid neotropics (Skutch 1985).

Prehistorically, the threat of nest predation for forest birds came primarily from native birds—the 'Io, an extinct harrier (Circus dossenus), stilt-legged owls (Grallistrix spp.), several Corvus spp., and perhaps even one or more extinct smaller passerines. Of these only the 'Io, Pueo (Short-eared Owl, Asio flammeus, likely a recent colonist), and 'Alalā continue to take nest contents today. Introductions of small mammals began with the Polynesian rat (Rattus exulans), a Polynesian introduction, and accelerated in the 1800s with the arrival of the cat (Felis catus), Norway rat (R. novegicus), black rat (R. rattus), and small Indian mongoose (Herpestes auropunctatus). The black rat poses the gravest threat to tree-nesting birds in rain forest because of its arboreal foraging and nesting behavior (VanderWerf 2001b;

this volume, Chapter 11), whereas feral cats are important predators in short-stature, subalpine dry forests on Mauna Kea (Chapter 23).

Of the eggs that escape predation and are incubated to term, some fail to hatch because they are infertile or otherwise inviable. Failure to hatch contributed to the loss of about 5% of nests or eggs on average (range 0–14%, n = 15). These rates are not unusually high for passerine birds: Koenig (1982) analyzed 155 nesting studies and found that a mean of 9.4% of eggs were unhatchable. There is evidence, however, that the level of 'Alalā fertility is unusually low, only about 63–64% (n = 57 eggs from 26 clutches; Banko, Ball, et al. 2002).

The montane forests of Hawai'i are some of the rainiest places on earth, receiving up to 7 m of rain annually (Juvik and Juvik 1998). Severe weather takes its toll on the nests of Hawaiian forest birds. The montane birds in particular are vulnerable to high winds that toss nests about, tear them from the treetops, and break branches or topple trees. In addition, heavy rains can impact a parent's ability to forage and provide for nestlings, as has been observed for Po'o-uli (Kepler et al. 1996). Twenty-two studies reported an average of 10% (0–25%) of nests lost or failed because of severe weather (see Table 8.4). Species that nest in treetops were vulnerable, including the 'Elepaio, Maui Parrotbill, 'Anianiau, Hawai'i Creeper, 'I'iwi, and 'Apapane. Laysan Finches, one of only two ground-nesting honeycreepers, lost nests to flooding in wet years (Morin 1992a). Droughts have been shown to be associated with lower rates of nest initiation and higher rates of nest predation in Puaiohi (Snetsinger et al. 2005) and dry forest species (Lindsey, Pratt, et al. 1997). Because weather influences nesting success, climate change with its accompanying storms and droughts may have important implications for Hawaiian forest bird demography (Chapter 25).

A few species may lose chicks to starvation, but this seems rare. Starvation of entire broods has been recorded in only Palila (9% of nests; van Riper 1978, 1980b), Hawai'i 'Amakihi (4%; van Riper 1978, 1987), and Maui 'Alauahio (2%; H. Baker and P. Baker 2000). Brood reduction (starvation of only some of the young in the nest) has been observed in Puaiohi, where pairs typically fledge two young in good years but fledge only one of their two young in poor years (Snetsinger et al. 2005; see also Tweed et al. 2006). Brood reduction may also occur in Laysan Finches (Morin 1992a).

FACTORS AFFECTING NESTING SUCCESS

Several studies indicate that within species, the higher an individual builds its nest, the better its chances of fledging young. Nest success is positively correlated with nest height in Puaiohi (Snetsinger et al. 2005), Hawai'i 'Elepaio in māmane-naio forest on Hawai'i (van Riper 1995), Palila (Pletschet and Kelly 1990), and Hawai'i 'Amakihi (Kilpatrick et al. 2006). There was no effect for 'Elepaio in 'ōhi'a-koa forest on Hawai'i or on O'ahu (Vander-Werf 1998a, VanderWerf and Smith 2002) or for 'Ōma'o (Wakelee 1996). However, we do not see the same pattern among species, that is, species that nest higher above ground or higher in a tree relative to tree height do not have nest survival rates greater than those that nest lower (Table 8.5; Figs. 8.2 and 8.3; regression for height: $P = 0.32$, df $= 19$; regression for nest height to tree height: $P = 0.23$, df $= 10$). This suggests that once a nest is high enough, other factors of nest site selection or parental behavior are more important, that is, differences among species "swamp" the effect of nest height. The few species that have been studied in more than one habitat showed no consistent tendency to have higher rates of nest success in particular habitats (see Table 8.4 and Nielsen 2000).

Two factors that might influence nest predation rates among species of Hawaiian forest birds are nest sanitation behavior and nest odor. Cardueline finches typically allow fecal material to accumulate around the rims of their nests toward the end of the nesting period (Newton 1972), a behavior that is still found in the Laysan Finch, Palila, Maui Parrotbill, and 'Ākepa. The other Hawaiian honeycreepers and all other Hawaiian forest birds, however, keep their nests clean through fledging. Although fecal sac accumulation could increase the detectability of nests by predators, the nests of species that allow fecal sacs to accumulate on their nests do not appear to suffer greater predation than those kept clean through fledging (daily nest survival rates were 0.972 ± 0.019, $n = 2$ species, versus 0.976 ± 0.008 SE, n $= 8$, respectively; the small sample size of species that allow fecal accumulation precluded statistical analysis; see Tables 8.2 and 8.4).

Many tropical birds have strong odors, but little is known about their function (Weldon and Rappole 1997). Most Hawaiian honeycreepers possess a distinctive and peculiar odor, rather like that of old canvas tents, that is not found in any other taxon (Pratt 1992). The origin or function of the odor is unknown, though Perkins (1893) and later workers suggested that it might serve to repel the avian predators that were common in prehistoric times (Pratt 2005). The nests of honeycreepers take on their scent, potentially making them more vulnerable to introduced mammalian predators. However, we found no evidence to suggest that nesting success is associated with the presence or absence of drepanidine odor (0.972 ± 0.010, $n = 7$ species, versus 0.984 ± 0.009, $n = 3$, respectively; the small sample of species without drepanidine odor precluded statistical analysis; see Tables 8.2 and 8.4). Studies of multiple species in the same habitat or experimental studies will be needed to elucidate what role nest sanitation or odor

Table 8.5. Nest sites of Hawaiian forest birds

Species	Nest Type[a]	Location	Tree Species[b]	Nest Height[c]	Canopy Height[c]	Source[d]
'Io (Hawaiian Hawk)	ST	Branch	'Ōhi'a	9.3 (23)	15.8 (23)	1
'Io	ST	Branch	'Ōhi'a > nonnative species	—	—	2
'Ō'ō 'ā'ā or Kaua'i 'Ō'ō	CA	Trunk	'Ōhi'a	10 (3)	20.5 (2)	3
'Alalā (Hawaiian Crow)	ST	Branch	'Ōhi'a > kōlea	12.7 (22)	Tree crown	4, 5
'Oahu 'Elepaio (OA)	ST	Tree branch	Various nonnative species	7.6 (32)	—	6
'Oahu 'Elepaio (OA)	ST	Branch fork	Various native and nonnative species	9.5 (86)[e]	—	7
Hawai'i 'Elepaio (Mauna Loa Strip, HI[f])	ST	Branch fork	'A'ali'i > koa	5.69 (138)	7.8 (139)	8
Hawai'i 'Elepaio (Hakalau, HI)	ST	Branch fork	'Ōhi'a > koa	12.0 (151)	15.9 (152)	9
Hawai'i 'Elepaio (Mauna Kea, HI)	ST	Terminal[g]	Māmane > naio	—	4.5–10.5 (34)	10
Hawai'i 'Elepaio (Hakalau, HI)	ST	Branch, grass	'Ōhi'a	9.8 (34)	15.1 (34)	5
Millerbird	ST	Branch, grass	'Āweoweo, bunchgrass	0.33 (35)	Shrubs, grass	11
'Ōma'o (Hakalau, HI)	SH	Trunk	'Ōhi'a and koa	7.5 (37)	23.5 (31)	12
'Ōma'o (Mauna Loa, HI)	SH	Trunk	'Ōhi'a = koa	9.2 (14)	—	12
'Ōma'o (Hakalau, HI)	SH			7.9 (13)	24.1 (13)	5
Puaiohi	SH	Cliff	n/a	4.2 (172)	9.5 (157)	13
Laysan Finch	ST > CA	Various	Bunchgrass > shrubs, rock cavities	<0.4 (53)	Grass, shrubs	14
Nihoa Finch	CA	Rock cavity	n/a	n/a	n/a	14
Palila	ST	Branch fork	Māmane > naio	5.2 (26)	6.9 (26)	15
Palila	ST	Branch fork	Māmane > naio	3.9 (85)	5.0 (85)	16
Greater Koa-Finch	ST	Terminal	Koa	Canopy	High	17
Maui Parrotbill (Hanawī, MA)	ST	Terminal	'Ōhi'a	11.8 (10)	13.2 (10)	5
Hawai'i 'Amakihi (Mauna Kea, HI)	ST	Terminal	Māmane > naio	4.6 (228)	6.5 (217)	18
Hawai'i 'Amakihi (Hanawī, MA)	ST	Terminal	'Ōhi'a	7.5 (22)	8.9 (22)	5
Hawai'i 'Amakihi (Hakalau, HI)	ST	Terminal	'Ōhi'a	16.6 (32)	22.4 (33)	5
O'ahu 'Amakihi	ST	Terminal	Various trees	—	—	19
Kaua'i 'Amakihi	ST	Terminal	'Ōhi'a	5.6 (23)	—	20
'Anianiau	ST	Terminal	'Ōhi'a	6 (24)	—	21
Lesser 'Akialoa	ST	Branch fork	Koa, in lichens	Canopy	High	22
'Akiapōlā'au	ST	Terminal	'Ōhi'a > koa	14.2 (13)	18.7 (10)	23
'Akikiki	ST	Terminal	'Ōhi'a	8.5 (3)	—	24
Hawai'i Creeper (Hakalau, HI)	ST > CA	Fork > cavity	'Ōhi'a > koa	13.6 (52)	23 (52)	25, 5
Hawai'i Creeper (Mauna Loa, HI)	ST > CA	Fork > cavity	Koa > 'ōhi'a	14.2 (9)	20.1 (9)	26

Species	Nest type	Nest placement	Plant species	Mean height (n)	Mean height (n)	Source
O'ahu 'Alauahio	ST	Branch fork	Naupaka kuahiwi, kukui	2.5, 7 (2)	—	27
Kākāwahie	ST	Terminal	'Ōhi'a > kāwa'u	3, 5 (2)	—	27
Maui 'Alauahio (Waikamoi, MA)	ST	Terminal	'Ōhi'a > others	8.67 (68)	~14.5	28
Maui 'Alauahio (Hanawī, MA)	ST	Terminal	'Ōli'i	11.4 (8)	12.9 (8)	5
'Akeke'e	ST	Terminal	'Ōhi'a	11.0 (5)	—	29
'Ākepa (MA)	ST?	Terminal?	'Ōhi'a	—	—	30
'Ākepa (HI)	CA	Branch	'Ōhi'a = koa	10.2 (54)	22.1 (33)	30
'I'iwi (KA)	ST	Terminal	'Ōhi'a	7.2 (17)	—	31
'I'iwi (Hanawī, MA)	ST	Terminal	'Ōhi'a	6.8 (2)	7.3 (2)	5
'I'iwi (Hakalau, HI)	ST	Terminal	'Ōhi'a	17 (144)	21.8 (145)	5
'I'iwi (Hakalau, HI)	ST	Terminal	'Ōhi'a	18.2 (183)	13.7 (183)	32
'Ākohekohe (Hanawī, MA)	ST	Terminal	'Ōhi'a	12.4 (49)	13.3 (49)	5
'Ākohekohe (Waikamoi, MA)	ST	Terminal	'Ōhi'a	15.3 (13)	—	33
'Apapane (KA)	ST	Terminal	'Ōhi'a	8.4 (38)	—	34
'Apapane (Hanawī, MA)	ST	Terminal	'Ōhi'a	11.6 (20)	12.1 (20)	5
'Apapane (Kīlauea, HI)	ST	Terminal	'Ōhi'a	13.2 (129)	15.3 (129)	35
'Apapane (Kona, HI)	ST	Terminal	'Ōhi'a	11.2 (81)	14.6 (81)	34
'Apapane (Hakalau, HI)	ST	Terminal	'Ōhi'a	16.2 (44)	19.8 (44)	5
'Apapane (Laysan)	ST	Various	Bunchgrass > 'āweoweo	<1?	Grass, shrubs	33
Po'o-uli	ST	Terminal	'Ōhi'a	8 (2)	15 (2)	36

Notes: Scientific names are listed in Appendix 8.1. "—" indicates that no data were available.

[a] Nest type: SH, nest on a shelf of a tree branch or cliff; ST, a statant nest, an open cup supported at the base by twigs or branches; CA, nest in a tree or rock cavity.

[b] Frequently used plant species: most frequently used species marked by ">." Scientific names: 'ōhi'a (Metrosideros polymorpha); kōlea (Myrsine lessertiana); a'ali'i (Dodonaea viscosa); koa (Acacia koa); māmane (Sophora chrysophylla); naio (Myoporum sandwicense); 'āweoweo (Chenopodium oahuense); bunchgrass (Eragrostis variabilis); naupaka kuahiwi (Scaevola gaudichaudiana); kukui (Aleurites moluccana); kāwa'u (Ilex anomala).

[c] Mean in meters followed by (sample size). Sample size is the number of nests from which data were collected.

[d] Sources: (1) Griffin et al. 1998; (2) Klavitter et al. 2003; (3) Sykes et al. 2000; (4) Banko, Johnson, et al. 2002; (5) USGS, unpubl. data; (6) Conant 1977; (7) VanderWerf and Smith 2002; (8) Sarr et al. 1998; (9) VanderWerf 1998b; (10) van Riper 1995; (11) Morin et al. 1997; (12) Wakelee and Fancy 1999; (13) Snetsinger et al. 2005; (14) Morin 1992b, Morin and Conant 2002; (15) van Riper 1980b; (16) Pletsche: and Kelly 1990; (17) Olson 1999a; (18) van Riper et al. 1993; (19) VanderWerf 1997; (20) Lindsey et al. 1998; (21) Lepson 1997; (22) Lepson and Johnston 2000; (23) Pratt, Fancy, et al. 2001; (24) Foster et al. 2000; (25) Lepson and Woodworth 2002; (26) Sakai and Johanos 1983; (27) P. Baker and H. Baker 2000; (28) H. Baker and P. Baker 2000; (29) Lepson and Pratt 1997; (30) Lepson and Freed 1997; (31) Fancy and Ralph 1998; (32) W. Kuntz, unpubl. data; (33) VanGelder and Smith 2001; (34) Fancy and Ralph 1997; (35) Nielsen 2000; (36) Kepler et al. 1996.

[e] Weighted average of heights of 41 failed and 45 successful nests (VanderWerf and Smith 2002).

[f] Study localities are given when more than one study is presented per species. Island abbreviations are given when more than one island is concerned: KA, Kaua'i; OA, O'ahu; MA, Maui; HI, Hawai'i.

[g] "Terminal" indicates a nest concealed in leafy twigs.

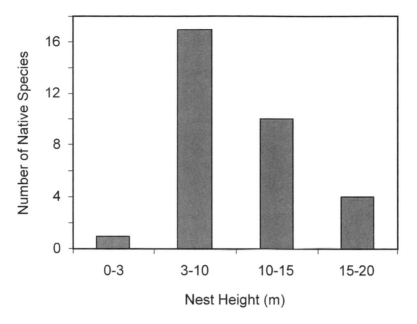

Figure 8.2. Distribution of nest heights for forest birds on the main Hawaiian islands. Historically, Hawaiian forest birds have nested high above the ground, either in trees or on cliff faces. The sole exception is the ground-nesting Pueo. The bars show the number of species nesting in each nest-height category. A species may be represented in more than one category. Data (from Table 8.5) are for all native bird species in the main Hawaiian islands for which nests are known (n = 23). Excluded are samples from nonnative vegetation and the predator-free Northwest Hawaiian Islands.

may play in the nest predation rates of Hawaiian honeycreepers.

ANNUAL REPRODUCTIVE SUCCESS

Annual productivity is determined not only by the success and productivity of individual nests but by the number of nesting attempts per pair per season, which is in turn a function of the length of the nesting cycle, the time required to renest, and breeding season length. Furthermore, not all pairs will attempt to breed in a given year, and the proportion breeding can be affected by both population density (VanderWerf 2004) and weather (Lindsey, Pratt, et al. 1997; Snetsinger et al. 2005). As seen

in Table 8.6, annual reproductive success (the mean number of young produced by a pair over the course of a breeding season) is known empirically for only a handful of species. Therefore, we supplemented empirical data with modeled estimates for seven species from Kilpatrick (2006). A complete description of the model structure may be found in Pease and Grzybowski (1995), and a full description of input parameters and results for each species may be found in Kilpatrick (2006). Because modeling parameters were obtained from the same field studies as the empirical estimates, empirical and modeling estimates should not be considered independent.

The propensity to renest after nest failure is widespread among Hawaiian forest birds, increasing the probability that a particular pair will be successful in a given season. Hawai'i 'Elepaio in montane koa woodland suffer high rates of nest predation from rats (40% of nests). Renesting, however, allowed 69% of pairs to be successful eventually and 5% to raise two broods of young (Sarr et al. 1998). 'I'iwi may fledge three broods (W. Kuntz, pers. comm.), and a Puaiohi pair fledged four

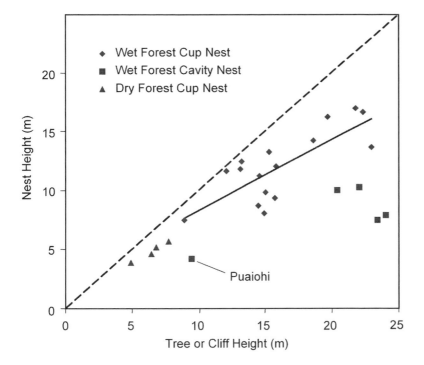

Figure 8.3. Nest height in relation to substrate height on the main Hawaiian islands. Hawaiian forest birds tend to nest as high as possible. Open cup nests are closest to the canopy top in smaller trees, as shown for dry forests and lower-stature rain forest trees. In tall-stature forest, nests are placed even higher, but proportionately lower in the canopy, perhaps because height is less important above a minimal threshold. Points for cavity nesters, including the cliff-nesting Puaiohi on the far left, are proportionately lower still. This low ratio of nest height to tree height for cavity nesters might be expected from the height distribution of potential sites in thick branches and trunks. Puaiohi must avoid ground predators climbing from both below and above. The dashed line represents the one-to-one relationship of nest height to tree canopy height. The solid line represents the correlation of cup nest location to canopy height. The points (data from Table 8.5) represent individual nesting studies for single species; 16 species are represented by 1–4 points (studies) each; n for each point ≥10 nests, except for Kaua'i 'Ō'ō and Po'o-uli, with 2 each.

broods in a single year (Snetsinger et al. 2005). Indeed, most of the species listed in Table 8.6 are known to double-brood at least occasionally; only 'Io, 'Alalā, Maui Parrotbill, 'Akiapōlā'au, Maui 'Alauahio,

and 'Ākepa are apparently restricted to a single successful nest per year (see Table 8.6). Maui 'Alauahio and 'Ākepa may be restricted to single broods because they have short breeding seasons relative to other Hawaiian forest birds (<100 days; see Fig. 8.1), whereas Maui Parrotbills and 'Akiapōlā'au may only be able to raise one brood because young remain dependent on the parents for extended periods (Simon et al. 2000; Pratt, Fancy, et al. 2001).

There is strong interannual variation in productivity within species. Van Riper (1995) found that Hawai'i 'Elepaio fledged an estimated 1.21 young per pair in 1974, compared with only about 0.29 young per pair the previous year. In better years, a higher proportion of birds nested, and birds bred over a longer period, fledging more young. Likewise, Puaiohi raised an average of 4.9 young per territory in the 1997 breeding season but only 0.4 young per territory in 1998, possibly because of an El Niño Southern Oscillation drought (Snetsinger et al. 2005). In that year, fewer pairs

Table 8.6. Productivity of Hawaiian forest birds

Species	Max No. Broods per Year[a]	Mean No. Chicks per Active Nest (n)	Mean No. Chicks per Successful Nest (n)	Percentage of Pairs Successful	Chicks per Pair per Year (empirically observed)[b]	Chicks per Breeding Pair per Year (modeled)[c]	Source
'Io (Hawaiian Hawk)	1	0.53 (88)	1.0 (47)	—	0.46 (102)	—	Klavitter et al. 2003
	—	—	—	—	0.41 (27)[d]	—	Griffin 1985
	—	—	—	—	0.72 (18)[d]	—	Clarkson and Laniawe 2000
	—	—	—	—	0.59 (39)[d]	—	Clarkson and Laniawe 2000
'Alalā (Hawaiian Crow)	1	0.75 (57)[e]	1.3 (32)	58%	0.77 (52)[e]	—	W. Banko and P. Banko 1980, in Banko, Ball, et al. 2002
	—	0.5 (16)[f]	1.1 (7)	—	0.66 (12)	—	Temple and Jenkins 1981, in Banko, Ball, et al. 2002
	—	0.9 (22)	1.4 (14)	—	—	—	Giffin 1983, in Banko, Ball, et al. 2002
O'ahu 'Elepaio	2	—	1.11 (63)	—	0.70 (74)[g]	—	VanderWerf and Smith 2002
					0.33 (54)[g]		
Hawai'i 'Elepaio	2	—	—	69% (58)	—	0.87[h]	Kilpatrick 2006
	1	—	1.18 (127)	53% (188)	—	—	Sarr et al. 1998
	2	—	—	—	—	—	VanderWerf 1998a, 2004
Millerbird	2	1.26 (19)	1.67 (15)	—	0.56 (—)[i]	0.29, 1.21[j]	van Riper 1995
	2	—	—	—	—	—	Conant and Morin 2001
'Ōma'o	2	0.70 (27)	1.50 (13)	—	—	—	Wakelee 1996
	—	—	—	—	—	1.80[h]	Kilpatrick 2006
Puaiohi	4	1.10 (94)	1.60 (66)	69%	2.30 (—)	—	Snetsinger et al. 2005
Laysan Finch	2	0.90 (165)	—	—	—	—	Morin 1992a
Palila	2	0.95 (20)	1.73 (11)	—	—	1.80[j]	van Riper 1978, 1980b
	—	0.65 (54)	1.35 (26)	—	—	—	Pletschet and Kelly 1990
	2	0.72 (280)	1.46 (138)	—	—	—	Pratt, Banko, et al. 1997
	2	0.71 (163)	1.27 (91)	—	—	—	Banko, Johnson, et al. 2002
	—	—	—	—	—	1.32[h]	Kilpatrick 2006
Maui Parrotbill	1	—	—	—	—	—	Simon et al. 2000
Hawai'i 'Amakihi	2	0.94 (79)	—	—	—	2.5[j]	van Riper 1978, 1987
	—	—	—	2.42	—	3.13[h]	Kilpatrick 2003, 2006
Kaua'i 'Amakihi		2.43 (21)	3.00 (17)	—	—	—	Eddinger 1970
'Anianiau		1.78 (27)	2.82 (17)	—	—	—	Eddinger 1970
'Akiapōlā'au	1	—	1.00 (—)	85.7%	0.86 (—)	—	Ralph and Fancy 1996
	1	—	—	14%	0.14 (—)	—	Pratt, Fancy, et al. 2001
	1	0.71 (7)[k]	1.00 (20)[k]	48%[k]	0.48 (20)[k]	—	Pejchar et al. 2005; Pejchar, unpubl. data

Species						Reference
Hawai'i Creeper	2	0.52 (33)	1.70 (7)	—	1.85[h]	Woodworth et al. 2001
		—	—	—	1.28[h]	Kilpatrick 2006
Maui 'Alauahio	1	—	—	—	—	H. Baker and P. Baker 2000
'Ākepa	1	—	—	—	—	Lepson and Freed 1997
		—	—	—	1.47[h]	Kilpatrick 2006
'I'iwi	—	1.07 (15)	2.00 (8)	—	—	Eddinger 1970
	3	—	1.44 (73)	67%	—	W. Kuntz, unpubl. data
		—	—	1.33 (63)	—	VanGelder and Smith 2001
'Ākohekohe	2	0.95 (—)	1.67 (11)	—	—	Simon et al. 2001
	2	1.10 (40)	1.47 (30)	—	—	Kilpatrick 2006
		—	—	—	2.47[h]	
'Apapane	—	1.92 (38)	2.70 (27)	—	—	Eddinger 1970
	—	1.90 (34)	—	—	—	Nielsen 2000

Notes: Scientific names are listed in Appendix 8.1. "—" indicates that no data were available.

[a]Refers to the maximum number of successful broods a pair will attempt in season (most species renest readily and multiple times if nests fail).

[b]Refers to all territorial pairs, whether or not they attempted to breed, except where noted. Because some pairs may not have attempted to breed in a given year, the number of chicks per breeding pair is typically higher than the number of chicks per territorial pair. Sample size (n) is number of pair-years.

[c]Modeled data include any seasonal fecundity values that were calculated from information on individual nest success rates, chicks per nest, and length of breeding cycle and breeding season rather than from directly observed fecundity of individual pairs. See footnotes h and j for more information on methods.

[d]Data are for chicks per breeding pair; includes only pairs that attempted to breed in a given year. Chicks per territorial pair per year were not reported.

[e]Chicks per breeding pair = 1.0 (n = 40).

[f]Duckworth et al. (1992: Table 2.3) report that chicks per active nest = 0.7 based on 50 nests for this data set. We follow the reanalysis of Banko, Ball, et al. (2002: Table 2) in this chapter.

[g]Productivity with and without predator control, respectively.

[h]Productivity was estimated using the model of Pease and Grzybowski (1995). In this model seasonal fecundity is estimated using data from the literature for each species, including nest success rates, chicks per successful nest, and length of nest building, egg laying, incubation, nestling, and postfledging periods. For information on model structure, see Pease and Grzybowski (1995). For information on the parameter values used to calculate seasonal fecundity estimates for each species, see Kilpatrick (2006: Table 1). The model of Pease and Grzybowski (1995) provides estimates in terms of female chicks per female per year, and these estimates were multiplied by 2 to get total chicks per pair per year for comparison with empirical values.

[i]Includes only pairs that laid eggs. Approximately 15% of pairs held territories but did not attempt to breed, and an additional 7% built nests but did not lay eggs (Vander-Werf 2004).

[j]Productivity was estimated by van Riper using this formula: Productivity = $(C \times B \times S)/N$, where C = clutch size, B = length of breeding season in days, S = breeding success (proportion of eggs laid that fledged young), and N = length of the nesting cycle in days. See van Riper (1978) for details. 'Elepaio productivity differed significantly between 1973 (0.29 chicks per pair) and 1974 (1.21 chicks per pair).

[k]"No. Chicks per Active Nest" was obtained from seven monitored nests, but "No. Chicks per Successful Nest," "Percentage of Pairs Successful," and "No. Chicks per Pair per Year" were inferred by "pair follows" of 25 territorial pairs over two years and identification of 24 successful breeding attempts based on the presence of fledglings. Because fledglings may perish shortly after leaving the nest and before detection by the researcher, these are minimum estimates. Note that although no pairs were seen accompanied by more than one fledgling in this study, one of the seven monitored nests fledged two chicks.

attempted to breed, the breeding season was drastically shortened, nest predation increased, pairs did not renest, and successful nests fledged one rather than two young (Snetsinger et al. 2005).

Despite relatively long breeding seasons and the propensity to renest, Hawaiian forest birds are limited in their productivity by small clutch sizes, long developmental periods, and the failure of about half of all nesting attempts on average. As a result, successful nests usually fledge only one young, and only three species—Puaiohi, Hawai'i 'Amakihi, and 'Ākohekohe—have been observed or predicted to fledge an average of two or more young in a season (see Table 8.6), although others such as 'I'iwi and 'Apapane may yet be found to do so. In contrast, a House Finch in the temperate zone may have less than four months to breed, but a single successful nest is all it needs to fledge a brood of four or five young (Hill 1993). The consequences of this limit on productivity for Hawaiian forest bird population dynamics are profound, as discussed later.

HUMAN-CAUSED CHANGES IN AVIAN DEMOGRAPHIC RATES

The natural world that Hawaiian forest birds inhabit has changed dramatically since prehuman times (Chapters 2 and 6). The demographic parameters we measure today are altered from historic levels, and life history characters evolved under different conditions. How have these new population drivers—habitat loss and modification, food competition, mammalian predation, and disease—impacted bird demography?

Effects of Habitat Loss and Degradation

Worldwide, most declining species persist not at the core of their historical geographical ranges, as one might predict, but rather in the portions of their range that are farthest from human-related disturbance

(Channell and Lomolino 2000). Declining species overall appear to undergo a directional range contraction away from anthropogenic threats and impacts. Hawaiian forest birds, restricted to high-elevation montane forests by rampant disease and habitat loss in the lowlands, have followed this pattern of range contraction (Chapter 2). This is significant because, as one moves from the core to the periphery of a species' geographical range, populations generally occupy less favorable habitats and exhibit lower and more variable densities (Gaston 1990, Brown et al. 1995). As a result, some populations remaining in Hawai'i today may occupy habitats that are not ideal and to which they are not well adapted. We know virtually nothing about how species' demographics might have been different in formerly occupied habitats. As only one example, there is evidence that populations of Hawai'i 'Amakihi in low-elevation forests breed over more extended periods than those at high elevations, which could be important for population persistence and evolution of disease resistance (Woodworth et al. 2005).

Effects of Introduced Mammalian Predators

Rates of nest predation in Hawai'i have likely been altered substantially in historic times by predator introductions and habitat change (Chapter 11). The types of nest predation have also changed since the arrival of humans in Hawai'i, from depredation entirely by birds (corvids and raptors) to that mainly by mammals. Increased nest predation has obvious short-term effects on productivity but may also cause evolutionary shifts in birds' nest site selection, behavior, and life history.

The current levels of nest predation are remarkably variable, with some species relatively unaffected, while for others the level of nest predation is high (see Table 8.4). These levels of predation can make the difference between populations producing enough young to persist or becoming locally extinct (Kilpatrick 2006). Hawai'i 'Elepaio in montane koa woodland lost 39.5% of their nest starts to predation, primarily by rats; 'Ōma'o lost 47.4% of nests; captive-bred Puaiohi in the Alaka'i Swamp, Kaua'i, lost 37.5% of nests to predation; and Maui 'Alauahio lost 40.4% (see Table 8.4). The best information on the impact of introduced rodent predators on forest bird demography comes from 'Elepaio in nonnative-dominated forests on O'ahu, where rodent control dramatically increased 'Elepaio reproduction by 112% and survival of female 'Elepaio by 66% (Vander-Werf and Smith 2002). Modeling suggests that in the absence of mammalian predation on nests and adults, many threatened populations of Hawaiian forest birds would attain stable or positive population growth rates (Kilpatrick 2006).

Predation can influence the organization of nesting birds such that each bird species hides its nest in a novel way (Martin 1988). Temperate continental forest bird assemblages subject to nest loss to small mammals, snakes, and other predators typically have suites of species nesting on the forest floor, in the understory, or in the tree canopy (Oniki 1985, Hansell 2000). One would expect that Hawaiian forest birds would also show a diverse dispersion of nesting sites among species, given former predation by raptors, owls, and crows in trees and by flightless rails and ibis on the ground, but this is not the case now. Hawaiian forest birds on the main islands today all nest in the tree canopy, except the Pueo, which nests on the ground (see Fig. 8.2). Unlike in other tropical forest bird communities, there is no nesting on the forest floor or in the shrub layer. The lowest average height at which native forest birds now nest is about 4 m above ground (Palila at 3.90 m; see Fig. 8.2) in subalpine dry forests on Mauna Kea, Hawai'i Island, where the forest is relatively short. The lowest nests in rain forest are from Hanawī, Maui (Hawai'i 'Amakihi

at 7.5 m). The potential variation of nest heights in rain forest is much greater because of the range of tree sizes.

The lack of diversity in nests and nest sites closer to ground level in present-day Hawaiian forests is likely the result of two processes: extinction and adaptation. First, now-extinct ground-living birds that were extirpated by either people or introduced mammals (Olson and James 1982b), especially the flightless species, presumably nested on the ground. The only living examples of ground-nesting honeycreepers, the Laysan and Nihoa Finches, prehistorically inhabited the main Hawaiian islands. These species nest primarily in grass tussocks and rock cavities, respectively, on their essentially treeless and predator-free islands, and they may have nested in similar situations on forested islands. Second, there appears to have been selection for the remaining species to nest as high as possible within the forest. Among several species, higher nests are more likely to fledge young (discussed earlier). Moreover, forest birds apparently avoid nesting in shrubland and low-stature dry forest, even though these habitats are visited for foraging. Exceptions include apparently breeding 'Ōma'o, Hawai'i 'Amakihi, and 'Apapane high above the timberline on Hawai'i Island. Study is needed to clarify these patterns.

In summary, evidence from studies of nesting success and nest site selection indicates that predation by small mammals currently appears to have a major influence on Hawaiian forest birds. We hypothesize that the long-term effects of nest predation on Hawaiian forest birds have been to exclude entire nesting habitats (shrubland and low-stature forest) and, within forests, to restructure nest dispersion so that ground and shrub layers are also excluded. Because cats and rats must reach nests from the ground by climbing upward, nest success increases with nest height, and nests are relatively safe only high in trees or on cliffs.

Effects of Introduced Disease

The introductions of avian malaria (*Plasmodium relictum*) and avian pox virus to Hawai'i (Chapter 9) have had far-reaching impacts on the demography of virtually all of Hawaiian forest birds. In aviary experiments, avian malaria causes significant mortality in many native species, with mortality rates ranging from 65 to 90% (Atkinson et al. 1995; Yorinks and Atkinson 2000; Atkinson, Dusek, et al. 2001). Birds that recover from acute infection apparently carry a chronic infection for life (Atkinson, Dusek, et al. 2001). The impact of disease on the survival rates of birds in the wild is expected to be greater than that observed in the aviary because of the effects of disease combined with climate, food limitation, and predation. The result cannot be overstated: introduced diseases have virtually eliminated native birds from low-elevation forests throughout the Hawaiian Islands (van Riper et al. 1986), and only two native species, the O'ahu 'Elepaio and Hawai'i 'Amakihi, appear able to persist in areas where malaria is endemic (VanderWerf et al. 1997, Shehata et al. 2001, Woodworth et al. 2005).

Although Hawaiian forest bird survival rates appear high, the majority of studies have been of populations in high-elevation montane rain forests outside of the range of the mosquito *Culex quinquefasciatus*, the primary vector of avian malaria. Thus, these rates represent survival rates in the absence of avian malaria. Even here, there is evidence that disease can impact survival rates, particularly for species that live in middle-elevation forests. Periodic outbreaks of avian malaria can wipe out entire cohorts of young 'Apapane (Chapter 9), and pox epidemics have pronounced impacts on the demography of O'ahu 'Elepaio (Vander-Werf 2001a, 2004).

Morbidity due to infection has been shown to negatively affect the reproductive success of some avian hosts, with important consequences for avian population dynamics (Hudson et al. 1998). Certainly acute infection, with its effect on the physiology and activity levels of birds (e.g., Yorinks and Atkinson 2000), could reduce or even preempt breeding attempts. However, once recovered from initial infection, chronically infected birds appear to be able to reproduce as well as birds that have never been infected (Hawai'i 'Elepaio, VanderWerf 2001a; Hawai'i 'Amakihi, Kilpatrick et al. 2006). Human-caused increases in predation are expected to interact with disease to limit populations: repeated nesting attempts due to nest predation may stress individual birds, leading to decreased immune response and relapse (Nordling et al. 1998), and predators may be more likely to depredate moribund birds (Yorinks and Atkinson 2000). Meanwhile, there is a renewed interest in the role of immunological competence as a life history trait (Ricklefs 1992, Sheldon and Verhulst 1996, Nordling et al. 1998), and this is a fertile area for research.

Life History, Demography, and Vulnerability to Extinction

Species vary in their susceptibility to different threats, leading to patterns in extinction across taxa (Bennett and Owens 1997). Understanding the processes of extinction is fundamental to designing effective conservation strategies. These factors may be ecological (e.g., increased predation) or demographic (e.g., population resilience or reproductive rate). Species attributes commonly associated with high extinction probability are small populations, restricted ranges, and/or low population density (Terborgh and Winter 1980, Diamond 1985, Foufopoulos and Ives 1999, Hughes 2004), body size (Brown 1971, Owens and Bennett 2000), population fluctuation (Karr 1982a, 1982b; Newmark 1987), low resilience (Pimm 1991), specialization on vulnerable habitats or patchy or seasonal resources (Terborgh and Winter 1980; Karr 1982a, 1982b; Foufopoulos and Ives 1999), and low dispersal ability (Brown and Kodric-Brown 1977). Life histories affect the vulnerability of species to extinction by influencing their demographic rates (Pimm 1991, Saether et al. 2002) and genetic diversity and the ability of populations to respond to selective pressures (Fisher 1958).

Among the birds of the world, island taxa have suffered the greatest losses—over 93% of avian extinctions have occurred on islands (King 1985). The same processes that operate on continental birds worldwide are magnified on islands (Diamond 1985; this volume, Chapter 1). Certain demographic traits within the Hawaiian avifauna may predispose a species to be vulnerable to endangerment and extinction, and this question has been investigated by several workers. By dividing the Hawaiian avifauna into eight "ecomorphs," Freed (1999) used extinctions on individual islands as replicates in a comparative analysis to examine factors causing vulnerability among the Hawaiian avifauna. He concluded that the species that went extinct had the most limited historical ranges, were restricted primarily to a single habitat, and had lower reproductive rates. In a separate analysis, Banko and Banko (Chapter 2) argue that habitat and food specialization have been the most important predictors of avian extinctions in Hawai'i in historic times. Importantly, Banko and Banko (Chapter 7) show that Hawaiian forest birds that are foraging specialists have lower reproductive rates than generalists (smaller clutch volume per weight and lower incubation efficiency), demonstrating how species' foraging ecology and life history traits may interact to influence their extinction risk.

In a broad review of bird taxa, Owens and Bennett (2000) demonstrated that species traits that predispose species to extinction vary depending on the mechanism of extinction. That is, different threats act in different ways on different traits and therefore threaten different species. In Hawai'i,

we anticipate that extinction risk from habitat loss would disproportionately affect those species that are specialized on lowland habitats (Freed 1999; this volume, Chapter 2), food resources (Pimm and Pimm 1982; Freed 1999; this volume, Chapter 7), or nest sites (cavity-nesting 'Ākepa, Freed 2001; cliff-nesting Puaiohi, this volume, Chapter 22). Introduced nest predators, on the other hand, would be expected to affect those species that nest and/or roost in vulnerable places such as on the ground, have conspicuous or odiferous nests, and have low levels of intrinsic reproduction. Disease would be expected to pose the greatest threat to species with low levels of immunity (Chapter 9) and genetic diversity (Jarvi et al. 2004), limited distribution away from disease transmission (i.e., lowland species such as O'ahu 'Elepaio, VanderWerf 1998a), lack of defensive behavior against mosquitoes (LaPointe, pers. comm.), and/or low reproductive potential (and therefore an inability to compensate for higher rates of mortality). Finally, introduced competitors would most affect those species that are highly specialized on certain food resources, particularly those foods used by common introduced species, such as the Japanese White-eye (Zosterops japonicus) (Mountainspring and Scott 1985), species of small body size, and those with poor behavioral competitive ability. Thus, three traits in particular—specialization on lowland habitat, food specialization, and low reproductive potential—make species vulnerable to multiple concurrent threats.

Adult survival rates are a driving factor in population dynamics, and population models show that they are the most important contributor to growth rates. Adult survival rates are especially important to population growth rates in long-lived species that mature late and have small clutch sizes (Saether and Bakke 2000). Thus, the impact of avian disease on the population dynamics of Hawaiian forest birds is likely to be profound. Despite their relatively long breeding seasons, propensity to renest, and relatively high rates of nest success, Hawaiian forest birds' productivity is limited by small clutch sizes, long developmental periods, and the failure of about half of all nesting attempts on average. Successful nests usually fledge only one young, and only a few species are observed or predicted to fledge two or more young in an average season (see Table 8.6). Populations that undergo the "double whammy" of both decreased survival rates due to disease and decreased fecundity due to predators are really struggling at the fringes of viability, and they may or may not be capable of persisting and adapting over evolutionary time. The continued decline in many vulnerable Hawaiian forest bird species indicates that populations are not stable (Chapter 5) and that mitigation of threats is crucial (Chapter 25).

Finally, although we have come a long way in the past 20 years, our knowledge of Hawaiian forest bird biology is still very incomplete. Despite a general sense of urgency surrounding some endangered species whose numbers are precarious, huge gaps remain in our knowledge of the Maui Parrotbill, 'Akiapōlā'au, 'Akikiki, and 'Ākohekohe. Even some relatively common species, particularly those restricted to the forests of Kaua'i, such as 'Akeke'e and 'Anianiau, remain virtually unstudied. Only by filling these gaps in our knowledge can we hope to fully appreciate the spectrum of life histories that have evolved among the Hawaiian avifauna and understand and address the threats that they face.

SUMMARY

The adaptive radiation of Hawaiian forest birds into various foraging niches was accompanied by diversification in their breeding behavior and life history traits. The selective forces resulting from living on a tropical island have led to an evolutionary shift toward longer lifespans and lower levels of fecundity among the native forest birds of Hawai'i. These birds display some of the highest survival rates of any avifauna studied thus far, and their clutch sizes are small and development times long compared to those of their temperate continental relatives. Most Hawaiian forest birds do not routinely breed until they are in their third year or later, and this postponement is often associated with strong sexual dichromatism and delayed plumage maturation. Most Hawaiian forest birds breed during periods when their food resources are most abundant (most breed in December–May) and have relatively long breeding seasons (180 days). Nesting success rates are variable among species but averaged about 50% of nests, and nest predation is probably the most significant source of nest failure, followed by hatching failure and severe weather. Even with predation rates elevated by introduced mammals, nest losses are much lower in this island situation than in continental tropics.

Thus, Hawaiian forest bird life histories seem to be typically tropical, except for the rather higher level of nesting success that they enjoy. This suggests that nesting success may not have been the driving force in the evolution of clutch size or development rate in Hawaiian forest birds or that other factors have overridden the selective force of predation on nests, although we admit having imperfect knowledge of historic predation rates in Hawai'i.

Human activities have impacted forest bird demography in multiple ways. Birds are restricted to only portions of their historical geographic range. Introduced disease decreases the survival of all native Hawaiian forest bird species. Because of the sensitivity of population growth rates to adult survival rates, we expect this to be a major driver in population dynamics for all species. Introduced nest predators not only affect demographic rates but influence the evolution of nest site selection and habitat selection.

Hawaiian forest birds are threatened by multiple mechanisms, and each species is vulnerable to a different suite of threats depending on aspects of its life history and behavior. Three factors that appear most important are habitat specialization, especially on lowland habitats, food specialization, and low reproductive rate. Finally, it is clear that further research is needed in many aspects of Hawaiian forest bird biology in order to determine how best to mitigate the extensive threats these unique birds face.

ACKNOWLEDGMENTS

This research was made possible in part thanks to support from the U.S. Geological Survey Invasive Species Program and Wildlife and Terrestrial Resources Program; the National Science Foundation (Grant DEB00-83944: Biocomplexity of Introduced Avian Disease in Hawai'i); and the U. S. Fish and Wildlife Service (support to Hakalau Forest Birds Project 1994–1999). We thank individuals who contributed unpublished data, as well as W. Kuntz, E. Morton, R. Ricklefs, E. VanderWerf, and an anonymous reviewer for providing critical feedback on the manuscript.

APPENDIX 8.1. FAMILY AND SCIENTIFIC NAMES, MEAN CLUTCH SIZE, INCUBATION PERIOD, NESTLING PERIOD, AND FEMALE BODY MASS FOR HAWAIIAN FOREST BIRDS

Family	Common Name	Scientific Name	Clutch Size[a]	Incubation Period (days)[b]	Nestling Period (days)[b]	Mean Female Body Mass (g)[c]	Source[d]
Accipitridae	'Io	Buteo solitarius	1.0	38	61	621	Clarkson and Laniawe 2000
Meliphagidae	Kaua'i 'Ō'ō	Moho braccatus	2.0	—	—	—	Sykes et al. 2000
Corvidae	'Alalā	Corvus hawaiiensis	2.8	20	42	485	Banko, Ball, et al. 2002
Monarchidae	O'ahu 'Elepaio	Chasiempis sandwichensis ibidis	2.0+	18	16	12.2	Conant 1977, VanderWerf and Smith 2002
	Hawai'i 'Elepaio	C. s. sandwichensis	1.9	18.4	16	14.4	van Riper 1995, Sarr et al. 1998
Sylviidae	Millerbird	Acrocephalus familiaris	2.2	16	—	18.2	Conant and Morin 2001
Turdidae	'Ōma'o	Myadestes obscurus	2.0*	16	19	49.9	van Riper and Scott 1979, Wakelee 1996, Wakelee and Fancy 1999
	Puaiohi	M. palmeri	2.0	13.5	18.3	39.1^	Snetsinger et al. 2005
Fringillidae	Laysan Finch	Telespiza cantans	3.2	15.7	24	32.2	Morin 1992a
	Nihoa Finch	T. ultima	2.8	15	24	—	Morin and Conant 2002
	Palila	Loxioides bailleui	2.0	16.6	25.3	37.6	Pletschet and Kelly 1990
	Maui Parrotbill	Pseudonestor xanthophrys	1.2[e]	14.5	20	19.7	Simon et al. 1997
	Hawai'i 'Amakihi	Hemignathus virens	2.5	14.1	16.8	13.3	van Riper 1978, 1987
	Kaua'i 'Amakihi	H. kauaiensis	3.1	14	18.8	16.4	Eddinger 1970
	'Anianiau	H. parvus	3.0	14	18	10.3	Eddinger 1970
	Lesser 'Akialoa	H. obscurus	2.0*	—	—	—	Perkins 1903

'Akiapōlā'au	H. munroi	1.0*	—	21	27.1	Pratt, Fancy, et al. 2001
'Akikiki	Oreomystis bairdi	2.0*	—	—	14.5^	Eddinger 1972b
Hawai'i Creeper	O. mana	2.1	16	18	14.7	Woodworth et al. 2001
O'ahu 'Alauahio	Paroreomyza maculata	2.0*	—	—	—	P. Baker and H. Baker 2000
Maui 'Alauahio	P. montana	2.0	16.5	18	12.3	H. Baker and P. Baker 2000
'Akeke'e	Loxops caeruleirostris	2.0	—	—	11.2	Eddinger 1972a
'Ākepa	L. coccineus	1.7^f	15	18	10.4	Lepson and Freed 1997
'I'iwi	Vestiaria coccinea	2.0	14	21.7	16.7	Eddinger 1970
'Ākohekohe	Palmeria dolei	1.7	16.8	21.5	23.6	Berlin and VanGelder 1999
'Apapane	Himatione s. sanguinea	2.8	13	16.1	14.1	Eddinger 1970
Po'o-uli	Melamprosops phaeosoma	2.0*	16	21	25.5^c	Kepler et al. 1996

Notes: Species for which no data are available other than body mass are excluded. Scientific names for Hawaiian forest birds mentioned elsewhere in this chapter but not in this table are Greater Koa-Finch (*Rhodacanthis palmeri*), Kona Grosbeak (*Chloridops kona*), Kaua'i Greater 'Akialoa (*Hemignathus ellisianus stejnegeri*), and Kākāwahie (*Paroreomyza flammea*).

a For clutch sizes marked with an asterisk (*), mean clutch size was not available, so we used the modal clutch size; if only a range of clutch sizes was given, we used the midpoint of the range. For those sources that listed more than one study, we calculated the mean clutch size for all studies weighted by sample size, as indicated by a +. Kaua'i 'Ō'ō and Lesser 'Akialoa clutch sizes are based on a single nest with eggs or chicks, respectively.

b Where available, incubation and nestling periods are mean values; otherwise they are the midpoint of the range recorded in the literature, excluding known outliers. "—" indicates that no data were available.

c Mean body mass of adult females in grams, unless sex-specific data were unavailable (marked by a ^). For Po'o-uli only, mass is that of holotype.

d The sources listed are for demographic data; data on body mass are from Birds of North America accounts (Poole and Gill 1992–2002).

e Maui Parrotbill clutch size based on pairs with fledglings: 26 pairs with single fledglings and 5 pairs with two fledglings.

f 'Ākepa clutch size based on number of chicks visible at entrance to nest cavity.

APPENDIX 8.2. FAMILY AND SCIENTIFIC NAME, MEAN CLUTCH SIZE, INCUBATION PERIOD, AND NESTLING PERIOD FOR NORTH TEMPERATE CONTINENTAL BIRDS FROM TABLE 8.3

Family	Common Name	Scientific Name	Clutch Size[a]	Incubation Period (days)[b]	Nestling Period (days)[b]	Source
Accipitridae	Red-shouldered Hawk	Buteo lineatus	3.1+	33	38.5	Crocoll 1994
	Broad-winged Hawk	B. platypterus	2.4+	29.5	38.5	Goodrich et al. 1996
	Short-tailed Hawk	B. brachyurus	2.0*	36.5	—	Miller and Meyer 2002
	Swainson's Hawk	B. swainsoni	2.4+	34.5	43	England et al. 1997
	White-tailed Hawk	B. albicaudatus	2.26	31.2	50.2	Farquhar 1992
	Red-tailed Hawk	B. jamaicensis	2.8	31.1	44	Preston and Beane 1993
	Ferruginous Hawk	B. regalis	3.6+	32.5	44	Bechard and Schmutz 1995
	Rough-legged Hawk	B. lagopus	3.4+	34	32.5	Bechard and Swem 2002
Corvidae	American Crow	Corvus brachyrhynchos	4.7+	18	32.8	Verbeek and Caffrey 2002
	Northwestern Crow	C. caurinus	4.0	18.2	31.7	Verbeek and Butler 1999
	Fish Crow	C. ossifragus	4.5	17.5	36	McGowan 2001
	Carrion Crow	C. corone	4.5*	—	—	Brazil 1991
	Jungle Crow	C. macrorhynchos	4.0*	—	—	Brazil 1991
	Chihuahuan Raven	C. cryptoleucus	4.9+	20.3	38.5	Bednarz and Raitt 2002
	Common Raven	C. corax	5.3+	21.7	38.5	Boarman and Heinrich 1999
Sylviidae	Aquatic Warbler	Acrocephalus paludicola	4.1	—	—	Halupka and Wróblewski 1998
	Sedge Warbler	A. schoenobaenus	5.2	—	—	Halupka 1996
	Great Reed Warbler	A. arundinaceus	4.5*	—	—	Brazil 1991
	Black-browed Reed Warbler	A. bistrigiceps	4.5*	—	—	Brazil 1991

Turdidae	Townsend's Solitaire	Myadestes townsendi	3.6+	12.2	11.8	Bowen 1997
	Veery	Catharus fuscescens	3.9	12	11	Moskoff 1995
	Gray-cheeked Thrush	C. minimus	4.0	12.5	11	Lowther et al. 2001
	Bicknell's Thrush	C. bicknelli	3.6	12	11.4	Rimmer et al. 2001
	Swainson's Thrush	C. ustulatus	3.7+	12	13	Evans Mack and Yong 2000
	Hermit Thrush	C. guttatus	3.4+	12	12	Jones and Donovan 1996
	Wood Thrush	Hylocichla mustelina	3.3+	12.7	13.5	Roth et al. 1996
	American Robin	Turdus migratorius	3.3+	13	13	Sallabanks and James 1999
	Varied Thrush	Ixoreus naevius	3.2	12	14	George 2000
Fringillidae	Gray-crowned Rosy-Finch	Leucosticte tephrocotis	4.5+	14	17.5	MacDougall-Shackleton et al. 2000
	Black Rosy-Finch	L. atrata	4.6	12.5	20	Johnson 2002
	Brown-capped Rosy-Finch	L. australis	4.2	13	18.5	Johnson et al. 2000
	Pine Grosbeak	Pinicola enucleator	3.8	13.5	—	Adkisson 1999
	Purple Finch	Carpodacus purpureus	3.9	12.5	14.5	Wootton 1996
	Cassin's Finch	C. cassinii	4.3	12	—	Hahn 1996
	House Finch	C. mexicanus	4.2+	13.5	15.5	Hill 1993
	Red Crossbill	Loxia curvirostra	3.0*	14	20	Adkisson 1996
	White-winged Crossbill	L. leucoptera	3.0*	—	—	Benkman 1992
	Common Redpoll	Carduelis flammea	4.7+	11.7	11.5	Knox and Lowther 2000a
	Hoary Redpoll	C. hornemanni	4.5*	12.3	12	Knox and Lowther 2000b
	Pine Siskin	C. pinus	3.8	13	15	Dawson 1997
	Lesser Goldfinch	C. psaltria	4.2	12.5	14	Watt and Willoughby 1999
	Lawrence's Goldfinch	C. lawrencei	4.9	12.5	13.5	Davis 1999
	American Goldfinch	C. tristis	5.2	13	12	Middleton 1993
	Evening Grosbeak	Coccothraustes vespertinus	3.5*	13.4	13.5	Gillihan and Byers 2001

[a]For clutch sizes marked with an asterisk (*), mean clutch size was not available, so we used the modal clutch size; if only a range of clutch sizes was given, we used the midpoint of the range. For those sources that listed more than one study, we calculated the mean clutch size for all studies weighted by sample size, as indicated by a +.

[b]Where available, incubation and nestling periods are mean values; otherwise they are the midpoint of the range recorded in the literature, excluding known outliers. "—" indicates that no data were available.

CHAPTER NINE

Ecology and Pathogenicity of Avian Malaria and Pox

CARTER T. ATKINSON
AND DENNIS A. LAPOINTE

Few examples of the effects of introduced pathogens and vectors on wildlife populations are as clear or compelling as the extensive declines in the abundance, diversity, and geographic distribution of Hawaii's native birds after the introduction of mosquitoes, avian malaria, and pox virus (Warner 1968, Dobson and May 1986, van Riper et al. 1986). Hawai'i once had the distinction of being one of only a handful of tropical areas in the world that were free of common biting arthropods such as mosquitoes (Culicidae), black flies (Simulidae), sand flies (Ceratopogonidae), and deer flies (Tabanidae) and the protozoan, helminth, bacterial, and viral pathogens that they can transmit. With the introduction of a highly competent mosquito vector to the islands in 1826 (*Culex quinquefasciatus*) and the presence of large populations of highly susceptible forest birds, the stage was set for the introduction of one or more vector-transmitted diseases. Two pathogens with broad host specificity—avian malaria (*Plasmodium relictum*) and avian pox virus (*Avipoxvirus*)—successfully made the jump from either migratory or introduced nonnative birds to native species, and, with few exceptions, native forest bird populations have been reeling from their impacts ever since.

In this chapter we summarize what is currently known about the epidemiology and pathogenicity of avian malaria and pox virus in Hawai'i. A clear understanding of the ecology and transmission of these parasites is essential for the development of conservation strategies to restore native species, yet the variability of the system across both temporal and spatial scales can make this difficult. In addition, new levels of complexity are continually being added to the system as invasive plants, invertebrates, and nonnative birds, insects, and small mammals are introduced in increasing numbers to the islands (Chapter 15). This rich mix of native, indigenous, and introduced species from throughout the tropics and subtropics can lead to dynamic and unpredictable interactions in the disease system, particularly when new reservoir hosts or vectors enter the picture (Chapters 14 and 17). All of these features make the Hawaiian Islands an exceptional place to investigate the dynamics and evolution of introduced disease organisms and endemic and introduced hosts, with important lessons to be learned about how invasive species can alter a landscape and its evolutionary future.

THE HISTORY AND ORIGINS
OF AVIAN DISEASE IN HAWAI'I

Studies by Richard Warner in the 1950s and Charles van Riper III in the 1970s were the first to clearly identify and establish the role that avian malaria and avian pox virus may have played in the decline and extinction of many of Hawaii's endemic forest birds (Warner 1968; van Riper et al. 1986, 2002). Avian malaria is a disease caused by intracellular, mosquito-transmitted protozoan parasites in the genus Plasmodium. These parasites have a worldwide distribution in multiple avian families but occur primarily in passerine birds (songbirds or perching birds) (Garnham 1966, Valkiūnas 2005). Although more than 40 species of Plasmodium have been described, their taxonomy is currently in a state of flux and surprisingly few have been studied in the field (Atkinson and van Riper 1991, Atkinson 2008). Avian pox, by contrast, is a viral infection caused by a large, double-stranded DNA virus that typically causes tumorlike swellings on exposed skin or diphtheritic lesions on the mouth, trachea, and esophagus of infected birds. It can be transmitted mechanically through contact with infected objects and also the mouth parts

of blood-sucking arthropods. Thirteen different species are currently recognized worldwide and are defined by host associations ranging from the genus to the family level (Tripathy 1993, van Riper and Forrester 2007). It is significant that both of these pathogens are capable of infecting multiple species of avian hosts. Their broad host range makes them particularly well suited for invading new areas and, in part, explains why they have had such broad impacts on highly susceptible native Hawaiian forest birds.

It is not clear precisely when or how both of these pathogens first reached Hawai'i. Despite its isolation, the Hawaiian Archipelago is regularly visited by migratory shore birds and waterfowl, but the role of these hosts in the movement and introduction of disease agents to the islands is not known. Avian malaria and pox virus have been reported from most of these migrants or their close relatives in other parts of the world (Bennett et al. 1982, Kreuder et al. 1999), but comprehensive surveys of parasite fauna in migrants to the Hawaiian Islands have never been done. Although it is possible that migrants may have first brought malaria to the islands, the absence of suitable vectors prior to European contact would have prevented spread of the infection into endemic forest bird populations. The picture is less clear for avian pox virus, because it can be transmitted by contact with infected surfaces. It is possible that migratory shorebirds or waterfowl first brought the virus to the islands, but differences in the host specificity of the virus and differences in habitat use by migratory waterfowl, shorebirds, and native forest birds make it unlikely that the virus became established in the islands through this pathway.

Perhaps the defining moment in the history of these diseases in Hawai'i was the introduction of Culex quinquefasciatus, the southern house mosquito. Historical records suggest that this mosquito was first released on Maui in 1826 from the whaling ship *Wellington* when sailors emptied fresh water casks containing larvae that had been taken on board in Mexico (Van Dine 1904, Hardy 1960). There are subsequent reports of mosquitoes on Oʻahu in the 1830s, and it is likely that the close association of this mosquito with human development aided its spread among lowland settlements and agricultural communities. Development of the sugar industry in the late 1800s may have further enhanced its expansion through the large-scale clearing and fragmentation of low-elevation native forests and the development of extensive irrigation systems (Chapter 6). The growth of the cattle industry in Hawaiʻi undoubtedly provided further larval habitat for this mosquito (Reiter and LaPointe 2007), and its range eventually expanded to include virtually all forested areas throughout the state.

The exact date when avian pox was introduced to the islands remains unknown, but it is clear from descriptions of tumor-like swellings on dead or moribund forest birds by late nineteenth-century naturalists working on Hawaiʻi Island (Henshaw 1902) and from recent amplification by polymerase chain reaction (PCR) of pox virus sequences from museum specimens collected during that period (Jarvi, Triglia, et al. 2008) that the virus was well established in native forest bird populations by that time. The first definitive diagnosis of avian pox in wild populations was made by Bureau of Animal Industry diagnosticians in 1902 from an ʻĀkepa (*Loxops coccineus*) (van Riper et al. 2002). The virus had been diagnosed a year earlier from domestic chicken flocks on Oʻahu, but otherwise we know virtually nothing about its point of entry or how rapidly it spread among the main islands. Perkins (1903) reported more frequent occurrence of poxlike lesions in wetter, windward forests, an observation that supports the role of *C. quinquefasciatus* as the primary vector of the virus. Most workers have long assumed that domestic poultry was the most likely route for introduction of the virus to Hawaiʻi, yet recent studies by Tripathy et al. (2000) and Jarvi, Triglia, et al. (2008) indicate that in native forest birds at least two genetically distinct forms of the virus are circulating that are distinct from fowlpox and have strong similarities to canarypox based on the sequence of the virus 4b core protein.

Our knowledge about how and when avian malaria spread into Hawaiian forest bird populations is even more fragmentary than what is known about pox virus. *Plasmodium* infections were first detected by blood smear in a Red-billed Leiothrix (*Leiothrix lutea*) and a Japanese White-eye (*Zosterops japonica*) collected at Hawaiʻi Volcanoes National Park in the 1940s by Paul Baldwin (Baldwin 1941, Fisher and Baldwin 1947). Richard Warner subsequently detected transmission of avian malaria in lowland habitats near Līhuʻe, Kauaʻi, in the 1950s by exposing Laysan Finches (*Telespiza cantans*), Kauaʻi ʻAmakihi (*Hemignathus stejnegeri*), ʻAnianiau (*Hemignathus parvus*), and ʻApapane (*Himatione s. sanguinea*) in unscreened cages. Most of his birds died from fulminating pox and malarial infections, and this study became the cornerstone for future investigations by Charles van Riper in the 1970s (van Riper et al. 1986). The failure to detect *Plasmodium* in native birds sampled by Baldwin and the relatively high prevalence of malaria in these species in the 1970s led van Riper to suggest that malaria was introduced into the islands later than pox. The most likely opportunity for this introduction was in the early twentieth century, when the Hui Manu or local bird clubs introduced nonnative passerines from around the world to replace low-elevation native birds that were vanishing from pox infection and other limiting factors (Chapter 13). More than 100 documented introductions of passerine birds from Southeast Asia, North and South America, and Africa were made between 1900 and the 1960s, with more than 50 species becoming established there. It is unlikely that any of these birds were screened for infectious diseases before being released into the wild, and in the absence of comprehensive surveys of native and nonnative species for pathogens, our knowledge is still fragmentary about how many species of parasites were introduced and became established (Alicata 1964, van Riper and van Riper 1985).

Because of taxonomic confusion in the classification of *Plasmodium* spp. and the fact that the erythrocytic or blood stages of these parasites have so few distinguishing morphological characteristics, there was considerable initial confusion about which species of *Plasmodium* had been introduced to Hawai'i or whether multiple species were present. Laird and van Riper (1981) conducted a detailed morphological study of blood smears from native and nonnative species across the state and concluded that only a single species of *Plasmodium* was present there and that it was most similar to *Plasmodium relictum*, subspecies *capistranoae*. The variability in parasite morphology and identifications of multiple species was attributed to the high levels of parasitemias in immature red blood cells that are common in susceptible honeycreepers. Interestingly, the type host of this subspecies is the Blue-breasted Quail (*Coturnix chinensis*) from the Philippines, a gallinaceous bird (heavy-bodied ground-feeding domestic or game bird) that was introduced to Kaua'i in 1910 and to other islands in 1922 (Moulton et al. 2001) but failed to become established. More recent studies suggest that even this potential origin may be in error, because live isolates of the parasite made from native birds in the 1990s are not infectious to gallinaceous birds (Valkiūnas et al. 2007) and the occurrence of malaria in wild game birds that are taxonomically most closely related to Blue-breasted Quail has never been definitively established in spite of extensive surveys on Hawai'i Island (R. Nakamura, pers. comm.). Recent efforts to identify the origins of Hawaiian *P. relictum* by examining the divergence of the cytochrome b gene have found that the mitochondrial lineage present in Hawai'i is extremely rare in the New World but more common in Old World passerines. This suggests that the importation and release of one or more Old World passerines may have been the source of the pathogen (Beadell et al. 2006).

THE ROLE OF POX AND MALARIA IN THE DECLINE AND EXTINCTION OF HAWAIIAN BIRDS

Direct evidence implicating avian pox and malaria in the extinction of any of Hawaii's endemic birds is actually very scant, even though these diseases are often cited as primary factors in the demise of Hawaiian forest birds. Van Riper and colleagues (1986, 2002) have suggested that two major waves of extinction that occurred in the late 1800s and again between 1960 and 1990 match closely the presumed times of introduction of avian pox and malaria. However, lack of detailed historical records from the late 1800s, field observations that are limited primarily to anecdotal observations by early naturalists, and only limited surveys for avian malaria in wild populations near the middle of the twentieth century make the matching of these events difficult to confirm.

Perhaps the best evidence that these pathogens have had a major impact on forest bird populations is their strong negative correlation with extant native birds. Scott et al. (1986) noted high densities of native forest birds and the restriction of endangered species to elevations above 1,500 m on Kaua'i, Maui, and Hawai'i islands and their near absence at middle elevations despite abundant, seemingly suitable habitat. This pattern is repeated throughout the islands, with almost complete loss of native species on O'ahu, Moloka'i, Lāna'i, and Kaho'olawe, where elevations do not exceed 1,500 m. Warner (1968) was the first to draw a parallel between declines in native bird populations at lower elevations and the presence of mosquitoes and the diseases they transmit. He drew arbitrary lines around profiles of the islands at the 600 m contour, suggesting that mosquito populations declined precipitously above this elevation. Later work by Goff and van Riper (1980), Scott et al. (1986), and van Riper et al. (1986) documented occurrence of the vector as high as 2,500 m but also supported Warner's basic idea by demonstrating a negative association between numbers of mosquitoes and native birds at elevations below 1,500 m.

Experimental studies have demonstrated the high level of susceptibility of native honeycreepers to avian malaria and pox. Both Warner's use of sentinel Laysan Finches and high-elevation Kaua'i 'Amakihi, 'Anianiau, and 'Apapane and the experimental studies by van Riper et al. (1986) with Hawai'i 'Amakihi (*Hemignathus virens*), 'I'iwi (*Vestiaria coccinea*), and 'Apapane strongly supported the role of disease in the extinction of native species. These studies were criticized at the time, however, because their sample sizes were small and because they failed to make convincing links with disease transmission and mortality in wild populations (Atkinson 1977). Missing elements from both studies were isolation of the disease agents from dead native birds, reproduction of the disease using routes of inoculation and dosages that are comparable to natural situations, and demonstration of demographic impacts (recruitment and survival) on wild populations. Later work in the 1990s filled many of these gaps (Atkinson et al. 1995, 2000; Atkinson, Lease, et al. 2001; VanderWerf 2001a). Although it will be impossible to reconstruct the historical spread and impacts of these diseases on forest bird populations, evidence presented here and elsewhere in this volume suggests that pox and malaria acted in concert with other limiting factors to eliminate endemic species, particularly honeycreepers, from low- and midelevation native forests.

THE PATHOGENICITY OF AVIAN MALARIA AND POX VIRUS IN NATIVE FOREST BIRDS

Malaria

Warner (1968) provided the first detailed descriptions of the pathogenic effects of malarial infection in Laysan Finches, Kaua'i 'Amakihi, 'Anianiau, and 'Apapane that he exposed to mosquitoes in lowland areas on Kaua'i. Key observations were the development of fulminating erythrocytic infections with parasitemias that exceeded 95% of the circulating erythrocytes, development of an associated acute anemia with packed cell volumes as low as 16%, and rapid deterioration of body condition and vigor, with almost 100% mortality between 5 and 15 days after exposure (Fig. 9.1). Van Riper et al. (1986) reproduced the disease in Hawai'i 'Amakihi, 'I'iwi, and 'Apapane under more controlled conditions using intramuscular needle inoculation of a standardized dose of erythrocytic asexual stages of the parasite. They observed 100% mortality in five Laysan Finches, 66% (4/6) mortality in high-elevation Hawai'i 'Amakihi from Mauna Kea Volcano, 20% (1/5) mortality in high-elevation Hawai'i 'Amakihi from Mauna Loa Volcano, 60% (3/5) mortality in 'I'iwi, 40% (2/5) mortality in 'Apapane, and no mortality in Red-billed Leiothrix, Japanese White-eyes, or Common Canaries (*Serinus canaria*) inoculated with the same dose of the parasite. The levels of parasitemias in native species infected by van Riper et al. by needle inoculation were lower than those reported by Warner (1968) but were still at high levels relative to nonnative species, with peak parasitemias at 30–35% of the circulating erythrocytes. Van Riper et al. (1986) also described the gross pathology seen in his experimental fatalities, including enlargement and discoloration of the liver and spleen—two characteristic pathological changes that occur in acute malarial infections—but provided few details about the course of experimental infections over

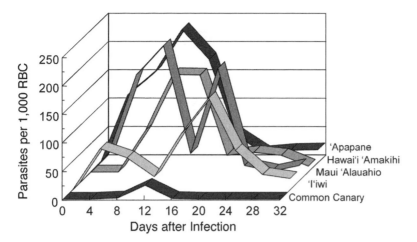

Figure 9.1. Acute phase malarial parasitemias in an 'Apapane, Hawai'i 'Amakihi, Maui 'Alauahio, 'I'iwi, and Common Canary. All birds were exposed to single infective mosquito bites. The number of parasites per 1,000 red blood cells (RBC) is much higher and more prolonged in native birds than in Canaries and other nonnative species, making them exceptional reservoir hosts for infecting mosquitoes and initiating and amplifying disease outbreaks during seasonal epizootics.

time, including the duration of the acute clinical phase.

These findings differ remarkably from what is known about natural and experimental malarial infections in passerine birds from areas where the parasite is endemic. There are relatively few reports of fatal, naturally acquired malarial infections in wild birds, and most continental populations are hosts to a suite of hematozoan parasites, including species of *Plasmodium*, *Haemoproteus*, *Leucocytozoon*, *Trypanosoma*, and filarial worms with few or no detectable demographic impacts of infection (Atkinson and van Riper 1991, Atkinson 2008). Analogous to the situation with Hawaiian birds, the few exceptions in which demographic impacts have been documented are instances in which hosts are moved outside of their normal range and/or exposed to vectors and parasites to which they have had no prior exposure. Some examples include the high mortality in penguin collections at zoological parks from local strains of avian malaria (Stoskopf and Beier

1979) and the introduction of *Leucocytozoon marchouxi* and its impacts on the Pink Pigeon (*Columba mayeri*) of Mauritius (Peirce et al. 1997, Swinnerton et al. 2005).

Atkinson et al. (1995, 2000, unpubl. data); Atkinson, Lease, et al. (2001); and Yorinks and Atkinson (2000) extended the work of Warner and van Riper et al. by studying a series of experimental malarial infections in the native 'Ōma'o (*Myadestes obscurus*), Hawai'i 'Amakihi, Maui 'Alauahio (*Paroreomyza montana*), 'I'iwi, and 'Apapane and in the nonnative Red-billed Leiothrix, Japanese White-eye, House Finch (*Carpodacus mexicanus*), House Sparrow (*Passer domesticus*), and Nutmeg Mannikin (*Lonchura punctulata*) using single and multiple mosquito bites as infective doses (Fig. 9.2). These experiments confirmed earlier work by van Riper and colleagues and used natural routes of infection to document differences in mortality among species of honeycreepers as well as significant differences in susceptibility between native and nonnative species (Table 9.1). Declines in food consumption and loss of body mass began approximately one week after infection in Hawai'i 'Amakihi, 'I'iwi, and 'Apapane that were exposed to single infective mosquito bites. These declines correlated closely with increases in numbers of erythrocytic parasites in the birds' peripheral circulation. The effects were much more severe in 'I'iwi; individuals of this

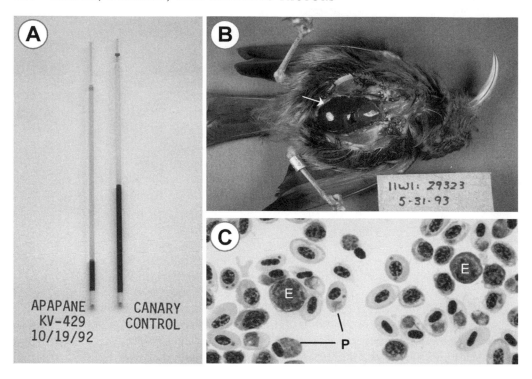

Figure 9.2. Pathological effects of acute phase malarial infections in Hawaiian honeycreepers. (A) Severe anemia and a low packed cell volume in a wild 'Apapane from Kīlauea Crater that was captured during the 1992 epizootic at Hawai'i Volcanoes National Park. (B) Gross lesions in an 'I'iwi with an experimental malarial infection. Note the enlarged, darkened liver (arrow), which is immediately evident when the body cavity is opened. (C) Giemsa-stained blood smear from an 'I'iwi with an acute malarial infection. Note the parasitized erythrocytes (P) and erythroblasts (E), which normally occur only in bone marrow. *Source:* Photos by Carter Atkinson, U.S. Geological Survey.

species had significantly higher levels of mortality (90%) and peak parasitemias at death. Gross and microscopic lesions were also more severe in 'I'iwi and included massive enlargement of the liver and spleen with extensive deposition of malarial pigment (see Fig. 9.2B), diffuse extramedullary erythropoiesis in the liver and kidneys as birds struggled to replace red blood cells destroyed by the infection, and profound alteration of the normal cellular makeup of the blood, with mature erythrocytes replaced almost entirely by immature erythrocytes and precursors that are normally found only in bone marrow (see Fig. 9.2C). Unlike 'I'iwi, up to one-third of Hawai'i 'Amakihi and 'Apapane exposed to the same doses and isolates of the parasite recovered and developed low-intensity chronic infections that stimulated their immunity to reinfection when challenged with multiple infective mosquito bites. These individuals became, in effect, outstanding reservoir hosts for the very disease that had almost killed them.

The development of serological tests (Atkinson, Dusek, et al. 2001) and molecular probes (Feldman et al. 1995, Jarvi et al. 2002) for diagnosing chronic, low-level infections in birds that recover from infection have helped to demonstrate the low sensitivity of blood smears for detecting chronic infections and the significantly higher prevalence of infection in wild populations than was documented by earlier workers (Atkinson et al. 2005). When the duration of infections in recovered individuals was followed over time in a

Table 9.1. Comparative mortality of native and nonnative species following exposure to infective mosquito bites

Species	Origin	Number of Infective Bites	Sample Size	Mortality (%)
'Ōma'o	Endemic	1	4	0
Hawai'i 'Amakihi	Endemic	1	20	65
Maui 'Alauahio	Endemic	1	4	75
'I'iwi	Endemic	1	10	90
'Apapane	Endemic	1	8	63
Red-billed Leiothrix	Introduced	5–10	5	0
Japanese White-eye	Introduced	5–10	5	0
House Finch	Introduced	1	5	0
House Sparrow	Introduced	1	5	0
Nutmeg Mannikin	Introduced	5–10	8	0

mosquito-proof aviary, Hawai'i 'Amakihi were still PCR-positive and infectious to mosquitoes up to four years after initial infection, and it is likely that such individuals remain infected for life (Jarvi et al. 2002; Atkinson, unpubl. data). These individuals maintained body masses that were comparable to those of uninfected birds, suggesting that the costs of controlling chronic infections were relatively low (Kilpatrick et al. 2006). Results from these series of experimental studies are in agreement with the current altitudinal distribution of Hawai'i 'Amakihi, 'I'iwi, and 'Apapane, with 'I'iwi rare or absent from elevations below approximately 1,200 m, where mosquito numbers and disease transmission increase.

The susceptibility of other groups of native Hawaiian birds to introduced malaria is less clear. 'Ōma'o and possibly O'ahu 'Elepaio (*Chasiempis sandwichensis gayi*) and Hawai'i 'Elepaio (*Chasiempis s. sandwichensis*) appear to have some resistance to malaria based on their current altitudinal distribution and occasionally high abundance in some mid- and low-elevation habitats where disease prevalence is high. This is particularly true on O'ahu, where O'ahu 'Elepaio occur almost exclusively in extensively modified low-elevation forests with high mosquito densities. 'Ōma'o also ap-

pear to be relatively resistant to malaria and fail to develop clinical signs of disease when exposed to single infectious mosquito bites (Atkinson, Lease, et al. 2001). This observation was unexpected because 'Ōma'o have undergone a major range reduction on the island of Hawai'i and during the first half of the twentieth century disappeared entirely from the leeward and northern sides of the island, where they had formerly been common (Wakelee and Fancy 1999). One of the leading theories for this range contraction is susceptibility to disease, yet midelevation populations on the windward side of the island have prevalences of infection that exceed 90% and yet appear to be stable (Atkinson, Lease, et al. 2001; this volume, Chapter 5). One possible explanation is that 'Ōma'o populations crashed after the introduction of malaria and then expanded as the birds developed resistance to the parasite. Alternatively, 'Ōma'o and other native thrushes that are now extinct were resistant to pox and malaria but lost their former ranges and/or went extinct because of other factors, including food resource limitation, predators, habitat degradation, and competition with introduced birds (Chapters 6, 7, 11, and 13). Fancy et al. (2001) successfully translocated 16 wild 'Ōma'o and 25 captive-reared 'Ōma'o to leeward Hawai'i, and the success of this technique may make it possible to test alternate hypotheses about range limitation.

Pox

We know relatively little about the pathogenesis of avian pox virus in Hawaiian forest birds, and the report by Warner (1968) still provides some of the best documentation about the course of infection and pathogenicity of the virus in native and nonnative birds. Warner exposed 24 Laysan Finches to mosquitoes in the downtown Honolulu area of O'ahu after keeping the cage wrapped in mosquito-proof cheesecloth for two months prior to the start of the experiment. Within two weeks after the protective cheesecloth was removed, six finches developed indurated swellings on the lores, tarsal, and wing joints, where mosquitoes had access to exposed skin. The swellings increased in size and then erupted into granular, tumorlike lesions that became necrotic, with accompanying secondary bacterial infections. As the tumors progressed in severity they tended to bleed, particularly from the foot lesions. By the end of one month of exposure, every finch had at least one lesion. Death typically occurred over a variable period of time after the birds lost body condition and the lesions became necrotic.

Van Riper et al. (2002) brought naturally infected Hawai'i 'Amakihi, 'Apapane, and House Finches with small lesions in early stages of development into captivity to observe the course of their infections. All 'amakihi succumbed to pox infection with significant declines in body mass, but van Riper et al. did not provide details about the course of lesion development, severity of lesions, or time to death or information about concurrent malarial infections.

More recently, Jarvi, Triglia, et al. (2008) documented differences in virulence between two genetically distinct isolates of pox virus that share similarities with canarypox. Both isolates caused lesions when inoculated into the footpads of Hawai'i 'Amakihi, but the size and severity of the lesions differed. Lesions associated with Pox Variant 1, isolated from a wild Hawai'i 'Amakihi from Kīlauea Volcano, were relatively small and self-limiting, while those associated with Pox Variant 2, isolated from a wild Hawai'i 'Amakihi from Ainahou Ranch at Hawai'i Volcanoes National Park, were large, bloody, proliferative, and ultimately fatal in four experimentally infected birds. The more virulent Pox Variant 2 was not infectious to either Japanese White-eyes or House Sparrows when inoculated into their footpads. Sample sizes in the experimental study were limited but suggest that 'amakihi that recover from infection with Pox Variant 1 have little or no immunity to subsequent exposure to Pox Variant 2. These findings raise some interesting questions about the epidemiology of the virus in wild birds and suggest that interactions between these genetically and biologically distinct pox viruses, their susceptible hosts, and the presence or absence of concurrent malarial infections may play important roles in determining the severity of epidemics.

Other than these limited experimental studies, information about the potential impacts of pox virus on Hawaiian forest bird populations is based on observations of poxlike lesions on captured wild birds (Fig. 9.3). These studies are limited by the presumptive nature of the diagnosis, because other bacterial infections can cause morphologically similar swellings. Based on the recovery of virus from 20 of 22 wartlike lesions of wild birds, van Riper et al. (2002) argued that most of these lesions are likely to be caused by pox virus. Using a presumptive diagnosis based on the presence or absence of tumorlike swellings and/or missing digits, VanderWerf (2001a) studied demographic effects of pox virus in Hawai'i 'Elepaio and found declines in the size of some breeding cohorts that were correlated with the occurrence of pox epizootics. Preliminary observations of O'ahu

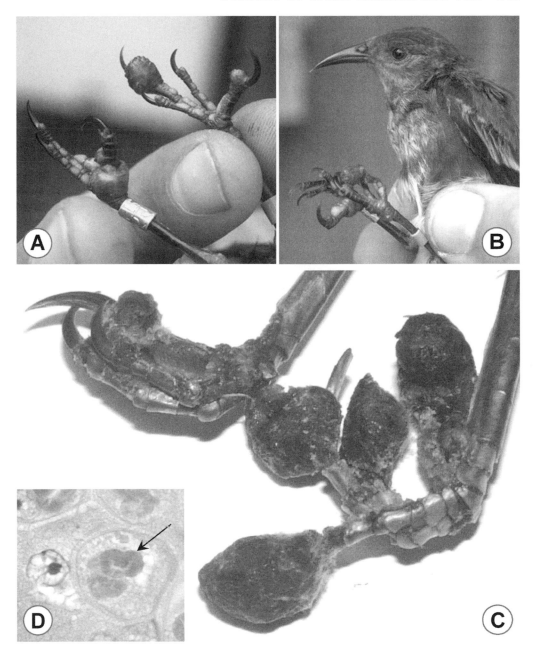

Figure 9.3. Avian pox lesions from wild 'Apapane (A, B) and Hawai'i 'Amakihi (C). The definitive diagnosis is isolation of the virus from the lesions and/or detecting the presence of viral inclusion bodies (Bollinger bodies) (D, arrow) in the epithelial cells of the lesions by histopathology. *Source:* Photos by Carter Atkinson, U.S. Geological Survey.

'Elepaio suggest that annual mortality of birds with active lesions is up to 40%, while birds with mild infections involving only one or more toes frequently recover (VanderWerf et al. 1997, VanderWerf 2001a). More recent work indicates that the prevalence of malaria in O'ahu 'Elepaio is also extremely high, with an average infection rate of 87% in birds that were sampled between 1995 and 2005 (Vander-Werf, Burt, et al. 2006). Pox and malaria infection appeared to be independent of each other in this study, whereas in other field studies in Hawai'i, birds with poxlike lesions have been found to be infected with concurrent malarial infections more frequently than expected by chance (van Riper et al. 1986, Atkinson et al. 2005). Although this finding suggests that the two diseases are interacting with each other, we know little about whether this apparent interaction is caused by simultaneous transmission of the diseases by the vector or by differential mortality among pox, malaria, or pox-malaria infected birds or is simply a reflection of the longer exposure that birds with chronic malaria would have to potential pox infections. The high frequency of concurrent pox and malaria infections makes it extremely difficult to tease out the demographic impacts of either agent alone.

PATTERNS OF MALARIA AND POX INFECTION IN NATIVE AND NONNATIVE FOREST BIRDS

Warner (1968) conducted limited surveys of wild birds in lowland habitats on Kaua'i and in mid- and high-elevation areas above 1,200 m along Saddle Road and in Hawai'i Volcanoes National Park in the late 1950s. He detected patent malarial infections in nonnative House Finches and Japanese White-eyes on Kaua'i and no infections in a limited number of native birds that were sampled on Hawai'i Island. The first large-scale field efforts to sample disease prevalence in wild populations were done in the 1970s on windward Mauna Loa and Kīlauea volcanoes (van Riper et al. 1986, van Riper et al. 2002). These studies demonstrated a high level of prevalence of avian malaria and poxlike lesions in native forest birds at elevations between 900 and 1,500 m and a rapid decline in prevalence above 1,500 m, where mosquito abundance

fell because of cool temperatures. Prevalence of malaria was higher in wet forests than in mesic and xeric habitats, with a seasonal peak in the late fall, immediately after Culex populations peaked and the number of birds with new infections was highest. Unlike Warner (1968), who found malarial infections only in nonnative species, van Riper et al. (1986) detected the highest prevalence in native species, with infection rates of up to 30% detected by microscopy in 'Apapane, and less than 3% in nonnative species, including Red-billed Leiothrix, Japanese White-eyes, and Northern Cardinals (Cardinalis cardinalis). This finding is somewhat counterintuitive given the belief that nonnative birds were the origin of the parasite that is now so widespread in the islands, but it has been supported by the use of more sensitive PCR and serological tests for detecting chronic, low-level infections (Feldman et al. 1995, Atkinson et al. 2005).

Several hypotheses could explain different patterns of malarial infection in native and nonnative birds. Malaria may have been introduced in a passerine or game bird species to which it was closely adapted and had low transmissibility to other nonnative passerines but high transmissibility to native species. After the initial spread of the disease to native forest birds, the source species may not have become established in the islands, leaving a parasite that was highly pathogenic in native birds but with low infectivity and pathogenicity to other nonnative species. Alternatively, the source host may still be in the islands but with a range that is more restricted than the current distribution of avian malaria. Several of the most common and widespread nonnative species—the Japanese White-eye, Red-billed Leiothrix, and House Finch—develop transient infections and are capable of infecting mosquitoes for only a brief period of time following acute phases of the infection (C. Atkinson, unpubl. data). Because of their resistance to infection, these nonnative hosts could act as "buffers" by absorbing infective mosquito bites that would otherwise have gone to susceptible endemic hosts. By contrast, House Sparrows develop parasitemias that are intermediate in intensity between those in the more resistant species and those in the native forest birds and remain chronically infected and infectious to mosquitoes for up to one year after acute infection. We feel that this is evidence that this host is better adapted to Hawaiian isolates of *P. relictum* and suggest that either House Sparrows or birds from another species of nonnative passerine with similar susceptibilities to the parasite may have been primary source hosts for introduction of the parasite to the islands. Recent genetic studies indicate that there is little diversity in the cytochrome b gene among geographically diverse isolates of malaria from the islands, which argues that few introductions of the parasite from few source hosts were successfully made to Hawai'i (Beadell et al. 2006).

Patterns of pox infection in Hawaiian forest birds mirror those of malarial infections, with the highest prevalence in native species, but overall prevalences are lower than those of malaria for any given sample of wild birds. This lower overall prevalence may reflect the ability of native and nonnative birds to completely recover from infection. Like that of malaria, the prevalence of poxlike lesions varies temporally across altitudes, ranging from <5% in some high-elevation habitats to up to 35% at elevations lower than 600 m, depending on host species and time of year.

VECTOR ABUNDANCE AND TRANSMISSION ACROSS ALTITUDINAL GRADIENTS

One of the key outcomes of field investigations by van Riper and colleagues (1986, 2002) was a model depicting the transmission of avian pox and malaria across an altitudinal gradient on Mauna Loa and Kīlauea volcanoes. Van Riper et al. (1986) depicted declining numbers of mosqui-

toes and increasing numbers of native birds between sea level and 1,800 m, with the highest incidence and prevalence of malarial parasites and pox virus infections where these two curves intersect (Fig. 9.4). This model places the highest rates of malaria and pox transmission in mid-elevation (1,200 m) forests, where highly susceptible hosts and vectors overlap. Disease transmission declines above this elevation, virtually disappearing at 1,500 m and effectively creating high-elevation "refugia" from avian diseases.

Scott et al. (1986) were the first to quantify the presence or absence of mosquitoes and their relationship to native bird distribution, and they found that virtually all of the threatened and endangered passerines were less abundant in areas with detectable mosquito populations. They observed "staggering" population declines for Maui 'Alauahio, 'I'iwi, and 'Ākohekohe (*Palmeria dolei*) in habitats with mosquitoes and noted that almost all native bird populations reached their highest densities at altitudes higher than the 1,500 m contour line. These observations agree remarkably well with aviary studies of the pathogenicity of malarial infections in Maui 'Alauahio and 'I'iwi, whose mortality exceeded 75% after exposure to single infective mosquito bites (Atkinson et al. 1995; Atkinson, Lease, et al. 2001). They also agreed well with laboratory studies by LaPointe (2000) of thermal constraints on the sporogonic development of *P. relictum* in experimentally infected *C. quinquefasciatus*, where sporogonic development of the parasite dramatically slowed at 15° C and ceased at 13° C. Interestingly, the 15° C isotherm in Hawai'i closely follows the 1,500 m contour line. This finding suggests that high-elevation refugia are a result more of thermal limits on malarial development in the vector than of the absence of suitable vectors.

At elevations below 1,500 m, the key factor driving the epizootics of pox and malaria is the seasonal and altitudinal distribution and density of the primary vector

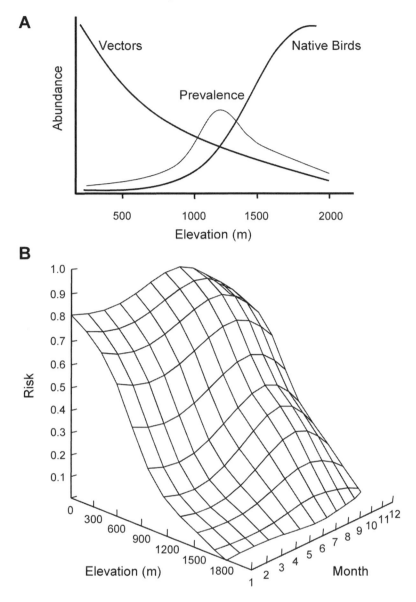

Figure 9.4. The van Riper model for understanding the altitudinal distribution of malarial infections (A) and our current understanding of how disease risk varies across both altitude and season on the eastern slopes of Mauna Loa and Kīlauea volcanoes (B). The primary factor driving the current high rates of malaria transmission at lower elevations is the presence of recently emergent populations of Hawai'i 'Amakihi that have evolved some resistance to the disease. *Source:* Top figure modified from Fig. 11 in van Riper et al. (1986).

of these diseases, *Culex quinquefasciatus*. The important role of this mosquito species in the epidemiology of malaria transmission has been consistently supported by identification of infected individuals in the wild during periods of peak disease transmission and by laboratory experiments that have demonstrated that *C. quiquefasciatus* is several orders of magnitude more susceptible to the parasite than other introduced mosquito species (van Riper et al. 1986, LaPointe et al. 2005).

LaPointe (2000) conducted studies of seasonal and altitudinal variations in vector populations in windward forests of Mauna Loa Volcano. Key findings were the seasonal nature of mosquito populations above elevations of 300 m, with peaks between June and August; the near absence of detectable populations above 1,500 m; their year-round presence in lowland forests; and their remarkably low density relative to many continental mosquito populations. LaPointe (2000) used attractive traps to monitor monthly trends in host-seeking and ovipositing *Culex* populations. In areas with abundant forest bird populations, carbon dioxide (CO_2)–baited traps were less effective in capturing host-seeking mosquitoes. Similarly, oviposition traps in areas with abundant natural and artificial oviposition sites were less effective in trapping ovipositing mosquitoes. In spite of these difficulties, the general trends in mosquito abundance across an increasing altitudinal gradient were found to be similar to the declining curve depicted by van Riper et al. (1986, 2002), with the highest prevalence and intensity of malarial infections in the vector population at elevations between 900 and 1,200 m, where the abundance of native forest birds began to increase. Of particular interest were the prevalence of malaria infection in up to 10% of host-seeking mosquitoes, the low numbers of mosquitoes per trap night, and the high rate of survivorship of mosquitoes in cool, midelevation habitats—all indicating that epizootic disease transmission in Hawaiian forests is efficiently maintained by a vector population that is low in density relative to other parts of the world (LaPointe 2000).

Other than the early reports by Warner (1968) that described pox and malaria transmission to caged native birds in lowland areas on Kaua'i and O'ahu, little information has been available anywhere in Hawai'i about the prevalence and transmission of these diseases in habitats below about 600 m. Shehata et al. (2001) and

VanderWerf et al. (1997) measured the prevalence of malarial and pox infections in lowland nonnative birds and in O'ahu 'Elepaio and O'ahu 'Amakihi (*Hemignathus flavus*) but did not study vector populations or measure transmission rates. The prevalence of malarial infections in O'ahu 'Amakihi captured in Mānoa Valley were surprisingly low, with only 1 of 16 birds (6.3%) found positive for malaria by serological methods and none of 42 individuals found positive by PCR (Shehata et al. 2001). The prevalence of infection in a sample of 268 nonnative birds from the same area was 11.6% by PCR, suggesting that the O'ahu 'Amakihi may have some resistance to infection. Without information about vector populations or transmission rates, however, it is difficult to rule out alternative explanations, including the possibility that the birds were immune survivors of a previous infection with parasitemias at levels too low to be detected by PCR (Jarvi et al. 2002) or that they were simply not exposed to infective mosquito bites. Van derWerf, Burt, et al. (2006) found high prevalences of malarial and pox infection in O'ahu 'Elepaio from nearby valleys on O'ahu and determined that acute, active pox infections may be having demographic impacts on the species' survival either alone or in combination with malarial infection (VanderWerf et al. 1997; VanderWerf 2001a; VanderWerf, Burt, et al. 2006).

In the Puna District of Hawai'i Island, recently emergent populations of Hawai'i 'Amakihi have been found in remnant patches of 'ōhi'a (*Metrosideros polymorpha*) forest at elevations below 300 m (Woodworth et al. 2005, Spiegel et al. 2006). The densities of these populations exceed those of Hawai'i 'Amakihi at elevations above 1,500 m on Mauna Loa Volcano in spite of remarkably high prevalences of infection, which in some areas exceed 80% of the 'amakihi population (Woodworth et al. 2005). These 'amakihi are resident, are reproducing, and appear to be expanding in range in spite of continuous exposure to pox and malaria. The abundance of these low-elevation 'amakihi and the low prevalence of infection in nonnative birds in the same habitat indicate that resident native birds are most likely supporting the high rates of disease transmission.

Given their abundance in areas with high levels of disease transmission, low-elevation 'amakihi appear to be more resistant to the pathological effects of avian malaria and may be evolving some resistance to the disease. This conclusion is supported by experimental studies with low-elevation 'amakihi that show they have a high susceptibility to malaria but greater physiological tolerance to the pathological effects of the disease (C. Atkinson, unpubl. data). Interestingly, these low-elevation birds also have unique nuclear and mitochondrial haplotypes that are not found in 'amakihi from high elevations (Foster et al. 2007, Eggert et al. 2008). Analysis of museum specimens of low-elevation 'amakihi collected between 1898 and 1948, just prior to the presumed introduction of malaria to the island of Hawai'i, identified the same unique haplotypes that are present today, suggesting that the recent resurgence of these birds originated from pockets of surviving individuals with some natural disease resistance rather than from recolonization of the lowlands by high-elevation birds (Foster et al. 2007, Eggert et al. 2008). Low-elevation Hawai'i 'Amakihi have also recently been recognized at 300 m elevation in areas of East Maui that are dominated by 'ōhi'a and almost at sea level in Pelekunu Valley on Moloka'i (S. Aruch, pers. comm.). Similar to the situation on Hawai'i Island, the prevalence of infection in low-elevation 'amakihi on Moloka'i exceeds 75% (C. Atkinson, unpubl. data), suggesting that selection for disease resistance has occurred on other islands as well.

Recent findings from low-elevation Puna and elsewhere suggest that the van Riper model of disease transmission needs modification to take into account both emergent low-elevation native birds and the highly seasonal nature of disease transmission at elevations above 900 m. This change in the model requires modification of the original parasite (disease prevalence), vector, and host curves into three-dimensional surfaces that vary continuously across both seasons and elevations (see Fig. 9.4). The highest incidence of disease will occur where all three surfaces intersect. In areas with native species and mosquitoes, the transmission interface ranges continually from year-round transmission in enzootic lowland areas with high rates of malarial infection in both hosts and vectors to more seasonal epizootic transmission in midelevation habitats, with peaks in the warmer months between July and October. Factors that affect the shape and slope of these surfaces include climatic variables such as rainfall and temperature, availability of larval habitat for Culex mosquitoes, and anthropogenic factors that affect the distribution and density of this habitat across the larger landscape (Chapter 14).

When highly virulent pathogens are introduced into naive populations, epizootics can spread rapidly and have major demographic impacts over wide geographic areas. Based on what we know about the high susceptibility of species of honeycreepers to malaria from both laboratory and field studies, it seems likely that malaria swept rapidly across all of the lower Hawaiian Islands after it was introduced, leaving few survivors. The cool, high-elevation mountains of Kaua'i, Maui, and Hawai'i not only provided the only refugia from the diseases but also fueled high rates of disease transmission at middle elevations by providing a continuous source of highly susceptible native birds. These conditions create a source-sink dynamic across the landscape, because disease transmission at middle elevations is stoked every fall by increasing mosquito populations and a seasonal influx of dispersing, nonimmune birds from higher elevations.

The recent emergence of spatially disjointed low-elevation 'amakihi populations suggest that the system is evolving at different rates because of variation in selective pressure across the altitudinal gradient (Woodworth et al. 2005, Foster et al. 2007). With transmission occurring year-round at lower elevations and low-elevation populations not being continually diluted by emigrating, highly susceptible juvenile birds from high elevations, it might be predicted that disease resistance would first appear at low elevations. This scenario suggests that disease resistance may subsequently spread upslope over the next few decades, with eventual recovery of mid-elevation populations of the more resistant species.

Although this predicted scenario is encouraging for the more common native species that currently have the widest altitudinal and geographic distribution, there is concern that rarer threatened and endangered species may not have enough remaining genetic variability to adapt to these diseases. 'I'iwi, one of the species of honeycreepers most susceptible to malaria, has virtually no variability in mitochondrial DNA haplotypes, yet their levels of diversity at 13 peptide-binding codons in Class II major histocompatibility complex (Mhc) loci were comparable to those of Hawai'i 'Amakihi (Jarvi et al. 2004). This suggests that exposure to avian malaria and/or other parasites is helping to select for diversity in Class II loci in 'I'iwi after what appears to have been a genetic bottleneck in this species, but it is not yet clear whether resistance or susceptibility to avian malaria can be directly linked to particular Mhc haplotypes or other nuclear genes.

THREATS FROM OTHER DISEASE AGENTS

Detailed parasitological and microbiological surveys of native forest birds were not done prior to the introduction of non-native passerines and game birds. Consequently, we know virtually nothing about the original microbiological and helminth fauna of the islands. Perkins (1903) reported an acanthocephalan (Apororhynchus hemignathi) and a cestode (Drepanidotaenia hemignathi) from an 'Akialoa (Hemignathus obscurus), but the host species is now extinct, and the parasites have not been observed or reported since. Van Riper and van Riper (1985) prepared a detailed host parasite checklist of island birds that, though limited in host range and geographic scope, is still the best source for baseline information on potential pathogens present in Hawai'i. It is clear that some of these parasites are endemic to Hawai'i and evolved with their hosts, for example, the proventricular worm Viguiera hawaiiensis (Cid del Prado Vera et al. 1985). Others that commonly occur in Hawaiian honeycreepers, such as the cestode Anonchotaenia brasilense (Vogue and Davis 1953), have a more cosmopolitan distribution and were possibly introduced via migratory birds or imported passerines.

Work and Hale (1996) reported sporadic mortality from trichomoniasis in the introduced Barn Owl (Tyto alba), yet there are no records of its occurrence in the 'Io (Hawaiian Hawk, Buteo solitarius), possibly because pigeons and doves, primary hosts for Trichomonas gallinae, are rarely taken for prey. Other diseases and disease organisms reported from wild birds include toxoplasmosis (Toxoplasma gondii) from captive-reared and released 'Alalā (Hawaiian Crows, Corvus hawaiiensis), Nēnē (Hawaiian Geese, Branta sandvicensis), and wild game birds (Work et al. 2000, 2002); erysipelas (Erysipelothrix rhusiopathiae) from captive-reared and released 'Alalā (Work et al. 1999); echinuriasis (Echinuria uncinata) from Laysan Ducks (Anas laysanensis) (Work et al. 2004); and sporadic outbreaks of avian botulism (Clostridium botulinum) from waterbirds on O'ahu and Hawai'i islands (National Wildlife Health Center, unpubl. data), but none of these diseases or parasites has had the population-level impacts of avian pox and malaria. Although there is currently no evidence that other protozoans, helminths, and bacteria are serious causes of mortality in native Hawaiian forest birds, the mixture of birds that were introduced to the islands, the wide range of potential native and nonnative intermediate hosts for parasites with multihost life cycles, and the favorable environmental conditions and host interactions that may encourage host switching from a nonnative to a native species make ongoing monitoring for new diseases important.

Hawai'i is at risk for introduction of West Nile virus (WNV) from the U.S. mainland, for this introduced pathogen has continued to spread across the hemisphere. Given the high susceptibility and mortality of mainland raptors, owls, corvids, and passerines to the virus; the widespread distribution of vectors in the Hawaiian Islands; the wide host range of the virus; and the devastating history of avian pox and malaria in Hawaiian forest bird communities, the introduction of WNV-infected hosts or vectors through natural migration or commerce could be a disaster for the remaining populations of endemic forest birds (LaPointe et al. 2009). The potential effects could be especially devastating for native species at lower elevations, particularly those that appear to be evolving some natural resistance to pox and malaria. Although pox and malaria appear to have had their most serious effects on endemic honeycreepers, WNV may have its most serious effects on other endemic species, particularly the Nēnē, 'Io, and 'Alalā—all species with close relatives on the mainland that are highly susceptible to the virus.

Development of WNV in the mosquito vector exhibits the same temperature dependence as malaria, with both a delay in development of disseminated infections and a decrease in the proportion of individuals that become infectious after taking a blood meal from a viremic host (Dohm et al. 2002). At ambient temperatures of 18° C, dissemination of the virus to the salivary glands where it becomes infectious can be delayed by up to 25 days, with fewer than 30% of engorged mosquitoes developing disseminated infections. This suggests that should WNV reach Hawai'i, the epidemiology of the virus may have similarities to that of avian malaria, with transmission dropping sharply at elevations above 1,200 m, where vector numbers and mean temperatures fall below 18° C. It is possible that existing high-elevation refugia on Kaua'i, Maui, and Hawai'i islands may continue to afford protection to remnant populations of endemic forest birds if the virus reaches Hawai'i, but only if climatic conditions remain stable.

GLOBAL WARMING AND ITS IMPACT ON HAWAIIAN DISEASE ECOLOGY

The diverse range of climates and rainfall across steep altitudinal gradients in the Hawaiian Islands makes the archipelago an outstanding place to investigate the impacts of global warming on ecological processes. As discussed earlier, the ecology of pox and malaria transmission in Hawai'i is dependent on climatic conditions— primarily seasonal changes in temperature and rainfall, which drive vector populations (Chapter 14). These environmental factors can explain much of the seasonal and altitudinal variation in the abundance of *Culex quinquefasciatus* in Hawai'i (Ahumada et al. 2004), and, in conjunction with thermal limits on the ability of *P. relictum* to complete its development to infectious stages in the vector (LaPointe 2000), these may be the primary factors responsible for the persistence of high-elevation refugia on the higher islands. Benning et al. (2002) recently modeled the effects of a warming of 2° C on the altitudinal location of the 13° C and 17° C isotherms. These isotherms delineate a high-risk zone for malaria transmission (below the 17° C isotherm), a transition zone with seasonal epizootic transmission (between the 17° C and 13° C isotherms), and a low-risk zone for transmission (at and above the 13° C isotherm) where thermal constraints inhibit the complete development of the parasite in the mosquito vector. Following a projected temperature rise of 2° C, the high-elevation forest habitat available in the low-risk zone declined by 57% at Hanawī Natural Area Reserve on Maui, where subalpine habitats are still intact and undisturbed, to as much as 96% at Hakalau Forest National Wildlife Refuge on Hawai'i Island, where cattle grazing has eliminated most forest cover above elevations of 1,800 m. The Alaka'i Wilderness Preserve on Kaua'i currently has no areas in the low-risk zone for malaria transmission because its highest elevation is less than 1,800 m, but it would experience an 85% loss of forested habitat where transmission is highly seasonal to conditions where transmission could occur through most of the year.

If climatic changes occur slowly enough, the tree line and forest bird habitat might be able to keep pace with warming temperatures by moving up the sides of the high volcanoes on Hawai'i and Maui and allowing the lower limits of the high-elevation refugia to continue to remain above the level of disease transmission (Benning et al. 2002). However, the tropical inversion layer may play a more significant role than temperatures in determining the tree line in the Hawaiian Islands (Giambelluca and Luke 2007). Often visible as a layer of clouds around higher peaks, the inversion layer forms as Hadley circulation moves air from equatorial regions to the northern and southern hemispheres (Giambelluca and Schroeder 1998). As this cool, dry air mass descends at latitudes between 15 and 20°, it meets warm, moist air that is driven upward by both convection currents and topographic relief in the Hawaiian Islands in the form of trade winds sweeping eastward into the high mountains of the archipelago. When these air masses meet, they form an inversion layer that caps moisture and cloud development between 1800 and 2400 m. Recent analyses of climatic data from Hawai'i indicate that although temperatures have shown a significant upward trend over the past 80 years, the inversion layer has remained relatively stable in height and increased in frequency of occurrence during the past 25 years (Cao et al. 2007, Giambelluca and Luke 2007).

Although it is difficult to predict whether this trend will continue, a rise in the base height of the inversion layer and subsequent effects on rainfall and forest bird habitat may ultimately have a substantial impact on the persistence of high-elevation disease refugia. Remaining high-elevation forest bird populations may be squeezed between expanding disease transmission from lower elevations and the upper limits of suitable habitat. These changes may push the remaining populations of threatened and endangered honeycreepers to extinction and cause declines in nonendangered species such as 'I'iwi and Maui 'Alauahio that exhibit high levels of susceptibility to avian malaria (Chapter 14).

SUMMARY

The accidental introduction of pox virus, malaria, and mosquito vectors to Hawai'i is an outstanding example of how invasive species and diseases can affect isolated populations of susceptible endemic hosts and how those hosts can respond to those threats through coevolutionary adaptation. Without question, the one factor that prevented widespread and rapid extinction of virtually all of Hawaii's endemic and highly susceptible honeycreepers after the introduction of these diseases was the presence of significant altitudinal gradients on Kaua'i, Maui, and Hawai'i, where susceptible native birds could maintain large populations in relatively disease-free refugia. While providing havens where rarer species can persist, these high-elevation refugia also set up conditions for maintaining high rates of epizootic transmission of the parasite in adjacent midelevation habitats through seasonal movement of susceptible, dispersing juvenile birds and through altitudinal movement of susceptible adults as they track nectar resources. In this sense, more than 70 years after their introduction to Hawai'i, pox and malaria still behave like emerging diseases, with high rates of transmission and mortality in native host populations. Nevertheless, there is evidence that at least some native species are evolving resistance to these diseases and expanding in numbers and geographic distribution in remnant patches of low-elevation native forest. Over the next 50 years it will be interesting to see whether, in lieu of effective control measures (Chapter 17), resistance spreads into midelevation habitats, where birds have been under less selective pressure and where extensive efforts to recover and manage native forest habitats are currently in progress. Preservation of the remaining low-elevation forests, where coevolutionary processes between remaining native species and introduced pathogens can continue to play out, may be the most cost-effective strategy for the long-term sustainability of more common endemic species, but this will do little or no good for the rarest species, which may have already lost the genetic haplotypes that may be needed for resistance to these diseases. The complexity of this system makes management of avian disease in these populations a staggering task for resource managers with few precedents for guiding and assessing the long-term success or failure of specific management actions. The sustainability of Hawaii's rarest birds may ultimately be dependent on vaccine development, captive breeding and propagation to select for disease resistance, and management actions to reduce or eliminate vector populations and expand the high-elevation habitats available.

ACKNOWLEDGMENTS

We thank the U.S. Geological Survey (USGS) Wildlife and Invasive Species programs, the USGS Natural Resource Protection Program, and the National Science Foundation (Grant DEB0083944) for the financial support that made this research possible. The collection of data for the various studies described here would not have been possible without the assistance and support of numerous technical staff, co-workers, and student interns, who spent many long days in the field and the laboratory under often difficult living and working conditions. We are grateful to all of them.

CHAPTER TEN

Genetics and Conservation

SUSAN I. JARVI, ROBERT C. FLEISCHER,
AND LORI S. EGGERT

Ecosystems in Hawai'i are unique because of their extreme geographic isolation, diversity, and high numbers of endemic species, and, regrettably, because of the large number of species that are teetering on the brink of extinction. Biological and genetic changes in populations and species are governed by evolutionary processes that are molded by the unique landscape of the Hawaiian Islands. Genetic studies play a major role in the comprehensive approaches required for effective conservation of endemic avian species. Although the primary focus of many genetic studies begins at the level of the gene itself—at the level of nucleotide variation—a better understanding of the biological or functional consequences of genetic change is especially critical in these often very rare Hawaiian species. This chapter summarizes our current understanding of the conservation genetics of Hawaiian forest birds and how genetic studies are being applied to the long-term sustainability of the native avifauna.

The primary role of genetics in early conservation studies of birds was in documenting the loss of genetic diversity in declining populations, for instance, in the Whooping Crane (*Grus americana*) (Glenn et al. 1999; Jarvi, Miller, et al. 2001) and Clapper Rail (*Rallus longirostris*) (Fleischer et al. 1995). Whereas early theoretical studies supported the idea that reduced genetic variability was associated with reduced population viability (Soulé 1980, Frankel and Soulé 1981), at that time there was little empirical evidence that loss of genetic variability contributed to the decline of wild populations. Since then, genetic diversity has been increasingly recognized as important to the persistence of wild populations. For Song Sparrows (*Melospiza melodia*) on Mandarte Island, British Columbia, Keller et al. (1994) showed that less inbred birds were favored by natural selection during a population crash. Similarly, a long-term study of Greater Prairie Chickens (*Tympanuchus cupido pinnatus*) in Illinois documented reductions in genetic variability and fitness that accompanied reductions in population and habitat size, demonstrating the link between genetic variability and vital demographic traits (Westemeier et al. 1998).

Estimating the loss of diversity in declining populations, as well as the extent of diversity in naturally evolving populations, is important, but with the availability of recently developed molecular genetic tools, conservation geneticists today can do so much more. Researchers can efficiently characterize not just genes or gene fragments but also entire genomes. Through studies of genome expression, researchers are beginning to evaluate the functional consequences of genetic change in natural populations over time. The rapid development and application of streamlined protocols from a growing molecular genetics toolchest has broadened and redefined the role of genetics in conservation.

Genetic data not only yield important insights into the historical and current levels of genetic diversity; they can also provide important information about migration and dispersal over periods of hundreds or thousands of years; assist in defining units of conservation, such as evolutionarily significant units (ESUs) (Moritz 1994); and help measure the effects of natural selection by agents such as newly arrived parasites and diseases. Perhaps more important, genetic studies allow characterization and assessment of biological change resulting from genetic change in natural populations. The information we obtain from genetic studies allows us to make more accurate assessments of the current status of a given species in the context of its natural history, and these assessments can influence management decisions for the species' long-term survival.

Many of the avian lineages in Hawai'i likely began through colonization by a small number of founders due to geographic isolation. Because of the relatively limited size of the islands, avian populations may have remained smaller on average than they might otherwise have been on continents. Frankham (1997) found that island populations generally have lower levels of genetic variation than mainland populations. Genetic studies show that island populations and species differ from those of their mainland counterparts because physical isolation affects gene flow and influences diversity among and within species (Grant 1998a; Johnson, Adler, et al. 2000).

Since the publication of Darwin's theory of evolution by natural selection (Darwin 1859), island ecosystems have long been viewed as natural laboratories for evolutionary studies because of their unique geography and diversity of species and habitats across relatively small spatial scales (Emerson 2002). Important influences on island evolution include extinction and colonization events. The extent and rate of species divergence on islands are dependent on rates of successful migration to and colonization of the islands and on extinction rates (MacArthur and Wilson 1967; Johnson, Adler, et al. 2000). Depending on the degree of species' isolation, founding populations of species may consist of a small number of individuals, and gene flow from the source population may be severely restricted or cut off entirely. Because these founders do not represent the full complement of alleles at the frequencies present in the source population, island populations are especially subject to genetic change due to both genetic drift (random changes in allele frequencies) and natural selection (Grant 1998a).

An important mechanism of genetic change is mutation, a heritable change in genes. When mutation occurs, the selective forces that act on populations can result in higher- or lower-than-expected frequencies of some alleles within populations. The most significant of these genetic changes are those that result in functional change and have biological significance. Mutations that are beneficial may persist in populations through processes of natural selection. Potent forces acting on natural selection can include climatic conditions, competitors, parasites, and disease. Island populations and species flourish or perish depending on whether they possess alleles that allow them to survive and reproduce under novel ecological conditions.

Whereas island populations as a group are physically isolated from mainland populations, Hawaiian bird populations and species also demonstrate island specificity and within-island elevation specificity, most likely because of the unique topography and wide range of habitats in the Hawaiian Archipelago (Woodworth et al. 2005; this volume, Chapter 1). The causes and consequences of inter- and intraisland genetic fragmentation are discussed later in this chapter and elsewhere (Frankham 1997; Johnson, Adler, et al. 2000; Johnson and Seger 2001). One consequence is that species of island birds differentiate rapidly. For example, the sedentary 'amakihi (*Hemignathus* spp.) have evolutionarily distinct species status among islands (Tarr and Fleischer 1993), and populations of high- and low-elevation Hawai'i 'Amakihi (*Hemignathus virens*) on Hawai'i Island appear to have evolved genetic and morphological distinctions between them that suggest that they, too, could eventually be considered biologically distinct populations (Foster et al. 2007, Eggert et al. 2008).

This chapter provides an update on the many genetic studies of native Hawaiian forest birds and how their results are applicable to the conservation of island birds. Detailed explanations of genetic methods and terms are provided in the genetics reference guide found in Box 10.1. We begin with an overview of the current understanding of evolutionary relationships of Hawaiian birds and discuss the use of genetics for determining ESUs. Next we describe the use of genetics to document changes in genetic variation associated with historic population declines and habitat loss. We then describe studies focusing on the major histocompatibility complex (*Mhc*), a diverse group of genes under intense selection pressure. The *Mhc* is potentially relevant to disease resistance in native birds and therefore to their conservation biology. Finally, we describe other molecular tools currently being used to assist with the recovery of Hawaiian forest birds. Our goal is to summarize present and past genetic research on endemic forest birds and to illustrate how these tools can ultimately contribute to the recovery and long-term management of native avifauna in Hawai'i.

TAXONOMIC AND PHYLOGENETIC STUDIES

The precontact forests of the Hawaiian Islands contained a diverse array of bird species (Chapter 1). Evidence that the Hawaiian honeycreepers are monophyletic (originating from a single ancestral source) was first provided by an analysis of allozyme variation in nine species of honeycreepers (Johnson et al. 1989). This evidence was later further supported by analysis of their mtDNA (mitochondrial DNA) sequence (Fleischer et al. 2001). It is thought that as few as 6 lineages of passerines (perching birds) colonized the Hawaiian Islands and subsequently evolved into at least 72 species (Fleischer and McIntosh 2001). Similarly, as many as 19 lineages of nonpasseriforms colonized the islands and evolved into at least 36 species (Fleischer and McIntosh 2001). Thus, adaptive radiation (the process of rapidly changing into diverse forms in response to natural selection) is a common mode of evolution of birds in Hawai'i, particularly

Box 10.1. **Genetics Reference Guide**

Many molecular techniques are based on the fact that DNA naturally occurs as two complementary strands. If these strands are separated (denatured) by heat or by the addition of chemicals, small fragments of single-stranded DNA can find their complements. This is the basis for the design of primers for use with the polymerase chain reaction(PCR), which can amplify specific gene regions from genomic DNA. This process is integral to many of the techniques described here. Some techniques are used to analyze multilocus markers, those that represent many loci throughout the genome. Others target a single informative locus, such as a mitochondrial DNA region or a single nucleotide polymorphism locus. Some of these techniques analyze dominant markers, some co-dominant markers. Co-dominant markers detect each allele at a locus, so one can readily distinguish homozygotes from heterozygotes. This allows direct assessment of allele frequencies and calculation of other population-level statistics (e.g., the major histocompatibility complex and microsatellites). Dominant markers (amplified fragment length polymorphism, restriction fragment length polymorphism) are scored as either present or absent (null), and heterozygotes cannot be directly distinguished from homozygotes. Statistical adjustments in analyses need to be applied, but in general, the sheer number of loci evaluated using dominant markers versus co-dominant markers offsets this distinction, and the end results appear comparable in many studies (Bensch and Akesson 2005).

Amplified Fragment Length Polymorphism (AFLP)

AFLP is a method that consists of four steps: (1) restriction digestion of genomic DNA with enzymes that cut DNA (like molecular scissors), (2) attachment of short DNA segments (adapters) to the ends of digested DNA fragments, (3) use of the PCR to amplify fragments using selective primers, one of which is labeled with a fluorescent dye, and (4) separation and visualization of amplified fragments by gel electrophoresis. The use of the PCR allows for analysis of samples with limited DNA concentrations. Because no prior knowledge of the DNA sequences is required, AFLP has been used across diverse taxonomic groups (Vos et al. 1995). This method produces a large amount of data: one primer set can yield as many as 50–100 bands, representing 50–100 loci. Our research group is currently using AFLP for analyses of native and introduced bird species including the Laysan Duck, ʻAlalā, Hawaiʻi ʻAmakihi, Hawaiʻi Creeper, ʻIʻiwi, ʻApapane, Japanese White-eye, and Red-billed Leiothrix (*Leiothrix lutea*).

Evolutionarily Significant Unit (ESU)

ESU is a term used to describe populations or groups of populations that have achieved a significant level of genetic isolation and appear to differ in adaptive variation from other such populations (Crandall et al. 2000). Although the criteria for defining ESUs remain controversial, they usually include measures of genetic differentiation of both mitochondrial and nuclear DNA (Moritz 1994).

F_{ST} (Fixation Index)

F_{ST} is a measure of genetic differentiation between populations or groups of populations. Although it can be calculated in a number of ways, all measures reflect the effect of population subdivision on

Box 10.1.   *Continued*

inbreeding (Frankham et al. 2002). It is the probability that two alleles drawn randomly (originating from one individual or from multiple individuals) are identical by descent. A low value of F_{ST} indicates high rates of gene flow among populations; as populations diverge and become inbred, the value of F_{ST} increases.

Major Histocompatibility Complex (Mhc)

The Mhc is a group of genes involved in immunity, including susceptibility or resistance to infectious disease. Approaches to detecting Mhc diversity include restriction fragment length polymorphism, single-stranded conformational polymorphism, and the sequencing of Mhc genes or gene fragments that are produced by PCR amplification. There have been numerous studies of Mhc genes in a variety of avian species, including Hawaiian honeycreepers (Jarvi et al. 2004).

Management Unit (MU)

MU is a term used to describe populations that exhibit substantial but incomplete genetic divergence from other such populations. For conservation purposes, such populations may be considered sufficiently independent to warrant separate monitoring and management (Moritz 1994).

Microsatellite Loci

Microsatellite loci consist of tandem repeats of short sequence motifs that range from one to six base pairs in length. PCR primers are designed using the sequences flanking the repeat region, and variation is typically the result of differing num-

bers of repeats, which are reflected in changes in the size of the amplified fragment. Microsatellite loci are distributed throughout eukaryotic genomes and have been used in genetic mapping (Goldstein and Pollock 1997). The large number of alleles (variants that differ in size) and the high level of variation found in individuals have made these loci excellent tools for fine-scaled studies of wild populations. Microsatellites have been developed in or applied to a number of Hawaiian taxa, including ʻAlalā, ʻElepaio, and honeycreepers (Tarr et al. 1998, Tarr and Fleischer 1999, Eggert and Fleischer 2004, Burgess and Fleischer 2006, Eggert et al. 2008).

Mitochondrial DNA (mtDNA)

MtDNA is the remnant genome of a bacterium that colonized early eukaryotic cells more than a billion years ago. In birds as in most animals, it is a circular, maternally inherited, nonrecombining molecule and is roughly 15,000–20,000 base pairs in length. MtDNA generally has a rapid rate of sequence evolution (relative to most nuclear genes) and is a useful marker for population genetics, phylogeography, and phylogeny reconstruction. Initially, variation in an mtDNA sequence was analyzed by purification of the entire molecule from tissues, followed by restriction fragment length polymorphism analysis. Currently, mtDNA is analyzed by PCR amplification and direct sequencing of individual mtDNA genes or regions; in some cases entire mtDNA genomes are sequenced. MtDNA has been shown to be useful for resolving the phylogenies of many Hawaiian birds, including terrestrial waterfowl (Cooper et al. 1996; Soren-

continued

Box 10.1. *Continued*

son et al. 1999; Paxinos, James, Olson, Sorenson, et al. 2002), raptors (Fleischer et al. 2000), rails (Slikas et al. 2002), and songbirds (Tarr and Fleischer 1993, Fleischer et al. 2001).

Molecular Sexing

Although morphological characters are useful to differentiate sex in some bird species, if the species does not exhibit sexual dimorphism or if variability of the characters between and among age classes confounds accurate gender determination, other methods become necessary. Molecular sexing in birds exploits genetic distinctions between the Z and W sex chromosomes, and karyotyping has shown that females are the heterogametic (ZW) sex and males are homogametic (ZZ). PCR-based approaches are based on several gene regions, including the CHD (chromo–ATPase/helicase–DNA-binding domain) gene (Ellegren 1996; Griffiths et al. 1996, 1998). The genetic differences between the W and Z copies of the CHD1 gene have been used for gender determination in a number of birds, including Hawaiian forest birds (Jarvi and Banko 2000, Jarvi and Farias 2006).

The Genetically Effective Size of a Population (N_e)

The N_e of a population differs from the actual numbers of individuals in the population or census size (N) (Wright 1931). The N_e is usually smaller than the number of breeding adults, as actual populations differ from the assumptions of an idealized population in sex ratio, in distribution of family sizes, in constancy of numbers in successive generations, and in having overlapping rather than discrete generations. The N_e is frequently about an order of magnitude lower than N (Frankham et al. 2002).

Nuclear Introns

Nuclear introns are noncoding portions of genes interspersed between coding regions (called exons). They appear to be largely unconstrained by natural selection; thus they tend to be much more variable than exons and can be informative in phylogeographic and phylogenetic analyses (Walsh and Friesen 2003). Like mtDNA, nuclear introns are amplified by PCR. Variation can be analyzed by direct sequencing, cloning and sequencing, or a visualization method such as single-stranded conformational polymorphism (Friesen et al. 1997). Variation in nuclear introns has been assessed for Hawaiian honeycreepers and shows that there is variation both within and among species (Foster et al. 2007).

Restriction Fragment Length Polymorphism (RFLP)

RFLP involves digesting either genomic DNA or selected gene regions with restriction enzymes that recognize particular sequences. If the individual possesses that sequence, the DNA is cut. The restriction products are separated by gel electrophoresis and either visualized directly by staining or transferred to a membrane and visualized by hybridization with a labeled probe (e.g., Southern blot). Individual RFLP profiles are then compared. RFLP has been used for the detection of *Mhc* variation in Hawai'i 'Amakihi and 'I'iwi (Jarvi, Atkinson, et al. 2001).

Box 10.1. *Continued*

Single Nucleotide Polymorphisms (SNPs)

SNPs occur when a single base of DNA sequence differs between individuals. They make up the majority of genetic variation between individuals and can be detected by PCR-based approaches (Brumfield et al. 2003), in which primers are designed for the region of the polymorphism. The PCR is carried out using selected labeled nucleotides that are terminating (only one base is allowed to attach), and the single base polymorphism is detected by gel electrophoresis. Multiple polymorphisms can be detected simultaneously, and this method has been found to be useful when many individuals are analyzed simultaneously. This technique has been used to detect variation in an avian malaria gene (Farias 2008, Farias and Jarvi 2009).

Single-Stranded Conformational Polymorphism (SSCP)

SSCP techniques involve PCR amplification of selected gene regions or use of digested genomic DNA. The DNA is processed (denatured) to become single stranded, then separated by gel electrophoresis. Single-stranded DNA will form unique conformations (secondary structures) under nondenaturing gel-running conditions based on its DNA sequences. Differences in DNA sequences can then be visualized by staining. SSCP has been used for detection of *Mhc* variation (Jarvi et al. 2004) and in the detection of malaria variants (S. Jarvi and M. Farias, unpubl. data).

---

for the Hawaiian honeycreepers (Freed et al. 1987). Although it appears that at least 10 of all the colonizing lineages radiated into 2 or more species, most of these radiations are now extinct. Late Holocene, pre–Western contact extinctions have been well documented elsewhere (James and Olson 1991, Olson and James 1991), and extinctions following Western contact are summarized in Chapter 2.

Recently, the ability to extract and sequence ancient DNA from museum specimens and subfossil (partly fossilized) bones has enabled us to learn much about the evolutionary history of these extinct taxa. These techniques have allowed us to document the closest mainland relatives of Hawaiian lineages, determine the phylogenetic relationships of species in Hawaiian radiations, and even determine the time frames in which the radiations occurred (Fleischer et al. 1998, Fleischer and Mc-

Intosh 2001). Pratt and Pratt (2001) reviewed how molecular and morphological studies have recently influenced the systematics of Hawaiian birds such that nearly all distinct populations are now classified at the species level.

An essential application of genetics to conservation is the determination of units of conservation for species of Hawaiian birds, usually framed in terms of ESUs (Moritz 1994, Fleischer 1998, Crandall et al. 2000). Determination of an ESU involves basic phylogeographic and population genetic analyses using molecular and nonmolecular markers. The results are used to determine whether the species should be divided into geographically separate populations that are on divergent evolutionary trajectories. Most definitions have required genetic indicators of separation, such as evidence of reciprocal monophyly of mtDNA sequences, where each ESU is a

Figure 10.1. A Laysan Duck, part of a group translocated to Midway Island to establish a second population. The Laysan Duck was historically restricted to Laysan and Lisianski islands in the Northwestern Hawaiian Islands chain. Ancient DNA methods applied to subfossil bones have uncovered extinct populations of Laysan Ducks on the main Hawaiian islands, indicating a once widespread range. This information helped justify the decision to reestablish populations of Laysan Ducks on islands where none historically occurred. *Source:* Photo courtesy of James H. Breeden Jr., U.S. Geological Survey.

monophyletic group (descended from a single common ancestor) that does not include any portion of the other and shows significant divergence of allele frequencies at nuclear loci (Moritz 1994). More recently, significant morphological or ecological differences have been considered equally important for ESU delineation. The use of genetics to determine ESUs has important political ramifications, because the Endangered Species Act requires definition of a discrete unit for preservation, whether species, subspecies, or "distinct population segment" (DPS). In some studies, DPSs have been equated with ESUs (U.S. Fish and Wildlife Service 1996a), although this notion has been criticized (Pennock and Dimmick 1997). A less restrictive concept, that of the "management unit" (MU), has been defined as a population that differs significantly from others in allele frequencies (Moritz 1994).

An increasing number of studies are being conducted to determine the subspecific or specific status and ESU delineation of Hawaiian forest birds. These studies have provided information about the extent and boundaries of conservation units and their priority for conservation, along with information about ancestral ranges and justification for reintroduction. An early study using ancient DNA methods on historical and extinct samples revealed that the Laysan Duck (*Anas laysanensis*) (Fig. 10.1) has not always been restricted to the Northwestern Hawaiian Islands but occurred more widely through the archipelago until just prior to Western contact (Cooper et al. 1996, Rhymer 2001). Information about the species' prior range has allowed the U.S. Fish and Wildlife Service to develop plans for management and reintroduction that now include areas within and outside of the Northwestern Hawaiian Islands.

Some authors have suggested that conservation priority should be given to taxa that are more phylogenetically unique than other taxa (Faith 1992, Krajewski 1994). One example is the Poʻo-uli (*Melamprosops phaeosoma*), a unique but potentially extinct member of the honeycreeper subfamily (Chapter 21). Genetic studies played a key role in defining the classification of the Poʻo-uli, whose phylogeny and systematics were left unsettled by morphological studies (Pratt 1992, Fleischer et al. 2001, Pratt 2001, James 2004). A concurrent phylogenetic study using both mtDNA sequences and osteological characters placed the Po'o-uli within the drepanidine clade but estimated that it had split from its closest relatives early in the drepanidine radiation, at approximately 3.8 ± 0.9 million years (Fleischer et al. 2001). Phylogenetic analyses of 13 nuclear chromo–ATPase/helicase–DNA-binding domain protein 1 (CHD1) sequences (Jarvi and Farias 2006) also supported a close yet distinct phylogenetic relationship of Po'o-uli with extant honeycreepers. Combined phenotypic and currently available genotypic data concur that *Melamprosops phaeosoma* has been a distinct and somewhat genetically and morphologically unique lineage within the Drepanidinae.

Genetic markers have also been used to reclassify some subspecies of honeycreepers as species. Based on mtDNA, the Kaua'i 'Amakihi (*Hemignathus kauaiensis*) and O'ahu 'Amakihi (*H. flavus*) are sufficiently differentiated from Hawai'i 'Amakihi on Mauinui and Hawai'i islands (*H. v. wilsoni* and *H. v. virens*, respectively) to be considered distinct species (Tarr and Fleischer 1993). The subspecies of Hawai'i 'Amakihi on Maui-nui and Hawai'i conform at the least to definitions of subspecies and ESUs. The reciprocal monophyly of Hawai'i 'Amakihi on Maui and Hawai'i is supported by analyses of mtDNA control region sequences. Other studies reveal differentiation among islands (Jarvi et al. 2004) and also provide evidence of genetic structuring by elevation within Hawai'i Island based on mtDNA data (Foster et al. 2007, Eggert et al. 2008). Another lineage that shows patterns of interisland differentiation is the 'Elepaio (*Chasiempis sandwichensis*) (Burgess 2005). In addition, sequences obtained from the bones of an extinct crow (*Corvus* sp.) on Maui appear to be from a sister taxon to the 'Alalā (Hawaiian Crow, *Corvus hawaiiensis*) of Hawai'i Island (Fleischer et al. 2003).

These genetic studies have confirmed that allopatry (speciation due to physical isolation) of closely related populations has allowed for speciation (or the beginning stages of it) in many lineages.

By contrast, a few species show no clear evidence of genetic structure based on mtDNA control region sequences either within or among islands. These include the 'I'iwi (*Vestiaria coccinea*) and 'Apapane (*Himatione s. sanguinea*) (Jarvi et al. 2004, Foster et al. 2007), species that also show no morphological variation among islands. They are vagile species that presumably have had less restricted gene flow.

DEMOGRAPHIC HISTORY AND CURRENT GENETIC STATUS OF POPULATIONS

Neutral genetic markers such as mtDNA, minisatellites (DNA repeats), or microsatellites (extremely short DNA repeats) are very useful for both determining standing genetic variation and documenting changes in variation over time and space. These markers can provide information about the demographic history of populations by averaging diversity estimates across various time periods based on their relative rates of mutation and fixation. A number of studies have used these markers to document genetic variation in Hawaiian birds and to interpret this variation in reference to historical population declines (bottlenecks), habitat loss, and population fragmentation.

Comparisons of Populations

One of the earliest studies to document the genetic variation of a population within a native Hawaiian bird species assessed variation at 33 allozyme loci in Laysan Finches (*Telespiza cantans*) from their source population (Laysan Island) and from translocated populations on Pearl and Hermes Reef (Fleischer et al. 1991). The study found evidence of a low level of genetic variation in the species but an unexpectedly higher level of genetic variation in the translocated rather than the source population. Only 5 of the 33 loci were variable, and this low number may have resulted in insufficient power to clearly resolve similarities and differences between the two populations. An F_{ST} estimate (an indication of relatedness between populations; see Box 10.1) based on these variable allozyme loci of 0.05 was considered relatively high, especially in light of the history of this population, and suggested that the founding events and small population sizes had resulted in considerable genetic drift.

A later study by Tarr et al. (1998) assessed variation at nine dinucleotide microsatellite loci and found that the degree of genetic variation, as measured by the average number of alleles and heterozygosity, was lower in the translocated populations. Oddly, each of the three Pearl and Hermes populations also showed significantly more heterozygosity than expected, while the Laysan populations did not. These authors speculated that this finding may have resulted from nonrandom mating or selection for heterozygotes in these small and perhaps more stressful environments. The F_{ST} values on all islands were significantly different from zero and were even higher for the microsatellite loci (~ 0.17) than for the allozymes. This differentiation suggests that genetic drift plays a major role in these small and mostly isolated populations. Interestingly, the Laysan and Pearl and Hermes Reef populations would be considered separate "management units" based on their divergence and current MU definitions (Moritz 1994), in spite of the fact that one population was formed from the other by translocation less than 40 years ago.

Neutral genetic markers have also been used to investigate the genetic structure of Palila (*Loxioides bailleui*), another endangered Hawaiian honeycreeper that is currently restricted to high-elevation habitats on Mauna Kea Volcano. Analysis of minisatellite DNA from two small and isolated populations on Mauna Kea revealed typical and relatively high levels of variation (Fleischer et al. 1994). Thus the recent population declines the species has suffered have not reduced its minisatellite variation. The two populations also had very similar minisatellite allele frequencies, with a mean F_{ST} of 0.026. This finding suggests that not much differentiation has occurred subsequent to the relatively recent fragmentation of this species. In spite of continued population decline, a recent comprehensive study of 171 Palila (including 24 complete families) by amplified fragment length polymorphism (AFLP) also suggests levels of variability that are comparable to those reported from other finch species (Patch-Highfill 2008).

The Genetics of Population Bottlenecks

The Nēnē (Hawaiian Goose, *Branta sandvicensis*) (Fig. 10.2) is another highly endangered native Hawaiian bird that has suffered a historical bottleneck. Nēnē are known from the Holocene subfossil record of most of the main Hawaiian islands (Olson and James 1991) but were found after Western contact only on Hawai'i Island. Estimates of their historical population size vary, but Nēnē were thought to have been extremely common on Hawai'i at the time of Captain James Cook's arrival in 1778 and shortly thereafter (Baldwin 1945, Kear and Berger 1980). The Nēnē population declined precipitously during the nineteenth and early twentieth centuries due to hunting, introduced predators, and habitat loss and was reduced to perhaps as few as 17 individuals by the 1950s (Elder and Woodside 1958, Kear and Berger 1980). Only through captive breeding programs in Hawai'i and England was the Nēnē able to be reintroduced to its prior ranges and escape extinction.

As might be predicted from their historical bottleneck, the level of genetic variation found in both captive and wild Nēnē populations is extremely low in compari-

Figure 10.2. Nēnē (Hawaiian Goose) pair. Nēnē exhibit a low level of genetic variation. Analyses of ancient DNA from subfossil bones and museum specimens suggest that much of the variation may have been lost during a population decline that occurred several hundred years ago, well before the species' historic bottleneck in the nineteenth century. *Source:* Photo © Jack Jeffrey.

son to that found in other species of geese based on minisatellite DNA, mtDNA sequences, and microsatellites (Rave et al. 1994; Rave 1995; Paxinos, James, Olson, Ballou, et al. 2002; Veillet et al. 2008). Rave's (1995) analysis of 75 wild Nēnē from six populations revealed no difference between wild and captive birds in genetic variation or distinct fragments at minisatellite DNA markers, suggesting that all or most of the current Nēnē population may be derived from captive birds. Either the wild population was extirpated prior to reintroduction or birds of wild origin have not contributed substantially to present-day wild populations. In captive Hawaiian Nēnē, Rave et al. (1998) also found evidence of inbreeding effects, including reduced fertility, hatchability, and total survivorship. Data from ten polymorphic microsatellite loci also suggest high levels of inbreeding in the Nēnē (Veillet et al. 2008).

A study by Paxinos, James, Olson, Ballou, et al. (2002) has shown that Nēnē from Hawai'i Island went from typical mtDNA variation prior to about 800 years ago to almost no variation by about 500 years ago. This lack of variation continues into present-day populations, leading these authors to propose that the Nēnē population on Hawai'i Island had undergone a major prehistoric decline in genetic variation sometime during that 300-year period. They suggested that the same factors that had caused the extinction of many other large waterfowl and ibises during that period had also caused extirpation of the Nēnē from the other main islands and a major population decline on Hawai'i Island (Paxinos, James, Olson, Sorenson, et al. 2002), as had Olson and James (1991). The authors of the later study speculated that Nēnē may have survived on Hawai'i because of kapu (taboos) placed on them by powerful Hawaiian ali'i (chiefs) or because of changing land use patterns caused by an increase in warfare.

The 'Alalā is an endangered Hawaiian crow of uncertain taxonomic affinities (James and Olson 1991; Fleischer and McIntosh 2001; this volume, Chapter 20). Historically found only on Hawai'i Island, it appears to have had a relatively large population as recently as the late nineteenth or early twentieth century (Scott et al. 1986, Duckworth et al. 1992). Its numbers began to decline severely after 1890, and by 1978 the species had been reduced in the wild to an estimated 76 ± 18 individuals (Scott et al. 1986). By 1992, the estimated population size was 11 individuals (Duckworth et al. 1992). Before 1986, 14 wild-caught birds were brought into the captive program, of which only 4 contributed to subsequent generations. These 4 produced 6 captive-bred young (Duckworth et al. 1992). Starting in 1993, 13 eggs from the nests of 3 wild pairs from adjacent territories on the McCandless Ranch were hatched and raised in captivity. Some of these captive-reared birds were released back into the wild population, and some were incorporated into the captive flock. Because the relationships among the original 4 founders and these later 6 founders were not known, genetic analyses were completed to estimate these relationships, to compare the observed pedigree with genetic evidence, and to evaluate the extent of genetic diversity in this small population using mtDNA sequences, microsatellites, and AFLP.

For the mtDNA control region, all sequences were identical among the 46 individuals that were tested (Fleischer 2003). 'Alalā had experienced mtDNA variation in the past, for three haplotypes among six sequences were found in 'Alalā museum specimens collected during the late nineteenth century (Fleischer et al. 2003). The current low level of variation in 'Alalā also does not appear characteristic of corvids; comparatively higher levels of variation were found in the same mtDNA region in the Common Raven (*Corvus corax*) (Tarr and Fleischer 1999, Omland et al. 2000).

Microsatellite results also indicated that 'Alalā have very low levels of genetic variability and that variation was lost during the population decline that occurred during the nineteenth and twentieth centuries (Fleischer 2003). Kinship and cluster analyses can depict known family relationships and were used to estimate the relatedness of birds of unknown relationship. These analyses were able to show that the McCandless Ranch birds and earlier founders were not closely related.

AFLP analyses of 41 individuals with seven sets of primers again revealed low levels of diversity in the extant 'Alalā population (Jarvi and Bianchi 2006). A total of 19 distinct banding patterns was distinguished among these 41 individuals, with one predominant pattern that included individuals originating from several families and lineages. Phylogenetic analysis revealed two main clusters, with the McCandless Ranch birds in one cluster. By all methods used to date, the level of genetic variation in 'Alalā appears relatively low.

EVOLUTION IN ACTION: CONSEQUENCES OF EXPOSURE TO INFECTIOUS DISEASE

Although multiple factors continue to influence the fate of Hawaiian birds, the relatively recent introduction of two infectious diseases, avian malaria (*Plasmodium relictum*) and avian pox virus (*Avipoxvirus* sp.), have significantly affected the distribution and abundance of many honeycreeper species. The genetics of host-parasite(s) interactions is complex and involves many innate and adaptive mechanisms of hosts' immunity. The Mhc is a group of genes with a significant role in hosts' immunity. The Mhc was first identified as the genetic locus responsible for allograft (tissue) rejection and is responsible for determining what is viewed as "self" and what as "nonself" by the immune system. Mhc Class I and Class II genes have been found in all well-characterized vertebrates (Klein 1986). Class I molecules are present on essentially all nucleated cells in the body (including erythrocytes in birds), whereas

Class II molecules are expressed on only certain cells of the immune system. The Mhc is polymorphic (multiallelic) and multigenic in most species investigated to date.

The genes of the Mhc, like many other genes, are subject to selective pressure by infectious diseases. Selection by infectious disease has long been thought to influence the genetic makeup of hosts (Haldane 1948, 1949; May and Anderson 1983); a well-known example in humans is the selective drive of malaria (Hill and Weatherall 1998). Mhc genes are thought to undergo positive or balancing selection, likely due to differential recognition of pathogen proteins (Kaufman and Salomonsen 1997, Hill and Weatherall 1998, Hughes and Yeager 1998, Hedrick and Kim 2000). Evidence of balancing selection includes high numbers of alleles and sequence polymorphism among Mhc genes, long-term retention of alleles (i.e., transpecies polymorphism), and relatively high rates of nonsynonymous nucleotide changes (d_N). These changes result in amino acid changes (phenotypic changes), whereas synonymous or silent substitutions (d_S) do not result in amino acid changes at the site of the Mhc-pathogen interaction or in the peptide-binding region (Hughes and Yeager 1998, Hedrick 1999). The use of disease to determine the effects of such selection on variation at the Mhc is addressed or reviewed elsewhere (Hill and Weatherall 1998, Jarvi et al. 2004).

The Hawaiian honeycreeper disease system is being used as a model for investigating the influences of parasite-driven selection on host genes and the evolution of virulence. Studies of the Mhc in honeycreepers have described extensive variation at the peptide-binding codons (PBCs, the Mhc codons hypothesized to be involved in pathogen interaction) in species that otherwise have very low levels of genetic diversity (Jarvi et al. 2004). The 'I'iwi, a species in which other genetic markers show extremely low levels of variation, has levels of nonsynonymous Mhc variation (indica-

tive of a functional change) that are roughly equal to those of the Hawai'i 'Amakihi, a species with higher levels of genetic variation. Both in honeycreepers and in Darwin's finches, there are also distinct clusters of sequences (likely representing loci) that have very low levels of variation (especially nonsynonymous levels) that are atypical of this Mhc region. These contrast with the main cluster (consisting of multiple loci) of sequences, which collectively show a high rate of nonsynonymous substitution and high levels of diversity that are typical of avian Mhc sequences (Jarvi et al. 2004).

There is evidence of balancing selection on the Mhc of honeycreepers in the finding of a d_N/d_S ratio of 8.6 at the PBC sites among variable loci. This d_N/d_S ratio in honeycreepers is notably (nearly three times) higher than those in the Darwin's finches $(d_N/d_S = 3.10)$, but the ratios are comparable at the other nonsynonymous sites (Jarvi et al. 2004). Elevated rates of nonsynonymous substitutions have been documented in island birds compared with their mainland counterparts at mtDNA genes (Johnson and Seger 2001), but the honeycreeper d_N/d_S ratio at the Mhc is high, even for island species, possibly due to exposure to introduced infectious diseases.

Jarvi et al. (2004) detected identical sequences, likely representing identical alleles, among honeycreeper species. The existence of identical alleles among species has been reported in many mammalian (Klein et al. 1993) and some avian species (Vincek et al. 1997). Alleles shared between species are presumed to have arisen by either convergent evolution or interbreeding of species, or they may have been present in the ancestral species and were retained, possibly by selection, throughout the process of speciation (transspecies hypothesis; Klein et al. 1993). The interdigitation of Mhc sequences from different honeycreeper species provides additional evidence of balancing selection (long-term retention of alleles) on the Mhc. These alleles are likely important if they transcend the processes of speciation and should be considered in any conservation strategy that seeks to manipulate gene pools.

Genetic comparisons between Hawai'i 'Amakihi and 'I'iwi (Jarvi et al. 2004) suggest that 'I'iwi are significantly less variable (P < 0.001) at Class II loci than Hawai'i 'Amakihi based on restriction fragment length polymorphism (RFLP) or Southern blot analyses (Jarvi, Atkinson, et al. 2001). However, there is no significant difference in Mhc sequence variability between 'I'iwi and Hawai'i 'Amakihi at the 13 PBC sites, and their d_S and d_N values and genetic distances did not differ overall (Jarvi et al. 2004). At the 32 non-PBC sites, the diversity in 'I'iwi appears consistently lower than that observed in Hawai'i 'Amakihi. Thus, the RFLP or Southern blot analyses of Class II genes may reflect the low variability in 'I'iwi at non-PBC or other adjacent or linked sites but cannot resolve the higher level of variability at the 13 PBC sites. Additional studies of the nuclear genes of individuals subjected to experimental infection with malaria and of those from natural populations are currently under way. These studies will allow us to more directly determine the relationship between particular Mhc variants and malaria resistance.

MtDNA sequences have thus far shown no evidence of variability within or among island populations of the 'I'iwi, whereas mtDNA variation in the Hawai'i 'Amakihi is substantial (Tarr and Fleischer 1995, Fleischer et al. 1998, Jarvi et al. 2004, Foster et al. 2007). In addition, allozyme variation in the 'I'iwi appears reduced relative to the level found in the Hawai'i 'Amakihi (Johnson et al. 1989). The relatively low levels of variation in these genes most likely resulted from a population bottleneck that we believe occurred in 'I'iwi populations sometime in the recent evolutionary past— that is, thousands or tens of thousands of years ago. We do not think this putative bottleneck occurred more recently because

the nuclear variation in the 'I'iwi as assessed by AFLP and microsatellite data is similar to that found in the Hawai'i 'Amakihi (discussed in Jarvi et al. 2004; microsatellite data are found in Eggert et al. 2008). The microsatellite DNA sequences and selected AFLP markers are more rapidly mutating than the allozyme loci and mtDNA and should return to higher or equilibrium levels more rapidly, but they still would need thousands of years to do so. The current malaria and pox epidemics did not start until after their mosquito vector was introduced to the Hawaiian Islands in the early 1800s. Because the Mhc mutation rate should not be higher than those found for allozyme loci, these results in combination suggest that some other, earlier factor has maintained Mhc variation in the face of an apparently major genetic bottleneck (Jarvi et al. 2004).

FURTHER TOOLS FOR RECOVERY

Molecular Sexing

Altered sex ratios can significantly influence the genetic structure of a population. In natural populations, changes in genetic structure, for example, those due to sex bias or other factors, influence the effective population size (N_e) (see Box 10.1). It is N_e, not the actual census size (N), that determines the genetic future of a population, and unequal sex ratios reduce the effective size of the population toward the number of the sex with fewer breeders (Frankham et al. 2002). Populations, especially those that are small and fragmented, or perhaps particularly those impacted by artificial manipulation, should be monitored for sex bias.

Previous studies that evaluated sex ratios in Hawaiian birds were limited and traditionally were based on morphological differences in sexually dimorphic species (Fancy et al. 1992, Jeffrey et al. 1993, Lindsey et al. 1995). Many species of Hawaiian birds, however, are sexually monomorphic.

Molecular-based tests have been indispensable for distinguishing sex in these species and can be especially important in studies involving juveniles. PCR methods based on the sex-linked CHD1 gene are capable of accurately identifying sex in the eight endemic species and eight introduced species tested so far (Jarvi and Banko 2000, Jarvi and Farias 2006).

PCR-based sexing also provides important information at the level of the individual bird. In 1998, attempts by three independent laboratories to establish the sex of the last three remaining Poʻo-uli by means of PCR-based sexing using DNA extracted from feathers produced inconsistent results. The results were especially difficult to interpret for an individual (HR2) brought into captivity in 2004 (Chapter 21). Based on the results of one laboratory, bird HR2 was reported as female (Groombridge, Massey, Bruch, Malcolm, Brosius, Okada, Sparklin, et al. 2004), but the PCR products were not sequenced for verification. Upon reevaluation by Jarvi and Farias (2006), the PCR results using three primer sets consistently indicated that the captive individual was male, matching the earlier results from two of the independent labs. The PCR results also confirmed the original determination that the bird was male based on its plumage (Baker 1998). The 13 CHD1-Z sequences obtained from this individual were also compared with other species to provide further insight into the relationship between the Poʻo-uli and other honeycreepers. So, although CHD1 gene sequences are important for determining birds' sex when morphological characters are inconclusive, because these genes are highly conserved, they are also useful for evolutionary and phylogenetic studies.

Disease Diagnostics and Parasite Coevolution

Species of *Plasmodium* and other avian blood parasites have frequently been used to test evolutionary hypotheses about hosts' sexual selection, immunocompetence, and evolution of virulence in birds and reptiles (Hamilton and Zuk 1982; Perkins et al. 1998; Westneat and Birkhead 1998; Jarvi, Atkinson, et al. 2001). These organisms are well suited to studies of the effects of infection on hosts' fitness because, after an initial acute phase of infection, their hosts develop chronic, low-level parasitemias that are regulated by the hosts' immunity. Avian hosts remain infected for life, undergoing periodic relapses that are controlled by complex interactions between the hosts' immune response, physiological stress, and endocrinology (Atkinson and van Riper 1991; Jarvi, Atkinson, et al. 2001; Jarvi et al. 2002; Atkinson et al. 2005; this volume, Chapter 9). Individuals with chronic infections are of particular interest because they have survived acute stages of the infection and may have some genetic resistance to the parasite (Jarvi, Atkinson, et al. 2001), making them desirable candidates for incorporation into captive breeding and translocation programs.

As is the case with any infectious disease, accurate diagnostics are essential for identifying disease outbreaks and implementing and measuring the effectiveness of control programs. The "gold standard" for malaria detection remains the visual detection of the parasite by microscopy of stained blood smears. However, the numbers of circulating parasites may become so low in chronic infections that they are virtually impossible to detect by microscopy. The extreme sensitivity inherent in PCR-based tests suggests that they should be highly sensitive for diagnosing chronic, low-level parasitemias. Jarvi et al. (2002) compared the sensitivity of two PCR-based methods with microscopy and a serological (antibody-based) technique for diagnosing avian malaria in experimentally infected Hawaiʻi ʻAmakihi. Microscopy was significantly less sensitive than either serology or nested PCR for detecting chronic malarial infections. Individually, none of the diagnostic methods was 100% accurate in detecting subpatent infections. Thus, although PCR by itself may underestimate true prevalence, when combined with other methods it provides a very reliable means for diagnosing chronic malarial infections in Hawaiian birds.

PCR-based approaches are essential for sequencing genes or gene fragments for the detection of variation in both *P. relictum* (Jarvi et al. 2002, 2003; Beadell et al. 2006) and *Avipoxvirus* (Jarvi, Triglia, et al. 2008). Mitochondrial and nuclear dihydrofolate reductase–thymidylate synthase (DHFR-TS) genes have been used to evaluate current lineages of avian malaria in Hawai'i and elsewhere to determine the geographic origins of the parasite (Beadell et al. 2006). The resulting data suggest that a single mitochondrial and DHFR-TS lineage exists in Hawai'i. Mitochondrial DNA is also being used to determine the approximate time of introduction of disease to the islands through analysis of museum specimens (R. Fleischer and S. Jarvi, unpubl. data). Other studies using nuclear gene region sequencing, RFLP, and single nucleotide polymorphisms suggest higher levels of variation and show that individual birds are infected by multiple variants of *P. relictum* based on analysis of the TRAP (thrombospondin-related anonymous protein) gene (Farias 2008; Jarvi, Farias, et al. 2008).

A recent study of *Avipoxvirus* 4b core polypeptide sequences in Hawaiian birds detected two variants that were distinct from fowl pox (Jarvi, Triglia, et al. 2008). In contrast to the situation in malaria infections, all of the 15 individual birds evaluated were infected with either one or the other variant; no evidence of multiple infections was found. The 4b core gene sequence data from an 'Elepaio collected in 1900 on Hawai'i Island clustered closely with one of the two variants, suggesting that this variant has been in Hawai'i for at least 100 years. The high level of variation found between the three 4b clusters provides evidence of multiple, likely independent, introductions of *Avipoxvirus* to Hawai'i

and does not support the hypothesis that native species were infected through the introduction of infected fowl. Experimental *Avipoxvirus* infections in native Hawai'i 'Amakihi suggest that the 4b-defined variants may be biologically distinct, with one variant appearing more virulent. Studies are currently under way to further evaluate how avian pox viruses interact with avian malaria and impact virulence through the modulation of hosts' immune responses.

The Development of Cell Culture Lines from Hawaiian Birds

The cloning of "Dolly" the sheep (*Ovis ares*) in 1997 (Wilmut et al. 1997) brought the subject of cloning back into mainstream discussions. This technology involved nuclear transfer to enucleated sheep ova, and in 2001 a mouflon sheep (*O. musimon*) was cloned using donor nuclei obtained from a dead mouflon (Loi et al. 2001). Although cloning has long been and will likely remain a provocative issue, it has potential as an important tool in the conservation of species. Uncertainties abound regarding the fitness of cloned individuals and the extent to which cloned individuals can contribute to the conservation of gene pools. However, the collection of cells for cell culture development should be a priority if subsequent cloning is to be considered as a means of contributing, along with other conservation strategies, to the retention of genetic variation in especially small populations (Ryder 2002).

The Hawaiian bird species from which viable cell cultures have been produced at this time are the Nēnē, 'Alalā, Maui Parrotbill (*Pseudonestor xanthophrys*), Hawai'i Creeper (*Oreomystis mana*), 'Ākepa (*Loxops coccineus coccineus*), Po'o-uli (Frozen Zoo®, Conservation and Research for Endangered Species, Zoological Society of San Diego; O. Ryder, pers. comm.), and Hawai'i 'Amakihi (C. Atkinson, pers. comm.). Cloning may be the only viable option in extreme cases such as that of the Po'o-uli, where other measures are infeasible and attempts to facilitate reproduction are not likely to succeed (VanderWerf, Burt, et al. 2006; this volume, Chapter 21).

CONCLUDING THOUGHTS ON GENETICS AND CONSERVATION OF HAWAIIAN FOREST BIRDS

Molecular technology is providing increasingly powerful tools that are essential for comprehensive conservation management. These tools allow accurate estimates of genetic diversity in small, fragmented populations and also provide the information needed to reduce their genetic impoverishment and risk of extinction. For instance, managers use genetic studies to increase their confidence in the status or relationships of populations under study. At another level, detection of declines or other changes in genetic diversity is important in management because reductions in heterozygosity or allelic diversity in especially small, fragmented populations greatly elevates their risk of extinction. A loss of genetic variation, especially the variation related to adaptive evolution, reduces the capacity of a population to evolve to respond to environmental change. A fundamental of Hawaiian forest bird conservation should be to preserve natural patterns of genetic variation so that evolution by natural selection and other processes can continue unimpeded. Unfortunately, the Hawaiian avifauna has been and continues to be greatly affected by a variety of human-related activities. A summation of the genetic studies of Hawaiian birds, major parasites, and mosquito vectors that are currently published or known to be in progress at this time is presented in Appendix 10.1.

The *Revised Recovery Plan for Hawaiian Forest Birds* (U.S. Fish and Wildlife Service 2006) covers 21 taxa in the main Hawaiian islands. Ten of these have not been observed in the past 10 years and may now be extinct. The remaining species are mostly found in fragments of wet forest at elevations above 1,200 m. In general, small populations tend to have less initial genetic diversity, and they tend to lose diversity at a greater rate than do large populations (Frankham et al. 2002). These small, fragmented populations are also much more susceptible to the effects of genetic drift and inbreeding, in part because selection is less effective in small populations. If selection is less effective, deleterious alleles are less likely to be removed by natural selection and may become fixed, leading to reduced fitness and increased risk of extinction. Thus, it is essential to evaluate levels of genetic diversity within and gene flow between these metapopulations and develop strategies toward the maintenance of optimal diversity.

For now, high-altitude populations are protected from the introduced parasites and diseases that were implicated in the disappearance of forest bird species from lower elevations, because the cooler temperatures at high altitudes limit the development of both the avian malarial parasite and its mosquito vector (Chapters 9 and 17). However, it is possible that this protection is temporary, because global warming will result in the transmission of malaria and pox at higher elevations (Benning et al. 2002; this volume, Chapter 9). Natural resistance to avian malaria is evidently rapidly evolving in at least low-elevation Hawai'i 'Amakihi (Woodworth et al. 2005), and in this case of "evolution in action" the 'amakihi is apparently flourishing. However, this rapid rate of evolution can simultaneously be a double-edged sword if such relatively small, sedentary populations under intensive selection for malaria resistance encounter another significant disease, potentially West Nile virus, for instance, or if limited gene flow drives them toward extinction (Frankham et al. 2002).

Another honeycreeper species, the 'I'iwi, has a very low level of natural resistance —approximately 2% (C. Atkinson, pers.

comm.). This low level is apparently not enough to allow natural expansion of this species back into habitats where it historically lived prior to the introduction of mosquitoes. However, this species has at least some natural resistance. Further research is needed to identify wild populations of as many species as possible that are capable of survival in habitats in which malaria is widespread (U.S. Fish and Wildlife Service 2006). Individuals from these populations are candidates for use in captive breeding and translocation programs. Perhaps more important, they can help us to better understand the genetic mechanisms of malarial resistance.

Although avian malaria and avian pox virus are the "diseases of the day," over evolutionary time parasites and pathogens have come and gone, and others are yet to arrive, such as West Nile virus and avian influenza. Efforts to understand the genetic basis for disease resistance to malaria and interactions with concurrent pox infections are important. It is critical to gain an understanding of the genetic mechanisms and the level of diversity for maximum genetic resiliency to allow for the long-term survival of native Hawaiian forest birds. This includes a better understanding of the biological consequences of the observed genetic changes in birds thriving or perishing in these unique Hawaiian Islands.

SUMMARY

Biological and genetic changes in Hawaiian populations and species are governed by evolutionary processes that are molded by the unique landscape of the Hawaiian Islands. Studies of genetic change help us define significant biological changes that, in turn, allow us to make more accurate assessments of the current status of a given species and more knowledgeable management decisions for its long-term survival. In this chapter we provide an overview of the many genetic studies of native Hawaiian forest birds and how their results are applicable to conservation of the avifauna.

Genetics research has played an important recent role in our understanding the phylogeny of Hawaiian birds. The mtDNA sequence, often including that of ancient DNA, has now been studied for many taxa and has provided independent evidence by which to evaluate phylogenies derived from studies of morphology of both living and long-extinct species. This research has led to improved classification of Hawaiian birds and to identification of units of conservation for forest birds. These conservation units are identified as geographically separate populations that are on divergent evolutionary trajectories, such as the sedentary 'amakihi, now recognized as three species endemic to single islands.

Genetic studies also add to our knowledge of the historical patterns of species distribution and can help us detect some of the diversity that has been lost. Ancient DNA methods applied to subfossil bones have identified extinct populations of species once more widespread, such as the Laysan Duck. Loss of genetic variation has been found in other species, such as the Nēnē, 'Alalā, and Palila. Studies of the genetic structure of translocated populations (e.g., the Laysan Finch) and geographically restricted species (e.g., the Palila) have documented previously somewhat unexpected patterns of diversity potentially important in long-term management strategies. Historical studies of parasites can help us to better understand the evolution of virulence and its impacts on native populations.

One consequence of the genetic isolation of the Hawaiian avifauna is its extraordinarily high level of susceptibility to introduced diseases. Genetic evaluation of immune system genes such as the *Mhc* has provided evidence of exceptionally high selection intensities in honeycreepers, likely due to their exposure to introduced infectious disease. The power of selective pressures in molding genetic and biological change in these Hawaiian species is evidenced by the maintenance of Mhc variation in the ‘I‘iwi, a species that has apparently undergone a major genetic bottleneck. To effectively study disease, accurate diagnostics are essential. Molecular-based approaches have been developed and are now being applied to detecting disease outbreaks and implementing and measuring the effectiveness of disease control programs.

Another type of genetic tool that is critical to conservation of the Hawaiian avifauna is PCR-based sexing of populations. Unequal sex ratios reduce the effective size of populations. PCR-based sexing is currently being used to evaluate sex bias in the Palila, ‘amakihi, and Japanese White-eye. Finally, although cloning has long been and will likely remain a provocative issue, it has potential for the conservation of species. Viable cell cultures have been produced from many species of Hawaiian birds, and in some cases, particularly that of the Po‘o-uli, may be the only viable option for species survival.

Molecular technology is providing increasingly powerful tools that are essential for preserving natural patterns of genetic variation so that evolution by natural selection and other processes can continue. Concerted efforts should be implemented to achieve a better understanding of the biological consequences of genetic changes and the level of diversity necessary to allow for the long-term survival of native Hawaiian forest birds.

ACKNOWLEDGMENTS

For technical assistance and sample collection, we thank S. Barwise, J. Beadell, K. Bianchi, M. Farias, J. Foster, A. Giannoulis, C. McIntosh, C. Pfifer, S. Skinner, C. Tarr, K. Weigand, the field crew, and other members of the working group for Biocomplexity of Introduced Diseases: Threats to Biodiversity of Native Forest Ecosystems. We appreciate the helpful comments and suggestions of C. Atkinson, M. Farias, L. Patch-Highfill, B. Holland, R. Zink, and two anonymous reviewers. *Mahalo i ka po‘e o Hawai‘i no ko lākou kōkua a me ko lākou aloha* (Thank you very much to the people of Hawai‘i for their help, and aloha). Much of this work was funded by NSF Grant DEB 0083944, the Biological Discipline of the U.S. Geological Survey, and the Smithsonian Institution.

APPENDIX 10.1. GENETIC STUDIES OF HAWAIIAN BIRDS, THE MAJOR PARASITES AVIAN MALARIA AND AVIAN POX VIRUS, AND MOSQUITO VECTORS

Method of Analysis	Topic	Reference
Bird Hosts		
AFLP	Diversity in ʻAlalā	Jarvi and Bianchi 2006
AFLP	Population diversity in Hawaiian honeycreepers	Jarvi et al., unpubl. data
Allozymes	Origin and relationships of Hawaiian honeycreepers	Johnson et al. 1989
Allozymes	Relationship of subpopulations of Laysan Finches	Fleischer et al. 1991
CHD1 nuclear gene	Sex determination in Hawaiian honeycreepers	Jarvi and Banko 2000
CHD1 nuclear gene	Molecular sexing of Hawaiian honeycreepers	Jarvi and Farias 2006
DNA fingerprinting	Diversity in Nēnē	Rave 1995
DNA–DNA hybridization	Relationships of Hawaiian honeycreepers	Sibley and Ahlquist 1982
Mhc fingerprinting	*Mhc* diversity in ʻamakihi and ʻIʻiwi	Jarvi, Atkinson, et al. 2001
Mhc sequencing	*Mhc* natural selection, diversity in ʻamakihi and Iʻiwi	Jarvi et al. 2004
Mhc sequencing and microarray	Population diversity in Hawaiian honeycreepers	Jarvi et al., unpubl. data
Microsatellite primers	Diversity in honeycreepers	Eggert and Fleischer 2004
Microsatellites	Diversity in ʻElepaio	Burgess and Fleischer 2006
Microsatellites	Population diversity in Nēnē	Veillet et al. 2008
Microsatellites	Evolution of microsatellites in Hawaiian honeycreepers	Eggert et al. 2009
Microsatellites	Population structure in Hawaiian honeycreepers	Eggert et al. 2008
Minisatellite	Genetic structure and mating system of Palila	Fleischer at al. 1994
Minisatellite	Reproductive success and genetics of Nēnē	Rave et al. 1998
mtDNA	Relationships within the ʻamakihi complex	Tarr and Fleischer 1993
mtDNA	Relationships of Hawaiian honeycreepers	Tarr and Fleischer 1995
mtDNA	Relationships of the Laysan Duck as determined using ancient DNA	Cooper et al. 1996
mtDNA	Calibration of rates of mtDNA sequence evolution	Fleischer et al. 1998
mtDNA	Relationships of an extinct eagle using ancient DNA	Fleischer et al. 2000
mtDNA	Systematics and biogeography of Hawaiian avifauna	Fleischer and McIntosh 2001
mtDNA	Relationships of the Poʻo-uli	Fleischer et al. 2001
mtDNA	Origins and diversity of Nēnē as determined using ancient DNA	Paxinos, James, Olson, Sorenson, et al. 2002
mtDNA	Genetic variation in an ancient population of Nēnē	Paxinos, James, Olson, Ballou, et al. 2002
mtDNA	Adaptive radiation of Hawaiian songbirds	Lovette et al. 2002
mtDNA	Phylogeny of Pacific rails	Slikas et al. 2002
mtDNA	Millerbird phylogeny and genetic variation	Fleischer et al. 2007
mtDNA	Phylogeny of Hawaiian thrushes	Millier et al. 2007
mtDNA	Relationships of extinct moa-nalos as determined using ancient DNA	Sorenson et al. 1999

Method of Analysis	Topic	Reference
mtDNA and microsatellites	Diversity in ʻAlalā	Fleischer 2003, Fleischer et al. 2003
mtDNA, mini- and microsatellites	Founder events and genetic variation in Laysan Finches	Tarr et al. 1998
mtDNA, mini- and microsatellites	Diversity in Mariana Crows	Tarr and Fleischer 1999
mtDNA/nuclear introns	Population structure in Hawaiʻi ʻAmakihi	Foster et al. 2007
nuclear/mtDNA	Phylogeny of Hawaiian honeycreepers	Reding et al. 2008
RAG1/mtDNA/ B-fibrinogen	Phylogeny of Hawaiian ʻōʻō (*Moho* spp.) and relatives	Fleischer et al. 2008
Parasites		
4b core *Avipoxvirus* gene	Diversity and virulence of *Avipoxvirus*	Jarvi, Triglia, et al. 2008
Comparative diagnostics	Comparative nested PCR, smear, ELISA in detection of *P. relictum*	Jarvi et al. 2002
mtDNA	PCR diagnostics using mtDNA	Beadell and Fleischer 2005
mtDNA	Avian malaria in the Common Myna (*Acridotheres tristis*)	Ishtiaq et al. 2006
mtDNA	Global relationships of *P. relictum*	Beadell et al. 2006
PCR-based diagnostics	PCR amplification of *P. relictum* 18S rRNA genes	Feldman et al. 1995
PCR-based diagnostics	PCR amplification of *P. relictum* in Oʻahu birds	Shehata et al. 2001
RFLP	Characterization of pox viruses	Tripathy et al. 2000
RFLP	Pox viruses in Nēnē	Kim and Tripathy 2006
SNP-TRAP gene	Rare allele detection of *P. relictum*	Farias and Jarvi 2009
TRAP gene	Mixed-genotype infections of *P. relictum*	Jarvi, Farias, et al. 2008
Mosquito Vectors		
Microsatellites	Development of microsatellites for *Culex*	Fonseca et al. 1998
Microsatellties, mtDNA	Population genetics of *Culex* in Hawaiʻi	Fonseca et al. 2000
Microsatellites	Population genetics of *Culex* in Hawaiʻi and the world	Fonseca et al. 2006
Microsatellites	Fine-scale population structure of *Culex* on Hawaiʻi Island	Keyghobadi et al. 2006

CHAPTER ELEVEN

Small Mammals as Predators and Competitors

GERALD D. LINDSEY, STEVEN C. HESS, EARL W. CAMPBELL III, AND ROBERT T. SUGIHARA

Alien small mammals can affect populations of insular forest birds through four processes: directly through (1) predation and indirectly through (2) competition (both scramble and contest types), (3) habitat degradation, and (4) subsidization of other predators (Ebenhard 1988, Moors et al. 1992). Small mammals are unnatural on many oceanic islands, and their accidental or deliberate introduction by explorers and settlers has been devastating to native insular biotas worldwide (Atkinson 1973, 1977, 1985; Merton 1977; King 1985; Stone and Loope 1987; Ebenhard 1988; Griffin et al. 1989; Campbell 1991; Moors et al. 1992; Steadman 1995). Commensal rodents first arrived aboard early sailing ships and escaped at many sites where these ships stopped. They now occupy 82% of the world's major islands and island groups (Atkinson 1985). Domestic cats (*Felis catus*), also carried aboard sailing ships, were given as gifts, were taken by island residents, or were lost or escaped (Baldwin 1980, King 1984), and they now occur on most islands worldwide (Apps 1983). Other small mammalian predators were deliberately introduced to islands in unsuccessful biological control programs, as was the small Indian mongoose (*Herpestes auropunctatus*) to control rats in Jamaica and Hawai'i (Baldwin et al. 1952, Tomich 1986) and mustelids (*Mustela* spp.) to control rabbits in New Zealand (King 1984, O'Donnell 1996). Islands where alien small mammals became established have experienced myriad economic, health, and ecological problems (Atkinson 1985, 1989; Tomich 1986; Buckle and Fenn 1992; Moors et al. 1992). In island ecosystems worldwide, introduced small mammals have had ecological impacts on avian, reptilian, mammalian, land mollusk, and plant populations (King 1985, Stone 1985, Morgan and Woods 1986,

Case and Bolger 1991, Hadfield et al. 1993, Towns et al. 1997). Ninety-three percent of the land and freshwater birds that have gone extinct worldwide since 1600 have been insular forms (King 1985). Worldwide, predation by alien wildlife has been considered second only to habitat loss as the leading cause of avian extinctions and declines on islands, with rats (*Rattus* spp., 56%) and domestic cats (26%) implicated in most avian extinctions caused by alien predators (King 1985). Introduced mammalian predators have been implicated in nearly 70% of Pacific island bird extinctions, and they continue to have profound effects on surviving avifauna (Griffin et al. 1989).

Predation and competition from alien small mammals have played important roles in the collapse of the Hawaiian avifauna along with other factors resulting from direct and indirect effects of human activities (Chapters 1 and 2). The Polynesian rat (*Rattus exulans*) arrived with Polynesian voyagers on remote islands throughout the central Pacific, including Hawai'i (Matisoo-Smith and Robins 2004). During the century following the discovery of the Hawaiian Islands by Captain James Cook in 1778, mammals such as the black rat (*R. rattus*), Norway rat (*R. norvegicus*), house mouse (*Mus musculus*), domestic cat, and small Indian mongoose were accidentally or intentionally introduced. Many island bird species that evolved in the absence of native mammalian predators had lost their ancestors' acquired antipredator responses and were not able to defend themselves against the onslaught of these alien invaders (Moors et al. 1992; this volume, Chapter 1).

A BRIEF HISTORY OF ALIEN SMALL MAMMALS IN HAWAI'I

The history, biology, and past impacts on native flora and fauna by introduced mammals in the Hawaiian Islands have been extensively covered in the literature (Stone 1985; Scott et al. 1986; Tomich 1986; Cuddihy and Stone 1990; this volume, Chapter 2). Here we present a brief history of the alien small mammals that have been or are a threat to Hawaiian forest birds.

The Black Rat

The black, roof, or ship rat evolved in Asia, but the rats taken to Hawai'i were derived from European stock (Tomich 1986). Although not documented in Hawai'i until 1899, this species probably arrived about 1870 (Atkinson 1977). It once occupied all main Hawaiian islands but now appears to be absent from Kaho'olawe (Lindsey, Atkinson, et al. 1997). The black rat inhabits dry and wet forests, sugar cane and macadamia nut plantations, and urban areas from sea level to about 3,000 m elevation. The most common rat in Hawai'i, the black rat is locally abundant in low- and mid-elevation forests and is sparsely distributed in dry forests (Tomich 1986, Sugihara 1997, Lindsey et al. 1999). In montane wet forests of Maui and Hawai'i islands, the rat species composition in snap trap samples was 64% and 61% black rats, 36% and 38% Polynesian rats, and 0% and 1% Norway rats, respectively (Sugihara 1997, Lindsey et al. 1999). The mean capture rate over 41 months of study on Hawai'i Island at 1,500–1,650 m elevation was 19.9 black rats per 100 live trap nights. In contrast, the capture rate within dry subalpine woodlands at 1,520–2,130 m elevation on the same island was 0.7 black rats per 100 trap nights (Amarasekare 1994).

Because of its arboreal lifestyle, the black rat (Fig. 11.1) is considered the most significant avian predator among the three rat species introduced to Hawai'i (Atkinson 1977, Ebenhard 1988). In wet forests, Lindsey et al. (1999) found that 99% of the rats captured in trees were black rats. In mesic and wet forest habitats, black rats nest in cavities of native 'ōhi'a-lehua (*Metrosideros polymorpha*) and koa (*Acacia koa*) trees and in the tops of hāpu'u tree ferns (*Cibotium*

Figure 11.1. A black rat depredating eggs in a bird nest.
Source: Photo © Jack Jeffrey.

glaucum), where they are a threat to nesting and roosting forest birds (Lindsey et al. 1999).

The Polynesian Rat

The Polynesian rat, originating in Southeast Asia, accompanied early Polynesian voyagers to virtually every island in the Pacific, becoming the third most widely distributed rat species on earth (Kirch 1982, Matisoo-Smith and Robins 2004). It is characteristically a lowland species, flourishing best in wet forests, agricultural lands, and adjacent waste areas (Tomich 1986). However, this rat occurs in native forests all the way to the timberline. In the main Hawaiian Islands, the Polynesian rat currently occupies Kaua'i, O'ahu, Moloka'i, Lāna'i, Kaho'olawe, Maui, and Hawai'i islands, but it has not been recorded from Ni'ihau (Tomich 1986). Because of its introduction to Hawai'i perhaps as early as 1,200 years ago, its major influences on the Hawaiian native biota are assumed to have occurred long before Europeans arrived (Blackburn et al. 2004). It has been documented preying on ground-nesting

seabird species, including the Laysan Albatross (*Phoebastria immutabilis*), Bonin Petrel (*Pterodroma hypoleuca*), Bulwer's Petrel (*Bulweria bulwerii*), and Wedge-tailed Shearwater (*Puffinus pacificus*) (Tomich 1986). This rat is an agile climber, but it is now seldom found in trees, probably due to competitive exclusion by the larger arboreal black rat (Lindsey et al. 1999). Stone and Loope (1987) noted that Polynesian rats became more arboreal in some areas of Hawai'i where black rats had been controlled.

The Norway Rat

The Norway rat arrived in Hawai'i in the late 1700s aboard European sailing ships. This species occupies primarily urban and agricultural environments where domestic and agricultural food sources are available in abundance (Tomich 1986, Tobin and Sugihara 1992). It is occasionally found occupying low- and midelevation forests. Lindsey et al. (1999) captured only 13 Norway rats out of 1,251 total rat captures in a montane wet forest during a four-year study on Hawai'i Island, and Sugihara (1997) did not capture any Norway rats in similar forest habitat on Maui. No Norway rats were captured in a three-year study within natural ecosystems ranging from open grassland savannas to closed rain forests in the

Hawai'i Volcanoes National Park (Tomich 1981a). This rat is not currently considered a threat to forest birds because of its limited distribution in forest habitats.

The House Mouse

The house mouse arrived in Hawai'i by 1816 and presently has the most widespread geographical and ecological distribution of any alien mammal in Hawai'i (Tomich 1986). This species occurs on nearly all the large and small Hawaiian islands, regularly occupying dry grasslands, scrublands, and dry and wet forests from sea level to 3,920 m and occasionally up to 4,200 m. Dense mouse populations occur regularly in dry grasslands and in sugar cane and pineapple fields, with populations sporadically reaching plague proportions during late summer and fall in the drier beach, grassland, scrub, and forest areas on O'ahu, Maui, Kaho'olawe, and Hawai'i (Stone 1985, Tomich 1986). Although predation by mice on Hawaiian forest birds has not been documented, it cannot be discounted. In lowland sugar cane fields and adjacent gulches, insects were a major food source for mice (Kami 1966). Mice may be important competitors with birds for insects, and they may also serve as periodically abundant food for mongooses and feral cats. For example, heavy seed fall in New Zealand beech (*Nothofagus* spp.) forests is known to trigger mouse irruptions that result in increased litter sizes and survival of young stoats (*Mustela erminea*) (O'Donnell 1996).

The Domestic Cat

The domestic cat has been present in Hawai'i since the early days of European contact, when feral populations soon became established, although the spread of this species in the wild has not been well documented (Tomich 1986). William Brackenridge made one of the earliest observations of feral cats while traveling through a remote wilderness area en route to Kīlauea Volcano in December of 1840 (Brackenridge 1841). Cats were reported to be abundant in forests of Lāna'i and O'ahu by 1892 (Rothschild 1893– 1900), and Perkins (1903) recorded predation by cats on forest birds in the late 1800s. Feral cats currently live throughout the main Hawaiian islands, ranging from relatively high densities near sea level, where pets are frequently abandoned and fed by people, to sparse populations in areas completely isolated from human subsidies, including montane forests and even alpine areas of Maui and Hawai'i islands (Simons 1983, Hu et al. 2001, Winter 2003).

Although diet studies of feral cats worldwide show that they prey primarily on small mammals, these studies also show that island birds can be an important part of feral cats' diet (Fitzgerald 1992; Snetsinger et al. 1994; Smucker et al. 2000; Hess et al. 2004, 2007). Ebenhard (1988) considered feral cats the most significant predators ever introduced by man, and he cited 38 known or probable cases in which they have seriously reduced the abundance of prey populations. In Hawai'i, cats are presently important predators on seabird colonies and terrestrial birds that nest on the ground, in understory shrubs, and in woodland trees (Stone 1985, Scott et al. 1986, Ainley et al. 1997, Kowalsky et al. 2002, Hess et al. 2004). Where trap-neuter-release colonies of feral cats have been maintained near nesting seabirds on O'ahu, substantial predation has been documented (Smith, Polhemus, et al. 2002). Feral cats have also played an important role in the loss of birds from high-elevation dry forests, where other predators are uncommon and avian diseases are limited.

The Small Indian Mongoose

The small Indian mongoose was deliberately taken to Hawai'i Island from Jamaica in 1883 and released along the Hāmakua coast by sugar planters to reduce the rat

Figure 11.2. A small Indian mongoose scavenging a dead 'Apapane. *Source:* Photo © Jack Jeffrey.

populations in sugar cane fields (Tomich 1986). Within the next few years, offspring from these animals or additional individuals from the West Indies were introduced to O'ahu, Moloka'i, and Maui. Tomich (1986) reported that mongooses occupied all habitats on these four islands from sea level to elevations of 2,100 m. On the west slope of Mauna Kea Volcano, 5 of 73 recent mongoose captures have been recorded above 2,500 m elevation in māmane (*Sophora chrysophylla*) woodland, with the highest capture at 2,750 m (R. Danner and P. Banko, unpubl. data). Mongooses have not been found on Lāna'i or Kaho'olawe but have been reported from time to time on Kaua'i (K. Gunderson, W. Pitt, and T. Telfer, pers. comm.). Although mongooses have been seen climbing trees on occasion (Baldwin et al. 1952), they are considered weak climbers and not a threat to canopy-dwelling birds.

In contrast to most rat species, which are predominantly nocturnal, mongooses are diurnal, a factor that may limit their ability to control rat populations. Mongooses are nonspecialized carnivores, feeding mainly on rodents and insects (Kami 1964), with

birds occurring in <6% of their scats examined (Baldwin et al. 1952, Kami 1964). Mongooses are a persistent, direct threat only to introduced game birds, nesting seabirds, and native forest and water birds that nest or forage on or near the ground, such as the Nēnē (Hawaiian Goose, *Branta sandvicensis*) and 'Elepaio (*Chasiempis sandwichensis*) (Banko 1992, VanderWerf 1998a, Hays and Conant 2007) (Fig. 11.2). Nonetheless, the decline in the Pueo (Hawaiian Short-eared Owl, *Asio flammeus sandwichensis*) may in part be related to the species' dietary overlap with mongooses, introduced Barn Owls (*Tyto alba*), and cats (Mostello 1996). Owl pellets from mongoose-infested areas contained fewer arthropods than those from mongoose-free areas, a pattern consistent with competition.

The European Rabbit

European rabbits (*Oryctolagus cuniculus*) were introduced to remote Lisianski and Laysan islands, in the Northwestern Hawaiian Islands, where they had particularly devastating impacts on the fragile island environments and avifaunas through excessive herbivory (Tomich 1986). On Lisianki, rabbits were presumably introduced between 1904 and 1909, compounding the effects of mice that had been present since 1846 (Olson and Ziegler 1995). Rabbits eliminated most of the island's vegetation by 1914, subsequently causing their own starvation by 1923. Laysan Rails (*Porzana palmeri*), of which 45 were translocated to Lisianski in 1913, disappeared at this time, as did the mice. During guano mining on Laysan Island in 1902, Captain Max Schlemmer released domestic rabbits and guinea pigs (*Cavia porcellus*) for meat canning and to amuse his children (Tomich 1986, Rauzon 2001). An expedition was sent in 1912–1913 to eradicate the rabbits by shooting, but this effort failed and was not completed until 1923, coinciding with the desertification of Laysan, the extinction of the Laysan Honeycreeper (*Himatione sanguinea freethii*) and Laysan Millerbird (*Acrocephalus f. familiaris*), and the last observations of the Laysan Rail on that island (Ely and Clapp 1973).

Near the main Hawaiian islands, rabbits once existed on Lehua, Ford, Mānana, Molokini, and Kaho'olawe islands (Swenson 1986; Tomich 1986; C. Swenson, pers. comm.). More limited releases have occurred on the main Hawaiian islands without lasting success. An incipient rabbit population was eradicated in Haleakalā National Park on Maui in 1990 by shooting, trapping, and snaring (Loope et al. 1992), and another on Kaua'i was controlled by trapping in 2003 (C. Martin, pers. comm.). No further discussion is presented here because rabbits are essentially herbivorous, and for the time being they are not an ecological problem in the Hawaiian Islands. Future releases of domestic rabbits could, however, readily lead to the establishment of the species on one or more islands.

ECOLOGICAL IMPACTS
ON FOREST BIRDS IN HAWAI'I

The evidence of the ecological impacts of small alien mammals on Hawaiian forest birds includes historic patterns of the birds'

declines and extinctions on islands, direct observations of predation, and predator diet analyses. However, the most robust evidence of the impact of predation on forest birds comes from experimental manipulations of predator populations (Box 11.1). Many predator control programs are conducted primarily as urgent conservation measures for critically endangered species. Monitoring these actions has not only provided valuable feedback on the effectiveness of conservation treatments; it has also yielded measures of population responses and other insights as to which avian life stages are most vulnerable to predators or amenable to protection by management.

It has been challenging to study the effects of mammalian predators in Hawai'i. From a historical perspective, little contemporary biological expertise and awareness existed at the time alien mammals were becoming established in Hawai'i, when their effects on birds were likely to have been most severe. Furthermore, although there is scant documentation or direct evidence that allows us to determine the relative importance of past impacts, it is likely that multiple factors in concert, rather than any single factor, were responsible for the extinction or reduction in population numbers of Hawaiian bird species during that period. We also note that the inherent difficulties of studying these secretive and often nocturnal mammals have been further complicated by the steep terrain, vegetation, and heavy rainfall of Hawai'i.

Predation

Predation by small mammals on insular avian species has been well documented (Moors 1985, Moors et al. 1992) because of the high level of attention that the worldwide decline of island birds has generated within the conservation and scientific communities. The primary effect of small mammals on forest birds is predation of eggs, nestlings, fledglings, and sometimes adults (Atkinson 1977) (Table 11.2). However,

Box 11.1. Small Mammal Control and Evidence of Increased Forest Bird Productivity in Hawai'i

Predator control can be a useful management tool for protecting native bird species in island ecosystems (Griffin et al. 1989; Table 11.1), but the importance of predation on a bird population must be weighed against losses from other causes (King 1984). Côte and Sutherland (1997) reviewed a selected number of predator control programs and found that in some cases these programs do not result in an increased breeding population of the protected species. Other factors affecting the species of concern, such as poor survival of juveniles, lack of habitat for additional breeding birds, another predator species filling the void left by the removal of the target predator, disease, or emigration of excess breeding birds outside the treated area, may potentially weaken or nullify the effect of the predator control program. Conversely, increased emigration as a consequence of increased juvenile survival may contribute to overall metapopulation size even if it does not enhance the local breeding population.

A number of literature sources discuss the effects of introduced predators on birds, but few are available on the ability of predator control to provide long-term increases in forest bird populations (Atkinson 1973, 1977, 1985; Johnstone 1985; King 1985; Griffin et al. 1989). The pioneering work of Innes et al. (1995) and some New Zealand and Cook Island research (Robertson et al. 1994, Clout et al. 1995, O'Donnell 1996, Dilks et al. 2003) has demonstrated dramatically increased nesting success and fledgling and parental survivorship following predator control. A rodenticide suitable to protect native plants and animals, diphacinone, was first registered for Hawai'i in 1994 (Conry 1994). Biologists and land man-

agers now have a tool to control rodents and test their effects on forest bird populations. Examples of predator control programs conducted in the Hawaiian Islands and effects on bird populations are presented next.

O'ahu 'Elepaio Study

During the second half of the twentieth century, the O'ahu 'Elepaio (*Chasiempis sandwichensis ibidis*) underwent a dramatic population decline (VanderWerf et al. 1997). Predation by rats and feral cats, among other factors, was considered responsible. Between 1997 and 1999, Vander-Werf and Smith (2002) used snap traps and diphacinone baits in bait stations to remove rats and protect breeding 'Elepaio, which nest in trees. Rat removal increased the number of fledglings per pair from 0.37 to 0.70, and the rate of nest success increased from 41% to 60% when compared with untreated populations. The population went from sharp decline to a slight increase with rat removal. The authors concluded that rat removal was a viable management tool to help prevent further declines in the 'Elepaio and other Hawaiian forest birds.

Hakalau Predator Control Project

Bait stations with diphacinone were used at Hakalau Forest National Wildlife Refuge to control rats on a 48 ha grid from 1996 to 1999 (Nelson et al. 2002). Although indexes of rat abundance were reduced by 58–90% each year after treatment, black rats accepted diphacinone baits more readily than did Polynesian rats, and therefore the abundance of Polynesian rats remained >10 times higher than that of black rats as indexed by snap

Box 11.1. Continued

Table 11.1. Small mammal control programs and their effect on select populations of Hawaiian forest birds

Bird Species	Locality	Predators	Control Methods	Effect on Population	Source
Nēnē	Hawai'i	Feral cats, mongooses	Poison baits, trapping	Not monitored	Hawai'i Department of Land and Natural Resources (1974)
'Alalā	Hawai'i	Black rats, mongooses, cats	Trapping, tree bands at nests	Reduced nest predation	D. Ball (pers. comm.)
O'ahu 'Elepaio	O'ahu	Black rats	Poison bait stations, trapping	Mean increase of 0.33 fledglings per pair	VanderWerf and Smith (2002)
Hawai'i 'Elepaio	Hakalau (Hawai'i)	Black rats, feral cats, mongooses	Poison bait stations, trapping	Increased young:adult ratio	B. Woodworth (pers. comm.)
Palila	Hawai'i	Black rats, cats	Poison bait stations, trapping	Reduced nest predation during trapping	Hess et al. (2004)
'I'iwi	Hakalau (Hawai'i)	Black rats, feral cats, mongooses	Poison bait stations, trapping	Increased young:adult ratio	B. Woodworth (pers. comm.)
'Apapane	Hakalau (Hawai'i)	Black rats, feral cats, mongooses	Poison bait stations, trapping	Increased young:adult ratio	B. Woodworth (pers. comm.)
Po'o-uli	Maui	Rats	Poison bait stations, trapping	No evidence of predation	W. Sparklin (pers. comm.)

trapping. Intensive efforts were made to monitor the demographic responses and nest productivity of birds during rodent control. However, because the reference and treatment areas were located in close proximity, approximately 10% of young birds were captured in both areas, making independent comparisons of ratios of young to adults problematic (B. Woodworth, pers. comm.). Nonetheless, in 1997 the ratios of young to adult Hawai'i 'Elepaio, 'Apapane (*Himatione s. sanguinea*), and 'I'iwi (*Vestiaria coccinea*) were greater in the treated area than in the reference area. Perhaps due to the close proximity and small sizes of the study areas, no differences could be found between treated and reference areas in nest success, and no differences were detected in bird abundance during point-count censuses or mist netting. Low nesting productivity associated with El Niño Southern Oscillation climate events in 1998 and 1999 and the small spatial scale

continued

Box 11.1. *Continued*

of the study also contributed to variability that limited strong inference.

Midway Atoll Study

On Midway Atoll, the accidental introduction of the black rat from within the Hawaiian Islands chain in 1943 (Johnson 1945) was responsible for the extinction of introduced populations of Laysan Rails and Laysan Finches and for drastic declines of the Bonin's Petrel population. Seto and Conant (1996) determined that the Bonin Petrel was most vulnerable to rat predation during the incubation stage of its nesting cycle. In 1993 and 1994, they used the rodenticide bromethalin in bait stations to successfully suppress rat numbers at three treatment sites within Bonin Petrel nesting colonies. The level of nest success increased from 46% in areas without rat control to 90% in areas with rat control. Rats were subsequently eradicated from Midway Atoll, providing permanent protection for seabirds from predation (J. Murphy, pers. comm.). Although mice remain abundant on the atoll, seabird populations have increased noticeably in the absence of rats (J. Klavitter, pers. comm.).

---

not every predator-bird interaction results in the continued decline or extinction of the bird species. Worldwide, coexistence of bird species with predators is a common situation (Atkinson 1985).

Among the first documented evidence of direct predation in Hawai'i were observations on Lāna'i Island by Perkins (1903), who found 22 native birds killed by feral cats within a two-day period, and he twice observed cats feeding on 'Ō'ū (*Psittirostra psittacea*). But even by the early 1900s, Henshaw (1902) and Perkins (1903) suspected avian diseases as a probable cause for falling Hawaiian native bird populations. Atkinson (1977) presented circumstantial evidence that a stepwise, island-by-island decline in endemic forest birds occurred following the invasion and establishment of the black rat. As wharves were constructed on each island to support shipping, black rats were able to escape from ships via mooring lines. Scott et al. (1986) suggested that the introduction of black rats along with numerous other factors may also have contributed to the decline in bird populations during that period.

Contemporary rates of predation by small mammals probably bear little resemblance to the predation rates that occurred during the historic declines and extinctions of forest birds (King 1984, Innes and Hay 1991). When predators first arrived in Hawai'i, the colonizing mammal populations may have initially reached extremely high densities, and forest birds naïve to mammalian predators were most likely easy prey. Atkinson (1977) recounted the invasion of black rats from a ship grounded at Lord Howe Island in 1918 and the subsequent extinction of five species of birds, mostly within two years. Another black rat invasion of Big South Cape Island, New Zealand, in 1964 resulted in the decline of three species and extinction of two other species within three years (Atkinson 1989). On Midway Island, Fisher and Baldwin (1946: 8) described an "extreme [black] rat infestation of 1943 and 1944," to which they attributed the abrupt local extirpation of the Laysan Rail and Laysan Finch (*Telespiza cantans*) in less than two years. Atkinson (1977) also noted that daylight observations of rats climbing trees are rare among contemporary biologists in Hawai'i but were frequently recorded during the 1890s, indicating that this was a period of population irruptions.

Table 11.2. Levels of evidence of small mammal predator on select Hawaiian land birds

Bird Species	Level of Evidence[a]	Predators	Stage of Life Cycle Affected	Effect on Population	Source
Nēnē	Observational data	Mongooses, feral cats	Eggs, goslings, incubating females	Contributed to decline	Banko (1992), Taylor (1994)
'Alalā	Observational data	Black rats, feral cats, mongooses	Eggs, nestlings, juveniles	Contributed to decline	Giffin (1983); this volume, Chapter 20
O'ahu 'Elepaio	Designed study	Rats, mongooses, feral cats	Eggs, nestlings, adults	Contributed to decline	VanderWerf (1997), VanderWerf and Smith (2002)
Hawai'i 'Elepaio	Observational data	Black rats, cats	Eggs, nestlings, adults	Coexisting with predator	Snetsinger et al. (1994), Sarr et al. (1998)
'Ōma'o (Myadestes obscurus)	Casual observation	Black rats	Fledglings	Coexisting with predator	Berger (1981)
Puaiohi (Myadestes palmeri)	Observational data	Black rats	Eggs, fledglings, adults	Contributed to decline	This volume, Chapter 22
Laysan Finch	Casual observation	Black rats	Unknown	Translocated population became extinct	Fisher and Baldwin (1946), Berger (1981), Tomich (1986)
'Ō'ū	Casual observation	Feral cats	Adults	Contributed to decline and possible extinction	Perkins (1903)
Palila	Observational data	Black rats, feral cats	Eggs, nestlings, adult females	Coexisting with predator	Pletschet and Kelly (1990); this volume, Chapter 23
Hawai'i 'Amakihi (Hemignathus virens)	Observational data	Black rats, feral cats, mongooses	Eggs, fledglings, adults	Coexisting with predators	van Riper (1978), Snetsinger et al. (1994)
'Apapane	Casual observation	Black rats	Eggs, incubating adults, fledglings	Coexisting with predator	Berger (1981)
Po'o-uli	Casual observation	Black rats, feral cats	Eggs, fledglings, adults	Contributed to decline?	Pratt, Kepler, et al. (1997); this volume, Chapter 21

[a]Categories as defined by McArdle (1996).

Blackburn et al. (2004) recently analyzed avian extinction patterns, biogeographic factors, and mammal presence on 220 oceanic islands throughout the world, including the six largest Hawaiian islands, and provided stronger evidence for the role of predators in island extinctions. They found that the total number of alien predator species, not the total number of alien mammal species or the presence of rats or cats, best explained the probability of avian extinctions from islands. Moreover, the effects of these predators were disproportionately greater on bird species that were endemic to island groups or single islands. This analysis also determined that the proportion of currently threatened species is statistically independent of the number of extant alien predators, suggesting that the most susceptible bird species have already been driven to extinction and that the threats have changed over time. This process, known as the "filter effect," was described conceptually by King (1984) and in more detail by Holdaway (1999). A major implication of this process is that the establishment of additional predator species on islands will lead to additional avian extinctions, and it is therefore critically important to prevent further predator introductions.

Humans and introduced mammals, birds, invertebrates, and plants have altered the environment in which both predators and birds now live. Island bird species that coexist with introduced mammals today may have developed behavioral means of avoiding predation, may simply have been eliminated from structural habitats where they are vulnerable to predation, or may have reached demographic equilibria with predators (Lande 1987). Moreover, alien predators may no longer encounter the superabundant food resources that once drove their population irruptions, or their populations may be limited by interactions with other alien predators (Atkinson 1977). In Hawai'i and on many other Pacific islands, mammals were introduced and quickly established ecological interactions with other native and alien biota. On many of these islands, introduced mammals assumed the roles of apex predators (feral cats) or became numerically dominant (black rats) due to the lack of other animals with comparable life histories, sizes, and behaviors (Boersma and Parrish 1998).

Direct observations of predation on birds implicate alien mammals but do not quantify their impacts at the population level. In Hawai'i, predation on 20 bird species or their eggs by introduced small mammals has been recorded (Berger 1981, Atkinson 1985), although this number probably represents only a portion of the total number of Hawaiian bird species upon which introduced small mammals actually prey. Most small mammals, particularly rodents and feral cats, hunt at night and leave little sign of their nocturnal activities. Recent advances in field research techniques, such as the development of improved miniature surveillance cameras, are now allowing biologists to document the importance of predation on forest birds. For example, in 16 Palila (*Loxioides bailleui*) nests monitored by video surveillance cameras, two nestling predation events were documented, both involving feral cats (Hess et al. 2004). One of these events, described in detail by Laut et al. (2003), was unusual in that it occurred during daylight hours but left little sign of the predator's presence at the nest (Fig. 11.3).

Factors that influence the frequency and severity of predation on particular bird species include the life histories, morphology, and behavior of birds; the abundance and behavior of their predators; and the general ecological conditions where birds and predators interact (Burger and Gochfeld 1994). Especially vulnerable bird species exhibit life history characteristics such as delayed maturity, reduced clutch sizes, extended incubation and nestling development periods, and highly specialized feeding behavior (Chapters 2 and 8).

Figure 11.3. Still frame from a video of a feral cat that killed but did not eat two Palila nestlings. Filmed during daylight hours on June 22, 1999. The cat has the head of a nestling in its mouth after both nestlings begged to be fed. *Source:* Photo courtesy of Pacific Island Ecosystems Research Center, U.S. Geological Survey.

Such species may have a reduced ability to sustain unnatural predation relative to species that are generalist feeders, reproduce at younger ages, and have larger clutches, with the ability to renest after losing broods (Chapter 2).

A bird species' morphological characteristics, such as body size, egg size, and eggshell thickness relative to predator body size, and its behavioral adaptations, such as its level of aggressiveness, nest placement and concealment, and roost site selection, influence the species' overall susceptibility to predation (Burger and Gochfeld 1994, Boersma and Parrish 1998, Bradley and Marzluff 2003). Important characteristics of predators include their temporal and spatial abundance, body sizes, and behaviors such as habitat use and diurnal activity patterns (Atkinson 1977, 1985). Finally, general ecological conditions and the lack of alternative food sources may cause nutritional stress that intensifies predation risk and makes birds spend more time in search

of food and less time in vigilance and nest defense (Martin 1992a). Disease infection can also predispose birds to predation by reducing their ability to escape (Yorinks and Atkinson 2000).

Predation pressure can be constant throughout the year or may occur only during certain periods of a bird's life cycle, such as nesting, fledging, or juvenile periods. Birds with breeding seasons that coincide with seasonal peaks in predator populations are generally more vulnerable to predation (Boersma and Parrish 1998). An additional factor that may have contributed to the vulnerability of the Hawaiian honeycreepers to predation is their characteristic odor (Perkins 1903, Atkinson 1977), which could attract small mammal predators to their nests and roosting birds.

The magnitude of predation thus depends upon the biology and behavior of both the predator and the prey species. The black rat is the most arboreal and agile climber of the three rat species (Atkinson 1985). Most nests and roosting forest birds are therefore vulnerable to predation by this rat. Predator species and population sizes can also differ between habitats. 'Alalā (Hawaiian Crows, *Corvus hawaiiensis*) in mesic

Figure 11.4. Palila nestlings depredated by a feral cat. Source: Photo courtesy of Pacific Island Ecosystems Research Center, U.S. Geological Survey.

forests of west Hawai'i Island are exposed to mongoose and feral cat predation during the first two weeks after the young leave the nest because the fledglings cannot sustain upward flight and often perch on the ground (Scott et al. 1986). On Mauna Kea, Hawai'i Island, rat populations reach peak levels in the montane wet forest (Lindsey et al. 1999) but are generally quite low in the montane dry forest (Amarasekare 1993). Within the montane dry forest, the endangered Palila, with its extended nesting season, strong odor, and open-cupped nests, is vulnerable to predation by feral cats (Banko, Johnson, et al. 2002; this volume, Chapter 23) (Fig. 11.4). In the Mauna Loa strip section of Hawai'i Volcanoes National Park, where rats are more abundant in the mesic forest, Sarr et al. (1998) found that the annual rate of nest success of Hawai'i 'Elepaio (*Chasiempis s. sandwichensis*) over two years was 28% (n = 31) and 59% (n = 27). Although 'Elepaio lack the characteristic drepanidine odor,

predation, most likely by black rats and feral cats, accounted for 34–57% of all nest failures. Nonetheless, because of their ability to renest up to eight times after failures, 65–74% of pairs were successful in fledging at least one chick. (For a review of nest losses in Hawaiian forest birds and a discussion of the likely selection of nest sites inaccessible to rodents, refer to Chapter 8.) Unfortunately, it is impossible to compare current rates of nest loss to those at any time before the introduction of small mammals to Hawai'i; such data do not exist.

Competition

Competition takes two forms: scramble competition (the winner is the one that consumes resources the fastest) and contest competition (the winner maintains exclusive control of resources whether they are consumed or not). Competition is common, although extremely difficult to document in the field, and direct evidence of this process is essentially lacking for Hawai'i. It can affect island fauna if alien small mammals usurp the limited resources of endemic species, such as food or nest sites (King 1985, Sugihara 1997). Rodents may limit the distribution and numbers of frugivorous, seed-eating, and nectivorous birds by removing preferred fruits, seeds, or nectar (Scott et al. 1986; this volume, Chapter 2). Rodents may also compete with insectivorous birds that have restricted diets or substrate foraging capabilities (Banko and Banko 1976), such as birds that forage on branches and trunks for burrowing larvae, pupae, and adult insects (Baldwin and Casey 1983, Scott et al. 1986). Rats are suspected of competing for shelter or nest sites with some cavity-nesting birds (Moors et al. 1992, Sugihara 1997). Lepson and Freed (1997) hypothesized that black rats competed with cavity-nesting 'Ākepa (Loxops coccineus) for the use of cavities for daytime den sites.

Our current understanding of competition between small mammals and forest birds in Hawai'i for food resources and nest sites (Table 11.3) is based primarily on circumstantial evidence and on what is known from the literature about these animals' apparent dietary overlap and reduction of principal foods (Banko and Banko 1976; this volume, Chapter 2). Proof of actual resource shortage is difficult to document. Direct competition would be likely between native birds that specialize on fruit and large conspicuous invertebrates that are active day and night (Stone 1985). Banko and Banko (1976) and Berger (1981) suspected that the black rat had significantly reduced 'Ō'ū and 'Alalā populations by competitive interactions, primarily through competition for fruit. Insect eggs, pupae, and many small invertebrates on and above ground level might also be vulnerable to both birds and rats. Stone (1985) suggested that there may be competition for invertebrate foods between rats and native birds such as thrushes and the Po'o-uli (Melamprosops phaeosoma). However, few rigorous studies have been conducted on the interrelationships of resource competition between small mammals and native bird species in Hawai'i.

Rats and house mice, particularly during population irruptions, may be important food competitors with birds by consuming invertebrates and plant resources and by nesting in cavities that are used by native birds (Chapter 7). Sugihara (1997) documented that ≥93% of Polynesian rat stomachs contained invertebrate material, and he suspected that rats may compete with Hawaiian forest birds for food. The effects of mice on native biota in Hawai'i are currently poorly known, but mice could exert strong ecosystem impacts through their sheer abundance, competing with native birds for arthropod and plant seed food resources (Stone 1985, Tomich 1986, Sugihara 1997).

Table 11.3. Levels of evidence of resource competition between small mammals and select Hawaiian land birds

Bird Species	Level of Evidence[a]	Predators	Resource	Effect on Population	Source
'Alalā	Casual observation	Black rats	Fruit, berries	Contributed to decline?	Banko and Banko (1976); this volume, Chapter 20
'Ō'ū	Casual observation	Rats	'Ie'ie fruit	Contributed to decline?	Perkins (1903)
'Ākepa	Casual observation	Black rats	Nest cavities	Unknown	Lepson and Freed (1997)
Po'o-uli	Casual observation	Black rats	Arthropods and mollusks	Contributed to decline?	Pratt, Kepler, et al. (1997); this volume, Chapter 21

[a]Categories as defined by McArdle (1996).

Habitat Degradation

Besides engaging in predation and competition, rats also damage the habitats of forest birds by consuming and destroying the flowers, fruit, seeds, bark, and foliage of native trees, as well as native invertebrates, particularly arthropods and native snails (*Achatinella* and *Partulina* spp.) (Gagné and Christensen 1985, Stone 1985, Sugihara 1997). In the cases of some important plant species used by birds as food, rats may destroy high proportions or entire crops of viable seeds, thereby interrupting reproduction entirely (Cuddihy and Stone 1990). In Tonga, for example, McConkey et al. (2003) found sites used by rats to strip husks from seeds; these sites contained 13,555 empty husks from at least 18 plant species, with only 165 viable seeds and seedlings remaining. Although rats destroy seeds with their strong teeth, frugivorous birds pass many viable seeds through their guts, a process that often increases the probability of germination. Some Hawaiian plants whose seeds fall prey primarily to black rats include hōʻawa (*Pittosporum* spp.), ʻieʻie (*Freycinetia arborea*), ʻiliahi (*Santalum* spp.), and loulu (*Pritchardia* spp.) (Fig. 11.5), all of which are or were important food plants for Hawaiian birds (Perkins 1903, Beccari and Rock 1921, Male and Loeffler 1997, Sugihara 1997). Black rats are also known to strip bark from koa, thereby inhibiting regrowth of this important forest species (Scowcroft and Sakai 1984). Cuddihy and Stone (1990) presented a more complete account of the plants affected by rats in Hawaiʻi.

Although seed predation does not occur among carnivores, feral cats and mongooses, as well as rodents, all prey on invertebrates. Kami (1964) found evidence of invertebrate prey in 69% of mongoose scat. Invertebrates appear to be a surprisingly important food source for feral cats. Hess et al. (2004, 2007) found invertebrates in 55% of feral cat stomachs from Mauna Kea and in 54% of stomachs from Kīlauea Volcano, which most frequently contained

Figure 11.5. Hōʻawa (*Pittosporum* sp.) seed capsules depredated by rats. *Source:* Photo © Jack Jeffrey.

Orthoptera. Sugihara (1997) also documented invertebrates in 85–100% of black and Polynesian rat stomachs depending on the time of year. The loss of ecological function provided by these invertebrates has not been quantified in Hawaiʻi but may include soil nutrient cycling, herbivory, interactions with other invertebrates such as predation, and, perhaps most important, pollination of native Hawaiian plants.

Predator-Prey Interactions

Interactions between two or more introduced mammal species, which may facilitate or suppress an increase in the population of an introduced predator species, can also impact native bird species (Ebenhard 1988, Moors et al. 1992). Small predatory mammals such as rodents may serve as abundant prey for larger predatory species such as feral cats or mongooses, thereby maintaining greater abundance of larger predators that also prey on native birds (Atkinson 1985). This effect is known as

Figure 11.6. Feral cat scat containing leg bands and bones of a depredated bird. A band can be seen in the scat at the center of the photo. *Source:* Photo courtesy of Pacific Island Ecosystems Research Center, U.S. Geological Survey.

hyperpredation (Courchamp et al. 2000). Another phenomenon, known as mesopredator release, may occur when the absence or removal of apex predators results in an abundance of smaller predators, such as rodents (Courchamp et al. 1999). In Hawai'i, evidence of these predator-prey interactions comes only from predator diet analyses, but more direct evidence comes from other oceanic islands where predator populations have been manipulated.

In cane fields and pastures of the Hāmakua District on Hawai'i Island, Kami (1964) found evidence of rodents in 72% of mongoose scats. He concluded that predation on rodents served a useful purpose, but its effectiveness in controlling rodent populations was unknown. He also found, however, evidence of birds in 6% of scats and evidence of invertebrate prey in 69% of scats. Although Hawaiian forest birds are not associated with these habitats, this example illustrates the potential for carnivores to subsist primarily on abundant species but also prey on birds.

Further upslope on Mauna Kea, in the habitat of endangered Palila, the grassy understory of the subalpine dry forest supports an abundant house mouse population, which in turn supplies a ready food source for feral cats. Snetsinger et al. (1994) recorded rodent remains in 87% of cat scats and bird remains in 59% (n = 87) collected from the dry māmane-naio (*Myoporum sandwicense*) forest on Mauna Kea. Smucker et al. (2000) analyzed an additional 17 cat scats from this dry forest and found 76% of the scats with house mouse remains and 53% with bird remains (Fig. 11.6). Stomach samples from 96 cats on Mauna Kea showed, however, that the greatest proportion of prey items in the diet of feral cats was bird remains (82%), represented primarily by passerines (songbirds or perching birds) (53%) (Hess et al. 2004). In this study, mice comprised a smaller proportion of items (43%) than insects (55%), and insects may be less likely to appear in scat samples.

Examples of ecological effects resulting from interactions between introduced mammals come mainly from islands other than Hawai'i. On Macquarie Island, halfway between Australia and Antarctica, the introduction of rabbits led to an increase in the abundance of feral cats, which, in turn, led to increased predation on and the extinction of the Macquarie Island Parakeet (*Cyanoramphus novazelandiae erythorotis*) (Taylor 1979, Johnstone 1985). The removal of one alien predator (e.g., the feral cat) through a control program may also allow a population increase in another alien predator (e.g., the rat), resulting in continued predation on the native species of concern (Côte and Sutherland 1997, Courchamp et al. 1999). On North West Island, Australia, the eradication of feral cats resulted in an increasing mouse population, which, in turn, required control by long-term baiting (Domm and Messersmith 1990). Similar increases in rat populations have been noted on Wake Atoll following the recent removal of cats (Rauzon et al. 2008). Years of successful stoat trapping on South Island, New Zealand, may have been related to larger-than-usual postseedfall irruptions of rats that were nearly as damaging to nesting Mohua (*Mohoua ochrocephala*) as were stoats (Dilks et al. 2003). Conversely, prey switching has also been documented when numbers of rodents have been reduced. At certain sites in one forest on North Island, New Zealand, successful rat-poisoning operations resulted in a higher level of bird consumption by stoats when their main food source was reduced (Murphy and Bradfield 1992, Murphy et al. 1998). These case studies provide insight on the potential relevance to Hawaiian forest birds of situations in which the abundance of one small mammal species is influenced by population changes in a second species.

Disease

The role of diseases in small mammals that are transmissible to Hawaiian forest birds is poorly understood. Pasteurellosis, caused by the bacterium *Pasteurella multocida* and usually transmitted by cats, is known to kill Pueo and introduced Barn Owls (Work and Hale 1996). Erysipelas, caused by the *Erysipelothrix rhusiopathiae* bacterium, can be found in decomposing rat carcasses. This organism, which has also been implicated

in die-offs of other species of wild birds, caused the death of one free-ranging 'Alalā (Work et al. 1999). The deaths of four other free-ranging 'Alalā were attributed to toxoplasmosis, a disease caused by *Toxoplasma gondii* and primarily associated with cats (Work et al. 2000). While 37% of 67 feral cats from Mauna Kea had antibodies indicating past exposure to *T. gondii*, about 7% had active infections and may have been capable of transmitting toxoplasmosis (Danner et al. 2007). 'Alalā may contract this disease by ingesting *T. gondii* oocysts from the feces of infected feral cats or tissue cysts from transport hosts such as infected rodents or birds or from invertebrates that ingest oocysts. Toxoplasmosis, as well as avian malaria (*Plasmodium relictum*), may have been a significant factor in the 'Alalā decline, increasing the birds' mortality rates directly or indirectly by enhancing their susceptibility to avian and nonnative mammal predators (Work et al. 2000, Yorinks and Atkinson 2000). The significance of these diseases among populations of Hawaiian forest birds is unclear, but as Work et al. (1999) recommended for the 'Alalā, continued vigilance and sustained effort to objectively document the causes of morbidity and mortality and the impacts of diseases carried by small mammals on forest bird populations appears warranted. As a case in point, the emerging avian influenza virus H5N1 has recently found new hosts among many species of felids (Kuiken et al. 2004). Cats, including domestic cats, contract avian influenza by eating infected birds and can also transmit the disease to other cats. The appearance of such new diseases in Hawai'i may be potentially devastating to the avifauna, particularly when numerous hosts or vectors already exist.

FUTURE RESEARCH AND MANAGEMENT NEEDS

Effective management policies for predator control programs must be based on an understanding of the ecology of predators, what controls the distribution and numbers of native bird species, and whether and where predation actually still limits wild bird populations (King 1984). The importance of predation or the rate of loss of individuals from a population can be understood only by simultaneously studying entire populations of predators and their prey. Decisions regarding control programs based on historic predation data may be misleading because factors currently affecting bird populations may have little to do with predators. For example, direct predation on birds is of no importance if those birds would not have lived to breed because of other limiting factors such as disease, competition, or lack of food or breeding habitat (King 1984). With an increased interest in the conservation of Hawaii's native forests and the availability of new and/or improved research tools, for instance, surveillance cameras, and control techniques such as the use of registered pesticides, data can be obtained on the importance of predators within these habitats. Experimental, manipulative studies to evaluate the contemporary effects of introduced small mammals on native forest bird species—effects such as direct predation, food resource competition, and nest site competition—can provide robust evidence to supplement observational studies and historic information. Data on the levels of predation on native Hawaiian forest birds by small mammals are generally lacking, with only a few exceptions (Chapter 8).

A better understanding of factors that influence the current distribution and abundance of introduced small mammal predators in Hawaii's native forests and the part they presently play in limiting forest bird populations will assist us in developing improved management strategies for preserving Hawaiian forest birds. A suite of studies (Tomich 1981b; Amarasekare 1993, 1994; Sugihara 1997; Lindsey et al. 1999) have provided relevant data on the life histories and biology of rodents in Hawaiian forests. Additional information on the life histories, home range sizes, and behavior of rodents, feral cats, and mongooses will be useful to improve the timing, spacing, and configuration of traps and toxicant delivery systems for more effective management efforts. Understanding factors that cause an unusually high abundance of predators, such as rodent population irruptions, will also allow reliable predictions of these events for more selective control efforts. Comprehensive knowledge of predators' ecology and life histories will allow the development of control strategies to reduce the threat of predators under a wide variety of environmental conditions.

SUMMARY

Alien small mammals have had substantial negative effects on the native forest birds of Hawai'i through predation, resource depletion, habitat degradation, and hyperpredation, as they have on many islands throughout the world. The role of mammals in the transmission of diseases is less well understood but may also be important. Because the avifaunas of most Pacific islands were originally naïve to predatory small mammals, the greatest impacts probably occurred soon after first contact with each new alien mammal species. Thus the main effects of Polynesian rats, which arrived with Polynesian voyagers some 1,200 year ago, are presumed to have occurred long ago. Although Polynesian rats are not now as detrimental to native Hawaiian forest birds as the black rats introduced by Europeans, Polynesian rats are known to prey on ground-nesting seabirds and could well have extirpated numerous forest bird species nesting on or near the ground. Through habitat alteration and competition, they could otherwise have greatly impacted these bird communities. The presence of Norway rats was noted as early as 1835, but black rats were not documented in Hawai'i until 1899. Black rats may have been responsible in large part for bird declines and extinctions of the 1890s due to their ability to climb trees and prey on bird eggs and nestlings. The impacts of house mice, established by 1816, are not as well known as those of larger rodents, but mice may compete with native birds for seeds, fruit, and arthropods. All of these rodents may also serve as abundant food resources for larger mammalian predators: feral cats and mongooses.

Domestic cats, often used to control rodents aboard ships, quickly became popular with island residents and established feral populations, probably by 1840. Feral cats are known to prey on colonial seabirds and terrestrial birds that nest on the ground, in understory shrubs, and in woodland trees. Moreover, feral cats carry diseases such as toxoplasmosis that are known to kill Hawaiian birds and may have major impacts on some critically endangered species. The small Indian mongoose, introduced in 1883 to control rats in sugar cane fields, soon became ineffective at its intended task but established populations in the wild on most islands except Kaua'i, Lāna'i, and Kaho'olawe. Although mongooses may climb trees on occasion, they are not considered a major threat to canopy-dwelling birds. Mongooses primarily threaten nesting seabirds and native forest and water birds that nest or forage near or on the ground, such as Nēnē and 'Elepaio. Alien mammals are now found throughout virtually all of the Hawaiian islands and continue to have pervasive effects on native forest birds, although they are perhaps not as destructive as they were during their early irruptive phases.

Predator control programs have produced some notable successes for the conservation of endangered forest birds. New Zealand has been a leader in this line of research, but there have also been many successful programs on the islands and atolls of Hawai'i. Landscape-scale rodent control programs have been developed for Hawai'i, but further research is needed to assess the effects of these programs on ecological interactions among alien species, such as competitive exclusion and predator-prey relationships, among other concerns.

ACKNOWLEDGMENTS

We gratefully acknowledge the assistance and dedicated conservation efforts of D. Ball, P. Banko, R. Danner, K. Gunderson, J. Klavitter, C. Martin, J. Murphy, W. Pitt, M. Rauzon, B. Sparklin, T. Telfer, E. Tweed, E. VanderWerf, and B. Woodworth. We thank C. M. King, P. Q. Tomich, and an anonymous reviewer for helpful comments and constructive criticism of the manuscript.

CHAPTER TWELVE

The Ecology and Conservation of Hawaiian Raptors

JOHN L. KLAVITTER

Hawaiian raptors have continued to soar, despite the winds of uncertainty that are buffeting the archipelago and causing declines and extinction in many of the other native Hawaiian forest birds (Scott et al. 1986). The 'Io (Hawaiian Hawk, *Buteo solitarius*) (Plate 1) is the only endemic raptor that has persisted over time, primarily because of its ability to cope with anthropogenic change (Olson and James 1991). The Pueo (Hawaiian Short-eared Owl, *Asio flammeus sandwichensis*) (Plate 1), although classified as an endemic subspecies, may actually be indistinguishable from North American and Asian forms. Its fossil history suggests that it may have colonized the archipelago only after the Polynesians created extensive grasslands and introduced Polynesian rats (*Rattus exulans*) to the islands (Olson and James 1991, Burney et al. 2001). The Barn Owl (*Tyto alba pratincola*) is a recent human introduction to Hawai'i from North America, and it has successfully occupied all the main islands (Berger 1981).

The 'Io is found only on the island of Hawai'i, while the other two species occur throughout the archipelago (Clarkson and Laniawe 2000). The 'Io is a small, broad-winged hawk, an adaptable forest nester often seen soaring over the forest's edge. It exhibits the most pronounced sexual size dimorphism of any member of the genus *Buteo*, with males averaging 441 g and females 606 g (Griffin 1985). In addition, 'Io occur in distinct light and dark color phases. The Pueo is predominantly diurnal or crepuscular, but nocturnal activity has also been recorded (Holt and Leasure 1993, Mostello 1996), whereas the Barn Owl is predominantly nocturnal. The Pueo inhabits grasslands, shrublands, montane parklands, and, less frequently, forested areas (Griffin 1989). The Barn Owl is more

293

often associated with forests, agricultural habitats, and urban settings and is frequently observed hunting near roadsides (Marti 1992).

Island raptors are unique compared to their continental counterparts, with relatively small population sizes and nonmigratory habits (White and Kiff 2000, Klavitter et al. 2003). When the *Buteo* ancestors of the 'Io arrived in the Hawaiian Islands, they found an environment missing their most important prey: rodents and other small mammals. To survive, the early *Buteo* were forced to subsist on a diet of birds, spiders, and insects. This was true for the 'Io, but its diet may have consisted of more nestlings and eggs and fewer mature birds (Perkins 1903). The 'Io positioned itself atop the terrestrial food chain and joined a few other, now extinct, raptors and the corvids as the only predators of birds (Olson and James 1991). With the introduction of mammalian predators to Hawai'i, the proportion of the species' original prey diminished, but the diversity of prey increased (including three species of *Rattus* and the house mouse, *Mus musculus*) and continued to expand with the introduction of nonnative bird species. The 'Io adapted to this shift in prey base, allowing it to survive. In addition, its ability to capture a diversity of prey in a variety of size classes, ability to nest in a variety of trees, and apparent resistance to avian diseases explain why it is the only forest bird in Hawai'i that has been considered for delisting from federally endangered status. The Pueo shares similar qualities, except as a ground nester it is susceptible to mammalian predators and it faces competition from a larger, introduced species, the Barn Owl. Habitat degradation, increased urbanization, mammalian predators, and West Nile virus (WNV) are the primary conservation concerns for both native species.

Raptors are an integral component of the ecosystem in Hawai'i. The 'Io and Pueo may function as both "flagship" and "umbrella" species for Hawai'i (Simberloff 1998). When habitat is protected for 'Io and Pueo, other native Hawaiian organisms found in the same habitat also benefit. 'Io and Pueo are charismatic vertebrates prominent in ancient Hawaiian culture, and today they are recognized by many people throughout the archipelago (Chapter 3).

In this chapter I present background information on Hawaiian raptors that includes their fossil record and systematics. I also summarize data on the 'Io, Pueo, and Barn Owl, encompassing their (1) life histories, (2) population status, and (3) role as predators of native passerines (songbirds or perching birds). Finally, I offer management recommendations and future research needs along with their implications for policy makers, managers, and planners.

RAPTORS IN HAWAI'I

Native Hawaiian Raptors

Raptors arrived on the Hawaiian Archipelago by long-distance dispersal (Carlquist 1966). This idea is reinforced by the frequent observation of vagrant raptors in Hawai'i each year (Pyle 2002). Several species are capable of long-distance migrations in the continental setting. Successful colonization of Hawai'i may have been enhanced by repeated arrivals, recently called "migration dosing" (Bildstein 2004), which would have supplemented small, fluctuating founder populations. Most likely, inexperienced juveniles constitute the majority of "dosed" founders, for they often migrate together, are particularly vulnerable to wind drift, and have the propensity to develop new migratory habitats (Kerlinger 1989, Viverette et al. 1996). The Swainson's Hawk (*Buteo swainsoni*), a close relative of the 'Io, is an example of a raptor that is physically capable of flying to Hawai'i. It is a long-distance, transequatorial migrant that breeds in the Nearctic and overwinters in the Neotropics. It migrates in groups of up to 1,000 individuals and regularly "doses" the Neotropics with hundreds or thousands of misguided hawks (Bildstein 2004).

In Hawai'i, ancient bird bones (fossil and subfossil) have been discovered in caves (lava tubes), limestone sinkholes, lakebeds, and dunes. These bones provide a window into the past and an opportunity to envision some of the now extinct raptors and other birds that once inhabited the islands. Prehistoric (extinct before Europeans arrived) raptors once inhabiting Hawai'i included a sea eagle, a harrier, and at least four species of long-legged owls (Olson and James 1982b, 1991; Burney et al. 2001). Their extinction can likely be attributed to Polynesian settlement through habitat destruction (clearing of lowland forest, primarily by fire), loss of avian and insect prey, and depredation by people. Judging from their morphology, the long-legged owls of the genus *Grallistrix* and the harrier (*Circus dossenus*) evolved as bird and insect predators, but the fossil record suggests that both became extinct prior to the arrival of the Pueo. Subfossil castings attributed to *Grallistrix* show that they ate Hawaiian honeycreepers. The Hawaiian sea-eagle is thought to be a recent colonist and was probably the White-tailed Eagle (*Haliaeetus albicilla*) from Asia, which feeds on fish and waterbirds (Fleischer et al. 2000). The 'Io and Pueo are all that remain of a past filled with a more diverse native raptor fauna.

Vagrant Raptors

A few vagrant raptors arrive in Hawai'i each year, some species annually and others more infrequently. They include the Osprey (*Pandion haliaetus*), Black Kite (*Milvus migrans*), Steller's Sea-Eagle (*Haliaeetus pelagicus*), Northern Harrier (*Circus cyaneus*), Gray Frog-Hawk (Chinese Goshawk, *Accipter soloensis*), Rough-legged Hawk (*Buteo lagopus*), Golden Eagle (*Aquila chrysaetos*), Merlin (*Falco columbarius*), and Peregrine Falcon (*F. peregrinus*) (Pyle 2002). These wandering raptors are mostly highly migratory arctic and boreal species.

Their presence suggests how ancestors of native Hawaiian raptors may have arrived on the islands.

THE 'IO: STATUS AND ECOLOGY

Fossil Record and Former Range

Based on the fossil record, the former range of the 'Io included the islands of Kaua'i, Moloka'i, and Hawai'i (Olson and James 1997, Burney et al. 2001). Given this documented distribution, it is likely that 'Io were also established on O'ahu, Lāna'i, and Maui, but no fossil material has been found yet on these islands. In several instances 'Io remains were found in association with Hawaiian archaeological sites, suggesting that Hawaiians may have eaten them and/or utilized their feathers. Remains of 'Io have also occasionally been encountered in lava tube openings on the western slopes of Hawai'i Island (Olson and James 1997). On O'ahu, fossils from a larger *Buteo* species are unusually common in excavated limestone sinkholes located within a Pleistocene lakebed at Ulu-pa'u Head (Olson and James 1997). These bones are not identifiable as *B. solitarius* and may represent a species that evolved into the modern 'Io or a species that became extinct in the islands. Because Kaua'i is the main island farthest from Hawai'i, the fossils from Kaua'i show that 'Io are capable of dispersing naturally to any of the islands in the main Hawaiian Archipelago. Although a handful of 'Io have been sighted briefly on Kaua'i, O'ahu, and Maui within the past 200 years, it remains a mystery why the current population is restricted to Hawai'i Island (Banko 1980–1987).

Systematics

Understanding how the 'Io is related to other raptors may aid its management and conservation. Based on the species' morphology and behavior, Griffin (1985) hypothesized that the 'Io is most closely related to the Swainson's Hawk. Recent mitochondrial DNA analysis indicates that 'Io are least divergent from the Short-tailed Hawk (*Buteo brachyurus*) and are part of a clade that includes the Short-tailed Hawk, Galápagos Hawk (*B. galapagoensis*), White-throated Hawk (*B. albigula*), and migratory Swainson's Hawk (Fleischer and McIntosh 2001, Riesing et al. 2003). It can be hypothesized that the ancestor of this clade was a long-distance migrant morphologically similar to the Swainson's Hawk, which could reach remote islands. The 'Io and the Short-tailed Hawk are most similar in appearance, life history, and ecology (Griffin 1985, Miller and Meyer 2002).

Population Status

The 'Io was included in extensive forest bird surveys conducted by Scott et al. (1986) from 1976 to 1981. These authors found 'Io distributed throughout Hawai'i Island from sea level to 2,600 m. Within this elevation range, 'Io were absent only from arid grasslands on the northwest side of the island, the Ka'ū Desert, the dry scrublands of Kapāpala, and the open pastures of Kahuku. The population size was not estimated because 'Io failed to meet many of the assumptions that underlay their density estimates.

From 1980 to 1983, Curtice Griffin studied the distribution, physical characteristics, habitat use, reproductive biology, foraging behavior, home range characteristics, and taxonomy of the 'Io (Griffin 1985, Griffin et al. 1998). He found 'Io distributed throughout the island of Hawai'i and estimated the population at 1,400–2,500 hawks, although this estimate was likely positively biased (Griffin 1985, Klavitter et al. 2003). The species' home range size was dependent upon its habitat, ranging from 48 ha in an agricultural area with papaya and guava orchards to 526 ha in a wet 'ōhi'a-lehua (*Metrosideros polymorpha*) forest with native and exotic understory (Griffin 1985). Some habitat modification has

benefited 'Io, at least in the short te increasing the densities of their pre as rats in orchards, and allowing 'I crease their home range size.

The 'Io was federally listed as an gered species in 1967 based on its distribution; a low population siz ceived threats to its native forest from agriculture, logging, and co cial development; and a lack of infor on the species' status (Oreinstein Berger 1981, U.S. Fish and Wildlife 1984a, Griffin 1985). Additionall tors (especially bird-eating and fish species) had been declining worl from contaminants such as DDT an drin, and the effects of these chemi 'Io populations were unknown (N 1979, 1998). In 1993, the U.S. Fis Wildlife Service (USFWS) began a of all 'Io data to ascertain whether a in federal listing status under the E gered Species Act was warranted. the reasons for down-listing the "threatened" status was "to ensure sustaining 'Io population in the rar 1,500–2,500 adult birds in the w distributed in 1983, and maintaine stable, secure habitat." The listing ued with this requirement: "For pur of tracking the progress of recovery, will be used as a target to reclass threatened status" (U.S. Fish and W Service 1984a: 25). No justificatior given for the target 'Io populations. fin's (1985) population estimate me criterion, but because the estimate nearly a decade old, the USFWS requ that Hall et al. (1997) complete ar dated 'Io population survey on Ha From December 1993 to January 1 these authors found 'Io distributed thro out the island of Hawai'i in both nativ nonnative habitats and estimated the ulation at 1,600 hawks. After a thorc review of all the information availabl USFWS withdrew its proposal to d list the 'Io (U.S. Fish and Wildlife Se 1994). In their survey Hall et al. coul

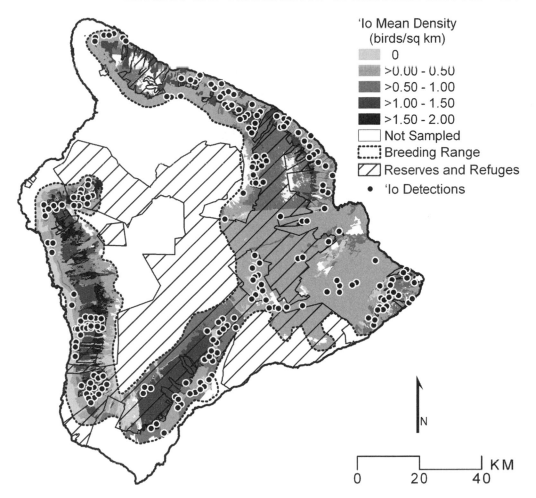

'Io Mean Density
(birds/sq km)
- ☐ 0
- ☐ >0.00 - 0.50
- ☐ >0.50 - 1.00
- ☐ >1.00 - 1.50
- ☐ >1.50 - 2.00
- ☐ Not Sampled
- ☐ Breeding Range
- ☐ Reserves and Refuges
- • 'Io Detections

Figure 12.1. 'Io (Hawaiian Hawk) distribution and density. Projected 'Io density in 2007 and the distribution of 'Io detected during the 1998 and 2007 surveys is shown in relation to state and federal reserves and refuges on Hawai'i Island. Territorial 'Io occur only in forest and nearby habitats. About 55% of the species' breeding range lies outside state and federal conservation land. *Sources:* Data are from Klavitter et al. (2003) and Gorresen et al. (2008).

ascertain the population status with confidence, and important factors remained unresolved, such as breeding pair productivity and fledgling survival.

In 1998, the USFWS once again considered a change in the species' listing status and requested that Klavitter et al. (2003) conduct a two-year 'Io study to address the unresolved population factors. Based on an intensive point-count survey, it was determined that 'Io were distributed throughout much of the island of Hawai'i (Fig. 12.1), as they had been in the 1980s (Griffin 1985, Scott et al. 1986), with a population size estimated at $1{,}457 \pm 176$ SE distributed over 6,144 km². Protected lands supported approximately 469 'Io (95% CI = 244–901). 'Io appeared resistant to avian diseases, and no evidence was found of negative impacts from contaminants. From following 30 radio-tagged adults and monitoring 113 nests, Klavitter et al. (2003) judged that the population appeared stable based on reproductive and survival rates, and they recommended down-listing to threatened status.

In 1998, the USFWS also formed the 'Io Recovery Working Group to make recommendations on 'Io reclassification. Based on a review of all the 'Io information, they recommended delisting the 'Io but prefaced the action with a recommendation for monitoring the population ('Io Recovery Working Group 2001). On August 6, 2008, the USFWS published a proposed rule to remove the 'Io from the federal list of endangered and threatened wildlife (U.S. Fish and Wildlife Service 2008a). As of February 2009, the USFWS had not published a determination as to whether to reclassify the 'Io (K. Marlowe, pers. comm.).

In 2007, Gorresen et al. (2008) completed another islandwide 'Io count survey of the locations originally sampled by Klavitter et al. (2003). Improving upon the survey methodology developed by Klavitter and Marzluff (2007), these authors developed a more rigorous model to calculate 'Io abundance from point-count surveys, and they more conservatively estimated the size of the 'Io breeding range at 5,755 km^2. They reanalyzed the 1998 survey data and estimated that the population was comprised of 3,239 hawks (95% CI = 2,610–3,868) in 1998 and 3,085 hawks (95% CI = 2,496–3,680) in 2007. No significant difference in densities was found among years at either regional or islandwide scales.

Adult and Juvenile Survival

Klavitter et al. (2003, Appendix 1) estimated the annual rate of survivorship by monitoring radio-tagged 'Io as 0.94 for adults (n = 34) and 0.50 for first-year birds (n = 26). Such a rate of adult survivorship is high for a raptor, especially compared to temperate, continental species such as the Common Buzzard (Buteo buteo, 0.76) and the Eurasian Sparrowhawk (Accipiter nisus, 0.64) (Ryttman 1994). 'Io probably have a lower mortality rate because they do not migrate and have fewer predators. Klavitter et al. (2003) also concluded that adult survivorship was the most im-

portant demographic parameter regulating 'Io population growth. With fecundity and first-year survival held constant, the adult survival rate needs to exceed 0.91 for a population to remain stable or to increase. This is consistent with published data on 49 bird species where the mean elasticity value (relative contribution of a demographic parameter to population growth rate) for adult survival rate was significantly larger than the mean elasticity for fecundity (Saether and Bakke 2000). The importance of adult survival may be pronounced in tropical, insular species such as the 'Io, which have low rates of mortality and reproduction (Murray 1985, Faaborg 1986). 'Io typically lay only a single clutch of one egg per season and live up to 17 years (Griffin 1985, Klavitter 2000). The rate of first-year 'Io survivorship (0.50) is similar to that of the Common Buzzard (0.48) and greater than that of the Eurasian Sparrowhawk (0.36) (Ryttman 1994). Juvenile 'Io probably have a lower mortality rate than most continental species, because they have an extended postfledging dependency, do not migrate or disperse over great distances, and have fewer predators.

Causes of Mortality and Population Threats

A number of mortality factors have been documented for 'Io (J. Klavitter, unpubl. data). Birds have been found shot, hit by automobiles, drowned in cattle watering troughs, poisoned, and killed by feral cats (Felis catus), feral dogs (Canis familiaris), and small Indian mongooses (Herpestes auropunctatus). All of the causes mentioned are density independent. In the early 1900s, farmers on the southwestern portion of Hawai'i Island reportedly shot 'Alalā (Hawaiian Crows, Corvus hawaiiensis) and in some instances were actually paid a bounty for each bird (Munro 1944). Malicious shooting of birds continues today at some low level (Griffin 1985, van Riper and Scott 2001, Thompson 2003); those shot include some 'Io that prey upon domestic chickens

(*Gallus gallus*) (J. Klavitter, unpubl. data). These chickens are raised and exported as gamecocks to the U.S. mainland and other parts of the world and can fetch as much as $250 each (Hoover 2003). Most juvenile 'Io deaths are attributed to starvation, especially after prolonged periods of rain (Klavitter 2000, Klavitter et al. 2003). 'Io do not fly during heavy rainfall and therefore would have a difficult time foraging (J. Klavitter, unpubl. data).

DISEASE AND EXTERNAL PARASITES

The two main diseases affecting native passerines, avian pox virus (*Avipoxvirus* sp.) and avian malaria (*Plasmodium relictum*), do not appear to be serious diseases for 'Io (Chapter 9). Only three 'Io have been recorded with pox lesions (n = 44, Griffin 1985; n =155, Klavitter et al. 2003). 'Io have not shown antibodies to avian malaria in their blood (n = 75, Griffin 1985; n = 7, Klavitter et al. 2003). 'Io have the ability to persist in areas where many native Hawaiian honeycreepers have been extirpated by pox and malaria (Jarvi, Atkinson, et al. 2001; Klavitter et al. 2003). Toxoplasmosis also does not appear to be a problem. Only one 'Io showed possible evidence of past exposure to *T. gondii* (n = 7, Klavitter et al. 2003). *T. gondii* caused mortality in 'Alalā (n = 5) in the same area in which 'Io were sampled (Work et al. 2000). Red-tailed Hawks on the U.S. mainland also show resistance to *T. gondii* (Lindsay et al. 1991).

In 1999, one juvenile female 'Io was found dead in the South Kona District on Hawai'i Island. Preliminary findings showed that the 'Io died as a result of trichomoniasis, caused by the protozoan *Trichomonas gallinae* (T. Work, pers. comm.). The protozoan forms a lesion, often in the oral cavity and surrounding areas such as the sinuses, pharynx, crop, or esophagus. The bird may have contracted trichomoniasis while foraging on wild doves or pigeons, known carriers of the disease in Hawai'i (Kocan and Banko 1974). Doves and pigeons are considered reservoirs for trichomoniasis in other areas of the world (Pokras et al. 1993). This is a common disease in falconry birds and wild raptors eating columbiformes (Krone et al. 2005).

Two adult 'Io were known to have died from aspergillosis (*Aspergilla* sp.) in 1997 (G. Massey, pers. comm.). These wild-caught birds were being held temporarily (~3 months) in an aviary with a gravel- and grass-covered floor in a cool, moist environment at a high elevation (1,460 m) when they contracted aspergillosis and died from the illness. Aspergillosis has not been detected in free-ranging, wild birds.

Ectoparasites do not appear to be a major limiting factor for adult or nestling 'Io. In one study, only five adult 'Io (n = 155) had minor infestations of nonnative hippoboscid flies (*Ornithoica vicina*), and infected birds showed no obvious ill effects (Klavitter et al. 2003). Tropical fowl mites (*Ornithonyssus bursa*) also infect nests and nestlings (4 of 113 nests, Klavitter et al. 2003) and may negatively affect young in the nest. An infected nestling in the study just mentioned suffered lethargy and swollen eyes but eventually fledged one week early.

Although 'Io are relatively resistant to diseases already present in Hawai'i, the susceptibility of 'Io to other diseases is unknown, and unfortunately the opportunity for new diseases to be introduced to the islands is high (Chapter 15). Currently, WNV is of great concern in North America for its lethal effects on birds and people. The virus affects hawks and many species of their avian prey (Eidson et al. 2001, Garmendia et al. 2001, Malakoff 2002). Because 'Io are closely related to Red-tailed Hawks, which are susceptible to WNV, one could predict that the 'Io population might also be affected.

ENVIRONMENTAL CONTAMINANTS

There is no evidence that environmental contaminants (organochlorines or heavy metals) currently limit the 'Io population (Berger 1981, Griffin 1985, Klavitter et al.

2003). Organochlorines were once used extensively in Hawai'i (Allen et al. 1997), but it is unclear how 'Io were affected. Salvaged 'Io eggs and a chick were tested for contaminants and found to have none or only trace amounts that would be unlikely to cause deleterious effects (Berger 1981, n = 1 egg; Griffin 1985, n = 3 eggs, 1 chick; Klavitter et al. 2003, n = 5 eggs).

HABITAT LOSS AND DEGRADATION

'Io breeding occurs throughout the island of Hawai'i (Klavitter 2000, Klavitter et al. 2003). Most nests are built in native 'ōhi'a trees (81%, n = 113), but some are in exotic trees (16%). Eucalyptus spp. are the exotic trees most often used for nesting (5%). 'Io may select 'ōhi'a trees because they are the most abundant trees in native forests (Jacobi 1990). The physical properties of 'ōhi'a (strong branches, great height, and thick foliage) may also make them preferred for nesting. Invasive, nonnative vegetation may outcompete 'ōhi'a trees and limit their recruitment as older trees die. 'Ōhi'a grow very slowly, potentially taking 200–300 years to reach a size that would be adequate for 'Io nesting. Nesting success (percentage of pairs raising an offspring to fledging) is slightly higher in native habitat compared to nonnative habitat (1998 and 1999 data pooled: 62% vs. 40%, n = 88, P = 0.04), indicating the importance of maintaining both 'ōhi'a trees and native forests for 'Io (Klavitter et al. 2003).

Today, 'Io appear limited primarily by habitat availability (Klavitter et al. 2003). They occur in nearly all habitats that have some large tree component (Banko 1980, Scott et al. 1986, Hall et al. 1997) and appear to exist at carrying capacity in mature native forests with a grass understory, lower-elevation forests, grasslands, and plantations with forest edges. These habitats have numerous accessible perch sites, an open understory, seldom-used dirt roads, and large numbers of introduced birds and small mammals (Clarkson and

Laniawe 2000, Klavitter et al. 2003). Logging and grazing in mature native forests have created this type of "edge" habitat (Chapter 6), which has increased the densities of Buteo elsewhere (Santana and Temple 1988, Preston and Beane 1993, Zelenak and Rotella 1997). Presently, 'Io densities may be twice as high in altered native forest as in relatively unaltered native forest (Klavitter 2000). 'Io coexist with moderate levels of anthropogenic forest manipulation (logging, grazing, nonnative tree plantings or invasions), but their numbers may become low in such areas where native tree recruitment does not occur. Human-altered forests may develop into grasslands that support lower densities of 'Io (Scott et al. 1986). Fire may also act in the same negative manner, transforming quality 'Io habitat into grassland. 'Io are also limited by low-stature forest, lava, and large expanses of orchards without windrows or forest edges. Extensive urban development and orchard monocultures do not provide nest trees or foraging areas.

The rapid spread of urban development threatens to drastically alter the remaining quality 'Io habitat. Extensively built-up areas support few 'Io and lack suitable nest trees (Klavitter et al. 2003). Urban areas are hostile environments for 'Io because of traffic, cats and dogs, nest disturbance, and human harassment. Human population growth and commercial development will increase on the island of Hawai'i (Hawai'i Department of Business, Economic Development and Tourism 1998), decreasing the amount of quality habitat for 'Io. As of 1998, 143,135 people resided on the island, an estimate expected to increase by approximately 2,850 per year (U.S. Bureau of the Census 1998). In 1998, 773 new family homes were built on the island (Hawai'i State Department of Business, Economic Development and Tourism 1998). These figures show modest growth, but it is difficult to predict future growth. If development continued to a point where all lands except those protected by reserves, parks, and refuges were converted to urban landscapes, 'Io numbers could fall below 500 (Klavitter et al. 2003).

'Io as Predators of Forest Birds

Before the arrival of humans in Hawai'i, 'Io foraged on native birds and insects (Perkins 1903, Munro 1944). With the introduction of nonnative mammals, insects, and birds, their prey base has expanded (Tomich 1971, Berger 1981, Griffin 1985). 'Io are both opportunists and generalists in their foraging endeavors. They easily utilize introduced prey, especially rats. In one study, black rats (*Rattus rattus*) constituted most of the 'Io diet during the breeding season (Griffin 1985).

'Io are documented to prey upon a variety of native forest birds including Hawai'i 'Amakihi (*Hemignathus virens*), 'I'iwi (*Vestiaria coccinea*), and 'Apapane (*Himatione s. sanguinea*) (Tomich 1971; Clarkson and Laniawe 2000; P. Hart, pers. comm.). The effect of 'Io on native bird populations is unknown but believed to be low (Griffin 1985), the exception being 'Alalā (discussed later). In an intact native ecosystem, predators and prey coexist. Predators serve an important role in weeding out sick, weak, and injured individuals and may be an important force in the evolution of prey characteristics (Newton 1979). Perhaps 'Io still play this role to some extent in intact native forests.

The ability of 'Io to kill and consume large prey is impressive. They occasionally take adult Kalij Pheasants (*Lophura leucomelanos*) (Clarkson and Laniawe 2000; J. Klavitter, unpubl. data), which greatly outweigh 'Io. 'Io have been observed attacking Nēnē (Hawaiian Geese, *Branta sandvicensis*) goslings and adults on the ground and in the air, although these were not likely foraging attempts (Rojek 1994; P. Banko, pers. comm.).

'Io have been documented killing and eating critically endangered 'Alalā, which are the same size as a male 'Io. Managers are now faced with a conservation conundrum —one endangered species preying upon another. Historically, 'Io undoubtedly preyed upon 'Alalā but most likely did not impact their population greatly. While the 'Io population seems to have benefited from some anthropogenic changes (increased "edge" habitat, introduced rats), the 'Alalā population has dwindled. Currently, the increased 'Io population in altered native forest may explain the higher rate of 'Io predation on 'Alalā. 'Alalā have also been negatively affected by habitat loss and alteration, shooting, disease, and introduced predators (Chapter 20). The 'Alalā population is extinct in the wild and currently numbers 60 birds in captivity (Chapter 19). As part of the 'Alalā Recovery Program, captively reared 'Alalā were released into the wild as fledglings from 1993 to 1998. Although their rate of survivorship was high from 1993 to 1994, it dropped dramatically beginning in 1995, and a significant portion of their mortality was attributed to 'Io (Banko, Ball, et al. 2002). 'Io now seem to be a formidable impediment to their recovery (Chapter 20).

THE PUEO: STATUS AND ECOLOGY

A Recent Colonist?

The Pueo is thought to be relatively new to the Hawaiian Islands, perhaps establishing itself after humans arrived in the islands (Olson and James 1991, Burney et al. 2001). The pre-Polynesian fossil record offers no evidence of Pueo as yet, suggesting that this owl's colonization was linked to the recent anthropogenic expansion of open habitat and introduction of diverse mammalian prey (the Polynesian rat; Norway rat, *Rattus norvegicus*; black rat; and house mouse) (Olson and James 1982b, Burney et al. 2001). Pueo are normally birds of open country; hence they would have benefited from forest clearing by humans. Even though Short-eared Owls are able to capture avian prey (Mikkola 1983, Snetsinger et al. 1994), it seems unlikely that they are well adapted to subsisting mainly upon an avian prey base. Furthermore, the now extinct Hawaiian long-legged owls were adapted for foraging on birds and may have played a role in preventing Pueo from earlier establishment.

The ancestry of the Pueo remains unresolved. The literature suggests that founders may have originated from either American or Asian stock, for these migratory populations of Short-eared Owls are capable of flying long distances and are occasionally seen far out at sea (Bryan 1903, Mostello 1996).

Systematics

Pueo are considered an endemic subspecies of the Short-eared Owl, a species distributed widely across North America and throughout much of the world (Holt and Leasure 1993). Recently, there has been some question as to the validity of the Hawaiian race (Burney et al. 2001). The Short-eared Owl has proven itself as an exceptional colonist, especially of islands. The Pueo is one of up to nine subspecies, five of which are island endemics (Holt and Leasure 1993). Morphologically, these owls appear similar, with slight differences in plumage. The minor plumage differences described for the Pueo require verification. All subspecies are typically associated with nonforest habitats (Holt and Leasure 1993).

Population Status

During the Hawaiian Forest Bird Survey (1976–1983), Pueo were found on all the main islands that were surveyed (Scott et al. 1986). These owls occurred in grasslands, shrublands, and montane parklands but were observed less frequently flying low over closed-canopy forests. Pueo have also been observed in agricultural areas (sugar cane and pineapple), meadows, mowed grass fields, wetland areas, and high-elevation forests (Mostello 1996). There have been sightings of Short-eared Owls for many of the Northwestern Hawaiian Islands, but these owls could have been vagrants of Asian or North American rather than of Hawaiian origin (Pyle 1981).

Although Pueo were recorded during the Hawaiian Forest Bird Surveys, their population size has never been estimated, and no distribution map has been created (Scott et al. 1986, Griffin 1989, Mostello 1996). The species was widespread on all the main islands in the 1890s but was most likely experiencing a decline due to hunting and habitat loss (Perkins 1903). Cats and small Indian mongooses were introduced to Hawai'i in the late 1700s and 1883, respectively (Tomich 1986). Both are known avian predators and must have also depleted the ground-nesting Pueo. This apparent decline continued through the 1930s and led the Hawaii Audubon Society to pass a resolution in 1941 requesting the discontinuation of hunting of several Hawaiian birds, including the Pueo (Bryan 1933, Hawaii Audubon Society 1941, Munro 1946). Pueo appear to have continued to decline, at least on Kaua'i and Maui, from the 1970s to the 1990s based on the Christmas Bird Counts (Mostello 1996).

The apparent population decline of the Pueo has raised few conservation alarms. In 1982, the Pueo was listed by the state of Hawai'i as endangered on O'ahu (Mostello 1996). On all the islands the Pueo also benefits at some level from protection under the Federal Migratory Bird Treaty and Lacey Act. There is currently no intention to federally list Pueo as threatened or endangered, perhaps because of the lack of data on its population status and trend and because this bird is distributed across all the main Hawaiian islands. If the Pueo were found to be indistinguishable from the continental Short-eared Owl, that might also weaken the justification for listing it.

Causes of Mortality and Population Threats

PREDATION AND NEST LOSS

Several studies have investigated the causes of Pueo mortality and nesting failure (Aye 1994, Mostello 1996, Work and Hale 1996). Pueo are ground nesters vulnerable to predation and disturbance from cats, small Indian mongooses, dogs, feral pigs (Sus scrofa), and other ungulates (Berger 1981, Mostello 1996). Mongoose predation was most likely responsible for two documented nest failures (one at the egg stage, one at the chick stage) that occurred while four nests were monitored in Lualualei Valley, O'ahu (Mostello 1996).

TRAUMA

Pueo deaths have also been documented to occur through trauma. Pueo are hit by automobiles and occasionally by aircraft (Aye 1994, Work and Hale 1996). Pueo that are compromised by starvation, disease, or contaminants may be more susceptible to such collisions than those that are healthy. Owls are probably attracted to foraging on or near roadsides and runways because pavement, especially adjacent to dense vegetation, offers an obstruction-free surface where they can easily detect and capture prey. Pueo were shot in the past during open hunting seasons (Mostello 1996), and today illegal shootings may occasional occur (Thompson 2003).

DISEASE

The term sick owl syndrome was coined to describe the numerous dead or live but lethargic and seemingly impaired Pueo and Barn Owls found along roadsides in Hawai'i from the 1960s (TenBruggencate 1992). Sick owl syndrome prompted two studies to examine sick owls collected near roadsides and at the Līhu'e Airport, Kaua'i. Work and Hale (1996) examined dead owls (n = 86), including five Pueo, found opportunistically by residents and agencies from 1992 to 1994 and determined that trauma (n = 3), infectious disease (n = 1, Pasturella multocida), and emaciation (n = 1) were causes of mortality in both Pueo and Barn Owls. The pathway and growth of the infectious disease was unclear, although emaciation, evidenced by pathology and blood samples from nearly dead owls, may have predisposed the birds to bacterial infection. Necrotic stomatitis, intestinal disease (coccidiosis), liver abscesses, stomach erosion, and hemorrhagic lesions were also documented as causes of mortality (Aye 1994). Aye (1994) theorized that injuries sustained in collisions with motor vehicles serve as the primary cause of sick owl syndrome. The high rate of emaciation and dehydration seen in these birds may be due to the inability of owls to hunt following injury or during periods of harsh weather. Also, rodent abundance is cyclic in Hawai'i (Tomich 1986), and sick owl syndrome may occur when rat and mice populations crash after peak abundance, leading to starvation in owls.

Avian pox and malaria have not been reported in Pueo (Mostello 1996; Work and Hale 1996, n = 5). WNV is lethal to Great Horned Owls (Bubo virginianus) in North America (Eidson et al. 2001), so there is some potential that Pueo will be affected if the virus spreads to Hawai'i. Because there are no baseline population estimates for Pueo, it would be difficult to quantify the impacts of WNV on Pueo.

ENVIRONMENTAL CONTAMINANTS

As in many agricultural environments, pesticides including organochlorines and organophosphates were heavily used in Hawai'i and continue to be used today (Mostello 1996; Mineau et al. 1999; L. A. Woodward, pers. comm.). These pesticides were responsible for the decline of many raptor species worldwide. The plight and recovery of the Peregrine Falcon and Bald Eagle (*Haliaeetus leucocephalus*) are well known and documented (Cade 1983, Barclay 1988). The effects of pesticides on Pueo in the past were not studied in detail but must have occurred at some level. A thorough review of the documented cases of raptor mortality in the United States, Canada, and the United Kingdom from 1985 to 1995 (Mineau et al. 1999) reported that Short-eared Owls and Barn Owls were found affected by cholinesterase-inhibiting pesticides. Despite these findings, at the present time there is no evidence that contaminants are affecting Pueo (Work and Hale 1996).

HABITAT LOSS AND DEGRADATION

Habitat loss and alteration pose major threats to Pueo. Intensive grazing in pastures limits the amount of nesting habitat available, because Short-eared Owls require the protective cover of grass and shrubs for nesting (Holt and Leasure 1993). Fire, urbanization, invasive nonnative plants, and intensive agriculture currently alter grassland habitats (Chapter 6).

Pueo as Predators of Forest Birds

Pueo and other subspecies of Short-eared Owls forage predominantly on small mammals such as mice, voles, and rats (Holt and Leasure 1993). Additionally, they capture and consume insects and avian prey. Pueo pellets from low-elevation areas on Kaua'i, O'ahu, Moloka'i, Lāna'i, and Maui (n = 163) and from two high-elevation sites rich with native passerines at Haleakalā National Park, Maui (n = 14), and Hakalau Forest National Wildlife Refuge (Hakalau), Hawai'i Island (n = 35), were analyzed to determine the prey proportions (Mostello 1996). House mouse remains were most frequently encountered (74%), followed by birds (29%), insects (33%), and rats (18%). Avian prey consisted of six species of nonnative birds (with Zebra Doves, *Geopelia striata*, most numerous). No native

birds were found in pellets, possibly because the majority (77%) of the samples were collected in low-elevation areas devoid of native passerines. A second study by Snetsinger et al. (1994) also examined Pueo pellets (n = 36). Avian prey were found in similar portions (36%) to those reported by Mostello (1996) but included native birds: Pacific Golden Plovers (*Pluvialis fulva*), 'amakihi, 'I'iwi, and 'Apapane. Many of the samples (48%) were from high-elevation areas where native passerines were present, which may explain why native birds were found in this study and not in Mostello's. There is some evidence of Pueo robbing nests of forest birds. A Pueo was observed snatching an endangered Palila (*Loxioides bailleui*) chick (T. Pratt, pers. comm.), and circumstantial evidence suggested that Pueo were taking 'Ākohekohe (*Palmeria dolei*) chicks (VanGelder and Smith 2001). Owls in Hawai'i may have also contributed to population declines in the endangered 'Akiapōlā'au (*Hemignathus munroi*), although more observations are needed to substantiate this notion (Fancy et al. 1996). Although Pueo do capture native passerine chicks and adults, at the present time it appears that their overall effect is negligible.

THE BARN OWL: STATUS AND ECOLOGY

History of Introduction

In 1958, the Hawai'i Commissioners of Agriculture and Forestry approved the introduction of the Barn Owl to Hawai'i as a biocontrol agent for rats and mice, which were causing heavy damage in sugar cane fields (Thistle 1959). From 1958 to 1963, approximately 86 owls from California and Texas were released on the islands of Kaua'i, O'ahu, Moloka'i, and Hawai'i (Berger 1981, Aye 1994), and by 1966 Barn Owls were breeding and had established a self-sustaining population (Au and Swedburg 1966). They had successfully colonized all the main Hawaiian islands by 1987 owing to their ability to disperse among the islands and to use a variety of habitats for nesting and foraging (Pratt et al. 1987). No studies have been initiated to determine how successful Barn Owls are at controlling rodents in agricultural areas in Hawai'i.

Systematics

The Barn Owl has proved able to use a broad array of habitat and prey (Marti 1992). This characteristic has allowed it to become one of the most cosmopolitan of all raptors. The species is widely distributed on all continents except Antarctica and is recognized by many people in the world today. Body size and coloration define the 35 recognized subspecies. The North American race (*Tyto alba pratincola*) was introduced to the Hawaiian Islands. Its status as nonnative is significant from a conservation perspective, because state and federal wildlife agencies are tasked with protecting and preserving mainly native species and the habitats on which they depend. Under this directive, the Barn Owl is not a species promoted in wildlife refuges and preserves. Nonetheless, the Barn Owl is protected by federal and state law.

Population Status

Since the introduction of Barn Owls to Hawai'i, there have been no studies of their abundance, survivorship, or reproduction. Their successful colonization in a relatively short period of time indicates that the population was increasing for several decades (approximately 1960–1990). Their ability to use a variety of prey (mice, rats, birds, insects), habitats (open grasslands, residential areas, wet and dry forests), and nest sites (tree cavities, buildings and other artificial structures, lava tube cave ledges) presumably has aided their rapid spread. The Barn Owl's high reproductive rate may have been a contributing factor in its success. Its mean clutch size ranges from 3.1

in the Galápagos Islands to 7.1 in Utah, with a clutch of 18 reported in Germany (Marti 1992). Two Barn Owl nests, both located in 'ōhi'a tree cavities, have been observed on Hawai'i Island (J. Klavitter, pers. obs.). One nest contained 5 eggs, the other 5 nestlings. Clutch size is probably related to food abundance (Marti 1992). Double broods per season are also typical in tropical Malaysia (Lenton 1984).

Causes of Mortality and Population Threats

PREDATION AND NEST LOSS

Population threats for the Barn Owl are poorly known but probably similar to those for Pueo but less severe. Nesting off the ground makes Barn Owls less exposed to predation from cats, small Indian mongooses, pigs, and dogs. 'Io rarely prey on Barn Owls. But the remains of a Barn Owl were found in an 'Io nest in 1999 at the Pu'u Wa'awa'a Wildlife Sanctuary (J. Klavitter, unpubl. data).

TRAUMA

Several anthropogenic factors have been documented as causes of Barn Owl mortality. Barn Owls have been hit and killed by automobiles and aircraft (Aye 1994, Mostello 1996, Work and Hale 1996). Most (46%, n = 81) of the owls examined from 1992 to 1994 died as a result of trauma (Work and Hale 1996). Because large birds such as the Barn Owl present a hazard for airplanes, they are legally shot around the Līhu'e Airport, Kaua'i (Work and Hale 1996). Although uncommon, strong winds may lead to Barn Owl mortality. For example, a Barn Owl was found dead in Volcano, Hawai'i, after strong winds apparently caused it to crash (J. Klavitter, pers. obs.).

DISEASE

Sick owl syndrome and disease have also been listed as causes of mortality for Barn

Owls. Eighty-one dead Barn Owls found opportunistically by residents and agencies were examined to determine their causes of death (Work and Hale 1996). A relatively high rate of disease prevalence (26%) was found, and most disease cases (95%) were identified as trichomoniasis. Barn Owls may contract trichomoniasis while foraging on Zebra Doves, known carriers of the disease (Kocan and Banko 1974, Mostello 1996). Other diseases noted (some concurrent with trichomoniasis) included *Pasturella multocida*, *Sarcocystis* sp., and *Aspergillus* sp. Pasteurellosis was probably a cat-induced trauma based on the location of a foot lesion and serotype. Starvation was the cause of death in 22% of mortalities. Avian pox and malaria were not present in owls examined by Work and Hale (1996) and have not been documented to occur in Barn Owls in Hawai'i (Mostello 1996). Aye (1994) reported necrotic stomatitis, intestinal disease (coccidiosis), liver abscesses, stomach erosion, and hemorrhagic lesions as additional causes of mortality. The establishment of WNV in Hawai'i would most likely be lethal to Barn Owls (Eidson et al. 2001).

ENVIRONMENTAL CONTAMINANTS

There is little information on the effects of pesticides on Barn Owls in the Hawaiian Islands. Barn Owls were found to have been affected in the past by cholinesterase-inhibiting pesticides (Mineau et al. 1999). Work and Hale (1996) tested 18 Barn Owls for pesticides and found only low levels of organochlorines and no carbamates or organophosphates. Gassmann-Duvall (1988) examined four dead barn owls from Kaua'i and Maui that exhibited sick owl syndrome and found evidence resembling a vitamin E deficiency. Gassmann-Duvall (1988) suggested that a chemical agent (vitamin E, selenium, or glutathione-peroxidase antagonist) was responsible for the deaths. In 1992, no evidence of toxins was found in 104 Barn Owls (87 from Kaua'i and 17 from O'ahu) that were subjected to a complete necropsy with supportive laboratory tests (Aye 1994). Thus, at present there is little evidence of contaminants as an important cause of mortality in Barn Owls in Hawai'i.

HABITAT LOSS AND DEGRADATION

Habitat loss and alteration pose some threat for Barn Owls. Their use of suitable habitat is limited by nest requirements such as hollow trees, cavities in cliffs and river banks, and some human structures (Marti 1992). Large expanses of grasslands and lava without suitable perch and nest sites are unusable by Barn Owls, as are highly urbanized areas.

Barn Owls as Predators of Forest Birds

Barn Owls were introduced to Hawai'i to help control rodents that were agricultural pests, especially to sugar cane. Their potential negative effects on native bird populations were not assessed prior to their introduction, and few studies to evaluate this issue have been initiated since the species' establishment. Barn Owls are strictly crepuscular and nocturnal. In North America, they forage mainly on voles and other small mammals active at night. Birds are usually eaten in low numbers, and most are small passerines that roost in the open (Marti 1992).

In Hawai'i, Barn Owls forage mostly on mice, rats, and insects (Tomich 1971, Snetsinger et al. 1994, Mostello 1996, Work and Hale 1996). In the Mostello (1996) study previously mentioned for Pueo, Barn Owl pellets from low elevations where native passerines were absent (n = 340) and from high-elevation sites with native passerines (Hakalau, n = 176; Alaka'i Swamp, n = 14) contained mostly house mouse remains (79%), followed by those of rats (41%), insects (25%), and birds

(10%). Avian prey represented seven species of nonnative birds (with Zebra Doves most numerous) and six species of native forest birds. Native bird prey (~30% juveniles) included 'Elepaio (*Chasiempis sandwichensis*), 'Ōma'o (*Myadestes obscurus*), 'amakihi, 'Aki-apōlā'au, 'I'iwi, and 'Apapane. In the Snetsinger et al. (1994) study in which the majority of pellets (93% of 301 pellets) were collected from high-elevation sites with abundant native birds, again introduced rodents were the major prey of Barn Owls, although remains of 'Ōma'o, 'amakihi, 'Apapane, and four nonnative bird species were also found. At Hanawī Natural Area Reserve, an important habitat for Maui forest birds, 98% of Barn Owl pellets (n = 60) contained rodent remains, and one contained bird remains, of an endangered 'Ākohekohe (Kowalsky et al. 2002). In the study by Work and Hale (1996), the stomach contents of 18 Barn Owls (64% of the 28 birds examined) contained mainly katydids and crickets, 5 (18%) contained mainly rodents, and the remaining 5 had mixtures of rodents and insects or grass. The finding of insects as a principal food source of Barn Owls by Work and Hale (1996) contrasted with the findings of Mostello (1996), Snetsinger et al. (1994), and Kowalsky et al. (2002) and may be due to differences in sampling techniques (studying stomach contents versus pellets) and populations sampled (salvaged birds versus presumably healthy ones). Investigations in Hawai'i show that Barn Owls forage mainly on rodents, and their predation on common native forest birds appears to be minor. However, Barn Owls do have the potential to impact endangered forest birds, because the loss of even one or a few individuals can be significant over time. Whether the possible role of the Barn Owl in reducing rodent populations is beneficial to native birds and outweighs occasional depredation on the birds themselves remains to be determined.

MANAGEMENT RECOMMENDATIONS AND RESEARCH NEEDS

'Io

'Io are extant on only one island, making their population more vulnerable to catastrophic and anthropogenic events. Events such as fires, alien plant invasions, introduced avian diseases, shooting, and changes in land usage such as urban development could lead to population declines or favor hawks able to coexist with people. Other large-scale changes in land usage could affect 'Io, such as the spread of *Eucalyptus* tree plantations. These plantations support few 'Io, because the quantities of prey are probably limited there and the trees are harvested before they are large enough to support an 'Io nest (J. Klavitter, pers. obs.).

The Hawaiian Hawk Recovery Plan emphasizes habitat protection, reducing human-related mortality, monitoring 'Io populations, and improving public awareness (U.S. Fish and Wildlife Service 1984a). Since the plan was published, additional considerations have arisen. Next I discuss the potential for 'Io translocation, improvements for monitoring, the importance of 'ōhi'a trees with respect to habitat management, implications of proposed large-scale control of rodents in forests, and the threat of WNV.

Translocating 'Io to other islands would seem a logical step. With proper reintroduction techniques, viable populations of 'Io probably could be established on other islands because of the species' adaptability. Some of the possible negative consequences resulting from 'Io translocation include the following: (1) funding may be diverted from more imperiled species, (2) 'Io could prey upon endangered birds at translocation sites, and (3) the action might limit the possibilities of translocating critically endangered 'Alalā to 'Io-free islands. To address such concerns, the reintroduction sites for 'Io to be considered first would be those with few or no endangered birds.

Although 'Io appear viable on Hawai'i Island, the population could decline. To detect a decline and alert managers to the need for recovery action, the Io Recovery Working Group (2001) and Klavitter et al. (2003) recommended that 'Io abundance and adult survivorship be monitored periodically. Recall that adult survivorship is the most important parameter regulating the population. To be statistically meaningful, survival should be monitored with a radio-tagging study at five-year intervals, tagging a minimum of 30 birds from various habitats around the island. The use of radio tags would avoid underestimating 'Io survival.

'Io range and abundance can be monitored islandwide with survey methods developed by Klavitter and Marzluff (2007) and Gorresen et al. (2008). Adult 'Io calls are broadcast at point-count stations, and a correction is made for the 'Io movement that occurs before detection by an observer. However, this 'Io survey technique produces rather wide confidence intervals and currently has the power to detect only major changes in the population. For example, the population would have to drop by 48% to detect a significant decrease with a 95% level of confidence. Klavitter et al. (2003) recommended that surveys be performed every five years and vegetation changes tracked at survey points and on a landscape level using satellite imagery and geographic information systems.

Although not the primary regulating parameter for this population, reproduction could also be monitored. Klavitter et al. (2003) recommended that a minimum of 30 nests be monitored every 10 years in a mixture of habitats dispersed around the island. If productivity studies occurred simultaneously with survival studies, adult females could be tracked directly to their nests. Because the level of fecundity is currently low, it may be one of the few parameters that can be enhanced by management. Because population regulation is most sensitive to adult survival, fecundity would need to be dramatically increased to have a major impact on the population. For increased fecundity to have an effect, juvenile survival would have to remain stable and there would need to be open territories for the "additional" birds to occupy.

Other recommendations by Klavitter et al. (2003) are the management of 'ōhi'a trees, implementation of public educational programs, and protection of additional habitat. Long-term persistence of 'ōhi'a trees and their recruitment are essential, because 'ōhi'a are the principal nest trees for 'Io. Invasive exotic vegetation, fire, and intensive grazing limit 'ōhi'a recruitment (Cuddihy and Stone 1990). Public educational programs would encourage the protection of habitat and nest trees, prevent harassment at nest sites, discourage the purposeful shooting of birds, and foster an appreciation for the 'Io. A large portion of 'Io habitat is currently protected, but few high-density areas fall within the boundaries of a reserve, refuge, or park (see Fig. 12.1) (Klavitter et al. 2003, Gorresen et al. 2008). Incentives that encourage landowners to manage these lands for 'Io would help. On Hawai'i Island, landowners are eligible to receive a land tax discount for lands in conservation status.

One of the long-term goals of land managers in Hawai'i is to control rats with rodenticides over large areas of native forest to assist native passerine populations and other native biota (Nelson et al. 2002). Managers contemplating the use of rodenticides should be aware that there is a risk to raptors that feed on poisoned rats (Stone et al. 1999), although the risk is believed to be low (Lindsey and Mosher 1994). Evidence suggests that hawks eviscerate small mammals prior to consumption, and this behavior probably greatly reduces their risk when scavenging poisoned prey (Schmutz et al. 1989). Also, because rats presently make up a substantial portion of the diets of 'Io and Pueo, these proposed control efforts will reduce the amount of rodent prey available for Hawaiian raptors. Native raptor populations may decrease if large numbers of rats and mice are removed from native ecosystems. The removal of rodents could also force raptors to prey upon larger numbers of native bird nestlings and eggs.

Monitoring 'Io reproductive success and survivorship in the rodent control areas beforehand would provide a baseline from which to assess how 'Io will be affected. 'Io will probably decline at least initially with a decrease in their main prey base, rodents. They may also shift to a diet mainly of insect and avian food as in prehistoric times. Because the predation effects from rodents would then be lifted from native passerines, more of these prey might become available to 'Io.

WNV could prove extremely detrimental to the 'Io population (Chapter 9). Strategies are being implemented to limit the introduction of this virus and other diseases to Hawai'i, and these strategies could be strengthened (Chapter 15). Because mosquitoes are a vector for spreading WNV and other diseases, mosquito control in both conservation and urban areas would reduce transmission (Chapter 17).

Pueo

Basic abundance and demographic information is lacking for Pueo. A study initiated to obtain these data would require extraordinary effort, with research needed on a variety of different habitat types on all the islands. An effective method for determining abundance that surveyed large expanses of habitat efficiently and reliably would need to be developed. Pueo could be outfitted with transmitters and monitored to determine their productivity, survivorship (juvenile and adult), and population trends.

Controlling predators such as dogs, cats, small Indian mongooses, and pigs would assist Pueo in conservation areas by limiting predation at their nest sites and perhaps by reducing interspecific competition. Large-scale rodent control is needed in conservation areas to preserve biodiversity, but Pueo population levels may drop initially due to the loss of a substantial portion of their prey base. The monitoring of Pueo abundance done simultaneously with rodent control would provide information on how the species responds. Pueo would potentially benefit from the creation of "hedge rows" and other habitat enhancements commonly used for game birds in grassland areas that would provide more nesting cover and foraging habitat. Care should be taken to reduce fire risk and to control fires.

Preventing the introduction of the small Indian mongoose to Kaua'i, Lāna'i, and Kaho'olawe may be the most cost-effective management strategy (Mostello 1996). Occasional sightings of mongooses on Kaua'i are alarming and indicate that greater effort is needed to prevent the spread of alien species within the Hawaiian Islands (Chapter 15).

How might Pueo respond to the short-term control of Barn Owls? This is an interesting research question that could shed light on competition between Pueo and Barn Owls and inform land managers regarding this potentially limiting factor.

In areas where Pueo pose aviation hazards around airports, nonlethal owl deterrent methods should be explored. These might include trapping Pueo and translocating them to conservation areas with low owl densities or using them for education in zoos and other settings.

Federal listing of the Pueo as threatened or endangered is debatable because so little is known regarding its conservation status and its taxonomic uniqueness. Nonetheless, federal listing would help garner research and management funding as well as provide additional legal protection. Listing the Pueo as endangered by the state of Hawai'i on islands beyond O'ahu may have similar effects. Additional phylogenetic and systematic work could clarify its endemic status. Regardless, the Pueo could be thought of as an "endangered species unit" (White and Kiff 2000). Finally, recovery planning for the Pueo would provide needed guidance by prioritizing research and management activities for this species in Hawai'i.

Barn Owls

Because the Barn Owl was brought to Hawai'i by humans, its management is treated differently than that of native raptors. Raptor research and management in Hawai'i focus primarily on assisting the 'Io, because the species is both native and endangered, and secondarily on assisting the more widespread Pueo. The role of Barn Owls as predators of native birds has been discussed. It may be useful for researchers to explore the possible effects of Barn Owls on the endangered Hawaiian Hoary Bat ('ōpe'ape'a, *Lasiurus cinereus semotus*), because this owl occasionally preys upon bats elsewhere (Speakman 1991). The level of Barn Owls' impact on Hawaiian seabirds may warrant investigation, because they may be an important cause of mortality for the threatened 'A'o (Newell Shearwater, *Puffinus auricularis newelli*) and other species (J. Sincock, pers. comm., in Conant 1980). When Barn Owls are known to depredate Hawaiian birds, including seabirds, managers may be justified in removing them.

SUMMARY

There is little space for top predators perched in short insular food chains, and consequently the original raptor community of the Hawaiian Islands was composed of few species: the 'Io or Hawaiian Hawk, a sea-eagle, a harrier, and four species of owls. All were endemic except possibly the sea-eagle, and all are now extinct except

the 'Io. Anthropogenic pressures most likely drove these extinctions. With the recent colonization by the Pueo or Short-eared Owl and introduction of the Barn Owl, only three species occupy the trophic niches once filled by a more diverse raptor fauna.

Living in an island system, Hawaiian raptor populations have always been small compared to those of their continental counterparts, making them more susceptible to the negative effects brought about primarily by humans and to a lesser extent by natural catastrophic events. The 'Io and Pueo have weathered recent changes but continue to survive in relatively small populations.

The endangered 'Io persists as a robust population, most recently estimated at 3,085 hawks distributed throughout Hawai'i Island, although its prehistoric range extended across the archipelago. It occupies a broad suite of habitats but requires trees for nesting, particularly native 'ohi'ā trees. Originally a predator of birds and insects, it has expanded its diet with the human introduction of rodents and alien birds and arthropods.

Although classified as an endemic subspecies, the Pueo scarcely differs from the continental Short-eared Owl, and it may someday be moved from endemic to indigenous status. It is distributed throughout the Hawaiian chain but is now rare and perhaps increasingly confined to open habitats. Surprisingly little is known about this species in Hawai'i despite its prominence in Hawaiian culture. No source of threat particularly stands out, although this ground-nesting species may suffer substantially from losses to introduced mammals, competition from Barn Owls, and conversion of its habitat.

All three raptors prey to some extent on native forest birds, but this predation does not seem to inflict substantial losses to any species except the 'Alalā. Because raptors are predators, managers are faced with a conservation conundrum when attempting to balance raptor recovery with that of native passerines, especially the rare and critically endangered species. A precarious balance must be achieved in maintaining viable populations of raptors while at the same time protecting rare native passerines on which they prey.

The key elements in conserving raptors in Hawai'i are as follows:

1. Maintain large tracts of native forest, especially the large 'ōhi'a trees important for nesting 'Io, by suppressing fire and grazing and preventing invasions of ecosystem-altering nonnative species.
2. Preserve large tracts of grasslands and shrubland for Pueo.
3. Prevent the establishment of new avian diseases in Hawai'i, particularly West Nile virus and control the vector (mosquitoes) responsible for spreading most avian diseases in Hawai'i.
4. Reduce populations of feral cats and small Indian mongooses that compete with, and in some instances prey upon, native raptors. Prevent the introduction of mongooses to Ni'ihau, Kaua'i, Lāna'i, and Kaho'olawe.
5. Promote educational programs and bird watching to increase awareness of Hawaiian raptors and to prevent the purposeful shooting of them.

6. Plan for and minimize urban development. This includes making urban areas usable by 'Io, for instance, by maintaining 'ōhi'a trees and controlling feral animals.
7. Reconcile rodent control in native ecosystems with recovery of the two native raptors.

Fortunately, the majority of the management actions needed for Hawaiian raptors are similar to those for other forest birds (Stone 1985, Scott et al. 1986, van Riper and Scott 2001). As is the case of so many conservation programs, appropriate funding, political support, and public education to increase knowledge and awareness of native ecosystems are the keys to the conservation of raptors in Hawai'i (Restani and Marzluff 2001, 2002a). With humans' respect and appropriate management, it is likely that Hawaiian raptors will soar well into the future.

ACKNOWLEDGMENTS

Special thanks are reserved for L. Laniawe, J. Marzluff, M. Vekasy, and all those who have contributed to the conservation of raptors in Hawai'i. L. Kiff, J. Marzluff, and an anonymous reviewer provided excellent comments that improved the manuscript.

CHAPTER THIRTEEN

The History and Impact of Introduced Birds

JEFFREY T. FOSTER

A quick stroll through any terrestrial habitat in Hawai'i reveals an abundance of birds —cooing doves, twittering white-eyes, and squawking mynas. Yet people living in lowland Hawai'i at the beginning of the twentieth century were faced with a much different auditory sensation: silence. In his 1854 book *Sandwich Island Notes by a Haole*, G. W. Bates wrote, "A traveler is more impressed with what there is not, than what he sees . . . the almost universal absence of singing birds" (in Berger 1974: 61). Disease, habitat loss, hunting, and introduced predators were among the multitude of factors that had decimated the native birds. Furthermore, the extreme isolation of the Hawaiian Islands has created few chances for natural colonizations by land birds from other areas. Human residents of the islands, many of them accustomed to a mainland dawn chorus of birds, aimed to replace the native birds that had been lost, and after a wide array of releases, a new community of birds took shape.

Hawai'i now hosts the most diverse assemblage of introduced birds in the world, among them pheasants, ducks, egrets, parrots, swifts, owls, and finches. A combination of intentional and accidental releases has allowed the number of birds introduced to exceed 170 species, and this number grows every year (Berger 1981, Long 1981, Pyle 2003). Not every importation has been successful, but 54 species have established breeding populations in the wild (Pyle 2002), along with at least 13 other potentially naturalized species. The result of these releases is that introduced birds are now the dominant avian element, in both numbers of species and individuals, in nearly all terrestrial habitats of Hawai'i. Due to the large numbers of releases of an assortment of species, patterns of introduction success have emerged and allowed for analysis of the factors determining what makes a successful invader. The Hawaiian Islands have thus become one of the premier natural laboratories in which to study avian invasion biology.

Yet, as quiet and barren as the Hawaiian lowlands appeared a century ago, native birds, plants, and invertebrates continued to survive, particularly in native montane forests. In a simplified view, two separate bird communities would emerge, native birds at high elevations and introduced birds at low elevations. Over time, however, introduced birds began invading native forests, and two disparate avian communities would meet, a native one replete with endemic species locally adapted over several million years versus a suite of worldwide invaders adapting to novel environments. These meetings have had profound consequences for the native flora and fauna and have become among the most challenging conservation issues. Do introduced birds compete with natives for food resources, or are they merely filling empty or vacated niches? With the loss of most native frugivores, are introduced birds replacing natives in the roles of seed dispersers? Do introduced birds serve as disease reservoirs, or, alternatively, do they buffer native birds from biting mosquitoes, reducing infection rates among natives? In essence, what are the impacts of these introductions? These and similar questions are at the core of current research on introduced birds, and the answers they provide will aid us in the maintenance of native ecosystems as well as enlighten us about fundamental ecological and evolutionary processes.

This chapter describes the history of bird importation into Hawai'i and the series of actions that enabled so many species of birds to be released. This history provides the context through which we can understand the succession of processes by which birds have reached Hawai'i. The chapter then summarizes the research on the factors determining introduction success from studies in Hawai'i and elsewhere, providing insight into why some releases succeed while others fail. Third, it describes what happens to those species that have established, including genetic and morphological changes and the population dynamics of these species as they invade new areas. Finally, it assesses the effects of introduced birds on native ecosystems and addresses methods of preventing or minimizing their impacts.

BRAVE NEW WORLD: INTRODUCED BIRDS AND HOW THEY GOT THERE

The true list of all birds introduced into Hawai'i will never be known. While writing *Exotic Birds of Hawaii* in the 1930s, Edward Caum encountered numerous difficulties in determining the numbers and identities of released birds. Often introductions were not recorded, and when lists were kept, they were incomplete. For example, a description of a release in 1921 merely stated, "75 pigeons, 100 doves, 100 quail" (Caum 1933: 33). Nonetheless, he and other authors have been able to assemble a long list of known releases (e.g., Long 1981, Lever 1987). The types of birds introduced have been taxonomically narrow, with releases and naturalizations (establishment of a wild breeding population) limited primarily to a few avian orders. Passerines (songbirds and perching birds), gallinaceous birds (chickens, pheasants, and related birds), and doves comprise nearly 80% of all introductions in Hawai'i (Moulton et al. 2001) (Fig. 13.1). More important, these groups represent nearly 90% of all naturalized species (Pyle 2002).

Introductions worldwide reflect the cultural interests and aesthetics of the people conducting the releases (Long 1981), and releases in Hawai'i have followed this pattern. Sometime after their arrival around A.D. 800, Polynesians introduced to Hawai'i the first exotic bird, the Red Junglefowl (*Gallus gallus*) (Caum 1933). No other in-

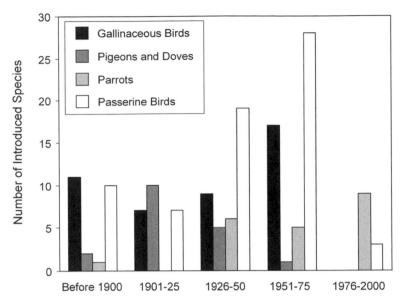

Figure 13.1. Changes in the pattern of avian introductions in Hawai'i over time. Introductions of species outside of these four groups constitute only 11% of introductions and are not graphed. Sources: Caum (1931), Berger (1981), Long (1981), and the SIGHTINGS database of the Bishop Museum, Honolulu.

troductions occurred until the late 1700s, when arriving Europeans brought domestic Rock Pigeons (Columba livia) that soon became feral (Walker 1967). Since then, there have been four main methods of introduction: escape or release of cage and domestic birds, deliberate introduction of songbirds, introductions of birds as biological control agents, and game bird releases. A fifth introduction method, "adventitious arrival," which is unintentional human-aided introduction, has not yet resulted in any naturalization. On O'ahu there have been sightings of the European Starling (Sturnus vulgaris) and Great-tailed Grackle (Quiscalus mexicanus), likely stowaways on cargo ships docking at Honolulu Harbor (Pratt 2002).

Cage Bird Escapes and Releases

Early immigrants to Hawai'i brought many cage birds commonly kept as pets or for food (Berger 1974). Spotted Doves (Streptopelia chinensis) were raised by Chinese immigrants for food and likely escaped captivity (Caum 1933). The Hwamei, or Melodious Laughingthrush (Garrulax canorus), is believed to have escaped in the Honolulu fire of 1900, and the Red Avadavat (Amandava amandava) likely escaped captivity in the early 1900s (Caum 1933). However, the majority of the cage bird releases or escapes have taken place in the past 40 years (Fig. 13.2). In the 1960s, there was a plethora of releases of finches common in the pet trade. Many of these species became naturalized, especially around Diamond Head, O'ahu, and Pu'u Wa'awa'a, Hawai'i Island. Passerine releases continue to this day and include the Black-hooded Oriole (Oriolus xanthornus), sighted on O'ahu in 2000 (Shapin 2000). Significantly more passerines have been successfully introduced to Hawai'i than members of either of the two other most successful orders, Galliformes and Columbiformes (Moulton et al. 2001), although the shift in releases has recently changed to parrots (Psittaciformes) (Pratt 2002).

Most parrot species common in the pet trade have been sighted free-flying in Hawai'i. Because parrots are long-lived and

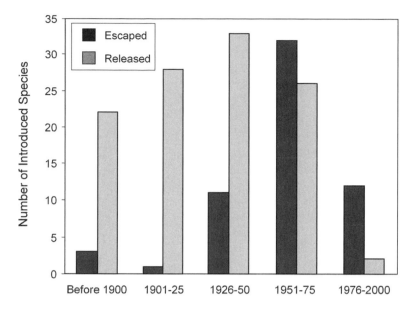

Figure 13.2. Shifts in avian introductions in Hawai'i over time. Such introductions have shifted from intentional introductions to escape of birds from captivity. Data from the same sources as Fig. 13.1.

populations may be continuously supplied by additional releases, determining their degree of naturalization is difficult. Only the Rose-ringed Parakeet (*Psittacula krameri*), Mitred Parakeet (*Aratinga mitrata*), and Red-crowned Parrot (*Amazona viridigenalis*) are currently listed as naturalized species (Pyle 2002), although several other species have likely established breeding populations. At least one parrot species that naturalized on Maui subsequently died out; Pale-headed Rosellas (*Platycercus adscitus*) were present at 'Ulupalakua from 1877 until at least 1933 (Caum 1933). O'ahu hosts most free-flying parrots in Hawai'i, particularly at sites in Honolulu. Recent sightings include the Salmon-crested Cockatoo (*Cacatua moluccensis*) and Sulphur-crested Cockatoo (*C. galerita*) (Pratt 2004).

Intentional Releases

Naturalization societies, government agencies, and private individuals all carried out the intentional release of birds. Much of the work conducted by the naturalization societies and individuals was based on the belief that lowland Hawai'i was a desolate land that required an imported flora and fauna to bring it to life (Berger 1974). Many of the releases by government agencies were of game birds for sport and of other birds such as the Cattle Egret (*Bubulcus ibis*), Barn Owl (*Tyto alba*), and Common Myna (*Acridotheres tristis*) for pest control in sugar cane and other crops. Legislation was passed to protect nonnative birds from trapping or killing so that they could become established (Act 78, S.L. 1953, Revised Laws of Hawaii 1945). Although the laws have been revised, most past releases of species that have naturalized were legal. Species such as the Ring-necked Pheasant (*Phasianus colchicus*), Japanese White-eye (*Zosterops japonicus*), and Northern Cardinal (*Cardinalis cardinalis*) were released by as many as four different organizations, highlighting the fact that many different groups have been responsible for bird releases. Private individuals, mostly wealthy landowners acting on their own accord, have also released numerous species, but the cohorts have usually been small, decreasing the likelihood of their establishment.

The Buy-a-Bird Campaign

The buy-a-bird campaign was initiated in 1928 by an agricultural official of the Territory of Hawai'i and then sponsored and organized by the Hilo Chamber of Commerce in 1929 (Locey 1937). Schoolchildren were encouraged to raise money for the introduction of the Hwamei, Red-billed Leiothrix (*Leiothrix lutea*), and Northern Cardinal to the island of Hawai'i. The importation of birds intended for release required approval from the Hawaiian Sugar Planters' Association and the Board of Agriculture and Forestry. In 1937, during a second campaign, 501 individuals of five species were released, including Varied Tits (*Parus varius*), Blue-and-White Flycatchers (*Cyanoptila cyanomelana*), Japanese White-eyes, Indigo Buntings (*Passerina cyanea*), and Painted Buntings (*P. ciris*) (Bryan 1937c). Of these species, only the Japanese White-eye is known to have established a breeding population.

Despite this enthusiasm from schoolchildren and the Hilo Chamber of Commerce, letters from agricultural officials indicated the hesitancy among the sugar companies regarding the widespread importation of birds (Pemberton 1936, Spalding 1936). Their rationale was economic; imported insectivorous birds would eat a range of insects, including insects imported by the sugar companies to control crop pests. For instance, it was feared that flycatchers would be more likely to eat the introduced dipteran parasite of the beetle borer than the beetle itself (Pemberton 1936). No further releases were conducted by the buy-a-bird campaign, but the mantle was picked up by another release society, the Hui Manu.

Hui Manu

When the name of the person who introduced the first songbird to the Islands is learned, a monument will be erected to commemorate that important event.

—Anonymous (1937: 2)

The Hui Manu, a bird acclimatization society, was the most successful private organization behind releases in Hawai'i, and many of the birds we see today are a result of their efforts. The society was first organized in 1929 as an offshoot of a ladies' gardening club whose sole purpose was to introduce songbirds to Hawai'i. The crusade was begun with aesthetic goals, to add song and color to the islands (Lewis 1937, Hogue 1957). Through the introductions of birds and plants, it was thought, a "composite masterpiece" could be created (Anonymous 1937).

By World War II, Hui Manu releases included now resident species such as the White-rumped Shama (*Copsychus malabaricus*), Japanese Bush-Warbler (*Cettia diphone*), and Northern Cardinal and nonnaturalized species including the Narcissus Flycatcher (*Ficedula narcissina*) and Japanese Red Robin (*Erithacus akahige*) (Berger 1974). The major obstacle to releases by the Hui Manu was that only the hardiest of species could survive transoceanic travel. The mortality rate was high during the lengthy time required to ship the birds to Hawai'i (Berger 1981).

By 1968, many of the goals of the society had been achieved, and the Hui Manu was disbanded due to stricter regulations on bird importation, an aging membership, and dwindling funds (Anonymous 1970). The Guam Swiftlet (*Aerodramus bartschi*) was the last species they introduced. Approximately 150 swiftlets were released on O'ahu in 1962, and an additional 200 were released in 1965 (Berger 1981). The species is still an established resident of Hālawa Valley.

Government Releases

There are many species in America, the most accessible source of supply, that are of great value, both from the aesthetic and the economic standpoint.

—Henshaw (1901: 133)

The aspirations of the territorial government were similar to those of the naturalization societies. Officials of the Board of Agriculture and Forestry desired to fill Hawai'i with any useful species that could survive (Fisher 1948). Their emphasis was on introductions to control agricultural pests, and the board was particularly interested in importing insectivores. Frugivores such as the Blue Jay (*Cyanocitta cristata*) were banned from importation because of their potential damage to fruit crops (Baldwin 1972).

City and county governments were also involved in bird releases. For instance, in the 1920s the city of Honolulu released Emerald Doves (*Chalcophaps indica*) and the counties of Maui and Hawai'i released Nicobar Pigeons (*Caloenas nicobarica*) and Diamond Doves (*Geopelia cuneata*), respectively (Caum 1933).

Biological Control

Three species introduced as biological control agents naturalized, but no species succeeded in the control of its target. First introduced in 1865 to control outbreaks of armyworms in pastures, the Common Myna was abundant in Honolulu by 1879 (Caum 1933), and by the 1900s it had become one of the most common birds in native forests (Henshaw 1902, Perkins 1903). Myna populations have drastically declined since then but are locally common along forest edges and in human-developed areas (Scott et al. 1986, Foster et al. 2004). The Cattle Egret was introduced from Florida to control flies (Breese 1959). It was released on five of the main islands, slowly spread, and became abundant in the lowlands (Berger 1981). The Barn Owl was imported from California to control rats that fed on sugar cane. They were initially released in 1958 on the island of Hawai'i, and subsequent releases occurred on that island and on Kaua'i, O'ahu, and Moloka'i until 1965 (Berger 1981). Like the Common Myna and Cattle Egret, the Barn Owl is currently resident on all the main Hawaiian islands.

By the 1950s, the threats posed by the import of birds that were generalist feeders and their lack of efficacy in controlling pests were well known. However, when Hillebrand introduced Common Mynas, the hazards were poorly recognized (Caum 1933). In what was to become the field of economic ornithology, promoters of birds for biocontrol stressed the economic value of bird importation, such as the introduction of insectivores to control sugar cane pests. They believed there was a specific need for birds to inhabit lowland agricultural lands and urban areas due to the absence of native insectivores (Henshaw 1901, Bryan 1912). However, some researchers were aware of potential nontarget effects and recommended that judicious care be taken not to import harmful species and to ensure government regulation (Henshaw 1901).

Game Bird Releases

The territory of Hawai'i maintained a game bird rearing facility on O'ahu in the 1930s and aimed to release 60,000 birds per year (Locey 1937). As of 1970, at least 80 species and subspecies of game birds had been released in the islands (Lewin 1971). Since then, the state of Hawai'i and game bird clubs have concentrated primarily on supplemental releases of naturalized species. There are currently 14 game bird species naturalized in Hawai'i. These are mainly gallinaceous species but also include three doves and a sandgrouse (Pyle 2002).

The Pu'u Wa'awa'a Ranch on Hawai'i Island was the site of the most intensive and best-documented game bird release program. At least 2,611 individual birds of 33 species were introduced (Lewin 1971). The only game bird released at Pu'u Wa'awa'a that is a primary resident of native forest is the Kalij Pheasant (*Lophura leucomelanos*), which was released in 1962. Within 20 years, kalij had dispersed and reached nearly all suitable habitat on the island (Lewin and Lewin 1984). Unlike passerine releases, successful game bird releases provide for sport, and hunting serves as a source of revenue for the state.

The Bird Trade

The state of Hawai'i currently has four categories for birds to be imported into the state: Conditional, Restricted A, Restricted B, and Prohibited (Hawaii Administrative Rules §4-71). All birds on the Conditional list can be imported with minimal restrictions. The Restricted A list allows importation by commercial bird dealers and private individuals, and the Restricted B list allows for importation for exhibitions or research, such as importation by zoos. Prohibited species are not allowed entry for any reason. Currently 60 bird species are prohibited from entry into the state of Hawai'i, and an additional 339 species are allowed for import on the Restricted B list. Species not on any list are denied entry by default.

As restrictions on the import of nonnative birds intended expressly for release have become more stringent, most new sightings have been the result of accidental or intentional release of cage birds (Pratt 2002). The current pool of potential invaders, excluding those illegally imported, reflects the availability and abundance of birds in the pet trade, as is the case in California (Hardy 1973). The naturalization of finches in the 1960s, seemingly the result of a concerted release effort, may simply be a reflection of the many thousands of finches that had been imported into Hawai'i at that time. As more and more birds were imported, inevitably some finches escaped captivity or were released by their owners. The dynamics of exotic bird markets have been modeled, indicating that short-lived, fecund species such as finches are prone to oversupply and that this surplus makes release more likely (Robinson 2001). It was also reported that during surges in parrot popularity in the United States, even expensive species were given to bird sanctuaries. Not all species are turned over to bird sanctuaries, and the abundance of released birds in areas such as Hawai'i, Florida, and southern California may reflect not only the suitable environmental conditions in those places (Hardy 1973) but also the likelihood of release into areas where people believe these birds can survive in the wild. Furthermore, an abundance of nonnative fruiting plants may sustain these released populations (Owre 1973, Pranty and Garrett 2003).

In 2003, in response to outbreaks of Newcastle disease, the state of Hawai'i imposed a ban on the importation of all birds from southern California and Nevada (Fujimori 2003). Other restrictions on bird imports had previously been imposed, including a complete ban on the importation of birds through the mail, due to the outbreak of West Nile virus in the continental United States. Despite increasing regulation and quarantine requirements, thousands of birds are imported into Hawai'i each year. In 1999, private individuals and commercial dealers officially imported 3,289 individual cage birds (Hawai'i Department of Agriculture database). Of these, 2,811 were parrots of 28 species and 478 were passerines of 23 species. The most commonly imported bird was the Budgerigar (*Melopsittacus undulatus*), with 2,408 individuals imported, followed by the Cockatiel (*Nymphicus hollandicus*), with 163 individuals, and the Peach-faced Lovebird (*Agapornis roseicollis*), with 110 birds. The Zebra Finch (*Taeniopygia guttata*) was the most commonly imported finch, with 154 imports. Recent escapees thus appear to be originating from breeders or current pet owners, because many species not currently imported have recently appeared in the wild.

Interisland Dispersal

The degree of interisland dispersal by introduced birds is unknown. However, it is not uncommon. Bryan (1958) suggested that Japanese White-eyes had colonized the island of Hawai'i from Maui, although the extensive efforts to release this species on Maui bring this idea into question. Nonetheless, white-eyes (*Zosterops* spp.) are island colonists throughout the Indo-Pacific region. They have been sighted at sea miles from land (Berger 1981), and interisland dispersal by other species in the Hawaiian Islands is not unprecedented. The Japanese Bush-Warbler, initially released on O'ahu in 1929 by Hui Manu, successfully colonized Lāna'i, Moloka'i, and Maui by 1980 (Carothers and Hansen 1982), Kaua'i by 1986 (Pratt 1988), and the island of Hawai'i by 1997 (Nelson and Vitts 1998). White-rumped Shama have only just colonized Moloka'i (Pratt 2004), potentially flying there from O'ahu. The Mourning Dove (*Zenaidura macroura*), a recent arrival on islands from Maui to Kaua'i, likely colonized from populations on Hawai'i Island. Thirteen nonnative bird species on uninhabited Kaho'olawe likely colonized on their own from the adjacent islands, although introduction by humans is also possible. These species include the Spotted Dove, Zebra Dove (*Geopelia striata*), Barn Owl, Sky Lark (*Alauda arvensis*), Japanese Bush-Warbler, Japanese White-eye, Northern Mockingbird (*Mimus polyglottos*), Common Myna, Red-crested Cardinal (*Paroaria cristata*), Northern Cardinal, House Finch (*Carpodacus mexicanus*), African Silverbill (*Lonchura cantans*), and Nutmeg Mannikin (*Lonchura punctulata*). Zebra Doves and Northern Mockingbirds appear to have colonized the island of Hawai'i from Maui (Schwartz and Schwartz 1949, Berger 1981). Thus, water barriers only delay the spread of many species. Interisland dispersal also suggests that the nonnative bird communities of the main islands will eventually come to mirror each other.

WHAT MAKES AN INVADER SUCCESSFUL?

The remoteness of the Hawaiian Islands has minimized the likelihood of natural colonization from the continents or other islands. In the past 5 million years, only six passerine species are known to have colonized the main Hawaiian islands unaided by humans (Chapter 1). As is evident from the preceding pages, settlement by humans drastically changed these odds, and as a result, there have been many opportunities for birds to gain entry into the Hawaiian avifauna. Once a species has been successfully introduced, it is likely to remain a resident; the birds that are there are there to stay. The majority (75%) of the nonnative bird species established in 1933 are still present (Pyle 1992).

Given the large number of successful introductions to island groups such as Hawai'i, it is tempting to assume that islands are more successfully invaded than continental areas. In a comparison of bird introductions to island and mainland areas, Sol (2000) concluded that invasibility by birds is not an intrinsic attribute of islands. However, although most species released in Hawai'i have not naturalized, roughly 30% of all species released have established viable populations in the wild (Pyle 2002). Because of the large numbers of independent releases on separate islands, the Hawaiian Islands have become a testing ground for theories on community assembly. Are there predictable patterns of species success or failure based on species-specific attributes? Do some species typically succeed while others usually fail?

Patterns from Hawai'i

Research on the factors influencing the introduction success of nonnative birds began with an assessment of the role of competition in determining the assembly of birds in the Hawaiian Islands (Moulton and Pimm 1983). Interspecific competition has often been invoked as a dominant force determining the invasibility of avian communities (Moulton and Pimm 1983, Lockwood et al. 1999). If competition is occurring, with some species competitively excluding others, three patterns in the extant avifauna may emerge. Naturalization should become increasingly difficult over time, as was demonstrated by an analysis of the introductions of finches to Hawai'i (Moulton and Pimm 1983). In addition, naturalized species should be more morphologically dissimilar (morphologically overdispersed) from one another than expected by chance (Moulton and Lockwood 1992, Lockwood et al. 1993). Furthermore, the bill morphologies of pairs of congeneric species that have successfully naturalized should be significantly different from one another compared with those of pairs whose introductions have failed (Moulton 1985). Moulton, Lockwood, and their coauthors have found all three patterns of competition among introduced finches in Hawai'i and other island groups.

Simberloff and Boecklen (1991) detailed an all-or-none pattern of success or failure suggesting intrinsic attributes of successful avian invaders in the Hawaiian Islands. They concluded that some species typically succeed, while others inevitably fail. Their work also indicated that birds' introduction history could be used to predict the likelihood of their introduction success. Might some species be competitively superior to others? Their conclusions have not had widespread support (e.g., Moulton 1993, Moulton et al. 2001). One explanation is that regardless of an invader's attributes, some releases fail due to factors such as environmental stochasticity or insufficient release numbers, and larger data sets are necessary to test such hypotheses.

Worldwide Patterns

The work from Hawai'i has spurred a series of publications testing the relative importance of the role of competition compared with other factors affecting introduction success worldwide. The major challenge has been accounting for the suite of confounding factors, particularly differences in release effort. In addition, humans have exhibited taxonomic, geographic, and character selectivity with regard to the species they introduce, and this nonrandom sampling has made statistical analysis challenging (Duncan et al. 2003). Although the patterns found in Hawai'i suggest a history of competition among nonnative species, other factors affecting introduction success are potentially more important. Duncan and his coauthors maintained that these results are more likely caused by differences in introduction effort and possibly by predator abundance (Duncan 1997, Duncan and Blackburn 2002). After evaluating all the current literature on bird introductions and from their own work in New Zealand, they further concluded that there is no compelling evidence that competition is a major determinant of establishment (Duncan et al. 2003). Their evidence suggests that factors unrelated to species' characteristics, such as release environment and introduction effort, are often the major determinants of introduction success. For instance, populations of nonnative species were more likely to be successful in areas where climate conditions matched those in their native range (Blackburn and Duncan 2001).

Introduction effort is often a significant factor in the establishment of introduced birds (Veltman et al. 1996, Duncan 1997, Green 1997, Duncan et al. 2001). In Hawai'i, naturalized galliforms were released in significantly greater numbers than species that had failed, but there are insufficient data to allow us to assess the importance of introduction efforts for passerines and doves (Moulton et al. 2001). Many of the successful invaders in Hawai'i were repeatedly released in large cohorts by naturalization societies such as the Hui Manu or the game bird release programs.

Other factors are clearly important as well. For instance, behavioral flexibility has been reported to increase the chances for bird establishment, likely through the ability to adapt to novel environments (Sol et al.

2002). Bird species with larger geographic ranges are more likely to be successfully introduced than those with small ranges, also suggesting that successful species are better able to adapt to the local environment (Duncan et al. 2001, Moulton et al. 2001, but see Moulton and Scioli 1986).

Regardless of the relative importance of factors affecting introduction success, competition between native and introduced birds has not been a factor in introduced bird establishment in Hawai'i. Habitat segregation (nonnative birds in the lowlands, natives in the montane areas) has meant that during the initial stages of introduction, introduced birds have had minimal contact with most native birds. There is a possibility of postestablishment competition between native and introduced birds, however, potentially influencing the distribution and abundance of both sets of species (see "Impacts on Native Birds and Flora," following).

Adaptive Shifts

Birds introduced to a new environment, coupled with genetic change, can lead to speciation events. This combined effect may be particularly true for insular introductions due to reduced population sizes, founder effects, and geographic isolation. Thus, populations of nonnative species on islands can allow for analysis of microevolutionary processes. Often these patterns can be observed through changes in morphological characters. Evolution of racial differentiation in birds is typically thought to occur over long time scales, perhaps thousands of years. It has been shown, however, that differentiation can occur rapidly. Within roughly 100 years, distinct genetic changes have occurred in Common Mynas introduced to Hawai'i from India, indicating that there has been rapid genetic differentiation due to population bottlenecks and genetic drift (Baker and Moeed 1987). Johnston and Selander (1971) demonstrated rapid differentiation

in introduced House Sparrows within 100 generations. Size differences in these sparrows corresponded with environmental gradients, suggesting that the response was adaptive. Similar clinal responses were seen among House Sparrows introduced to New Zealand (Baker 1980). Compared with North American and many European populations, House Sparrows introduced to Hawai'i have unique plumage characteristics likely due to their extended isolation from mainland populations (Johnston and Selander 1964). In addition to plumage characters, morphometric measurements from populations in their introduced range can be compared with those from ancestral stock. Eurasian Tree Sparrows (*Passer montanus*) introduced to the U.S. mainland were significantly smaller than the ancestral populations and exhibited significant morphological change in their bills (St. Louis and Barlow 1991). In terms of invasion success, the ability of introduced birds to rapidly adapt to novel environments may be more important than phenotypic plasticity or their ability to tolerate a wide range of environmental conditions (Lee 2002).

However, not all changes in exotic birds are adaptive, nor are they all genetically based. Male House Finches in Hawai'i are predominantly yellow or orange rather than the typical red of most mainland populations. Grinnell (1912) found the Hawai'i population so different from mainland populations that he considered it a separate species. But this change is a reflection of a lack of carotenoid pigments in the diets of House Finches in Hawai'i rather than an example of evolutionary change (Hill 1993).

Population Dynamics of Introduced Birds

Many populations of nonnative birds have a lag phase in their introduction. After their initial releases, populations of nonnative birds slowly grow and then suddenly explode. For some species, this lag is explained by the lag period of a simple exponential growth curve. Many currently naturalized species were not found breeding or were not even sighted for many years following release, such as White-rumped Shamas (Anonymous 1953), Barn Owls (Tomich 1962), and Guam Swiftlets (Donagho 1970). Populations of the Common Myna and Red-billed Leiothrix in Hawai'i both experienced a slow building period followed by rapid growth. Alternatively, rapid evolutionary adaptation may account for the emergence of nonnative species from such lag phases (Mack et al. 2000, Lee 2002).

Species whose populations have exploded do not always remain abundant, however. Common Myna populations irrupted to almost countless flocks in 1900 but had declined appreciably by 1905 (Perkins and Swezey 1926). Nonetheless, mynas were a formidable force in native forests for at least a decade. Currently, they remain common at lowland sites, particularly in urban areas, but are rare in native forests. The Red-billed Leiothrix has also experienced periods in which it has been among the most abundant birds. Leiothrix established on Kaua'i, but Richardson and Bowles (1964) reported them rare in the 1960s, and by the 1990s they were likely extirpated (Male and Snetsinger 1998). The Japanese or Varied Tit (*Parus varius*) apparently established breeding populations on Kaua'i and O'ahu, although both populations subsequently died out (Caum 1933). Population crashes are common among naturalized species, and disease is often implicated, although frequently unproven (Simberloff and Gibbons 2003). Yet freedom from disease and parasites is one reason that many nonnative species may thrive upon release (Torchin et al. 2003). The presence of multiple populations of introduced species on different islands but in similar habitats may allow for assessment of the factors regulating their populations. Such studies will not only provide insight into nonnative species biology but may also suggest which fac-

tors are limiting for native populations, such as food availability, disease, and nest predation.

IMPACTS ON NATIVE BIRDS AND FLORA

We ought to proceed with caution before loading ourselves with too many species of birds that some day we may be sorry for.

—Spalding (1936: 1)

Many introduced birds are likely to remain dominant members of forest bird communities on one or more islands of Hawai'i (Box 13.1). Previous releasers of birds have rarely considered the ecological effects these species would have on native communities in Hawai'i. Although some members of the Hui Manu and other organizations were concerned about negative impacts, often the primary concern was the potential threat to sugar production.

What have been the consequences of bird introductions? Although it is an essential aspect of invasion biology and conservation, quantifying the impact of nonnative species and determining the full extent of their effects have proven challenging. Among the many potential effects of introduced birds are competition with native species, predation, transmission of disease, and seed dispersal. Prominent among conservation concerns are the effects of nonnative birds on endangered birds. Even minor effects could contribute to the extinction of these species, particularly those with small populations. For instance, introduced birds may be competing with native birds, and discerning which invaders have the most impact is essential. Introduced birds may also influence native ecosystems through the dispersal of native and nonnative plants. Thus, assessing the full range of their effects is essential.

Competition

Introduced birds may compete with native birds for food, territories, or nest sites. The distance between the introduction sites of

Box 13.1. **The Big Seven**

Here seven introduced bird species (with family names in parentheses) are described that are common in native forest, are on at least one of the Hawaiian islands, and are potentially deleterious to native ecosystems. See Plates 10 and 11 for photo illustrations. Other introduced birds, such as the House Finch and House Sparrow, may serve as reservoirs for avian disease, but the extent and significance of these interactions are not yet known. All of these species may compete with native birds for food, disperse nonnative plant seeds, or both, and they are listed in order of increasing perceived threat. Unless otherwise indicated, distribution and natural history information are based on Berger (1981), Scott et al. (1986), and Pratt (2002); diet information is from Foster (2005) and Foster and Robinson (2007).

Kalij Pheasant (Phasianidae)

The only gallinaceous bird among the big seven, kalij are established throughout montane forests on the island of Hawai'i, although a few wild birds were discovered recently on O'ahu (S. Mosher, pers. comm.). Kalij feed primarily on the ground but will also forage in shrubs and trees. Their eclectic diet includes quantities of fruit and seeds of many native and introduced species. The ecological effects of kalij foraging have recently been studied (K. Postelli, pers. comm.), demonstrating that this pheasant's digestion kills most seeds and that its seed dispersal is limited to a few species of plants.

Red-vented Bulbul (Pycnonotidae)

This is the more common of two bulbul species present on O'ahu. It has not yet naturalized on any other island. It is abundant in montane native forests, low-elevation nonnative forests, and urban areas. It feeds primarily on fruit but will eat invertebrates and small reptiles as well (Blanvillain et al. 2003). Its primary threat is as a disperser of nonnative seeds.

Japanese Bush-Warbler (Sylviidae)

With its extremely loud vocalizations and high population densities, the bush-warbler seems to be the only species in some forests. Bush-warblers reside in native and nonnative forests from low to high elevations on all the Hawaiian islands, apparently requiring only a dense understory. Males are highly territorial toward conspecifics, but aggression toward other species has not been documented. Bush-warblers eat primarily arthropods, creating possible competition with native insectivorous birds.

White-rumped Shama (Turdidae)

A showy bird with a flutelike song, the shama is common in mesic to wet forests on O'ahu and Kaua'i, with a new population establishing on Moloka'i. They are primarily insectivorous and may compete for food resources with the native Puaiohi (*Myadestes palmeri*) (Snetsinger et al. 1999).

Hwamei (Timaliidae)

This species is the largest introduced passerine. It resides in native and nonnative forests from low to high elevations on Hawai'i, Maui, O'ahu, Kaua'i, and Moloka'i, but its specific habitat requirements are unknown. Hwamei are omnivorous and forage on the ground and understory. Arthropods dwelling in the leaf litter are common in their diet. They eat a variety of nonnative fruit, but

continued

Box 13.1.   *Continued*

their impact as seed dispersers is minimal due to their low population densities.

Red-billed Leiothrix (Timaliidae)

This species is locally abundant and resides primarily in mid- to high-elevation native forests on Hawai'i, Maui, Moloka'i, and O'ahu. It can occasionally be found at low elevations. Like the white-eye, the leiothrix eats a wide variety of arthropods, fruit, and snails. Its diet shifts primarily to fruit when that food is abundant. At times arthropods can comprise nearly all food items for individual birds. The leiothrix is another major disperser of nonnative plants. During the non-breeding season, leiothrix and white-eyes both range widely, increasing their potential spread of seeds.

Japanese White-eye (Zosteropidae)

Since its introduction, the Japanese White-eye has become the most abundant and ubiquitous bird in the Hawaiian Islands. It thrives in a wide range of habitats on all islands, from low-elevation urban areas to montane rain forests. It is omnivorous, feeding on arthropods, other invertebrates, nectar, and fruit. Due to its flexibility in habitat and diet in addition to its high densities, the white-eye has been implicated as a potential competitor with native birds. Perhaps more significant is its effect as the primary seed disperser of many nonnative fruiting plants. Despite its small size (~10 g), it is capable of ingesting seeds as large as 3–4 mm in diameter.

introduced birds and the habitats of most native forest birds has made interspecific competition unlikely during the establishment phase. Yet as introduced populations of some species grow and begin to colonize new habitats—particularly away from urban areas and into native habitats—they have an increasing potential to enter into conflict with native species. Despite the large number of releases of exotic birds, surprisingly few fare well in native forests. Habitat segregation between native and introduced birds is a common occurrence at introduction sites throughout the world (Diamond and Veitch 1981, Case 1996). Much of this segregation can be explained by the fact that introduced species are often adapted to human-disturbed areas, whereas many native species are not (Sax and Brown 2000). Of the 54 species considered naturalized in Hawai'i, only 17 are permanent residents of forests, and many of these species inhabit primarily forest

edges or highly disturbed sites (Table 13.1). Species such as the House Finch and Nutmeg Mannikin will nest along forest edges but forage and reside in grassy areas such as pastureland and roadsides. Furthermore, many of the introduced species that have successfully invaded native forests decrease in abundance as one enters increasingly intact native forests (Scott et al. 1986, Foster et al. 2004).

Confounding this relationship between human disturbance and invasion success is the fact that the native forest bird community generally increases in abundance and diversity in more intact forests. Thus, one wonders if the avoidance of native forests by many nonnative species is habitat-related or if species within the native bird community resist invasion by competitively excluding nonnative birds. A few introduced species, however, appear to have higher densities in native forests, such as the Red-billed Leiothrix (Camp et al. 2003)

Table 13.1. Successfully introduced species commonly found in native forest

Common Name	Scientific Name	Year First Introduced
Red Junglefowl	*Gallus gallus*	After A.D. 800
Nutmeg Mannikin	*Lonchura punctulata*	1865[a]
Common Myna	*Acridotheres tristis*	1865[a]
House Finch	*Carpodacus mexicanus*	Before 1870[a]
Spotted Dove	*Streptopelia chinensis*	Before 1900[a]
Hwamei or Melodious Laughingthrush	*Garrulax canorus*	1900[a]
Red-billed Leiothrix	*Leiothrix lutea*	1918[a]
Zebra Dove	*Geopelia striata*	1922[a]
Northern Mockingbird	*Mimus polyglottos*	1928[a]
Japanese White-eye	*Zosterops japonicus*	1929[a]
Japanese Bush-Warbler	*Cettia diphone*	1929[a]
Northern Cardinal	*Cardinalis cardinalis*	1929[a]
White-rumped Shama	*Copsychus malabaricus*	1931[a]
Kalij Pheasant	*Lophura leucomelanos*	1962[b,c]
Red-whiskered Bulbul	*Pycnonotus jocosus*	~1965[c]
Red-vented Bulbul	*Pycnonotus cafer*	~1965[c]
Yellow-fronted Canary	*Serinus mozambicus*	~1965[d]

Sources: [a]Caum (1933).
[b]Berger (1981).
[c]Long (1981).
[d]Lever (1987).

and Japanese Bush-Warbler (Foster 2005). This observation suggests that a series of factors rather than one mechanism governs the habitat selection and success of introduced birds in Hawai'i.

Interspecific competition can be proved only if stringent requirements are met, from simple distribution patterns and degree of resource overlap to individual and population-level effects (Wiens 1989). Studies of native and introduced bird distributions in montane native forests of Hawai'i suggest that introduced birds may be competing with native species for food resources (Mountainspring and Scott 1985, Ralph and Noon 1988). However, the only introduced species implicated in native bird declines has been the Japanese White-eye, an abundant and opportunistic forager. In particular, correlation analysis of native and introduced bird distributions suggests that white-eyes may compete with two native species, the 'Elepaio (*Chasiempis sandwichensis*) and 'I'iwi (*Vestiaria coccinea*)

(Mountainspring and Scott 1985). In an analysis of foraging behaviors and bird densities, Ralph and Noon (1988) concluded that there are negative interactions between white-eyes and native species. Detailed comparisons of native and nonnative diets have not been published, and the diets of most nonnative birds in Hawai'i remain poorly described. Freed et al. (2008) reported that white-eyes were causing the decline of the Hawai'i 'Ākepa (*Loxops c. coccineus*), but their results are controversial and conflict with other results of forest bird research in the islands (U.S. Fish and Wildlife Service 2008b). The Japanese White-eye may compete with native species, but thus far studies have been only suggestive, and a more thorough analysis of the competitive effects of white-eyes and other introduced birds is needed.

Recent research suggests that there is considerable overlap in the diets of the native Maui 'Alauahio (*Paroreomyza montana*) and those of three nonnatives, the Japanese

Bush-Warbler, Japanese White-eye, and Red-billed Leiothrix (Foster 2005). However, the existence of an overlap in diet or foraging niche does not indicate that competition is occurring, particularly if food is abundant. For example, in the Bonin Islands, a study of the impact of the introduced Japanese White-eye on the endemic Bonin Island White-eye (*Apalopteron familiare*) "detected almost no negative ecological relationship between Z. *japonicus* and A. *familiare* and suggests that competition is not a significant threat to either species" (Kawakami and Higuchi 2003: 583). Thus, we need to first evaluate if food is a limiting resource. We can measure food resources directly, as is currently being done by the U.S. Geological Survey Pacific Island Ecosystems Research Center (R. Peck, pers. comm.). Additionally, we can evaluate the potential effects of food limitation on clutch size, nesting success, annual number of nesting attempts, or survival. Food limitation does not appear to affect any measured variables of the reproductive success of bush-warblers, white-eyes, or leiothrix, but research suggests that it may decrease white-eyes' survival in high-elevation forests (Foster 2005). The potential effects of limited food supplies on juvenile survival are not yet known but are possible.

Predation

Competition is not the only direct effect of an introduced bird on a native one. Two introduced species are bird predators. The Barn Owl is the only naturalized avian predator of adult birds (Chapter 12). Snetsinger et al. (1994) documented evidence of native and introduced birds in the pellets of Barn Owls. The Common Myna is a known predator on the eggs and young of other birds, but because at present it rarely forays into the forest interior, it is most likely a problem along fragmented and forest-edge habitats. In Tahiti, mynas are nest predators of the endangered Tahiti

Flycatcher (*Pomarea nigra*) and, along with the Red-vented Bulbul (*Pycnonotus cafer*), have likely contributed to flycatcher nest desertion through harassment (Blanvillain et al. 2003), indicating potential negative interactions between bulbuls and native species in Hawai'i.

Avian Disease

The devastating impact of introduced parasites and disease on native birds has been well documented (Berger 1975; van Riper et al. 1986; van Riper and Scott 2001; this volume, Chapter 9). Avian malaria (*Plasmodium relictum*) and avian pox virus (*Avipoxvirus* sp.) were likely brought into Hawai'i through the import of infected birds. Because of the ecology of these diseases and their mosquito vectors (*Culex quinquefasciatus*), populations of most native forest birds were decimated in lowland habitats but remained at high densities within montane forests.

Despite their seeming resistance, as evidenced by high nonnative bird densities in areas where disease and its vectors are present, introduced birds can be affected as well. Pox has a high prevalence in House Finches (van Riper et al. 2002), and roughly 12% of nonnative birds sampled at low elevations on O'ahu tested positive for malarial parasites (Shehata et al. 2001). Malaria susceptibility ranges widely among introduced birds. House Sparrows are relatively susceptible to infection, while Japanese White-eyes are predominantly resistant (Chapter 9). Nonetheless, infection does not always lead to disease, and although introduced birds can be hosts for pox and malaria, the impacts of these diseases on their populations appear to be minimal. The presence of malaria-resistant nonnative birds in a community may actually help reduce the rates of transmission to highly susceptible native birds if the resistant birds absorb infective mosquito bites that would otherwise have been received by native species (Chapter 14). A high level of disease resistance in most introduced birds likely allows them to inhabit a much larger geographic range than most native birds. Among other disease threats to Hawaiian birds, West Nile virus is the most prominent and has recently spread across the mainland United States. Its potential arrival in Hawai'i through the importation of infected birds could be devastating. Introduced birds also potentially spread other protozoan and helminth parasites, but we have only limited records about such parasite introductions to the Hawaiian Islands and whether they have made the jump to native species (van Riper and van Riper 1985, van Riper and Scott 2001).

Nonnative Plant Dispersal

Introduced birds have tremendous potential to affect plant communities through fruit consumption and seed dispersal. In Hawai'i, native birds have traditionally played the crucial role of dispersing many fruit-bearing plants (Price and Wagner 2004). In the absence of extinct or depleted native frugivores, that role has been partially filled by many introduced birds with similar diets (Kjargaard 1994, Medeiros 2004, Foster and Robinson 2007). Nonnative birds invading new environments often feed on the fruit of nonnative plants and thus can significantly aid the dispersal of these plant species. In Hawai'i, 37% of the major weed species are dispersed by introduced birds (Smith 1991). The increased spread of nonnative seeds may thus have ecosystem-level effects. Japanese White-eyes are the major dispersers of the invasive tree *Morella faya* (previously *Myrica faya*), which threatens native forests through its nitrogen fixation and development of monotypic stands (Vitousek et al. 1987, Woodward et al. 1990).

The gape sizes of introduced birds affect the range of seed sizes that are dispersed. Due to the current absence of large frugivores, either native or nonnative, the fruits of many medium- to large-seeded plants

are not adequately dispersed. Numerous nonnative horticultural plants, such as certain palms, may suddenly become invasive if this dispersal bottleneck is breached by the naturalization of larger frugivores (Medeiros 2004), for instance, the Hill Myna (*Gracula religiosa*), a fruit-dove, or toucan.

In disturbed areas, introduced birds likely contribute more to the dispersal of nonnative than native seeds. For example, Lewin and Lewin (1984) found that Kalij Pheasants eat the seeds of invasive plants such as banana poka (*Passiflora mollisima*), guava (*Psidium guajava*), and thimbleberry (*Rubus rosaefolius*), in addition to the seeds of many native plants. Gut content analysis in their study revealed that not all seeds are crushed in the gizzard and that the Kalij Pheasant likely is capable of spreading a variety of nonnative plants.

Through positive interactions between introduced birds and nonnative plants, introduced birds may be sowing the seeds of their own success by spreading desirable food resources (Mandon-Dalger et al. 2004). This seed dispersal may also affect native birds by means of the subsequent changes in plant communities. However, the loss of native frugivores also suggests that introduced birds may play a positive role in the maintenance of native plant communities. Many of the islands in the Hawaiian Archipelago have lost all or most of their native frugivores (Perkins 1903), and many fruiting plants, particularly in the understory, now depend on introduced birds for dispersal. In some habitats, nonnative game birds are significant dispersers of native plants and may contribute to the reseeding of degraded native habitats (Cole et al. 1994). The extinction or local extirpation of nearly all native frugivores may be offset by the addition of several introduced frugivores. Not all introduced frugivores are necessarily beneficial. Parrots have strong, sharp bills made to crush seeds and are unlikely to be dispersers. Some species such as lorikeets feed on flowers, and their effect on native 'ōhi'a-lehua trees (*Metrosideros polymorpha*) and other native flowering plants could be devastating. Introduced birds may therefore have the potential for both positive and negative effects on plant communities. Determining the prevalence of these two countervailing processes will aid us in estimating their relative impacts. The results of such study will have important management and restoration implications.

FUTURE RESEARCH AND CONSERVATION

Introduced birds are a conservation conundrum, both biologically and politically. Removal of well-established nonnative bird populations may not be feasible. Thus, our knowledge that introduced birds may be competing with native birds for food resources or are the major vectors of nonnative plant dispersal is a challenge to conservation management. Ethical concerns surrounding bird removal must also be considered, and lethal methods to control charismatic species such as parrots are often not publicly supported. Nonlethal capture and housing of large numbers of birds is expensive and difficult. Therefore, eliminating nascent populations of birds is likely the only way to prevent the spread and establishment of nonnative birds.

Several species in Hawai'i have populations that are small and appear controllable. For instance, the population of Mitred Parakeets on Maui has slowly grown to more than 100 birds, and its removal seems feasible from a logistical standpoint. Successful removal is not unprecedented. In the mid-1970s, state officials eliminated a breeding population of five potentially invasive Red-billed Magpies (*Urocissa erythrorhyncha*) (Berger 1981).

Stemming the release of nonnative birds starts with improving import restrictions. Ensuring the health of imports prevents the invasion of new pathogens and parasites. The recent ban on the mail-order importation of birds into Hawai'i has added to the control of birds and pathogens entering the state. Further improvements in the quarantining of Hawai'i from bird importations is part of the much larger and very complex issue of regulating the traffic of animals and plants to the state (see Chapter 15).

Although the likelihood that any release will be successful is only slight, there are many species whose successful introduction could be disastrous (Loope et al. 2001). Nonnative nectar-feeding species, such as lorikeets, would be serious competitors with native nectarivores. Other threats include the introduction of corvids, an intelligent and adaptable family whose members are known as adept nest predators. The introduction of avian brood parasites such as cowbirds (*Molothrus* spp.) and Cuckoos (*Cuculus* spp.) could also have pervasive effects on naïve native birds.

Research on the impacts of introduced birds on native and nonnative flora and fauna is vital to protecting native ecosystems in Hawai'i. The positive interactions of introduced birds, nonnative plants, and potentially nonnative insects, would allow introduced birds to have a greater effect in degrading native communities. Given the current trajectories in bird populations, it appears that some species of native birds will decline and many nonnative birds will increase. Understanding the future complexion of bird communities aids our understanding of the threats they pose. One challenge is that the introduction of just one bird species may have dramatic effects, and it is not known which species will be introduced in future years.

SUMMARY

Hawai'i has had more avian introductions than any other place of comparable size: more than 170 species have been introduced, resulting in the establishment of

54 species. How did so many nonnative species arrive in Hawai'i? Before 1945, nearly all releases were intentional, made by government agencies, naturalization societies, or individuals motivated to fill a void created by the loss of native birds. For example, the Hui Manu, an acclimatization society, released many of the songbirds commonly seen today. After 1945, legislation started to be passed to restrict the release of birds, and new releases mainly consisted of escapees from captivity and game bird releases. The majority of introductions since then have reflected the species commonly found in the bird trade, destined for people's homes as cage birds. Of the 23 species naturalized since 1950, 13 are common cage finches or parrots. Thus, the avenues for introduction have changed radically over time. Understanding the history of introductions enables us to discern the succession of processes that have led to nonnative bird naturalization and will allow us to identify future threats that seem likely to come from the cage bird trade. Indeed, one key factor has been the sheer volume of introductions, whether intentional or accidental, which has likely allowed so many species to successfully invade.

A persistent question is why some introduced species have thrived while others have failed. Assessing the factors determining introduction success has become an integral part of invasion biology, and patterns of insular bird introductions have been used to test theories of community assembly. Hawai'i has played a prominent role in this research due to its isolation, large number of introductions, and relatively well-documented invasion history. The introductions in Hawai'i suggest that interspecific competition among invaders is a primary force determining establishment. However, results from other island groups indicate that other factors, such as climate and introduction effort, may be more important in determining introduction success.

Due to patterns of disease and habitat loss, most native songbirds in Hawai'i are restricted primarily to upper-elevation forests, while the majority of introduced birds reside in nonforested habitats in the lowlands such as agricultural fields and urban areas. Thus far, 17 introduced bird species are reliably encountered in native forests, the primary forest type of native birds, but only 7 are major threats. The introduced birds found in native habitats have the potential to do widespread ecological damage. Primary concerns are the effects of introduced birds on the native flora and fauna through competition, introduction and transmittal of diseases to native birds, and the spread of seeds of nonnative plants. In particular, introduced birds have a tremendous potential to change plant communities through seed dispersal, ultimately affecting native birds dependent on native ecosystems.

Exotic bird introductions have presented an excellent opportunity to better understand basic ecological processes. At the same time, they have provided prime examples of the perils of these introductions. Previous ornithological work in Hawai'i has concentrated primarily on the study and conservation of its endemic birds. Nonetheless, the impact of introduced birds on the population dynamics and fitness of native birds is largely unstudied, although it is one of the most important practical reasons for studying these species' interactions. We need to identify negative impacts on endangered species and ecosystems caused by introduced birds in order to develop methods of mitigation. More important, however, is preventing future unwanted introductions so we do not continue this grand ecological experiment. In the universal aviary that Hawai'i has become, birds sing once again, but most have different voices than the birds of earlier times. We are only beginning to understand the ramifications of past introductions to Hawai'i, and protecting the islands' unique avifauna depends on a detailed knowledge of the threats to its existence.

ACKNOWLEDGMENTS

This work was part of my doctoral research on the ecological impacts of introduced forest birds. I thank the U.S. Geological Survey Pacific Island Ecosystems Research Center, which hosted me during my work in Hawai'i. Funding was provided in part by an Environmental Protection Agency STAR (Science to Achieve Results) Fellowship. I thank the following reviewers of earlier drafts: J. Lockwood, M. Moulton, and an anonymous reviewer.

CHAPTER FOURTEEN

Modeling the Epidemiology of *Avian Malaria and Pox*

JORGE A. AHUMADA,
MICHAEL D. SAMUEL,
DAVID C. DUFFY, ANDREW P. DOBSON,
AND PETER H. F. HOBBELEN

Interest in wildlife diseases has increased greatly in recent years, but most of our knowledge is based on anecdotal events or the highly visible epidemics. Three overlapping but frequently separate disciplines based on medical, epidemiological, and ecological approaches have been used to study wildlife diseases (Smith et al. 2005, Wobeser 2006). As a result, we know a great deal about the pathology of diseases in individuals, the characteristics of disease agents, or major outbreaks affecting large numbers of animals, but we have only limited understanding of the ecological dynamics of wildlife diseases and their contribution to the dynamics of host animal populations. Historically, interest in wildlife disease ecology has been academic, but this issue has become increasingly important because zoonoses have emerged as human health issues (Cleaveland et al. 2001) and diseases can significantly impact biological conservation, biodiversity, and wildlife populations (Daszak et al. 2000, Harvell et al. 2002). West Nile virus, Nipah virus, avian influenza virus, toxoplasmosis (*Toxoplasma gondii*), and severe acute respiratory syndrome are recent examples (Rappole et al. 2000, Dubey and Odening 2001, Bell et al. 2004, Ferguson et al. 2004, Food and Agriculture Organization 2004, Hsu et al. 2004), while other zoonoses such as sylvatic plague (*Yersinia pestis*), hantavirus, and Lyme borreliosis (*Borrelia burgdorferi*) represent persistent or reemerging problems (Bacon 1985, Barbour and Fish 1993, Nichol et al. 1993).

Koch's postulates involving the presence and role of pathogens in hosts were developed as criteria to establish a causal relationship between a pathogen and a host disease (Rivers 1937), but there are no similar guidelines to establish the importance of disease at the host population level. Direct measurements of diseased and healthy populations often cannot be made at appropriate temporal or spatial scales to assess disease impacts because they may require decades or centuries and hundreds or thousands of square kilometers (Kitron 1998). Direct measurements are also not available to assess the consequences of the initial arrival of a disease on "virgin soil" or in a previously unexposed population (Crosby 1974, Stokstad 2004). Furthermore, when dealing with rare wildlife species or zoonoses affecting human health, there are limited opportunities for trial-and-error responses to disease, and there is a critical need to "get ahead of the curve" in developing effective strategies to respond to disease threats such as avian influenza and Usutu virus (Weissenbock et al. 2002, Ferguson et al. 2004).

Since the late 1970s, there has been an increasing interest in using mathematical approaches to model complex disease processes (Anderson and May 1978, May and Anderson 1978). Mathematical modeling of host populations and their interaction with parasites, diseases, and vectors offers a means of assessing the importance of disease agents on wildlife populations and the potential effectiveness of alternative disease management strategies (Aaron and May 1982). Models can summarize what we know about a disease system and the relative importance of different parts of the system (Grenfell et al. 1995). Models serve not to duplicate nature but to capture its essential features or to narrow the window of possible ways in which a system can work (Kingsland 1995). Recently, modeling has become an increasingly sophisticated tool for integrating the medical, epidemiological, and ecological approaches for understanding and predicting the dynamics of disease processes such as those of rabies virus in carnivores (Murray et al. 1986; Smith, Lucey, et al. 2002), foot-and-mouth disease (Aphthovirus) in domestic cattle (Woolhouse et al. 2001, Keeling et al. 2003), or measles virus (Grenfell and

Bolker 1998, Grenfell et al. 2001) and HIV/AIDS in humans (Wodarz and Nowak 2002). Our chapter presents a modeling approach and preliminary assessment of the parasite–vector–multiple host system involving avian malaria (Plasmodium relictum) in Hawai'i.

Avian malaria may have been brought to Hawai'i in the early twentieth century (van Riper et al. 1986; for a detailed biological account of this disease system, see Chapter 9). This disease, together with the previously introduced avian pox virus (Avipoxvirus sp.), posed a major new threat to immunologically naïve native birds. An effective vector of both diseases, the southern house mosquito (Culex quinquefasciatus), was introduced as early as 1826 (Halford 1954, Hardy 1960). Van Riper et al. (1986) concluded that avian malaria was responsible for a wave of extinctions of native bird species during the 1920s and 1930s, and subsequently native birds below 1,500 m elevation were at continual risk of malaria infection. Above that elevation mosquitoes were rare, so the risk was greatly reduced, and as a result native forest bird populations survive mainly at higher elevations. This elevational pattern of malaria infection may also vary because forest birds differ in their movement patterns, from nomadic species such as 'I'iwi (Vestiaria coccinea) and 'Apapane (Himatione s. sanguinea), which travel from high- to lower-elevation forests in search of seasonal or ephemeral food resources (especially nectar), to sedentary and mainly insectivorous species like Hawai'i 'Amakihi (Hemignathus virens) and the nonnative Japanese White-eye (Zosterops japonicus), which usually remain in their home ranges throughout the year.

Atkinson and colleagues (Atkinson et al. 1995, 2000; Atkinson, Dusek, et al. 2001; Atkinson, Lease, et al. 2001), using improved surveillance methods, reported a greater prevalence of malaria infection than previously described, differences in susceptibility among native species and between native and nonnative species, and low-intensity, chronic infections in several native species, which made them life-long reservoirs for avian malaria. Epidemiologically, the result is that most endemic Hawaiian species are highly affected by *Plasmodium relictum*, are effective disease transmitters, and are long-term reservoirs of disease. In contrast, malaria has minimal impact on the survival of nonnative birds, which are capable of effective disease transmission for only a limited period.

The last Poʻo-uli (*Melamprosops phaeosoma*) died in November 2004 with a concurrent infection of avian malaria (Chapter 21). The loss of this species reemphasizes that for eight decades malaria has played a major role in the population dynamics and conservation of native Hawaiian birds. Modeling may be the only way to understand the complexity of this disease, how low-elevation populations of some species may be evolving disease resistance (Woodworth et al. 2005), or how climate change may reduce mosquito-free, high-elevation avian refugia (Benning et al. 2002). In addition, modeling may help reveal how avian malaria plays different roles on each island, with their different arrays of native bird hosts and varied landscape patterns. This chapter presents an overview of our modeling efforts to understand the avian malaria system in Hawaiʻi. These efforts and the biological data required to determine model parameters are the result of a multidisciplinary project, Biocomplexity of Introduced Avian Diseases in Hawaiʻi (see acknowledgments at end of chapter).

COMPLEX INTERACTIONS: A MODEL OVERVIEW

The persistence and dynamics of avian malaria and pox in Hawaiian bird communities are influenced by a complexity of interrelated factors that produce unexpected patterns. These factors include endogenous components of the disease system (vector and host abundance) and exogenous components that drive the system (weather, food abundance, and human influence on the environment), as well as landscape components (especially elevation) that influence the rates of biological processes and distribution of species. Interactions among these components are complex and nonlinear and often include feedback mechanisms. To develop a model that captures this complexity, it is useful to conceptualize the avian malaria-pox system in a flow diagram that summarizes the main interacting components (Fig. 14.1).

We have separated the system into five different but highly interrelated components: (1) pathogen-vector-host dynamics, (2) the biotic structure, (3) the biotic community, (4) the weather, and (5) human influences. The first component describes dynamics of the disease system as influenced by host and vector dynamics, transmission between vector and host, and within-host and vector parasite dynamics. It is primarily the behavior of this component (our "response variable") that we want to predict and understand as a function of its interaction with other parts of the system. The biotic structure includes features such as food abundance, forest cover, and number of vector oviposition sites that might limit populations of vectors and hosts. In contrast, the biotic community includes alternate hosts that can act as reservoir species for the parasite or predators that have the potential to affect avian hosts. The weather component includes variables such as temperature and rainfall that affect vector and host populations, potential interactions between parasite and vector, food availability, and other structural components of the system. Finally, there are three important human influences: the introduction and management of feral pigs (*Sus scrofa*), ranching activities, and land-use changes. Each of these can affect biotic structures such as the abundance of vector oviposition sites, host nesting sites, forest cover, and food. The human

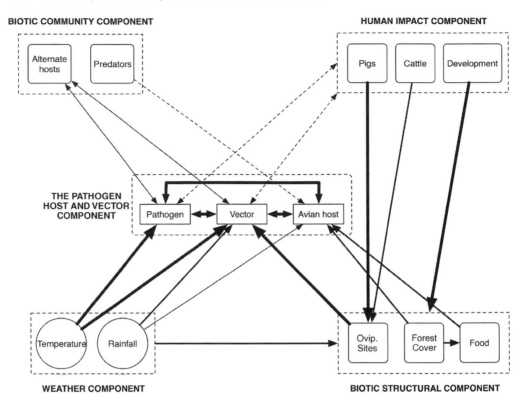

BIOTIC COMMUNITY COMPONENT

HUMAN IMPACT COMPONENT

THE PATHOGEN HOST AND VECTOR COMPONENT

WEATHER COMPONENT

BIOTIC STRUCTURAL COMPONENT

Figure 14.1. Conceptual model of the malaria-pox disease system in Hawai'i. Unidirectional arrows from one component (A) to another (B) denote "A influences B." Bidirectional arrows refer to reciprocal interactions between components. Solid arrows are hypothesized relationships between components; broken arrows are tentative relationships. The thickness of the arrows represents the relative importance assigned to the relationships in influencing critical system dynamics. Ovip., oviposition.

component can also provide an alternate system for disease persistence, for instance, transmission and maintenance of disease in domestic fowl or House Sparrows (*Passer domesticus*).

A General Model Structure

Our simulation incorporates the different components outlined in Fig. 14.1 in a spatially explicit framework (Fig. 14.2). The spatial unit of the simulation is a cell of 1 km^2, and the local dynamics of vectors and hosts are modeled within each cell.

As in many disease systems, pathogens are modeled implicitly by dividing hosts and vectors into susceptible, latent, infectious, and in some cases, recovered classes. We formulated these local population models as a series of continuous-time ordinary differential equations (ODEs) to facilitate sensitivity analysis of model parameters. Environmental and biological variables within each cell can vary spatially and temporally across the landscape based on temperature, rainfall, food abundance, or breeding sites. Predators and alternate or reservoir hosts can be included in the disease cycle as additional equations. All these variables provide input for the population models (see the following sections).

A crucial component of the simulation model is the ability to incorporate movement of individuals between cells in space. Hosts and vectors can move from one cell to the next by following an environmental or biological gradient (food abundance or hosts). We can use simple decision algo-

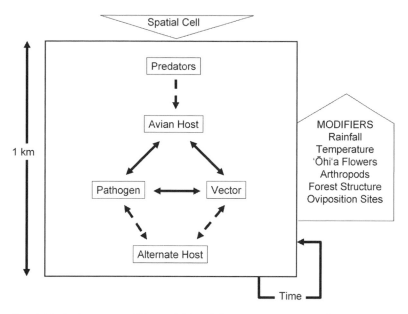

Figure 14.2. Basic structure of the model. Local dynamics are modeled in explicit space at the cell level. Biotic and abiotic components may modify cell variables and influence within-cell parameters.

rithms to drive this exchange of individuals between cells; for example, some avian hosts may track flower abundance through time and space. Mosquito vectors may also actively or passively disperse, although it seems that a scale of 1 km is appropriate for most vector movements (Chapter 17).

We selected a time step of one day for model simulations. Although this is a rather short time for the hosts, it is appropriate for the vectors, which have a much shorter lifespan. Daily estimates of temperature and rainfall are derived from weekly averages and used to drive the population models. Stochastic simulation or computationally simpler discrete-time equations can be used when exploring main model features and behaviors. However, for model prediction and assessment, stochastic simulations may be conducted by choosing values from the statistical distribution describing specific parameters (distribution of rainfall or temperature). Results from particular combinations of parameter values are then summarized by pre-

senting their mean and 95% confidence envelopes.

A crucial issue in the modeling process is the estimation of key parameters that drive the interactions and dynamics of the model. Our approach has been to use the best observational and experimental evidence available to estimate these model parameters. This is possible for parameters related to host and vector demography, but estimation of other parameters, such as transmission or force of infection, are much more elusive. We recently conducted laboratory and field studies as part of the previously mentioned Biocomplexity project to estimate important parameters or develop ways to calculate these parameters indirectly. In general, parameter estimation can be a complex problem, and the technical details are outside the scope of this chapter.

The Malaria-Pox Pathogen System, Mosquito Vector, and Avian Hosts

The disease system is the core component or the "response component" of our model. It is composed of the pathogens for malaria and pox, the mosquito vector (the southern house mosquito), and several species of native and nonnative Hawaiian forest birds that represent the avian hosts. In the case of avian malaria, the vector is required to transmit the disease to the host and vice versa. Avian pox, however, can be transmitted either by indirect contact through a vector that has bitten an infected individual or by direct contact between an infected and a susceptible host (van Riper et al. 2002), although the relative importance of these alternate routes of transmission is not well known. Because pox transmission is more complex, we will restrict most of our discussion to malaria and refer to pox when appropriate.

The life cycle of avian malaria is well described, and details can be found in van Riper et al. (1994) and in Chapter 9. Our modeling approach is to develop a system of equations similar to those developed for classical human malaria, which describes the dynamics of susceptible, latent, and infected vectors and a single host. For avian malaria, this approach is extended to several host species to describe disease dynamics at a local scale (Dobson 2004). Many factors that drive mosquito population or avian host dynamics depend on other components of the model (weather, food abundance, and oviposition sites). For example, the daily temperature directly influences the development rate of mosquito larvae and the rate of malaria parasite development within an infected mosquito.

Much about the crucial dynamics of disease transmission within Hawaiian forest birds can be captured by considering birds in various disease classes and how these demographics are affected by seasonal, annual, and elevational factors. All hatched birds of both native and nonnative species are susceptible to avian malaria and pox. Once a susceptible bird (either immature or adult) has become infected with malaria, it will either die from the disease or survive the infection. During infection, native birds develop a much higher level of parasites circulating in their blood than do nonnative birds, which facilitates transmis-

sion to biting mosquitoes (Yorinks and Atkinson 2000; this volume, Chapter 9). Many native birds are likely to die from malaria infection, whereas most nonnatives are likely to survive (Chapter 9). Native birds that survive malaria develop a lifelong immunity to future infection, but they also maintain a sufficient number of circulating parasites to infect a mosquito. Thus, native birds remain in the infectious class and function as an important reservoir for disease transmission to uninfected mosquitoes. In some years and at elevations with lower levels of disease transmission, more susceptible birds (both immature and adults) will survive until the following year. If several years of lower disease transmission occur sequentially, the increase in the abundance and proportion of susceptible birds in the population can set the stage for epizootic events with substantial mortality in native bird populations.

Host species differences and dynamics associated with disease transmission; bird abundance, survival, and recruitment; spatial movement patterns; and seasonal climate patterns can be integrated into the overall modeling systems by using submodels to represent native and nonnative species. Host submodels interact with the vector submodel to simulate disease transmission between vectors and hosts, and these processes are influenced by spatial and environmental drivers that determine seasonal changes in life history dynamics (recruitment or movement), relationships with habitats, habitat patch sizes, and elevational distribution. Host submodels include four bird species that represent the different characteristics of disease susceptibility, movement patterns, and life history dynamics found among Hawaiian avifauna. These species are the ʻIʻiwi, a nomadic native species with a high level of disease susceptibility; the ʻApapane and Hawaiʻi ʻAmakihi, nomadic and sedentary native species, respectively, with moderate disease susceptibility; and the Japanese White-eye, a sedentary nonnative species with low susceptibility to avian malaria and pox. 'I'iwi and 'Apapane are important nectar-consuming species that can track seasonal and elevational changes in flowering of the dominant rain forest tree, 'ōhi'a-lehua (*Metrosideros polymorpha*). Seasonal movement of high-elevation birds like 'I'iwi to lower elevations in search of 'ōhi'a nectar can put these highly susceptible birds at almost certain risk of malaria infection and mortality. Each host submodel is developed to incorporate variation among species and endemic disease patterns.

The Biotic Structural Components

The biotic structure refers to features such as habitat, food, or nesting sites that are complex biological systems in themselves but also affect population growth rates or local carrying capacities of the vector and host species. We briefly describe three factors that are crucial in modeling this system: (1) the number and characteristics of mosquito oviposition sites, (2) the forest structure and cover for birds, and (3) the abundance of food for birds, primarily consisting of 'ōhi'a flowers and arthropods (see Fig. 14.1).

Female mosquitoes oviposit in many types of cavities throughout the wet areas of Hawai'i, including artificial cavities provided by human residential and agricultural activities, such as abandoned drums and tires, cemetery vases, roadside puddles, and ditches. In forest habitats, mosquitoes oviposit mostly in water-filled cavities in tree ferns, hāpu'u (*Cibotium* spp.). Feral pigs create these tree fern cavities when feeding on the fallen trunks of the hāpu'u. Therefore, the presence of pigs likely increases the abundance of mosquitoes and the potential infection of susceptible avian hosts.

Forest structure is another important biotic element affecting vector persistence. Mosquitoes are unevenly distributed from lowland dry 'ōhi'a scrub to high-elevation 'ōhi'a and koa (*Acacia koa*) forests (van Riper et al. 1986, LaPointe 2000, Ahumada et al. 2004). Forests provide a humid environment that enhances the survival of adult mosquitoes (Dow and Gerrish 1970) and their breeding habitats. However, information is limited on how forest types affect mosquito distribution and abundance. Based on current information, we assume that mosquitoes require minimal forest cover and have higher recruitment rates in closed-canopy wet forests.

Forest characteristics and food abundance are also important in determining host abundance and distribution. Bird and habitat surveys conducted from the mid-1970s through the late 1980s provided information on the distribution and habitat preferences of most birds (Scott et al. 1986). A recent study by Camp et al. (2003) on Hawai'i Island found that each native and nonnative bird species has unique requirements and distribution based on vegetation characteristics, species ecology, and life history characteristics that should be useful in estimating the local (cell) carrying capacity of different bird species. Spatial and temporal changes in 'ōhi'a flowers and arthropod abundance also affect host distribution, movement, population size, and breeding season throughout the east side of Hawai'i. Nectar from 'ōhi'a flowers is a major food source for many native species throughout the year and likely influences breeding at high elevations (Ralph and Fancy 1994b). However, 'ōhi'a tree morphology and flower production is a highly variable function of elevation and substrate type and changes seasonally (Scott et al. 1986, Ralph and Fancy 1995, Berlin et al. 2000). The phenology of arthropod communities has been studied only at high elevations (>1,800 m) (Ralph and Fancy 1994b; Banko, Oboyski, et al. 2002), where their abundance is relatively constant. Our consideration of food abundance in the model will rely on information gathered by the Biocomplexity project on the flower production of 'ōhi'a trees and on arthropod abundance along an elevational gradient along the east side of Mauna Loa.

The Biotic Community Components

Mosquitoes may also feed on alternate hosts that then affect the malaria-pox disease system. Alternate hosts for malaria and pox can spatially or temporally extend disease transmission beyond the native forest bird–vector pathogen cycle by acting as disease reservoirs or, conversely, if they are not infected, by diluting transmission. However, if the alternate host has a lower rate of infection than native species owing to, for example, defensiveness to mosquito bites, its presence might have little dilution effect on the disease cycle. Some of these theoretical issues are explored in the following section on modeling transmission. Because many nonnative birds are more resistant to malaria infection than native Hawaiian species, they probably play a minor role in disease transmission. However, House Sparrows appear to be infectious and may be an important reservoir species (Chapter 9).

In addition to being alternative hosts, predators may directly affect the population dynamics of native hosts (Chapter 11). Introduced black rats and domestics cats are nest predators of native forest birds, and there is limited evidence that native birds infected with malaria may be more prone to predation than uninfected birds (Yorinks and Atkinson 2000). Currently, there is not sufficient information on the effects of predators on native bird populations to allow us to incorporate this factor into a modeling framework.

The Weather Component

Weather, especially rainfall and temperature, can have direct effects on the physiology and development of vectors and the development of the malaria parasite, as well as indirect effects on resources important for vectors and hosts. Temperature and rainfall in Hawai'i are highly variable across different temporal and spatial scales. Rainfall occurs primarily between October and March (Giambelluca and Schroeder 1998). Temperature fluctuates seasonally, with warmer conditions typically occurring during the middle of the year. Within these seasonal patterns, both rainfall and temperature vary depending on elevation, aspect, and slope (Chapter 6).

At higher elevations, malaria transmission is typically restricted to the warm season, when environmental conditions are suitable for completion of the mosquito life cycle and extrinsic development of the parasite. Because temperature and rainfall are consistently favorable at lower elevations, transmission occurs throughout the year, with only minor seasonal fluctuations. In midelevation forests, transmission occurs primarily during the fall and early winter, when temperature and rainfall support abundant mosquito populations and development of malaria parasites. Annual variation in malaria infection in birds occurs primarily in midelevation forests, likely caused by bird population susceptibility and annual shifts in weather patterns that drive vector abundance. These annual disease patterns vary from full-blown epizootics that infect nearly all of the susceptible population to lower-intensity events that infect approximately 50% of the susceptible population (Chapter 9).

On the island of Hawai'i, moisture-laden trade winds blow from the northeast, forming clouds that rise along east-facing slopes to a point where a stable dry air mass is reached (the inversion layer at about 2,000 m). This situation creates wet conditions at middle and low elevations on the island's windward side (Fig. 14.3). Annual and seasonal rainfall also can vary substantially between wet and dry years (see Fig. 14.3), depending on El Niño Southern Oscillation events or droughts (Giambelluca and Schroeder 1998).

The effects of rainfall on vector and host population dynamics can be complex. For

1990—Wet Year   1995—Dry Year

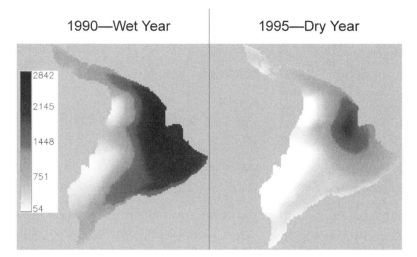

Figure 14.3. Map of the eastern side of the island of Hawai'i showing spatial variation in annual rainfall (mm) for two contrasting years (1990—wet year and 1995—dry year). Rainfall interpolation was achieved using the thin-plate spline-smoothing method of Hutchinson (1990).

mosquitoes, a lack of rain can increase larval mortality in breeding cavities (Alto and Juliano 2001), but adult mortality can increase during extreme rainfall events (>200 mm per day) (Hayes and Downs 1980, LaPointe 2000). Both of these effects can influence the dynamics of mosquito vectors in Hawai'i (Ahumada et al. 2004). Rainfall may indirectly affect host demography by influencing the location, timing, or abundance of food resources. Arthropods and 'ōhi'a flower nectar constitute the major diet staples for Hawaiian honeycreepers (Chapter 7), but it is unclear how rainfall affects the distribution and abundance of these resources in Hawai'i.

Temperature decreases predictably with elevation (dry adiabatic lapse rate) at 9.78°C per km, and this decrease slows the development rate of mosquito larvae (Rueda et al. 1990). Lower temperatures can also produce a minor increase in the mortality rate of adults (Oda et al. 1999), although this effect has not been studied in Hawai'i. Perhaps the most important temperature effect on avian malaria is

through control of the parasite incubation period. Following a blood meal from an infectious host, 6 to 28 days are required for a mosquito to produce salivary gland sporozoites and become infectious. This incubation period is inversely dependent on temperature (LaPointe 2000), which restricts the elevational distribution of infectious mosquitoes and thus disease transmission in Hawai'i (Benning et al. 2002).

Human Impacts

Human activity also has the potential to affect the avian malaria-pox disease system. For simplicity, we describe three human activities that are treated independently: feral pig management, ranching and agriculture, and urban development. These and other human impacts can be considered activities that indirectly affect the disease system by influencing its biotic structure or by providing alternate hosts (see Fig. 14.1). For example, feral pig management affects mosquito breeding sites through the disturbance of tree ferns in midelevation forests. The disease cycle could also be extended by providing nesting places for House Sparrows, which act as an additional reservoir for disease. Human activities might also provide dead-end hosts that do not transmit malaria, such as domestic animals. Many of the human-associated effects are difficult to include in a modeling framework because their relative importance and strength are unknown or difficult to measure. For example, mosquitoes tend to associate closely with humans, using artificial cavities for breeding, but the impact of human development on mosquito populations is largely unmeasured in Hawai'i. Other impacts, such as the effects of deforestation on bird populations or ranching activities, which provide water holes for mosquito breeding, may be easier to model. Experiments and field studies that measure the effects of these human activities would provide additional insights on the avian malaria-pox system.

MODEL DEVELOPMENT

In the following sections we describe a mathematical formulation of many of the model components presented above. We focus on ODEs that model the dynamics of vectors and hosts and their interaction through disease transmission. We use these equations to model the rate of change in the number of individuals during daily (instantaneous) time steps instead of modeling the numbers of individuals over a discrete time period (a month or year). In these equations, positive terms increase the state variables (number of individuals), while negative terms decrease them (an example follows). We also incorporate the effects of weather on vector and malaria parasite dynamics. Table 14.1 contains a complete list of variables and parameters used throughout the equations.

The Disease System

We can describe general vector dynamics by considering ODEs for a single host species and mosquitoes:

$$\frac{dS_M}{dt} = bM - \lambda(I_H, H)S_M - \mu S_M,$$

$$\frac{dL_M}{dt} = \lambda(I_H, H)S_M - \partial L_M - \mu L_M, \text{ and}$$

$$\text{[Eqs. 1, 2, 3]}$$

$$\frac{dI_M}{dt} = \partial L_M - \mu I_M,$$

where S_M, L_M, and I_M are susceptible, latent (without mature sporozoites), and infectious mosquitoes; M is the total number of mosquitoes; the force of infection for susceptible mosquitoes, $\lambda(I_H, H)$, is determined by the number of infectious hosts and host density or by the rate at which they acquire the disease from infected hosts (I_H); H is the number of hosts; b is the per capita fertility of mosquitoes; ∂ is the rate at which latent mosquitoes become infectious (sporogony rate); and μ is the per capita mortality rate.

Understanding the biology behind these equations is relatively easy. For example, Equation 1 describes the rate of change in the number of susceptible mosquitoes (the state variable) per unit time, and it has one positive term and two negative terms. The positive term is the rate at which new mosquitoes enter the population and therefore results in an increase in the number of susceptible mosquitoes. There are two negative terms: the rate at which susceptible mosquitoes become infected (force of infection times the number of susceptible mosquitoes) and the rate at which mosquitoes die (mortality rate times the number of mosquitoes). These terms decrease the magnitude of the state variable. Equations for the dynamics of the host are similar:

$$\frac{dS_H}{dt} = fH - \theta(I_M, M)S_H + \gamma I_H - dS_H, \text{ and}$$

$$\text{[Eqs. 4, 5]}$$

$$\frac{dI_H}{dt} = \theta(I_M, M)S_H - \gamma I_H - dI_H - \nu I_H,$$

where S_H and I_H are the number of susceptible and infectious hosts, H is the total number of hosts, θ is the force of infection

Table 14.1. Variables and parameters used in equations throughout the text

Symbol	Definition	Units
S_M, L_M, I_M	Number of susceptible, latent, and infectious mosquitoes, respectively	Number of individuals
M	Total number of mosquitoes	Number of individuals
b	Fertility of mosquitoes	Number of individuals per individual per day
μ	Mortality rate of adult mosquitoes	Number of individuals per individual per day
∂	Rate at which mosquitoes become infectious	Number of individuals per individual per day
λ	Force of infection on susceptible mosquitoes	Number of individuals per host per day
H	Number of hosts	Number of individuals
F	Fertility of hosts	Number of individuals
f	Fertility of hosts per unit time	Number of fledglings per individual per day
γ	Recovery rate of host	Number of individuals per individual per day
d	Background mortality rate of host	Number of individuals per individual per day
v	Disease-induced mortality rate in host	Number of individuals per individual per day
θ	Force of infection on susceptible hosts	Number of individuals per mosquito per day
χ	Infectious contact rate between susceptible and pox-infected hosts	Number of infectious contacts per day
a	Probability that a infectious bite transmits the disease to a susceptible host	No units
β	Number of mosquito bites per host	Number of bites per host per day
A'	Probability that a bite transmits the disease to a susceptible mosquito	No units
β'	Number of bites by a mosquito per unit time	Number of bites per mosquito per day
R_0	Basic reproductive number of a disease	Number of new infections per infected host per day
C	Abundance of host species C	Number of individuals
B	Abundance of host species B	Number of individuals
c	Length of the infectious period for host C	Number of days
b	Length of the infectious period for host B	Number of days
a_i	Relative attraction of species i to mosquitoes	No units
A_i	Abundance of host species i	Number of hosts
η_i	Length of the infectious period for host species i	Number of days
K	Carrying capacity of mosquito larvae at a breeding cavity	Number of individuals per area
$d\tau$	Development rate of mosquito larvae as a function of temperature	Number of individuals per day
τ	Daily temperature	°C
τ_{min}	Minimum temperature for mosquito larvae development	°C
μ_L	Daily mortality rate of mosquito larvae	Number of individuals per day
L	Density of mosquito larvae	Number of individuals per area

continued

Table 14.1. Continued

Symbol	Definition	Units
M	Density of adult mosquitoes	Number of individuals per area
DD	Degree-days (number of days to develop at 1°C above τ_{min})	Number of days
$d^*\tau$	Development rate of the parasite inside the mosquito	Number of days
R_t	Rainfall at time t	mm rainfall
δ_t	Accumulated number of dry days at time t	Number of days
i_t	Indicator variable = 1 when rainfall falls below some threshold	No units
ψ	Threshold rain below which mosquito breeding sites begin to dry	mm rainfall
ω	Shape parameter controlling the sensitivity of mosquito larvae to rainfall	No units
g	Shape constant that ensures $\mu_L(R)$ starts at 0	No units
ϕ	Indicator variable = 1 if $R_t > 300$ mm	No units
h	Increase in adult mosquito mortality due to extreme rainfall events	No units

for susceptible hosts, γ is the recovery rate of infected hosts, f is the fertility rate, and d and ν are the natural and disease-induced mortality rates, respectively. We do not include a latent class because the time required for a newly infected host to become infectious is short compared to the lifetime of the host. The force of infection for hosts depends on the proportion of infected mosquitoes in the population as well as the number of susceptible hosts available to bite. Later we explore some functional forms for this term. Although many nonnative host species might recover from the infection ($\gamma > 0$) and return to the susceptible class, native species typically are infectious to susceptible mosquitoes for the remainder of their lives ($\gamma \approx 0$) (Atkinson et al. 2000; Atkinson, Dusek, et al. 2001; Atkinson, Lease, et al. 2001; this volume, Chapter 9). This persistence of infection can easily result in a large pool of infectious native hosts that serve as a reservoir for avian malaria. A diagram summarizing mosquito and host interactions is shown in Fig. 14.4.

The dynamics of pox can be described by a similar system, except that infection might also result from environmental transmission between susceptible and infected hosts. An equation for the rate of change of infected hosts with pox can be of the form

$$\frac{dI_H}{dt} = \theta(I_M, M)S_H + \chi S_H I_H - \gamma I_H - dI_H - \nu I_H, \qquad [\text{Eq. 6}]$$

where θ is the force of infection from pox-bearing mosquitoes, γ is the recovery rate, and d and ν are the natural and pox-induced mortality rates, respectively. The term $\chi S_H I_H$ describes the rate at which susceptible individuals become infected by indirect transmission from infected individuals. Although our knowledge of avian

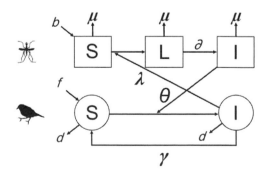

Figure 14.4. Diagrammatic representation of the vector–host model. See the text (especially Table 14.1) and Equations 1–5 for an explanation of the symbols used.

pox dynamics in Hawai'i is limited, current evidence suggests that birds may recover from pox infection and become susceptible to infection later (Jarvi, Triglia, et al. 2008).

Transmission

Transmission is the essential component that couples the dynamics of diseased and susceptible hosts with infectious and susceptible vectors. The function $\theta(I_M, M)$ describes transmission from infectious mosquitoes to susceptible hosts, and $\lambda(I_H, H)$ denotes transmission from infectious hosts to susceptible mosquitoes. A functional form for the force of infection on susceptible hosts is

$$\theta(I_M, M) = \frac{a\beta I_M}{M}, \qquad [Eq. 7]$$

where a is the probability that an infectious bite causes an infection in the host and β is the number of bites received by a host per unit time. This equation shows that the rate at which susceptible hosts become infected is the product of three processes: the probability that a bite from an infectious mosquito will result in an infection (a) of a susceptible bird, the biting rate (β), and the proportion of infected mosquitoes (I_M/M). The rate at which new infections are acquired by susceptible mosquitoes when biting infectious hosts is

$$\lambda(I_H, H) = \frac{a'\beta' I_I}{H}, \qquad [Eq. 8]$$

where a' is the probability than an infectious host will transmit the disease to a susceptible mosquito per bite; β' is the number of times a mosquito bites per unit time, or $\beta' = \beta/M$; and I_H/H is the proportion of infected hosts.

But what happens when mosquitoes feed on multiple species of avian hosts? We can address this question by examining the

conditions under which disease might be established using several simplifying assumptions, such as constant mosquito density. The potential for an outbreak of a vector-transmitted pathogen, such as malaria or pox, will be determined by R_0, the basic reproductive ratio (number) of the pathogen (Anderson and May 1982, Dieckmann et al. 1990). The methods developed by Dieckmann et al. (1990) can be used to derive a general expression for R_0 for a vector-transmitted pathogen with n host species (Hasibeder and Dye 1988, Dobson and Foufopoulos 2001). Consider a vector-transmitted pathogen with two host species, C and B, that occur at abundances C and B, respectively. Each host is bitten at a rate ($\bar{a}$) determined by the mosquito biting rate and by the host's relative abundance in the community, for example, the relative abundance of species C = C/(C + B). If birds of species C are n times more attractive than birds of species B, the proportion of bites received by C is nC/ (nC + B). We also assume that each host species has a relative infectious period determined by $1/(v + d + \gamma)$, where v, d, and γ are the disease mortality, natural mortality, and disease recovery rates, respectively (Equations 4 and 5). We designate the infectious period c for species C and b for species B. A short infectious period indicates that the host is infectious for a short period of time, dies, or quickly recovers. In addition, we assume a constant mosquito population density of M and a lifespan of $1/\mu$ days.

Using these simplifying assumptions we can develop a simple transmission matrix (T) to project the number of infections produced when an infected host is introduced into the population. The first row of this matrix includes disease transmission from mosquitoes to other mosquitoes and hosts (because mosquitoes cannot directly transmit malaria to other mosquitoes, this term is zero). The second row includes transmission from host B to mosquitoes and other hosts (also zero because malaria is transmitted to hosts only by vectors), and the third row is for species C:

$$T = \begin{bmatrix} 0 & \dfrac{\beta B}{\mu(B+C)} & \dfrac{\beta C}{\mu(B+C)} \\[2ex] \dfrac{MbB}{B+C} & 0 & 0 \\[2ex] \dfrac{McC}{B+C} & 0 & 0 \end{bmatrix}.$$

The basic reproductive number for the pathogen can be determined from the dominant eigenvalue of this matrix (Dobson 2004, Gandon 2004):

$$R_0 = \frac{\beta}{\mu(B+C)}\sqrt{\mu M(bB^2 + cC^2)}. \quad [\text{Eq. 9}]$$

This approach can readily be generalized for n species of hosts:

$$R_0 = \frac{\beta}{\mu\sum\limits_{i=1}^{n} a_i A_i}\sqrt{\mu M \Sigma \eta_i (a_i A_i)^2}, \quad [\text{Eq. 10}]$$

where a_i is the relative attraction of species i to mosquitoes, A_i is the abundance of species i, and η_i is the infectious period of host i $(1/(v_i + d_i + \gamma_i))$. Equation 10 can be used to evaluate how the number of species, the length of the infectious period for hosts, and their attractiveness to mosquitoes affect the persistence of the disease. Figure 14.5 illustrates the impact on R_0 of adding new primary host species with differing infectious periods to the community. We can also consider host species that have either longer or shorter infectious periods and are more or less abundant. In general, the addition of more host species reduces R_0 (see Fig. 14.5A and 14.5C–D), except when the new species both are more abundant and have a long infectious period (see Fig. 14.5B). However, when the new hosts are more abundant and have a short infectious period, we see a dilution

(or buffering) effect that reduces R_0 below the outbreak threshold (<1.0) (see Fig. 14.5A). The other cases may reduce R_0 but not sufficiently to consistently prevent an outbreak.

These simple equilibrium models provide broad insight into the potential ways disease transmission and host diversity may affect the long-term dynamics of malaria. However, spatial and temporal heterogeneity in vector and host abundance will likely have a strong influence on transmission dynamics. Observed patterns of avian malaria infection vary based on elevation and seasonal weather patterns, because transmission is strongly influenced by environmental variables and host and vector density.

The Importance of Climate in Vector and Parasite Dynamics

Annual and seasonal weather patterns play a crucial role in determining the spatial and temporal distribution of vector-transmitted diseases (Aaron and May 1982). A discrete-time mosquito model incorporating temperature and rainfall effects on mosquito demography in Hawai'i was developed by Ahumada et al. (2004). Here we present a compact ODE model that also incorporates the effect of daily temperature variation on the extrinsic incubation period of the parasite.

We model mosquito populations using a system of equations describing the daily dynamics of immature and adult mosquitoes and incorporating density-dependent effects for immature mosquitoes:

$$\frac{dL}{dt} = bM\left(\frac{K-L}{K}\right) - (d(\tau) + \mu_L)L, \text{ and}$$

$$\frac{dM}{dt} = d(\tau)L - \mu M, \quad [\text{Eqs. 11,12}]$$

where L represents the density of immatures, M is the density of adults, b is the per capita birth rate of adults, μ_L is the mor-

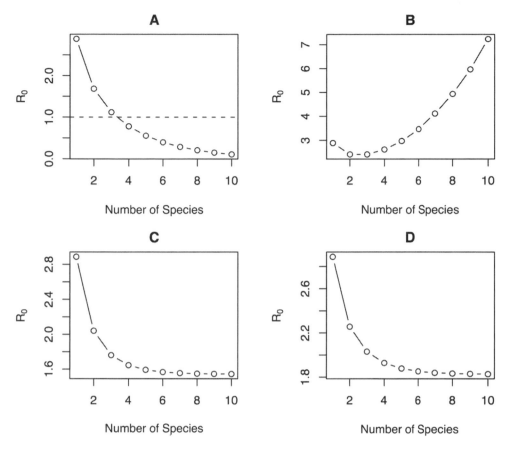

Figure 14.5. Effect on R_0 (the basic reproductive number for the pathogen) of adding species under various abundance and disease viability regimes. Species that are more viable for the disease have a longer infectious period than species that are less viable. (A) Sequential addition of species that are 50% more abundant than the previous species but 50% less viable. (B) Sequential addition of species that are 50% more abundant and 50% more viable. (C) Addition of species that are 50% less abundant and 50% less viable. (D) Addition of species that are 50% less abundant and 50% more viable.

tality rate of immatures, μ is the mortality rate of the adult mosquitoes, and K is the carrying capacity of the immatures.

We incorporate temperature (τ) by assuming that immatures develop at a rate proportional to the daily temperature, $d(\tau)$. This development rate is the inverse of the number of days from the first immature stage to the adult stage, at a constant temperature (τ). For example, if $d(\tau) = 0.1$, an immature develops into an adult in 10 days. Based on data from Rueda et al. (1990), the development rate for *Culex* mosquitoes is of the form

$$d(\tau) = \frac{DD - 1}{\tau - \tau_{min}}, \qquad \text{[Eq. 13]}$$

where DD is the number of degree-days a mosquito needs to develop from egg to adult (one DD is the number of days at 1°C above the minimum development temperature) and τ_{min} is the minimum development temperature (10°C for mosquitoes in Hawai'i). We could also make mosquito birth and mortality parameters (b, μ_L, and μ) dependent on temperature, but for simplicity we assume that they are constant.

For a simple case in which the daily temperature is constant, we can calculate the number of adult mosquitoes at equilibrium.

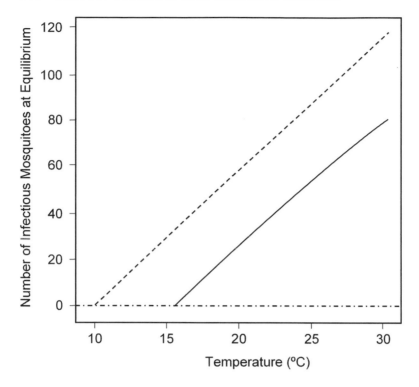

Figure 14.6. Number of infectious mosquitoes expected at equilibrium as a function of temperature. The figure shows the equilibrium values when the incubation period of the parasite is built into the model (solid line) and when it is ignored (broken line, Equation 18). These functions are the result of solving Equations 14–16.

For mosquito populations to persist, the rate of adult recruitment must be larger than the product of the average lifespan of the adults $(1/[d(\tau) + \mu_L])$. As expected, the abundance of infectious mosquitoes at equilibrium increases approximately linearly with temperature (Fig. 14.6). More complicated cases could also be considered, such as temperature effects on adult survival or fecundity.

PARASITE INCUBATION

We can also consider the importance of temperature on the development of *Plasmodium relictum* within a mosquito by creating an additional time lag for the transition of latent mosquitoes to an infectious state. If malaria is endemic, with the proportion of infectious hosts approximately constant, we can write ODEs for susceptible, latent (infected but not yet infectious), and infectious adult mosquitoes:

$$\frac{dS_M}{dt} = d(\tau)L - \lambda(X)S_M - \mu S_M,$$

$$\frac{dL_M}{dt} = \lambda(X)S_M - d^*(\tau)L_M - \mu L_M, \text{ and}$$

[Eqs. 14, 15, 16]

$$\frac{dI_M}{dt} = d^*(\tau)L_M - \mu I_M,$$

where $d^*(\tau)$ is now the development rate of the parasite inside the vector and $\lambda(X)$ is the force of infection for susceptible mosquitoes. Based on data from LaPointe (2000) for *Culex* mosquitoes in Hawai'i, this function can take the form

$$d^*(\tau) = -6.8 \times 10^{-4}\tau^2 + 0.042\tau - 0.49.$$

[Eq. 17]

Considering a simple case in which the temperature is constant, we can determine the expected number of infectious mosquitoes at equilibrium (by setting equations 14–16 equal to zero and solving for the equilibrium value of I_M):

$$I_M^* = \frac{d^*(\tau)\lambda(X)K[bd(\tau) - \mu(d(\tau) + \mu_L)]}{b\mu(\lambda(X) + \mu)(d^*(\tau) + \mu)}. \qquad \text{[Eq. 18]}$$

Infectious individuals persist when $I_M^* > 0$, which occurs when

$$d^*(\tau)\lambda(X) > 0.$$

Thus, the persistence of infectious mosquitoes depends on whether temperature is high enough to produce mosquitoes at a rapid rate relative to their lifespan and to allow the development of the malaria parasite inside infected mosquitoes. This framework allows us to compare the potential growth rates of adult and infectious vectors across geographic regions with varying temperature. The expected number of infectious mosquitoes increases with temperature, and fewer infectious mosquitoes are produced when the extrinsic incubation period of the parasite is included in the model (see Fig. 14.6).

We can also evaluate the seasonal effects of temperature on vector and parasite dynamics in Hawai'i. Numerical solutions for the ODEs were evaluated for several temperature regimes (seasonal, seasonal with white noise, and pure white noise) shown in Fig. 14.7. For the first two regimes, when the temperature is below the threshold for parasite development (broken horizontal line in the temperature graphs and horizontal thick lines in the vector graphs), infected mosquitoes do not persist. If the temperature falls below the threshold for immature development, the mosquito population crashes. A random (white noise) temperature regime does not have long consecutive periods of "favorable" or "unfavorable" temperatures. Therefore, the population hovers around a quasi-equilibrium determined by the long-term average temperature.

RAINFALL EFFECTS

Because mosquito larvae develop in water pools, they can also be affected by rainfall (Ahumada et al. 2004). We incorporate the effects of rainfall on larval mortality by first calculating the number of dry days since time t:

$$\delta_t(R_t) = (\delta_{t-1} + i_t)i_t, \qquad \text{[Eq. 19]}$$

where δ_t is the number of accumulated dry days at time t, R_t is the rainfall at time t, and i_t is an indicator variable equal to 1.0 if the rainfall is below a threshold value ψ or zero otherwise (Ahumada et al. 2004). The mortality rate of larvae μ_L can be estimated as a logistic function of the accumulated number of dry days δ_t:

$$\mu_L(R_t) = 1 - \frac{(1 - \mu_L)\exp(-\omega\delta_t(R_t))}{g + \exp(-\omega\delta_t(R_t))}, \qquad \text{[Eq. 20]}$$

where ω is a shape parameter controlling the sensitivity of immatures to dry conditions and g is a constant to ensure that the function starts at 1.0 when the number of dry days is zero (Ahumada et al. 2004). This expression for larval mortality can be inserted into the ODE describing larval dynamics (Equation 11).

Adult mosquitoes are susceptible to catastrophic rainfall events that cause extensive mortality. This effect can be incorporated by writing a new equation for the adult mortality rate (μ):

$$\mu(R_t) = (1 - \phi(R_t))\mu + \phi(R_t)\mu h, \qquad \text{[Eq. 21]}$$

where $\phi(R_t)$ is an indicator variable with the value of 1.0 if $R_t > 300$ mm rainfall in a day and zero otherwise and h is the increase in adult mortality rate (about 90%) due to this event. A more detailed analysis

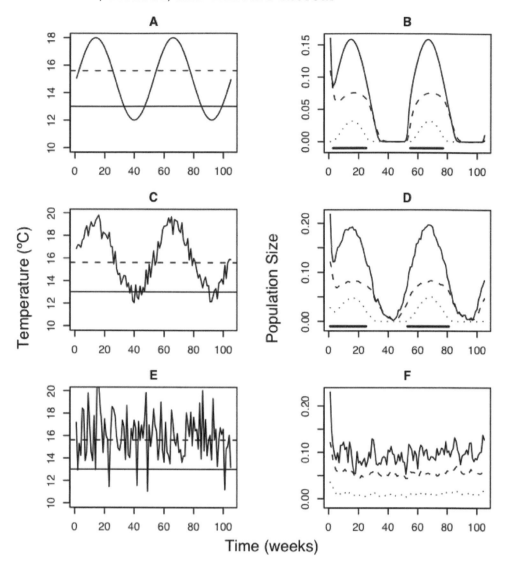

Figure 14.7. Numerical solutions of Equations 14–16 under different temperature regimes: a regular seasonal regime (A and B), stochastic seasonal regime (C and D), and random regime (E and F). The left-hand graph in each row shows temperature as a function of time (solid curve), the threshold temperature for the development of the mosquito (solid horizontal line), and the threshold temperature for the development of the parasite (broken horizontal line). The right-hand graph in each row shows the relative population size of susceptible (solid line), latent (broken line), and infectious (dotted line) mosquitoes through time. The thick horizontal lines indicate periods when the environmental temperature is greater than the threshold temperature for the development of the parasite.

of this model can be found in Ahumada et al. (2004). Model predictions show that wet years (e.g., 1990) produce a greater abundance of mosquitoes during fall and winter than do dry years (1995), except when catastrophic rainfall events occur and dramatically reduce the numbers of mosquitoes (Fig. 14.8).

GLOBAL WARMING

Global warming is predicted to expand the distribution of disease vectors, especially for diseases in which a significant

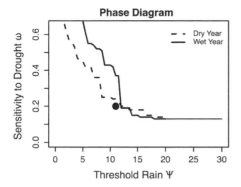

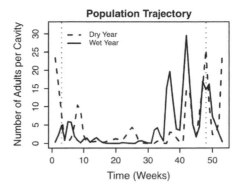

Figure 14.8. Effects of contrasting rainfall on the population dynamics of mosquitoes. Left-hand graph: isoclines of population under different combinations of threshold rain and larvae sensitivity to drought. Mosquitoes can persist to the right of each line and go extinct otherwise. Each isocline represents predicted population effects in two different years: 1990 (an unusually wet year) and 1995 (a dry year). The dot is the combination of parameters chosen to illustrate the effects of a rainfall regime on the population trajectories of mosquitoes (right-hand graph). The vertical dotted lines represent rainfall events greater than 300 mm per day.

portion of the life cycle is controlled by environmental conditions (Harvell et al. 2002). In Hawai'i, both mosquito population dynamics and parasite development rates respond positively to increased temperatures. As a result, we know that higher temperatures will increase the abundance of mosquitoes and the rate at which latent mosquitoes become infectious and therefore will increase the rate of infection in birds. How will global warming affect the dynamics of avian malaria in Hawaiian forests? Although detailed predictions of the future climate remain uncertain, an increase in temperature of 2°C may be likely during this century (Benning et al. 2002).

We used our simulation model to evaluate the potential impacts of increases of 1°C and 2°C in daily temperatures on bird populations at different elevations in Hawaiian forests. We considered three native birds, 'Apapane, 'I'iwi, and Hawai'i 'Amakihi, which are affected by malaria, and an introduced bird, the Japanese White-eye,

which is not affected. We evaluated bird populations' response after 25 years of increased temperatures at three (low, middle, and high) forest elevations using bird, mosquito, and habitat data collected for our Biocomplexity project (Fig. 14.9). For midelevation forests, we also evaluated two different study sites with high and low amounts of habitat for mosquito larvae.

Our simulations indicate that global warming will have little impact on any of the bird species found in low-elevation forests (see Fig. 14.9, 36 m). Low-elevation forests already have warm temperatures throughout the year, which favors fast development of mosquitoes and parasites, and this fast development facilitates rapid infection of nearly all the susceptible birds. Vulnerable native species such as the 'I'iwi and 'Apapane are rare or absent in many of these forests because high rates of disease transmission cause considerable mortality. Higher temperatures only increase the rate at which susceptible bird species are infected in low-elevation forests. In midelevation forests, the rate of disease transmission to native birds is highest during fall and early winter, when favorable temperatures cause an increase in mosquito populations and parasite development rates. At midelevation sites with abundant habitat for mosquito larvae (see Fig. 14.9, 1,024 m), global warming may have a minimal impact on 'Apapane, 'I'iwi, or Hawai'i 'Amakihi because high levels of mosquito abundance and rates of malaria transmission already produce high rates of infection

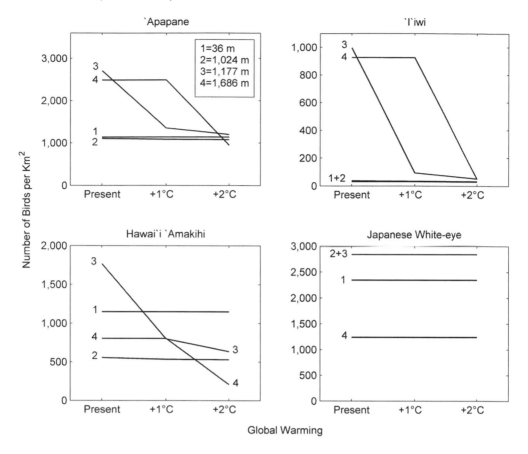

Figure 14.9. Predicted densities of forest birds at different elevations and levels of global warming. Shown are the simulated densities of 'Apapane, 'I'iwi, Hawai'i 'Amakihi, and Japanese White-eye at low- (36 m), mid- (1,024 m and 1,177 m), and high-elevation (1,686 m) forest sites under the present climate conditions and at increases of 1° C and 2° C in daily average temperatures for each site. Bird densities are estimated as average annual abundance after a 25-year simulation based on the bird, mosquito, and habitat conditions at each study site.

in susceptible native birds. However, at midelevation sites with limited habitat for mosquito larvae (see Fig. 14.9, 1,177 m), global warming can increase the rate of larval and parasite development, thus increasing the abundance of infectious mosquitoes to levels that will increase malaria transmission to birds and substantially

reduce the abundance of 'Apapane, 'I'iwi, and Hawai'i 'Amakihi. In high-elevation forests, global warming should have minimal implications until temperatures increase by >1°C (see Fig. 14.9, 1,686 m). At higher temperatures, increased mosquito populations and parasite development are expected to produce infectious mosquitoes that will successfully transmit malaria to birds, causing a dramatic decline in the population of native species. Japanese White-eyes are not susceptible to avian malaria; as a result, their abundance should not be affected by global warming.

In summary, global warming is likely to produce a significant increase in malaria transmission throughout most mid- and high-elevation forests in Hawai'i. Malaria transmissions in midelevation forests may shift from seasonal epidemics to continuous year-round transmission, as is currently found in low-elevation forests. Simultaneously, high-elevation forests will likely develop seasonal epidemics similar to those now occurring in midelevation forests. General trends are for 'Apapane and 'I'iwi populations to decline to levels similar to those currently found in low-elevation forests; however, Hawai'i 'Amakihi maintain larger populations in low-elevation forests due to the malaria resistance that has apparently developed in this population. To further exacerbate these problems, mosquito populations and successful parasite development will likely increase in elevational distribution with global warming (Benning et al. 2002; this volume, Chapter 9), further restricting the disease-free refugia for native Hawaiian birds to a few high-elevation sites. This combination of factors will likely produce further reductions and extinctions of native Hawaiian bird species.

MOSQUITO CONTROL

Strategies to control mosquito-borne disease transmission have usually focused on reducing the longevity or abundance of vectors, thus removing the transmission link of pathogen to bird host (see the discussion on mosquito control in Chapter 17). One of the most effective ways to reduce mosquito abundance is to remove larval habitat. However, reduction of man-made or naturally formed larval habitat in Hawaiian forests requires diligence and poses substantial challenges. In some forests, significant larval habitat reduction may require a combination of strategies, including pig removal, destruction of breeding sites, and chemical or biological control.

We used our simulation model to evaluate the potential effects of reducing mosquito larval habitat at the same forest sites where we considered the impacts of global warming (discussed earlier) (Fig. 14.10). At high-elevation sites, larval habitat reduc-

tion had little impact on the abundance of 'Apapane, 'I'iwi, or Hawai'i 'Amakihi because mosquito populations are small and little disease transmission occurs at these sites (see Fig. 14.10, 1,686 m). In contrast, low-elevation forests have a large amount of larval habitat and/or temperatures that are favorable for rapid development of larvae, producing a large and continuous abundance of adult mosquitoes. In these low-elevation forests, nearly complete removals of larval habitat may be required to substantially reduce mosquito abundance and thus increase 'Apapane or 'I'iwi populations (see Fig. 14.10, 36 m). The resistance of low-elevation Hawai'i 'Amakihi to malaria infection means that complete removal of larval habitat will have little impact on populations of these birds (see Fig. 14.10, 36 m). In midelevation forests, the challenges and potential benefits of larval habitat removal may depend on the amount of available habitat present. When larval habitat is abundant, substantial reductions are required to cut mosquito abundance and disease transmission (see Fig. 14.10, 1,024 m). However, if reductions can be sustained, increases in populations of 'Apapane, 'I'iwi, and Hawai'i 'Amakihi may be possible (see Fig. 14.10, 1,024 m). In midelevation forests with a small abundance of larval habitat, mosquitoes and disease transmission are already limited. Thus, further reduction in larval habitat may have a minimal impact on native bird populations.

Our results suggest that control actions that reduce mosquito larval habitat will be most successful in midelevation forests where larval habitat is abundant. Unfortunately, these sites will likely require a permanent reduction of nearly all (>80%) of the larval habitat to produce a significant improvement in native Hawaiian birds' abundance. In many cases, control efforts that substantially reduce larval habitat without reaching critical thresholds may result in the waste of effort and resources and

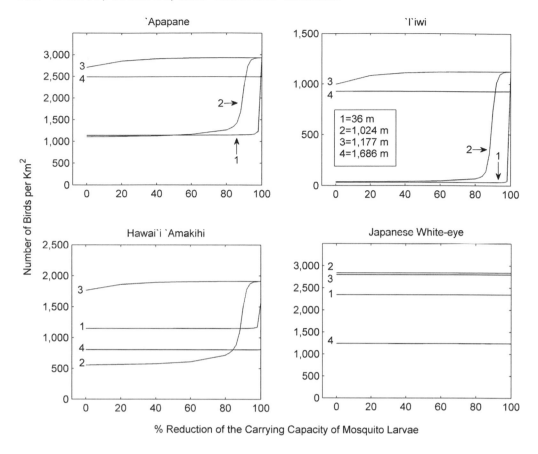

Figure 14.10. Predicted densities of forest birds at different elevations and levels of proportional reduction in mosquito larvae habitat at each site. Shown are the simulated densities of 'Apapane, 'I'iwi, Hawai'i 'Amakihi, and Japanese White-eye at low- (36 m), mid- (1,024 m and 1,177 m), and high-elevation (1,686 m) forest sites. Bird densities are estimated as average annual abundance after a 25-year simulation based on the bird, mosquito, and weather dynamics of each study site.

will not likely reduce disease transmission or increase bird populations.

FUTURE CONSERVATION MANAGEMENT

The most immediate goals for conserving the Hawaiian forest birds are to prevent further declines in populations, range reductions, and extinctions. Longer-term goals are to enhance population levels and to restore extant species to their historical distribution. How can modeling contribute to achieving these goals?

Benefits of Modeling to Conservation Management at the Strategic Level

Models can play a key role in formally describing the basic components, linkages, and dynamics present in complex ecosystems like the Hawaiian forest bird disease system. Models can assist us in developing hypotheses about system linkages and in evaluating the potential impacts of management intervention or conservation actions on the system. In many complex ecological systems, models can project the potential consequences of different actions or the effects of changes in conditions on system behavior. Without a modeling approach, it would be difficult to determine whether the system has reached equilibrium or is continuing to evolve in both predictable and unpredictable ways. Without modeling it would be difficult to predict system dynamics spanning several biotic levels ranging across genetic diversity and disease resistance or virulence, landscape patterns and elevational gradients, bird and mosquito demographics, and seasonal or annual environmental changes. An important goal of the modeling work is to identify weak links in the disease system and determine the ecological scale necessary for intervention to achieve the desired conservation goals. Models can also be used to understand the relative importance (or sensitivity) of different processes in affecting the system dynamics and patterns. However, because all models involve different levels of abstraction and simplification from the complex biological world, their ability to predict general and detailed patterns of system dynamics should be carefully considered (McCallum 2000).

Our analytical models can be useful in determining how broad-scale changes in biological processes affect system dynamics and potential equilibrium levels. For example, model assessments indicate that temperature plays a crucial role in determining the rate of mosquito population growth and therefore potential disease transmission. At the higher temperatures found in low-elevation forests throughout the year, mosquito populations grow at such a rapid rate that control efforts focused on the mortality of adults and/or larvae may be nearly futile. In contrast, at higher elevations, cooler annual and seasonal temperatures slow mosquito population growth and parasite development rates. Thus, at higher elevations, mosquito control by increasing adult or larval mortality may have limited benefits.

In many situations, detailed representation of the biological complexity of a system exceeds the basic level of problems that can be easily solved using analytical methods. In the Hawai'i situation, numerical simulation is also needed to eval-

uate the complexities of disease transmission, vector demographics, and bird demographics over a variety of landscape scales and to identify key factors that contribute to disease transmission. A modeling approach can accommodate important biological complexities, including seasonal and annual changes in weather, random variation in model parameters, and biological attributes of individuals, such as the age, sex, or genetic resistance of birds.

Of particular interest in assessing future risks to Hawaiian forest birds is determination of the impact of malaria and pox on population levels and species distribution patterns. For example, global warming with predicted temperature increases and more frequent wet cycles may increase both the abundance and the elevational distribution of infected mosquitoes, increasing their disease transmission in midelevation forests and decreasing the high-elevation mosquito-free refugia for sensitive Hawaiian species (Benning et al. 2002). How would the introduction of additional mosquito-borne diseases like West Nile virus (Stokstad 2004) or Usutu virus (Weissenbock et al. 2002) impact the forest bird communities? How would the changes caused by a new disease impact the dynamics of avian malaria and pox? How would the introduction of a new non-native bird or mosquito vector with different temperature tolerances influence the dynamics of avian malaria and pox transmission to endemic Hawaiian birds? Modeling can help us answer these important questions.

Benefits of Modeling to Conservation Management at the Tactical Level

Managers have considered a variety of conservation strategies to enhance or restore native Hawaiian forest bird populations facing avian malaria and pox (Chapter 17). Efforts focused on birds populations include reducing their susceptibility by using vaccines or creating controlled infections, translocating wild birds that are genetically resistant to infection, or reducing the abundance of nonnative species, such as House Sparrows, that might act as disease reservoirs. Computer modeling can be used to help us evaluate the potential success of these conservation actions by considering how biological factors, including infection rate, survival rate, reproduction, and number of released or treated birds, might interact to affect the outcome. Suggested programs to control vector populations (Chapter 17) include removal of feral pigs from forested areas to reduce the creation of mosquito breeding cavities, habitat modification or chemical control of adult mosquitoes and larvae in agricultural or human housing areas, and genetic alteration of mosquitoes. Computer modeling can assist us in evaluating the effects of various treatment levels or combinations and the scale of treatment needed to affect the dynamics of avian malaria transmission and the mortality of Hawaiian forest birds. Finally, computer simulations can help us assess the relative effectiveness of alternative actions, identify optimal timing for control actions, and predict the potential effectiveness of programs that integrate several different control strategies.

SUMMARY

Introductions of avian malaria and a competent mosquito vector have significantly contributed to the loss of diversity and abundance of native Hawaiian birds, especially the unique and highly visible honeycreepers. Despite our general awareness of the potentially devastating impacts of malaria on these important species, our knowledge about the dynamics of disease transmission and its demographic impacts on Hawaiian birds has remained limited. Recent research and modeling studies have considerably advanced our understanding of the ecology, evolutionary adaptation, and epidemiology of the Hawaiian forest bird–avian malaria system. The information currently available indicates that many factors, including climate, topography, habitat, nonnative birds, and human influences, affect the dynamics of vectors, hosts, and malaria transmission. These factors have combined to produce patterns of disease that vary by elevation, year, and species of bird. The general avian disease patterns are characterized by (1) high levels of malaria transmission in low-elevation forests, with little seasonal or annual variation in infection; (2) episodic levels of transmission in midelevation forests, with dramatic seasonal and annual variation depending on climatic conditions and their effects on mosquito dynamics; and (3) disease refugia in high-elevation forests, with only a slight risk of infection occurring during summer, when climatic conditions are favorable for pathogen and mosquito development.

Mathematical models provide a quantitative but simplified description of the complexity of avian malaria and pox in Hawaiian forest birds. We present a framework for modeling the dynamics of avian hosts, mosquito vectors, and disease in the Hawaiian system that links these biotic components through disease transmission between birds and mosquitoes, incorporates climatic (temperature and rainfall) and elevational drivers of these dynamics, and considers how other ecological factors (global warming and mosquito control) might affect system processes and behavior. Our models are the first to explicitly include the effects of environmental variables on the population dynamics of vectors and link this effect to the population dynamics of hosts in the Hawaiian system. These models link disease, ecology, climate, and epidemiology to provide an integrated evaluation of system processes, an understanding of how different components contribute to disease dynamics, and an identification of key processes on which additional scientific research would be beneficial. They also provide the framework for a systematic evaluation of factors influencing disease transmission and how transmission might be altered to control disease cycles in Hawaiian bird species. Using a modeling approach to understand these complex and interacting processes facilitates our evaluation of alternative conservation programs that might be used to restore native bird populations and our predictions of how future human development and climate may affect disease cycles and bird populations.

ACKNOWLEDGMENTS

This chapter is the result of a project supported by National Science Foundation Grant DEB-00-83944; the U.S. Geological Survey (USGS) Wisconsin Cooperative Wildlife Research Unit, University of Wisconsin, Madison; and the USGS National Wildlife Health Center, Madison. We are grateful to C. Atkinson, R. A. M. Fouchier, P. Hart, D. LaPointe, and B. Woodworth for useful discussions on the avian disease system in Hawai'i.

Plate 1. Hawaiian Forest Birds—Natives

Hawk, Owl, and Crow

'Io (ee-oh)
Hawaiian Hawk. Found only on Hawai'i Island. Feeds on small birds, rodents, and insects. *Source:* Photo © Jack Jeffrey.

Pueo (poo-eh-yo)
Hawaii's only surviving native owl. Diurnal as well as nocturnal. Found on all main Hawaiian islands. *Source:* Photo © Jack Jeffrey.

'Alalā (ah-lah-LAHH)
Hawaiian Crow. Hawaii's only living crow species. Native to Hawai'i Island but now lives only in captivity. *Source:* Photo © Jack Jeffrey.

Plate 2. Hawaiian Forest Birds—Natives
Monarch Flycatchers

Kaua'i 'Elepaio
(kau-wah ee eh-leh-PAH-yo)
A monarch flycatcher endemic to Kaua'i Island.
Captures insects in flight and gleans from leaves
and bark. *Source: Photo © Jack Jeffrey.*

O'ahu 'Elepaio (oh ah-hoo eh-leh-PAH-yo)
Endangered. Endemic to O'ahu Island. *Source: Photo
© Jack Jeffrey.*

Hawai'i 'Elepaio
(hah-wai ee eh-leh-PAH-yo)
Endemic to the island of Hawai'i. No 'Elepaio
are known from the islands of Moloka'i,
Lāna'i, and Maui. *Source: Photo © Jack Jeffrey.*

Plate 3. Hawaiian Forest Birds—Natives
Warbler and Thrushes

Millerbird
The only Old World Warbler in Hawai'i. These insect feeders are known from Laysan (extinct) and Nihoa (endangered) in the remote Northwestern Hawaiian Islands. *Source:* Photo © Jack Jeffrey.

'Ōma'o (ohh-MAH oh)
A thrush native to the island of Hawai'i. A berry feeder and important disperser of native plant seeds. Shown here in pilo. Related species are now extinct on the other main Hawaiian islands. *Source:* Photo © Jack Jeffrey.

Puaiohi (poo-ai-OH-hee)
An endangered thrush found only on Kaua'i. Unique for its habit of nesting on steep ravine banks, as shown here. *Source:* Photo © Jack Jeffrey.

Plate 4. Hawaiian Forest Birds—Natives
Hawaiian Honeycreepers

Laysan Finch
Endangered. Only on Laysan Island, but introduced to Pearl and Hermes Atoll in the Northwestern Hawaiian Islands. Feeds on seeds, insects, and seabird eggs. A male is pictured. *Source:* Photo © Jack Jeffrey.

Nihoa Finch (nee-hoh-ah finch)
Endangered. Only on tiny Nihoa Island in the Northwestern Hawaiian Islands. Fossils of both Nihoa and Laysan finches have been found on the main Hawaiian islands. Feeds on seeds, insects, and seabird eggs. A male is pictured. *Source:* Photo © Jack Jeffrey.

Palila (pah-lee-lah)
Endangered. Now found only in the dry māmane forests of Mauna Kea Volcano, Hawai'i Island. Feeds almost exclusively on the green seeds in the bean pods of the māmane tree, shown here. A male is pictured. *Source:* Photo © Jack Jeffrey.

Plate 5. Hawaiian Forest Birds—Natives
Hawaiian Honeycreepers

Maui Parrotbill (mau-wee parrotbill) Endangered. Now only on the east slope of Haleakalā Volcano, Maui Island. With its powerful bill it opens bark and wood in search of insects. A female is pictured. *Source:* Photo © Jack Jeffrey.

'Akiapōlā'au
(ah-kee-ah-pohh-LAHH-AU) Endangered. Endemic to Hawai'i Island. Note the unusual bill. Its feeding behavior is very unusual; it pecks bark with its lower bill, then uses its upper bill to probe for insect larvae in the openings. Also drills sap wells. A male is pictured. *Source:* Photo © Jack Jeffrey.

Plate 6. Hawaiian Forest Birds—Natives
Hawaiian Honeycreepers

Hawai'i 'Amakihi
(hah-wai ee ah-mah-KEE-hee)
A honeycreeper common throughout the island chain except on Kaua'i and O'ahu. Forages mainly for insects and flower nectar. Here in golden māmane flowers. A male is pictured. *Source:* Photo © Jack Jeffrey.

O'ahu 'Amakihi
(oh ah-hoo ah-mah-KEE-hee)
Like the Hawai'i 'Amakihi, this species seems increasingly tolerant to bird diseases. Has moved down into foothills and suburbs of Honolulu. Here in nonnative rose apple flowers growing at St. Louis Heights. A male is pictured. *Source:* Photo © Jack Jeffrey.

Kaua'i 'Amakihi
(kau-wah ee ah-mah-KEE-hee)
Has a larger bill than other island 'amakihi and is known to peel bark from dead trees in search of insects. Here in native lobelia flowers. A male is pictured. *Source:* Photo © Jack Jeffrey.

Plate 7. Hawaiian Forest Birds—Natives
Hawaiian Honeycreepers

'Akikiki (ah-kee-KEE-kee)
Also known as the Kaua'i Creeper.
Rare and declining on Kaua'i. Forages for
insects while creeping over tree trunks and
along branches. *Source:* Photo © Jack Jeffrey.

Hawai'i Creeper (hah-wai ee creeper)
Endangered, but more numerous than the previous
species. Endemic to Hawai'i Island. Creeps along
tree trunks and branches in wet and mesic forests
searching for insects. *Source:* Photo © Jack Jeffrey.

Maui 'Alauahio
(mau-wee ah-lau-wah-HEE-oh)
Now restricted to Maui. Travels
in small family groups creeping
on tree trunks and branches while
foraging for insects. A regular
cooperative breeder. *Source:*
Photo © Jack Jeffrey.

**Plate 8. Hawaiian Forest
Birds—Natives**
Hawaiian Honeycreepers

ʻAnianiau (ah-nee-ah-NEE-au)
A tiny honeycreeper endemic to Kauaʻi Island.
It feeds on nectar and arthropods. Here with
ʻōhiʻa lehua flowers. Photo © Jack Jeffrey.

ʻAkekeʻe (ah-keh-KEH eh)
Endemic to Kauaʻi Island. Feeds on insects in the crown
foliage of ʻōhiʻa trees. With its crossed bill, forces open leaf
buds in search of insects. *Source:* Photo © Jack Jeffrey.

ʻĀkepa (ahh-KEH-pah)
Endangered on Hawaiʻi Island and likely extinct on Maui.
A close relative of the ʻAkekeʻe. This is the only known
cavity-nesting honeycreeper. A male is pictured. *Source:*
Photo © Jack Jeffrey.

Plate 9. Hawaiian Forest Birds—Natives
Hawaiian Honeycreepers

ʻIʻiwi (ee ee-vee)
Once common to all the Hawaiian islands, it is now regularly found only on the highest forest slopes of Kauaʻi, Maui, and Hawaiʻi islands. It is a nectar feeder and important pollinator for many native plants. Here in ʻōhiʻa lehua flowers. *Source*: Photo © Jack Jeffrey.

ʻĀkohekohe
(ahh-KOH-heh-KOH-heh)
Endangered. Endemic to Maui. Dominates other nectar feeders where flowers are plentiful and feeds mainly at ʻōhiʻa lehua flowers. *Source*: Photo © Jack Jeffrey.

ʻApapane (ah-pah-PAH-neh)
The most common and widespread honeycreeper. Found on all of the forested main Hawaiian islands. Somewhat resistant to bird disease. A generalist that feeds on nectar and insects, it mainly visits ʻōhiʻa lehua flowers. *Source*: Photo © Jack Jeffrey.

Plate 10. Hawaiian Forest Birds—Introduced Species

Kalij Pheasant
Originally from the Himalayas. Established on Hawaiʻi Island and recently seen on Oʻahu. A male is pictured. *Source:* Photo © Jack Jeffrey.

Japanese Bush-Warbler
From Japan. Now on all of the main Hawaiian islands. *Source:* Photo © Jack Jeffrey.

Red-vented Bulbul
From Southeast Asia. On Oʻahu only. *Source:* Photo © Jack Jeffrey.

White-rumped Shama
From Southeast Asia. On Kauaʻi, Oʻahu, and Molokaʻi. A male is pictured. *Source:* Photo © Jack Jeffrey.

Plate 11. Hawaiian Forest Birds—Introduced Species

Hwamei
From Southeast Asia. On all forested main Hawaiian islands except Lāna'i. Photo © Jack Jeffrey.

Red-billed Leiothrix
From Southeast Asia. On all forested main Hawaiian islands except Kaua'i and Lāna'i. Source: Photo © Jack Jeffrey.

Japanese White-eye
From Japan. On all main Hawaiian islands. Source: Photo © Jack Jeffrey.

House Finch
From North America. On all main Hawaiian islands. A male is shown. Source: Photo © Jack Jeffrey.

Plate 12. Hawaiian Forest Birds—Species That Have Recently Disappeared

ʻŌʻōʻāʻā (ohh OHH ahh AHH) Kauaʻi ʻŌʻō. This was the last of an enigmatic group of five Hawaiian forest birds. This species was last seen in the mountains of Kauaʻi in the mid-1980s. *Source:* Photo © Rob Shallenberger.

ʻŌʻū (ohh OO) This fruit-eating honeycreeper was once abundant on all the forested Hawaiian islands. It disappeared from Kauaʻi and Hawaiʻi islands in the late 1980s. *Source:* Photo © Rob Shallenberger.

Poʻo-uli (poh oh-OO-lee) This honeycreeper was rare when first discovered on Maui in 1973. The last known bird died in captivity in 2004. *Source:* Photo courtesy of Paul E. Baker, U.S. Geological Survey.

Plate 13. Hawaiian Forest Birds—Extinct Species

Hawaiʻi Mamo (hah-wai ee mah-moh)
Plate "Drepanis pacifica" of Wilson and Evans
(1890–1899).

ʻUla-ʻai-hāwane (oo-lah ai hahh-WAH-neh)
Plate "Ciridops anna" of Wilson and Evans
(1890–1899).

Hawaiʻi ʻŌʻō (hah-wai ee oh OH)
Plate "Moho nobilis" of Rothschild
(1893–1900).

Kauaʻi Greater ʻAkialoa
(kau-wah ee ah-kee-ah-LOH-ah)
Plate "Hemignathus procerus" of Rothschild
(1893–1900).

PENNULA ECAUDATA

Plate 14. Hawaiian Forest Birds—Fossil Birds

The flightless Hawaiian Rail or **Moho** (moh-hoh), endemic to Hawai'i Island. This rail, which perished in the late 1800s, represents a diverse extinct fauna of flightless Hawaiian birds composed of waterfowl, ibises, and rails. Plate "Pennula ecaudata" of Wilson and Evans (1890–1899).

Subfossil remains of an undescribed extinct goose from Hawai'i Island. Biologist Jon Giffin is shown with a skeleton he found in a lava tube cave. *Source:* Photo © Jack Jeffrey.

Plate 15. Hawaiian Forest Bird Habitats

Top: Coastal shrub community on Nihoa Island, habitat of the endangered Millerbird and Nihoa Finch. This vegetation type was formerly more common on the main Hawaiian islands. Photo courtesy of Jack Jeffrey, U.S. Fish and Wildlife Service.

Bottom: Subalpine dry forest of māmane and naio on Mauna Kea Volcano, Hawai'i Island. Resident birds include the endangered Palila and several common species. *Source:* Photo by Thane K. Pratt, U.S. Geological Survey.

Plate 16. Hawaiian Forest Bird Habitats

Koa-'ōhi'a mesic forest on the island of Hawai'i. On the "Big Island," this forest type supports the greatest diversity of native bird species and is the main focus of habitat protection and expansion. Formerly forested lands that were historically converted to pasture can be rapidly reforested with koa, a fast-growing leguminous tree that attains enormous size, as seen here. *Source:* Photo by James D. Jacobi, U.S. Geological Survey.

'Ōhi'a wet forest on Moloka'i Island. Sheltered in climatic zones too wet and cold for agriculture, the montane wet forests are the most intact remaining Hawaiian forest type. 'Ōhi'a-lehua, a tree in the myrtle family, is the community dominant and is of vital importance to native forest birds for nectar and arthropod food and particularly for nest sites. *Source:* Photo by James D. Jacobi, U.S. Geological Survey.

Plate 17. Hawaiian Forest Bird Habitats

Top: Lava flows create mosaics of habitat on Kīlauea Volcano, Hawaiʻi Island. Flows of different ages and composition appear as shades of gray in this picture. A kīpuka—an island of vegetation—is in the lower left corner. These habitat patches are often quite large and can support small populations of native birds. In time, kīpuka also serve as a source of plants and animals to colonize the surrounding lava flows. It takes hundreds of years for forest to reclaim the flows. Snow-capped Mauna Kea lies in the distance. *Source:* Photo courtesy of James Kauahikaua, U.S. Geological Survey.

Bottom: A lava flow enters the forest on Kīlauea Volcano, Hawaiʻi Island. *Source:* Photo courtesy of James Kauahikaua, U.S. Geological Survey.

Plate 18. Hawaiian Forest Bird Habitats

Top: Cattle in a once forested pasture. Vast areas of dry forest and mesic forest have been cleared by cattle grazing. *Source:* Photo © Jack Jeffrey.

Bottom: Pasture with remnant ʻōhiʻa trees and a young koa plantation (right) on Kamehameha School land, Keauhou, Hawaiʻi Island. Note the thick canopy of the koa stand. This habitat supports high densities of endangered ʻAkiapōlāʻau, which seek insects in koa bark and wood. *Source:* Photo by Thane K. Pratt, U.S. Geological Survey.

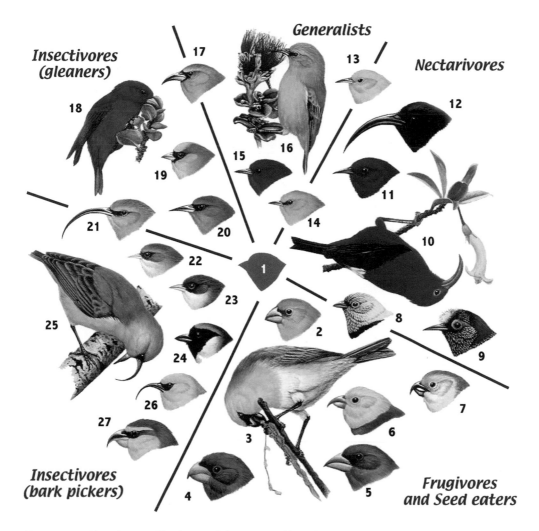

Plate 19. Adaptive radiation of the Hawaiian honeycreepers
(*Fringillidae: Drepanidinae*)

Shown are 26 of the 32 species known from the historic period. An additional 18 species have so far been described from subfossil bones. At the time of human arrival in the islands, the subfamily total must have well exceeded 50 species. *Source: Painting © H. Douglas Pratt.*

1	Hypothetical ancestral finch	10	ʻIʻiwi	19	ʻAkekeʻe
2	Laysan Finch	11	ʻApapane	20	Greater ʻAmakihi
3	Palila	12	Hawaiʻi Mamo	21	Lesser ʻAkialoa
4	Kona Grosbeak	13	ʻAnianiau	22	Hawaiʻi Creeper
5	Greater Koa-Finch	14	Maui ʻAlauahio	23	ʻAkikiki
6	ʻŌʻū	15	Kākāwahie	24	Poʻo-uli
7	Lānaʻi Hookbill	16	Hawaiʻi ʻAmakihi	25	ʻAkiapōlāʻau
8	ʻUla-ʻai-hāwane	17	Kauaʻi ʻAmakihi	26	Kauaʻi Nukupuʻu
9	ʻĀkohekohe	18	ʻĀkepa	27	Maui Parrotbill

Plate 20.

Ruling Chiefs. Painting © 2008 Herb Kawainui Kāne. Kāne's caption (1997: 32) reads: "A paramount chief, or king (*ali'i nui*), wears a feathered helmet (*mahiole*) and cloak ('*ahu'ula*). Full length cloaks were worn ceremonially; in battle, shorter capes were worn. The feathered malo worn around the waist and over the left shoulder signifies investiture as a king. Both figures wear carved whalestooth pendants on neckpieces of finely braided hair of ancestors (*lei niho palaoa*). An abstraction of a tongue, it signifies that the wearer speaks with authority. She wears a voluminous *kapa* wrap, a feathered head *lei*, a boar's tusk bracelet, and carries a small feathered *kāhili*, a fly whisk that evolved as a chiefly symbol.

"The temple image at left is carved in a style typical of Hawai'i Island. Beneath the image the *kahuna nui*, high priest of Kū (patron spirit of the chiefs), holds a feathered spirit image wrapped in *kapa*. Beside the *kahuna nui* stands the *kālaimoku* (prime minister and chief diplomat), holding a stalk of *tī*, a sign of truce or peace. At right stands a servant, a guard bearing a feathered standard (*kāhili*) of a height that warns commoners of the king's approach, and an armed bodyguard. As in Europe, Hawaiian rulers often recruited sons of district chiefs for training in their court, a practice calculated to insure loyalty." (With permission from H. K. Kāne.)

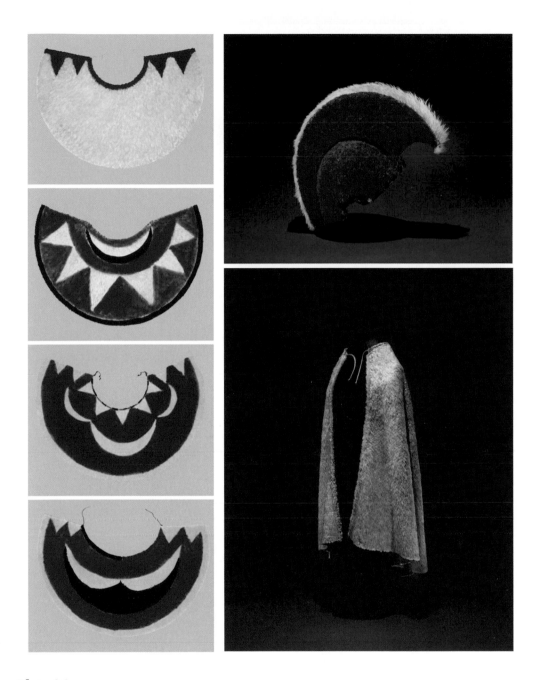

Plate 21.

Examples of Hawaiian feather work made of red ʻIʻiwi feathers and black and yellow ʻōʻō and Hawaiʻi Mamo feathers. (Left) Four feathered capes, ʻahuʻula. (Top right) A feathered helmet, mahiole, that is believed to have been given by Kamehameha the Great—Kamehameha I—to King Kaumualiʻi of Kauaʻi (photo by Seth Joel). (Bottom right) The cloak of Kamehameha the Great, made from the golden feathers of an estimated 80,000 Hawaiʻi Mamo (photo by Ben Patnoi). *Source:* Photos courtesy of Bernice P. Bishop Museum.

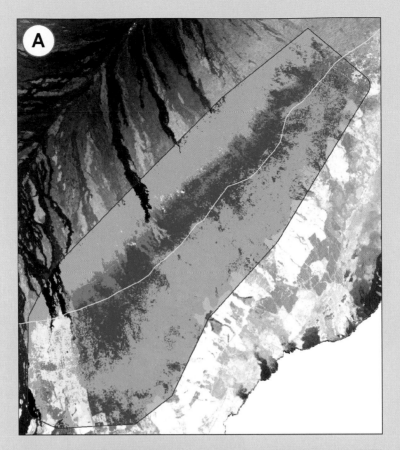

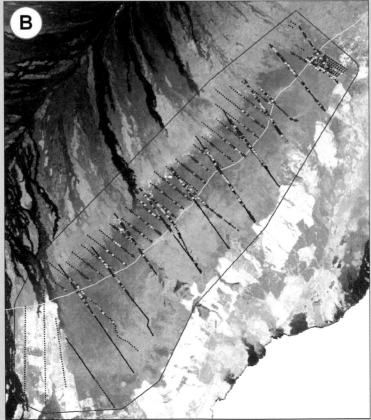

Plate 22.

Predicted population distribution and abundance of 'I'iwi in Ka'ū, Hawai'i Island. Maps were generated from a habitat association model that combines multiple biotic and abiotic factors and station-specific density estimates in a geographic information system (adapted from Gorresen et al. 2007). *Opposite: A:* Satellite imagery showing general land cover types (forest and woodland–green, grass and shrubland–yellow, barren and lava–brown, gray, and black). The study area is outlined in light red. *Opposite: B:* Observed 'I'iwi densities (black dot, 0 birds per ha; blue dot, >0–3 birds per ha; green dot, >3–6 birds per ha; yellow dot, >6–9 birds per ha; red dot, >9 birds per ha) at sampling stations. *Above: C:* 'I'iwi distribution and density as predicted from a species-habitat model; density increases from 0 birds/ha (blue) to >14 birds per ha (red). The yellow line indicates the 1,500 m contour.

Plate 23. Key to Plant Communities *(see Plate 24, opposite, above)*

Alpine Communities. Dry plant communities cover the upper slopes of the high mountains of Maui and Hawai'i. Alpine plant communities are sparsely vegetated dry shrublands of pūkiawe (*Leptecophylla tameiameiae*) and 'ōhelo (*Vaccinium* spp.) that extend upslope from the subalpine zone. The 'Ōma'o on Hawai'i is the only forest bird species resident in the alpine zone.

Subalpine Dry Forest and Shrubland. Dry forests of the subalpine zone are 'ōhi'a lehua (*Metrosideros polymorpha*) on western Hawai'i and māmane (*Sophora chrysophylla*) and naio (*Myoporum sandwicense*) trees on eastern Hawai'i and east Maui. Resident bird species are Hawai'i 'Amakihi and in certain seasons 'I'iwi and 'Apapane. The endangered Palila is dependent on the māmane forests of Mauna Kea.

Wet Forest and Shrubland. *Lowland areas:* Wet shrublands of 'ōhi'a are stunted versions of lowland wet forest that occur on windward ridges and island summits. Lowland wet forests of 'ōhi'a and the matted fern uluhe (*Dicranopteris linearis*) are common on the steep windward slopes of most islands and on young substrates of Hawai'i Island. Forests of 'ōhi'a and 'ōlapa (*Cheirodendron* spp.) occur in the cloud zone, and 'ōhi'a shares dominance with koa (*Acacia koa*) at a few sites on Kaua'i, Maui, and Hawai'i. Forests of 'ōhi'a and lama (*Diospyros sandwicensis*) are rare on 'a'ā lava substrates of windward Hawai'i. Typical bird species are 'Apapane, 'Elepaio, and (on Hawai'i) 'Ōma'o. *Montane areas:* Wet shrublands of 'ōhi'a are dwarf forests that grow on ridges and summits, particularly on Moloka'i and west Maui. Mixed shrubs and ferns grow on steep mountain slopes and valley ridges. Montane wet forests of 'ōhi'a are the most common forest of Kaua'i, Moloka'i, Maui, and Hawai'i. On Hawai'i, hāpu'u tree ferns (*Cibotium* spp.) typically form a dense understory in this forest. Wet koa-'ōhi'a forests are limited to Maui and Hawai'i; these have a diverse understory of native trees and shrubs. This is the primary habitat of most extant forest birds.

Mesic Forest and Shrubland. *Lowland areas:* Mesic shrublands are found on the lower leeward slopes and ridges of all main islands. Lowland mesic forests have the same dominant trees as dry forests: 'ōhi'a, lama, koa, and rarely olopua

(*Nestegis sandwicensis*). Diverse mesic forests with no dominant species are relictual, and the extent of their original distribution is unknown. Component birds are the 'Elepaio, three species of 'amakihi, and the 'Apapane. *Montane areas:* Mesic communities of the montane zone are forests of 'ōhi'a or mixed 'ōhi'a-koa found on Kaua'i, Maui, and Hawai'i. Diverse forests of 'ōhi'a-koa-a'e (*Sapindus saponaria*) or olopua are very rare on Hawai'i and Maui. Mesic forests support rich bird assemblages of 'Elepaio, 'amakihi, 'I'iwi, 'Apapane, and endangered species such as 'Akiapōlā'au and 'Ākepa.

Dry Forest and Shrubland. *Lowland areas:* Dry shrublands predominated on the lower leeward slopes of all the main islands. In lowland regions with higher levels of rainfall and soil development, there are dry forests of 'ōhi'a, lama, and less commonly koa or olopua. Wiliwili (*Erythrina sandwicensis*) forests grow in dry, rocky areas and often contain 'ohe makai (*Reynoldsia sandwicensis*) trees. Common birds are the three species of 'amakihi and the 'Apapane. *Montane areas:* Above 2,000 m, montane dry shrublands are typically 'a'ali'i (*Dodonaea viscose*) shrubs with 'ōhi'a trees. Montane dry forests are composed of 'ōhi'a on upper leeward slopes and mixed koa-māmane trees on windward slopes. The bird species of dry montane communities are 'Elepaio, 'amakihi, 'I'iwi, and 'Apapane.

Lowland Dry Shrubland and Grassland. In very dry areas, pili (*Heteropogon contortus*) was formerly the dominant cover; this grass has been largely replaced by alien species. Small, remnant communities of 'a'ali'i, 'akoko (*Chamaesyce* spp.), or alahe'e (*Psydrax odorata*) shrubs and mixed grasses are extant on Kaua'i, O'ahu, and Maui. No native forest bird species remain in these habitats.

Coastal Plant Communities. Shrublands and forest once fringed the coasts of all islands (not shown on map). Today, low-stature, diverse strand vegetation grows on the Northwestern Hawaiian Islands and on a few sandy beaches and dunes on the main islands. Bird species that inhabit coastal plant communities of the Northwestern Hawaiian Islands are the Millerbird, Laysan Finch, and Nihoa Finch.

Plate 24. *(opposite, above)*

Plate 25. *(opposite, below)*

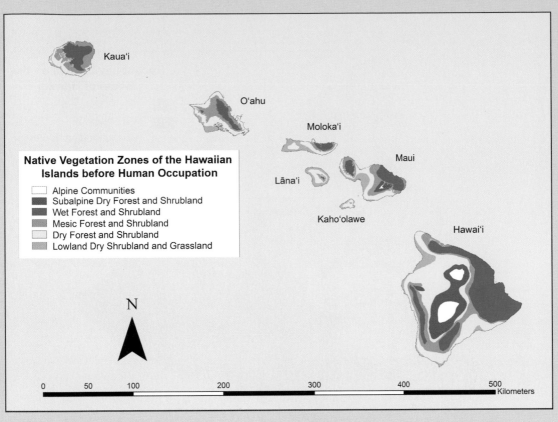

Native Vegetation Zones of the Hawaiian
Islands before Human Occupation

Alpine Communities
Subalpine Dry Forest and Shrubland
Wet Forest and Shrubland
Mesic Forest and Shrubland
Dry Forest and Shrubland
Lowland Dry Shrubland and Grassland

Kaua'i
O'ahu
Moloka'i
Maui
Lāna'i
Kaho'olawe
Hawai'i

N

0   50   100   200   300   400   500
                                                        Kilometers

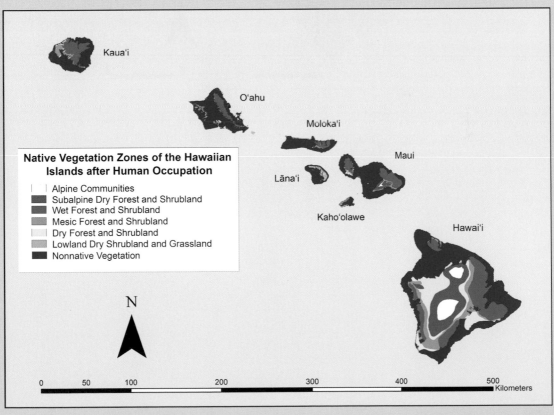

Native Vegetation Zones of the Hawaiian
Islands after Human Occupation

Alpine Communities
Subalpine Dry Forest and Shrubland
Wet Forest and Shrubland
Mesic Forest and Shrubland
Dry Forest and Shrubland
Lowland Dry Shrubland and Grassland
Nonnative Vegetation

Kaua'i
O'ahu
Moloka'i
Maui
Lāna'i
Kaho'olawe
Hawai'i

N

0   50   100   200   300   400   500
                                                        Kilometers

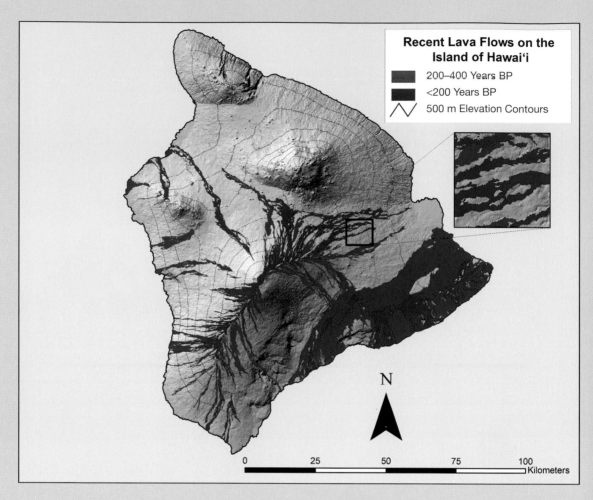

Recent Lava Flows on the Island of Hawai'i

■ 200–400 Years BP
■ <200 Years BP
/\/ 500 m Elevation Contours

N

0   25   50   75   100
Kilometers

Plate 26.

Recent lava flows on the island of Hawai'i. The inset shows the landscape dissected into small islands of vegetation (kīpuka) by lava flows. Volcanic activity creates a biogeographically dynamic landscape of forest patches of different ages. Protection of forest bird habitat should take into account this shifting patchwork of successional vegetation. Substrate information adapted from Wolfe and Morris (1996).

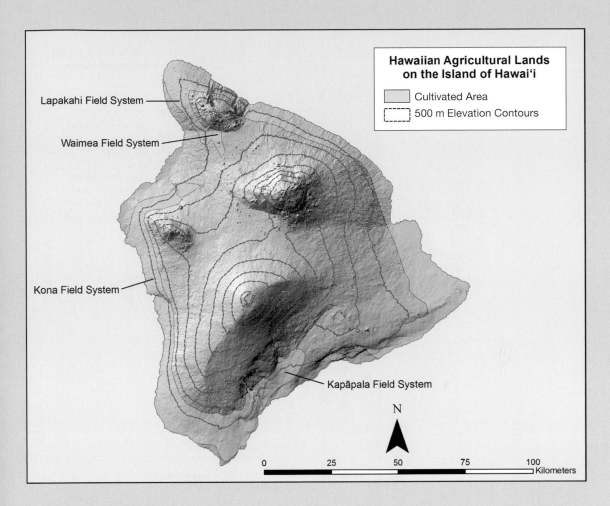

Plate 27.

Areas suitable for agriculture on the island of Hawai'i prior to Western contact. Prehistoric settlement occurred along the coast, leaving montane forests relatively undisturbed until modern times. Based on data from Newman (1972) and Kirch (1985). Field system locations from Kirch (1985).

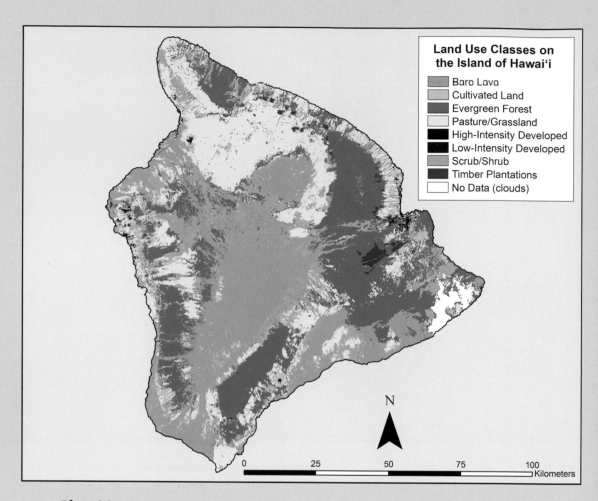

**Land Use Classes on
the Island of Hawaiʻi**

Baro Lava
Cultivated Land
Evergreen Forest
Pasture/Grassland
High-Intensity Developed
Low-Intensity Developed
Scrub/Shrub
Timber Plantations
No Data (clouds)

N

| 0 | 25 | 50 | 75 | 100 |

Kilometers

Plate 28.

Current land use on the island of Hawaiʻi. Most of the remaining wet forest is protected, and some potential bird habitat historically used for pasture is being reforested. Adapted from National Oceanic and Atmospheric Administration, Coastal Change Analysis Program (2000).

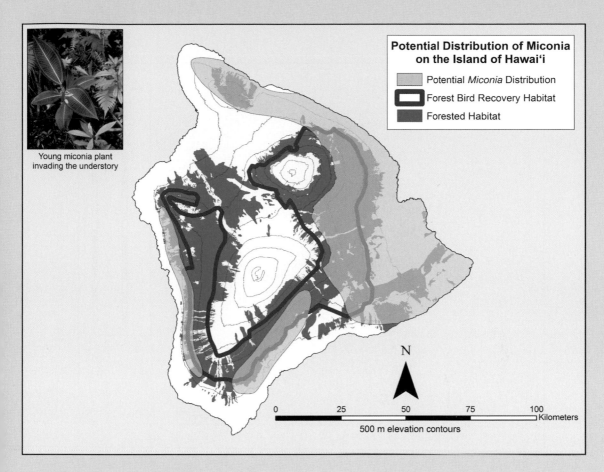

Potential Distribution of Miconia on the Island of Hawai'i

Potential *Miconia* Distribution

Forest Bird Recovery Habitat

Forested Habitat

Young miconia plant invading the understory

N

0   25   50   75   100 Kilometers

500 m elevation contours

Plate 29.

Potential distribution of the invasive tree miconia with respect to forest bird habitat on the island of Hawai'i. Very large areas of native forest are likely to be converted to monospecific stands of miconia if current efforts to control the species fail. Photo courtesy of Arthur Medeiros, U.S. Geological Survey.

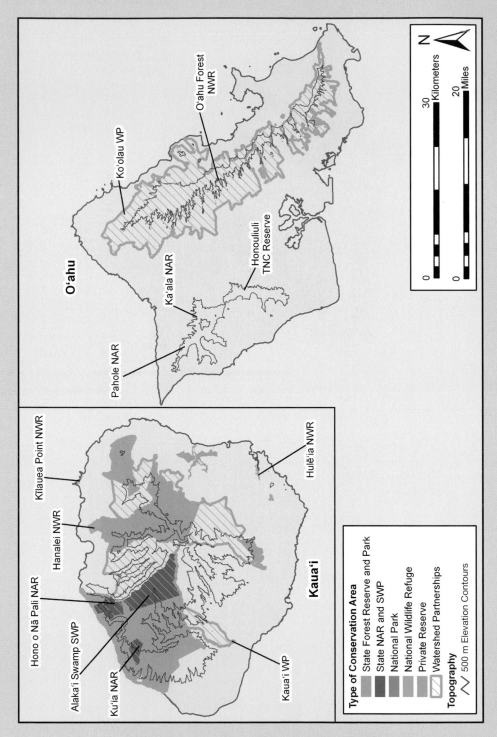

Plate 30. Major conservation areas—Kaua'i and O'ahu. Abbreviations: NP, National Park; NWR, National Wildlife Refuge; FR, state Forest Reserve; NAR, state Natural Area Reserve; SWP, State Wilderness Preserve; TNC, The Nature Conservancy of Hawai'i; WP, Watershed Partnership.

O'ahu

Ko'olau WP

O'ahu Forest NWR

Pahole NAR

Ka'ala NAR

Honouliuli TNC Reserve

N

0   30
Kilometers

0   20
Miles

Kaua'i

Kīlauea Point NWR

Hanalei NWR

Hono o Nā Pali NAR

Alaka'i Swamp SWP

Ku'ia NAR

Kauai'i WP

Hulē'ia NWR

Type of Conservation Area
State Forest Reserve and Park
State NAR and SWP
National Park
National Wildlife Refuge
Private Reserve
Watershed Partnerships

Topography
500 m Elevation Contours

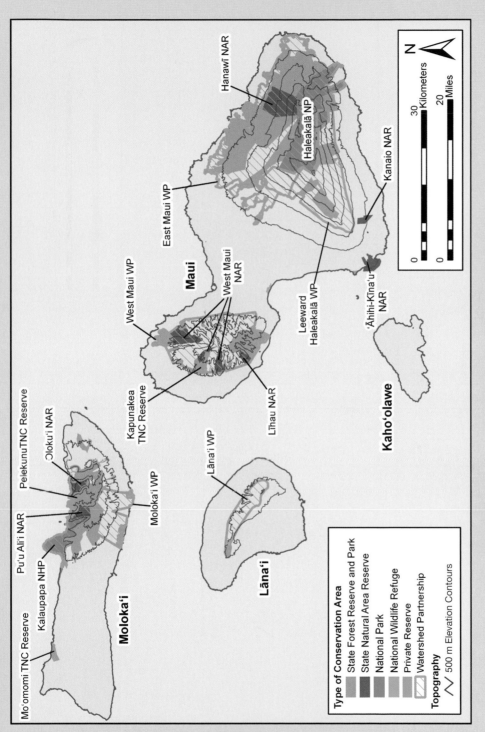

Plate 31. Major Conservation Areas—Moloka'i, Lāna'i, and Maui. See Plate 30 for abbreviations.

Moloka'i

Mo'omomi TNC Reserve

Kalaupapa NHP

Pu'u Ali'i NAR

Pelekunu TNC Reserve

'Ōloku'i NAR

Moloka'i WP

Lāna'i

Lāna'i WP

Kapunakea TNC Reserve

West Maui WP

West Maui NAR

Līhau NAR

East Maui WP

Maui

Hanawī NAR

Haleakalā NP

Leeward Haleakalā WP

Kanaio NAR

'Āhihi-Kīna'u NAR

Kaho'olawe

N

0   30
Kilometers

0   20
Miles

Type of Conservation Area

State Forest Reserve and Park
State Natural Area Reserve
National Park
National Wildlife Refuge
Private Reserve
Watershed Partnership

Topography

500 m Elevation Contours

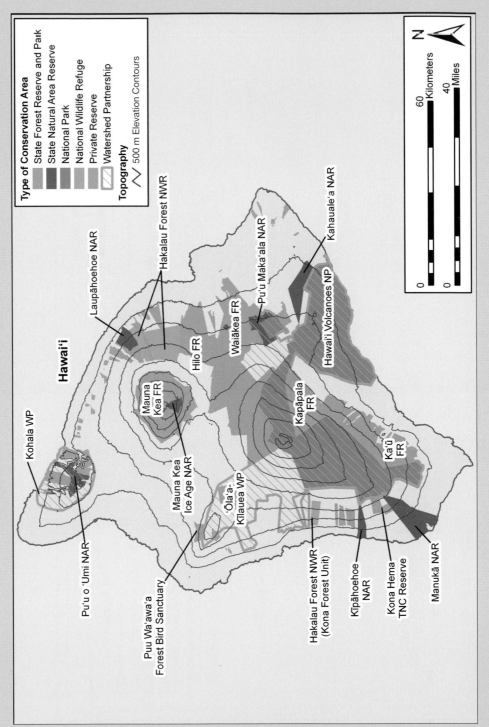

Plate 32. Major Conservation Areas—Hawai'i Island. See Plate 30 for abbreviations.

PART 3

Applying Research to Management

CHAPTER FIFTEEN

Preventing Establishment and Spread of Invasive Species

Current Status and Needs

LLOYD L. LOOPE AND FRED KRAUS

Invasive species constitute the largest category of threat to Hawaiian forest birds (Chapter 2). Existing and future invasions affect all aspects of forest bird ecology: habitat change, disease, predation, and competition. The nature and magnitude of these impacts are often unexpected. The effects of the accidental introduction of the brown treesnake (*Boiga irregularis*) to Guam were unpredicted (Savidge 1987), but inference suggests that other generalist snakes could have similar effects if allowed to establish in Hawai'i (Kraus and Cravalho 2001, Loope et al. 2001). Establishment and rapid spread of insectivorous coqui frogs (*Eleutherodactylus coqui*) within the past 10–15 years has introduced a new and potentially formidable competitor for Hawaiian forest birds (Kraus et al. 1999, Kraus and Campbell 2002), and the incipient establishment of veiled chameleons (*Chamaeleo calyptratus*) has brought a new potential predator (Kraus 2004). However threatening to forest birds, the arrival of newly invasive species is a much larger issue that affects all native biota and ecosystems, agriculture, and the quality of living of Hawaii's people (Box 15.1). For example, newly emergent plant pathogens and insect pests are posing direct threats to Hawaii's dominant tree species and the ecosystems they support (Box 15.2).

The future of all native biodiversity of Hawai'i depends primarily on how well the government and the public cooperate to prevent establishment and spread of new introductions and on how well natural resource managers control established invasive alien species. Given the high rate at which new species are successfully entering the state (Eldredge and Evenhuis 2003), the record in Hawai'i to date for stemming biological invasion is not good, although recent efforts may surpass those in most

Box 15.1.   **A Worsening Future?**

The Brown Treesnake

Dramatic, cascading ecological, economic, and quality-of-life effects of the invasive brown treesnake (BTS) (Fig. 15.1) in Guam have been well documented (Savidge 1987, Rodda et al. 1997, Fritts and Rodda 1998). There is no reasonable doubt that comparable effects would occur in Hawai'i if the BTS were to establish there (Burnett et al. 2006), and most birds would be eliminated in Hawaii's lowlands. It is not entirely clear, however, how comprehensively the BTS would affect Hawaii's endemic forest birds, which primarily persist at high elevations. The impact of the BTS on them would depend primarily on the cold tolerance of the snake population established.

No experiments have been conducted to determine the critical thermal limits of the BTS (Rodda et al. 1999). The highest record of a BTS is from an elevation of 1,375 m in Papua New Guinea (McDowell 1984). The species occurs as far south as the northern suburbs of Sydney,

Australia, at 36° S latitude, where freezing temperatures occur at least occasionally in winter. The source population for the BTS on Guam is probably the Admiralty Islands (Rodda et al. 1992), the highest point of which is merely 702 m. The BTS population on Guam (highest point, 407 m), present for more than 50 years to date, has likely never been challenged by temperatures as low as 10°C. One can only speculate on potential effects of genetic founder effects and microevolution away from cold tolerance on Guam, as well as on the potential for the acclimatization and microevolution of the BTS toward cold tolerance should a population become established in Hawai'i. It is doubtful that temperatures under Hawaii's high-elevation rainforest canopies ever reach as low as those in the BTS habitat around Sydney, Australia. If the BTS reached Hawai'i, it might be limited to elevations below 1,000 m (and have

Figure 15.1. Brown treesnake. *Source:* Photographed in New Guinea by Fred Kraus.

Box 15.1.   *Continued*

only modest effects on Hawaii's forest birds), but might just as plausibly reach 2,000 m under forest canopy (and eliminate most forest birds).

West Nile Virus

The mosquito-borne West Nile virus (WNV) can infect humans, birds, horses, and many other animals. This flavivirus, widespread in Africa and Eurasia, first reached the northeastern United States in the New York City area in 1999. By the end of 2006, it had spread across most of North America, 1,008 humans had died from WNV infection, and 317 bird species had been found infected (Kilpatrick et al. 2007). WNV had not been detected in Hawai'i as of early 2009, though two invasive mosquito species (*Culex quinquefasciatus* and *Aedes albopictus*) are widely abundant in Hawai'i and able to transmit the disease. The same mosquitoes transmit avian malaria (*Plasmodium relictum*) and avian pox virus (*Avipoxvirus* sp.), diseases believed to be responsible for a major decline of Hawaiian forest birds. A third species capable of transmitting WNV, *Aedes japonicus*, was first discovered in the Hawaiian Islands on Hawai'i Island in 2004 (Chapter 17).

Based on the relatively severe effects of WNV on North American birds, it is expected that WNV infection would produce a high level of mortality in many bird species endemic to Hawai'i. Biodiversity concerns, along with concerns for human health and tourism, provide strong motivation for preventing WNV from establishing in Hawai'i.

There are three potential routes for WNV to reach Hawai'i: importation of infected animals, importation of infected mosquitoes, and migratory birds. Fortunately, viremia levels in WNV-infected humans are insufficient to transmit the disease (Kilpatrick et al. 2004). Continuation of the embargo on bird shipments (Hawai'i Department of Agriculture 2002b) to Hawai'i addresses the importation issue, although the process could be refined and expanded. Measures are urgently needed to prevent WNV-infected mosquitoes from arriving in the cargo holds and cabins of aircraft from North America (Kilpatrick et al. 2004). Migratory birds are believed to present a low risk of WNV transmission due to the physical rigor of the long-distance over-water migration necessary to reach Hawai'i (Kilpatrick et al. 2004).

---

other states, such as Hawaii's comparatively stringent regulations and enforcement against vertebrate introductions (Kraus 2003). What should be Hawaii's main line of defense against alien invasions—geographic isolation—has been broken by the forces of globalization.

Because of its special vulnerability to invasion, Hawai'i needs prevention and management efforts much greater than those of the rest of the country. This greater

level of need, combined with Hawaii's small population and limited tax base, means that optimal protection against invasions is not being provided. This chapter gives an overview of current authorities and programs for preventing establishment and spread of new invasions in Hawai'i, the limitations of current programs, and some recent innovations in Hawai'i and elsewhere that offer hope for improved protection.

Box 15.2. **Ecosystem Killers**

The loss of ecologically crucial foundation tree species due to outbreaks of invasive pathogens or insect pests—hemlock woolly adelgid, white pine blister rust, and many others—is rampant in the United States (Campbell and Schlarbaum 2002, Ellison et al. 2005). Two potential ecosystem killers were newly discovered in Hawai'i in early 2005—the erythrina gall wasp (EGW, *Quadrastichus erythrinae*) and 'ōhi'a rust (*Puccinia psidii*) (Heu et al. 2006, Killgore and Heu 2007). Two of Hawaii's dominant trees, wiliwili (*Erythrina sandwicensis*) and 'ōhi'a (*Metrosideros polymorpha*), are now under attack. This juxtaposition of events has brought about the realization that Hawaii's dominant native plant species must be proactively protected from plant pests, just as sugar cane, pineapple, coffee, and orchids have been protected by Hawai'i quarantine regulations for a century.

The EGW (Fig. 15.2) is a newly invasive, tiny, previously undescribed eulophid wasp species that apparently originated in Africa. First noted and collected on La Réunion Island in the Indian Ocean in late 2000, it reached Singapore and Taiwan in 2003 and was described as a new species a year later (Kim et al. 2004). By 2005, it had arrived in Hawai'i (via an as yet undocumented pathway) and was severely infecting and deforming cultivated *Erythrina* spp. and the important endemic tree of dryland forests, wiliwili, in lowland areas throughout Hawai'i. The Hawai'i Department of Agriculture (HDOA) rapidly implemented a biocontrol program in the hope of saving the wiliwili from extinction.

'Ōhi'a rust (Fig. 15.3) arrived with an already notorious reputation due to a dramatic host jump in its native Brazil from guava (*Psidium guajava*) to large, nonnative eucalyptus plantations, where it

Figure 15.2. Erythrina gall wasp galls on the native Hawaiian *Erythrina sandwichensis* or wiliwili. *Source:* Photo © Jack Jeffrey.

inflicted damage on young leaf growth, and due to its broad host range within the Myrtaceae family (Coutinho et al. 1998). Recognized as the most serious threat to eucalyptus cultivation worldwide, it is also considered among the foremost of the newly emerging tree diseases (Slippers et al. 2005). It has been present in Florida since the 1970s, and its establishment as an outlying population in Hawai'i—bringing the Neotropical rust to a hub of transportation 8,000 km nearer—stimulated much concern in Australia, home to more than 1,500 Myrtaceae species (Glen et al. 2007). There was also much concern in Hawai'i when the rust, first found on a young 'ōhi'a plant on O'ahu, spread within a few months throughout the main Hawaiian islands. However, its ecosystem-level effects were not on 'ōhi'a but on the invasive rose apple (*Syzygium jambos*), which by early 2008 was starting

Box 15.2.   Continued

Figure 15.3. *Puccinia psidii* is a fungal pathogen that recently arrived in Hawai'i, where it is known as 'ōhi'a rust. It is shown in this photo defoliating rose apple trees. *Source:* Photo courtesy of Pacific Island Ecosystems Research Center, U.S. Geological Survey.

to undergo extensive mortality. The Hawaiian conservation community has encouraged the HDOA to keep out new strains of *P. psidii*, given that the damage to 'ōhi'a is modest to date and that *P. psidii* has a reputation for multiple strains, only superficially documented and vaguely defined by the suite of host plants it infects. The decimation of rose apple dramatizes the potential virulence of this rust species complex for non-Neotropical Myrtaceae.

In its first three years in Hawai'i, no sexual stage of this rust has been detected (J. Uchida, pers. comm.), and preliminary DNA analysis has not yet detected genetic variation, which is compatible with the single-strain concept (Zhong

et al. 2008). Shipments to Hawai'i by the flower and foliage industry of California are believed to have been the source of the existing *P. psidii* strain in Hawai'i and to be a likely threat for vectoring additional strains (Loope and La Rosa 2008). In August 2007, after *P. psidii* had been repeatedly intercepted on Myrtaceae foliage from California, Hawaii's Board of Agriculture approved an interim rule empowering restriction of movement into the state of plants and plant products in the Myrtaceae family from areas infested with *P. psidii* to prevent the arrival and establishment of new strains of the rust. An effort was under way as of February 2009 to develop a long-term rule and strategy to prevent the importation of not only new strains of *P. psidii* but also numerous other invasive pests—both insects and fungi—that are moved by the flower and foliage trade and the eucalyptus industry and that attack Myrtaceae (Wingfield et al. 2001, Cortinas et al. 2004).

THE HISTORICAL CONTEXT

The uniqueness of Hawai'i and other high Pacific islands lies in their isolation from continental landmasses and their great topographic and climatic diversity (Carlquist 1974). Natural colonization and sustained establishment in these islands by foreign species has taken place very infrequently, perhaps once every 2,500 years (Loope 1998). Evolution in isolation over time has led to the generation of species and higher lineages unique to particular islands (Price and Clague 2002). This isolation has resulted in the evolution of approximately 10,000 species found nowhere else on earth, out of a total native biota of approximately 20,000 native species (Eldredge and Evenhuis 2003). Topographic variability, interacting with regional climate, has also resulted in a mix of habitats that can place tropical rain forests within a few miles of baked desertlike conditions, creating climatological gradients that would occur over much greater distances in continental situations. For these reasons, Hawai'i holds a significant portion of the United States' heritage of unique biological wealth.

With the breakdown of natural geographic isolation by human activities, Hawaii's native biota and ecosystems have been overwhelmed by establishment of more than 5,000 species of alien plants and animals in the past 200 years—a rate of successful colonization of new species on the order of 50,000 times the estimated natural rate over the past 5 million years (Loope 1998). This pattern shows no sign of abatement, and from 1995 to 2003 the Hawai'i Biological Survey documented an average of 89 additional alien species in Hawai'i each year (Miller and Eldredge 1996, Eldredge and Evenhuis 2003). Under these circumstances, Hawaii's ecological meltdown is not unexpected and can be depicted in a number of ways.

The first is loss of native species. Hawai'i has lost hundreds of species to extinction,

currently has 322 species (26% of the U.S. total) recognized as endangered or threatened by the U.S. Fish and Wildlife Service (USFWS), and has hundreds more that deserve protection but remain unlisted (M. Bruegmann, USFWS, pers. comm.). Most Hawaiian forest birds are now extinct, 24 are federally listed as endangered (including 11 that may be extinct), and only 14 are extant and not classified as endangered (Chapter 5). Most terrestrial endangered species, including forest birds, are currently endangered primarily by alien organisms (M. Bruegmann, pers. comm.). Alternatively, if one considers the landscape, Hawai'i has lost a massive percentage of its native habitats (fewer than 10% are undisturbed), especially the dry and mesic forests that once supported the greatest diversity of forest birds (Chapter 6). The large majority of this habitat loss is due to either replacement of native vegetation by invasive plants or destruction of vegetation by alien ungulates.

Although no well-researched effort has been undertaken to address the question, we estimate that of Hawaii's 5,175 established alien species (Eldredge and Evenhuis 2003), approximately 300–500 are causing ecological damage. This group would include approximately 20–40 vertebrates, 150–200 plants, and an unknown but large number of invertebrates and pathogens. Especially damaging are certain species groups such as mammals, social insects, fire-promoting grasses, and many trees that alter ecosystem function or community structure.

Broad-based concern over introduction of nonnative species in Hawai'i goes back to 1888–1890, when action was taken to protect coffee and other agriculture from insect pests and diseases (www .hawaiiag.org/hdoa/pi_pq.htm). Nondomestic animal invasions were recognized as a potential problem by 1905, when 14 snakes were seized. Later, when air transport was in its infancy, the Hawaiian Sugar Planters' Association established a quarantine station on Midway Island to prevent alien pests originating in Asia from reaching Hawaiʻi. Yet these early and continued efforts were much less successful at preventing accidental introductions than were similar efforts elsewhere. McGregor (1973: 19), a consultant to the U.S. Department of Agriculture (USDA), recognized the point that innate characteristics of Hawaiʻi seem to make quarantine there more difficult: "[For insects and mites] in the period 1942–72 the rate of colonization per thousand square miles was 40 species, 500 times the rate of continental United States," in spite of a larger quarantine force in relation to the volume of commerce. McGregor (1973: 19) speculated on the reason: "Although there is much greater diversity of crops and habi-

tats within the continental United States, these are dispersed over a vastly larger land area. In Hawaii . . . the various habitats are more readily accessible from the principal port of entry. The more moderate and stable climate is also more favorable for an invading species than is the climate over much of the United States."

The Hawaiʻi Department of Agriculture (HDOA) began reviewing all animal and micro-organism import requests through a subcommittee in 1965. (The complex mix of governmental and nongovernmental organizations involved in alien species prevention in Hawaiʻi can be difficult to comprehend. To help the reader, we have listed most of the organizations in Table 15.1, along with their abbreviations.) Changes in state law in 1975 formalized

Table 15.1. List of agency abbreviations used in text, corresponding agency names, and their roles in invasive species prevention

Abbreviation[a]	Agency	Role
Federal Agencies		
DHS	Department of Homeland Security	Inherited most of the APHIS-PPQ role of intercepting listed pests at the national borders as of March 2003
FAA	Federal Aviation Administration	Regulates the nation's airports and aircraft
GAO	U.S. Government Accountability Office (formerly General Accounting Office)	An arm of the U.S. Congress investigating federal programs and expenditures
OTA	Congressional Office of Technology Assessment	The now defunct research and reporting arm of the U.S. Congress investigating technical issues
USDA	U.S. Department of Agriculture	Promotes U.S. agricultural products, manages the nation's national forests and rangelands; until recently provided border inspection to exclude listed pests
APHIS-PPQ	Animal and Plant Health Inspection Service, Plant Pest Quarantine Branch	An arm of the USDA responsible until recently for inspecting imported goods for targeted pest species
USFWS	U.S. Fish and Wildlife Service	Responsible for exclusion of listed invasive animals, protection of endangered species, and management of wildlife refuges
USGS	U.S. Geological Survey	Provides scientific research on the nation's natural resources
HEAR	Hawaiian Ecosystems at Risk Project	A web-based project providing online information on invasive species in Hawaiʻi

continued

Table 15.1. Continued

Abbreviation[a]	Agency	Role
State Agencies		
BOA	Board of Agriculture	Sets policy for HDOA; oversees petitions for deliberate introductions of alien plants and animals to Hawai'i
HISC	Hawai'i Invasive Species Council	A government body with the authority to provide funding and set state policy for invasive species
HDOA	Hawai'i Department of Agriculture	Manages Hawaii's agricultural resources and importation rules for alien organisms
PQ	Plant Quarantine Branch	Regulates alien organisms, inspects for targeted pests from mainland and foreign sources, and certifies some exported materials
HDOT	Hawai'i Department of Transportation	Regulates Hawaii's ports and highways
Private and Quasi-governmental Agencies		
CGAPS	Coordinating Group on Alien Pest Species	A statewide coalition of government and private agencies designed to improve the government's response to invasive species threats in Hawai'i
ISC	Invasive Species Committee (generic term for the following groups)	An island-specific coalition of government and private agencies to provide rapid-response programs to eradicate or contain newly invasive species prior to their irreversible establishment
BIISC	Big Island Invasive Species Committee	
KISC	Kaua'i Invasive Species Committee	
MISC	Maui Invasive Species Committee	
MoMISC	Moloka'i Invasive Species Committee	
OISC	O'ahu Invasive Species Committee	
MAC	Melastome Action Committee	An informal coalition of agencies active from 1991 to 1997 to stop the spread of miconia and other invasive melastomes in Hawai'i
NRDC	Natural Resources Defense Council	A national conservation organization with shared responsibility for an early report warning of the danger of invasive species in Hawai'i
TNCH	The Nature Conservancy of Hawai'i	A national conservation organization long active in promoting programs to rectify the invasive species problem in Hawai'i
International Organizations		
WTO	World Trade Organization	An international body setting rules for trade between nations, including rules regarding living organisms
Foreign Agencies		
MAF	New Zealand Ministry of Agriculture and Forestry	Responsible for executing New Zealand's biosecurity programs
MQS	MAF Quarantine Service	Regulates importation of alien organisms, inspects arriving goods, and controls or eradicates targeted invasive pests

[a]Agency abbreviations indented from the left are branches or programs of the agency immediately above.

and added to this process by requiring that import requests be reviewed by a Plants and Animals Committee and by the Board of Agriculture to assess potential impacts on agriculture, human health, and the environment. In the 1980s, the profound worldwide effects of transport and establishment of alien species were beginning to be recognized (Mooney and Drake 1986, Drake et al. 1989). In 1992, The Nature Conservancy of Hawai'i (TNCH) and the Natural Resources Defense Council (NRDC) produced a comprehensive report on Hawaii's alien-species problem, "The Alien Pest Species Invasion in Hawai'i: Background Study and Recommendations for Interagency Planning" (hereafter, TNCH/NRDC). Among their conclusions (p. ii) was this: "The silent invasion of Hawai'i by pest species—weeds, disease organisms, predators, insects, etc.—has far-reaching consequences for the State's people, economy and natural environment." Although motivated by concern for biodiversity loss, the report effectively employed the concept of "mainstreaming" alien species prevention—recognizing the current and future damage caused by invasive pests to agriculture, the economy, human health, and the quality of life.

Alien species arrive through a variety of pathways that may be grouped into two major categories: intentional and unintentional introductions. Examples of the former include releases from the pet trade (legal or illegal), garden escapes, and biocontrol organisms; examples of the latter include stowaways in cargo and equipment, contamination of nursery stock, and seed contaminants. The importance of the pathway varies by taxon (Ruiz and Carlton 2003). Some groups, such as mammals, birds, reptiles, amphibians, and vascular plants, are primarily introduced purposely because someone perceives an amenity or use value for the species (for a review of pathways of bird introductions, see Kraus [2003] and this volume, Chapter 13). Others, such as the BTS, insects, pathogens,

and some weeds, are usually unintentional and unwanted introductions. This primary distinction is important, because efforts to address invasive species problems must recognize multiple pathways, and a comprehensive program requires that all important pathways be addressed.

CURRENT POLICY AND PRACTICE IN THE UNITED STATES AND IN HAWAI'I

Prevention: Federal Efforts

Federal agencies are charged with preventing the arrival of alien pests from foreign countries. Until very recently, most of the responsibility to prevent unintentional introductions has rested with the U.S. Department of Agriculture, Animal and Plant Health Inspection Service, Plant Protection Quarantine (USDA-APHIS-PPQ). However, on March 1, 2003, APHIS-PPQ's inspection operations at all U.S. international ports of entry merged with the operations of other agencies into the Department of Homeland Security (DHS). APHIS-PPQ retains the responsibility for preventing pests established in Hawai'i and Puerto Rico from reaching the U.S. mainland. Its work is guided by federal laws and rules aimed at protecting major U.S. agriculture with minimum interference to trade (Baskin 2002, Campbell and Schlarbaum 2002). The mandate of APHIS-PPQ (and now DHS as well) has never explicitly addressed species capable of harming natural ecosystems or those introduced deliberately (Congressional Office of Technology Assessment 1993). Consequently, the species covered by APHIS-PPQ quarantine measures are primarily invertebrate and pathogen pests of crops and livestock and important crop weeds and their seeds. Efforts to exclude these species consist primarily of a complex set of laws and regulations in concert with port-of-entry inspections of a sample of arriving goods and cargo deemed at high risk of harboring pests. Given the tremendous volume of trade arriving from overseas, inspectors are in practice faced with sampling no more than a small fraction of total arriving goods and cargo (National Research Council 2002).

USDA-APHIS-PPQ has a mandate to protect the U.S. mainland from introduction of certain pests present in Hawai'i using point-of-departure inspections for passengers and their possessions. This mandate began with the Plant Quarantine Act of 1912 and is now codified in the federal Plant Protection Act of 2000 (Public Law 106-224, Title IV) and the U.S. Code of Federal Regulations (CFR), Chapter 7, Sections 301 and 318. There is no reciprocal mandate for protecting Hawai'i from undesirable species from the U.S. mainland.

DHS has a mandate or policy at the Port of Honolulu virtually identical to that at other international ports in the United States. Preventive actions are taken based on an official list of prohibited organisms for which specific legal authority is deemed to exist. In practice, the actionable list has little to do with organisms that would affect Hawaii's agriculture or native biota. An exception involves a special USDA quarantine action policy for ants moving into or through the state of Hawai'i. This policy, first instituted in 2002, places any ant species not already present in Hawai'i (and those present for which active control efforts are under way) on an "actionable list" of species targeted by DHS or USDA inspectors (Krushelnycky et al. 2005).

With respect to intentional introductions, the USFWS bans the introduction without permit of 11 genera and one species of mammal, four species of bird, one genus of fish, and the BTS (50 CFR 16.11–16.15). It also prohibits release into the wild of any terrestrial vertebrate unless the release is conducted or approved by a state wildlife conservation agency. Federal public health regulations (42 CFR 71.52–71.53) further prohibit importation without a permit of primates as pets, turtles smaller than four inches or their eggs, and bats.

To better coordinate federal efforts and identify and correct weaknesses at the national level, identified in part by a Congressional Office of Technology Assessment study (Congressional Office of Technology Assessment 1993), in February 1999 President Bill Clinton issued Executive Order 13112, "Invasive Species," which required a National Invasive Species Council to produce a National Management Plan (NMP) for Invasive Species. In January 2001, the council issued the first NMP (www.invasivespecies.gov/council/nmp.shtml) to focus attention on invasive species and coordinate a national control effort involving the 36–40 federal agencies that are responsible for various aspects of managing them.

Still, as of February 2009, federal border protection against invasive species at Hawaiian ports of entry appeared not to have improved significantly and may have declined. For example, no improvement was evident in addressing many problems the National Plant Board had recognized in 1999. As the board reported then: "The APHIS-PPQ risk assessment process does not adequately describe the uncertainty present in the process, nor does it list assumptions on which subjective judgments or models are based. Many aspects of the process are overly simplistic" (National Plant Board 1999: x). The National Plant Board (1999: 51–54) had raised special concern over the inadequacy of APHIS's handling of risks involved with nursery stock. The U.S. Department of Agriculture, Animal and Plant Health Inspection Service (2004), presented its alarming dilemma involving inadequate safeguards for preventing introduction of plant pests via nursery stock and expressed concern (p. 71739) over the fact that its "ability to quickly apply new scientific research and information has been hampered by ... lack of resources." In 2005, the U.S. Government Accountability Office (2005b: 7) reported that on a nationwide basis, since DHS took over border inspections, "agricultural inspections at ports of entry . . . have declined over the past two years at a time when imports have increased."

Prevention: State Efforts

The Hawai'i Department of Agriculture, Plant Quarantine Branch (HDOA-PQ), is the state entity mandated to regulate importation and movement of all plants, nondomestic animals (vertebrate and invertebrate), and micro-organisms to prevent introduction of harmful insects, plant diseases, illegal animals, and other pests into Hawai'i. However, its authority extends primarily to materials coming from the continental United States. For protection from organisms in materials entering Hawai'i from international source areas, it relies on referrals from DHS, Customs and Border Protection, Agriculture Quarantine Inspection (formerly USDA-APHIS-PPQ and U.S. Customs), and the law enforcement branch of USFWS. Federal "preemption," codified by the Plant Protection Act of 2000 (Public Law 106-224), restricts Hawaii's latitude in defending its borders by negating Hawaiian quarantine rules for some invasive species and deterring HDOA from promulgating rules for others (Fox 2005).

Broad operating policies within HDOA and evolving lists of organisms approved or prohibited for import are set by a 10-member, governor-appointed Board of Agriculture (BOA) (Hawai'i Revised Statutes [HRS], Chapters 142 and 150A; Hawaii Administrative Rules [HAR], Title 4, Chapters 70–72). Legislation directs the BOA to maintain three lists for animals: "conditionally approved" (permit required for importation), "restricted" (permit required for both importation and possession), and "prohibited." Any animal not on the first two lists is also prohibited by law but may be added to either list by petition to the BOA. The BOA also maintains lists for micro-organisms. Whenever the BOA proposes a list revision, it must pro-vide public notice, a hearing, and a comment period before adoption. In practice, HDOA bans most animal species except for a relatively small number (many hundreds) allowed for pet trade sale, research, public exhibition, and private pet ownership. Potential penalties for violating prohibitions on animal importation are substantial (up to $200,000 in fines and three years' imprisonment), but to date such penalties have rarely been imposed (Kraus and Cravalho 2001).

The Hawai'i Legislature mandates that the BOA maintain a list of restricted plant species that may be imported by permit only (HRS 150A; HAR 4-70). Three plant species, giant salvinia (*Salvinia molesta*), dwarf salvinia (*S. minima*), and water lettuce (*Pistia stratiotes*), are prohibited. A few other species may be imported only under certain conditions. All other plant species are allowed into the state.

Individuals requesting the import of any animal species must file an application with HDOA-PQ. If an applicant requests the import of an animal not on the conditionally approved or restricted list, the species must be added to the appropriate list by the BOA before it may be imported. When such a request is made, a technical review is conducted within one of the BOA's five subcommittees (Land Vertebrates, Invertebrates and Aquatic Biota, Entomology, Micro-organisms, and Plants) comprised of researchers and other technical experts in these fields. A second review by an advisory committee examines potential negative impacts versus potential benefits. The BOA makes the final decision. When a permitted organism is brought into Hawai'i, it is verified at the port of entry. In some cases, payment of a bond may be required for entry. Postentry inspections to monitor certain importations may be required.

Passengers and crew of arriving aircraft and ships are asked to complete a mandatory HDOA Declaration Form and submit any imported agricultural items for inspection. In practice, this is largely an honor system, but HDOA does have a few trained beagle dogs to inspect checked baggage and hand-carried items. Potential penalties range from $100 to $200,000 in fines (HRS 150A-14). An amnesty provision exempts from penalty persons who voluntarily surrender, prior to the beginning of an investigation or seizure action, a prohibited animal or a restricted animal for which they have no permit. Nonetheless, illegal animals have been found loose in cabins of arriving airplanes, and smuggling of animals through airports is suspected of being an important pathway of invasive pest introduction (Box 15.3).

Shippers of domestic cargo must notify HDOA of arriving goods requiring inspection. Low staffing levels have made it impossible to inspect all cargo, but shipping manifests provide the ability to prioritize inspections based on the perceived risk of the imported commodity. Intensive HDOA sampling of air cargo at Kahului Airport, Maui, during 2000–2001 provided information on the risk levels of various im-

---

Box 15.3. **Lack of Credible Deterrence**

Loope et al. (2001) raised the consideration that it is not only the BTS that threatens Pacific birds; several hundred of the world's snake species, some of which are repeatedly smuggled as pets, might have similar impacts on native birds if transported to Pacific islands. Hawai'i has the most restrictive laws in the country to exclude invasive animals; all but a few hundred species of vertebrates are prohibited legal entry (HRS 150A-6.2, HAR 4-71-6). Notwithstanding these laws, a large number of alien animals, mostly vertebrates, are smuggled into the state each year. Many are lizards and snakes, but animals as large as cougars (*Puma concolor*) and gray wolves (*Canis lupus*) and as diverse as centipedes, tarantulas, piranhas, and pythons have been confiscated from local homes (D. Cravalho, HDOA, pers. comm.). This activity has contributed greatly to the large increase in reptiles established in Hawai'i since the 1940s (Kraus 2002).

Kraus and Cravalho (2001) provided an analysis of specific gaps in inspection and law enforcement programs that allow the smuggling of animals into Hawai'i. They noted that smuggling by passengers is easy because port-of-entry inspection relies upon voluntary declaration, a system sidestepped by smugglers. Inspectors cannot legally search passengers, mail, or most baggage, nor can they arrest violators if they are discovered. Furthermore, enforcement against owners of illegal animals will remain difficult as long as the numbers of enforcement personnel and specific training in the ramifications of invasive species importation are limited. Likewise, without vigorous prosecution of such cases, enforcement efforts will remain constrained by lack of credible deterrence.

Kraus and Cravalho (2001) noted that effective reduction of animal smuggling would require increasing the authority of inspectors and establishing a single point of exit at airports. There, passengers and their baggage would be screened by detector dogs and x-ray machines, providing HDOA inspectors with the "probable cause" necessary to stop and search persons entering the state or traveling between islands. Providing HDOA with trained enforcement personnel to pursue reported smuggling violations would also be required. To date, none of these improvements has been made.

ported commodities (Hawai'i Department of Agriculture 2002a).

HDOA has the authority to conduct inspections of goods shipped between islands, but due to limited resources, inspections have focused on the highest-risk pathways, such as shipments of nursery stock. When an invasive species establishes on one island, it commonly spreads to other islands (Wagner, Herbst, et al. 1999).

First-class mail is a major, highly problematic pathway. Federal laws have traditionally protected domestic first-class private and commercial mail against "unreasonable searches" (Congressional Office of Technology Assessment 1993). In 1990, however, the USDA and the U.S. Postal Service began a trial program in Hawai'i using trained dogs to detect outgoing packages (directed to the U.S. mainland) containing illegal agricultural products. Detection by dogs is deemed adequate evidence on which to obtain a warrant to open a package to determine whether the contents are illegal. The program continues and can be effective but is cumbersome, requiring a trip to a judicial officer (miles away through Honolulu traffic) each time a dog detects something suspicious in a package. In the early 1990s, Congress passed a law specific to Hawai'i that allows similar inspection of mail entering Hawai'i (Congressional Office of Technology Assessment 1993). The protocol is for the U.S. postal inspector to take packages deemed suspect into a room where the state inspector and dog check it. If the dog shows a positive response, the state may apply for a search warrant through the USDA to have it opened. This must occur within 24 hours or the package is released (N. Reimer, HDOA, pers. comm.). HDOA has not pursued this legal option but has, beginning in 2002, worked out a cooperative agreement with the U.S. Postal Service for an embargo on the shipping of birds to Hawai'i via the U.S. mail as a measure to prevent West Nile virus from reaching the state (Hawai'i Department of Agriculture 2002b).

HDOA's Animal Industry Division has an Inspection and Quarantine Branch that focuses on arriving cats, dogs, and other carnivores with an emphasis on keeping Hawai'i free of the rabies virus. It also has a Livestock Disease Control Branch that focuses on inspecting arriving livestock for diseases.

RAPID RESPONSE

HDOA has the authority (HRS 141, 152; HAR 68, 69A) to mount eradication efforts for plants and a very few animal species, including the right to enter private property after properly notifying the property owner. HDOA must demonstrate reasonable cause (an obvious infestation) to enter the property. There are enough caveats in interacting with landowners to make eradications very difficult, although strong public support, when it exists, helps officials in gaining access to private property. Response to new infestations is frequently delayed by inadequate staffing, conflicting jurisdictions, or lack of public cooperation (Kraus and Campbell 2002), sometimes allowing pests to establish and spread beyond control even when a threat is widely recognized (Box 15.4).

CONTROL

Hawai'i has a "Noxious Weed Control" law that establishes criteria and procedures for designation of plants as state noxious weeds and for cooperative agreements with landowners to facilitate eradication or control. In practice, most mechanical and chemical invasive plant control is done for species targeted by the staff of island invasive species committees (see below) or in natural areas by natural area managers.

Biological control, ideally involving state-federal cooperation, is also a potentially important tool in limiting invasions (Chapter 16). Largely as a result of reaction to

Box 15.4. **The Dangers of Delayed Response**

Miconia

More than $2 million is being spent each year in Hawai'i ($1 million on Maui alone) for containment of miconia (*Miconia calvescens*) (Fig. 15.4), a tree species brought to the islands in about 1960. In Tahiti, which is ecologically similar to Hawai'i, miconia was introduced (two plants) in 1937 and had spread to cover two-thirds of upland forests of that island by the 1980s, to elevations up to ca. 1,200 m (Meyer 1996). The tree has very large leaves (1 m in length), and light levels reaching the forest floor under its canopy are very low. It thus shades out all other species of plants and their associated animal communities. In Tahiti, 40–50 of the 107 plant species endemic to the island are believed to be on the verge of extinction primarily because of miconia invasion (Meyer and Florence 1996). The threat of miconia to watersheds in Hawai'i is also widely recognized (Plate 29); if miconia were to reach dominance, erosion and landslides will be likely on steep slopes with very high rainfall (up to 8,000+ mm annually) because of miconia's shallow root system.

Miconia was recognized in 1971 by Dr. Raymond Fosberg (pers. comm.) of the Smithsonian Institution as "the one plant that could destroy Hawaii's forests." Fosberg informed colleagues in Hawai'i of this threat, but no state or federal agency took action. By the 1970s, miconia was well known to have been planted on Hawai'i Island, and individuals and organizations mounted sporadic, local "eradication" campaigns; however, no control effort was sustained in Hawai'i until miconia was discovered 8 km from Haleakalā National Park in 1990. Individuals and agencies on Maui conducted volunteer efforts to remove miconia (Hur-

Figure 15.4. Young plants of miconia, a highly invasive tree from the Neotropics. *Source:* Photo © Jack Jeffrey.

ley 1991, Gagné et al. 1992). The Melastome Action Committee (MAC), formed in August 1991, raised the issue until public funding support was forthcoming to initiate action.

Significant public funding for miconia control in Hawai'i was first received from Maui County in 1995 and soon after from the State via the East Maui Watershed Partnership (Conant et al. 1997, Medeiros et al. 1997). After 1997, MAC's effort was continued by the Maui Invasive Species Committee (MISC). Despite persistent efforts by MISC and considerable funding from Maui County, funding was always chronically inadequate (though increasing) to contain miconia because each fruiting tree can annually produce millions of bird-dispersed seeds, making population growth rapid. During 2002–

Box 15.4.   *Continued*

2005, Haleakalā National Park provided substantial assistance to MISC in containing miconia on Maui, recognizing the threat of this invasive tree to the park's rain forests of Kīpahulu Valley, arguably the most biologically diverse and intact rain forest in the United States. When federal assistance was reduced in 2006, Maui County stepped in with additional funding necessary to maintain the intensive level of surveillance and removal effort required. For the present, containment efforts for miconia on Maui, Oʻahu, and Kauaʻi are effective, though costly; the prognosis for Hawaiʻi Island may be much less positive. Efforts for biological control of miconia are under way; potential for positive results is good but still uncertain (Chapter 16).

Coqui

In 1997, it was discovered that two species of small, direct-developing (no tadpole stage) Caribbean frogs of the genus *Eleutherodactylus* were established in Hawaiʻi (Kraus et al. 1999). The coqui (*E. coqui*) (Fig. 15.5) is especially noisy and has captured the public's attention. These frogs hitch-hiked into the state in potted nursery plants in the late 1980s or early 1990s and have since established hundreds of populations via dispersal of ornamental plants from infested nurseries (Kraus and Campbell 2002). Populations are established on the four largest islands but are concentrated on Hawaiʻi Island, with an estimated 2,000 ha infested with hundreds of thousands or millions of frogs in the Hilo area and Puna District alone (M. Wilkinson, DLNR, pers. comm.). In its native habitat in Puerto Rican rainforest, the coqui occurs at densities of over 20,000 animals per hectare and removes an average of 114,000 prey items per hectare per night; the species has been shown capable of exceeding this density threefold in Hawaiʻi (Woolbright et al. 2006). A variety of detrimental ecological and societal effects were predicted to occur if the frogs were allowed to spread (Kraus et al. 1999, Kraus and Campbell 2002), and

Figure 15.5. A coqui frog in Hawaiʻi. *Source:* Photo © Jack Jeffrey.

continued

Box 15.4.   *Continued*

the growth of these problems has since been documented (Kraus and Campbell 2002, Beamish 2004, Kaiser and Burnett 2006).

To effectively control the coqui, it was imperative that action be taken early in the invasion process (within 2–3 years), because the frogs' rapid intrinsic growth rate (Townsend and Stewart 1994, Kraus et al. 1999) provided only a narrow window of opportunity before populations exploded beyond human capability for containment. Instead, unhindered spread of frogs in contaminated nursery materials and deliberate spread of frogs by private individuals made the problem grow exponentially. Meanwhile, effective government response was delayed beyond the point at which hundreds of populations were established (Kraus 2007, Fig. 2). As a result, costs to now control the vast populations on Hawai'i Island are prohibitively expensive. It is conservatively estimated that, on average, $5,000 per infested acre per year is needed for frog eradication (M. Wilkinson, DLNR, pers. comm.). Thus, control costs in the Hilo-Puna region alone would likely total more than $25 million per year. This level of monetary support is not politically attainable and, clearly, control efforts came too late to be more than a palliative for most infestations.

In contrast, Kaua'i and O'ahu have had only a few established populations, and efforts to eradicate these have proven promising, though expensive. The effort to eradicate the single nine-acre population at Wahiawa, O'ahu, consisting of only a few hundred calling individuals, cost more than $400,000 from June 2003 through February 2006 (M. Wilkinson, DLNR, pers. comm.), approximately three times the average cost estimate provided earlier. That effort is nearly complete, but it seems feasible that populations on Kaua'i and O'ahu can be eradicated and that these islands (and Moloka'i and Lāna'i) can be kept frog-free if effective inter-island quarantine measures are in place to prevent reinfestation from Hawai'i Island nurseries. As of early 2007, such measures remained lacking, though pending. In summary, a serious pest that could have been eradicated in the late 1990s was out of control by the early 2000s, requiring the expenditure of millions of dollars for control, with no feasible hope for eradication of the large infestations on Hawai'i Island.

past indiscretions in the practice of biocontrol, Hawai'i has been a battleground of advocates and opponents of biological control; local policy concerning release of biocontrol agents has evolved to become more restrictive than probably anywhere else in the world (Lockwood et al. 2001). Nevertheless, almost everyone agrees that biocontrol is desperately needed for certain insect threats to biodiversity and for the most damaging weeds (for example, the EGW; see Box 15.2). Pemberton (2002) argued that safe biocontrol of weeds can be successfully achieved but that, given the existing funding constraints, limited quarantine space, and low number of biocontrol researchers, only a small portion of invasive weeds can be subjected to well-managed biological control programs.

LIMITATIONS TO EFFECTIVE PREVENTION

In general terms, the entire United States suffers severely from overarching structural limitations. First, there is no comprehensive approach to prevention or detection of invasive alien species and timely response. The current patchwork of programs is reactive rather than proactive and focused on a small array of primarily agricultural pests. Their adequacy even for protecting agriculture is questionable (National Plant Board 1999). This limitation is important to note because the national and international border protection quarantine systems were designed by and for agricultural interests, and natural areas suffer correspondingly (Van Driesche and Van Driesche 2000).

Second, deliberate introductions via the horticulture, agriculture, and pet industries are rarely addressed. Roughly 20,000 species of vascular plants have proved invasive somewhere in the world (Randall 2002, www.hear.org/gcw), but U.S. federal noxious weed control law currently prohibits only 91 species and five genera, most of which are well-documented threats to agriculture. Even so, some of the listed species have been spread daily in commerce, usually without challenge, including the giant salvinia that had been known since 1999 to be illegally traded in Hawai'i. In 2003, this species covered Lake Wilson on O'ahu, provoking a concerted and expensive response and resulting in abundant press coverage and attention from the public and politicians. Most invasive plants in Hawai'i were imported deliberately. These were introduced via botanical gardens and arboreta, nurseries, garden club and horticultural society seed exchanges, the seed trade industry, and government agencies (Wester 1992, Reichard and White 2001, Baskin 2002). Among woody invaders in the United States, 82% have a background as landscape plants (Reichard and Hamilton 1997). The pet trade is similarly a major and increasing source of invasive animals (Devick 1991; Kraus 2003, 2008). Sale using the Internet provides an efficient mechanism for disseminating species from diverse parts of the world. This problem is compounded by individuals' intentional release of alien pets to establish wild populations (Kraus and Cravalho 2001; Kraus 2003, 2008).

Unfortunately, Hawai'i continues to receive some of its most damaging pests unexpectedly from foreign countries, not only without risk assessments but without warning. Normally, the species is discovered upon its establishment, and the pathway is rarely known. Even if intercepted, most of these newly arrived pests would not have been considered quarantine pests by the U.S. system of border protection. Furthermore, once a damaging quarantine pest from foreign shores establishes in another part of the United States, it is no longer considered a quarantine pest by federal agencies.

IS IT POSSIBLE TO BUILD EFFECTIVE PREVENTION FOR HAWAI'I?

The challenges we have enumerated are formidable, but for over 15 years there has been sustained effort by individuals in Hawaii's conservation community to work with responsible institutions to limit further introduction and spread of invasive species. The TNCH/NRDC report (1992: v) called for an interagency planning effort to develop a comprehensive pest prevention and control system to close the many "gaps and leaks for pest entry and establishment." It recognized the need for strong public support through public education and recognition that the strategy for prevention and control is "compelling and practical." Finally, the report (p. v) suggested that "to succeed, [the planning process] should be guided by a simple, clear policy statement identifying the standard of excellence that Hawai'i aspires to in this field (e.g., 'Hawai'i will develop a pest prevention and control system that is the most effective in the world,' or '. . . that reduces the influx of new pest species into the State to ten percent of present levels by the year 2000')." This has not happened, and given the scope of the problem and challenges to progress, one might wonder whether a comprehensive and effective prevention system is even possible.

Averting our gaze for a moment from Hawai'i and the United States, the answer would appear to be "Yes." Good models for such a venture are available in the largely successful preventive systems in place in Australia and New Zealand. In Australia, a major review (the 1996 "Nairn Review") carried out by an independent team of experts commissioned by the Australian government in response to serious quarantine breaches in the early 1990s was crucial in establishing a more comprehensive focus, with emphasis on protecting the total environment, including but not limited to agriculture (Tanner and Nunn 1998, Hollingsworth and Loope 2007). In 2000, Australia set up "Biosecurity Australia" within its Ministry of Agriculture, Fisheries and Food, charged with the development of biosecurity policy and provision of technical assistance to the country's export market separate from the operational work of the Australian Quarantine and Inspection Service (www.affa.gov.au/ biosecurityaustralia and www.aqis.gov.au). Responsibility for biosecurity policy is now shared with the Ministry of Environment and Heritage to ensure that environment and biodiversity issues are appropriately addressed in biosecurity policy (www.environment.gov.au/biodiversity). Australia and some of its states have a politically and ecologically successful system for screening proposed plant introductions that has been used to evaluate thousands of plant species since its inception, with minimal challenge from affected industries (Walton et al. 1999; Baskin 2002; Randall 2002; R. Randall, Western Australia Department of Agriculture, pers. comm.). New Zealand has adopted a similar system.

The island nation of New Zealand provides an inspiring model for Hawai'i by illustrating what is possible (Williams 2000, Baskin 2002, Loope 2004). New Zealand has problems with biological invasions that rival those of Hawai'i but currently exhibits remarkable determination to reverse trends of ecological degradation through restoration (Veitch and Clout 2002). It simultaneously attempts to prevent continuing invasions with a strong border protection quarantine system. Border protection quarantine and surveillance have good public, legislative, and financial support (Baskin 2002). A striking current observation of visitors to New Zealand has to do with the relatively high level of public "quarantine literacy" and support for stringent quarantine measures (Hollingsworth and Loope 2007).

Based on the experience in New Zealand, Warren (2003) prepared a concise document titled "Elements of an Ideal Biosecurity System for Biodiversity Protection Purposes" (www.hear.org/articles/warren 200305). As part of her consultancy to the Galápagos National Park, she provides an excellent checklist of considerations beyond recommendations for potentially improving biosecurity in Hawai'i.

COLLABORATIVE EFFORTS TOWARD MOVING FORWARD

Whereas Hawai'i has not yet been able to attain the aspirations of the 1992 TNCH/ NRDC report, important progress within the state has ensued since that time.

The Coordinating Group on Alien Pest Species (CGAPS) and Hawai'i Invasive Species Council (HISC)

An interagency planning process following the publication of the 1992 TNCH/ NRDC report and led by TNCH resulted in the formation in January 1995 of CGAPS, a statewide coalition of government agencies and private organizations. The intention was to encourage the responsible agencies to work together to rectify the protection gaps identified by the report. At a major United Nations meeting on alien species held in Norway in July 1996, Holt (1996: 157–158), then chairman of CGAPS, reported on the first 18 months of progress by CGAPS by stating that it faced two significant challenges: "First, the launch of CGAPS coincided with the sharpest cutbacks in government budgets since statehood. This . . . left key members with insufficient funding and personnel to pursue the desired alien species management actions. Second, many of the individuals sitting on CGAPS as agency representatives are unable to make major commitments for their agency. CGAPS can develop excellent strategy and resolve problems that require little new funding and no major legislative work. Major improvements, however, require political leadership of the highest level, and this depends on widespread public support." In late 1996, the need for widespread public support led to a major public relations campaign highlighted by a booklet in glossy color titled "The Silent Invasion" (Coordinating Group on Alien Pest Species 1996).

The "Silent Invasion" booklet was well received and likely had a strong influence in educating individuals within the Hawaiian conservation community and in the media. The broader public's education came more slowly, but news reports on the most visible newly developing invasions have gradually achieved awareness by keeping invasive species issues in the public eye. The state of Hawai'i budget progressively declined through the 2003 session, and state agencies lost rather than gained funding and positions. Problems identified by Holt (1996) largely persist, but the budget situation improved beginning in 2004, accompanied by positive trends toward progress that continued until the threat of serious economic disruptions became apparent in 2008.

CGAPS has been effective in providing some coordination of effort among agencies and has pursued constructive efforts such as weed risk assessment (Daehler et al.

2004). CGAPS has diligently attempted to identify and support the programs, legislation, funding, and public buy-in required to mitigate the invasive species problem in the state. Hawai'i Act 085, passed by the 2003 Hawai'i Legislature, established the HISC and directed state agency chairs and department heads to address gaps in Hawaii's invasive species prevention and response measures. CGAPS, with a decade-long track record of deliberation and achievement, serves as an effective advisory group for the HISC, an important aspect of which has been annual funding of $2–4 million to provide seed monies for innovative work in prevention, response and control, research and technology, and public outreach.

Invasive Species Committees (ISCs)

In 1991, agencies and individuals on the island of Maui formed an interagency working group, the Melastome Action Committee, to combat invasion by the weed tree miconia (*Miconia calvescens*), which threatened Maui's watersheds. In 1997, this group decided to expand its scope and formed the Maui Invasive Species Committee (MISC), whose purpose was to eradicate or at least contain incipient populations of invasive species before they spread as far as miconia had. MISC partners included a variety of federal, state, county, and private entities. In 1998, MISC developed a plan that established categories (exclusion, eradication, containment, large-scale management) and set priorities and responsibilities for pest management. In 1999–2000, an action plan focused on eradication and containment of invasive plants was launched (funded by $700,000 raised from federal, state, county, and private sources) to employ a crew to combat top-priority species.

MISC's efforts inspired the formation of similar coalitions on other islands. Currently, there are ISCs not only on Maui but on Kaua'i (KISC), O'ahu (OISC), Moloka'i

(MoMISC), and the Big Island (BIISC). Increasingly, funding requests for the ISCs have been coordinated under the umbrella first of CGAPS and later of the HISC. These committees serve as successful models of local cooperation to address one aspect of the invasive species threat in Hawai'i. The ISCs have been successful in implementing some level of containment, and occasionally eradication, at the individual island level by reducing many populations of an array of invasive species, mostly plants. The efforts to date cannot be viewed as comprehensive because of the large standing crop of incipient invasive species in Hawai'i and the continual arrival of new invasions, but CGAPS-HISC and the ISCs are starting to develop and implement meaningful visions, including moving toward an increasingly comprehensive program of early detection of invasive plant targets (Loope et al. 2004). By 2008, MISC's total annual funding had grown to nearly $2 million per year, with half coming from Maui County, but half of the $2 million is devoted to containment of miconia (see Box 15.4).

Kahului Airport Record of Decision, Mitigation, and Progress

Arguably, the Kahului Airport expansion over the past two decades has had the effect of providing significant impetus for better quarantine of Hawai'i (Fox and Loope 2007: Table 1). Donald Reeser, Haleakalā Natonal Park (NP) superintendent, repeatedly articulated the need for vastly improved quarantine measures if federal dollars were to be used to expand and internationalize the Kahului Airport on Maui (Reeser 2002). Reeser's argument centered on the relative biological intactness of Haleakalā NP and the island of Maui, the high susceptibility of the island of Maui to invasions, and the feasibility of preventing invasions through better quarantine measures. In early 1998, the U.S. Department of the Interior asked the President's Council on Environmental Quality to coordi-

nate interdepartmental discussions to determine how to address this issue (Reeser 2002). A federal-state "Memorandum of Understanding regarding the Prevention of Alien Species Introduction through Kahului Airport" was signed by three federal departments and four state agencies and became part of the Record of Decision approving the final Environmental Impact Statement for expansion of Kahului Airport. In the decade since the agreement was signed in 1998, more than 20 meetings have been held to discuss potential improvement measures to prevent biological invasions.

The Federal Aviation Administration funded a risk assessment for passengers and cargo at Kahului Airport that resulted in a series of recommendations by HDOA to improve its quarantine program at the airport (Hawai'i Department of Agriculture 2002a). The Hawai'i Department of Transportation completed construction of an enclosed air cargo facility, partially funded with federal funds, in January 2008. The report of the Kahului Airport Pest Risk Assessment (Hawai'i Department of Agriculture 2002a), based on results of inspection blitzes in 2000–2001, has provided an important portrait of the rate of arrival of new species, the relative importance of pathways, and strategies for statewide prevention. Remarkable for its transparency and use of interception data to support its recommendations, this report may ultimately serve as a watershed development or "tipping point" toward upgrading HDOA's ability to protect Hawai'i from harmful new invasions (Hollingsworth and Loope 2007).

The BTS Control Committee

Perhaps the most successful example of an integrated containment, prevention, rapid response, and research initiative protecting Hawai'i and other Pacific islands is the interagency BTS prevention program based on Guam and its supportive research program based in Fort Collins, Colorado. Nine forest bird and three seabird species present on Guam in 1945 were eliminated by the alien BTS, which reached Guam soon after the close of World War II (Savidge 1987, Fritts and Rodda 1998) (see Box 15.1.). Concern about the ecological and economic effects that would result if the snake were to spread to other islands from Guam surfaced in the mid-1980s, eventually resulting in the creation of this program. The program consists of comprehensive (though voluntary for commercial shippers) inspection on Guam of outbound cargo and vessels and reduction of snake populations in port areas (Brown Treesnake Control Committee 1996). Since its implementation in 1995, the incidence of BTS export to other jurisdictions, including Hawai'i, has declined dramatically (Colvin et al. 2005). This program could serve as a model for other species-specific prevention programs throughout the Pacific. Despite a lack of base funding, the track record of this program indicates that comprehensive, coordinated responses to specific, high-impact invasive species threats can be successful.

The Hawai'i Ecosystems at Risk (HEAR) Project

In early 1994, responding to an initiative by the U.S. Department of the Interior, a group of scientists and managers in Hawai'i identified the need for an information clearinghouse on invasive species to serve land managers, partnerships, and the public. This initiative began in 1996, with the U.S. Geological Survey (USGS) and the University of Hawai'i as coordinators. It came to be known as the HEAR project and is best known for its Web site, www.hear.org. A midcourse review of this project (Van Driesche 2002: 39) praised its grassroots approach: "The purpose of an information sharing network is to support and enhance the conservation community as it exists—not to replace it with some ideal of how it

ought to be." HEAR has been a means of providing public information on Hawaii's invasive species problems and programs and linking those programs to relevant activities in other states and countries. Beginning in 2001, HEAR became the invasive species component of the more broadly focused USGS Pacific Basin Information Node.

SUMMARY

Hawaii's endemic forests birds, already challenged by avian disease and mammalian predation, face a future in which their survival may depend on keeping out West Nile virus, snakes, pathogens, insects that attack native trees, and other as yet unknown but similarly catastrophic invasive species. Containment of established habitat-displacing plants such as the miconia tree and potential competitors like the coqui frog may also prove crucial. There are many important needs for maintaining significant tracts of native biodiversity in Hawai'i, but we think it is clear that preventing establishment and spread of invasive species is paramount.

Hawaii's isolation, which for millennia has been its main line of defense against biotic invasions, has been destroyed by the forces of globalization. Although existing regulatory barriers have offered Hawaiian forest birds some degree of protection from invasion by alien species, new and damaging invasive species continue to enter the state to the great detriment of bird populations.

Invasions continue at a high rate and pose overwhelmingly the greatest current threat to Hawaii's endemic biodiversity, including its bird life, while also jeopardizing the state's economy, agriculture, health, and quality of life. For well over a decade there has been much interest and effort among conservationists toward rectifying this problem. The strongest progress to date has arguably been in the realm of public education, but this education has resulted largely from the press reports of the deteriorating situation caused by such new problems as miconia, frogs, chameleons, and giant salvinia. The Kahului Airport Pest Risk Assessment has provided the Hawai'i Department of Agriculture (HDOA) with an opportunity to float its promising vision of how gaps in Hawaii's prevention system could be filled.

To summarize current quarantine policy and practice, the main federal concern regarding Hawai'i involves protecting major agricultural crops in California and other states from fruit flies and other pests that reached Hawai'i long ago. Federal inspectors at Hawaii's airports screen baggage and hand-carried items of passengers bound for the U.S. mainland but provide no reciprocal protection of Hawai'i from pests of the mainland. The quarantine protecting Hawai'i from mainland U.S. pests is funded and implemented not by the federal government but by the state, which has limited jurisdiction and a relatively small tax base. Further, federal inspection of international arrivals focuses on essentially the same pests of concern at all U.S. ports: pests of major agricultural crops on the mainland. Consequently, federal quarantine activities in Hawai'i have been largely irrelevant to protecting the state from ecological pests that can threaten Hawaiian forest birds. The state, on the other hand, has limited authority to inspect international arrivals and, largely because of an inadequate quantity of state inspectors (though legal issues may pose a problem as well), there has been limited coordination among federal and state agencies in quarantine inspection. But increasing support by state legislators for invasive species concerns is leading to improved resources for HDOA inspection, beginning with a doubling of the number of HDOA inspectors (to 112) and budget (to $5.8 million) in the 2006 legislative session. A mechanism for funding improvements in state inspection, passed by the 2008 Hawai'i Legislature, seemed to promise hope for overcoming many of the federal-state coordination issues, enhancing diagnostic capacity, and developing a much improved statewide biosecurity system using the model HDOA had forged for Maui. However, for economic reasons the state's governor has opposed implementation of the fee of 50 cents per 1,000 pounds of freight brought in by air or sea (Lauer 2008), and its future was in question as of early 2009.

The state of Hawai'i is well positioned to make strong arguments to the federal government regarding the need for assistance in prevention of invasions not only from foreign shores but also from the U.S. mainland. Risk assessments of invasion pathways, with special attention to protecting Hawaii's dominant native trees and shrubs and most species-rich endemic plant radiations from plant pests, would provide a good and necessary way of documenting threats and need. The newly created Hawai'i Invasive Species Council has an excellent opportunity to help catalyze a response to challenges, many of which were identified by the TNCH/NRDC report (1992). The most pressing needs identified in this review are (1) a state funding mechanism for increasing the number of HDOA inspectors, identifiers, and support staff statewide to meet the need to inspect and monitor all high-risk pathways; (2) some form of major federal help toward achieving the level of prevention necessary to sustain meaningful biodiversity; (3) a way to stop proliferation of more and more invasive plant species and their spread into natural areas; and (4) credible deterrence to the smuggling of illegal organisms into Hawai'i.

ACKNOWLEDGMENTS

We thank P. Warren and two other reviewers for their helpful comments on an early draft of this chapter.

CHAPTER SIXTEEN

Protecting Forest Bird Populations across Landscapes

JONATHAN P. PRICE,
JAMES D. JACOBI, LINDA W. PRATT,
FREDRICK R. WARSHAUER,
AND CLIFFORD W. SMITH

Landscape is a term used in ecology to refer to a large geographical area within which plant and animal species and other elements of their ecosystem interact (Bastian 2001). From a conservation perspective, the size and shape of landscape units may strongly influence the ability of certain species or communities to maintain ecological stability within a region over time (Sanderson et al. 2002). Mobile species such as birds need large, contiguous habitat units or an expansive regional complex of smaller habitat units that contain adequate resources for foraging, breeding, and other vital functions (Askins 2000). Problems may arise when a landscape becomes so small or fragmented that it can no longer support important species, resulting in a collapse of structural or functional components of an ecosystem (Rolstad 1991, Simberloff 1995b, Wiens 1995).

Prior to the arrival of humans in the Hawaiian Islands about 1,200 years ago, Hawaiian forest bird habitat stretched nearly continuously from high elevations to the coast (Zimmerman 1965; this volume, Chapter 6), and most forest bird species were distributed across many different ecosystems (Perkins 1903; this volume, Chapter 2). Since the first Polynesian settlers arrived, and continuing after Western contact with the arrival of Captain James Cook in 1778, the composition and distribution of Hawaiian vegetation has been dramatically altered (Cuddihy and Stone 1990; this volume, Chapter 6). In addition to the direct degradation or loss of habitat through conversion, the islands experienced the fragmentation of native forests, which resulted in smaller tracts, often separated by substantial distance. With these changes to the Hawaiian ecosystems, many native plant and animal species were depleted or became extinct, no longer able to survive in altered eco-systems and subject to stresses from alien species. Many of the historically documented taxa of endemic Hawaiian forest bird species are now extinct, and many of the remaining taxa are listed as endangered or threatened (U.S. Fish and Wildlife Service 2006; this volume, Chapters 2 and 5). These declines can be tied to factors that can act both individually and in combination, including the aforementioned habitat loss (which occurred either directly through the clearing of habitat or indirectly through the impacts of invasive species), avian disease, predation, and resource competition. These factors are topics of other chapters in this book and will be discussed here within the context of landscape conservation.

Here we explore the nature of landscapes as they pertain to the conservation of Hawaiian forest birds. Beginning with an overview of why conservation at the landscape scale is important and the challenges it faces, we then appraise the present state of bird habitat in the Hawaiian Islands. A historical perspective of landscape-scale conservation is followed by a present-day assessment of lands designated for conservation in some capacity. Subsequently, we evaluate the management practices currently used on lands designated for conservation, including control of various threats and restoration. Finally, we present a vision for long-term conservation at the landscape scale, including consideration of reserve design, priorities for management, and restoration of both habitats and bird populations.

THE IMPORTANCE OF
LANDSCAPE-LEVEL CONSERVATION

The centrality of Hawaiian forest birds as key agents in ecological processes clearly calls for conservation of bird species within the context of functioning ecosystems. Seventeen percent of native Hawaiian plant species are primarily bird-pollinated (J. Price, unpubl. data), with many of these having evolved from insect-pollinated ancestors that originally colonized the islands (Price and Wagner 2004). Forty-nine percent of native flowering plant species are dispersed via ingestion of fleshy fruits by birds (J. Price, unpubl. data), and many of these plants have evolved from ancestors with capsular fruits (Price and Wagner 2004). In total, 61% of native flowering plant species are either pollinated by birds or dispersed through ingestion by birds (J. Price, unpubl. data), and these in return offer a diversity of food resources to their consumers. Insectivorous birds may be important in structuring arthropod communities by consuming midlevel predators, especially spiders, and may promote community stability through selective predation on insect outbreaks (Perkins 1903; Gruner 2004, 2005). Therefore, the presence of native bird species may be pivotal in maintaining various ecosystem processes.

Many bird species, especially those consuming plant-based resources, require large areas for foraging because the plant species they depend on vary temporally and spatially in providing a given resource. For example, māmane (*Sophora chrysophylla*), the primary food resource for the seed-eating Palila (*Loxioides bailleui*), exhibits variation in the timing of peaks in flower and fruit production along an elevation gradient (Banko, Oboyski, et al. 2002). Two nectar-feeding birds, the ʻIʻiwi (*Vestiaria coccinea*) and ʻApapane (*Himatione s. sanguinea*), migrate seasonally to different elevations in response to fluctuations in the flowering of ʻōhiʻa-lehua (*Metrosideros polymorpha*), their primary nectar resource (Baldwin 1953, Ralph and Fancy 1994b). Different fruit-bearing plant species may also exhibit seasonal peaks in production that vary by site, although there have not been sufficient studies to discern this (but see Berlin et al. 2000). Taken together, existing studies strongly suggest that a given bird species requires enough habitat to encompass areas with different timing of peaks in resource availability so that food resources are available throughout the year (Askins 2000; this volume, Chapter 7).

Large conservation areas are also necessary to maintain species that are distributed as metapopulations, essentially "populations of populations" in which a species has a patchy distribution (Levins 1969, 1970). According to this concept, populations in different areas exhibit different properties, such as reproduction or survival rates, and individuals may disperse between populations at different rates depending on the distances between populations. In some cases, a population may not be self-sustaining over time and may require occasional dispersal from other populations in order to maintain a target number of individuals, a situation referred to as the rescue effect (Brown and Kodric-Brown 1977). Metapopulations also require periodic genetic communication to remain viable. If small population units become too isolated without receiving occasional genetic input from other populations, they may become inbred and unhealthy and may ultimately collapse (Lande 1988, Rolstad 1991, Simberloff 1995b). Considering the nature of metapopulations, it is therefore crucial not only that multiple bird populations be protected but that populations continue to interact.

CHALLENGES TO
LANDSCAPE-LEVEL CONSERVATION

When dealing with large contiguous areas, the challenges associated with achieving conservation goals are as sizeable as the landscapes themselves. A primary challenge is to reconcile the needs of natural resources, such as forest birds, with the consequences of human tenure and demands on the landscape. In many areas of the world, this conflict has led to habitat loss, and in Hawai'i many previously forested areas have indeed been converted to

or designated for other uses (Chapter 6). Today, however, outright habitat conversion is largely limited to the island of Hawai'i.

There are numerous ways lands can be protected from habitat conversion, ranging from the discretion of a private landowner to a mandate of the federal or state government. The remaining areas of Hawaiian forest bird habitat are owned or managed by a wide array of interests. Even when two neighboring landowners share the goal of conserving bird populations, they may have contrasting philosophies, strategies, or organizational rules, so conservation could be practiced quite differently on their lands. Therefore, another major challenge to conservation at the landscape scale, which inevitably involves numerous landowners and managers, is to pull together these interests, reconcile differences, and maximize the efficiency of various habitat management practices across landscape units.

Even after large areas are secured against wholesale conversion to intensive urban or agricultural use, residual threats complicate the maintenance of viable forest bird habitat across whole landscapes. Not surprisingly, most of these threats are embodied in invasive alien species. All of the islands with forest bird habitat—the six largest Hawaiian islands—are inhabited by and subject to continued introductions of nonnative species. In contrast, New Zealand has numerous uninhabited islands with forest bird habitat that can be managed to provide isolated refuges from the impacts of nonnative species, particularly predators and ungulates (Butler and Merton 1992). The New Zealand offshore islands have proven invaluable as reserves where many bird species have been able to persist or be reintroduced. Unfortunately, in Hawai'i forest bird conservation is squarely challenged with having to address the threats of invasive species in nearly all areas, only a few of which are naturally protected from such invasion.

Both direct and indirect threats challenge Hawaiian forest birds. Direct threats result in the mortality of native birds sometime within their life cycle. Unlike in many parts of the world, in Hawai'i these threats are not now generally attributed to ongoing habitat loss and direct exploitation (Birdlife International 2000). Instead they come overwhelmingly from species brought in either deliberately or accidentally by humans, species that kill adult birds, nestlings, and eggs. Mammalian predators occupy most habitats (Chapter 11), while avian disease is largely restricted to lower elevations (Chapter 9).

Indirect threats, while not explicitly killing native birds, result in the loss of resources necessary for their survival and cause long-term or even rapid declines in avian populations. Arguably, the most dramatic and spatially widespread indirect impacts come from introduced ungulates, which modify vegetation, spread nonnative plant species (Cuddihy and Stone 1990; this volume, Chapter 6), and promote disease vectors (Chapter 9). Nonnative plant species are similarly widespread. Although many areas with native-dominated vegetation remain, expansion of alien weeds continues to degrade and displace these habitats (Chapter 6). Introduced insect predators and parasitoids (particularly ants and wasps) compete for universally important caterpillar food resources. Rats compete with some native forest bird species for seed, fruit, and insect resources (Chapter 11), simultaneously reducing native plant regeneration, while introduced forest birds overlap many native forest birds in resource use (Chapter 13). Each of these threats may be found even in native vegetation that might appear to be largely intact forest bird habitat.

Another distinct challenge is posed by the occurrence of disturbances that may complicate various human-related threats. Prior to the arrival of humans, bird populations likely endured periodic natural disturbances such as lava flows, fires, hurricanes, and large-scale tree canopy dieback (Mueller-Dombois 1983; Jacobi 1993; this volume, Chapter 6). Today, however, these natural disturbances have a greater capacity to impact bird species negatively. Where populations have been fragmented by anthropogenic factors, a single lava flow, fire, or widespread canopy dieback event can eliminate habitat a given species may ill afford to lose.

Confounding these challenges is the prospect that different threats reinforce each other or combine to form more serious synergistic impacts (Warshauer 1998). For example, fire or invasive alien native plant species alone may threaten native bird habitat. However, when fire occurs on islands with introduced fountain grass (*Pennisetum setaceum*), native forest communities can be rapidly converted to alien grasslands with no inherent habitat value for forest birds (Tunison and Leialoha 1988). As another example, by creating water-collecting hollows in tree ferns and on the ground, feral pigs (*Sus scrofa*) provide breeding habitat for mosquitoes that are the vectors of deadly avian malaria (*Plasmodium relictum*) and pox virus (*Avipoxvirus* sp.) (Chapter 9). There are arguably cases in which successful abatement of one type of impact (such as fire suppression or control of ungulates) may reduce other types (such as alien plants and vector-borne disease, respectively).

One future consideration is the potential for the spatial distribution of threats to change relative to that of bird species as a function of global climate change. For example, climate change could permit the upslope expansion of avian disease vectors that would threaten present bird habitat, while geology or land use, such as cattle grazing, could limit the ability of forest habitat to expand upslope (Benning et al. 2002).

Ultimately the widespread occurrence of any single type of threat can effectively fragment a bird population and threaten its interaction as a metapopulation. Such fragmentation may relegate individuals to small, isolated populations, each of which has a higher probability of being extirpated by disturbance events, a disease outbreak, drought, or the local outbreak of any of several threats. In addition, fragmentation may result in reduced rates of gene flow and increased occurrence of inbreeding, resulting in an overall decline in birds' genetic health (Lande 1988, Simberloff 1995b). Because threats vary in spatial extent and often produce a combined, synergistic effect, spatial coordination of various threat abatement programs is likely the best method to maintain large tracts of viable forest bird habitat. The real challenge, then, is to properly incorporate this coordination within the system of established protected areas and to expand protected areas where native bird populations are viable, or at least have the potential to be viable in the future.

THE PRESENT STATUS
OF FOREST BIRD HABITAT

Today, habitat supporting forest birds is found on the six islands with upland habitats. Not surprisingly, the three islands supporting the largest number of forest birds,

Kaua'i, Maui, and Hawai'i, also have the largest areas of native vegetation remaining. Figure 16.1 depicts a summary of native forest bird habitat distribution modeled as part of the Hawai'i Gap Analysis Project (HI-GAP) (Gon et al. 2006). For individual species on each island, the appropriate vegetation cover types, mapped from satellite imagery, were selected along with other criteria, such as the elevations and regions of each island where birds have been observed. The resulting habitat projections are a predictive snapshot of which habitats birds may currently be using. A few disease-resistant bird species, such as 'amakihi (Hemignathus spp.) and 'Apapane, can exist in a wide range of habitats at various elevations, but the majority are

Figure 16.1. Areas of potential bird habitat. The habitat for each bird species was determined for the Hawai'i GAP by selecting areas of each island with the appropriate land cover type, elevations, and biogeographic regions. Areas shown in gray represent potential habitat for up to four bird species and extend widely across most islands. Areas shown in black represent potential habitat for five or more species. These "hotspots" are restricted to upper-elevation forests on Kaua'i, Maui, and Hawai'i.

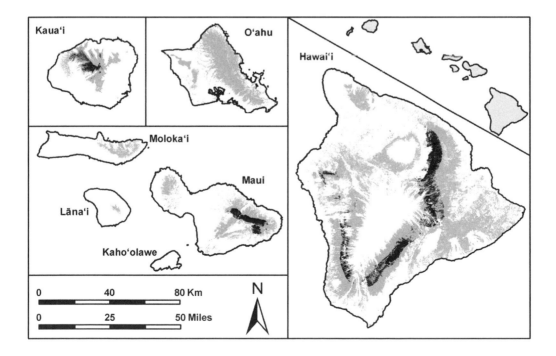

currently restricted to a narrow range of habitats and elevations. To summarize these distributions, some areas harbor only the hardiest species, whereas other high-elevation areas include these species in addition to concentrations of rare species (Chapter 5). Diversity hotspots include the Alaka'i Plateau of Kaua'i, windward East Maui, and a discontinuous belt of high-elevation forest on the island of Hawai'i. This pattern reflects the large areas at high elevations on these three islands and the fact that lowlands have been subjected to more habitat conversion and are also hot-beds of disease and other threats. None-theless, Hawai'i Island still has extensive forest bird habitat at very low elevations (essentially down to sea level) locally con-taining populations of 'Elepaio (*Chasiempis sandwichensis*), 'Ōma'o (*Myadestes obscurus*), Hawai'i 'Amakihi (*H. virens*), and 'Apapane (Reynolds et al. 2003). Other islands sim-ilarly contain substantial areas of appro-priate native vegetation at lower elevations, some of which may support some bird species.

Traditional landscape metrics, such as patch size and connectivity (Forman and Godron 1986), have been difficult to ap-ply in Hawai'i for several reasons, includ-ing the complexity of the vegetation and the lack of detailed vegetation maps until recently (Jacobi 1989, Gon et al. 2006). Nonetheless, the data available permit some general statements about changes in the amount and pattern of forest bird habitat (see Fig. 16.1). First, on all major islands, the amount of habitat available has declined over the past few decades as many plant communities have become increasingly dominated by invasive plant species. In ad-dition, the amount of habitat on the island of Hawai'i has further declined during this time as development, logging, and con-version to pasture have reduced native forest cover. With the exception of Hawai'i Island, the islands tend to have forest bird habitat consolidated into a contiguous area in each topographically discrete upland

(O'ahu and Maui have two separate upland regions each). These upland habitat regions tend to have a complex shape (with high ratios of perimeter to area) as a function of their topography and the degree of interdigitation of native with nonnative vegetation. On Hawai'i Island, a more com-plex pattern is evident; large, spatially dis-crete habitat regions occur on (1) the Ko-hala Mountains; (2) windward Mauna Kea, Mauna Loa, and Kīlauea volcanoes; (3) the southeast flank of Mauna Loa (Ka'ū Dis-trict); and (4) the western slopes of Mauna Loa and Hualālai volcanoes (Kona districts). Windward Mauna Loa and Kīlauea rep-resent a naturally discontinuous habitat as a function of relatively young lava flows with pioneer vegetation that dissect the forest habitat there. The forest habitat on the western side of the island is also dis-sected into patches; however, in this case it is due to fragmentation resulting from habitat conversion that has occurred in recent decades. Future work may resolve more specific patterns for individual bird species in order to evaluate the structure of metapopulations.

A Historical Perspective on Recognition of Lands for Conservation

Native Hawaiians divided each of the is-lands into large landscape units that they called ahupua'a. These were narrow land divisions that generally extended from the shoreline up into the forested inland re-gions of the islands, and they have left their stamp on land ownership up to this day. Ahupua'a were designed to contain the full range of resources that the people in these areas needed to survive. This range in-cluded heavily utilized near-shore and marine resources (kahakai) and lower-elevation lands for habitation and cultiva-tion (wao kanaka); upper-elevation regions (wao nahele, wao ma'u kele) with re-sources such as medicinal plants, birds, and religious sites to which people needed less extensive access; and little-used wilderness regions (wao akua). Resource use within an ahupuaʻa was regulated by a designated land manager (konohiki) to ensure appropriate land use and allocation of resources (Kamakau 1964, Minerbi 1999, Mueller-Dombois and Wirawan 2005).

The Hawaiian land tenure system, particularly the ahupuaʻa concept, began to change following Western contact in 1778. However, interest in protecting the remaining forests of the Hawaiian Islands was still strong during the latter days of the Hawaiian Kingdom as evidenced by the creation of the Bureau of Agriculture and Forestry (Kuykendall 1967). Even without government support, private landowners and owners of sugar plantations set aside forests and planted trees upslope of fields to assist watershed management and to ensure water supply for the plantations (Hall 1904). After Hawaiʻi became a U.S. territory, a bureau survey recommended the immediate establishment of a reserve system to protect the remaining native forests and provide watersheds for the current and future agriculture (Hall 1904). The first government forest reserve was created on the island of Oʻahu in 1904 (Hosmer 1959), and within 10 years 37 forest reserves had been established containing 323,890 ha (Bryan 1961). The second territorial forester, Charles Judd, proposed using native trees in forestry and recognized the pest potential of introduced species in native forest systems (Judd 1918). During Judd's tenure, native "protection forests" were valued for water production, and the reserve program's emphasis was on fencing, animal removal, and tree planting (Judd 1927a, 1931).

The Forest Reserve System grew to 65 reserves by 1952 (Hosmer 1959) and included private lands that comprised a third of the protected acreage of >400,000 ha (Lennox 1948). After World War II, however, in a shift to a multiple-use philosophy for forest reserves, recreational uses and commercial timber production were considered of equal value to watershed pro-

tection (Frame and Horwitz 1965, Rodwell 2001). This change altered the perception of introduced ungulates from that of forest pests to be reduced or eliminated (Bryan 1947) to that of wildlife resources valuable for recreational hunting. More recently, most private lands were removed from the Forest Reserve System in the 1960s (Little and Skolmen 1989), reducing the total reserve area to approximately 260,000 ha by the end of the twentieth century. Nonetheless, the State Forest Reserves remain today the largest area of protected forest in Hawaiʻi.

Although the state Forest Reserve System has declined in both scope and importance relative to conservation in the past 50 years, several other land categories have been introduced to better address this need. In the early 1970s, the state also developed the Natural Area Reserve System (NARS), which was designed to recognize and to preserve and protect representative samples of Hawaiian biological ecosystems and geological formations. To date, there are 15 NAR units encompassing approximately 44,000 ha on five islands. A parallel program, Natural Area Partnerships (NAP), promotes conservation on private lands, where designated state funds assist management and administration. In many cases, the management of biological resources on private lands is supplemented by funding through the Hawaiʻi Forest Stewardship Program and through various federal cost-sharing programs. Additionally, the U.S. Forest Service (USFS) funds, and the State of Hawaiʻi runs, the Hawaiʻi Forest Legacy Program for the purchase of limited conservation easements and the fee ownership of selected private lands to prevent conversion of these lands to non-forest uses.

Although most forest bird habitat is on state land, conservation activities are also being conducted on other lands. Three major national parks (Hawaiʻi Volcanoes, Haleakalā, and Kalaupapa) were established starting in the early twentieth century and later expanded, with forest bird conservation recently a major goal. Two national wildlife refuges (NWRs) (Hakalau Forest NWR and Oʻahu Forest NWR) were created specifically for forest bird protection and are managed to that end. Several private reserves on lands owned or controlled by The Nature Conservancy of Hawaiʻi, Kamehameha Schools, and Maui Pineapple Company are also managed for the benefit of forest birds, often funded by the state's NAP program. The network of major conservation areas on each island is identified in Plates 30–32.

Conservation Designations at the Landscape Level

As seen from the preceding discussion, the value of preserving large, contiguous areas has been a central theme of conservation in Hawaiʻi for more than a century. This focus has resulted in the recognition of several types of protection status to complement ground-based management activities under state, federal, or private jurisdiction. Three related approaches to implementing conservation in Hawaiʻi include (1) designation of areas for conservation, (2) conducting active management programs on the ground, and (3) utilizing administrative processes (such as the listing of threatened and endangered species by the state or federal government) that can support the previous two approaches.

The identification and designation of areas for conservation, such as forest reserves, is a crucial first step leading to management, indicated with a "D" in the "Category" column of Table 16.1. Depending on the type of designation, however, the resulting management may be focused on protecting natural resource values (native biota and ecosystems) or may be directed to other issues (timber production, watershed protection, recreation, etc.). At the present time, areas such as forest reserves and state parks are not specifically designated to receive on-ground management

(mainly abatement of threats) for protecting their native biological resources.

Areas indicated with an "M" in Table 16.1 have been designated for conservation and also receive some level of active management on the ground (as opposed to those marked "D," indicating conservation without active management). These types of lands include state NARs, national wildlife refuges and national parks, and some private land reserves, for example, The Nature Conservancy's preserves and Maui Pineapple Company's Puʻu Kukui Watershed, all NAP reserves.

A third group of programs, indicated with an "A" in Table 16.1, relate to administrative designations, especially those under either the state or federal endangered species laws. These include areas with an administrative designation, for example, critical habitat designation under the federal Endangered Species Act, as well as areas that are planning targets for conservation actions, especially recovery areas for forest birds (U.S. Fish and Wildlife Service 2006) and priority conservation areas identified in the state of Hawaii's Comprehensive Wildlife Conservation Strategy (Mitchell et al. 2005). In addition, the state conservation zones apply to both public (not federal) and private lands and relate to the types of activities that are allowed to be conducted. Four conservation subzones are recognized, with one of these, the P subzone, relating to native species–dominated habitats that are important for the conservation of forest bird species. These areas may overlap with the other two designations. For example, areas designated as critical habitat or recovery area by the U.S. Fish and Wildlife Service include state, federal, and private lands. Some of these are either designated as conservation areas ("D" in Table 16.1) or are actively managed ("M" in Table 16.1). In some cases, recognition of an area as critical habitat or recovery area may facilitate both designation and management of that area for conservation values.

Table 16.1. Summary of categories of areas designated for conservation on the main Hawaiian islands

Description	Category	Island					
		Ka	Oa	Mo	La	Ma	Ha
State Programs							
Forest reserves	D	✓	✓	✓		✓	✓
Conservation districts	A	✓	✓	✓	✓	✓	✓
State parks	D	✓	✓	✓	✓	✓	✓
Priority conservation areas (CWCs)	A	✓	✓	✓	✓	✓	✓
Wilderness preserve	M	✓					
Wildlife sanctuaries	M	✓	✓	✓		✓	✓
Natural area reserves	M	✓	✓	✓		✓	✓
Natural Area Partnership Program areas	M			✓	✓	✓	
County Programs							
Tax incentive programs for protecting forest habitats	A						✓
Federal Programs							
Critical habitats	A		✓				✓
Recovery areas (formerly essential habitat)	A	✓	✓	✓	✓	✓	✓
National wildlife refuges	M	✓	✓	✓		✓	✓
National parks	M		✓	✓		✓	✓
Private Programs							
The Nature Conservancy of Hawai‘i preserves	M	✓	✓	✓	✓	✓	✓
Local management areas (e.g., Kīlauea Forest, Ulupalakua)	M	✓				✓	✓
Landscape and Watershed Partnerships	M	✓	✓	✓	✓	✓	✓

Notes: A, an administrative designation relative to state or federal endangered species laws; D, designated for conservation with little or no on-ground management; M, designated for conservation and on-ground conservation management; Ka, Kaua‘i; Oa, O‘ahu; Mo, Moloka‘i; La, Lāna‘i; Ma, Maui; Ha, Hawai‘i Island.

In addition to considering the total size of protected areas, how these correspond to actual bird habitat is perhaps a better measure of the completeness of existing reserves relative to the focus of this volume. Table 16.2 indicates the amount of area of potential bird habitat (from Fig. 16.1) in square kilometers designated as "D" or "M" on each island. There is quite a large amount of variability in how much of the landscape on each island is designated and/or managed for conservation. Overall, 48% of forest bird habitat has some sort of conservation designation ("D" or "M"). In nearly all cases, these lands protect the majority of habitat for endangered species and the core habitat for more common species. However, only 19% of forest bird habitat is actually designated for active management ("M"), and only a subset of these areas are actu-

ally subject to intensive management actions such as ongoing ungulate control and/or weed control. The recent history (1975–2005) of bird habitat protection at the landscape scale is characterized by an expansion of areas designated ("D") and managed ("M") for conservation (Figs. 16.2 and 16.3).

In the 1970s, with the inception of the Endangered Species Act, there was a clear need for studies to guide various aspects of conservation in Hawai‘i. The Hawai‘i Forest Bird Survey (HFBS) conducted from 1976 to 1983 sought to assess bird ranges, populations, and habitat throughout the state to assist managers with recovery planning for the listed endangered species. The resulting work by Scott et al. (1986) concluded that there was a gap in habitat protection for endangered forest birds on Hawai‘i Island, ultimately leading to the

Table 16.2. Areas of potential bird habitat either designated (D) or managed (M) for conservation, by island

Island	Number of Extant Forest Bird Species	Number of Endangered Forest Bird Species	Area of Potential Bird Habitat	Area of D (% of bird habitat)	Area of M (% of bird habitat)
Kaua'i	9	4	447	205 (46)	55 (12)
O'ahu	4	1	783	106 (14)	33 (4)
Moloka'i	4	0	88	27 (30)	24 (27)
Lāna'i	2	0	11	0 (0)	0 (0)
Maui	7	2	446	154 (35)	127 (28)
Hawai'i	10	5	5,932	1,746 (29)	1,237 (21)
Total, all islands	19	13	7,707	2,238 (29)	1,476 (19)

Note: Area given in km^2.

creation of Hakalau Forest National Wildlife Refuge. Another area, the Kona Forest Unit, was added to this refuge in 1997. The O'ahu Forest Wildlife Refuge was created in 2000. National parks similarly expanded during this period: in 1980 Kalaupapa became a National Historical Park, in 2000 Ka'āpahu Ranch was added to Haleakalā National Park, and in 2003 Kahuku Ranch was added to Hawai'i Volcanoes National Park. The state of Hawai'i paralleled this trend by creating the first natural area reserve (NAR) in 1973, which was followed in the next decade by the creation of many more NARs, of which the most important for forest birds was Hanawī NAR on Maui. In addition to state and federal lands receiving such designations, many private lands were also designated for conservation over time, largely facilitated by The Nature Conservancy of Hawai'i and as a direct result of coordinated conservation planning using findings of the HFBS (Scott et al. 1987).

More recently, watershed partnerships have placed tracts of both private and public land under cooperative management for large-scale protection of watershed and natural resource values. These partnerships include both public and private landowners, as well as organizations such as the U.S. Geological Survey and USFS, which assist management by conducting applied research and monitoring in support of efforts to maintain watershed and natural resource values on the landscape. An important strategy within these partnerships is to deal with regional issues, for example, control of invasive species, across large landscapes that transcend property boundary lines among the participating landowners. Today there are nine watershed partnerships in Hawai'i covering more than 300,000 ha (750,000 acres) of important landscapes on six of the islands.

The existing system of conservation areas has been the result of responses of various agencies to both conservation needs and opportunities, although there has been only limited use of landscape ecology theory. One well-known example was the establishment of the Hakalau Forest NWR as a result of the original GAP analysis (Scott et al. 1986). Other aspects

Figure 16.2. Areas designated or managed for conservation, 1975–2005. The expansion of conservation areas is chronicled at 10-year intervals for Kaua'i and O'ahu. Areas designated for conservation but with no active management (D) are shown in gray; those designated and managed for conservation (M) are shown in black. Watershed partnership management areas, which are also designated to be managed for conservation, are indicated by diagonal lines. A clear trend of expansion of M designations is evident.

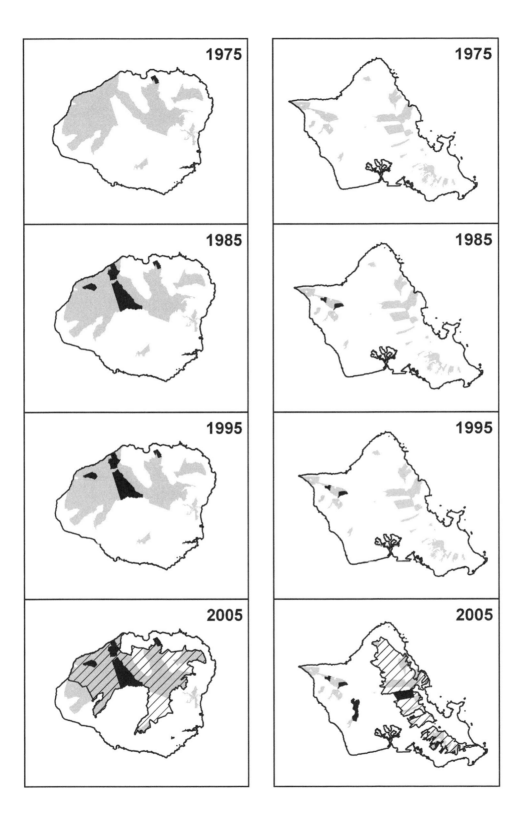

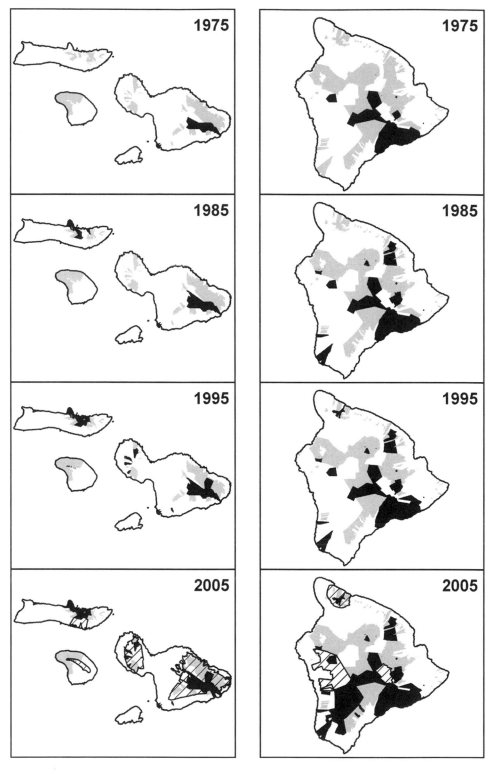

Figure 16.3. Areas designated or managed for conservation, 1975–2005. The expansion of conservation areas is chronicled at 10-year intervals for Moloka'i, Lāna'i, Maui, and Hawai'i. See Fig. 16.3 for an explanation of the symbols used.

of landscape ecology theory, such as consideration of patch dynamics and metapopulations, have not been incorporated into reserve design per se, although the increasing detail in the population data available (Chapter 5) and studies of habitat use may elucidate additional areas requiring conservation designation. Beyond conservation designation, principles of landscape ecology may also help target active management activities within existing conservation areas.

Management at the Landscape Level

The fact that approximately half of all bird habitat is designated for conservation in the broadest sense ("D" or "M" in Table 16.2) should not obscure the challenges that remain in managing these important landscapes for the benefit of maintaining native forest bird populations. As mentioned previously, only 19% of forest bird habitat is currently receiving some form of active management ("M"). Technically speaking, however, these various conservation designations address only one or two threats, namely conversion of habitat to other land use for "D" units and the addition of ungulate management for some "M" units. Although considerable progress has been made over the past decades in the expansion of areas identified for conservation, other threats are continuing to expand in both intensity and geographic extent. Thus, the mere designation of a landscape for conservation management does not ensure that a habitat is insulated from the impacts of other factors, particularly the actions of invasive species. Instead, different threats are only abated as a result of deliberate management actions. Implementation of such actions at biologically relevant scales where threats are controlled over a large area can promote self-sustaining bird populations (Chapter 5). Following is a summary of the progress to date in abating each major type of threat within recovery habitat for Hawaiian forest birds.

INTRODUCED UNGULATES

Control of ungulates involves the construction of fences designed to keep large mammal species out of an area, combined with removal measures such as shooting or trapping. In some areas, ungulates have been eradicated entirely. There are now large (>5,000 ha) ungulate-free areas in Haleakalā and Hawai'i Volcanoes national parks, Hakalau Forest NWR, and the 'Ōla'a-Kīlauea Partnership Area as a result of combined fencing and control programs for feral cattle (*Bos taurus*), pigs, and goats (*Capra hircus*). Although ungulates have been entirely removed from significant areas on some islands, these areas comprise a small fraction of bird recovery habitat (Table 16.3). For example, ungulates have been eliminated from about one-quarter of the forest bird habitat on Maui but have been eliminated from only a small area of Kaua'i. Moreover, the recent expansion of extremely mobile and difficult-to-contain ungulate species, specifically axis deer (*Axis axis*) on Maui and Moloka'i, black-tail deer (*Odocoileus hemionus*) on Kaua'i, and mouflon sheep (*Ovis gmelini musimon*) on Hawai'i, indicates that even areas that are presently free of ungulates can be invaded when these high-jumping species clear low fences intended for feral pigs and goats.

In some cases, local eradication is not possible, either because adequate fences are not available or are too expensive to construct or because sustained hunting or trapping programs cannot be implemented due to fiscal, administrative, social, or other concerns. Although some studies have attempted to quantify ungulate damage at low population densities, there are few data comparing areas where ungulate populations are subjected to moderate hunting pressure to areas where ungulates have been eradicated or where there is no population control whatsoever. As a result, there is no clear evidence to show that partial reduction of ungulate populations has produced significant conservation benefits

Table 16.3. Areas of potential bird habitat where ungulates have been removed or will be removed in the near future

Island	Area of Potential Forest Bird Habitat	Ungulate-Free Area (% of total potential forest bird habitat on the given island)	Area Where Eradication Is Planned (% of total potential forest bird habitat on the given island)
Kaua'i	447	0.4 (0.1)	35.0 (7.8)
O'ahu	783	2.4 (0.3)	9.7 (1.2)
Moloka'i	88	6.1 (6.9)	8.5 (9.7)
Lāna'i	11	0.0 (0.0)	1.2 (11.0)
Maui	446	106.0 (24.0)	307.0 (69.0)
Hawai'i	5,932	151.0 (2.5)	1,117.0 (19.0)
Total, all islands	7,707	266.0 (3.5)	1,478.0 (19.0)

Source: Data were compiled by the Nature Conservancy of Hawai'i and adjusted by J. Price.
Notes: Areas are given in km². Ungulate-free areas that do not include forest bird habitat, such as Haleakalā Crater, are not counted toward these totals.

to native plants or animals in Hawaiian ecosystems. Simple simulation models indicated that semiannual removal of 30–40% of pigs would be required to maintain them at half their equilibrium density in Hawaiian forests (Barrett 1983). This level of removal is likely far greater than that presently achieved by sport hunting, which requires a sustained yield. Therefore, any effort to control ungulates to some low population level would require a continuous shooting or trapping program to remove at least 40% of the animals from an area twice per year for as long as the habitat was to be managed. This approach could prove either extremely expensive or logistically infeasible. On the other hand, in an area where total eradication is achieved, the only sustained actions needed to maintain the area ungulate-free are fence upkeep and occasional control of animals that breach broken fences.

INVASIVE PLANTS

Invasive plants present a particularly difficult case for control at the landscape level. The spread of species exhibiting dense growth can rapidly modify ecosystems by excluding native species and even by changing the structure of the vegetation

(Chapter 6). There are several management actions that address invasive plants. The most effective action may well be preventing invasive plants from establishing on a given island in the first place (Chapter 15). The U.S. Department of Agriculture and the state of Hawai'i have produced lists of noxious weeds whose transport into the state and among islands is prohibited. However, the number of species with the potential to invade and modify bird habitat extends far beyond those recognized thus far. Furthermore, many of the species on the noxious weed list are already widespread in Hawaiian ecosystems. Reduction of the threats imposed by these invasive plants still requires some form of active control.

Manual control strategies involving hand weeding, chopping down, or otherwise physically removing invasive plant species, most often combined with local application of herbicides, have exhibited some success, particularly if incipient populations can be eliminated before they become widespread. For example, a 10-year program to eradicate fountain grass throughout Hawai'i Volcanoes National Park using the centripetal approach (working from the outside in) has reduced the population to fewer than 100 plants (Tunison et al.

1994). Such efforts require a long-term management program with sufficient funding and a highly focused, dedicated workforce. A much more ambitious, statewide attempt to contain, if not eliminate, miconia (*Miconia calvescens*) has been partially successful, but the program's viability is subject to a continuous struggle for funding (Meyer and Smith 1998). Widespread manual control programs are not common in Hawai'i, probably due to prohibitively expensive labor and herbicide costs.

Where a weed is widespread, associated native species are sensitive to disturbance or herbicides, and access is difficult or impossible, biological control (or biocontrol) may provide some level of a long-term solution to the problem. Biocontrol entails the purposeful introduction of an organism that will attack and kill or reduce the vigor of the targeted invasive species (Lincoln et al. 1998). To achieve this result, biocontrol agents are typically invertebrates or pathogens that specialize in a narrow set of plant species that includes the target invasive but not native species or species of agricultural importance. When successful, the biocontrol agent spreads widely, and the target organism is controlled at a biologically relevant spatial scale with no further involvement of managers. Because there is usually a density-dependent relationship between the agent and the target species, biocontrol rarely results in the complete elimination of a target, but control may be deemed successful if the invasive species is reduced to the point that it has a significantly reduced impact on native species or ecosystems.

To date, only a few species of plants that invade Hawaiian bird habitat have been the subjects of biological control programs. Examples include banana poka (*Passiflora tarminiana*) and Koster's curse (*Clidemia hirta*). The USFS started working on banana poka, a weedy vine, because of its intensive damage to native koa (*Acacia koa*) forests (Warshauer et al. 1983). Koa is a highly prized timber species as well as an important for-

aging substrate for many endemic birds. The fungal pathogen *Septoria passiflorae* was introduced and now causes acute defoliation of banana poka in some areas, for example, at Hakalau Forest NWR (Trujillo et al. 2001). Trujillo (1985) pioneered the use of pathogens as biological control agents in the Hawaiian Islands, successfully introducing a leaf spot fungus (*Colletotrichum gloeosporioides* f. sp. *clidemiae*) that causes defoliation of the invasive shrub *Clidemia hirta* in forested areas, but its impact has not been fully assessed (Trujillo et al. 1986).

INTRODUCED MAMMALIAN PREDATORS

Control of feral cats (*Felis catus*), rats (*Rattus* spp.), and small Indian mongooses (*Herpestes auropunctatus*) has proven feasible so far only at very local scales and most effectively at coastal sites and on small offshore islands (Chapter 18). The difficulty in dealing with mammalian predators in large forested landscapes lies in both successfully targeting and effectively excluding them, for appropriate fencing is currently infeasible. The efficacy of predator-trapping efforts is difficult to define spatially, but measurable reduction in predator populations has been achieved only at very local scales. To be effective, a control program would need to extend beyond the immediate area of interest due to the likelihood of constant predator influx from surrounding areas (Nelson et al. 2002). Poison bait, particularly diphacinone, has been used to eliminate rats on islets supporting seabird colonies, and its effectiveness has been tested in forest bird habitat as well (Spurr et al. 2003a, 2003b). Typically, bait is administered in enclosed bait stations. Although aerial dispersal from helicopters could allow widespread distribution of bait across the entire landscape, as yet there is no widespread use of diphacinone, because it is registered for use only in bait stations. Currently there is a coordinated effort between state, federal, and private organizations to finalize the registration of an aerial distribution system for diphacinone to be used over large areas of habitat (Chapter 18).

Although most management efforts have focused on ungulate and invasive weed control, over the long term such strategies would benefit from additional efforts to reduce or eliminate other factors that impact forest bird species at both the regional and local levels. These efforts might include reduction of the impacts of avian disease through vector control (Chapter 17), local reduction of populations of certain alien social insects (specifically ants and social hymenoptera) that compete directly with birds for food resources (Chapter 7), and reduction of disturbances such as fire.

Restoration Efforts

In addition to abatement of various threats, one management action that has received much attention recently is ecological restoration. Certain types of native habitat are now very restricted or have been badly degraded, even after being stabilized through subtractive management actions. Furthermore, many areas designated for management where the full suite of actions can be implemented do not presently consist of high-quality forest bird habitat. Therefore, restoration may be a way to increase the forest bird habitat available and maximize the potential of areas that have been designated for management.

The simplest form of restoration involves allowing native species to regenerate in the reduced presence of nonnatives. An early example of this passive style of restoration can be found on Mauna Kea, where reduction of introduced ungulates has been followed by a rebound in growth of māmane trees, a pivotal resource for the Palila, apparently stabilizing the bird's population (Banko et al. 2001). However, the primary focus for restoration of forest bird habitat has been the re-creation of upland koa forest (Pejchar and Press 2006, U.S. Fish and Wildlife Service 2006). In many pasture areas formerly supporting koa forest, koa seeds persist in the soil, and regeneration can be greatly facilitated through scarification (using a bulldozer to scrape the surface soil) to promote seed germination (Pearson and Vitousek 2001). In badly degraded areas, natural regeneration may be too slow, or there may be few species locally available to provide seeds for sufficient recruitment. In such areas, planting important canopy trees and understory species may be necessary.

Koa forest restoration can be seen at Hakalau Forest NWR, where extensive outplanting of koa, 'ōhi'a, and other species known to be important to forest birds is taking place in former pasturelands that had few native trees before restoration began. Dense pasture grass cover was scarified in linear strips that stretched from the lower forest up nearly 300 m in elevation to the upper boundary of the refuge. Koa was intensively planted in these strips in order to create lines of vegetation leading up from the forest. Eventually, understory species were also planted in these areas to provide for more diversity. Observations of various forest birds utilizing some of these newly planted areas, which are adjacent to forested areas with healthy bird populations, confirm the potential for restoration to expand the margins of the habitat available (Banko et al. 2001). Another example of active restoration is at Auwahi, a remnant mesic forest on leeward East Maui. There, extensive planting of a diverse suite of native tree species has resulted in recolonization by Hawai'i 'Amakihi (A. Medeiros, pers. comm.).

In other habitat restoration programs, conservation of bird species is not the direct objective. For example, over the past 20 years several areas of logged forest have been replanted or managed to encourage the growth of koa as a silviculture species (Pejchar and Press 2006). These types of habitats can develop into areas that are used by native forest bird species, including some rare species such as the ʻAkiapōlāʻau, Hawaiʻi Creeper (*Oreomystis mana*), and ʻĀkepa (*Loxops coccineus*), as has been seen in portions of Kamehameha School's Keauhou Ranch that were logged in the 1970s and 1980s and in grazed forest at the state's Kapapala Koa Management Area on Hawaiʻi Island (Plate 18) (Pejchar et al. 2005). The Nature Conservancy of Hawaiʻi is also experimenting with methods to restore former pasture and previously logged areas on private lands by turning them into landscapes that can provide better habitats for forest birds as well as some economic gain for the landowner through the harvesting of forest products, especially koa timber. Restoration of koa forest encircling the leeward flanks of Haleakalā Volcano on Maui is the goal of the Leeward Haleakalā Watershed Restoration Partnership. However, conversion of existing high-quality koa forest habitat that already supports native forest birds into plantation forests, even using native tree species, will likely reduce the overall conservation value of the habitat both for birds and for other native species that depend on a relatively intact ecosystem.

A modification of the Hawaiʻi Endangered Species Act in 1998 now allows private and public landowners to enter into Safe Harbor Agreements with both the state of Hawaiʻi and the U.S. Fish and Wildlife Service (Act 380, Session Laws of Hawaiʻi, 1997). This agreement encourages landowners to take actions that may enhance habitats for endangered or threatened species without being penalized if future actions, for example, logging the site, result in the incidental taking of endangered species above baseline populations determined at the outset of the agreement. However, this arrangement must provide for a "net benefit" to the species within the project area.

The most intensive and speculative form of habitat restoration is the complete reconstruction of plant communities where little native vegetation currently lives nearby and where it may even be unclear what vegetation is natural to the site. Many small-scale restoration projects involve outplanting native species to lowland sites that are now far removed from viable forest bird habitat and thus have little application to forest bird conservation.

THE VISION FOR LONG-TERM CONSERVATION AT THE LANDSCAPE SCALE

Identification of Priority Areas for Additional Conservation Management

Despite an existing network of reserves, doubt remains as to whether their size and configuration are appropriate to meet the long-term conservation goals for Hawaiian forest birds. Given the rapid pace of historic ecological change, conservation planning recognizes that in the future viable habitat for forest birds may not persist in areas without continuous management of the major limiting factors—ungulates, selected invasive plants, avian disease, and disease vectors. Therefore, it is critical to consider whether areas both designated and managed for forest bird conservation are adequate to provide self-contained regions of viable habitat. To see how the conservation landscape might look for Hawaii's forest birds in 2035, see Box 16.1.

The *Revised Recovery Plan for Hawaiian Forest Birds* provides a strategy for the recovery of 21 taxa of endangered Hawaiian passerines, or perching birds (U.S. Fish and Wildlife Service 2006). In addition to outlining the causes of each species' decline and the specific management actions needed to reverse these trends, it identifies areas necessary to promote viable and stable populations. This recovery habitat area

Box 16.1. **Hawaiian Forest Birds in 2035:**
What Does the Conservation Landscape Look Like?

In 2005, the amount of area designated for management (M) varied by island; Maui had the largest portion of its habitat (29%) in this category (see Table 16.2). In our vision for 2035, all islands have a comparable proportion of their forest bird habitat designated for management, and they receive management actions intensively so that they function as core habitats. Remaining habitat areas function as matrix habitats (Table 16.4).

Core Habitats

In 2035, the total amount of bird habitat is essentially the same as in 2005: about 7,700 km^2 or half of the total land area of the state (see Table 16.4). While some habitat has been lost due to continued conversion and degradation, some has also been added through restoration. Of this total, core areas largely coincide with the recovery area (U.S. Fish and Wildlife Service 2006; see also Fig. 16.3). These consist of habitats that are critical to the long-term survival of forest bird populations and are managed specifically to maximize their conservation value. These core areas support self-sustaining bird populations and include contiguous parcels actively managed for forest bird conservation. These core areas include relatively intact high-

quality habitats as well as some restored habitats.

Matrix Habitats

Areas within which the core habitats are embedded may support activities other than conservation, including limited logging, hunting, ecotourism, and other recreational uses. These matrix habitats make up the remaining forest bird habitat. Despite their lower intensity of conservation management and consequently their lower level of habitat quality, matrix areas nonetheless may be utilized by birds to some degree. For example, native tree plantations that have an economic use also provide habitat to some bird species. Matrix habitats may not support self-sustaining populations of many bird species. However, birds may utilize these areas opportunistically, resulting in populations that are effectively subsidized from nearby core habitats. Matrix areas also provide important connectivity among the core habitats necessary for metapopulation function. Matrix habitats may support self-sustaining populations of hardy species such as 'amakihi and 'Apapane. Finally, these areas provide a buffer between core habitats and areas of intense human use, where new outbreaks of alien species often emerge.

Table 16.4. A conceptual conservation landscape for forest birds in 2035

Island	Total Area of Potential Bird Habitat	Core Habitat (⅓ of total potential habitat)	Matrix Habitat (⅔ of total potential habitat)	Recovery Area
Kaua'i	447	149	298	95
O'ahu	783	261	522	268
Moloka'i	88	29	59	65
Lāna'i	11	4	7	0
Maui	446	149	297	408
Hawai'i	5,932	1,977	3,955	2,969
Total, all islands	7,707	2,569	5,138	3,805

Note: Areas are given in km^2.

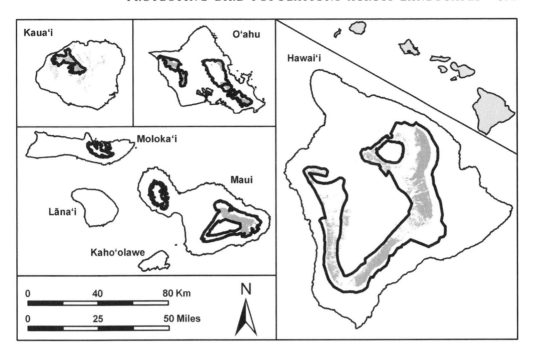

Figure 16.4. Endangered species recovery areas (polygons with dark outline) relative to modeled distribution of forest bird habitat. Recovery areas were mapped in the planning process for the *Revised Recovery Plan for Hawaiian Forest Birds*. Recovery areas closely match the projected distribution of endangered forest birds.

covers much of the upper elevations of the larger islands, and indeed closely matches the projected distributions of endangered forest birds (Fig. 16.4). The landscape scale considered by the recovery plan takes into account the desired size and distribution of bird populations, the limitations to distribution posed by avian disease, and the availability of existing and restorable habitat. These considerations are addressed on a species-by-species basis and then combined into a single strategy for each island. However, because the recovery plan deals with endangered species, nonendangered and more widespread species that may be critical to ecosystem function, such as 'amakihi and 'Apapane, must be considered elsewhere (compare Figs. 16.1 and 16.4, Box 16.2).

The Hawai'i GAP was designed to assess the adequacy of the existing conservation landscape for maintaining viable ecosystems and populations of native plants and animals by modeling the distributions of all native bird species and major plant communities and comparing the model to the layout of protected areas across the state (Gon et al. 2006). The relatively limited extent of managed areas ("M" in Table 16.1 and Fig. 16.2) suggests that there is an urgent need for more extensive and intensive habitat conservation, particularly relative to existing forest bird habitats. First, some areas that currently have no designation may need to be secured against habitat conversion. This is especially true for the Puna and South Kona regions on Hawai'i Island, where rapidly expanding real estate development threatens the largest remaining areas of lowland forest bird habitat in the state. Second, many areas may require an "upgrade" in their designation (from "D" to "M" in Table 16.1), particularly through expanded management of invasive species of plants and animals. Hawai'i Island, with extensive areas of remaining native forest, presents such an opportunity in the Kapāpala, Ka'ū, Upper Waiākea, and

---

Box 16.2.   **Hawaiian Forest Birds in 2035:
How Did the Conservation Landscape Come to Be?**

Planning

Changes to the conservation landscape in the 30 years from 1975 to 2005 largely consisted of the addition of new areas with some type of conservation designation (see Fig. 16.2). In the 30 years between 2005 and 2035, the changes have been in achieving management goals. The *Revised Recovery Plan for Hawaiian Forest Birds* (U.S. Fish and Wildlife Service 2006), the Hawai'i GAP project (Gon et al. 2006), and the Comprehensive Wildlife Conservation Strategy (Mitchell et al. 2005) all served as important guides that allowed us to strategically identify and secure important habitats.

Cooperation

With different venues available for cooperation, many different landowners, management agencies, and other organizations have been able to "buy in" according to their own needs in order to participate at various levels of conservation decision making. Examples of such venues include watershed partnerships, the Hawai'i Conservation Alliance, and Safe Harbor Agreements. In addition, an effective science-policy interface promotes an active reciprocal relationship between land managers and researchers.

Mainstreaming

A basic familiarity with Hawaiian forest birds and their conservation needs has become commonplace among the general public. Both regrettable losses and successful recovery programs have made the plight of Hawaiian forest birds well known. With the economic impacts of invasive species now well known, the public increasingly understands the issues and supports comprehensive control and prevention.

Hilo watershed forest reserves. These areas not only include considerable forest bird habitat that is of high quality and yet unmanaged, but they strategically connect existing conservation areas across recent and future areas of lava flows and could transform the region into a single conservation landscape. There are similar opportunities on Maui, particularly through rehabilitation of the mesic and xeric habitats that originally formed a nearly continuous forest around the western and southern slopes of Haleakalā Volcano. These areas are included in the Leeward Haleakalā Watershed Restoration Partnership area and, when restored, could provide important habitat for all Maui forest birds, especially the Maui Parrottbill (*Pseudonestor xanthophrys*).

Continued Management Action

There remains considerable need for expansion of numerous management actions despite the chronic dilemma of funding shortfalls for projects encompassing large areas or requiring expensive control methodology, particularly ungulate fencing, predator control, and biocontrol of weeds (Box 16.3). There is a major trade-off between management actions that are costly in the short term but need be applied only once and those that are less expensive but require continued implementation. Continually employing manual and chemical control of nonnative plants can be weighed against short-term investments in biocontrol. The latter is an expensive option, because considerable

---

Box 16.3. **Hawaiian Forest Birds in 2035:
How Is the Conservation Landscape Managed?**

Coordination

In core habitats, management agencies regularly and strategically pool their resources and target their actions, including not only habitat stabilization after ungulate removal but also control of direct threats such as predators. This action effectively maintains robust bird populations in core habitat areas. In matrix habitats, management actions are occasionally implemented, especially monitoring and targeting control of emergent invasive species. Forest bird conservation management is also coordinated with that for other organisms. For example, control of ungulates and rats also protects native plant species, and widespread control of predators is also beneficial to nesting seabirds.

Invasive Species Management

Controlling invasive species at different scales has been key in stabilizing and improving forest bird habitats. At the statewide scale, close regulation prevents new threats from entering Hawai'i or from spreading between islands. At the scale of individual islands, major nonnative threats have been reduced through a combination of biocontrol and strategic containment. New threats, while perhaps inevitable, appear infrequently, and well-established methodologies are in place to combat them when they appear on an island. The most widespread and persistent invasive species (especially social insects and predators) require continued intensive control in core habitat areas.

Adaptive Management

Between 2005 and 2035, few conditions have remained constant. Some threats have been greatly reduced (through biocontrol, for example), requiring research on how to best exploit the opportunities. Other threats have appeared suddenly and spread widely. Understanding the response of forest bird populations to these changing circumstances has been critical.

---

exploratory research and work on host specificity have to be undertaken by the interested agencies in special quarantine facilities prior to the release of potential biocontrol agents. Markin et al. (2002) estimated that each control agent successfully introduced to the Hawaiian Islands in recent years has required approximately $2 million in funding. Where the economic consequences of biological control programs have been evaluated elsewhere in the world, their cost-effectiveness has been variable, though generally favorable (van Wilgen et al. 2004). Not all invasive species are amenable to this approach, because suitable host-specific agents are not available for some. The more closely related the target and native species, the more difficult it is to find an agent that will not attack the native species (Pemberton 2000). When suitable biocontrol agents are available, the success rate to date in significantly reducing the impacts of the target invasive species has been about 50% (Julien and Griffiths 1998). However, the rate is increasing as better technologies, culturing and rearing strategies, and release mechanisms are developed. Even with the present success rates of biocontrol, the transformation of a serious habitat modifier to a more innocuous weed represents a clear improvement over costly and essentially perpetual manual control strategies.

One other thing resource managers have come to realize is that the concept of multiple use does not apply to certain activities in Hawai'i as it does in continental ecosystems. For example, it is not feasible to overlay sustained recreational hunting areas for ungulates with areas designated primarily for natural resource conservation because of the major impact on the latter objective. The fact that Hawaiian ecosystems evolved over millions of years without the presence of large mammals resulted in a biota that is highly susceptible to severe damage from these types of species. Most programs that focus on conservation of the native biological resources—those involved with natural area reserves, national parks, national wildlife refuges, and The Nature Conservancy preserves—have recognized this issue and have striven to eliminate ungulate populations from their management areas. A regional landscape conservation strategy needs to take this issue into account to ensure that core areas for conservation and game management are physically separated from one another.

The management for Hawaiian forest bird habitat in general may well be assessed and applied in at least two tiers. The first tier would involve overall habitat stabilization, such as eradication of ungulates and selected invasive plant species. The second tier would target more specifically other factors limiting the stability of forest bird populations, such as arresting nest predation, reducing disease vector populations, and preventing depletion of food arthropods during critical nestling and fledgling periods. Many areas are already slated to receive the first tier, for example, those areas where ungulate eradication is planned (see Table 16.3). The second tier, for maximum effect, would need to be applied within those areas where the first tier has been achieved and where potentially expensive and labor-intensive second-tier actions are likely to directly promote resources for forest birds or directly impede threats. Because all forest bird species co-occur in upper-elevation habitats (see Fig. 16.1), such a spatially targeted management scheme would offer benefits simultaneously to the full set of species.

Habitat Restoration Needs

In some extensive areas, restoration would enhance or add to existing habitat, to the benefit of forest birds. The best opportunities for this type of restoration are montane mesic regions formerly covered in koa forest on East Maui and Hawai'i. As threats such as ungulates and alien weeds are abated, the regeneration of native species should ideally result in communities that are resilient to further invasion and provide critical resources to birds.

In order to implement habitat restoration needs and fully take advantage of opportunities, conservation planners and managers will need to incorporate solutions to several stumbling blocks. For example, a streamlined permit and approval process could make it easier for landowners and conservation agencies to follow through on conservation plans. Similarly, engaging other parties would address the difficulties in depending on only variable, year-to-year government support for long-term funding, perhaps through the creative use of endowments, bequests, and "friends of" groups. Success in these and other tactical and strategic areas would go far toward developing habitat conservation and management for Hawaiian forest birds as part of the larger economic milieu of the islands. As has been found elsewhere, such a viewpoint more fully engages the public, which then concerns itself with the costs and benefits of more active management processes.

Expansion of Bird Populations

With new areas of habitat being improved or restored, native forest bird populations may have the natural capacity to expand their distributions. Another means of expanding bird populations may be through deliberate reintroductions into habitats they formerly occupied. Both the Puaiohi (*Myadestes palmeri*) and Palila have been successfully introduced from existing wild populations or by using captive-bred birds (Chapters 22 and 23). An apparently unsuccessful pilot translocation in 1996 sought to reestablish a population of 'Ōma'o on the west side of Hawai'i Island, where it had formerly occurred (Fancy et al. 2001). 'Alalā (*Corvus hawaiiensis*) that were reintroduced into another forest on western Hawai'i failed to reestablish for various reasons (Chapter 20). Future reintroduction programs for such species may exhibit more success if vigorous efforts to improve habitat quality are coupled with additional measures, such as predator control.

A bolder and potentially more controversial type of introduction would involve moving bird species among islands. The recent fossil record of Maui indicates that an extinct thrush once lived there (James and Olson 1991). To fill its place, extant related species, the 'Ōma'o and Puaiohi, could be considered for introduction to Maui. The importance of either frugivore in seed dispersal within native forests is clear, and Maui presently has no native frugivore. Also, a reserve population on Maui might ensure against a hurricane or similar catastrophic event. Obviously, any such interisland translocation would need substantial research to weigh the benefits and potential impacts. Nonetheless, the potential benefits to the bird species and to forest ecosystems in general are sufficient for the possibility to be considered.

Future Directions

Despite the dramatic decline of native plants and animals and their habitats since the arrival of humans in Hawai'i, much of this unique biota remains, and we now have better tools, mandates, and public support to halt, if not reverse, this declining trend. One of the most important steps is to prevent the introduction of new, problematic invasive species into these islands. Additionally, to be successful in the long term, conservation actions (both designation and management) need to be applied at the landscape scale, management actions have to be both effective and efficient, and restoration strategies need to be integrated and coordinated, both between organizations and on the ground.

The recent development of landscape or watershed partnerships throughout the state provides a unique opportunity to achieve conservation goals across large areas less restricted by property or jurisdictional boundaries. The model these efforts present—collaboration among landowners, managers, and researchers to enhance native biological resources across large landscapes—is already showing success by substantially increasing the size of managed habitats vital to the survival of Hawaiian forest birds. Unfortunately, at this point funding for most of the activities under the partnerships is provided on a year-to-year basis, primarily by the state's Natural Areas fund and short-term federal cost-sharing programs. It is extremely important that the area and funding support for these landscape partnerships be expanded and incorporated into the mainstream of conservation management in Hawai'i.

The continual refinement of management tools, such as control methods for invasive species and restoration techniques for depleted species, will be critical for success. It is essential that research and management efforts be closely integrated in the field. This strategy of adaptive management will allow for direct feedback from research and monitoring into management strategies to maintain and restore forest birds and their remaining habitats. Ultimately, however, the footprint of conservation actions in Hawai'i must include both large landscapes and a mix of "metahabitats" that are capable of supporting the diversity of both rare and common bird species within naturally evolving habitats. We hope that, through the development of sound conservation strategies and their implementation at both species and ecosystem levels, the native forest bird species that are a vital component of Hawaiian ecosystems will persist and increase in both distribution and abundance over the next century.

SUMMARY

The term landscape refers to large areas within which species and other elements of their ecosystem interact. Here we explore the importance of landscapes to the conservation of Hawaiian forest birds. The pre-human landscape of the Hawaiian Islands contained a diversity of forested environments housing many bird species that were adapted to unique aspects of island ecosystems. The arrival of humans initiated the widespread change of this landscape, with catastrophic consequences for the bird fauna. Conserving bird populations across broad areas is not only pivotal to their own survival but also bears on other aspects of conservation, for birds are central to many aspects of ecosystem function. However, maintenance of viable bird populations at this scale faces several challenges. In addition to administrative hurdles involving designation of lands for conservation and coordination among landowners and managers, landscape-level conservation must contend with numerous widespread threats, particularly habitat conversion and introduced species.

At present, owing to disease and other factors, forest birds may be found primarily in upland habitats on the six main islands, although notable lowland habitats also exist. Most habitats include some form of threat, and many rare bird species are restricted to those areas where threats are presently limited. Starting with the Hawaiian ahupuaʻa (land tenure) system, conservation efforts have progressed through the historical period, resulting in the designation of large portions of forest bird habitat for some sort of conservation. In recent years, a subset of these areas has been the focus of management actions intended to abate threats to forest bird populations. A statewide assessment of these areas indicates that, although conservation designation and management have expanded greatly in recent decades, much forest bird habitat remains unprotected.

Despite notable progress in managing habitats, dealing with many types of threats has proven difficult, expensive, and of limited efficacy. Management actions including control of mammalian predators and habitat-degrading ungulates, abatement of invasive plant species, and habitat restoration will all likely require implementation at larger spatial scales in order to sustain or improve bird populations. We present a vision of how to accomplish this task within the context of an expanded network of managed habitats. A functioning conservation landscape for the future would (1) consist of important core habitats for native birds embedded in matrix habitats that serve as buffers and provide connectivity; (2) result from comprehensive planning, interorganizational cooperation, and mainstreaming of conservation issues to promote public support; and (3) function through strategic and adaptive coordination among agencies managing forest birds and other biological resources.

ACKNOWLEDGMENTS

We acknowledge many people who have helped to shape our thinking on landscape management in Hawaiʻi. These include P. Banko, K. Bio, B. Gagne, S. Gon, R. Kennedy, L. Loope, L. Mehrhoff, S. Miller, S. Montgomery, D. Mueller-Dombois, T. Pratt, C. Rowland, and the late W. Gagne and L. Stemmermann.

CHAPTER SEVENTEEN
Managing Disease

DENNIS A. LAPOINTE,
CARTER T. ATKINSON,
AND SUSAN I. JARVI

The extinction and decline of Hawaiian avifauna due to introduced mosquito-borne disease has become a classic example of the impact of introduced diseases on naïve wildlife populations (Warner 1968; this volume, Chapter 9). Along with rinderpest virus and rabies virus, avian malaria (*Plasmodium relictum*) and avian pox virus (*Avipoxvirus* sp.) stood for decades as rare cases of invasive pathogens in wildlife species (Gulland 1995). Today, however, with ever-increasing globalization, emergent and invasive human and wildlife diseases are on the rise (Daszak et al. 2000, Gubler 2001, Friend et al. 2004). Once an academic curiosity to conservation biologists outside of Hawai'i, avian malaria and pox and their impact on Hawaii's native forest birds now have continental relevance with West Nile virus's sweep across North America and its potential threat to endangered species populations (Marra et al. 2004, Kilpatrick et al. 2007).

The Hawaiian Islands, like other isolated oceanic islands, experienced limited natural colonization by terrestrial biota. These founders brought few, if any, parasites and pathogens with them (Torchin et al. 2003; this volume, Chapter 1). For parasites with complex life cycles, alternate hosts were absent, as were the insect vectors of pathogens, such as mosquitoes, black flies, biting midges, and others. This all changed with the arrival of Westerners to the Hawaiian Islands in 1778 and the subsequent introduction of mosquitoes and mosquito-borne avian disease. Introduced mosquito-borne avian disease is considered a major factor limiting Hawaiian forest bird populations and an obstacle to the restoration of the islands' avifauna (U.S. Fish and Wildlife Service 2006). Future efforts to protect remaining Hawaiian forest bird species and to restore their populations will rely on the development of disease management strategies that can be applied at the landscape level. The long history documenting the control of vector-borne human disease (Harrison 1978) suggests that there will be no simple solutions and that an integrative and diligent approach will be necessary (Rose 2001).

In this chapter we (1) provide a brief overview of the mosquitoes in Hawai'i and the pathogens they vector to birds (see Chapter 9 for a more detailed account of disease biology); (2) outline possible management practices and evaluate them in the context of endemic Hawaiian bird conservation, from single-species captive populations to intact communities across broad landscapes; and (3), in synthesis, suggest management strategies to minimize the impact of vector-borne disease in Hawaiian forest birds.

KNOW THINE ENEMY

Mosquito Vectors of Avian Pathogens in Hawai'i

Six biting species of mosquito have become established in the Hawaiian Islands since the nineteenth century: *Culex quinquefasciatus* (established by ca.1826), *Aedes aegypti* (ca. 1892), *Aedes albopictus* (ca. 1902), *Aedes vexans nocturnus* (ca. 1962), *Wyeomyia mitchellii* (ca. 1981), and *Aedes japonicus* (ca. 2003) (Hardy 1960, Joyce and Nakagawa 1963, Shroyer 1981, Larish and Savage 2005) (Fig. 17.1). The first three species are known vectors of human pathogens and have been widely distributed throughout the tropic and subtropic regions by Western commerce (LaPointe 2007). By contrast, *A. v. nocturnus* and *A. japonicus* are competent laboratory hosts of encephalitis viruses but are not documented vectors of human or wildlife pathogens (Turell et al. 2001, Sardelis et al. 2003). *W. mitchellii* is not known to transmit any vertebrate pathogens (Shroyer 1981).

Only *C. quinquefasciatus, A. albopictus, A. japonicus*, and *W. mitchellii* have been found in native forest bird habitats, and only *C. quinquefasciatus* is presently common above 900 m in elevation (Goff and van Riper 1980, LaPointe 2000). *Culex quinquefasciatus* is a known vector of avian malaria in Hawai'i (LaPointe et al. 2005) and is the most likely vector of avian pox virus (van Riper et al. 2002). Only a few individuals of *A. albopictus* and *W. mitchellii* were found to support sporogany of *Plasmodium relictum* in the laboratory (LaPointe et al. 2005). Preliminary trials with the newly established *A. japonicus* suggest that this species does not support sporogony in the laboratory (LaPointe, unpubl. data). Avian pox virus does not require a specific vector and may be transmitted by any mosquito or biting arthropod. Although these three mosquito species may be more or less opportunistic in host selection (Tempelis et al. 1970, Edman and Haeger 1977, Tanaka et al. 1979) and abundant in some lowland Hawaiian forests, their marginal susceptibility to avian malaria and limited altitudinal distribution make them unlikely to be important vectors of either avian malaria or pox. Their possible role as minor, incidental vectors of avian malaria or pox is still unknown. Thus, due to its high level of vector competence and altitudinal distribution, *C. quinquefasciatus* can alone account for the current distribution and prevalence of avian malaria and pox in forest bird communities.

Hawaiian Landscapes, Feral Pigs, and Mosquito Abundance

Larval mosquitoes are aquatic and can be found in a wide range of temporary and permanent waters. *C. quinquefasciatus* larvae occur in natural and artificial containers, ditches, puddles, irrigation channels, cesspools, and the margins of ponds and flowing streams. This species is adapted to eutrophic waters heavily enriched with organic matter. Although generally not found in forest ground pools or open bogs in Hawai'i (Fig. 17.2), *C. quinquefasciatus* larvae have been recovered from ground pools and wallows where fecal

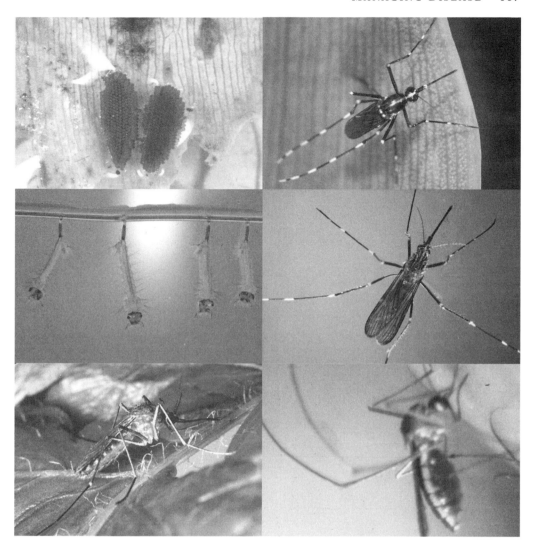

Figure 17.1. Common mosquito species found in forest bird habitats on Hawai'i Island. Left side, top to bottom: Egg rafts, larvae, and adult female *Culex quinquefasciatus*, the main vector of avian disease in the Hawaiian Islands. Right side, top to bottom: *Aedes albopictus*, *Aedes japonicus*, and *Wyeomyia mitchellii*. Source: Photos by Dennis LaPointe, U.S. Geological Survey.

matter from livestock or feral ungulates may have enhanced the microhabitat (D. LaPointe, pers. obs.).

Mosquitoes do not typically occur in many natural areas in the Hawaiian Islands because of temperature constraints on their development or the absence of suitable larval habitat. Adult and larval mosquitoes were rarely encountered in windward for-

ests above 1,500 m in elevation on Hawai'i Island (Goff and van Riper 1980, LaPointe 2000). Intermittent and ephemeral streams, however, may be important larval mosquito habitats in some Hawaiian landscapes. Surveys in Kīpahulu Valley on Maui documented *C. quinquefasciatus* larvae in rock pools of intermittent streambeds (Aruch et al. 2007).

In stark contrast, the younger volcanic landscapes of the east flank of Mauna Loa, Hawai'i Island, are all but devoid of permanent surface water. *C. quinquefasciatus*, however, are abundant in these wet forests, where their larvae rely primarily on rain-water-filled cavities in the native tree fern,

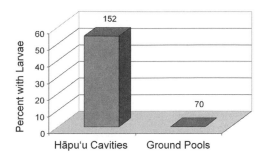

Figure 17.2. Larval *Culex quinquefasciatus* occupancy of available aquatic habitats in windward Mauna Loa forests, Hawai'i Island. The numbers over the columns represent the total number of individual habitats of the habitat type sampled for mosquito larvae that were encountered in 4.5 ha of forest. Although prevalent in these forests, ground pools do not appear to support mosquito larvae (LaPointe 2000).

hāpu'u (*Cibotium* spp.) (Fig. 17.3). These cavities are formed by the feral descendants of domestic pigs (*Sus scrofa*), which feed on the starchy core of the tree fern. After extracting the starch, a cup- or bowl-shaped cavity remains that will collect rainwater and leaf litter, thereby providing a favorable habitat for larval mosquitoes. Although the geological and hydrological nature of the Mauna Loa landscape precludes the production of mosquitoes, the occurrence of hāpu'u cavities throughout

large tracts of forest undermines this natural protection (LaPointe 2000) (see Fig. 17.2).

Feral pigs, however, are not the only culprits. The numbers of *C. quinquefasciatus* are much greater in suburban and agricultural areas than in natural areas. Conservation areas in Hawai'i often abut residential and agricultural communities that can produce high densities of mosquitoes through the creation of larval habitat and the presence of abundant hosts for blood meals (Mian et al. 1990, Reisen et al. 1990, Reisen et al. 1992). The rural community of Volcano Village, located just outside the boundaries of Hawai'i Volcanoes National Park, serves as an excellent example. Mosquito capture rates in the village are nearly three times greater than capture rates within the nearby forest (Reiter and LaPointe 2007). Household water storage in residential areas contributes to local mosquito abundance, but the impact of artificial containers and impoundments may be several times

Figure 17.3. Water-filled cavity in a hāpu'u tree fern (*Cibotium glaucum*) trunk created by feral pig feeding. Such cavities are the main breeding sites for mosquitoes in some native forests. Removal of feral pigs would reduce the number of breeding sites, which in turn would potentially reduce disease transmission among birds. *Source:* Photo by Daniel Lease, U.S. Geological Survey.

greater on agricultural lands. Cattle operations, in particular, create favorable habitats for C. quinquefasciatus in the form of stock ponds, troughs, cisterns, settlement ponds, and the old tires and tarps commonly used to cover feed (Reiter and LaPointe 2007).

Because the native tree ōhiʻa-lehua (*Metrosideros polymorpha*) still dominates many of these residential and agricultural landscapes, birds such as ʻApapane (*Himatione s. sanguinea*) and ʻIʻiwi (*Vestiaria coccinea*) move readily between the forests and suburban or agricultural areas where the likelihood of exposure to malaria and pox may be greater. Mosquitoes from these areas may also be dispersing into forests, thereby augmenting forest populations. Female C. quinquefasciatus were recaptured at the 3 km trap boundary during mark-release-recapture experiments in a closed-canopy Hawaiian forest. After accounting for trap density, 1.6 km would be a conservative estimate of the average dispersal of C. quinquefasciatus in 10 days (LaPointe 2008). Mainland studies in urban environments showed that C. quinquefasciatus can disperse up to 12.6 km (Reisen et al. 1992). The dispersal ability of this species has serious implications for disease management in fragmented areas.

On Mauna Loa's east flank, forest mosquito populations disappear during droughts but return after sufficient rain, suggesting that residential or lowland areas may serve as the source of these rebounding populations (LaPointe 2000). Low-elevation mosquito populations outside of forest bird habitats might hypothetically contribute to avian malaria and pox transmission if carried by wind to higher elevations (Scott et al. 1986).

POTENTIAL DISEASE MANAGEMENT STRATEGIES

Manipulation of Susceptible Hosts by Chemotherapy and Vaccination

The successes in vaccine development and the use of antimicrobial agents in the control of infectious disease in humans and domestic animals are well documented. However, these approaches have not been as successful in wild animal populations. Chemotherapy of wildlife populations is generally limited to very local situations, and immunizations of wildlife are dependent on novel methods of vaccine delivery (Wobeser 2002). Most wildlife vaccines are directed at mammalian reservoirs of human disease (Slate et al. 2005), livestock disease (Wilkinson et al. 2004), or critically endangered populations (Hastings et al. 1991). There are no examples of vaccine use in wild birds.

ANTIMALARIAL AGENTS

Chemotherapy of birds has been used only in captive or closely managed flocks. Common antimalarial agents used to treat human malaria, such as chloroquine and primaquine, were first evaluated in birds and may work as a prophylaxis, as supportive treatment, or as a radical cure for malaria in captive birds (Hewitt 1940, Stoskopf and Beier 1979). For example, chloroquine and primaquine were used to treat captive-reared, endangered ʻAlalā (Hawaiian Crows, *Corvus hawaiiensis*) that became infected with malaria while acclimating in outdoor aviaries prior to release. The crows recovered completely and were ultimately released (Massey et al. 1996).

The value of antimalarial agents in the treatment of wild birds is limited by the difficulties of delivery. Dosing wild birds via artificial nectar feeders is impractical and likely to fail at the landscape level because dosages cannot be regulated, and low doses could lead to rapid selection for drug-resistant parasites (White 2004). Similarly, deliberate exposure of birds to malaria and chemotherapeutic control of the acute phases of infection might allow individuals to mount an immune response that would ultimately be able to regulate and control chronic infections. However, this approach will also ultimately select for drug resistance in the parasite population. In addition, chronic infections may relapse or recrudesce when birds are immunologically stressed (Valkiūnas 2005).

AVIAN MALARIA VACCINE

Vaccines against protozoan pathogens have been notoriously difficult to develop, but they can avoid some of the problems associated with chemotherapy and drug resistance (Jones and Hoffman 1994). Attempts to develop subunit vaccines for human malaria have been unsuccessful because of the complex life cycle of the parasite, coinfection of multiple strains, and the ability of Plasmodium to change its antigenic, surface proteins (Desowitz 2000). However, recently there has been renewed interest in using irradiated sporozoites as a live, attenuated human vaccine. In laboratory trials with rodents and human subjects, this approach provided protective immunity for at least 10 months (Nussenzweig et al. 1967, Clyde et al. 1975, Hoffman et al. 2002). Its potential for use in Hawaiian birds has yet to be evaluated.

A technically more sophisticated approach is the use of DNA vaccines based on short sequences of specific genes (Doolan et al. 1996). Recently, a DNA vaccine using the circumsporozoite protein (CSP) of P. relictum afforded moderate but short-lived protection in canaries (McCutchan et al. 2004). In a similar CSP vaccine trial, the prevalence of malaria in a captive flock of penguins was reduced from 50% to 17% following vaccination. The vaccination of part of the captive flock also protected unvaccinated individuals by decreasing the proportion of susceptible individuals (Grim et al. 2004). DNA vaccines have clear benefits for captive populations, but the short duration of the immunity afforded limits their effectiveness in wild populations.

AVIAN POX VIRUS VACCINE

Attenuated live pox virus vaccines have been available for domestic birds since the 1940s and are commonly used in high-risk areas where birds and/or mosquitoes occur in high densities (Ritchie 1995). The vaccine is typically delivered by injection into the wing web, but reimmunization is often necessary, limiting its usefulness in wild birds. Another major concern is recombination of the attenuated virus in the vaccine with other viral strains or pox viruses, resulting in more virulent strains. Vaccine contaminants are also of concern. Reticuloendotheliosis virus causes immunosuppressive disease in domestic fowl and is a common contaminant of commercial fowlpox vaccine (Garcia et al. 2003). Finally, immune responses to avian pox can be very host-species specific, and additional information on the diversity and pathogenicity of avian pox in Hawai‘i is needed before an effective vaccine can be developed (Tripathy et al. 2000; Jarvi, Triglia, et al. 2008).

Even in the event of the development of a successful malaria or pox vaccine, difficulties in the method of delivery remain a formidable obstacle to the immunization of large wild populations. But technological advances offer some hope. It is now theoretically feasible to have a ubiquitous and potentially benign virus, such as fowl pox, vector recombinant DNA for the antigenic proteins of any number of potential pathogens (Paoletti 1996). Because pox viruses are easily transmitted by mosquito, a nonpathogenic pox virus or viral strain might serve as a vector for a Plasmodium or arboviral gene. A fitting irony would be use of the mosquito as the ultimate delivery system of this engineered vaccine.

Naturally Malaria-Tolerant Birds

There is evidence that some lowland populations of Hawai‘i ‘Amakihi (Hemignathus virens) may be evolving an innate tolerance to avian malaria (Woodworth et al. 2005). Translocation of tolerant birds to new sites may introduce this presumably heritable trait to currently vulnerable populations. However, translocated birds may introduce pathogenetically different strains of malaria or new pathogens into the recipient population and should be properly screened and/or treated before release (Atkinson and van Riper 1991, Cunningham 1996). It has also been suggested that tolerant individuals be brought into captivity as a breeding population for propagation and future release into the wild.

Whether translocated or captive-bred birds are used, the presumed heritable traits of disease tolerance may not become fixed in the wild populations if similar selective pressures—especially perennial malaria transmission—do not prevail at the release site or if gene flow from distant populations overcomes the introduced resistant genes. The latter is most likely to occur with highly vagile species such as 'I'iwi and 'Apapane (Fancy and Ralph 1997). For these reasons, the translocation of disease-tolerant individuals may have its greatest value in facilitating the lowland expansion of naturally evolving, disease-tolerant populations.

Manipulation of the Vector

Reduction of vector abundance and longevity remains the central control strategy for mosquito-borne disease today. Whether the pathogen is dependent on the vector to complete transmission (as in the case of *Plasmodium* and arboviruses) or the vector merely facilitates mechanical transmission (as with avian pox), the mosquito is the crucial link between infected and susceptible host. Remove the link, and transmission ceases. The early successes in human malaria and yellow fever control were both brought about through the control of the mosquito vector.

Many characteristics of the Hawaiian avian disease system would seem to make control through vector reduction an unattainable goal (Chapter 9). First, avian malaria and avian pox are pervasive in the Hawaiian Islands, and the prevailing climate favors transmission throughout most of the year. Second, the combined effects

of a highly competent vector, a favorable climate, and an ample reservoir of hosts make for efficient transmission even at low densities of *C. quinquefasciatus* relative to mainland areas. Third, native birds with chronic malarial infections are reservoirs for life (Jarvi et al. 2002), and although there are no estimates of species longevity, there are records of individually banded native passerines' living in excess of 10 years (Lindsey et al. 1998). Finally, the land needing control is enormous; approximately 25% of the state's land area (405,000 ha) is under some form of natural resource protection (Loope and Juvik 1998). These limitations suggest that management of avian disease in Hawai'i will require control methods that are exceptionally effective, cost-efficient, environmentally safe, and indefinitely maintained.

SOURCE REDUCTION OF LARVAL MOSQUITO HABITAT

The most effective way to limit mosquito numbers is to remove larval mosquito habitats, a practice referred to as source reduction. From the time of the Roman Empire to recent times, great efforts were made to drain swamps and lowlands where human malaria was common (Harrison 1978). These efforts were successful, and source reduction was practiced widely throughout the world until the environmental movement of the 1970s raised the alarm that wetlands were invaluable ecosystems of great productivity and biodiversity (Mitsch and Gosselink 1993). Although swamps and wetlands are no longer drained in the name of mosquito control, source reduction of artificial habitats or altered natural habitats remains a valuable component of most mosquito abatement programs.

Perhaps the greatest potential use of source reduction in Hawai'i lies in the residential and agricultural communities that encroach on natural areas. Catchment systems for household water are common in rural Hawai'i and are often poorly maintained. Artificial containers, particularly cemetery vases, refuse containers, discarded tires, and construction materials, may harbor hundreds of mosquito larvae in a relatively small volume of water. Unfortunately, control of mosquito production by means of source reduction requires considerable cooperation and vigilance from the public, and without strong incentives (or penalties) community-based campaigns are seldom successful. Source reduction campaigns are more likely to succeed on government-owned lands where environmental management practices can be mandated. On a small scale, such as in an atoll refuge, mandated source reduction has the potential to eliminate vectors entirely.

In Midway Atoll National Wildlife Refuge, removal of refuse containers and tarpaulins associated with environmental mitigation work and the decommissioning of military infrastructure was followed by a decrease in the occurrence of avian pox in nestling seabirds (J. Hale, U.S. Fish and Wildlife Service, pers. comm.). Unfortunately, C. quinquefasciatus and A. albopictus are still present on Midway Atoll, and in 2005 an avian pox outbreak occurred in nestling Laysan Albatrosses (Phoebastria immutabilis) (J. Klavitter, pers. comm.). The creation of artificial wetlands for the restoration of Laysan Ducks (Anas laysanensis) on Midway (U.S. Fish and Wildlife Service 2004b) may have inadvertently increased mosquito populations.

Similarly, in Hawai'i Volcanoes National Park, the feasibility of breaking avian malaria transmission through the source reduction of larval mosquito habitat was tested at a 100 ha former cattle ranch where the larval mosquito habitat was associated with ranch infrastructure. Though unpublished, the study demonstrated that the removal and treatment of troughs and cisterns virtually eliminated mosquitoes (D. LaPointe, unpubl. data). Unfortunately, the mosquito population rebounded when a new larval habitat was created by a nearby construction project. These examples emphasize that for source reduction to succeed, mosquito control must be a priority whenever alterations to refuge lands are considered.

Source reduction need not be entirely limited to human infrastructure. As mentioned earlier, feral pigs create larval mosquito habitats throughout the wet forests of Hawai'i. Source reduction of hāpu'u cavities by elimination of the feral pig is feasible but not without great cost and controversy. The costs of pig eradication in Hawai'i Volcanoes National Park during the 1980s were estimated to be $24,000 per km for fencing, $25,000 per year for fence maintenance and inspection on 71 km of fence line, and an average of $95 per pig for hunting (Hone and Stone 1989, Katahira et al. 1993). However, the cost of feral pig control is not the only obstacle. Feral pigs are a favored game species, and proposals to control or eliminate them from public lands have been met with opposition from some hunters. After years of dialogue between conservationists and hunters, the controversy over the control of feral pigs on some public lands continues (Tummons 1997).

CHEMICAL CONTROL OF MOSQUITO VECTORS

The use of insecticides to control mosquitoes has saved countless lives worldwide from the ravages of vector-borne disease. Chemical approaches to mosquito control also have had a negative impact on the environment. Over the past century, the chemical control of mosquitoes has shifted from treating for larvae (larviciding) with coal oil (kerosene) (Van Dine 1904) and the chlorinated hydrocarbon DDT to targeting adults (adulticiding) with organophosphates and pyrethroids (Nakagawa 1964, Mulla 1994, Rose 2001). In the past three decades, concern over the impact of pesticides on human health and the environment has brought about a more integrated approach to mosquito control, combining source reduction and biological control with novel, environmentally sound larvicides and ultralow-volume (ULV) adulticiding (Mulla 1994, Rose 2001).

When source reduction is inappropriate and larval mosquito habitats are accessible, larviciding is often the preferred method of control. Since the 1970s, artificial insect growth regulators and surfactants have been used in natural and man-made waters for control of mosquito larvae (Mulla 1994). Insect growth regulators such as methoprene (Altosid®) interfere with pupal molting by mimicking a hormone common to all insects and crustaceans, whereas surfactant monomolecular films alter the surface tension of natural waters and affect surface-respiring invertebrates. Although some adverse nontarget effects have been reported, these chemicals are essentially nontoxic to vertebrates and most invertebrates when applied at recommended rates (Nayar and Ali 2003, Pinkney et al. 2005). These larvicides can be effectively used in suburban and agricultural settings, but larval mosquito habitats in Hawaiian forests are too widely dispersed and inaccessible to allow employment of conventional larviciding.

When environmental conditions are not suitable for larviciding or when faced with a public health emergency, mosquito control professionals rely on adulticiding. Modern ULV adulticiding applies a small amount of organophosphate or pyrethroid insecticides per hectare and, using aerial applications, can treat vast areas in a short time. Under ideal conditions, ULV adulticiding can achieve 90% control without detectable nontarget effects (Mount 1998, Jensen et al. 1999, Zhong et al. 2003). Still, ULV adulticiding is no more likely to aid in the control of avian disease in Hawai'i than is conventional larviciding. ULV applications are not as effective in dense vegetation (Mount 1998) and therefore are unlikely to penetrate a closed-canopy Hawaiian forest. Furthermore, adult

mosquito numbers often rebound quickly after adulticiding. In Hawai'i, *C. quinquefasciatus* has multiple cohorts throughout the year and would require continuous applications to suppress adults for any length of time. Although chemical control may be effective in large open wetlands or in suburban or agricultural environments, it is difficult to imagine a cost-efficient and effective landscape-level broadcast of insecticides over native Hawaiian forests.

Aside from the economic and logistical constraints, there are greater concerns regarding the use of chemical pesticides. Evolving insecticide resistance makes chemical control subject to complete failure. Attempts to manage resistance may postpone or even prevent failure of chemical control, but such management would require exceptional coordination with the agricultural community (Brogdon and McAllister 1998). The final factor weighing against chemical control is the actual and perceived adverse impact of insecticides on human health and the environment. Mainland monitoring for nontarget effects associated with ULV adulticiding has been largely limited to a few indicator species (Mount 1998). Little is known about the long-term effects of adulticiding on biodiversity. Hawai'i has more than 5,000 endemic arthropod species, including more than 20 species proposed for federal listing as endangered (Howarth and Mull 1992). The regulatory hurdles required just to initiate adulticiding in Hawaiian natural areas would be daunting and likely insurmountable. Given the nonselective toxicity of these chemicals and the already fragile status of Hawaii's native fauna, it would be extremely problematic to consider chemical control in Hawaiian forest bird habitats.

CLASSICAL BIOLOGICAL CONTROL OF MOSQUITO VECTORS

Classical biological control can be defined as the control of a pest species by introduced natural enemies. The earliest attempts to control mosquitoes in Hawai'i included the establishment of the wrinkled frog (*Rana rugosa*) and the dart-poison frog (*Dendrobates auratus*). Both species failed to control mosquito numbers (Oliver and Shaw 1953).

A number of poeciliid fish have been identified as predators of mosquito larvae and have been used extensively in Hawai'i (Nakagawa and Ikeda 1969). The first species imported for mosquito control were the western mosquito fish (*Gambusia affinis*), sailfin molly (*Poecilia latipinna*), guppy (*Poecilia reticulata*), green swordtail (*Xiphophorus hellerii*), and southern platyfish (*Xiphophorus maculatus*) (Van Dine 1907, Brock 1960). The western mosquito fish is extremely adaptable and effectively reduces mosquito numbers. Unfortunately, it also displaces native fish species (Meisch 1985). The guppy is more effective in polluted waters than mosquito fish and has been used in Hawai'i to control *C. quinquefasciatus* in poultry waste runoff (Bay 1985). Both species were released in natural areas as well as artificial impoundments, so today these fish are ubiquitous in lowland waters of the state. Introduced poeciliid fish have been implicated in the population loss of native *Megalagrion* damselflies in Hawai'i (Englund 1999). Due to their proven negative impact on native aquatic biodiversity, introduced fish should not be considered for mosquito control in natural areas.

Over the past century, a number of invertebrate predators have been considered for mosquito control (Lacey and Orr 1994). *Toxorhynchites* is a genus of tree hole–inhabiting mosquitoes characterized by large predacious larvae that prey on smaller mosquito larvae. The iridescent blue-green adults are notably larger than other mosquitoes and do not feed on blood (Steffan 1968). Two species, *T. brevipalpis* and *T. amboinensis*, were released between the years 1950 and 1959 (Bonnet and Hu 1951, Nakagawa 1963, Steffan 1968). Although both species are probably established on all major islands, they have not reduced the populations of *A. albopictus* or any other container-inhabiting species (Nakagawa 1963). The failure of *Toxorhynchites* as a self-sustaining biocontrol agent in Hawai'i is consistent with other attempts made worldwide; only through inundative releases of this predator has control been achieved (Lacey and Orr 1994).

Copepods in the genus *Mesocyclops* have shown some promise for the control of container-inhabiting mosquitoes (Riviere and Thirel 1981). Marten (1984) observed that in Hawai'i *A. albopictus* larvae were readily fed upon by a naturally occurring copepod, *M. leuckarti pilosa*. Under experimental conditions, these copepods provided complete control of the mosquito. Similar success using various *Mesocyclops* spp. has been reported for other container-inhabiting *Aedes* spp. (Gorrochotegui-Escalante et al. 1998, Kay et al. 2002, Nam et al. 2005). *Mesocyclops* are less effective at controlling *C. quinquefasciatus*, although some species show promise. A few copepod species have been adventitiously established in the Hawaiian Islands, including a species commonly used in control, *Mesocyclops aspericornis* (Nishida 2002). Copepods are frequently encountered in streambed rock pools even at high elevations (1,800 m), but their significance in limiting mosquito populations is unknown (D. LaPointe, unpubl. data). The value of *Mesocyclops* for control of mosquitoes in natural areas is greatly minimized by its susceptibility to desiccation and limited dispersal ability. As in the case of other invertebrate predators, periodic augmentation of copepod populations would be necessary to achieve lasting control (Lacey and Orr 1994).

Microbial pathogens of mosquitoes include protozoa, fungi, bacteria, and viruses. Many of these agents do not recycle or reproduce and must be reapplied in the manner of a chemical agent. They are commonly referred to as biopesticides. The most extensively used agent is the bacterium *Bacillus thuringiensis israelensis* (B.t.i.) (Margalit and Dean 1985). It produces a toxin very selective for mosquitoes and other nematoceran flies such as midges, blackflies, and biting midges. No significant adverse nontarget effects have been reported from numerous laboratory and field trials (Lacey and Mulla 1990). However, there is evidence that the long-term use of B.t.i. may alter aquatic invertebrate communities (Hershey et al. 1998). In Hawai'i, there is great concern that microbial agents will impact already threatened native species such as *Megalagrion* damselflies (Howarth 1991), but there is little evidence of direct or indirect toxicity of B.t.i. to odonates (Painter et al. 1996). The major disadvantages of B.t.i. use are its reduced efficacy in polluted water, its nonresidual nature, and difficulties associated with its application in dense vegetation.

Bacillus sphaericus came into commercial use later than B.t.i. and is particularly effective against *Culex* species (Singer 1985). Unlike B.t.i., *B. sphaericus* provides good residual control, in part due to some natural recycling in the environment (Lacey et al. 1987). Some *C. pipiens* complex populations have developed resistance to *B. sphaericus*, but there is no evidence of cross-resistance with B.t.i. Thus, simple rotation of control agents in operational use should delay or prevent resistance development (Zahiri and Mulla 2003). No acute toxicity to or toxic trophic effects on nontarget species have been reported (Aly and Mulla 1987).

The most recently discovered microbe showing some promise for controlling *C. quinquefasciatus* populations is a baculovirus found naturally infecting *C. nigripalpus* (Becnel et al. 2001). This baculovirus has undergone limited testing to date and appears to infect only *Culex* species. It is a likely candidate for further development as a biopesticide (Andreadis et al. 2003).

In recent decades, the practice of biological control has been reevaluated as a growing body of evidence has clearly demonstrated the loss of native biota following the introduction of biological control agents (Simberloff and Stiling 1996). Some of the most compelling evidence comes from the Hawaiian Islands, where early agricultural researchers introduced hundreds of generalist predators and parasitoids, to the detriment of native insects (Howarth 1991, Brenner et al. 2002). Certainly, the early introductions of fish for mosquito control have had an impact on the native aquatic biota of the islands (Englund 1999). It would appear that neither vertebrate nor invertebrate predators are suitable for control of mosquitoes in Hawaiian natural areas, because these generalists might depredate an already naturally depauperate aquatic fauna. Short-lived, selective biopesticides pose the least environmental hazard and have their greatest value in artificial impoundments, although they may prove useful in streambeds or constructed wetlands. Both B.t.i. and *B. sphaericus* are currently used by Hawai'i Department of Health (HDOH) Vector Control for mosquito control in artificial impoundments. *B. sphaericus* has been approved for use in natural wetlands.

THE STERILE MALE TECHNIQUE
AND CYTOPLASMIC INCOMPATIBILITY

Perhaps the most efficient and elegant form of mosquito control would make the mosquito the agent of its own demise. Autocidal control methods have been considered since the 1940s and employ radio- or chemosterilization of males, genetic translocations, or bacterial symbiotes to interfere with insect reproduction.

The sterile male technique (SMT) has been employed numerous times against both agricultural pests and vectors of human or livestock disease (Asman et al. 1981, Klassen 2003). However, most attempts have had limited success in suppressing or eliminating populations, and many of those that have been successful have targeted island or incipient populations (Vreysen et al. 2000, Koyama et al. 2004). Perhaps the most successful attempt to suppress a mosquito population with SMT was conducted on Seahorse Key in Florida in 1969. The daily release of as many as 18,000 chemosterilized males for 10 weeks effectively eliminated C. quinquefasciatus from the key (Patterson et al. 1970). Later attempts by Patterson and co-workers (1977) were not as successful, and their failure was attributed to reduced sexual competitiveness in radiosterilized males and immigration into the target area by inseminated females.

Cytoplasmic incompatibility (CI) was first observed in the 1950s when researchers reported that some crosses of mosquito populations in the C. pipiens complex resulted in aborted embryonic development (Laven 1959). Originally thought to be a genetic effect, this phenomenon was later discovered to be caused by a maternally inherited bacterium, Wolbachia pipientis (Yen and Barr 1971). CI can occur when an infected male mates with an uninfected female or when infected males and females of different crossing types mate (Bourtzis and O'Neill 1998). The potential use of CI for the control of mosquitoes was demonstrated by Laven (1967), who was able to eradicate a wild population of C. quinquefasciatus in Burma through the repeated releases of incompatible males. Despite the apparent success of this early trial, CI has been largely ignored as a control strategy.

Successful suppression or eradication of mosquitoes using SMT or CI requires that certain biological and logistical criteria be met: (1) target species should display female monogamy and male polygyny, (2) target populations should be limited in number and isolated from immigration from native inseminated females or mosquitoes of different CI crossing types, (3) sterile or Wolbachia-infected males should exhibit competitive mating behavior, (4) the technique must achieve a rate of sterility >95 %, (5) there must be adequate facilities and proven techniques for the mass production of sterile or Wolbachia-infected males and the 100% exclusion of females, and (6) a reliable delivery system must be in place (Asman et al. 1981, Townson 2002).

C. quinquefasciatus populations have been suppressed in isolated villages and islets using SMT and CI, but is either approach applicable to the control of avian disease in the Hawaiian Islands? U.S. Department of Agriculture facilities and programs for the mass rearing and sterilization of tephritid fruit flies have been in Hawai'i for some time, and many of the logistical considerations about the rearing and release of these insects could be adapted for mosquitoes. An archipelago-wide survey for Wolbachia infection in C. quinquefasciatus needs to be conducted to determine if multiple crossing types are present or if an incompatible synthetic strain could be developed.

The size and landscape heterogeneity of the larger Hawaiian Islands would be insurmountable obstacles to mosquito eradication by SMT or CI. These techniques are more appropriate to smaller islands and atolls. Midway Atoll is geographically isolated, small, and accessible by air transport, making it an excellent location for an SMT or CI project. Complete eradication of mosquitoes on Midway Atoll would open the way for the translocation of Northwestern Island endemic passerines such as Nihoa Finches (Telespiza ultima) or Laysan Finches (T. cantans).

TRANSGENIC OR GENETICALLY MODIFIED MOSQUITOES (GMMS)

SMT and CI approaches to mosquito suppression have been largely eclipsed in recent years by genetic engineering, and the theoretical replacement of vector populations with a refractory transgenic mosquito (Box 17.1) has become the new Holy Grail of malaria control (Crampton et al. 1990, Morel et al. 2002). The main thrust of this research has been the development of molecular tools for the stable genetic transformation of mosquitoes (Catteruccia et al. 2000, Allen et al. 2001), identification of refractory effector genes that will block parasite development (Niare et al. 2002), and the development of mechanisms to drive the refractory genotype into wild mosquito populations (Rasgon and Gould 2005). Much of this work has focused on the Anopheline vectors of human malaria, most notably *Anopheles gambiae*. The recently completed genome map of *A. gambiae* may aid the development of a refractory transgenic mosquito (Holt et al. 2002), but finding an efficient genetic drive mechanism remains the major technical obstacle and may take several more years to resolve (Morel et al. 2002).

There are many concerns associated with the transgenic approach to malaria control, ranging from the ethical to the ecological. General criticism focuses on the redirection of available malaria research dollars away from integrated control strategies and toward the transgenic approach. Some biological concerns include the outright failure of this approach to reduce transmission, the variability in phenotypic expression of the transgene, the transgene's impact on fitness, inadvertent enhanced virulence, and a breakdown in efficacy over time, followed by a pandemic resulting from a loss of population immunity (Spielman 1994, Scott et al. 2002, Tabachnick 2003).

What is unanimously agreed upon is that any responsible trial release of a transgenic mosquito should be done in such a way that any negative consequences would be limited in both geographical extent and impact on the target population. The avian malaria system in the Hawaiian Islands has many of the recognized criteria of a responsible (from an anthropocentric perspective) test location (Clarke 2002), and the first step in developing a transgenic C. *quinquefasciatus* has already been taken (Allen et al. 2001). It remains to be seen if further consideration of the Hawaiian avian malaria system in transgenic research will materialize. Transgenic techniques might have greater immediate value in the development of markers or conditional lethal genes for mosquitoes used in CI or SMT programs (Benedict and Robinson 2003). Even if this long-shot technique were feasible, changing the vector competence of a mosquito population for malaria would have no impact on the mosquito's ability to serve as a mechanical vector of avian pox.

PROTECTING ENDANGERED BIRD POPULATIONS

A number of high-elevation refugia with intact endemic forest bird communities were identified on the islands of Kaua'i, Maui, and Hawai'i (Scott et al. 1986). Only some of this land was already under federal, state, or private management (Chapter 16).

Box 17.1. **Transgenic Mosquitoes 101**

In genetic engineering, genes from one organism are incorporated into the genome of another organism. The resulting offspring that have successfully received the transgene are said to be transformed or transgenic. Initial transformation is achieved by direct injection of the target gene and a transposable element into the recipient eggs with a microsyringe. The transposable element is a segment of DNA that moves genes about—the molecular equivalent of cutting and pasting. Refractory effector genes encode for mechanisms that prevent the development of the parasite or pathogen in a mosquito, and they are the target genes for malaria vector transformation. A genetic drive mechanism is a method by which the refractory genes are spread throughout a vector population.

In 1985, land was set aside for the Hakalau Forest National Wildlife Refuge (NWR), and in 1986 the Hanawī Natural Area Reserve was established on Maui. Preservation of high-elevation forest has continued with the acquisition of the South Kona subunit of Hakalau Forest NWR (1998) and the new Kahuku unit of Hawai'i Volcanoes National Park (2003). Studies of malaria prevalence in birds at high elevations (above ~1,500 m) suggest that local transmission is at most a very rare event (Feldman et al. 1995; Atkinson et al. 2005; this volume, Chapter 9) constrained by low-temperature effects on vector and parasite development (LaPointe 2000, Ahumada et al. 2004).

Maintaining Transmission-Free Refugia

Acquisition and management of transmission-free high-elevation habitat is crucial to the preservation and restoration of native Hawaiian forest birds but is valuable only if it can be maintained as such well into the future. Barring the possible effects of global warming or a land use change that would create a warmer microclimate, these refugia should remain avian malaria transmission-free zones even in the presence of increasing mosquito numbers. Global warming could increase the altitudinal distribution of avian disease and would particularly threaten present-day refugia where the geology or land use, such as cattle grazing, might impede the upslope expansion of suitable forest habitat (Benning et al. 2002). Securing deforested and pasture land adjacent to protected refugia and managing it for forest growth is the best long-term contingency plan against a warming scenario.

Of more immediate concern is the nature of land use change after cattle ranching operations are phased out. Should land use in these adjacent areas instead shift to a more intensive agricultural or residential use, the effect of infrastructure and water impoundment on transmission could be significant. Vector densities would increase as available larval habitat became more abundant, and temperature constraints on parasite development might be circumvented by infrastructure microclimates (Garnham 1948). Artificial structures absorb radiant energy, creating microclimates that foster parasite development. Careful consideration must be given to the construction of water impoundments and infrastructure so as not to create an environment conducive to high-elevation transmission of disease.

Even some techniques employed for game bird management, conservation, and habitat restoration work can backfire to support pathogen transmission. The use of black plastic tubs, barrels, or pond liners to provide water for game birds, outplantings, or bait for feral ungulates can support mosquito populations in areas where natural water sources or favorable water temperatures (through absorbed solar radiation) do not exist (D. LaPointe, pers. obs.). The construction of artificial wetlands for native waterfowl, such as Koloa (Hawaiian Ducks, *Anas wyvilliana*), is particularly risky and should be weighed against the potential harm to the overall avian community of the area. Approval of constructed wetlands should be contingent on routine monitoring for mosquito production and control when necessary.

Creating Small Disease-Free Refugia

In the Hawaiian avian disease system, very low vector numbers support high rates of transmission. Therefore, a mere reduction in vector numbers may not be sufficient to prevent disease. Whenever feasible, control measures should strive for local eradication. Although SMT and CI strategies might be viable options for eradication of mosquitoes on smaller islands (Ni'ihau, Lāna'i, and Kaho'olawe) and atolls (Midway), source reduction is the best strategy

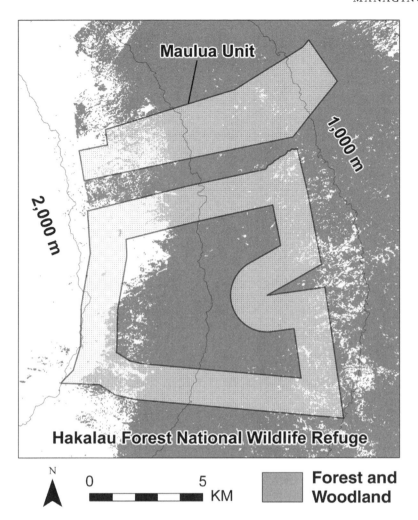

Forest and
Woodland

Figure 17.4. A model of mosquito ingression at the Hakalau Forest National Wildlife Refuge. The stippling indicates the zone of mosquito penetration from unmanaged lands outside the refuge and is based on the conservative estimate of *Culex quinquefasciatus* dispersal at 1.6 km mean dispersal distance. In this scenario, mosquitoes have the potential to inundate the entire area of the detached northern Maulua Unit and reduce the effective transmission-free area of the main refuge by 60%. This figure emphasizes the importance of refuge configuration to avoid unmanaged inholdings and to realize the significance of unmanaged adjacent lands in disease transmission. *Source:* Adapted from LaPointe 2008: 607, Fig. 5, with permission from The Entomological Society of America.

for the main islands. Source reduction will most likely succeed in habitats that are already marginal for vectors.

Unlike wet forests, where natural and feral pig–associated larval mosquito habitats are likely to be abundant, xeric and mesic forests may produce mosquitoes only through human activities that accidentally or purposely impound water. Source reduction and/or treatment of impoundments can be accomplished with far less effort and environmental harm than use of such strategies in more natural habitats. Eliminating local production of mosquitoes will not be sufficient, however, if there are substantial nearby populations that can disperse into the refuge. Kīpuka—islands of forest surrounded by lava flows—would make ideal sites for future refuges. Their natural isolation by younger lava flows would provide a maintenance-free buffer against dispersing vectors. Contiguous habitat would require the creation of a buffer to protect the core habitat from dispersing vectors. The extent of this buffer should be no less than 1.6 km, preferably 2–3 km, and refuge design should strive for a low ratio of perimeter to volume to minimize edge effects (Fig. 17.4). As a final consideration, planned refuges should not be in alignment with or adjacent to residential areas, roadways, or unmanaged wet forests along the path of prevailing winds. These areas might provide an abundance of vectors or corridors for wind-aided dispersal.

DISEASE MANAGEMENT IN LANDSCAPE-SCALE CONSERVATION UNITS

The prevalence of avian malaria and its impact on host populations vary greatly over the altitudinal range of forest bird habitat, so native Hawaiian forest birds are conspicuously missing from suitable habitat. On the high islands in particular, large tracts of native forest still extend the breadth of their elevational range, from coast to timberline, yet native bird densities and diversity are severely depleted at lower elevations (Scott et al. 1986).

The Landscape of Mosquito-Borne Avian Disease

Although pathogen transmission is virtually absent in high elevation forests (>1,500 m elevation), it is common seasonally in native birds in midelevation forests, often resulting in fatal epizootics. Most intriguing, however, is the situation in low coastal forests, where disease prevalence is extremely high and yet at least one group of honeycreepers, the 'amakihi

(Hemignathus spp.), appears to have evolved tolerance. The year-round transmission that is characteristic of these coastal forests could be driving evolved tolerance. Low-elevation, disease-ridden forests may not be the endemic bird wasteland once supposed; instead they may serve as the principal habitat for coevolution of introduced pathogens and native birds (Woodworth et al. 2005).

What seems clear is that the restoration of native Hawaiian forest birds will depend on the preservation of the full extent of native forest habitat, from the high-elevation refugia through the epizootic zone at middle elevations down to the coast, where coevolutionary adaptation toward disease tolerance can occur. Ideal conservation landscapes would emulate the traditional Hawaiian unit of land division, the ahupua'a, extending from mountains to seashore, and efforts to control disease should focus on midelevation forests.

Most of the large, midelevation forests remaining in the Hawaiian Islands are wet to mesic forests where C. quinquefasciatus and malaria are prevalent. The windward side of the main Hawaiian islands receives heavy rainfall (2,500–11,000 mm annual precipitation) (Giambelluca and Schroeder 1998). However, precipitation is not the sole predictor of a landscape's potential for production of vectors or transmission of disease. Climate, geology, surface hydrology, and land use are strong determinants of disease prevalence. Currently, eradication of mosquitoes is unlikely in wet forests, but some strategies may reduce their overall abundance and limit their populations during years of favorable conditions.

Mosquito Control over Broad Landscapes

Although high-tech solutions such as vaccines and GMMs should not be ruled out, there are proven traditional approaches to mosquito control that are applicable and immediately available. Elimination of feral pigs and source reduction of artificial larval mosquito habitats should be considered the first step in eliminating or reducing vector numbers. On windward Mauna Loa, for example, mosquito populations in large tracts of wet forests could be virtually eradicated through indirect source reduction in the form of feral pig removal. Though this management strategy is expensive and politically challenging, its potential overall benefit to the restoration of forest bird populations is unparalleled.

Due to the dispersal capability of C. quinquefasciatus, however, the success of feral pig removal in reducing transmission will be dependent on the size and shape of the managed land and on adjacent land use. Conservation units are better protected from immigrating mosquitoes when they are shaped to minimize edge effects and large enough to include a minimum dispersal buffer. The creation of narrow conservation units along altitudinal contours is to be avoided, because such units are more vulnerable to upslope, wind-enhanced mosquito dispersal.

For residential and agricultural areas adjacent to conservation units, integrated methods may be the best strategy. Initial survey and control measures could be conducted by the HDOH Vector Control staff and community volunteers while training landowners to identify and eliminate mosquito production on their personal property. Simple yard sanitation will suffice in many cases. However, domestic and agricultural impoundments that cannot be removed or covered may have to be treated with fish or a biopesticide such as B.t.i. Care should be taken when using fish to ensure that accidental introduction into natural waterways does not occur, and the continued use of biopesticides should include some rotation of agents to avoid resistance. The success of such efforts will depend on annual outreach, assistance from HDOH Vector Control, and the strength of neighborhood organizations.

In landscapes where intermittent streams produce the majority of vectors, there are few realistic options for control. It remains to be determined if streambeds can be successfully treated by aerial application of biopesticides, and the cost is expected to be great. Aerial applications may be limited to years of low rainfall, when streambeds make significant contributions to mosquito numbers. Additionally, C. quinquefasciatus production in riparian landscapes may be reduced by protecting streams from fecal contamination by feral and domestic ungulates.

LESSONS FROM 100 YEARS OF HUMAN MALARIA CONTROL EFFORTS

Since the early 1900s, many nations have been engaged in malaria control. The early successes of Gorgas in reducing yellow fever in Panama, Soper's eradication of A. gambiae in Brazil, and Italy's success in the Pontina and Sardinia fueled an optimism that with enough financial resources and manpower, malaria could be eradicated (Harrison 1978). With the introduction and use of DDT, a formal worldwide malaria eradication campaign was launched by the World Health Organization (WHO) in 1956 (Desowitz 1991, Spielman et al. 1993).

The key strategy of the campaign centered on the use of a residual pesticide, DDT, to kill infected females and in turn break transmission long enough for existing cases of malaria to self-cure, a period estimated at five years. Additionally, active infections were treated with chloroquine, the most effective chemotherapeutic agent at the time. By 1967, a handful of European and Caribbean nations claimed eradication, and India, Pakistan, and Sri Lanka had greatly reduced malaria transmission. However, resistance to DDT and chloroquine quickly turned the tables. Five years later, WHO formally abandoned the goal of global eradication and advocated instead for malaria control, first by new antimalarial drugs and later by an elusive malarial vaccine that has never materialized (Desowitz 1991). By the time of the malaria eradication campaign's demise, malaria had undergone a global resurgence and the campaign had cost the United States alone $790 million (Desowitz 1991).

The failure of the Global Eradication of Malaria Project was due to many factors. Poor communication from the field to policy makers led to an illusion of success, and a strict and unrealistic time frame derailed project funding before success could be achieved (Desowitz 1991, Spielman et al. 1993). For some regions, such as sub-Saharan Africa, the project failed because planners grossly underestimated the variability and plasticity of the vector and pathogen (Spielman et al. 1993). As pesticide and drug resistance spread around the world, the fallacy of reliance on a single control strategy was made clear. One of the most obvious errors of the program was the diversion of financial resources to control efforts with an almost complete abandonment of research and training (Spielman et al. 1993). When the old techniques failed, there were few alternative control methods to try, and few individuals were trained to take over the fight.

Although world health agencies are still lured by modern technology, pursuing elusive vaccines and the ultimate engineered mosquito, it is unlikely that a panacea for malaria will ever be found. Low-tech solutions, such as the use of insecticide-treated bed nets, have resulted in modest reductions in transmission, but when combined with traditional mosquito control and pharmaceutical treatment they could have a greater and lasting impact (Vogel 2002). Local, conservatively sized projects using integrated strategies are more likely to be sustainable and effective in malaria control (Spielman et al. 1993).

Though carried out at a much smaller scale, management actions to control mosquito-borne avian disease in Hawai'i (Table 17.1) would do well to heed these lessons. Eradication is realistic only for geographically isolated areas where mosquitoes cannot reinvade and where source reduction can completely eliminate larval habitat. For larger, more heterogeneous landscapes, an adaptive and sustainable management approach using integrative control strategies coupled with an active research program will ensure the selection of locally effective methods. Technologically sophisticated approaches such as transgenic refractory mosquitoes may be a risky investment but offer some potential. Finally, it is important to recognize that control efforts may require years of investment before significant results become evident and bird populations begin to recover.

SUMMARY

In the past century, Hawaiian forest birds have undergone extinctions and steep population declines largely due to introduced, mosquito-borne avian pox and malaria. Today, while a few bird species appear to be coevolving to exist with these pathogens, most populations of endemic species continue to be severely limited by them. Disease is most prevalent in wet forests below 1,500 m, where the climate is favorable to both the introduced mosquito vector and avian malaria. Feral pigs have a key role in this disease system, for they often create the only or most favored larval habitat. As we begin to understand the dynamics and impacts of these diseases, it is clear that control of avian disease will be crucial to the recovery and restoration of native Hawaiian birds.

Due to the difficulties of mass administration or immunization, the use of antimalarial drugs and vaccines is limited to the protection of captive birds. Resistant bird populations may seem a ready source of individuals for restoration, but these populations and the epidemiological conditions that maintain them are relatively rare. Therefore, the options for avian disease control in wild birds must target vec-

Table 17.1. A summary of management strategies for the control of avian disease through vector manipulation and potential obstacles to their use

Management Recommendation	Technology Available?	Research or Technological Development Needed?	Politically Sensitive?	Environmentally Sensitive?
Improve communication with HDOH Vector Control	Yes	No	Maybe	No
Develop and disseminate outreach materials	Yes	Yes	No	No
Mandate source reduction and mosquito-proof design of refuge infrastructure	Yes	No	No	No
Consider vector movements during refuge acquisition and design	Yes	No	No	No
Eliminate feral pigs from conservation lands	Yes	No	Yes	No
Attempt eradication of mosquitoes on Midway Atoll using SMT or CI	Yes	Yes	No	No
Monitor and control mosquitoes in refuge wetlands	Yes	Yes	Maybe	Yes
Evaluate efficacy and nontarget impacts of B.t.i. control in natural streambeds	Yes	Yes	Maybe	Yes
Support disinfection of aircraft and container cargo	No	Yes	Yes	No
Develop and release a GM refractory mosquito	No	Yes	Yes	Maybe

tor abatement. Traditional pesticides and growth regulators offer little applicability in conservation areas where larval habitats are difficult to locate and treat and where native wildlife and invertebrates would be affected by nonspecific toxicity. Historically, classical biological control of mosquitoes in the Hawaiian Islands has failed to be self-sustaining or, worse, has led to the decline of native species. Even long-term augmentation and inundative releases of most biocontrol agents are unlikely to succeed given the vast area to be covered and the low density of the target prey. Only the crossover biopesticides are sufficiently biologically selective and cost-effective to be considered for some riparian and wetland areas in Hawai'i.

Source reduction and mosquito-proof design of refuge infrastructure is the first step in avian disease management. Direct source reduction through removal and modification of artificial containers and impoundments and indirect source reduction through the elimination of feral pigs can successfully reduce and potentially eliminate vector populations. SMT may work to eradicate mosquitoes from small islands and atolls, but the transgenic mosquito approach is too immature for immediate consideration.

It is unlikely that there will be any magic bullets for abating avian malaria and pox in Hawai'i. The most productive conservation strategy would be to keep high-elevation refugia disease free, while the main management emphasis would focus on vector suppression at the middle elevations. We recommend investing resources into integrative mosquito control and the development of long-term disease management strategies. The successes and failures of the Global Malaria Control Program demonstrate the need to adopt and adapt the old, proven technologies, such as source reduction, while keeping a hand in developing technologies. Avian disease management is essential to the preservation of the Hawaiian avifauna, and with sound research and a commitment from land management agencies, the impact of disease can be minimized.

ACKNOWLEDGMENTS

We thank the U.S. Geological Survey (USGS) Invasive Species program, the USGS Natural Resource Protection Program, and the National Science Foundation (Grant DEB0083944) for the financial support that made this research possible. The collection of data for the various studies described here would not have been possible without the assistance and support of numerous technical staff, co-workers, and student interns, who spent many long days in the field and laboratory under often difficult living and working conditions. We thank J. McAllister, W. Reisen, and two anonymous reviewers for their comments on earlier drafts of this chapter.

CHAPTER EIGHTEEN

Controlling Small Mammals

STEVEN C. HESS, CATHERINE E. SWIFT,
EARL W. CAMPBELL III,
ROBERT T. SUGIHARA,
AND GERALD D. LINDSEY

Conservation of the unique and spectacular biota of Hawai'i and the Pacific islands requires the management of a number of alien mammalian predators and competitors. The control of small mammals on a landscape scale is especially important for the recovery of avian species (Stone and Loope 1987, Banko et al. 2001, U.S. Fish and Wildlife Service 2006). Although small mammals generate significant negative social and economic impacts through agricultural crop damage and the spread of zoonotic diseases worldwide, they are particularly devastating to the native biota of oceanic islands. In Hawai'i and on other remote islands throughout the world, the effects of novel mammalian predators and competitors have repeatedly caused the extinction and extirpation of insular bird species (King 1985; Stone 1985; Moors et al. 1992; this volume, Chapters 1, 2, and 11). Of all the factors that currently limit the native Hawaiian avifauna, the effects of alien small mammals are potentially the easiest to manage.

Small mammals have both direct and indirect impacts on native birds in Hawai'i (Chapters 8 and 11). Two carnivores, the feral cat (*Felis catus*) and small Indian mongoose (*Herpestes auropunctatus*), prey on eggs, nestlings, juveniles, and adults of ground- and tree-nesting land birds as well as burrowing seabirds. Feral cats have had tremendous impacts on insular vertebrate fauna worldwide and are directly responsible for the extinction of at least 33 bird species (Nogales et al. 2004). Feral cats also host diseases such as toxoplasmosis, which is known to kill endangered Nēnē (Hawaiian Geese, *Branta sandvicensis*) and 'Alalā (Hawaiian Crows, *Corvus hawaiiensis*). Rodents not only prey on birds at all life history stages; they also compete with native birds by preying on seeds of key plant foods and, along with other carnivores, on invertebrates. Rodents interrupt reproduction in important plants used by birds as food, such as hōʻawa (*Pittosporum* spp.), ʻieʻie (*Freycinetia arborea*), ʻiliahi (*Santalum* spp.), and loulu (*Pritchardia* spp.) by preying on seeds and seedlings. The effects of small mammals as predators and competitors of the Hawaiian avifauna are treated in greater detail in Chapter 11.

THE COMMUNITY OF SMALL MAMMALS IN HAWAIʻI

The contemporary structure of communities of small mammals on remote central Pacific islands can be relatively simple, with few mammalian predator species and few alternative prey species available to these predators. Prior to the arrival of humans, there were only two species of terrestrial mammal in Hawaiʻi, the Hawaiian hoary bat (*Lasiurus cinereus*) and an undescribed bat known only from prehistoric bone deposits (Ziegler 2002). Currently, four species of rodents, feral domestic cats, and mongooses are found widely throughout the Hawaiian Islands. The Polynesian rat (*Rattus exulans*) arrived in Hawaiʻi with the early Polynesian settlers (Kirch 1982, Olson and James 1982b). Three species of rodents, the domestic cat, and the mongoose did not arrive in Hawaiʻi until after European contact in 1778. Rodents that were introduced at this time included the house mouse (*Mus musculus*), the Norway rat (*R. norvegicus*), and the black rat (*R. rattus*). Atkinson (1977) provided an arrival chronology and an extensive discussion of black rats, which may not have had an opportunity to reach the Hawaiian Islands until after 1860. Although the earliest authenticated specimen of a black rat in the islands was not collected until 1899, Norway rats probably became established in Hawaiʻi before this time. However, their impacts may have been more limited, because they lack the developed tree-climbing abilities of black rats. House cats, employed as "mousers" on European sailing ships, were introduced early after European discovery and may have preceded the arrival of black rats by decades (Tomich 1986). Another carnivore that was introduced at a later date, in 1883, was the small Indian mongoose, from Jamaica (Baldwin et al. 1952, Tomich 1986), which was subsequently introduced to other Pacific islands such as Fiji. This species was intentionally introduced to Hawaiʻi by sugar planters to control rodents but soon became an important predator of ground-nesting birds.

DISEASES CARRIED BY SMALL MAMMALS IN HAWAIʻI

In addition to preying on and competing with the native avifauna of Pacific islands, small mammals also carry several diseases that are communicable to humans, domestic mammals, and native wildlife. The bacteriological diseases murine typhus and bubonic plague caused by the organisms *Rickettsia typhi* and *Yersinia pestis* are hosted by many mammal species and vectored primarily by the cat flea (*Ctenocephalides felis*) and the oriental rat flea (*Xenopsylla cheopis*), respectively (Tomich et al. 1984). These diseases have a long history of causing human illness and mortality in Hawaiʻi. Although plague has not occurred in the archipelago since 1957 (Tomich et al. 1984), murine typhus outbreaks still occur, with 47 confirmed human cases as recently as 2002 (Manea et al. 2001, Sasaki et al. 2003).

Leptospirosis, caused by the spirochete *Leptospira interrogans,* is one of the most widespread, sometimes fatal, zoonoses worldwide (Middleton et al. 2001). In Hawaiʻi, it has a mean annual incidence of 1.29 per 100,000 people (Katz et al. 2002). Leptospirosis is often transmitted when cuts or abrasions to the skin contact water or soil contaminated by mammal urine and is frequently associated with rodents and mongooses (Sasaki et al. 1993). Dogs, cattle, and swine may also serve as reservoirs (Middleton et al. 2001). Other diseases associated with rodents and carnivores, such as cryptosporidiosis, giardiasis, and salmonellosis, pose persistent and serious public health problems (Sasaki and Ikeda 2000, Katz et al. 2002).

Feral cats harbor diseases, including bovine tuberculosis and toxoplasmosis, which can be transmitted to humans, livestock, pets, and native wildlife (Ragg et al. 1995, Dubey and Beattie 1998). Cats are the definitive host of toxoplasmosis, caused by the coccidian protozoan *Toxoplasma gondii* (Wallace 1973). In humans, women infected with toxoplasmosis during pregnancy may experience spontaneous abortions, and although surviving infected fetuses rarely display obvious symptoms at birth, learning disabilities, impaired vision, or developmental delays may be manifested later in life (Wallace et al. 1969, Dubey and Beattie 1998).

Toxoplasmosis has been implicated in the mortality of captive Nēnē, wild Red-footed Boobies (*Sula sula*), and critically endangered 'Alalā, as well as introduced Erckel's Francolins (*Francolinus erckelii*) (Work et al. 2000, 2002). In one of the few studies of feral cat seroprevalence in Hawai'i, 37% showed evidence of past exposure, and 7% were currently or recently infected with *T. gondii* on Mauna Kea Volcano (Danner et al. 2007). Ground-feeding species such as the endangered Nēnē and the Erckel's Francolin may become infected after consuming food contaminated with sporulated oocysts (Work et al. 2002). *Toxoplasma* oocysts are known to enter marine environments through municipal sewage or stormwater runoff, sporulate in seawater, and infect a wide variety of marine mammals, including seals and dolphins (Holshuh et al. 1985, Migaki et al. 1990, Lindsay et al. 2003). Fatal toxoplasmosis infection was reported in an endangered Hawaiian Monk Seal (*Monachus schauinslandi*) on Kaua'i (Honnold et al. 2005). In cases in which colonies of alien mammals are subsidized by refuse or intentional feeding, high densities of animals may vector diseases such as toxoplasmosis, leptospirosis, and murine typhus, facilitating quick transmission through populations and increasing the risk to humans, wildlife, and domestic animals in the vicinity.

LAWS REGULATING SMALL MAMMAL CONTROL

A number of societal, environmental, and legal issues surround the control of small mammals in Hawai'i. Federal legislation regulates both the humane treatment of vertebrates and the use of toxicants to control small mammals. The Animal Welfare Act (P.L. 89-544, 1966, as amended in P.L. 91-579 and P.L. 94-279), promulgated by the U.S. Secretary of Agriculture, is the chief legislation that mandates the humane treatment of vertebrates. Although its original intent was to regulate the care and use of animals held for teaching and in laboratories, it has been amended several times to become the only federal law in the United States that regulates the treatment of animals in research, exhibition, transport, trade, or other activities supported by federal grants. Other laws, policies, and guidelines may cover additional species or specifications for animal care and use, but all refer to the Animal Welfare Act as the minimum acceptable standard.

For research activities, institutional animal care and use committees usually oversee compliance with the standards of the Animal Welfare Act. Established at institutions such as federal animal research facilities and universities, these committees are intended to minimize the use and ensure the humane treatment of vertebrates in research. Some of the considerations for testing techniques for the control of small mammals include level of pain caused by lethal toxicants and euthanasia; stress caused by handling; access to food, water, and dry

Figure 18.1. A small Indian mongoose captured in a live trap intended for feral cats. Captures of nontarget species greatly reduce the efficiency of live traps. *Source:* Photo courtesy of the U.S. Department of Agriculture, National Wildlife Research Center, Hawaii Field Station.

bedding during trapping; length of time that animals may be held in traps; and the type of traps that may be used. Two valuable resources for ethical treatment standards in field settings are the American Society of Mammalogists (1998) guidelines for the capture, handling, and care of mammals and the American Veterinary Medical Association (2001) report on euthanasia.

Rodent control programs rarely generate as much public concern about humane treatment as do programs for the lethal control of feral cats. In many cases, there is little public awareness of the problems caused by feral cat populations in natural areas and their impacts on native wildlife. Therefore, there is little public support for controlling cats in natural areas (Ash and Adams 2003). Several public groups concerned with animal welfare have demanded very high standards for the humane treatment of predators during control efforts in Hawai'i (Stone 1995). For example, the strenuous objections of cat advocates have narrowed the range of control techniques

used in Hawai'i primarily to live trapping (Fig. 18.1). Alternatives preferred by feral cat advocate groups, such as trap-neuter-release programs, do not address the fundamental problem of predation on native wildlife. Furthermore, some groups in Hawai'i currently feed colonies of feral cats adjacent to conservation lands containing endangered birds despite the fact that supporting these colonies may violate the Migratory Bird Treaty Act and the Endangered Species Act as well as local animal abandonment laws (Smith, Polhemus, et al. 2002; Winter 2003). Public activism and legal requirements for the humane treatment of animals are thus often at odds with provisions of the Endangered Species Act.

The use of vertebrate toxicants for the control and eradication of introduced mammals in native ecosystems has become widespread globally since their first use in New Zealand in the 1970s (Towns and Broome 2003, Galván et al. 2006). In the United States, vertebrate toxicants are regulated by the Federal Insecticide, Fungicide, and Rodenticide Act (FIFRA), which controls pesticide distribution, sale, and use. All pesticides sold in the United States must be registered by the Environmental Protection Agency (EPA) to ensure that when used in accordance with label specifications they will achieve the claimed benefits with a low risk of harm to human health and the environment. The registration process is lengthy, and applicants must submit data that demonstrate the efficacy of the product in controlling target species and also demonstrate that when used in accordance with the label it will not have unreasonable adverse effects on the environment, including soil and water, nontarget species, and humans.

The majority of household and agricultural pesticides are registered under Section 3 of FIFRA, which regulates a product's distribution, sale, and use throughout the United States, including its territories and possessions. In addition to EPA approval, Section 3 products are subject to registration by individual states' pesticide regulatory agencies. FIFRA has three provisions that permit special applications of pesticides in different situations. Under Section 24(c), states can request additional uses of a federally registered pesticide to meet a special local need (SLN). Natural resource managers in Hawai'i were the first in the United States to apply for and be granted SLN registrations for the use of vertebrate toxicants for conservation purposes. Section 18 of FIFRA authorizes the EPA to exempt state and federal agencies from certain provisions of FIFRA when emergency conditions exist, such as significant risk to threatened or endangered species. However, the Section 18 applicant must make progress toward registration of the product through Section 3 or Section 24c, generally within three years. A number of rat eradications undertaken on U.S. islands were granted emergency exemptions prior to the first conservation registrations under Section 3, and they were justified on the grounds of preventing imminent seabird extirpations.

Research to support the development of new registrations requires that an experimental use permit (EUP) be issued by either the U.S. EPA or a specific state's pesticide regulatory office under Section 5 of FIFRA. Detailed studies are usually required for registration of new pesticides. Research typically is conducted under good laboratory practice standards, which require a written protocol, meticulous record-keeping, and strict standards for quality assurance (Fagerstone et al. 1990, Jacobs 1992, Poché 1992). As a result, the cost of new pesticide registrations for small mammal control is often very high.

THE HISTORY OF SMALL MAMMAL CONTROL IN HAWAI'I

Its Roots in Agriculture

Many strategies have been developed in Hawai'i to control small mammals, primarily rodents, in association with agricultural situations, and these have formed the basis for later conservation applications (Table 18.1). One of the earliest strategies, a type of biological control, was the introduction of the small Indian mongoose to the island of Hawai'i in 1883 to control rodents in sugar cane plantations (Atkinson 1977, Tomich 1986). Atkinson (1977) discussed observations that mongooses appeared to reduce rat damage to sugar cane for a five-year period after their introduction, and he suggested that they may have been effective in controlling Norway rats prior to the arrival and spread of black rats in Hawai'i. Mongooses were also introduced to Maui between 1885 and 1888. Despite their inability to control rodents in sugar cane, as omnivores they thrived, later colonizing much of the land area of the Hawaiian Islands with the exception of Kaua'i, Lāna'i, and Kaho'olawe. Mongooses are believed to have contributed to the decline of ground-nesting native birds, particularly Nēnē (Tomich 1986, Stone and Loope 1987, Banko 1992, Hays and Conant 2007). Numerous sightings of mongooses on Kaua'i since the late 1960s indicate that they may be established on the island, although intensive trapping efforts have not yet produced definitive proof of

Table 18.1. Methods used to control small mammals in Hawai'i

Species	Method	Application	Context	First Source
Rattus spp.	Mongoose introduction	Biological control	Agricultural	Atkinson 1977, Tomich 1986
Rattus spp.	Thallium sulfate	Perimeter baiting	Agricultural	Doty 1938
Rattus spp.	Anticoagulants	Perimeter baiting	Agricultural	Doty 1945
All mammals	1080 (Sodium flouracetate)	Predacide baiting	Conservation	Hawai'i Department of Land and Natural Resources 1974
Rattus spp.	Zinc phosphide pellets	Aerial broadcast	Agricultural	Hilton et al. 1972
Carnivores	Trapping	Live capture	Conservation	Simons 1983
Rattus spp.	Fumarin / zinc phosphide	Bait stations / broadcast	Conservation	Stone and Loope 1987
Rattus spp.	Brodifacoum/bromethalin	Bait stations	Conservation	Ohashi and Oldenburg 1992
Mongoose	Diphacinone	Bait stations	Conservation	Stone et al. 1995
Rattus spp.	Diphacinone	Bait stations	Agricultural	Campbell et al. 1998
Rattus spp.	Diphacinone	Bait stations	Conservation	Nelson et al. 2002
Rattus spp.	Diphacinone	Experimental hand broadcast	Conservation	Spurr et al. 2003a
Rattus spp.	Diphacinone	Experimental aerial broadcast	Conservation	Spurr et al. 2003b

a breeding population (K. Gundersen, pers. comm.).

Early efforts to control rodents in sugar cane plantations used acute toxicants, including compound 1080 (sodium fluoroacetate), barium carbonate, strychnine alkaloid, thallium sulfate, and red squill (Doty 1938, Sugihara 2002). These compounds kill quickly with single, relatively small doses. Compound 1080 was also used as a general predacide to protect Nēnē from feral dogs, feral pigs (Sus scrofa), mongooses, and feral cats in the wild, although the benefits to Nēnē were not quantified (Hawai'i Department of Land and Natural Resources 1974). To reduce the possibility of consumption by 'Io (Hawaiian Hawks, Buteo solitarius), 1080 was injected into chunks of meat and then deposited in crevices near Nēnē nests and throughout sanctuaries. Single-dose, acute toxicants have fallen out of favor because of their limited efficacy due to target animals' bait shyness and the substances' high risk to nontarget species (Lund 1988, Prakash 1988, Sugihara 2002).

Multiple-dose anticoagulant baits such as warfarin, fumarin, and pival have also been used to reduce sugar cane damage by rats since the 1950s but have been restricted to bait stations placed on the ground along field perimeter access trails and roads, and they were extremely labor-intensive to maintain (Doty 1945, Sugihara 2002). Anticoagulants act by inhibiting the normal synthesis of vitamin K–dependent clotting factors in the liver, where they are gradually metabolized over a period that depends upon the initial amounts ingested and subsequent exposure. Individuals may survive an acute poisoning, because substantial proportions of large doses can pass through an animal without entering the vitamin K_1 cycle (Ren et al. 1977, Beasley and Buck 1982, Yu et al. 1982, Gilman et al. 1985, Freedman 1992, Timm 1994, Swift 1998). Chronic exposure to these first-generation anticoagulants (so called because they were

primarily developed in the 1950s, while a second group of single-dose anticoagulant compounds followed in the 1970s) greatly increases their toxicity. Thus, the total amount of a first-generation anticoagulant required for mortality is much lower when consumed in discrete doses for a number of days than is the amount needed to cause death with a single dose. Baits are formulated with low concentrations of the active ingredient and must be consumed daily for at least three to five days in order to accumulate a lethal dose (Marsh 1986, Lund 1988, Savarie 1991, Timm 1994). These characteristics significantly lower the risk of poisoning nontarget species, making the first-generation anticoagulants the most widely used class of vertebrate pesticides.

Although bait aversion was not a problem with the anticoagulants, they still did not adequately reduce rat damage in Hawaiian sugar cane. Laboratory tests confirmed that warfarin-treated rolled oats were not effective against Polynesian rats without consistent daily bait consumption (Bonnet and Gross 1951). When Lindsey et al. (1971) used biological tracers to evaluate the effectiveness of perimeter bait stations, they found that black rats living in non-crop vegetation on the edges of sugar cane fields could be controlled, but field-dwelling Norway and Polynesian rats encountered and consumed bait at a lower rate. They concluded that broadcast application of baits within the fields would be more effective in controlling these species. Experiments in the aerial broadcast of nontoxic bait with biological tracers further showed that fields and gulches adjacent to sugar cane should be baited for more effective control of Polynesian rats (Nass et al. 1971).

Zinc phosphide (an acute, nonanticoagulant toxicant) became the first toxicant for in-field, aerial broadcast approved by the EPA under an SLN in October 1970, vastly improving application efficiency (Hilton et al. 1972). Baits were formulated either as pellets or applied to oats and broadcast by fixed-wing aircraft at the rate of 5.6 kg per ha, with a maximum of four applications (22.4 kg per ha) per crop cycle. Damage to sugar cane was reduced by 50–60%, but the effectiveness of this method declined with repeated and prolonged use. Furthermore, the abundance of rats shifted from mainly Polynesian to Norway rats due to these species' differential susceptibility to the rodenticide (Sugihara et al. 1995). Zinc phosphide toxicants still retain Hawai'i SLN registration for use in sugar cane under the brand name Prozap®, but in the 1970s the sugar cane industry's research focus shifted to diphacinone, a first-generation anticoagulant. Experiments using diphacinone-treated oats (0.025%) in plastic bags broadcast into perimeter areas demonstrated a high level of control of Norway rats (Teshima 1976). However, sugar cane production began to decline to a small fraction of its former volume for economic reasons, and the pursuit of additional rodenticide registrations for use in Hawaiian sugar cane ceased.

The Hawai'i Department of Health, Vector Control Branch, also maintains an SLN registration for a 2% zinc phosphide bait to control commensal (associated with humans) rodents for public health purposes, usually to prevent mouse irruptions from causing murine typhus outbreaks. The registration allows for application of the toxicant in bait stations and by hand and for aerial broadcast in noncrop areas surrounding residential and resort areas and in rangeland.

Rodents heavily damage macadamia (*Macadamia integrifolia*) orchards, and much research was conducted to mitigate these effects, offering suitable applications for other forested environments. Perimeter baiting with zinc phosphide–treated oats and pellets reduced rat populations as much as 85% (Pank et al. 1978). Tobin et al. (1996) found, however, that rat activity occurred primarily in tree canopies during the night, and the authors recommended

that baits be placed in trees or in underground burrows rather than adjacent noncrop areas. Placebo baits with a biological marker indicated that placement in trees produced the highest degree of acceptance by rats and the lowest risk to nontarget species (Tobin, Sugihara, et al. 1997). Because crop damage later in the season resulted in the heaviest losses at harvest, an integrated approach to managing rodent damage with a combination of cultural, mechanical, and timed toxicant applications maximized returns on treatment costs (Campbell et al. 1998). These lines of research led to Hawai'i SLN registrations in 1998 for the use of diphacinone in macadamia orchards in bait stations placed on the ground or in trees and ultimately paved the way for its use in natural areas.

The Move from Agricultural to Natural Areas

Using New Zealand's spectacular successes as both model and inspiration (Towns and Broome 2003), Hawai'i has been at the forefront of U.S. efforts to adapt agricultural and commensal techniques of rodent control to native ecosystems. Developing rodenticide application techniques and obtaining registrations for them in Hawai'i has been pursued for the purpose of conserving native plants and animals while allowing restoration or recovery of species impacted by introduced rodents. This approach has substantially reduced rodent populations in valuable native ecosystems on the main Hawaiian islands and eradicated them from uninhabited offshore islands and remote atolls. Beginning in 1990, the U.S. Department of Agriculture, Animal and Plant Health Inspection Service, Wildlife Services (WS), eradicated rats from four remote Pacific atolls where these rodents were having devastating impacts on seabird colonies. The first of these efforts, conducted with the U.S. Fish and Wildlife Service and the Samoan Department of Wildlife and Marine Resources, targeted

Polynesian rats on uninhabited Rose Atoll, American Samoa. A second-generation anticoagulant, brodifacoum (WeatherBlok®, 0.005% brodifacoum), was used in bait stations spaced 50 m apart over the entire 6.3 ha island, along with live traps and snap traps (Morrell et al. 1991, Ohashi and Oldenburg 1992). Brodifacoum was selected because of the successful use of this toxicant in New Zealand (Innes and Barker 1999, Towns and Broome 2003). However, the initial application, although it substantially reduced rat numbers, did not result in eradication. Although it was unclear why the first attempt failed, a subsequent treatment with Vengeance® (an acute neurotoxin, 0.01% bromethalin) was successful (Murphy and Ohashi 1991).

In 1993, building upon the experience gained in the previous eradications, WS and the Hawai'i Department of Land and Natural Resources (DLNR) eradicated Polynesian rats from 129 ha Green Island, Kure Atoll, in the Northwest Hawaiian Islands using techniques similar to those used in the Rose Atoll operation (J. Murphy, pers. comm.). The following year, the U.S. Navy, U.S. Fish and Wildlife Service, and WS eradicated black rats from Eastern and Spit Islands of Midway Atoll (J. Murphy, pers. comm.). Intensive trapping and baiting of 134 ha Eastern Island using the same techniques previously employed on Rose and Kure atolls was completed within three months. Bait stations were maintained on the island for some months following, and a little over a year after the operation commenced, no evidence of rats was found (Murphy 1997a). The eradication of rats from 1 ha Spit Island was accomplished within a month using only live traps (Murphy 1997a).

The successful eradication of rats from the two smaller islands of Midway Atoll, combined with evidence of the devastating impacts of rats on a key seabird species, the Bonin Petrel (*Pterodroma hypoleuca*) (Seto and Conant 1996), persuaded the U.S. Navy to fund rat eradication on the final island of the atoll, Sand Island. In July 1996, the entire 486 ha island was overlaid with two 50 m grids, one for bait stations and one for live traps (Murphy 1997b). The last rat sighting was reported in October 1997. Sand Island remains the largest island and the only permanently inhabited island in the United States from which rats have been removed. The dramatic growth of the Bonin Petrel population, from an estimated 32,000 nesting birds (Seto and Conant 1996) to a number too great to estimate within just a few years of rat eradication (N. Hoffman, pers. comm.), provides compelling evidence of the enormous benefits of rat eradications. On Midway, native vegetation such as *Scaevola taccada* and *Tribulus cistoides* also became noticeably more dense and abundant. With the removal of rats from Sand Island, the mice on that island are the only small mammals remaining in the Northwest Hawaiian Islands.

Unfortunately, the next attempted eradication of small mammals from a Pacific atoll by WS—of black rats from Palmyra in the equatorial Line Islands in 2001—was not successful. This was by far the most complex eradication attempted by Hawai'i-based wildlife managers, involving approximately 275 ha and 52 islets, some of which were densely vegetated with coconut palms (*Cocos nucifera*), *S. taccada* bushes, and *Pisonia grandis* trees (Ohashi 2001). Numerous factors contributed to the failure, among them a three-dimensional habitat that resulted in smaller foraging ranges on the ground than the 50 m interval at which bait stations were spaced and the high level of bait take by the ubiquitous land crabs *Cardisoma carnifex*, *Coenobita brevimanus*, and *Coenobita perlatus*. Efforts to eradicate rats from Palmyra have continued. After the biological factors likely behind the failure were carefully evaluated in August 2004, a pilot eradication on several small islets using the hand broadcast of brodifacoum was conducted in July 2005. Although successful, the application rate of 90 kg per ha was too high to be practical. An assessment of Wake Atoll in the fall of 2007 identified the same issues regarding the application rates appropriate for an area with high rodent densities where land crabs also compete for bait.

The successes of rat eradication on remote islands have also brought about efforts for the restoration of offshore islets of the main Hawaiian islands. In 2002, the Offshore Islet Restoration Committee was formed to restore selected islets around the Hawaiian Islands. To date, black rat eradication using diphacinone in bait stations has been successful on tiny Mokoli'i near O'ahu, although reinvasion appears to have occurred (D. Smith, pers. comm.).

Feral rabbits (Oryctolagus cuniculus) were eradicated from 117 ha Lehua Island in 2006 using trained dog and hunter teams from Island Conservation (a nonprofit organization dedicated to protecting insular biota). Aerial broadcast of diphacinone for the eradication of Polynesian rats was conducted in January 2009 (C. Swenson, U.S. Fish and Wildlife Service, pers. comm.). Lehua will be monitored for two years to confirm eradication of rats.

In the 1980s and 1990s, a wide range of toxicants and delivery methods were tried in the main Hawaiian islands to develop a means for rodent control in native ecosystems that would be safe, effective, and economical. Both acute and chronic toxicants, including fumarin, zinc phosphide, and brodifacoum for use in bait stations as well as zinc phosphide for use in broadcasting, were tested in Hawai'i Volcanoes National Park (HAVO) and other areas on the island of Hawai'i. These tests included separate trials of fumarin and zinc phosphide in bait stations and trials of hand-broadcast zinc phosphide–treated pellets on 0.8 ha areas for two-week periods in HAVO (Stone and Loope 1987). The maximum rat reductions were only 32% with fumarin. The use of brodifacoum in bait stations was also tested for seven weeks in 1997–1998 in the 'Ōla'a Forest, HAVO. Of 81 bait stations with WeatherBlok®

(0.005% brodifacoum), only two were visited by rats, suggesting that reluctance to enter the bait stations rather than acceptance of the baits resulted in the low level of effectiveness of this method (G. Lindsey and T. Smucker, unpubl. data).

Diphacinone was ultimately selected as the preferred rodenticide for use in conservation areas of Hawai'i because of its effectiveness in controlling rats (Tobin 1992), its relatively low risk to nontarget species (Kaukeinen 1982), and its limited persistence in the environment (Lund 1988). Another advantage of anticoagulants such as diphacinone is that toxicosis is delayed until rodents ingest a lethal dose, thus circumventing the bait aversion that may be learned with repeated applications of acute toxicants (Lund 1988, Prakash 1988). Using the safety and efficacy data from existing agricultural and commensal registrations, as well as a Hawai'i-based laboratory efficacy bioassay (Tobin 1992), in 1994 the Hawai'i Division of Forestry and Wildlife applied for and received approval of an SLN registration for Eaton's Bait Blocks® rodenticide with molasses and peanut butter flavorizer (0.005% diphacinone; J. T. Eaton and Company, Twinsburg, Ohio) (Conry 1994). The registration, for bait station use in forests, on offshore islands, and in other noncrop outdoor areas to protect Hawaiian native and endangered plants and animals, was the first rodenticide registration ever issued in the United States to control rodents for conservation purposes.

The success of this registration led to the formation of the Toxicant Registration Working Group with the goal of obtaining other registrations for conservation use in Hawai'i. Group members included the many federal, state, and private researchers, managers, and conservationists who had assisted with and supported the registration effort. Members were the Hawai'i Division of Forestry and Wildlife; the Hawai'i Department of Agriculture (HDOA); Kamehameha Schools; the U.S. Department of

Figure 18.2. A bait box with diphacinone baits. Baits are secured with a skewer, and the box lid is locked to prevent nontarget species from dislodging the baits. Target species—rats in this case—enter the box through a hole of appropriate size. *Source:* Photo © Jack Jeffrey.

Agriculture (USDA), Animal and Plant Health Inspection Service (APHIS), Wildlife Services (WS); the USDA-APHIS-WS National Wildlife Research Center (NWRC); the U.S. Fish and Wildlife Service (USFWS); the U.S. Geological Survey Biological Resources Division; the National Park Service; the U.S. Army; the University of Hawai'i; The Nature Conservancy; and the Maui Land and Pineapple Company, among others.

The HDOA and the EPA required additional field testing in support of the initial J. T. Eaton and Company SLN registration to address efficacy and risks to nontarget species under the label's use conditions. These studies, along with other laboratory and field trials (T. Casey, unpubl. data; Lindsey and Mosher 1994; Nishibayashi 1995; Swift 1998), led to two additional SLN registrations, for Eaton's Bait Blocks® with fish flavorizer in 1997 and for Ramik® Mini Bars (0.005% diphacinone; Hacco, Madison, Wisconsin) in 1998. These registrations were for the same use patterns as the first SLN registration but included the important addition of the small Indian mongoose to the list of target species. Unfortunately, in 2005 J. T. Eaton and Company discontinued all field uses of their products, including Hawaii's conservation uses, citing the high cost of additional recordkeeping resulting from the EPA-required change of classification to restricted use.

Hawaii's registrations for conservation bait stations have been essential to protect native species, especially endangered forest birds, from rat predation. Diphacinone bait stations have been used at Hakalau Forest National Wildlife Refuge (Nelson et al. 2002), at Hanawī Natural Area Reserve (Chapter 21), in areas of Keauhou Ranch privately owned by Kamehameha Schools (T. Casey, pers. comm.), in Mauna Kea Forest Reserve to protect captive released and translocated Palila (*Loxioides bailleui*) (P. Banko, pers. comm.), and on the Alaka'i plateau, Kaua'i Island, to protect wild and captive released Puaiohi (*Myadestes palmeri*) (Chapter 22) (Fig. 18.2). Within the range of the Po'o-uli (*Melamrosops phaesoma*) at Hanawī, rats were controlled with 430 diphacinone bait stations (Malcolm et al. 2008).

The use of diphacinone bait stations and snap trapping has demonstrated major benefits for some species, reducing predation of artificial nests in Oʻahu ʻElepaio (*Chasiempis sandwichensis ibidis*) habitat by 45–55% (VanderWerf 2001b), increasing the survival rate of nesting females from 50% to 83%, and increasing the survival rate of fledglings per nest from 33% to 82% between nontreated and treated sites (VanderWerf and Smith 2002).

Developing Broadcast Methods for Landscapes

Deploying diphacinone in bait stations over large conservation areas requires frequent, repeated applications and is thus labor-intensive and expensive. Bait stations require nearly continuous operation during the annual nesting cycle of birds, which can last four months or more. Although rodent populations recover after baiting ceases, there may be a lasting effect of reduced densities from previous years' treatments (Nelson et al. 2002; E. VanderWerf, pers. comm.). The cost of equipment and supplies is roughly $7,000 per km^2 during the first year of operation but decreases to about $2,000 per km^2 in subsequent years (Nelson et al. 2002). This makes baiting impractical for use in large, rugged, remote areas. Furthermore, although Polynesian and black rats accepted hand-broadcast diphacinone baits equally in another study (Dunlevy et al. 2000), Polynesian rats at Hakalau did not accept bait in stations to the same degree as black rats, possibly due to neophobia (aversion to novel objects), competitive interactions between rat species (Nelson et al. 2002), or the rats' feeding only on small, sublethal portions during quick visits to bait stations. A similar pattern occurred at Hanawī, where Polynesian rat abundance remained high despite the near-absence of black rats after baiting (Malcolm et al. 2008). Thus, broadcast techniques were determined to be more effective over large areas and where multiple species of rodents coexist.

In 1994, the Toxicant Registration Working Group decided to pursue registration of a rodenticide for hand and aerial broadcast, especially broadcast by helicopter over large areas of native Hawaiian forest bird habitat. Diphacinone was ultimately chosen as the preferred rodenticide for use in conservation areas in Hawaiʻi. After review of available toxicants and discussions with manufacturers and others familiar with the registration process, two 0.005% diphacinone products, Eaton's Bait Pellet Rodenticide with Fish Flavorizer® and Hacco's Ramik® Green pellets (also fish flavored) were selected as candidates for registration. These bait products were selected because of the large amount of data on diphacinone and on the products themselves that already existed as a result of their agricultural and commensal registrations. Both products could be made into pellets of sufficient weight to penetrate dense forest canopy to the ground below, and Ramik® Green was particularly durable under extreme conditions of heat and moisture. The fact that both companies agreed to assist with the registration was also crucial.

The data requirements for a new use pattern, aerial broadcast over nonagricultural land such as watersheds, as well as concerns about public opinion and the risk to nontarget species, particularly feral pigs and native birds, necessitated that numerous studies be conducted in support of the registration application. A series of laboratory bioassays conducted under good laboratory practices determined that the minimum exposure time and the amount necessary for Ramik® Green to meet the standard EPA efficacy requirement of 80% mortality was seven days and 37.5 g of bait for a Hawaiian black rat, while six days and 30.0 g achieved 90% mortality of Polynesian rats (Swift 1998). Both species also consumed lethal doses in a laboratory choice test (Swift 1998). Next, a field ef-

Figure 18.3. An experimental aerial broadcast operation on Hawai'i Island in 2001. The flow of bait from the bucket towed beneath the helicopter is controlled by the pilot. *Source:* Photo © Jack Jeffrey.

ficacy trial using hand broadcast of non-toxic Ramik® Green pellets coated with a biological tracer compared three potential sowage rates. The trial determined that the optimal aerial sowage rate needed to maximize the pellets' exposure to rats while minimizing excess bait usage was 22.5 kg per ha (Dunlevy et al. 2000).

Broadcast trials conducted under Good Laboratory Practices with Ramik® Green were used in support of registration. An environmental assessment was completed, and experimental use permits were obtained in 1999 for broadcasts of Ramik® Green by hand and by helicopter in HAVO. Two hand-broadcast applications of 11.25 kg per ha conducted one week apart achieved 98–100% population reduction of rats for two to four weeks in wet and mesic forest study areas, each 4 ha in size (Spurr et al. 2003a). A separate study that tested field efficacy with actual aerial broadcast in HAVO confirmed 100% reduction of rats and bait disappearance within one month (Spurr et al. 2003b). Several commercially available bait disper-

sal buckets manufactured in New Zealand were also evaluated to determine if they met the desired criteria for aerial bait distribution in Hawaiian forests (Spurr 2002). These features included retractable legs to reduce bait breakage, an agitator rather than an auger to reduce clogging, a self-contained power pack, and compatibility with the Hughes 500D helicopter (Fig. 18.3). To ensure precise and complete coverage during applications, this report also strongly recommended the use of a differential global positioning system.

The success of the conservation registrations in Hawai'i and the need for broader labels applicable to islands throughout the U.S. jurisdiction led to three Section 3 registrations for conservation uses of brodifacoum and diphacinone. These products can now be used on islands throughout the United States and its territories and possessions to eradicate or control existing introduced rodent populations and combat new introductions; they can be applied in bait stations, in burrows, and in the crowns of coconut palms, and they can be broadcast by hand and from aircraft. These registrations would not have occurred without the work conducted in Hawai'i and on other Pacific islands such as the California Channel Island, Anacapa, where the first aerial application of a rodenticide

Figure 18.4. Aerial broadcast operation on Mokapu Islet near the island of Molokaʻi in February 2008. *Source:* Photo by Heather Eijzenga, courtesy of the U.S. Fish and Wildlife Service.

to eradicate rats from a U.S. island was conducted (Howald et al. 2005).

Under the newly approved Section 3 label for conservation uses of diphacinone, a joint project of the USFWS, Hawaiʻi DLNR, and WS eradicated Polynesian rats from the 7 ha Mokapu Islet off Molokaʻi in 2008 (P. Dunlevy, pers. comm.). This project represented two major milestones: it was the world's first rat eradication using the aerial broadcast of diphacinone, and it was the first use of a rodenticide in the United States for conservation purposes under Section 3 (Fig. 18.4).

Addressing Effects on Nontarget Species

Numerous studies were conducted to assess the risks of diphacinone baits to nontarget species in Hawaiʻi. A field study using infrared-triggered automatic cameras with placebo versions of Eaton's® and Ramik® Green diphacinone baits deter-

mined that the primary toxicity risks to nontarget wildlife were quite low, with rodents comprising 99% of recorded visits (n = 20,994) to the pellets (Dunlevy and Campbell 2002). Few alien and no native birds were observed at the bait-monitoring stations (P. Dunlevy, pers. comm.). A study conducted with Canada Geese (*Branta canadensis*) to assess the risk to Nēnē found that the birds would not ingest the bait voluntarily (Witmer 2002).

The risk of secondary exposure to native Hawaiian birds of prey was found to be small because most diphacinone-poisoned rats died underground and were therefore not accessible to other species. A study with unpoisoned rat carcasses showed that ʻIo and Pueo (Short-eared Owls, *Asio flammeus*) would not likely scavenge rats should they die aboveground (Lindsey and Mosher 1994). Spurr et al. (2003b) found a low risk of secondary toxicity due to the inaccessibility of more than 90% of diphacinone-poisoned rat carcasses during an aerial broadcast trial. Massey et al. (1997) fed diphacinone-poisoned rats to captive American Crows (*Corvus brachyrhynchos*) serving as surrogates for the 'Alalā. No mortalities occurred, although symptoms of anticoagulant intoxication (blood in the nares, prolonged clotting times) were observed in the group receiving the highest dosage, an exposure level unlikely under actual landscape broadcast conditions. The risk of mortality to Po'o-uli from consuming contaminated slugs and snails was estimated to be quite low (Johnston et al. 2005) by means of theoretical calculations based on actual diphacinone residue levels obtained from invertebrates feeding on diphacinone baits under natural conditions.

Assessing Risks to Humans from Broadcast Rodenticide Use

All of these studies, along with other published and unpublished research on diphacinone and its human pharmaceutical counterpart Dipaxin, were compiled and analyzed in a hazard assessment that comprised the centerpiece of Hawaii's registration application (Eisemann and Swift 2006). This hazard assessment examined the risks to humans from drinking water derived from native forest watersheds, ingestion of game species that might consume the rodenticide, and risks to native terrestrial and aquatic species. The techniques examined included worst-case scenarios for evaluation rather than scenarios most likely to occur. These conservative analyses indicated that the greatest hazard would be to pregnant women drinking untreated water from a watershed broadcast with rodenticide under drought conditions in which all of the toxicant applied would dissolve into the stream water, which would serve to concentrate the diphacinone at maximum levels. Even under this risk scenario, the level of sublethal effects, such as increased clotting times, would be low. This assessment also concluded that the risks to terrestrial and aquatic nontarget species would be very minimal, even under the most conservative risk scenarios.

The History of Carnivore Control

Although feral domestic cats have historically been widely distributed as human commensals throughout the Hawaiian Islands, until recently small Indian mongooses were restricted to the islands of O'ahu, Moloka'i, Maui, and Hawai'i. The development of control methods for mongooses has been particularly important in the conservation of ground-nesting birds such as Nēnē and in preventing the establishment of mongooses on additional islands.

In the 1970s, numerous sightings of mongooses on Kaua'i, including a documented road kill of a lactating female in 1976 (Telfer 1977), prompted the USFWS to contract the USDA APHIS Animal Damage Control Unit to develop a method for mongoose control. Both acute and chronic toxicants, including thallium sulfate, zinc phosphide, warfarin, and diphacinone, were found to be effective against mongooses in laboratory bioassays. Diphacinone was also found to be highly effective against mongooses in low doses (0.18 mg per kg), which minimized hazards to nontarget species (Keith et al. 1989). In 1988, an experimental use permit was obtained to test the use of 0.00025% diphacinone in lean, raw hamburger in HAVO and at James Campbell National Wildlife Refuge (Stone et al. 1995). In 1991, an SLN registration was approved for the use of 0.1% diphacinone concentrate in raw hamburger in bait stations designed for mongooses. The design specified a 4-inch-diameter PVC pipe in the shape of a T, with entrances in the arms of the T and bait placed in the supporting arm. This technique proved effective but expensive due to the cost of the bait, the labor involved in mixing the bait, bait station construction, and installation and maintenance in remote areas; therefore, it was impractical to apply to large conservation areas (Stone et al. 1995), and the registration was allowed to expire. Standard rodent bait stations containing Eaton's Bait Blocks® with fish flavorizer were later found to be effective in controlling mongooses in small areas (Smith et al. 2000).

Lethal trapping of mongooses was conducted in remote high-elevation areas on Maui within home ranges of the Poʻo-uli. In this area, traps designed for small carnivores were baited with sponges soaked in cod liver or fish oil. In this circumstance, the technique was considered effective against mongooses (B. Sparklin, pers. comm.). Current research on mongoose control is focused on developing better baits and lures with a large call distance (distance of effective attraction) to stand out in prey-rich environments (W. Pitt, pers. comm.).

Options for feral cat control in Hawaiʻi have been limited. Lethal trapping has been restricted due to animal welfare concerns, and state laws that prohibit the discharge of firearms after dark restrict hunting. To date, feral cat control has consisted of live trapping in several locations, including on Mauna Kea (Fancy et al. 1997; this volume, Chapter 23) and in Haleakalā National Park (Simons 1983, Stone and Loope 1987, Natividad-Hodges and Nagata 2001), and in HAVO live trapping was conducted for long periods for both mongooses and cats (Stone and Loope 1987; D. Hu, pers. comm.).

Outside of the main Hawaiian islands, feral cats have been eradicated from several of the central Pacific islands, including Baker Island in the 1960s, Howland Island in 1986, and Jarvis Island National Wildlife Refuge in 1990 (Rauzon and Flint 2002). Experimentation with several methods on Jarvis Island led to the combination of hunting, trapping, poisoning, and, to a lesser extent, the introduction of feline panleucopaenia virus (FPL) (Rauzon 1985). The eradication resulted in the recolonization of five extirpated seabird species. Feral cat eradication has also recently been completed on Wake Atoll; how-

ever, the abundance of Polynesian rats increased in the absence of cats (Rauzon et al. 2008). Successful programs to eradicate cats on 48 islands worldwide have relied on a variety of techniques, including combinations of trapping, hunting, poisoning, and introducing lethal pathogens (Nogales et al. 2004). Of these islands, only 10 were ≥10 km^2 in area; the largest of these, Marion Island (290 km^2) in the subantarctic Indian Ocean, was more than twice the area of the Hawaiian island of Kahoʻolawe (117 km^2). FPL was also released on Marion Island, resulting in an annual population decrease of 29% (van Rensburg et al. 1987). The 15-year successful eradication program continued with hunting and disease monitoring and culminated in trapping and poisoning (Bester et al. 2002).

FUTURE DIRECTIONS IN SMALL MAMMAL CONTROL IN HAWAIʻI

New techniques for small mammal control are needed that will be effective on a very large scale for natural areas throughout Hawaiʻi and the Pacific. They must also be humane and environmentally safe and pose a low risk to nontarget species. For rodents, this will primarily mean the registration of additional toxicants, such as chlorophacinone in bait stations, and the refinement of rodenticide application techniques, especially the broadcasting of bait from the air over larger or topographically complex landscapes. For carnivores on larger islands, improved bait and delivery systems, live trap technology, and toxicant delivery techniques will most likely be the primary focus of future research. Other important areas warranting further investigation include repellents, barriers, and fertility control, although to date there has been little research on fertility control in Hawaiʻi. Some of these applications may be highly resource-intensive and suitable only to protect critically endangered species in relatively small areas, while other applications may be developed for use at the landscape scale. Cooperation and communication with other national and international groups and agencies conducting these types of research and control programs will be essential in developing the most effective and environmentally safe techniques for Hawai'i.

Rodent Control

Further research is needed to refine the methodology and increase the scope and scale of rodenticide applications for Hawai'i. Future research in rodenticide applications will include investigating the spatial configuration of bait stations, developing large-scale broadcast application methods, monitoring genetically based resistance to rodenticides, and registering new rodenticides and methods for conservation areas to effectively target mice and Polynesian rats.

The current label for bait stations specifies that they are to be placed in a gridlike fashion. Although not stated, the intention is for the user to deploy many stations over a large area. This ensures that the outer stations eliminate individuals migrating into the grid, creating a core area free of rodents around the native species to be protected. Often, however, managers desire to protect isolated rare plants or small bird territories, so only a few bait stations are placed in the immediate vicinity of the protected species due to difficult terrain or lack of staff. Studies are needed to determine if this configuration is capable of controlling rodents in the targeted area to the extent that predation on the native species can be reduced to acceptable levels. Generally, controlling rodents in small areas is less effective than large-scale control because edge effects are greater in smaller areas (the ratio of perimeter to area is greater in smaller areas). Therefore, larger treatment areas will have lower reinvasion rates and longer-lasting effects from ro-

denticide applications. For example, in 4 ha trials of hand-broadcast diphacinone bait in Hawai'i, rats were effectively eliminated for only two to four weeks (Spurr et al. 2003a), whereas rat populations were still reduced six months after an aerial broadcast trial in 45 ha of nearby habitat (Spurr et al. 2003b).

Although the impacts of house mice in Hawai'i are primarily predation on arthropods and seeds of rare plants, these can be quite significant for some species, such as the endangered coastal shrub *Sesbania tomentosa* (D. Hopper, pers. comm.), and may have broad ecological consequences. Therefore, more attention needs to be given to developing mouse-specific control techniques. The current diphacinone bait station registration needs to be revised to effectively target mice. The minimum bait station spacing of 25 m may be too great for the foraging range of a mouse. Studies are needed to determine if this technique can effectively control mice and to find the optimal spacing needed between stations.

Alternative techniques for mouse control, including registrations of other types of rodenticides, need to be explored to provide effective solutions for conservation areas. Large-scale rat-poisoning operations in New Zealand have resulted in rapid increases of mice three to six months after the initial elimination of more than 90% of rats (Innes et al. 1995). Similarly, on Buck Island in the U.S. Virgin Islands, house mice, which were apparently present but suppressed by black rats, increased in abundance after rats were eradicated with diphacinone bait in 2000 (Hillis-Starr and Witmer 2004). During an aerial broadcast trial in Hawai'i, mouse numbers initially decreased but then recovered to pre-treatment levels three months after baiting and increased to three times pre-treatment levels six months after baiting (Spurr et al. 2003b).

Additional registrations also provide alternatives for use in case a company discontinues a product or behaviorally or genetically based resistance to particular formulations develops. The development of genetically based resistance to the toxicant (Prakash 1988, Greaves 1994) is a concern with prolonged use of single toxicants and inadequate control from bait stations. No assessments of resistance have been done in areas where long-term baiting with diphacinone has occurred in Hawai'i. With additional data, the current zinc phosphide label for controlling mouse irruptions for public health purposes could eventually be used in support of a conservation use registration. Data in support of registration for the bait station use of chlorophacinone are also needed. These toxicants may prove more effective against mice.

Under the current SLN registration for diphacinone broadcast in Hawai'i, the user and/or landowner must ensure that feral pigs from the treated area cannot be hunted for 28 days to prevent human consumption of any contaminated meat. This generally means that the area needs to be fenced, placing a serious constraint on conservation efforts for forest birds, because rodent control is excluded from large areas of habitat needed for bird conservation and many such areas will continue to have hunting programs. The only study of feral pig behavior during an aerial broadcast trial in Hawai'i was unable to assess the risk to pigs due to numerous deviations from the experimental label (Pitt et al. 2005). This study will need to be replicated under the more rigorously controlled conditions that will prevail under the approved label. A study currently under way at the NWRC field station in Hilo will measure diphacinone residues to determine if cooking degrades the compound in pig meat.

Carnivore Control

Wildlife managers in Hawai'i need a much broader range of control techniques for carnivores to effectively protect endangered birds, including improvements in detection capability and toxic bait delivery systems. The possibility of mongoose establishment on Kaua'i, where many ground-nesting birds live, increases the urgency of developing a rapid response strategy to detect and eradicate this incipient population. Highly trained, well-bred dogs have been used on islands of Northwest Mexico to track, flush, and locate the presence of feral cats (Wood et al. 2002). Finding such locations and den sites for mongooses can greatly increase the efficiency of placing traps and toxicant bait stations.

New strategies to control feral cats in Hawai'i will primarily involve improvements to the techniques for live trapping, including improved techniques for attracting cats to traps with lures and attractants that require infrequent refreshing or maintenance; development of multiple capture devices and trap-alerting devices that notify managers when traps contain animals; adaptive strategies for managing feral cat populations in a variety of habitats in the Pacific; and documentation and interpretation of the impacts of feral cats on native wildlife to help managers prioritize management, support requests for increased predator control, and inform decision makers and the public of the value of control programs in protected areas.

A major need of upcoming research will be the development of more efficient capture methods for feral cats, specifically methods to increase the capture rate per unit effort, and the design of reliable trap-alerting devices to immediately notify field staff when animals are captured, thereby reducing the effort involved in the daily maintenance of traps while providing adequate welfare for captured animals. These types of devices may employ a radio transmitter to monitor trap status and notify staff to check the appropriate trap when it is sprung (Benevides et al. 2008). Cellular telephones may also be employed to verify trap functionality and notify staff when captures occur through a system that calls a remote telephone with caller I.D. (Larkin et al. 2003). Because many cellular telephones are equipped with digital photograph capability, it may also be possible to use a motion detector to activate dialing and send an image of the trap to another telephone. This would allow staff to observe the actual trap contents and discriminate sprung traps and target versus nontarget captures. Ideally, remote mechanisms would be used to immediately release nontarget captures and reset sprung traps, reducing the need to visit remote trap locations.

Audio and synthetic olfactory lures are needed to increase the effectiveness of attracting carnivores into traps and also to reduce the need for frequent bait replacement. Audio lures may consist of microchip recordings of nestling birds, kittens, and rodents to attract predators or stimulate curiosity. Concentrated baits may contain chemically synthesized volatile odors or consist of dried calamari, fish, shrimp, chicken, and processed meats or mixtures of urine and feces (Algar et al. 2002). Key characteristics of good lures for predators are attractiveness over time, protection from invertebrate scavengers, and volant molecules carried by the air indicating that food is near. Fresh raw egg was an effective attractant for mongooses in Mauritius compared to a suite of other potential lures. However, volant odors in eggs that are attractive to an actively foraging predator break down relatively quickly, requiring frequent refreshing (S. Roy, pers. comm.). A synthesized chemical attractant impregnated in a durable matrix would provide a longer-lasting, more cost-effective lure. The NWRC Hawai'i Field Station systematically screened a suite of artificial and natural lures, attractants, and baits for mongooses to improve capture success. Food baits performed better than food and animal scents, and of the food baits, fish was the most effective, whereas beef scraps, hot dogs, and coconut performed well (W. Pitt and R. Sugihara, unpubl. data).

Because trapping is extremely effort-intensive, adaptive management strategies and prioritization for long-term control of predator populations are needed in habitats where carnivore impacts are most severe. Control strategies should consider timing, location, effort, and criteria for reducing or pausing trapping. These strategies should take into consideration different habitats, the vulnerability and status of prey species, existing carnivore densities, immigration from the landscape reservoir (Hansen et al. 2007), and the cost-effectiveness of continued trapping based on the rate of effort and benefit to native wildlife. Tests of lures and attractants in different locations would provide site-specific recommendations for the most effective combinations. Radio telemetry would be useful in determining ecological criteria for trap placement to increase the probability of capture (Veitch 1985). For example, knowledge of home range sizes, den and scat sites, and travel routes would provide useful information for locating traps and determining necessary buffer zones (Goltz et al. 2008).

Barriers and Repellents

Barriers and repellents are devices that can consist of fences, shields, electrified cables, or chemical odors that deter predators. These devices are typically used to intensively protect small areas, such as the immediate vicinity of nesting birds, from particular predator species. Once implemented, they often require frequent maintenance but may dramatically reduce or eliminate the need for the repeated use of traps, toxicants, or other forms of lethal control. The construction of ungulate fences and the subsequent removal of ungulates in natural areas in Hawai'i is an excellent example of the use of barriers that have led to the dramatic recovery of forest bird habitat within the state (Stone et al. 1992). Other physical and electrical barrier technologies designed to control the dispersal of small mammals (McKillop and Sibley 1988) hold promise for the future reduction of mammalian predation on native forest birds.

Tree bands are sheaths of sheet metal wrapped around the boles of trees to prevent small mammals from scaling them to access nests. In tests of the effects of tree bands on predation rates of artificial nests in HAVO, nests in banded trees were significantly less likely to be depredated than those in unbanded trees (Nielsen 2000). The rate of nest success of 'Apapane (*Himatione s. sanguinea*), although not statistically significant, was 30% higher, and Mayfield estimates of daily survival were 23% higher in banded trees than in unbanded trees. All trees with interconnected canopy need to be banded in order to prevent access, but tree banding in conjunction with other control methods, such as trapping or toxicant use, may be more effective than either method alone. In the Rotoiti Nature Recovery Project on the South Island of New Zealand, individual Kaka (*Nestor meridionalis*) nest trees were protected from introduced predators, primarily stoats (*Mustela erminea*), by sheathing a section of their trunks with sheet metal in addition to trapping and poisoning the stoats (Moorhouse et al. 2003). This combination of techniques significantly increased nesting success at managed sites, producing twice the number of fledglings per nest as the most productive site without predator control. Although tree banding may be relatively inexpensive in material costs, the cost of the initial deployment effort may be quite high, especially for larger areas.

Electrical barriers, consisting of networks of charged bare galvanized cable around tree boles, have been highly successful in repelling brown treesnakes (*Boiga irregularis*) from Mariana Crow (*Corvus kubaryi*) nests on

the island of Guam (Aguon et al. 2002). Electrical barriers, however, are effort-intensive and subject to grounding by vegetation. In test situations, electrical barriers were found to be 100% effective in stopping snake movement. However, care must be taken to ensure that predators of concern are removed from a tree above the electrical barrier prior or subsequent to barrier construction.

As the remaining suitable habitat for some critically endangered species continues to shrink, the value of predator-free, intensively managed areas becomes greater. Fencing provides the ability to completely exclude and eliminate mammalian predators from the enclosed area. To be effective in excluding predators, fence designs generally include an underground skirt and an outward extension from the top, commonly called a "hat," as well as a very high wall to prevent animals from digging under, climbing, or jumping over the fence. Although predator-proof fences have been used in Hawai'i only to protect semi-captive populations of Nēnē and 'Alalā during releases, they are used to enclose several larger locations in New Zealand, including the Karori Wildlife Sanctuary in Wellington, costing between US$52 and $188 per linear meter. Eradication of enclosed alien mammals is an additional expense subject to the limitations of current control methodology. Initial trials of a multispecies, predator-proof barrier in Hawai'i manufactured by the Xcluder Pest Proof Fencing Company of Cambridge, New Zealand, demonstrated efficacy against most small mammals (J. Burgett, USFWS, pers. comm.). Mice were the only species able to breach the barrier on bare lava substrates. However, once the edge of the mesh skirt was sealed to the underlying rock using a dry mortar mix tamped and compressed through the mesh, mice were unable to pass under the barrier.

Predator-proof fences require vigilance to keep intruders out and frequent maintenance for inadvertent breaches caused by falling vegetation. Alternatively, a wide swath may be cut around the fence perimeter, but this adds tremendous expense and may not be possible in many steep and rugged areas of Hawai'i. Nonetheless, barren lava flows on some of the younger Hawaiian islands may be ideal sites for construction and maintenance of predator-proof fences, particularly for alpine-dwelling seabirds. In this case, however, designs must be developed to minimize potential entanglement of nocturnal species such as 'Ua'u (Hawaiian Petrels, *Pterodroma sandwichensis*) (R. Swift, pers. comm.).

Chemical and auditory repellents may have the potential for short-term protection of native species from predators. Some animal advocate groups favor repellents because they may reduce the necessity of lethal control (Liss 1997). Because mammals tend to habituate to averse stimuli, the long-term usefulness of these techniques needs to be evaluated. Although in preliminary research on the repellent effects of mongoose urine and feces on Polynesian and black rats these substances were not proven to be effective in laboratory trials, fewer rats were captured in traps that had previously been used to capture mongooses (Tobin, Koehler, et al. 1997). However, in another experiment in a laboratory arena, wild Hawaiian black rats tended to avoid mongoose feces as well as similar synthesized chemical odors, apparently because they recognized them as familiar predator odors (Burwash et al. 1998). These lines of research have yet to produce field applications. However, even though landscape-scale delivery techniques are far from development, they could provide supplementary methods to protect critically endangered birds from depredation until more permanent, alternative solutions are found.

Biological Control

Biological control usually involves the release of self-propagating pathogens, natural enemies, or other biological means of limiting populations, such as virus-vectored immunocontraception (Courchamp and Cornell 2000). At lower densities, predators may affect native species less or be more easily controlled by auxiliary methods. Predator introductions themselves are a form of biological control, although these have generally resulted in negative consequences for insular biota.

Pathogenic biological control techniques are unlikely to be developed for use on vertebrates in the Hawaiian Islands due to their limited long-term efficacy at larger landscape scales, the high cost of development, unclear regulatory requirements, humane treatment concerns, and the possibility of affecting domestic animals and pets. These techniques are likewise impractical on smaller uninhabited islets because of the potential to accidentally transmit vertebrate biological control agents to inhabited islands.

Most biological control agents typically exert a massive selection pressure on a population. They lower the population levels of vertebrate pests for a finite period of time but do not usually result in eradication without auxiliary methods. Normally, populations rebound from individuals that are resistant to the biocontrol agent. For example, in Australia the introduction of myxomatosis to control rabbits resulted in the development of a resistant population of rabbits (Saunders et al. 1999, Kerr 2002). New strains of biological control agents are being continuously identified and released to manage rabbit populations in Australia as part of a larger, coordinated suite of control methods.

SUMMARY

Small mammals have caused population declines and extinctions of the native biota of Hawai'i and the oceanic islands of the Pacific through the complex interaction of predation, competition, habitat degradation, and the spread of diseases. Introduced mammals such as feral cats are predators of young and adult birds, while rodents and mongooses prey on eggs and on young and adult birds. In addition, rodents consume the arthropod prey of birds and also destroy the seeds of important native food plants and inhibit regeneration of forest species. Diseases carried by small mammals, including murine typhus, leptospirosis, and bubonic plague, have caused a long history of human illness and mortality in Hawai'i, but other diseases such as toxoplasmosis, hosted by feral cats, have also been fatal to endangered birds such as Nēnē and 'Alalā. Of all the factors that currently limit the native Hawaiian avifauna, small mammals are potentially the easiest threats to manage.

Many techniques have been developed and used in Hawai'i to control small mammals in agricultural settings and to reduce the risks of diseases carried by small mammals to humans. Several acute toxicants were initially employed to reduce crop losses to rodents, but many of these fell out of favor because rodents' bait shyness limited their efficacy. Some acute toxicants also posed an unacceptably high risk to nontarget species when applied for conservation purposes. Chronic toxicants such as anticoagulants overcame the problem of bait shyness. Both ground-placed bait stations and broadcast techniques using cereal grain baits were developed for the distribution of toxicants in agricultural settings such as sugar cane fields, and these methods were also well suited to treating rodents in natural areas.

Hawai'i has been at the forefront of U.S. efforts to adapt agricultural and commensal methods of rodent control to native ecosystems. Developing rodenticide application techniques and obtaining registrations for them in Hawai'i have been pursued to protect native biota with two goals: (1) to substantially reduce rat populations in valuable native ecosystems on the main Hawaiian islands and (2) to erad-

icate them from uninhabited offshore islands and remote atolls. By means of combinations of traps and toxicants, rats have already been eradicated from three remote Pacific atolls where they had devastating impacts on seabird colonies: Polynesian rats from uninhabited Rose Atoll (American Samoa) and Green Island (Kure Atoll) and black rats from Midway Atoll. Polynesian rats were also eradicated from the islet Mokapu (off Moloka'i) in the world's first island eradication with aerial broadcast of diphacinone. A consortium of conservation and management agencies—the Toxicant Registration Working Group—was formed in 1994 to advance the registration of toxicants such as diphacinone and brodifacoum for use in conservation areas throughout Hawai'i and the islands of the Pacific. Diphacinone, a first-generation anticoagulant, was identified as a promising chronic toxicant for natural areas. The second-generation anticoagulant brodifacoum had also been successfully used to eradicate small mammals in New Zealand.

Detailed studies on dosage, nontarget effects, sowage rates, efficacy, and human risk were conducted to support registration applications in natural areas, first for ground-placed bait stations and then for broadcast efforts. Many of these studies were conducted under rigorous standards during field trials of hand and aerial broadcasts for registration under the Federal Insecticide, Fungicide, and Rodenticide Act. Approval of a special local need (SLN) application was granted in 1994 for Eaton's Bait Blocks® rodenticide with 0.005% diphacinone for use in bait stations. Currently, there are two Hawaiian registrations for Ramik® products to protect native species, one for bait stations and one for hand and aerial broadcast. Three national registrations for conservation use of diphacinone and brodifacoum have been approved by the Environmental Protection Agency. These products can be applied by a variety of methods to islands throughout the United States and its territories and possessions to control or eradicate existing introduced rodent populations and prevent new introductions. These registrations would not have occurred without the work conducted in Hawai'i.

Toxicant applications were also developed for mongoose control in Hawai'i. An SLN application was approved in 1991 for 0.1% diphacinone concentrate in raw hamburger for use in bait stations specifically designed for mongooses. Although this registration subsequently expired, standard rodent bait stations containing Eaton's Bait Blocks® with fish flavorizer were also found to be effective in controlling mongooses over small areas, and SLN labels for two diphacinone bait products added the mongoose as a target species. The control of another carnivore, the feral cat, has primarily been restricted to live trapping due to concerns regarding humane treatment. Because feral cats are wary of traps, it has been difficult to reduce feral cat predation on native bird populations.

New small mammal control techniques are needed that are humane and environmentally safe, present a low risk to non-target species, and, most important, are effective on a very large scale for natural areas throughout Hawai'i and the Pacific. Techniques that hold the most promise for small mammal control include the further development and registration of toxicants for large-scale broadcast; technological advances in live trap alerting systems for carnivores; toxicants, improved baits, lures, and attractants for carnivores; and barriers that block a broad suite of mammals from protected natural areas.

ACKNOWLEDGMENTS

We gratefully acknowledge the assistance and dedicated conservation efforts of P. Banko, T. Casey, P. Dunlevy, J. Eisemann, J. Groombridge, K. Gundersen, N. Hoffman, D. Hopper, D. Hu, J. Murphy, W. Pitt, M. Rauzon, S. Roy, D. Smith, B. Sparklin, S. Stephens, C. Swenson, R. Swift, and M. Tobin. We thank K. Campbell, G. Howald, E. Spurr, and an anonymous reviewer for constructive criticism of earlier drafts of this chapter.

CHAPTER NINETEEN

Captive Propagation

ALAN A. LIEBERMAN
AND CYNTHIA M. KUEHLER

Loss of habitat to the encroachment of introduced plants, birds, insects, mammals, and diseases has led to the precipitous decline of avian diversity in the Hawaiian Archipelago (Chapter 2). Long-term, holistic programs to protect and restore habitat are required to preserve the remaining natural areas and ensure the survival of Hawaii's avifauna (Chapter 16). Unfortunately, measures to enhance and protect habitats take time and may not be initiated quickly enough to prevent the extinction of some bird species (Chapter 16). In such cases, alternative, immediate actions involving hands-on manipulation of wild and captive individuals can prove valuable as conservation tools. This chapter reviews the development of avian captive management programs for Hawaiian forest birds and describes the benefits of these programs to ongoing recovery efforts.

Captive management of birds for species restoration requires four basic steps: *research, maintenance, breeding,* and *release. Research* provides the information on a species' natural history necessary to guide the management of the species in captivity and its release back into the wild. Captive *maintenance* is the provisioning by humans of a bird's basic needs, such as food, shelter, and security (pens, aviaries, etc.)—the traditional "care and feeding." Captive *breeding* goes beyond simple maintenance and encompasses identifying sexes, determining compatible pairs, establishing a stable social structure, provisioning suitable nest sites and materials, monitoring health, and developing artificial incubation and neonatal care procedures. Captive breeding incorporates aspects of captive management at the population level designed to conserve the genetic and demographic integrity of the captive flock. *Release* of captive reared birds into suitable habitat is the last and often a technically complex step in the captive management process. Although it may be debated whether release is part of captive management, clearly captive propagation for species recovery without the requisite planning for release will leave the arrow short of its recovery target. It is important to plan, monitor, and evaluate all four aspects of captive management in an adaptive management framework to create a feedback loop for success and to avoid duplication of failure.

Captive breeding programs for reintroduction have proven a valuable recovery strategy for numerous species of endangered birds. Notable examples outside Hawai'i include programs for the Peregrine Falcon (*Falco peregrinus*) (Cade and Burnham 2003), California Condor (*Gymnogyps californianus*) (Snyder and Snyder 2000), Bearded Vulture (*Gypaetus barbatus*) (Fry 2002), Andean Condor (*Vultur gryphus*) (Lieberman et al. 1993), Mauritius Kestrel (*Falco punctatus*) (Jones et al. 1995), Pink Pigeon (*Columba mayeri*) (Swinnerton et al. 2000), and Whooping Crane (*Grus americana*) (Meine and Archibald 1996). Arguably, without captive propagation, several of these species would now be extinct. The most internationally recognized case in Hawai'i has been the program for the captive breeding and reintroduction of the Nēnē (Hawaiian Goose, *Branta sandvicensis*), which in concert with predator control and habitat management has led to the species' dramatic rebound from near extinction in the mid–twentieth century (Kear and Berger 1980, Black and Banko 1994). Many other species across a wide spectrum of diverse taxonomic groups—kiwis, cranes, ducks, rails, pheasants, raptors, parrots, pigeons, and passerines (Soorae and Seddon 1998, 2000) —with captive breeding and reintroduction programs under way will prove successful (or not) in the coming years.

Although there are scores of avian species currently in captive management programs (according to the International Species Information System, ISIS), only a few species of passerines (perching birds) have reached a level of sustainability in captivity that could support a reintroduction program. Two recent examples are the Bali Mynah (*Leucopsar rothschildi*) and San Clemente Loggerhead Shrike (*Lanius ludovicianus mearnsi*) (Captive Breeding Specialist Group 1990, Kuehler et al. 1993, Grant and Lynch 2004). The relative lack of captive management science for passerines initially proved an impediment to the development of such programs for Hawaiian forest bird species, the large majority of which are passerines. It is therefore our hope that this chapter reviewing the captive management of Hawaiian forest birds will have applications to passerine conservation beyond Hawai'i.

ALTERNATIVES TO CAPTIVE MANAGEMENT

Considering Captive Management versus Alternatives

Long-term propagation of birds in captivity is labor-intensive and costly and is not an effective recovery tool for all species (Conway 1986, 1995; Hutchins and Wemmer 1991; Derrickson and Snyder 1992; Rahbeck 1993; Magin et al. 1994; Hutchins et al. 1997; Snyder et al. 1997a, 1997b). Before captive propagation is considered as a recovery option for any endangered species, this question should always be asked: Is there something that can be done to solve the problem without removing birds (or eggs) from the wild? A review of the guidelines and policies established by government agencies and international conservation organizations is of benefit (Beck 1992, Ellis et al. 1992, Kleiman et al. 1994, Wilson and Price 1994, Conway 1997, Seddon 1998, U.S. Department of the Interior 2000, International Union for Conservation of Nature 2003). Such guidelines offer a matrix of conditions under which captive management should or should not be initiated, such as the security of long-term funding; the favorability of logistical conditions; the availability of avicultural expertise; the level of knowledge of the species' natural history; the efficacy of alternative conservation actions, including the reduction or elimination of limiting factors; the level of difficulty in keeping and breeding the species in question; and the availability of sufficient genetic material to establish a viable captive flock.

Given the aforementioned concerns, not all endangered species will make good candidates for captive breeding programs. For those that do not, other manipulation techniques such as field support, translocation, or cross-fostering may be more appropriate recovery strategies. In general, these alternative strategies may be more cost-effective than captive propagation and should be considered as recovery options prior to initiating a captive breeding program. However, recovery strategies involving translocation and cross-fostering require special conditions and programmatic considerations and may not always be feasible (Griffith et al. 1989, Butler and Merton 1992, Serena 1995, Wolf et al. 1996).

Field Support

Field support is the process of intensive management of free-living individuals to boost their productivity. This strategy has been utilized in the conservation and recovery of the Kakapo (*Strigops habroptilus*), Black Robin (*Petroica traversi*), and Seychelles Magpie-Robin (*Copsychus sechellarum*) (Butler and Merton 1992, Merton et al. 1999, Bell and Merton 2002). It may involve such activities as the health monitoring of all individuals, manipulation of female weight gains to increase the proportion of female young produced, supplemental feeding to support breeding birds and/or chicks in the nest, and physical protection and close monitoring of nests with the capability of intervening as necessary.

Translocation

Translocation, as defined here, is the capture of birds from one site and their subsequent release at a distant site. For some island endemics, such as the Ultramarine Lorikeet (*Vini ultramarina*), the Seychelles Brush-Warbler (*Acrocephalus sechellensis*), the Seychelles White-eye (*Zosterops modestus*), and numerous New Zealand species, translocation to predator-free habitats on another island or to disjunct habitats may be the best immediate option to save the species (Flack 1978, Komdeur 1991, Butler and Merton 1992, Armstrong and McLean 1995, Lovegrove 1996b, Kuehler et al. 1997, Lieberman et al. 1997, Armstrong 2000, Boyd and Castro 2000, Jamieson et al. 2000, Rocamora 2002). This approach has repeatedly been advocated for the endangered passerines of the Northwestern Hawaiian Islands: the Laysan Finch (*Telespiza cantans*), Nihoa Finch (*T. ultima*), and Millerbird (*Acrocephalus familiaris*) (U.S. Fish and Wildlife Service 1984b; Morin et al. 1997; Morin and Conant 2002, 2007). Translocation of the Laysan Finch to Pearl and Hermes Reef and Midway Atoll succeeded (although the Midway population succumbed when rats arrived), whereas introductions of the Nihoa Finch to French Frigate Shoals failed. Translocation still has great potential for recovering these three species (Chapter 25).

Although controversial, translocations of species to fill the ecological niches of similar but extinct species has been considered in Hawai'i. Elsewhere, this strategy has been applied to recover the Seychelles Fody (*Foudia sechellarum*) and Seychelles Brush-Warbler via translocation to Denis Island, Seychelles, outside the natural distribution of either species (Bristol 2005). In Hawai'i, extralimital possibilities are translocation of the Puaiohi (*Myadestes palmeri*) and 'Ōma'o (*M. obscurus*) from Kaua'i and Hawai'i islands, respectively, to Moloka'i or Maui to replace the likely extinct 'Oloma'o

(M. lanaiensis); translocation of the Nihoa Millerbird (Acrocephalus familiaris kingi) to Laysan to fill the niche of the extinct Laysan Millerbird (A. f. familiaris); or even introduction of the 'Alalā (Corvus hawaiiensis) on Maui to fill the niche left vacant by a now extinct Maui crow (Olson and James 1982a, James and Olson 1991).

Recent translocation efforts on the main Hawaiian islands have had limited success. In 1995, an experimental program by the U.S. Geological Survey Pacific Island Ecosystem Research Center and The Peregrine Fund evaluated the translocation of wild 'Ōma'o compared with the introduction of captive-reared 'Ōma'o as a potential recovery option for the endangered Hawaiian species of thrushes. This study demonstrated similar survival rates for both release groups of 'Ōma'o. However, captive-reared birds remained at the release site, whereas the translocated birds tended to disperse (Kuehler et al. 2000, Fancy et al. 2001). The release effort could also well serve the 'Ōma'o in the future if interest arises in reestablishing the species in its former range on the leeward side of Hawai'i Island. In contrast, an initial translocation of Palila (Loxioides bailleui) from the west side to the east side of Mauna Kea and subsequent efforts to translocate Palila from the west side to the north side of Mauna Kea have demonstrated homing behavior, with most of the wild birds returning to their capture site shortly after release (Fancy et al. 1997; this volume, Chapter 23).

Future conservation options for the 'Ākohekohe (Palmeria dolei) may include translocation. Because the 'Ākohekohe is a difficult species to maintain in captivity due to its pugnacious character, maintaining a large captive breeding flock would be impractical. In the late 1990s, a translocation project was organized using the 'I'iwi (Vestiaria coccinea) as the surrogate model for the 'Ākohekohe. 'I'iwi were captured in the forests of East Maui and released in West Maui to assess capture, acclimation, and release procedures and postrelease site fidelity (Hawai'i Division of Forestry and Wildlife 2001). The majority of the 'I'iwi apparently survived the translocation (G. Massey, pers. comm.; S. Fretz, pers. comm.). These initial successes with 'I'iwi indicate that translocation may be a low-risk, high-return recovery strategy for the 'Ākohekohe. Similarly, experimental translocation of Maui 'Alauahio (Paroreomyza montana) over short distances was attempted as preparation for translocating Po'o-uli (Melamprosops phaeosoma) (Chapter 21).

These studies indicate that translocation holds potential as a recovery option for Hawaiian birds if one takes into account the natural history of the species, the accessibility of capture and release sites, the sensitivity of the species to capture, the potential for the species to remain at its release site, the need for captive maintenance for a very short time (time to travel from capture site to release site), and finally the translocated species' ability to establish itself in the area of release.

Cross-Fostering

Cross-fostering is a technique by which one species acts as a surrogate parent for another species (Butler and Merton 1992, Nagendran et al. 1996). This can be especially useful in situations in which the surrogate foster parent species is common and the fostered species is extremely rare. Cross-fostering can help maximize production in rare species by inducing pairs to renest following the collection of eggs and/or chicks. Geographic and taxonomic proximity are required. If one examines potential donor-host pairings from each Hawaiian island, it quickly becomes apparent that there are too many dietary and behavioral incompatibilities and that few if any cross-foster possibilities exist. To date this technique has not been seriously promoted for any Hawaiian forest bird.

DECIDING ON CAPTIVE MANAGEMENT

Once the options of field support, translocation, and cross-fostering have been considered and discounted, captive management may be considered as a conservation option. Conservationists must establish a logical path of inquiry, asking why, when, and how such management is to be accomplished. The motives for captive propagation must be asked in a multidisciplinary forum that includes these questions: (1) Is a captive propagation and reintroduction program necessary to recover the species, or can alternative, more cost-effective strategies restore the species in the wild? (2) Do captive propagation and reintroduction have a reasonable chance of succeeding? (3) Will the program be part of an integrated landscape-level recovery

---

Box 19.1.   **Captive Management Strategies**

When captive propagation is being considered as an option for recovery, a critical analysis of the endangered avifauna must be completed to prioritize the species regarding when each species should become part of the captive program and which captive propagation strategy is most appropriate to prevent extinction and promote recovery (see Table 19.1). The criteria to consider for prioritization are presented here, in no particular order.

- Taxonomic uniqueness and recovery priority on lists of threat (International Union for Conservation of Nature–BirdLife International, U.S. Fish and Wildlife Service, state of Hawai'i, etc.).
- Urgency or degree of threat and current level of protection (legal and otherwise).
- Known cause of decline in the wild and resources available to rectify limiting factors.
- Knowledge of the species' natural history, including population size and distribution.
- Status of current field research and habitat management and potential for collaboration.
- Practical considerations (long-term availability of funding, expertise, labor, aviaries).

- Avicultural history and known (or presumed) degree of propagation difficulty.
- Release history and known (or presumed) degree of difficulty of release and availability of suitable release sites (healthy habitat and its management).
- Private and public landowner partnership agreements (Hawai'i Conservation Plan, Safe Harbor, etc.).
- Known value to the integrity of the ecosystem.
- Cultural and educational value.

The weighting of these criteria will depend on the species in question, the environment in which they are found, the resources available to dedicate to recovery, and the personnel and agencies involved.

This list for establishing the priorities for captive propagation is not comprehensive, and it may be controversial in concept. Arguably, all endangered species must be considered as equals in their need for assistance. However, if these criteria are applied only to establishing priorities for captive propagation as a subset of the overall recovery strategy, one can view these issues from a more subjective, practical perspective.

Table 19.1. Captive management strategies and priorities

Species	Captive Propagation Program Strategies	Captive Management Priorities
"On-the-brink" species; (e.g., Poʻo-uli)	8, then 4 or 5 or 6 or 7	1
Puaiohi	5	1
ʻAlalā	6, then 7	1
ʻĀkiapōlāʻau	9, then 5	2
Palila	5	2
Nēnē	7	2
ʻAkikiki	9, then 5	2
ʻAkekeʻe	9, then 5	2
Maui Parrotbill	5	2
Oʻahu ʻElepaio	1, then 3, then 4	3
Hawaiʻi ʻĀkepa	4 or 5	3
Hawaiʻi Creeper	4 or 5	3
ʻĀkohekohe	3, then 4	3
Hawaiʻi ʻElepaio	9	4
ʻIʻiwi	9	4
ʻŌmaʻo	4	4

Source: This table is modified from Table 13 in the *Revised Recovery Plan for Hawaiian Forest Birds* (U.S. Fish and Wildlife Service 2006).

Notes: All of the following captive management options are proposed to complement ongoing native habitat management with the goal of recovering the species in the wild. The three Northwestern Hawaiian Islands passerines (Millerbird, Laysan Finch, Nihoa Finch), all endangered, were not included in the original analysis. However, their recovery plan prescribes translocations for which captive management would be an important component (U.S. Fish and Wildlife Service 1984b). Given the critically endangered status of the Millerbird and Nihoa Finch (BirdLife International 2000), these two species could be ranked Priority 1, whereas the Laysan Finch could be ranked Priority 2 or 3.

Key: Strategy 1: no captive management necessary for recovery; Strategy 2: field support management (management of free-living individuals); Strategy 3: translocation; Strategy 4: rear and release; Strategy 5: captive breeding for immediate release; Strategy 6: captive breeding for self-sustaining captive population; Strategy 7: captive breeding for production; Strategy 8: emergency search and rescue; Strategy 9: development of captive technology; Priority 1: species in critical need of captive management; Priority 2: species in great need of captive management; Priority 3: species in need of captive management, but other conservation options are superior; Priority 4: species useful as a surrogate for other more critical species.

effort incorporating habitat management, research, and environmental education? (4) How much time will be required to prepare secure release sites? Objective analysis of these questions and the issues they raise (Box 19.1, Table 19.1) will identify the best role of captive propagation in a conservation environment of shrinking and ever more elusive fiscal resources.

It should be recognized from the outset of any recovery effort that captive breeding is not necessarily the first answer to an extinction problem and should not be viewed as the complete solution. Captive propagation and reintroduction must be part of an overall integrated conservation and recovery strategy that includes habitat management, field research, and public education. With or without captive breeding, recovery efforts that do not effectively address limiting factors affecting the wild population will ultimately fail.

THE EARLY HISTORY OF CAPTIVE PROPAGATION IN HAWAI'I

Captive breeding of endemic birds in Hawai'i began with the Nēnē recovery effort. The Nēnē program was not only the beginning of captive propagation as a conservation strategy in Hawai'i but was one of the first captive propagation programs in the world that focused on the recovery of an endangered species. Kear and Berger (1980) present a dramatic narrative of the near extinction and miraculous rebound of this island goose—the result of the combined efforts of energetic and forward-thinking conservationists. The Nēnē recovery effort was spearheaded by the Wildfowl Trust in England led by Sir Peter Scott, the Shipman Estate on the Big Island of Hawai'i, and Mr. Ah Fat Lee, supported by the territorial (later the state of Hawai'i) authorities. Beginning with the construction of the state's Pōhakuloa Endangered Species Propagation Facility (PESPF) in 1949 and continuing with the intensive management of the captive flock to the present, the Nēnē has enjoyed an increase from a wild population of no more than 31 birds in 1951 to present estimates of over 1,300 Nēnē on four islands: Kaua'i, Moloka'i, Maui, and Hawai'i (U.S. Fish and Wildlife Service 1983a, 2004a; A. Marshall, pers. comm.). In addition to working to preserve the Nēnē, the Pōkahuloa facility completed the conservation waterfowl troika with a similar captive propagation effort for the Koloa (*Anas wyvilliana*) and the Laysan Duck (*A. laysanensis*).

Despite its waterfowl emphasis, the PESPF also played a key role in the early captive management of the 'Alalā. It was not until the early 1970s that 'Alalā fledglings considered at risk were salvaged from wild nests in an effort comprising the initial attempts to keep 'Alalā in captivity for conservation purposes. All of the 11 wild birds salvaged were eventually taken to the PESPF from 1976 to 1981 (Banko, Ball, et al. 2002), and their management helped form the initial captive strategy—bring chicks in from the wild, establish the breeding program in Hawai'i, and avoid the stress of transport to the mainland (U.S. Fish and Wildlife Service 1982a). Several 'Alalā hatched under the parents at the PESPF in 1977 (Kuehler and Lieberman 2005), but none survived. In 1981, three 'Alalā eggs were sent to the Honolulu Zoo, where the "K Brothers" (Keawe, Kalani, and Keli'i) were hatched and reared and later sent back to the PESPF (Fig. 19.1).

Because the PESPF was considered deficient (Temple 1984) due to its proximity to ordnance demolition in the Pōhakuloa Training Area, as well as deteriorating aviaries and enclosures, disturbance by activities in the adjacent Mauna Kea State Park, and the generally dry climate of the area, the state relocated the captive program to the remodeled and converted Olinda Prison on Maui (Temple 1984, Duvall and Conant 1986), which was renamed the Olinda Endangered Species Propagation Facility (OESPF). This relocation took place over 1986–1989. The move to the OESPF included 9 adult 'Alalā, 20 Nēnē, and 17 Koloa but did not include the Laysan Ducks, which were sent to the Smithsonian National Zoological Park Conservation and Research Center at Front Royal, Virginia (Giffin 1989a). In addition to the physical change in venue, an effort was made to improve the inconsistent results of the first 10 years of the propagation program for the 'Alalā. In contrast, the propagation efforts for the Nēnē and Laysan Duck were doing very well, with 1,699 Nēnē goslings reared in 1949–1978 (Kear and Berger 1980), and there was similar success for the Koloa. The propagation efforts for the Laysan Duck were abandoned due to poor breeding records and the

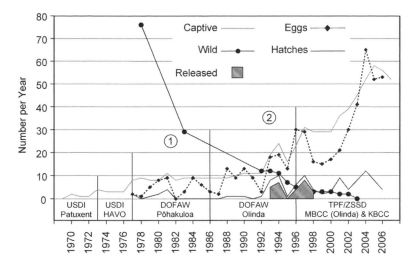

Figure 19.1. History of the 'Alalā in captivity. From 1968 to 2008, the care and breeding of the 'Alalā has passed through no fewer than five different management phases. Circle 1 indicates the three eggs laid at the Pōhakuloa Endangered Species Facility in 1981 that were hatched at the Honolulu Zoo, and Circle 2 indicates the 18 eggs collected from the wild in 1993 and 1994, from which hatched 13 chicks. The captive 'Alalā population gradually increased, while the wild population collapsed and became extinct, despite efforts to assist the wild birds. Note that the level of egg production has been high, but their hatchability has remained problematic. USDI, U.S. Department of the Interior; HAVO, Hawai'i Volcanoes National Park; DOFAW, Division of Forestry and Wildlife; TPF, The Peregrine Fund; ZSSD, Zoological Society of San Diego; MBCC, Maui Bird Conservation Center; KBCC, Keauhou Bird Conservation Center. Source: Graphic drafted by Nancy Harvey.

possibility of hybridization in captivity (Reynolds and Kozar 2000, U.S. Fish and Wildlife Service 2004b).

Following the review of the 'Alalā program's progress by outside consultants and the publication by the National Research Council of *The Scientific Basis for the Preservation of the Hawaiian Crow* (Duckworth et al. 1992), several additional programmatic changes were set in motion: the hiring of a program director and a full-time veterinarian for the OESPF and, perhaps most important of all, the settlement of a pending lawsuit against the U.S. Fish and Wildlife Service (USFWS) that allowed biologists to manip-

ulate the remaining three or four pairs of breeding wild birds on private lands in South Kona, Hawai'i Island. This latter action brought the Zoological Society of San Diego (ZSSD) and The Peregrine Fund (TPF) into the 'Alalā captive management program in 1993, with a one-year federal contract to incubate, hatch, and rear the wild 'Alalā eggs and to release and monitor captive-reared juveniles. Although TPF had earned a reputation as a world-class conservation organization that focused on endangered raptors, it took on a new role in Hawai'i with the captive management of passerines. Within one year following the publication of the National Research Council's report, TPF had hired staff, subcontracted the San Diego Zoo's Avian Propagation Center, designed a successful propagation protocol, built release aviaries, hatched seven wild-collected 'Alalā eggs and reared the chicks, and released the first cohort of five juveniles into the native forests of South Kona (U.S. Fish and Wildlife Service 2003; this volume, Chapter 20) (see Fig. 19.1).

The successful hatching of wild, partially incubated eggs of 'Alalā and rearing of the young in 1993 helped resurrect interest in captive propagation as a conservation option for other endangered Hawaiian passerines. The original recovery plans for these species had recommended actions that included captive propagation (U.S. Fish and Wildlife Service 1982b, 1983b, 1984b, 1984c, 1986). Two newly formed advisory teams, the Hawaiian Forest Bird Recovery Team and the Captive Propagation Recovery Working Group, convened their first meetings in 1994. The old recovery plans were revisited, and a new, comprehensive plan (minus the Northwestern Hawaiian Island passerines) was developed that included captive propagation strategies for many species (U.S. Fish and Wildlife Service 2006).

Although 11 "on-the-brink" species were the first to be targeted for captive intervention, several challenges quickly became apparent. Unfortunately, the wild populations of eight species that potentially could have benefited from captive management had declined precipitously during the 1970s and 1980s and could no longer be located. These were the Kaua'i 'Ō'ō (Moho braccatus), Kāma'o (Myadestes myadestinus), 'Oloma'o, 'Ō'ū (Psittirostra psittacea), Kaua'i and Maui Nukupu'u (Hemignathus lucidus hanapepe, H. l. affinis), O'ahu 'Alauahio (Paroreomyza maculata), Kākāwahie (P. flammea), and Maui 'Ākepa (Loxops coccineus ochraceus) (Chapter 5). As discouraging as it was to contemplate the potential success of captive management in saving the Kaua'i 'Ō'ō and other recently disappeared species from extinction, recovery biologists are inherently optimistic and quickly turned to the surviving endangered species. These species were reviewed by the two new recovery teams ('Alalā and Forest Birds), two working groups (Captive Propagation and Nēnē), and the managers of the captive program and were prioritized using the criteria presented in Box 19.1 and Table 19.1. Fortunately, there were common Hawaiian species that could be used as surrogates in developing avicultural protocols and procedures, as well as some previous institutional experience with certain Hawaiian thrush and honeycreeper species in captivity. In addition, surrogate models were developed using several species of mainland corvids for inclusion in the design of release strategies for the 'Alalā (Marzluff et al. 1995).

The Honolulu Zoo (HZ) has maintained several of the endemic Hawaiian species in its collection from 1969 to the present, establishing the care and feeding regimes for the Kaua'i 'Elepaio (Chasiempis sandwichensis sclateri), 'Ōma'o, Puaiohi, Laysan Finch, Palila, Hawai'i 'Amakihi (Hemignathus virens), Kaua'i 'Amakihi (H. kauaiensis), 'I'iwi, and 'Apapane (Himatione s. sanguinea) (P. Luscomb, pers. comm.). Perhaps the most significant contribution made by the HZ was the previously mentioned incubation and rearing of three 'Alalā from eggs laid by a pair of birds at the PESPF. Two of these 'Alalā later bred at the OESPF and the Hawai'i Endangered Bird Conservation Program. Most recently, HZ has focused on the maintenance and breeding of 'Apapane. Although none of the Hawaiian species bred in sufficient numbers to allow them to establish self-sustaining populations, much of the information gathered from the avicultural experience benefited later efforts to keep and breed these species, especially information on dietary requirements; heat, stress, and disease issues; aviary dimensions; and social and demographic considerations (Carlstead and Hunt 2002, Shepherdson et al. 2004).

Also, in the 1980s and early 1990s, a consortium of member zoos of the American Zoo Association (AZA) mounted several trips to collect the more common Hawaiian species. Although there have been several captive hatchings of 'Ōma'o, 'I'iwi, and 'Apapane in AZA institutions (Derrickson 1998), none of these species has established a viable captive population. Currently no mainland zoos are holding Hawaiian passerines (according to the International Species Information System— ISIS) other than the Hawai'i 'Amakihi, which now has a captive population of 12 males and 13 females in zoological institutions and has bred to a second generation (A. Oiler, pers. comm.). It is from the initial efforts of this AZA consortium that two of the more significant avicultural challenges presented by the Hawaiian avifauna have been confirmed: heat- and stress-related sensitivity (P. Luscomb, pers. comm.). Hypersusceptibility to fungal air-sacculitis (*Aspergillus fumigatus*) was one of the primary considerations (B. Rideout and P. Morris, pers. comm.) in designing a captive breeding program that originated with wild-collected eggs rather than wild, stress-prone adults. The threat of air sac infections has since been addressed with drug therapy, prophylaxis, and the modification of captive protocols to reduce stress through the careful management of birds. But the potential for fungal outbreaks remains a veterinary specter and is an ever-present consideration in designing protocols for birds' physical handling, their relocation from aviary to aviary, and their transportation from site to site.

In the early 1990s, the OESPF began working with the Maui 'Alauahio as a surrogate to refine the avicultural techniques used to rear, maintain, and release endangered species. Rather than collecting wild eggs, this effort focused on taking nestlings to avoid the labor-intensive period of incubating eggs and rearing day-one hatchlings. This project was successful in rearing and maintaining individuals of the species, but releases were hampered by the rat predation of several of the birds while they were being acclimatized in field aviaries prior to their release (P. Shannon, pers. comm.).

The Avian Propagation Center at the San Diego Zoo and several other AZA member institutions also provided the documentation and hands-on experience needed for the incubation and neonatal care of a wide variety of passerine species (Kuehler and Good 1990). Because passerines are altricial (helpless at hatch), this experience was invaluable for the establishment of a captive program of Hawaiian songbirds from wild-collected eggs that required artificial incubation and hand rearing.

THE HAWAI'I ENDANGERED BIRD CONSERVATION PROGRAM

In 1993, prompted by a recommendation from a blue-ribbon committee sponsored by the National Academy of Sciences (Duckworth et al. 1992), TPF entered into a 20-year cooperative agreement with the USFWS to design, build, and operate a second captive propagation facility (the OESPF had been the first) dedicated to the captive management of the 'Alalā and other endangered Hawaiian forest birds. The new facility, built with funds appropriated by the U.S. Congress and located adjacent to Hawai'i Volcanoes National Park on Kamehameha Schools' land, was named the Keauhou Bird Conservation Center (KBCC) and was opened on March 1, 1996. At the same time, the responsibility for the operations of the OESPF was transferred from the state of Hawai'i to TPF, and the facility was renamed the Maui Bird Conservation Center (MBCC). The captive propagation program—two facilities and the field component—was named the Hawai'i Endangered Bird Conservation Program (HEBCP). Finally, in 2000 the HEBCP transitioned from TPF to the ZSSD, with which it resides today. Operating funds for the HEBCP are provided through contracts, grants, and contributions from the USFWS, Hawai'i Division of Forestry and Wildlife (DOFAW), ZSSD, and private donors.

TPF set goals that would allow for the establishment of a successful captive propagation and reintroduction program: (1) build appropriate facilities, (2) establish the avicultural techniques required to build a captive flock of selected endangered Hawaiian species, (3) train staff in the requisite avicultural skills, and (4) establish

Figure 19.2. A "forest bird barn" at Keauhou Bird Conservation Center. Each of the two aviary buildings is divided into 19 units, each housing a pair of forest birds (Puaiohi or honeycreeper). However, each pair of 'Alalā is housed in a smaller, isolated aviary away from other pairs due to their aggressive territorial behavior when breeding. *Source:* Photo by Alan Lieberman, Zoological Society of San Diego.

working partnerships with government and nongovernment agencies and private landowners to promote recovery of the target species. The first goal was accomplished with the construction of the KBCC and the transfer of the OESPF to the HEBCP in 1996. Both facilities are sited away from urban areas to avoid unnecessary disturbance (noise, smoke, vandalism, etc.) but are in proximity to resources that support bird care (food supplies, staff needs, utilities, etc.). Consideration was also given to locations with climatic conditions suitable for maintaining a captive flock but out of harm's way in terms of volcanic activity. A matrix of program requirements and physical conditions was considered when planning the new captive propagation facility (U.S. Department of the Interior 1994). Both the KBCC and the MBCC were

designed to maintain and breed native Hawaiian passerines and Nēnē. Currently, the combined facilities have 55 forest bird (small passerine) aviaries, 34 'Alalā aviaries, and 12 Nēnē pens (Fig. 19.2). In addition to the breeding aviaries, the facilities have several independent kitchens designated for hatchlings, juveniles, and breeding stock. The infrastructure includes quarantine facilities, clinics, workshops, storage buildings, and on-site residences for staff, who provide security as well as round-the-clock bird care if required.

The second programmatic goal was to develop the avicultural techniques needed to build a captive flock for reintroduction. Although aviculture has a long and anecdotal history, there is very little analytical documentation of the avicultural experience required to keep, breed, incubate, and rear small passerine species in captivity (Roots 1970; Lieberman et al. 1989, 1990; Vince 1996). The work of Eddinger (1969, 1970, 1971) and Berger (1972, 1980) and personal communications with the personnel from the HZ and other members of the AZA Hawaiian Bird Consortium have been the main sources of information on the captive management of the Hawaiian avifauna (Giffin 1989b). In 1994–1999, several common species of native Hawaiian birds were incubated and reared by the HEBCP in order to establish the incubation and rearing protocols for use with related taxa of endangered Hawaiian species (Kuehler et al. 1994, 1996, 2000, 2001). The techniques established during this phase of surrogate work accomplished the second goal of developing the techniques necessary for the anticipated work on endangered Hawaiian forest birds.

The third goal of the HEBCP was to train staff in the techniques required to maintain the long-term continuity of the program. One of the greatest challenges of any propagation program is maintaining the integrity of the program through the normal dynamics of changes in staff. The solution lies in the mentoring of new staff by the experienced staff and the maintenance of accurate and complete records. From its inception, the HEBCP has taken great pains to pass avicultural skills from one staff person to the next and to keep detailed records of the incubation parameters of every egg, the feeding records of every chick, and the breeding records of every pair of birds in the program, as well as extensive necropsy and veterinary records and demographic and pedigree analysis of the captive inventory. These records are constantly reviewed.

The fourth and final goal was to establish and cultivate diverse partnerships in order to promote recovery. This has been accomplished through the formation of recovery teams and working groups whose members include agency biologists and administrators, representatives of nongovernment organizations, private citizens, and captive propagation staff. In particular, a Hawai‘i Captive Propagation Partnership was formed in 1996 by members of the USFWS, DOFAW, and TPF (now ZSSD) to establish and review programmatic priorities and action plans coordinated with other agency actions (Zoological Society of San Diego 2001).

THE CHALLENGE OF CAPTIVE MANAGEMENT IN HAWAI‘I

Captive management of birds for conservation purposes incorporates three distinct activities: project administration, aviculture, and release of birds. The history of the institutional involvement in Hawai‘i has been presented in the previous section, but the actual challenge of aviculture is at the heart of captive management. Having begun with the early avicultural experiences managing endemic Hawaiian species by the HZ, DOFAW, and AZA, followed by TPF and the ZSSD, captive management of native avian taxa now has a solid foundation with a proven record. From 1993 through 2007, the HEBCP handled 2,521 eggs from 15 endemic Hawaiian bird taxa, hatched 925 chicks, and reared 789 chicks to independence (Table 19.2) (Kuehler et al. 1994, 1995, 1996, 2000, 2001; Zoological Society of San Diego 2007). Significant contributions have now been made toward the recovery of several species of endangered Hawaiian forest birds: the Nēnē, ‘Alalā, Puaiohi, Palila, Maui Parrotbill (*Pseudonestor xanthophrys*), Hawai‘i Creeper (*Oreomystis mana*), and Hawai‘i ‘Ākepa (*L. c. coccineus*).

Basic avicultural knowledge can be described in terms of the challenges of the various life stages: eggs (handling and incubation), chicks (neonatal care or brooding), adults (capture, acclimation, and maintenance), reproduction (pairing, breeding, egg laying, and parental care), and finally reintroduction (transport, release, and monitoring). Although many of these challenges have been faced by bird fanciers and keepers throughout history, the forest birds of Hawai‘i present a number of unique challenges, including biological and behavioral unknowns, disease and stress susceptibilities, small egg and chick sizes, small wild populations, and, in the case of at least one species, the ‘Alalā, a high level of inbreeding (Ellis et al. 1992; this volume, Chapter 10).

Table 19.2. Summary of eggs and chicks hatched and reared in captivity in the Hawai'i Endangered Bird Conservation Program, 1993–2007

Species	Years	Total Eggs Collected or Laid	Eggs Viable at Collection	Chicks Hatched from Viable Eggs[a]	Chicks Survived to Independence	% Hatch from Viable Eggs[a]	% Survival of Chicks
Nēnē	1998–2007	666	431	365	346[b]	85	95
'Alalā	1993–2007	440	182	90	75	49	83
Hawai'i 'Elepaio	1995–2003	33	16	11	10	69	91
'Ōma'o	1995–1996	36	29	27	25	94	93
Puaiohi	1996–2007	1,001	311	261	202	84	77
Palila	1996–2007	135	96	76	54	79	71
Maui Parrotbill	1997–2007	41	18	16	13	89	81
Hawai'i 'Amakihi (Hawai'i Island)	1994–1995	38	26	21	21	81	100
Hawai'i 'Amakihi (Maui Island)	1997–2000	11	1	1	1	100	100
'Ākiapōlā'au	2001	1	0	0	0	0	0
Hawai'i Creeper	1997–2007	42	19	17	14	89	82
Hawai'i 'Ākepa	1998–2007	49	29	21	15	72	71
'I'iwi	1995–2001	15	12	11	6	92	55
'Ākohekohe	1997	6	6	6	5	100	83
'Apapane	1997	7	2	2	2	100	100
Totals		2,521	1,178	925	789	79	85

[a]Viable eggs are those that are fertile, intact, and not otherwise damaged in the nest in a way that would cause embryonic death prior to hatching.
[b]Nēnē goslings are precocial and, unlike those of the passerine species in this table, independent at hatch. For the purposes of this metric, "independence" for the Nēnē is considered 30 days of age.

In order to avoid the removal of reproductive adults from the wild, managers in Hawai'i have generally opted to collect only wild eggs for management purposes. This strategy obviously entails the problems associated with handling and transporting eggs and/or the rearing of newly hatched (day-one) chicks. The smallest eggs hatched thus far in the HEBCP have been from the Hawai'i 'Ākepa. The eggs and newly hatched chicks of this species weigh about 1.5 g and 1.0 g, respectively, and are among the smallest avian eggs and chicks successfully hatched and hand reared in captivity to date anywhere in the world (R. Wilkinson, pers. comm.). In contrast, 'Alalā eggs (~15 g) and newly hatched chicks (~12 g) are relatively huge. Although modern poultry incubation techniques have played a major part in the development of the protocols for aviculture, each Hawaiian species is considered unique and is managed individually as techniques are refined from one egg to the next. Such careful and circumspect treatment of each egg ensures that the proper temperature, humidity, and incubator and hatcher environments are managed in a constant state of review and adjustment (Kuehler and Good 1990; Kuehler et al. 1994, 1996, 2000, 2001).

As in the case of eggs, chicks are raised following poultry management practices as refined by exotic species aviculture. Although there are many similarities in diet and treatment among species, the dietary regimen of each Hawaiian species is distinct based on the diet of the wild adult. The spectrum ranges from obligate nectivory ('Ākohekohe) to obligate insectivory (Maui Parrotbill) to obligate semnivory (Palila). At hatch, the dietary regimes of small passerines are basically the same (bee larvae, crickets, and eggs, plus supplements). This basic diet, then, must be gradually adjusted as the chicks mature until they begin to eat the unique diet of the adult (Kuehler et al. 1994, 1996, 2000).

When possible, chicks are always raised with conspecifics to imprint them on their own kind. If no crèche mates are available, other species may be used, or, lacking these, a mirror is set in the brooder so the chick will imprint on itself. Additionally, in the case of the 'Alalā, puppets are used to help maximize conspecific imprinting and avoid imprinting on the human handler (Valutis and Marzluff 1999). Fortunately, we have always had multiple hatches of 'Alalā, so brood mates for each chick were available for imprinting (Kuehler et al. 1995). Taped recordings of adult vocalizations are played for all species from before hatch until weaning. Once the embryo pips the shell, adult recordings are played to motivate the chick to hatch. One of the first confirmations of the success of this early training (recordings, congener rearing, etc.) was the documentation of released, captive-reared Puaiohi breeding with wild Puaiohi (Kuehler et al. 2000; Tweed et al. 2003; Tweed et al. 2006; this volume, Chapter 22) and a captive-reared male Palila breeding with a wild female Palila on Mauna Kea (Chapter 23; C. Farmer, pers. comm.).

Fledged young are typically socialized in aviaries either with conspecifics or in mixed flocks, as would be found in the wild. As they mature, there is a selection of the birds to be released (if habitat is available) and those that will become part of the breeding program. This selection is based on sex, genetic background, behavior, and physical condition. This critical evaluation, bird by bird, pair by pair, has become one of the most important aspects of the 'Alalā program in an effort to maximize production and minimize inbreeding.

It has been found that most Hawaiian species can be managed in pairs, but one must always be prepared for seasonal aggression between the sexes that may require the separation of birds. The 'Ākohekohe, for example, is so aggressive that opposite-sex birds can be introduced to each another only for short periods and must be watched continuously to prevent injury or even death. Similar management is employed with the 'Alalā during the breeding season. Each pair of birds is continuously monitored via closed-circuit television to determine their compatibility, nest attentiveness, copulations, and egg laying.

Following the guidelines developed by the International Union for Conservation of Nature (1995), the birds selected for release are socialized in flocks in preparation for reintroduction. Sex, general health, physical fitness, and behavioral history are also considered in the selection of release candidates. To date, released 'Alalā, 'Ōma'o, Puaiohi, Palila, Hawai'i 'Amakihi, and 'I'iwi have been transported to release sites in cohort groups after receiving a health examination and being screened for diseases (International Union for Conservation of Nature 1995, Morris 2002, Armstrong et al. 2003), as well as treated with prophylactic antifungal medications to avoid stress-related infections. Disease screening is especially important to avoid any possibility of introducing novel pathogens to a native environment via the reintroduction of captive-bred birds. Prior to release, all candidate birds are given a physical examination to identify any problems that might prevent them from being successfully released. Fecal flotation is used to identify intestinal parasites, and a blood sample is taken to identify blood parasites as well as blood chemistries that might indicate infections or other conditions that could pose a threat to native birds or indicate that the candidate is not healthy enough to release (International Union for Conservation of Nature 1995). Depending on the species and the need to develop site fidelity, wild food recognition, and good body conditioning, the flock is acclimated at its release site in a field aviary for a period of up to 7 days for 'Ōma'o and Puaiohi (Kuehler et al. 2000), 14 days for Hawai'i 'Amakihi (Kuehler et al. 1996), 21 days for Palila (Lieberman and Hayes 2004), and more than 60 days for 'Alalā (Kuehler et al. 1995). All release cohorts are supported after release by supplemen-

tal food and are monitored using microtransmitters to determine the movement and fate of each individual. Postrelease monitoring is essential when evaluating the efficacy of the release (Tweed et al. 2003; this volume, Chapter 22).

OVERVIEW OF THE HAWAI'I CAPTIVE PROPAGATION STRATEGY

Through numerous discussions in various venues, including the 'Alalā Recovery Team (USFWS), Hawai'i Forest Bird Recovery Team (USFWS), Captive Propagation Recovery Working Group (USFWS), Nēnē Recovery Action Group (USFWS, DOFAW, ZSSD), and Hawai'i Captive Propagation Partnership (ZSSD, USFWS, DOFAW), as well as ad hoc discussions with knowledgeable individuals experienced in field biology, natural history, and aviculture, a matrix of proposed strategies was developed for the conservation of native Hawaiian forest birds from the main islands. This sequence of conservation actions is a dynamic process of planning and continues to be reviewed and modified. The most recent review of the proposed captive propagation priorities and strategies was incorporated into the *Revised Recovery Plan for Hawaiian Forest Birds* (U.S. Fish and Wildlife Service 2006), and a modified version is presented in Table 19.1. It is important to emphasize that this species priority list is not necessarily a list of species prioritized from most endangered to least endangered but rather a prioritization of species that can most benefit from captive management or intervention techniques in the field. What follows is the array of management strategies and how they have been or may still be applied to the endangered Hawaiian avifauna.

No Captive Management or Hands-On Manipulation Required

The decision *not* to establish a captive management effort for a species is not, by default, a decision to "do nothing" but rather a decision made in favor of some other management strategy. For example, the low priority ranking of the Oʻahu ʻElepaio (*C. s. ibdis*) and the decision against captive management for the species is appropriate because habitat management and change of land use practices offer the greatest chance for the conservation and recovery of this species. VanderWerf and Smith (2002) showed that Oʻahu ʻElepaio populations quickly rebounded where rat populations were reduced. When such options exist, it is clearly in the best interest of the species to work at the habitat level rather than to involve a labor-intensive, expensive captive breeding program.

Rear-and-Release

The rear-and-release strategy requires the collection of wild eggs for artificial incubation and hand rearing of young with the intention of immediately releasing the juveniles to the wild. This technique requires that wild nests be easily located and accessible to provide eggs, that the species be still fairly numerous, and that there be secure habitat for reintroduction. Rear-and-release is not necessarily more efficient or cost-effective than establishing a long-term captive breeding program. Assembling nest-searching teams, establishing and staffing incubation and rearing facilities, and possibly providing helicopter time for transporting eggs and later birds are expensive. If the target species breeds readily in captivity, it may be more cost-effective to develop a short-term captive breeding program designed for the immediate release of young. However, if the species does not breed readily in captivity, rear-and-release may be the preferred strategy as long as nests are accessible and the numbers of young can be easily hand reared. In many cases, rear-and-release offers the quickest and most cost-effective way to enhance wild populations because it avoids natural fledgling mortality (Newton 1998, 1999) and, in the case of poorly known species,

bypasses the uncertainties associated with developing captive breeding procedures and protocols. In New Zealand, such a rear-and-release strategy has been applied successfully in the recovery of the Takahe (*Porphyrio mantelli*) and Black Stilt (*Himantopus novaezelandiae*) (Maloney and Murray 2000, Bell and Merton 2002).

In Hawaiʻi, a model of an effective rear-and-release effort was the pilot project to incubate, rear, and release the Hawaiʻi ʻAmakihi (Kuehler et al. 1996). Twenty viable wild eggs were collected, and 16 chicks were released. Although the ʻamakihi is not a threatened species, it was selected as a model for this pilot study in order to develop avicultural and release techniques utilizing a native Hawaiian passerine.

The ʻŌmaʻo, Hawaiʻi Creeper, Hawaiʻi ʻĀkepa, and ʻĀkohekohe are considered good candidates for a rear-and-release strategy for recovery because their nests are easily located and accessible and artificial incubation and hand-rearing protocols for these species are well established. Although Hawaiʻi Creepers and Hawaiʻi ʻĀkepa have recently been maintained and bred in captivity, a rear-and-release program for these species clearly would be more expedient and cost-effective because key resources (aviary space and routine care) could be redirected to species requiring a more extensive propagation program. In the case of Oʻahu ʻElepaio, if habitat becomes available in which to reintroduce this flycatcher to a new area, rear-and-release might be preferable to translocation to avoid the likelihood that the wild birds would return to their capture site. If and when these forest birds require more intensive recovery efforts that might include captive propagation, the technology is in place to initiate a rear-and-release program.

The principal cost associated with the rear-and-release strategy is that for the fielding of personnel for locating nests and collecting and transporting eggs. For a few species, like the ʻAkiapōlāʻau (*Hemignathus munroi*), rear-and-release is not practical.

Their nests are extremely difficult to locate and are invariably constructed in an inaccessible location on the terminal tips of the flimsiest of 'ōhi'a lehua (*Meterosideros polymorpha*) branches. Collection of adults is much more feasible than collection of eggs, and the 'Akiapōlā'au makes a better candidate for a full captive breeding program founded on a captive population of adults.

Captive Breeding for Immediate Release

Maintaining a small flock of breeding birds for the release of young is the strategy most applicable to the recovery of the greatest number of Hawaiian forest bird species. The goal is the collection of wild eggs to establish a small captive flock that incorporates some of the genetic diversity of the wild population. This strategy is possible only if the aviary capacity is sufficient. Juveniles reared as progeny from even the first generation of captive-hatched birds could be released back to the wild. Each season, a few offspring would be retained to build the genetic and demographic stability necessary for a captive flock designed to produce birds for release. This option requires the maintenance of fewer captive animals than are required for a self-sustaining population intended to safeguard a substantial proportion of species' genetic diversity.

The two Hawaiian forest bird species being captive bred for immediate release are the Puaiohi and Palila (Chapters 22 and 23). The Puaiohi captive program originated with 15 wild-collected eggs from which 12 birds contributed to the genetic variability of the captive flock. This small, managed flock produced more than 250 chicks from 1998 to 2007, from which 176 birds were released back to the Alaka'i Wilderness Preserve (Fig. 19.3). Although the genetic variability of the captive flock has been predictably falling over the 10 years of propagation, the released birds are now pairing with wild birds that will provide an increase in heterozygosity in the resulting wild progeny. However, it is readily recognized that it will be necessary to someday improve the genetic diversity of the captive flock through the introduction of new genetic material (eggs) collected from the wild.

The Palila program, also originating from a small flock of wild-collected eggs, initially yielded 10 potential breeding adults. The program has produced more than 50 captive-propagated chicks since the year 2000. From this flock, 18 captive-bred birds and 3 adults that originated as wild eggs were released at Pu'u Mali on Mauna Kea in 2003, 2004, and 2005 (Lieberman and Hayes 2004, Zoological Society of San Diego 2007). Because these birds were reared in captivity, it is hoped they will exhibit a greater fidelity to the release site than those wild birds that have been translocated in the past which tended to return to their original home ranges (Chapter 23). Fidelity to the release site may prove one of the greatest advantages of captive breeding when managers are attempting to establish new populations of endangered passerines. Like the Puaiohi, captive-reared, released Palila have bred successfully with wild conspecifics (C. Farmer, pers. comm.) and have demonstrated the behavioral repertoire necessary to integrate into the wild flock.

Other Hawaiian forest bird species that would be appropriate candidates for this management strategy would be the Maui Parrotbill, 'Ākiapōlā'au, 'Akikiki (*Oreomystis bairdi*), Hawai'i Creeper, 'Akeke'e (*Loxops caeruleirostris*), and Hawai'i 'Ākepa (see Table 19.1). Space and resources in the captive facilities are currently overriding considerations when determining which species will most benefit from captive propagation and intensive manipulation for recovery. To date, only the Palila and the Puaiohi have merited this priority, but a Maui Parrotbill program has begun, and the 'Ākiapōlā'au, 'Akikiki, and 'Akeke'e may well be the species next established in captivity. As stated previously, this will depend on the resources that can be dedicated to finding

Figure 19.3. A Puaiohi chick hatched at the Keauhou Bird Conservation Center. The hand-reared chick either will become a member of the captive breeding flock or will be released into the wild on Kauaʻi Island. *Source:* Photo courtesy of Joop Kuhn, Zoological Society of San Diego.

nests and collecting eggs and on the availability of aviary space in which to establish a small but productive breeding flock.

An interesting variation of the traditional aviary scenario for rear-and-release, untested in Hawaiian forest birds, is the field aviary strategy, in which a pair or pairs of birds are kept in cages within their native habitat (Smales et al. 2000, Munkwitz et al. 2005, Woolaver et al. 2005). Any progeny that are reared by the parents can be released directly into the habitat without the inherent risks of improper imprinting, transporting juveniles, and unfamiliarity with conditions at the release site. Such a strategy could be effective for the Hawaiian passerines mentioned here.

Captive Breeding for a Self-Sustaining Captive Population

Captive breeding of birds to create a self-sustaining captive population is an option that can be considered a conservation "bank account" or a hedge against future "species genetic bankruptcy" for species that are about to fail in the wild. It is this option that was considered of highest priority in the early 1990s and was considered imperative to halt the decline, indeed prevent the extinction, of several species known to be at the point of disappearing forever from the Hawaiian landscape. Species that fall into this category would be maintained in captivity but not reintroduced until their habitat was safe and ready to receive them. Management of a self-sustaining captive population protects the genetic and demographic health of the species for many generations—a common target is the preservation of 90% genetic diversity for 100 years (Foose et al. 1986, Soulé et al. 1986)—especially if further recruitment of additional genetic material from the wild is not considered a viable option.

The ʻAlalā program is the only example of this strategy in Hawaiʻi (U.S. Fish and Wildlife Service 2003; this volume, Chapter 20). This species is extinct in the wild, and all individuals (60 in 2009) are now in captivity (Fig. 19.4; see also Fig. 19.1). The current goal is simply to produce enough birds so that genetic redundancy will provide genetically "surplus" birds with which to reinitiate releases into managed native

Figure 19.4. Monitoring the nesting behavior of captive 'Alalā using closed-circuit television. This technique allows surveillance of multiple nests simultaneously without disturbing the birds. *Source:* Photo by Alan Lieberman, Zoological Society of San Diego.

habitat following the genetic guidelines established by the International Union for Conservation of Nature (1995). This strategy recognizes that there is limited genetic diversity in the captive flock (Chapter 10) and that every effort should be made to safeguard this diversity. Therefore, careful genetic management is one of the principal goals of this captive breeding strategy (American Zoo Association Small Population Management Advisory Group 2000, Frankham et al. 2002).

Captive Breeding for Production

Captive breeding for production can be considered the "factory" option for captive propagation. After the avicultural questions have been answered, facilities built, personnel trained, and habitat for reintroduction is available, full-scale production of birds can be implemented to produce many birds for release into areas in need of infusions of new and/or additional birds.

This option would be considered only for endangered species that justify the expense of many cages and maximum labor for the production of as many birds as possible. The Nēnē program represents this strategy (U.S. Fish and Wildlife Service 2004a), and the 'Alalā recovery program will eventually reach this stage.

Emergency Search and Rescue

If one of the captive breeding programs described earlier is to have a reasonable chance of success, it needs to be established before a species is reduced to a critically low number. But if such a program is not established in time, the last-chance option of search and rescue applies. Species reduced to only a few individuals are removed from the wild and brought into captivity to save the species from imminent extinction. The search-and-rescue strategy should be considered only if captive propagation has a greater probability of recovering the species than translocation and/or habitat management. Although we may be saving the last few individuals by removing them from their natural habitat, we are losing an opportunity to study and protect the species in the wild, and there are no guarantees that captive propagation with such limited options will be successful.

Unfortunately, the opportunity that existed in the 1970s and early 1980s to do the most good for eight forest bird species most in need (the Kaua'i 'Ō'ō, Kāma'o, 'Oloma'o, 'Ō'ū, Kaua'i and Maui Nukupu'u, O'ahu 'Alauahio, Kākāwahie, and Maui 'Ākepa) has been lost with the extinction of these birds. Only for the Po'o-uli has this strategy been exercised. An effort was made to capture the remaining Po'o-uli (Chapter 21) in the hope that they would survive capture and acclimation, establish compatible pairs, lay fertile eggs, hatch chicks, and survive long enough to establish a captive flock and/or be released back to the wild. Sadly, only one bird was captured, and the remaining two known wild birds were never resighted. Although the first Po'o-uli died after only 79 days in captivity, apparently due to complications related to its advanced age, the bird acclimated well to captive conditions and its diet and maintenance regime.

Technology Development

The goal of technology development is to develop the captive propagation and release expertise necessary for any of the preceding strategies or in anticipation that some day captive propagation of a species will be needed. Many of the artificial incubation and hand-rearing techniques appropriate for Hawaiian forest birds have already been developed by the HEBCP through its selection of common endemic Hawaiian species as models for endangered endemic Hawaiian taxa: the Hawai'i 'Elepaio as a model for populations of the species on other islands; 'Ōma'o for native thrushes; and Hawai'i 'Amakihi, 'I'iwi, and 'Apapane for honeycreepers. In the future, this strategy would be chosen primarily for those species that still require development of captive breeding or release technology. The Hawai'i Island subspecies of 'Elepaio is the only surrogate being managed in captivity at this time in anticipation of the potential need for future captive management of its endangered counterpart, the O'ahu 'Elepaio.

Consideration is also being given to the experimental management of a species in captivity for its own conservation. Captive populations of Hawai'i 'Ākepa and Hawai'i Creepers have been established at the HEBCP to determine captive parameters in advance of future population declines should these occur. This action is a proactive approach to this classic question: When do we initiate a captive propagation program—when the species is so rare that we face few options to prevent extinction or when the species is rare but still common enough that we can take risks to explore more options and develop the skills necessary for captive management?

CAPTIVE MANAGEMENT AND DISEASE CONSIDERATIONS

One aspect of captive management that may potentially be incorporated into any of the previously described strategies is managing a species for disease resistance. A cautionary distinction must be made between selectively breeding captive birds for resistance to nonnative pathogens and selecting for captive management birds that are taken from wild populations evolving a natural resistance to disease.

Conservationists in Hawai'i have often speculated on the possibility of breeding birds for resistance against diseases that have ravaged the endemic bird population for the past 200 years. The concept of artificial selection for genetic traits is not considered responsible captive management by the managers of the American Zoo Association Small Population Management Advisory Group (2000), in recognition of the fact that selecting for one trait can often result in unwittingly causing or amplifying other deleterious traits. As Lacy states (1994: 81), "Even when universally undesirable genes have been identified, genetic damage can be done by artificial selection programs. Selection cannot remove some alleles from the population without incidentally removing others."

Captive management, however, may be employed to take advantage of naturally occurring disease resistance in wild populations of endemic forest birds, for example, Hawai'i 'Amakihi (Woodworth et al. 2005). This could be accomplished by collecting individuals from those unique populations that have demonstrated some level of natural resistance to infection, most notably avian malaria (*Plasmodium relictum*) and avian pox virus (*Avipoxvirus* sp.). Eggs could be collected from birds breeding in areas with a high incidence of these diseases, with captive propagules released in a new area or kept and bred to build a captive population with inherent resistance to disease.

SUMMARY

Captive management forms one of the cornerstones of forest bird conservation in Hawai'i. To date, 15 Hawaiian species have been raised in captivity, almost exclusively in island facilities. The goal has been to learn how to meet species' requirements in captivity and to maintain, raise, breed, and release native birds with the purpose of bolstering or reintroducing populations of endangered species. In rare instances, species have been brought into captivity while they were going extinct in the wild; examples include the 'Alalā and Po'o-uli. The path to saving forest birds by means of captive propagation has been made especially difficult because of the overall lack of comparable work on passerine birds elsewhere, necessitating the development of technology and expertise locally, and because of the great cost of building and staffing suitable aviary centers in the islands. Perhaps for these reasons, the use of captive propagation has come late to

Hawai'i and was not under way to save forest birds until the late 1980s, by which time many species most in need of such assistance had become extinct.

A variety of captive management options have been practiced in Hawai'i, and these serve as models for future work on endangered forest birds. Which species should be tackled first and what strategy would work best have been determined by a host of considerations, including primarily each species' biology and the particulars of its conservation situation. In lieu of captive management, field support of dwindling wild populations and translocations of endangered birds to safer locations have been or could be preferable (for example, in the case of Northwestern Hawaiian Island passerines). The captive management options appropriate for Hawaiian forest birds include (1) rear-and-release of chicks from wild-collected eggs (suitable for a variety of species but generally not used as yet), (2) captive breeding for immediate release (used for Puaiohi and Palila), (3) captive breeding for a self-sustaining captive population (Nēnē and 'Alalā), (4) captive breeding for production ('Alalā eventually), (5) emergency search and rescue (used for Po'o-uli), and (6) preemptive technology development for many species.

The Hawai'i Endangered Bird Conservation Program, run by the Zoological Society of San Diego, manages the two major avicultural centers in the state, one at Olinda, Maui Island, and the other at Keauhou, Hawai'i Island. Since 1993, the program has established the avicultural techniques necessary to incubate, hatch, maintain, breed, rear, transport, acclimate, and release all taxa of endemic Hawaiian birds it has attempted to recover to date.

Captive management of endangered Hawaiian bird species, when appropriate, is an integral component of species recovery that maximizes birds' reproductive potential and reestablishes their populations in habitat where limiting factors have been successfully managed. Captive management of endangered birds is now a proven technology that will play a role in avoiding future extinctions and assist managers in the recovery of endangered species in Hawai‘i. It serves as a conservation model for bird recovery programs elsewhere in the world, especially for passerine species.

ACKNOWLEDGMENTS

We gratefully recognize the commitment, dedication, hard work, and patience of the staff at the Maui Bird Conservation Center and the Keauhou Bird Conservation Center. Since the initiation in 1993 of the Hawai‘i Endangered Bird Conservation Program (HEBCP), nearly 40 biologists or aviculturists have been involved in the daily care of what is perhaps the most endangered captive bird inventory in the world. In addition, more than 40 interns have made an invaluable contribution, assisting with the work and taking with them the experience that can be gained only through hands-on participation. Also acknowledged are the leaders of The Peregrine Fund, who took the risk of committing to a conservation effort for which there was no precedent. The professionals of the Zoological Society of San Diego have been supportive throughout the history of the HEBCP and continue to set the standard for zoos that participate in species recovery and restoration programs.

We gratefully acknowledge the biologists and administrators of the U.S. Fish and Wildlife Service Pacific Ecoregion, the Hawai‘i Division of Forestry and Wildlife, and the U.S. Geological Survey Pacific Island Ecosystems Research Center, who have been outstanding partners, providing financial, biological, and moral support. Finally, to the community of donors, private landowners, and other good-will supporters of our efforts we extend a heartfelt *mahalo*. We hope that we have lived up to your trust and faith in our ability to help conserve the natural heritage of Hawai‘i.

Mahalo to D. Merton (New Zealand) and two anonymous reviewers for making valuable suggestions and for providing additional information and perspectives that have greatly improved this chapter.

PART 4

Recovery Programs

CHAPTER TWENTY

'Alalā

PAUL C. BANKO

The extinction in the wild of a species of crow must surprise nearly everyone. That a crow—a universal symbol of resourcefulness and adaptability—should live now in Hawai'i only in captivity demonstrates the desperate condition of Hawaiian ecosystems and species. Nevertheless, reports of wild 'Alalā (Hawaiian Crows, *Corvus hawaiiensis*) have not been substantiated since 2002, and a captive population of 60 individuals as of March 2009 is all that remains of Hawaii's historically largest and most charismatic songbird (Plate 1). With the recent presumed extinction of the Po'o-uli (*Melamprosops phaeosoma*) and other species (Chapter 5), the 'Alalā has become Hawaii's rarest bird and the only species whose recovery depends entirely on captive breeding. Building the captive flock and restoring habitats to support the eventual reintroduction of 'Alalā to the wild will signify a milestone in Hawaiian conservation.

Many other reintroduction programs indicate that 'Alalā recovery will be slow and difficult. Recovery programs that have depended solely on captive propagation, for example, that for the California Condor (*Gymnogyps californianus*), have required many years of reintroduction, sustained funding, and local support to be successful (Beck et al. 1994). Especially challenging will be preparing naïve captive birds for life in the wild, because the habitat conditions in which the 'Alalā evolved have changed dramatically since the islands were peopled many centuries ago. To achieve recovery, therefore, we must delve deeply into our knowledge of the species and its habitat as they were historically reported and should consider the evolutionary circumstances that shaped the species' ecology.

In this account I summarize aspects of 'Alalā ecology that seem most relevant to restoration: the biogeography of 'Alalā and related species; the species' habitat use and feeding ecology; its behavior, breeding, and demography; and patterns and causes of population decline. Reviewed next is a history of 'Alalā conservation and a discussion of major challenges and opportunities for the restoration of wild populations: extinction of the wild population, captive propagation programs, and releases of captive-bred birds; refuges and habitat management; and recovery planning.

ECOLOGICAL DIMENSIONS OF 'ALALĀ RESTORATION

The Biogeography of Hawaiian Corvids

The crows (*Corvus* spp.) that evolved in the Hawaiian Islands represent an extraordinary diversification of their kind for such a small land area, surpassing that of even the four crow species of the Caribbean (Goodwin 1986, Madge and Burn 1994). At least five Hawaiian species have been distinguished (four only from the fossil record, of which two have been described), but it is not known whether they radiated from a single colonization or from multiple colonizations by the same or different ancestors (James and Olson 1991). 'Alalā are more closely related to the Common Raven (*Corvus corax*) than to other corvids (Fleischer and McIntosh 2001). Subfossil evidence indicates that one or more species of crow once occupied four of the six largest islands in the archipelago: O'ahu, Moloka'i, Maui, and Hawai'i (James and Olson 1991; Banko, Ball, et al. 2002). As many as three species were found in the same region. The smallest and perhaps least morphologically specialized species, the 'Alalā was the only crow to have survived in Hawai'i into historic times. We can only wonder how much of the Alala's historical range and behavior reflect interactions with these other crow species. Neither can we fully understand 'Alalā ecology without knowing more about the many other extinct birds, invertebrates, and plants that have disappeared since human contact about 1,200 years ago.

'Alalā Distribution, Habitat, and Feeding Ecology

Now that the 'Alalā have disappeared from the wild, our understanding of their ecology is limited to inferences from ancient bones found in lava tubes, historical accounts, surveys since 1940, and an assortment of field studies (Banko, Ball, et al. 2002). Although 'Alalā were frequently collected and observed in the past, habitat relationships are difficult to interpret from these early records because much of their geographic range has been transformed by agricultural and other activities. Nevertheless, 'Alalā seem to have inhabited only mesic or dry forests on Hawai'i Island (Fig. 20.1), and subfossil bones of 'Alalā or a closely related species were found in relatively dry habitat on Maui. Similarly, evidence of other Hawaiian corvids has been found only in the drier, leeward regions of the islands, where bird bones are best preserved. Early naturalists found 'Alalā, or were informed of their presence by residents, only in Kona and Ka'ū districts. Further suggesting that 'Alalā populations never lived in windward regions is the fact that other bird species persisted in the extensive remaining wet forests long after their dry forest habitats became highly degraded (Chapter 2). 'Alalā that were occasionally reported from wet, windward regions of Hawai'i during the past 30 years were probably vagrants that dispersed as leeward populations disintegrated (W. Banko and P. Banko 1980).

Despite their historical distribution in mesic and dry habitats, 'Alalā were strongly associated with a native vine, the 'ie'ie (*Freycinetia arborea*), that was widespread and abundant at low and middle elevations in both mesic and wet habitats (Henshaw 1902, Perkins 1903, Rock 1913, Hillebrand 1981). 'Alalā relied heavily on 'ie'ie fruit and moved seasonally in response to

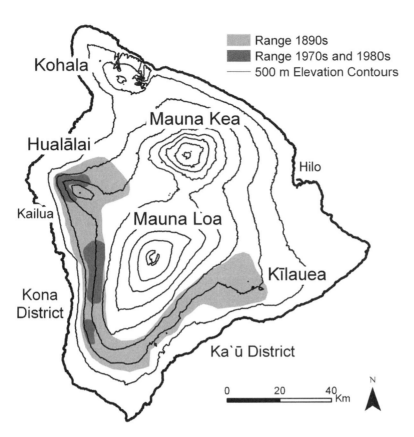

Figure 20.1. Historic and recent distribution of 'Alalā on Hawai'i Island. The species was distributed historically in mesic and dry habitats, where the levels of native shrub and tree diversity were high. Windward regions of the island were avoided. The distribution of 'Alalā contracted and fragmented considerably, and the last birds disappeared from central Kona in 2002. *Source:* Map adapted from U.S. Fish and Wildlife Service (2009: 1-5, Fig. 1).

its availability (Perkins 1903). Why, then, did 'Alalā not inhabit wetter forests, where 'ie'ie were abundant? 'Alalā are among the most arboreal of crows, and they ate more than 30 species of fruits (Perkins 1903, Sakai et al. 1986). The botanist Rock (1913) and others observed a greater variety of fruit-bearing species in dry and mesic leeward regions than in the wetter windward regions, and this disparity possibly influenced 'Alalā distribution more than the availability of 'ie'ie alone. Fruit is likely to be more available where the level of plant species diversity is high, because

different species tend to produce fruits at different times of year (Perkins 1903).

Arthropods and small bird eggs and nestlings were also important foods of 'Alalā, especially the young (Sakai and Carpenter 1990). These animal resources are likely to be more abundant in areas where the levels of plant species richness and abundance are high; therefore, 'Alalā habitat quality should be related to vegetation structure and composition. Bird and arthropod communities, and therefore the food available for 'Alalā, are now affected by introduced avian diseases and alien arthropod predators and parasites, such as ants and wasps (Chapter 2). Moreover, during historic times, highland habitats have supported a greater diversity and abundance of birds and arthropods because the impacts of habitat destruction and invasive species have been more severe in the lowlands (Perkins 1903, 1913; Scott et al. 1986). Partly for this reason, in the late 1900s 'Alalā were increasingly associated primarily with mid-elevation, mixed koa (*Acacia koa*) and 'ōhi'a-lehua (*Metrosideros polymorpha*) forests, where they foraged for fruits in the understory and hunted for arthropods and small bird nests in the trees (Giffin et al. 1987). Although 'Alalā exploited an assortment of fruits and animal prey, their stout bill and extensive use of 'ie'ie suggest a moderate degree of specialization (Chapter 7).

Behavior, Breeding, and Demography

'Alalā were never reported in large flocks (≥25 individuals) but instead formed tight family units in which juveniles remained close to their parents until the following nesting season (Banko, Ball, et al. 2002). In an area where a few pairs nested, families coalesced with their offspring, making for noisy "nurseries" to which parents flew with food. Offspring from the previous year also joined these gatherings, although it is not known whether they helped feed their new siblings. Pairs, families, and unmated birds also congregated shortly prior to nesting.

'Alalā were highly territorial during nesting, defending their areas against all other individuals, including offspring from previous years. Males proclaimed their territories at dawn with long bouts of loud calling (Banko, Ball, et al. 2002). Their vocal repertoire was stunningly rich, and this characteristic, together with their diverse behavioral displays and great size, made 'Alalā easily the most charismatic of Hawaiian forest bird species.

Maturing at two years of age, 'Alalā laid their eggs mainly during April and May, with a replacement clutch, if any, produced into June (Banko, Ball, et al. 2002). Clutches contained three eggs on average (n = 13 nests). Although there is little information that allows us to compare the breeding of 'Alalā with other crows endemic to islands, their 20-day incubation period is slightly longer than those of continental crow species, and their six-week nestling period is much longer (Chapter 8). Typical of other crows, 'Alalā nestlings usually could not fly effectively for a week or more after leaving the nest, spending hours perched quietly near the ground. The foraging skills of fledglings developed slowly, and parents continued to feed their offspring until they reached at least eight months of age. Slow reproduction suggests energetic constraints and is typical of Hawaiian forest bird species, especially those with specialized feeding habits (Chapters 7 and 8). Conservative reproduction also is common among long-lived species (Chapter 8). 'Alalā seem capable of living many years (18 years in the wild and 25 years in captivity), and they apparently maintain long-term, if not life-long, pair bonds and territories (Banko, Ball, et al. 2002).

Pattern and Causes of Population Decline

Historically, 'Alalā were widespread and relatively common in Kona and Ka'ū districts on the island of Hawai'i. They were readily collected by naturalists during 1887–1902, but by the 1940s, after decades of almost no collecting and little field reporting, 'Alalā were obviously missing or uncommon in portions of their historic range (Baldwin 1969, W. Banko and P. Banko 1980). The fact that most people are able to recognize such a conspicuous species as a crow prompted W. Banko (U.S. Fish and Wildlife Service) to interview many long-time residents of Kona and Ka'ū during the late 1960s and early 1970s. He learned that many considered the 'Alalā much less common than in earlier decades, and he undertook extensive field surveys to determine their distribution and abundance. These findings were unsettling, because most other species of crows are abundant and occupy a variety of habitats. By the mid-1970s, it was clear that fewer than 100 birds were scattered throughout Kona, and the islandwide forest bird surveys that soon followed confirmed the low abundance and restricted distribution of 'Alalā (Scott et al. 1986). Although some believed that the steep decline of 'Alalā was mainly due to avian diseases (Jenkins et al. 1989), habitat destruction and loss of fruiting plants in the forest understory due to ranching, logging, and feral ungulates would have reduced the species' food availability, with potential consequences for nest productivity and juvenile survival (Duckworth et al. 1992; Banko, Ball, et al. 2002).

Relict populations in Kona consisted mainly of established breeders who held their territories despite rapid habitat degradation. In fact, several pairs continued to nest in the same area even after their territories were disrupted by bulldozing (W. Banko and P. Banko 1980). Unfortunately, a number of birds were shot, particularly at Pu'u Wa'awa'a (Giffin 1983, Giffin et al. 1987). Given the formidable threats to both the species and its habitat, it is not surprising that few juveniles established territories of their own, although a low level of recruitment is typical of such long-lived birds (Chapter 8). While experienced pairs produced fledglings in the wild during the 1970s, the level of recruitment of new birds into the breeding population was low. Of 18 fledglings banded in the 1970s, only 2 were determined to be breeding more than 10 years later. The population probably would have crashed even sooner had it not been for the longevity of some birds in areas where they were not shot.

The last wild remnant of the species, consisting of 11 adult birds, was located on a private ranch in central Kona in 1992 (Banko and Jacobi 1992). The population included a 15-year-old female who had been banded as a fledgling in 1977 about 2 km away from her newly discovered nest in Kalāhiki territory. From this nest came the last wild fledgling. Of the 11 adults in the central Kona population, 8 were paired and occupied four breeding territories, which were separated by about 1–3 km (Banko, Ball, et al. 2002). Nesting was

recorded within each territory in at least two years during 1992–2002 (Table 20.1). The Keālia pair, which built nests every year until their disappearance in 2002, laid eggs in perhaps only four years, indicating that the female had become senescent or physiologically impaired after 1997. The female in Ki'ilae territory may also have been old or in poor condition, judging from her few abnormally shaped eggs. She was last seen in mid-1994, about seven months after her mate disappeared. The Kalāhiki female was the most productive bird, laying viable eggs and incubating reliably each year from 1992 to 1994. However, she disappeared about seven months after her mate vanished in 1995, possibly having been shot when wandering in the lower forest (Anonymous, pers. comm.). The Ho'okena female laid eggs in most years during 1993–1996, but her clutches were typically small and infertile, and incubation was usually disturbed by her mate or intruders in the territory. The Ho'okena territory became vacant after 1997. Breeding attempts within the four territories yielded 30 nests, which produced at least 35 eggs (16 nests), at least 7 chicks (5 nests), and 1 fledgling (see Table 20.1).

These birds exhibited no symptoms of avian malaria (*Plasmodium relictum*) or avian pox virus (*Avipoxvirus* sp.) during the 1990s, although both diseases affect all native forest birds and disease prevalence is relatively high within the recent range of the 'Alalā (Jenkins et al. 1989; Atkinson et al. 2005; this volume, Chapter 9). Biologists were not allowed to monitor wild birds using radio telemetry or leg bands, but mated pairs were frequently observed in their territories, and they never seemed sick. Some wild birds may have died of factors related to old age or from attacks by 'Io (Hawaiian Hawks, *Buteo solitarius*), themselves an endangered species. 'Io outnumbered 'Alalā in central Kona by at least two to one, and they harassed pairs near their nests or elsewhere throughout the year (Chapter 12). Disturbance by hawks

Table 20.1. Nesting activity of wild 'Alalā pairs in central Kona, 1992—2002

Territory	Year	Nest Attempts	Eggs	Chicks	Fledglings	Eggs Brought into Captivity
Kalāhiki	1992	1	ND	≥1	1	
	1993	1	4			4
	1993	2	≥3	≥2		
	1993	3	3			Salvaged 3
	1994	1	3			3
	1994	2	≥1	≥1		
Keālia	1992	1	≥2	2		
	1993	1	3			3
	1993	2	≥1	≥1		
	1994	1	3			3
	1995	1	ND			
	1995	2	0			
	1996	1	0			
	1997	1	1			1
	1997	2	ND			
	1997	3	ND			
	1998	1	0			
	1999	1	0			
	1999	2	0			
	2000	1	0			
	2001	1	0			
	2002	1	ND			
Ki'ilae	1992	1	ND			
	1993	1	1			1
	1993	2	0			
Ho'okena	1993	1	0?			
	1994	1	1			Salvaged 1
	1995	1	≥1			
	1995	2	≥1			Salvaged 1
	1996	1	4			Salvaged 3
	1996	2	≥3			Salvaged 2

Notes: In four nesting territories, pairs produced at least 35 eggs, about 7 chicks, and 1 fledgling in 30 nesting attempts. Of the 35 eggs, 25 eggs were intentionally taken for captive propagation as a contingency against imminent loss due to nest abandonment or disturbance. "Salvaged" indicates that the eggs were believed to be in danger of failing, not that they had already failed. Of the eggs removed from wild nests, 15 were fertile, 14 hatched, and 13 were recruited into the captive flock or were released to the wild. ND indicates no data.

may have contributed to some nest failures, but worse was the loss of territorial birds. The Kalāhiki male was observed being struck but not killed by an 'Io about six months before he disappeared. The disappearance of the recently widowed females of Kalāhiki and Ki'ilae suggests that solitary birds may have been more vulnerable to attack due to either wandering or loss of a vigilant companion. Additionally, three unpaired adults that frequented the Ho'okena area disappeared in 1994, and the Kalāhiki fledgling, who could never be definitively associated with a mate or territory, apparently died during 1997–1999. However much the 'Io may have hastened the demise of the 'Alalā in its final years in the wild, a variety of forces unleashed by humans long ago inexorably eliminated the wild population.

RESTORATION AND MANAGEMENT

Overview and Human Dimensions of Recovery

Recovering the 'Alalā is a complex process due to the many factors that may have contributed to its extinction in the wild. In contrast, the Mariana Crow or Aga (*Corvus kubaryi*), the only other endangered Pacific island crow, was decimated on Guam by one overwhelming factor, the introduced brown treesnake (*Boiga irregularis*). Recovery planning for the Aga is thus relatively straightforward, even if difficult to implement (U.S. Fish and Wildlife Service 2005). 'Alalā recovery is also challenged by human factors that transcend the formidable suite of biological agents threatening this and other Hawaiian forest birds. Longstanding animosity or apathy toward crows, particularly by Western cultures (Marzluff and Angell 2005), may have dampened enthusiasm for taking vigorous and expensive steps toward recovery sooner. At the same time, people's general familiarity with common continental crows may foster a perception that 'Alalā recovery should not require substantial or unusual conservation action. Additionally, the geographic range of the 'Alalā includes private lands valuable for agriculture and settlement at low and middle elevations and for ranching and logging at higher elevations. The majority of state-owned lands within its range are managed for multiple use, including the sport hunting of habitat-damaging ungulates. Commercial interests, multipurpose management policies, and personal views about the stewardship of natural resources have often influenced recovery planning and implementation (Johnston 2000; U.S. Fish and Wildlife Service 2009). Recovery depends more heavily on cooperation with private and public landowners in the case of 'Alalā than in that of any other Hawaiian species, and this consideration now permeates recovery planning, if not always

its implementation. Perhaps because the field of players in Hawai'i is relatively small, individual efforts, especially when motivated by personal beliefs or agency affiliations, have sometimes strongly affected outcomes—in both positive and negative ways (Walters 2006). At the same time, management action is often delayed or thwarted as government agencies struggle to provide funding and leadership for a cause that, by local standards, may seem hopelessly expensive, technically difficult, controversial, and distracting from other urgently needed conservation. Although expensive recovery programs have been mounted for many continental bird species, programmatic funding for Hawaiian birds and other island species falls far short of the need and priority (Restani and Marzluff 2001, 2002a, 2002b, 2002c; this volume, Chapters 24 and 25).

'Alalā are in some ways unusual among Hawaiian forest birds, yet they have influenced recovery planning for other species. For example, the Hawaiian forest bird propagation program was launched in response to the decline of 'Alalā numbers and range. Lawsuits and the threat of legal action also have marked recovery planning and funding. Although lawsuits have heavily influenced levels of support for a few high-profile continental species (Restani and Marzluff 2001, 2002a, 2002b, 2002c), 'Alalā and Palila (*Loxioides bailleui*) are the only Hawaiian species to have prompted legal action. Protecting critical habitat was the goal of lawsuits concerning Palila, whereas critical habitat has not been designated for 'Alalā or other endangered Hawaiian birds except the O'ahu 'Elepaio (*Chasiempis sandwichensis ibidis*) (Chapter 23). As populations declined during the 1980s, restoration increasingly depended upon access to the few birds remaining on private lands in central Kona. Refusal of access precipitated a lawsuit by conservation groups to compel the U.S. Fish and Wildlife Service (USFWS) to engage in recovery and to oblige the landowners to allow entry onto their lands (Tummons 1991c, 1991d). The lawsuit was settled out of court, but only recovery that had nothing to do with habitat studies or management was initiated. Momentum was lost, however, when the release program (described later) ended in 1999, funding to protect habitat from ungulates was forfeited, and negotiations for access to the refuge failed.

At about the same time that the lawsuit was being settled, the USFWS requested that the National Research Council commission an independent committee of biological experts to evaluate recovery options (Duckworth et al. 1992). After extensive interviews and analyses, the committee proposed integrating the management of wild and captive populations and recommended improvements to the captive propagation program, active management of the remaining wild birds, establishment of a recovery team with diverse membership, and other measures (Duckworth et al. 1992). New dimensions in Hawaiian bird recovery planning were introduced by considering the merits of taking wild eggs and nestlings for propagation, fostering captive nestlings to wild nests, releasing captive-bred juveniles to the wild, and other techniques used in restoration programs elsewhere. The committee's advice and concern about further controversy prompted commitments from federal and state agencies to support avicultural facilities and staff and to consider new approaches to solving longstanding problems. As a bonus, the 'Alalā initiative helped to launch other forest bird recovery programs, including those for translocating wild Palila and releasing captive-bred Palila, releasing captive-bred Puaiohi (*Myadestes palmeri*), translocating Po'o-uli, and breeding other species in captivity (Chapter 19).

Captive Propagation and Wild Nest Manipulations

The first few 'Alalā were taken into captivity in the 1970s in response to finding young fledglings that appeared to be sick, but the idea of breeding them in captivity soon followed (Banko, Ball, et al. 2002). The first two fledglings salvaged by W. Banko in 1970 appeared sick with avian pox virus and possibly malaria. Lacking the facilities or expertise to properly rehabilitate the sick birds, Banko sent the sibling birds by air freight to Patuxent Wildlife Research Center in Maryland to be tended by endangered species aviculturalists. Apparently because their health was compromised, one bird died soon after its arrival at Patuxent, and the other died about three years later. This inauspicious beginning to the captive propagation of the species was in some ways an omen for the decades of poor results that followed.

Three more fledglings were brought into captivity during 1973 because they seemed debilitated by disease or parasites. Later it became clear that fledglings normally cannot sustain upward flight for one to two weeks after leaving the nest and that they often perch near the ground within easy reach of predators for hours at a time (Banko, Ball, et al. 2002). The three birds salvaged in 1973 were the nucleus of a captive breeding program intended to reintroduce birds into their historic range within Hawai'i Volcanoes National Park. However, the Hawai'i Division of Forestry and Wildlife (DOFAW) requested the fledglings, and the USFWS transferred them to their care at the Pōhakuloa breeding facility in early 1976. Eight more wild fledglings were added to the flock during 1977–1983. After struggling with limited resources, staff capacity, and program guidance, the state moved the flock to a renovated facility at Olinda, Maui, in 1986 (Tummons 1991b, Duckworth et al. 1992). In 1996, shortly after modern facilities were built in Volcano, Hawai'i Island, the captive flock was divided between Olinda and Volcano, and the DOFAW turned the propagation program over to The Peregrine Fund. In 2000, the Zoological Society of San Diego (ZSSD) assumed responsibility for propagation at both sites (Tummons 1996, 2000a; this volume, Chapter 19).

During the decade at Pōhakuloa (1976–1986), the flock usually consisted of fewer than 10 birds; 0–9 eggs were laid annually (46 total), and 7 eggs hatched, including 3 that were shipped to the Honolulu Zoo, where the only fledglings of the era were produced (Chapter 19). After the flock was moved to Olinda, their egg production increased, but the hatching rate remained low. However, a breakthrough came in 1993 and 1994, when 18 eggs were taken from wild nests (see Table 20.1) to be artificially incubated in captivity (Kuehler et al. 1994, 1995). Of these, 14 were fertile, 13 hatched, and 12 survived to fledge (7 eggs removed during 1995–1997 were infertile). Also during 1993–1994, 4 fledglings were produced from 19 eggs laid in captivity. These, together with 8 other juveniles captive-reared from wild eggs, were released to the wild in 1993 and 1994, and the remaining 4 offspring from wild eggs were retained in the captive flock. This infusion of stock and the subsequent offspring of captive birds resulted in a doubling of the breeding pairs from 4 in 1996 to 8 in 1999 at Olinda and the new facility in Volcano, and pairs were recombined to increase their productivity (Harvey et al. 2002). Egg laying increased, but the hatching rate remained low. During 2000–2008, when no birds were released, only 46 birds fledged from 375 eggs, adding an average of 5.1 (range 2–10) young birds to the captive flock each year (ZSSD, unpubl. data). The flock had grown to 60 birds by March 2009.

To improve the chance that eggs would hatch in captivity, wild nests were monitored from blinds at ground level, and eggs were taken after being incubated partially to term. Females often laid replacement clutches, but these were left to be parent reared unless the nest appeared to be failing. Seventeen eggs were taken from the nests of two females (see Table 20.1), underscoring the potential conservation benefit of this technique even when few nests are available. Once viable wild eggs became unavailable, the goal of propagation switched to building up the captive flocks at Volcano and Olinda to provide stock for future releases and to increase the genetic diversity of offspring (U.S. Fish and Wildlife Service 2009).

Releasing Birds to the Wild

With the hope that captive-bred birds would mate with solitary wild birds and among themselves, 27 juvenile 'Alalā were released in the vicinity of the last wild population during 1993–1999 (Fig. 20.2). Five birds were released in 1993, followed by 7 more in 1994. By the end of 1996, 8 of the 12 birds were still alive, and restoration seemed to have started in earnest. Nevertheless, optimism for recovery soon dimmed, despite the release of 8 new birds in 1997. Six birds died in 1997, leaving a surviving population of only 10 captive-bred birds in the wild by year's end. When 8 more birds died in 1998, hopes for a smooth road to recovery were dashed. Although 7 additional 'Alalā were released after 1997, birds continued to die, and the 6 surviving birds were returned to captivity for safety and propagation.

Survival declined with each cohort released into central Kona (Fig. 20.3). Two birds in Cohort 1 (January 1993) survived five years before being captured and returned to captivity. In contrast, only one bird in Cohort 2 (October 1994) was alive three years after release. Birds in Cohort 3 (January 1997) and Cohort 4 (September 1997) died within one to two years of their release. The deaths of four of seven birds in Cohort 5 (January 1998) and Cohort 6 (January 1999) within a year of their release precipitated the decision to return all surviving birds to captivity.

Increased mortality of birds within successive cohorts suggested that predators were becoming more effective at killing these relatively naïve, noisy birds. In fact, monitoring the radio-tagged and banded birds indicated that 'Io (or possibly owls) were implicated in 7 of 13 cases in which bodies of both healthy and sick birds were

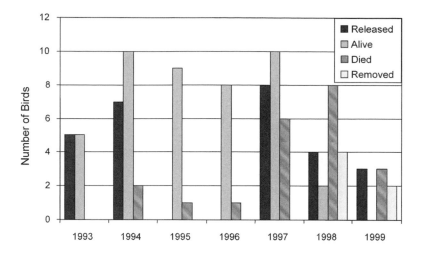

Figure 20.2. Numbers of captive-bred 'Alalā released and their fate over time. A total of 27 'Alalā were released in central Kona during 1993–1994 and 1997–1999. The number of birds surviving in the wild declined slightly after 1994, but the population increased briefly when additional birds were released in 1997. The population plummeted during 1997–1998, when 14 birds died. In response, the 6 surviving birds were captured and returned to captivity during 1998–1999.

recovered. A feral cat (*Felis catus*) or other small mammal apparently killed another bird (U.S. Fish and Wildlife Service 2009). Necropsies and health checks indicated that a protozoan (*Toxoplasma gondii*), two bacteria (*Erysipelothrix rhusiopathiae* and *Escherichia coli*), and a fungus (*Phlebia ludociviana*) compromised or killed the other birds (Work et al. 1999, 2000). As in the case of wild 'Alalā, avian malaria and avian pox virus seemed not to have afflicted captive-bred birds, which were exposed to mosquitoes while being held in conditioning aviaries months prior to their release but were treated with prophylactics as a precaution against their loss (Massey et al. 1996). Perhaps contributing to their survival in an area where disease prevalence was high was the abundance of nutritious food in their aviaries (Banko, Ball, et al. 2002). Although one bird developed a pox-like lesion, it seemed unimpaired.

Released cohorts did nothing to supplement the wild population in the long term, and these young birds, unaccompa-nied by parents, may have instead created a disruptive social environment that inhibited breeding. Cohorts also tended not to mix with one another often, and, just as released birds were usually chased by the socially dominant wild 'Alalā, older cohorts tended to push younger birds into less favorable habitats at lower elevations. Breeding behavior was most evident in the first release cohort: two birds formed a pair bond and initiated rudimentary nests, and a female was forming a pair bond with an unmated wild male before being killed (apparently by an 'Io) in her fourth year.

Managing Habitat for Recovery

Restoring the density of understory and canopy vegetation, where 'Alalā primarily forage for fruit, arthropods, and bird nests, is a critical but unfulfilled goal of habitat management (Giffin et al. 1987; U.S. Fish and Wildlife Service 2009). Browsing and rooting by domestic and feral ungulates have removed food plants in the understory, reduced tree cover, and facilitated the spread of weeds, especially grasses that increase the risk of damaging fires, and other invaders that suppress native vegetation. Although ungulate control is a prerequisite of restoring habitat, removals should be strategically planned to discourage the dangerous buildup of fine fire fuels or the sudden release of highly

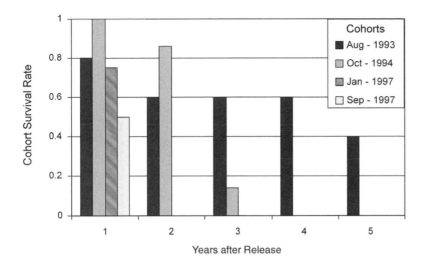

Figure 20.3. Survival rate of released captive-bred 'Alalā. The survival rate progressively declined with each of the six cohorts released in central Kona during 1993–1999. Two birds from Cohort 1 survived five years before being captured and returned to captivity. Only one bird from Cohort 2 was alive three years after release. Birds from Cohorts 3 and 4 died within one to two years of their release. Birds from Cohorts 5 and 6 died or were captured and returned to captivity within a year of their release. Birds were released in August 1993 (Cohort 1: 5 birds), October 1994 (Cohort 2: 7 birds), January 1997 (Cohort 3: 4 birds), September 1997 (Cohort 4: 4 birds), January 1998 (Cohort 5: 4 birds), and January 1999 (Cohort 6: 3 birds). Cohorts 5 and 6 are not shown in the graph.

competitive weeds. Forest restoration also could be accelerated by planting species of trees and shrubs that would otherwise recruit slowly and by reducing rodent populations to safeguard seedling recruitment. As major seed dispersers, 'Alalā themselves will help increase the density of understory vegetation, which will also reduce the suitability of habitat for 'Io (Klavitter et al. 2003). Research to guide vegetation management has focused on the diet composition and habitat use patterns of wild 'Alalā (Sakai et al. 1986, Sakai and Carpenter 1990). Detailed analyses of the responses of birds released in the past (U.S. Fish and Wildlife Service 2009) are needed to improve management planning.

The first 'Alalā recovery plan proposed habitat restoration in two areas on Hualālai

Volcano and three areas in Kona and Ka'ū on Mauna Loa (U.S. Fish and Wildlife Service 1982a). The revised recovery plan, however, is less specific about designating areas for restoration (U.S. Fish and Wildlife Service 2009), but vegetation structure and composition were recently assessed to evaluate the potential suitability of 'Alalā reintroduction areas (Price and Jacobi 2007). As responses of released birds to habitat conditions within their historic range on Hawai'i Island are understood in more detail, assessing habitat and food availability within the species' potential prehistoric range on Maui may be warranted where small areas of dry and mesic forest remain.

Two sanctuaries have been established for 'Alalā restoration, yet the habitats there have received little or no management. In 1984, Pu'u Wa'awa'a Wildlife Sanctuary on Hualālai was the first natural area designated for protecting 'Alalā and other endangered birds and plants (Tummons 1991a, 2000b; U.S. Fish and Wildlife Service 2009). This state-owned land, consisting of 1,542 ha near the dry northwestern edge of the species' historical range, had once been important to 'Alalā. Several pairs nested in or near the sanctuary during the 1970s, but the last bird disappeared in the early 1990s (W. Banko and P. Banko 1980; Banko, Ball, et al.

2002). Although cattle were removed and the native vegetation began to recover, feral ungulates remained, alien grasses and other weeds flourished, a fire burned a portion of the area, and the sanctuary became increasingly isolated as ranching continued in the surrounding region (Blackmore and Vitousek 2000). DOFAW is currently removing the remaining feral ungulates.

After lengthy negotiations, a second sanctuary for 'Alalā was purchased in 1999. Designated the Kona Unit of Hakalau Forest National Wildlife Refuge, it consists of 2,145 ha of seasonally wet forest in central Kona (U.S. Fish and Wildlife Service 2009). The last wild birds were declining rapidly here, and captive-bred birds released to bolster their numbers were dying frequently from predation and diseases. Releasing captive-reared birds and monitoring the population absorbed most of the funding allocated, leaving little for predator removal and none for ungulate removal and habitat restoration (U.S. Fish and Wildlife Service 2009). However, even the limited program to reduce the numbers of introduced mammalian predators along roads and in nesting territories was stopped when the refuge office closed and staff dispersed following the collapse of negotiations over terms for overland access to the refuge (Tummons 2003a, 2003b). A more tragic consequence of the stalemate over access was the forfeiture of $1 million in private funding for fencing and ungulate removal (U.S. Fish and Wildlife Service 2009). Following protracted and contentious legal negotiations, unlimited access to the refuge was finally gained in 2005 (Tummons 2005b). Long-deferred habitat restoration plans are being implemented as resources become available, although it will likely be years before conditions will be suitable for 'Alalā (U.S. Fish and Wildlife Service 2009).

Lessons Learned and Plans for Recovery
To recover the 'Alalā, we must learn from our shortcomings in protecting habitat,

preventing the extirpation of the wild population, and reestablishing a free-living population of captive-bred birds. At the core of recovery planning is the need for leadership to secure adequate funding and talent, guide the development of adaptive recovery strategies, and build strong partnerships between organizations, landowners, and the public. Even now, years after losing the wild population, steps to raise funds and other support, manage habitats, formulate new release strategies, and increase the captive population seem slow and tentative. If the past few decades are an indication, leadership for recovery may have to come from outside government agencies, although their energetic participation is crucial to success.

Without vigorous, sustained management across large, contiguous areas of habitat, 'Alalā recovery will fail for lack of resources and protection. Because 'Alalā moved seasonally with changing food availability, preferred sites for reintroducing captive-bred birds to the wild should span a wide range of elevations and include different habitat types. The largest and least disturbed habitat available for 'Alalā conservation lies within Ka'ū, although working there will be logistically challenging. Birds released there will respond to environmental conditions along substantial gradients of elevation and rainfall, and patterns of their habitat and resource use should help guide management in Kona. However, until landscape-scale management can be provided in the fragmented and disturbed forests of Kona, smaller units there can be intensively managed for reintroduction of the 'Alalā and for developing management strategies and methods. Although threats can be reduced more efficiently within small areas, supplemental feeding or other measures may be needed to prevent birds from wandering when food becomes locally scarce. Techniques for the small-scale restoration of vegetation have been proposed by Mueller-Dombois

(2005) that might lend themselves to this approach.

Again considering conservation at the landscape scale, we should integrate 'Alalā recovery with other initiatives to conserve native bird communities and ecosystems and take advantage of economies of scale and management synergy (U.S. Fish and Wildlife Service 2009). Restoring the size and health of native forest communities and watersheds can only improve 'Alalā habitat while stimulating landowner partnerships and public support. Similarly, conservation of other native forest birds would generally benefit 'Alalā. For example, reintroducing another native fruit-eating bird, the 'Ōma'o (Myadestes obscurus), into its former range in Kona could help guide 'Alalā restoration in similar habitats while hastening forest recovery through seed dispersal and adding one more species on which crows might prey.

The release of captive-bred birds demonstrated that wild birds are not easily replaced, and this lesson may also be relevant to other endangered Hawaiian birds. Nevertheless, even in the near absence of habitat and predator management, some released birds survived five years before being returned to captivity. Moreover, a few birds exhibited behavior suggesting that they would have bred had they not died or been disturbed by later cohorts. In the future, releasing cohorts farther apart and releasing birds in different social groupings might accelerate breeding. For example, proven pairs might be released after they have become well represented genetically in the captive flock. Not only would they possess the minimal social skills required for breeding, but pairs might be less vulnerable to attack by 'Io in that they would have at least some investment in the welfare of their mates. However, birds held longer in captivity might require more prerelease conditioning. Additionally, mixed-age release groups might also shorten the time to recruitment (White et al. 2005). Any release strategy will pose

trade-offs, but the key is to identify factors that contribute most to success. However, testing release methods must wait until more birds are available and their habitats are further restored. Progress on both fronts has been slow.

'Io may have killed 30% of released 'Alalā, but we do not know if they became more adept at killing crows as successive cohorts were released or if diseases or more open habitat made birds more vulnerable to attack. Nevertheless, predator aversion training might improve the awareness and response of released birds to the threat of raptors (White et al. 2005). Reducing rodents and other nonnative prey, such as Kalij Pheasants (Lophura leucomelanos), might reduce 'Io densities and, together with feral cat and mongoose control, could help protect 'Alalā.

When it is again possible to release birds to the wild, every effort should be made to evaluate the different qualities of the birds and the habitats where they will live to identify management options that will result in greater survival and breeding. Past monitoring was invaluable, and future monitoring must take advantage of new technology while ensuring that keen human observers are available to interpret behavior and events in the field. Among the many issues bearing on 'Alalā restoration that monitoring should address are how the diversity and abundance of fruiting plants affect 'Alalā survival, breeding, and movements; how tree cover influences hawk-crow interactions; how disease prevalence varies between habitat types, over time, and in response to management; and how rearing methods and social development in captivity affect the survival and behavior of released birds.

SUMMARY

The last of five Hawaiian corvid species, 'Alalā, are now extinct in the wild, and their recovery depends entirely on the success of establishing wild populations from captive-bred birds. 'Alalā historically occupied a range of mesic and dry habitats in Kona and Ka'ū districts that suffered tremendously from human settlement and agriculture and from invasive species. 'Alalā fed heavily on 'ie'ie and a variety of other fruits, arthropods, and small birds, making them especially vulnerable to the destruction of understory vegetation by ungulates. 'Alalā society seemed organized primarily by family units composed of pairs apparently mated for life and their most recent offspring. Pairs were strongly territorial and reluctant to move even when habitats were being rapidly degraded. A long-lived species with slow reproduction, 'Alalā dwindled with little notice during the early and mid-1900s, and the severity of their decline was not appreciated until the 1970s. Although pairs were generally productive from the 1970s to the 1990s, few young birds seemed to survive after their first year, and fewer still established breeding territories, suggesting that habitat quality, predators, and diseases may have limited the recruitment of breeding pairs.

More by happenstance than by design, a program of captive propagation slowly developed as the wild population rapidly dwindled. Propagation improved when eggs were brought into captivity from the last wild pairs in central Kona, and 27 captive-bred birds were released during 1993–1999 to bolster the wild population. Despite the relatively high rate of survival of the first few release cohorts, subsequent birds died at high rates, and the last 6 birds were returned to captivity for safekeeping and breeding. Predation by 'Io and several abnormal diseases, not including avian malaria and avian pox virus, were implicated in many deaths of released birds; however, habitat quality may have indirectly contributed to their mortality. Although 2 released birds formed a pair bond, no eggs were laid. Wild pairs disappeared one by one as individuals apparently fell victim to old age or 'Io attacks. The last wild pair was seen in 2002, marking the low point of 'Alalā conservation. Although two refuges were created for 'Alalā, little has been accomplished to restore their habitats. The slowly growing captive population, comprised of 60 birds in March 2009, offers a second chance for recovery, but vigorous leadership, increased funding, and strong public support are needed to restore habitat, control threats, and reestablish birds in the wild.

ACKNOWLEDGMENTS

This research was made possible by support from the U.S. Fish and Wildlife Service (USFWS), National Park Service, U.S. Forest Service, World Wildlife Fund, Hawai'i Division of Forestry and Wildlife (DOFAW), and U.S. Geological Survey wildlife research programs. Permits and access to study sites were provided by the USFWS, DOFAW, and many private landowners. I am deeply indebted to D. Ball, J. Klavitter, K. Clarkson, K. Shropshire, M. Sherman, and many other staff and volunteers who carried out work in the field. F. Benevides provided valuable technical assistance with monitoring equipment. 'Alalā were reared in captivity and released to the wild by staff of DOFAW, The Peregrine Fund, and the Zoological Society of San Diego. A. Lieberman provided unpublished information on the captive breeding program. I thank members of the 'Alalā Recovery Team for many stimulating discussions and ideas. I especially thank W. Banko for his inspiration, encouragement, and guidance during our memorable days together in the field and during long discussions of Hawaiian bird population decline and conservation. Earlier drafts of the chapter were improved by comments and suggestions from F. Duvall, J. Marzluff, P. Vitousek, and an anonymous reviewer.

CHAPTER TWENTY-ONE

Poʻo-uli

JIM J. GROOMBRIDGE

The Poʻo-uli (*Melamprosops phaeosoma*) (Plate 12), a species endemic to Maui Island, was discovered in 1973, the same year the Endangered Species Act was enacted (U.S. Fish and Wildlife Service 1975). The existence of this species on the brink of extinction has presented a challenge to conservation biologists ever since. When the species was discovered by undergraduate members of a University of Hawaiʻi expedition, only nine individuals of this unusual-looking bird could be found, of which two were subsequently taken as voucher specimens. Sadly, the Poʻo-uli was last seen in 2004, and the species may now be extinct. However, because the bird has just recently disappeared and may yet reappear, it has not been declared extinct and is still considered extant by government agencies attempting to save it. During the 31-year period of active fieldwork with the Poʻo-uli, the bird was very difficult to detect in the field, and this scarcity of observations means that much of the basic knowledge of avian biology that we take for granted for other Hawaiian forest birds remains unknown for the Poʻo-uli. The short history of this species and our lack of basic information on its biology have made recovery efforts challenging. A popular book on the Poʻo-uli was recently published and gives a fast-paced account of the bird's history and the human endeavor to save it (Powell 2008).

This chapter reviews the recovery efforts for the Poʻo-uli in the context of what we know—and do not know—about the life history and ecology of this species. It then describes the substantial progress as well as the setbacks experienced by conservation biologists working to save the Poʻo-uli and discusses the lessons that might be applied from successful recoveries of other critically endangered bird populations.

WHAT IS THE PO‘O-ULI?

When the Po‘o-uli was first discovered, its distinctive appearance earned the species its own monotypic genus *Melamprosops* (Casey and Jacobi 1974), a status that has helped set the Po‘o-uli apart from other Hawaiian honeycreepers (Pratt 1979, 2001; Berger 1981). Two *Xestospiza* honeycreepers most closely resemble the Po‘o-uli, but both species are known only from sub-fossil bones (James and Olson 1991).

The extreme rarity of the Po‘o-uli and the intensive efforts to recover this species have generated interest in the distinctiveness of the Po‘o-uli in relation to other Hawaiian forest birds. H. D. Pratt (1992) performed a detailed study to determine whether the Po‘o-uli rightly belonged with Hawaiian honeycreepers (Fringillidae: Drepanidinae). Among other characters, he reviewed evidence on plumage patterns, tongue morphology, and the presence or absence of a distinctive odor that is shared by almost all drepanids. Pratt (1992: 179) declared that the Po‘o-uli is phenotypically very different from any drepanid, stating that the Po‘o-uli "does not look, smell, act, or sound like a Hawaiian honeycreeper," a finding that he subsequently corroborated by analysis of a wider selection of phenotypic characters (Pratt 2001).

In contrast, analyses of osteological characters and molecular DNA sequences by Fleischer et al. (2001) provided a different perspective, placing the Po‘o-uli firmly within the drepanidine clade. Significantly, Fleischer et al. (2001) were able to quantify the amount of genetic distinctiveness contributed by each species to the honeycreeper group as a whole. They determined that the Po‘o-uli contributed more evolutionary diversity to the clade than any other member, and on this basis they recommended that the Po‘o-uli become a high priority for conservation efforts. In summary, despite a lack of consensus on the evolutionary origin of the Po‘o-uli, most evidence indicates that this species is a unique Hawaiian forest bird.

CURRENT STATUS AND HISTORICAL DISTRIBUTION

There is little doubt that the Po‘o-uli population has declined in the 36 years since the species' discovery, although the accuracy we can credit to field survey results is limited by the legendary difficulty of detecting Po‘o-uli in the field. What we do know is that Po‘o-uli numbers declined from an estimated 76 per km^2 in 1975 to 15 per km^2 in 1981 and further still to 8 per km^2 in 1985 (Scott et al. 1986, Mountainspring et al. 1990). In 1980, the population was estimated at approximately 140 ± 280 birds (Scott et al. 1986), but this estimate (with a somewhat nonsensical 95% confidence interval) must of course be treated with caution because of this species' cryptic behavior.

Surveys in 1994–1995 visually confirmed six Po‘o-uli at four locations, whereas surveys in 1997–2000 located only three of these birds (Reynolds and Snetsinger 2001). No other Po‘o-uli have been located since the three birds were color banded in 1997 and 1998 (Baker 1998, 2001). The last evidence of any breeding in the wild was documented in 1995 (Reynolds and Snetsinger 2001), and the last wild sightings were made in 2004 (U.S. Fish and Wildlife Service 2006).

Since discovery of the Po‘o-uli, all field records of this species have been confined to a 1,300 ha area of wet montane forest on the northern and eastern slopes of Haleakalā Volcano, East Maui (Engilis 1990, Mountainspring et al. 1990). The three remaining birds occurred within the Hanawī Natural Area Reserve (Hanawī NAR), a 3,035 ha fenced forest reserve, and each resided within a separate, nonoverlapping home range between 1,500 and 1,950 m in elevation (Fig. 21.1). Po‘o-uli were restricted to this high-elevation habitat of wet,

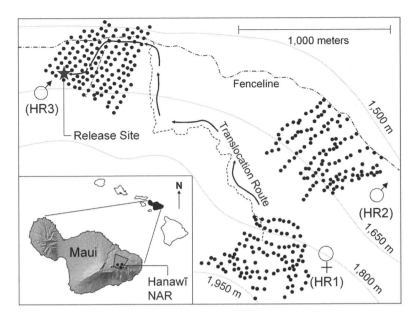

Figure 21.1. Home ranges of the last three Po'o-uli. The home ranges of the three banded individual Po'o-uli are indicated as HR1, HR2, and HR3. In the translocation, the female in HR1 was carried along the route shown and released in HR3. The black dots indicate the predator control grid system in each home range. *Source:* This figure has been adapted from Groombridge, Massey, Bruch, Malcolm, Brosius, Okada, Sparklin, et al. 2004: 367, Fig. 2.

mixed shrub–montane forest, and their presence there was associated with low levels of disturbance to vegetation by feral pigs (*Sus scrofa*), whose damaging effects on the habitat are thought to have been an important cause of the decline in Po'o-uli numbers (Mountainspring et al. 1990). The range of the remaining individuals coincides with high population densities of other endangered honeycreepers, and this distribution is thought to be delimited by habitat suitability and the disease-bearing mosquitoes prevalent at elevations below 1,500 m (Scott et al. 1986). Notably, the endangered Maui Parrotbill (*Pseudonestor xanthophrys*) lives at high densities in the Po'o-uli range, and its abundance there presumably reflects the good quality of the habitat for an insectivorous excavator (Mountainspring 1987, Simon et al. 1997). However, the current distribution of the Po'o-uli may not represent its optimal habitat in view of the dry, leeward, and lower elevations (300 m) on Maui where subfossil Po'o-uli have been found (James and Olson 1991).

A HISTORICAL PERSPECTIVE

The Po'o-uli was undoubtedly rare when first discovered, but might it have always existed in low numbers? Whether we are dealing with a species that has always been rare or one that has only recently crashed to low numbers can be important from an evolutionary and population genetic perspective when judging how well an island species might recover from a severe bottleneck. A long history of small population size may imply the species' robustness to the deleterious genetic effects of inbreeding (Bataillon and Kirkpatrick 2000), whereas a larger population, or one with historically high levels of genetic diversity that crashes to a few individuals, may be more vulnerable (e.g., Groombridge et al. 2000, 2001).

We are lucky to have insight into the historical populations of some Hawaiian honeycreepers through various lines of evidence. In contrast with colorful species whose feathers were collected in ancient and early historic times by native Hawaiians, the drab Po'o-uli is entirely absent from Hawaiian history and the anthropological record (Chapter 3). It is conspicuously absent, too, from the historical museum collections that were assembled for almost all other Hawaiian forest birds before its discovery in 1973. Perhaps, as I have said, Po'o-uli populations naturally had relatively low numbers, or perhaps their population was already restricted to its present location, remote from the collecting sites of the early Western naturalists. Indeed, another explanation is that the Po'o-uli simply went unnoticed due to its cryptic brown coloration, which provides little contrast with the low understory vegetation (Engilis et al. 1996). Surprisingly, however, the paleontological record tells a different story. Subfossil bones identified as those of Po'o-uli have been found at elevations as low as 300 m on the dry, leeward slopes of Maui, suggesting an islandwide distribution at least until prehistoric human settlement (James and Olson 1991). Hence, whether the ancestral Po'o-uli population was large or small remains a mystery.

BEHAVIOR AND FEEDING ECOLOGY

The Po'o-uli feeds by probing into dead wood and moss-covered bark, gleaning from it small insect larvae and other invertebrates (Pratt, Kepler, et al. 1997). However, what sets the Po'o-uli apart from other Hawaiian forest birds is its consumption of forest snails, mainly the endemic *Succinea* species, which it searches for and extracts from understory vegetation (Baldwin and Casey 1983). The tongue of the Po'o-uli is spoon-shaped and robust, which is relatively unusual among Hawaiian passerines (perching birds) but is presumably adaptive for extracting this mollusk prey (Bock

1978). Crucially, population densities of the *Succinea* snails are believed to be at risk from predation by the introduced garlic snail (*Oxychilus alliarius*), which is a serious threat to endemic island snails and is found across the present distribution of Po'o-uli (Cowie 1998). Consequently, the snail diet of the Po'o-uli has intrigued conservation biologists; snail densities may affect Po'o-uli abundance and distribution. So far, experimental plots and surveys of snail populations within the Po'o-uli range have been unable to clearly define the problem or prescribe a solution to snail decline (Maui Forest Bird Recovery Project, unpubl. data). These surveys have shown that the distribution and density of succineid snails are extremely patchy and localized, but more work is needed to clarify their role, if any, in determining the movement patterns and distribution of Po'o-uli (Groombridge et al. 2006).

Po'o-uli are often seen within their home ranges as part of mixed feeding flocks, commonly consisting of Maui Parrotbills and Maui 'Alauahio (*Paroreomyza montana*). This flocking behavior of Po'o-uli is not fully understood (not all honeycreepers participate in mixed-species flocks), but the innate willingness of the Po'o-uli to interact with these other species might prove a useful trait to facilitate successful releases of captive-raised birds.

THE PO'O-ULI IN THE CONTEXT OF CONSERVATION TODAY

The status of the Po'o-uli as one of the most endangered birds in the world has generated a great deal of conservation interest and controversy, in equal measure (Conry et al. 2000). This attention is not surprising because the possibility of extinction attracts a wide media, but never more so than in cases like this one, in which humans might have only a small chance of saving the species. The regional and worldwide media have viewed the critical rarity of the Po'o-uli and attempts to recover it very differently. On Maui, public hearings organized by the U.S. Fish and Wildlife Service (USFWS) and the Hawai'i Department of Land and Natural Resources (DLNR), which have been given the mandate to work toward this species' recovery, produced a mixed response to their announcement that they intended to save the Po'o-uli from extinction (Hawai'i Department of Land and Natural Resources 1999). Many thought that habitat protection for forested watersheds on Haleakalā, where the few remaining Po'o-uli lived, was the most important priority, whereas others viewed direct recovery action to save the Po'o-uli itself as an equal—if not more worthy—cause. The question of how to approach recovery of the Po'o-uli has always produced varied (and usually conflicting) opinions (Conry et al. 2000), but despite these differences, a common theme has been a sense of urgency to act without delay, tempered by a measure of caution to get things right. In contrast to the situation in the region, the status of the Po'o-uli and news of decisions and progress toward its recovery have attracted continual international interest, ensuring that the plight of the Po'o-uli is not ignored.

How does the rarity of the Po'o-uli compare with that of other endangered bird species in Hawai'i? Progress continues to be made in many bird recovery programs across Hawai'i. Ongoing recovery programs for the 'Alalā (Hawaiian Crow, *Corvus hawaiiensis*), Puaiohi (*Myadestes palmeri*), and Palila (*Loxioides bailleui*) are the main examples described in this volume, and these programs have adopted a diverse array of recovery techniques, ranging from habitat restoration to intensive captive breeding and reintroduction (Chapter 19). The Po'o-uli remains the rarest of these species, and it is set apart further still in that there are no individuals in captivity, where recently developed husbandry

techniques can offer additional recovery options (Chapter 19).

The most important factor that distinguishes the Po'o-uli from other species is the absence of a breeding pair. However, success stories elsewhere indicate that if this obstacle were overcome, recovery of the Po'o-uli could become feasible. The Mauritius Kestrel (*Falco punctatus*) was recovered from a single pair in 1974 to more than 600 individuals today (Jones et al. 1995), and the Black Robin (*Petroica traversi*) of Chatham Islands was restored from one pair to several hundred birds (Butler and Merton 1992). Given these successes, it is no surprise that much of the attention given to recovering the Po'o-uli has focused on the immediate creation of a single breeding pair, as well as wide acknowledgement of the small chance of success.

PHYSICAL AND BIOLOGICAL CHALLENGES TO RECOVERY

The task of restoring any population from two nonbreeding birds stands as a classic exercise in population recovery. However, additional physical and biological factors have increased the difficulty of achieving this for the Po'o-uli. Spatially, the home ranges of the two Po'o-uli individuals selected for pairing were separated by approximately 1.5 km. This distance between the home ranges was entirely feasible to access on foot, but the topography on the northeasterly slope of Haleakalā limits the field operations that can be achieved there. Logistically, access by helicopter was preferable, but the trade winds create dense cloud cover over the three home ranges, and these conditions can hamper activities that require continual field presence and support.

Upon arrival in the field, locating the Po'o-uli in preparation for any recovery action was not easy. Po'o-uli were notoriously difficult to detect in the field as a consequence of their low numbers, a preference for foraging in dense understory, and their infrequent vocalizations, which can closely resemble the repertoires of other honeycreepers (Pratt, Kepler, et al. 1997). An example illustrates their capacity to avoid detection. One Poʻo-uli individual remained undetected within its home range for several weeks during intensive mist-netting operations by an experienced field team and was finally detected only when caught (Groombridge, Massey, Bruch, Malcolm, Brosius, Okada, Sparklin, Fretz, et al. 2004). This elusiveness limits the accuracy with which a negative result (its absence) can be interpreted from a field survey, which becomes important when neighboring areas are surveyed for new birds. Even simple confirmation that the remaining Poʻo-uli were still alive each year remained a considerable year-round effort (Hawaiʻi Department of Land and Natural Resources, unpubl. data).

The key decisions about recovering the Poʻo-uli have demanded information on its breeding biology, but very little information on reproduction is available for this species. All reproductive data on the Poʻo-uli originated from a single breeding pair that was discovered in 1986. This pair consecutively built two nests and laid eggs in both. The first nest produced nestlings but failed, whereas the second nesting attempt fledged a 21-day-old offspring (Kepler et al. 1996). Valuable observations at the nest provided the first tentative opportunity (and, sadly, the only one) to attribute plumage and behavioral differences to each sex (Engilis et al. 1996). Following the discovery of this nesting pair, two conclusions proved useful in shaping options for a Poʻo-uli recovery. First, knowing when Poʻo-uli actually breed (March–June, $n = 1$) has allowed managers to plan for specific recovery actions. Second, knowledge that the pair can naturally recycle after a first failed nesting attempt is positive news for managers considering removal of eggs for captive rearing.

Accurate determination of the sexes and ages of known Poʻo-uli individuals also posed a major challenge. Established morphological criteria are available for many endangered Hawaiian forest birds but not for the Poʻo-uli. The first live-caught Poʻo-uli was described by Baker (1998) as an adult male by comparison of its plumage with that of the holotype specimen (an immature male) and field descriptions of the pair at the only reported nest (Engilis et al. 1996). The three wild birds that have been monitored since 1996 were caught and banded by 1998 and considered adults, but knowledge of their sexes had become critical for planning a recovery strategy. Consequently, their sexes were determined from the DNA of feathers taken at the initial capture of the birds for banding. Three laboratories attempted the analyses between 1997 and 1999: University Diagnostics Limited (UDL, now the Laboratory of the Government Chemist, United Kingdom), Avian Biotech International, and the genetic laboratory of the National Museum of Natural History at the Smithsonian Institution. Of these, UDL pronounced a definitive sex for each bird (two were females, one a male) using the method of Griffiths et al. (1996), whereas other workers experienced technical difficulties or ambiguities in the sexing results for one or two of the three individuals. The need to act on the best available information had become critical by early 2000, at which time the birds were thought to be at least five or six years old. Consequently, the sex results of UDL were adopted. However, when the first live-caught bird was finally recaptured and brought into captivity, further tests from blood samples using polymerase chain reaction techniques and ultimately a necropsy proved that it was a male, reversing the UDL determination that it was a female (Jarvi and Farias 2006).

A REVIEW OF RECOVERY ACTIVITIES

Elimination of Feral Pigs and Habitat Recovery

More than a decade of ecosystem management on Maui had already benefited the restoration of Po'o-uli habitat prior to 2000. Initial creation of the Hanawī NAR began in 1986 (the same year a single breeding pair of Po'o-uli was discovered), and since then restoration activities have followed a logical progression (Groombridge, Massey, Bruch, Malcolm, Brosius, Okada, Sparklin, Fretz, et al. 2004). The subsequent fencing of 800 ha of forest that included part or all of the three known Po'o-uli home ranges was completed by 1996 (Hawai'i Department of Land and Natural Resources 1988, 1996, 1999). Total removal of feral pigs from inside the fenced enclosure was achieved by 1997 (Hawai'i Department of Land and Natural Resources and U.S. Fish and Wildlife Service 1999), and adjacent forest areas became protected through the management of neighboring lands by the National Park Service and the East Maui Watershed Partnership (Hawai'i Department of Land and Natural Resources 1996).

Predator Control

An additional recovery initiative implemented in Hanawī NAR was to control introduced mammalian predators, including feral cats (*Felis catus*), small Indian mongooses (*Herpestes auropunctatus*), and two species of rat (*Rattus rattus* and *R. exulans*). Initial efforts began in 1994 in the Kūhiwa Unit of Hanawī NAR following concentrated sightings of Po'o-uli there. By 1997, a grid system of rat traps and Fenn™ traps for mongooses and cats was installed and regularly serviced in two of the three Po'o-uli home ranges. These trapping grids were supplemented by a system of rodenticide stations baited with diphacinone—currently the only rodenticide approved for conservation use in Hawai'i by the Environmental Protection Agency. By the end of 1998, rodenticide and trapping grids were in continual operation for all three home ranges (Hawai'i Department of Land and Natural Resources and U.S. Fish and Wildlife Service 1999). Servicing of these grids throughout a year routinely involved an expensive investment in labor, and the inaccessibility of Hanawī has restricted the regularity of their maintenance.

Despite these constraints, monitoring of the efficiency of these methods in reducing mammal densities in Hanawī did provide some useful information. For example, these control methods can reduce rat densities significantly, but comparisons of rat density inside and outside the treated Po'o-uli home ranges indicated that there is both temporal and spatial variation at a fine-scale resolution. Trends of decline in rat density through space and time are often different for *R. rattus* and *R. exulans* in Hanawī (*R. exulans* are less affected by baiting), suggesting that there are localized sources and sinks of rat population growth throughout the Hanawī NAR (Malcolm et al. 2008).

These long-term, managed reductions in rat density are likely to be of benefit to the Po'o-uli by reducing the risk of direct rat predation on adults and nests and by increasing the densities of one of the bird's foods, *Succinea* snails. Although maintaining this strategy in Hanawī is costly and labor-intensive, the advantages to the ecosystem may become felt more widely, particularly when delivery techniques such as aerial broadcast of rodenticide baits become a more feasible alternative.

Environmental Assessment

Intensive search efforts (1994–1997) to confirm the locations of any additional birds unfortunately proved fruitless (Baker 2001; Reynolds and Snetsinger 2001; U.S.

Geological Survey, unpubl. data). This impressive search involved more than 729 person-days and at least 6,000–8,000 search hours and was devoted to an 800 ha region of forest that extended from the known Po'o-uli home ranges to neighboring habitat and beyond. Nevertheless, the surveys did serve to focus attention on the three birds that remained, culminating in an *Environmental Assessment for Proposed Management Actions to Save the Po'o-uli* (Hawai'i Department of Land and Natural Resources and U.S. Fish and Wildlife Service 1999). This environmental assessment was drawn up between the two main agencies involved, the USFWS and DNLR, which together formed a partnership with experts in captive breeding and land managers to decide on the best recovery action (Vander-Werf, Groombridge, et al. 2006). By 2000, a full-time implementation team, the Maui Forest Bird Recovery Project, had been established by the DNLR and USFWS. This field team had already been monitoring the presence of the three birds for three years and was therefore ready to carry out any recovery action.

The alternative management actions outlined in the environmental assessment for recovering the Po'o-uli consisted of (1) continuing the habitat management regimes already in place (feral pig and predator control in Hanawī) but not manipulating the three birds, (2) removing the birds to some form of captivity (either in situ field captivity or an ex situ propagation facility), or (3) attempting to initiate a wild breeding pair by translocating one bird to another of the opposite sex. These alternatives represented a full spectrum of possible levels of intervention and manipulation. Unsurprisingly, regional and international experts who were consulted for advice returned a similar diversity of opinion. However, following a strengthening of personnel and expertise in the Maui Forest Bird Recovery Project during 2000, it became clear that a decision about what

to do with the three remaining Po'o-uli was then critical to progress.

By September 2000, all possible alternatives had been considered by representatives of the contributing agencies, including the USFWS, the DNLR, The Peregrine Fund, and the Zoological Society of San Diego (ZSSD), which in 2000 took over from The Peregrine Fund the management of the captive propagation program in Hawai'i. The alternatives were compared in terms of their relative merits and potential pitfalls, alongside the best information available on existing sex ratios and reproductive parameters, as well as professional opinions on the likelihood of success of the three options and their operational feasibility under the harsh environmental conditions in Hanawī (Vander-Werf, Groombridge, et al. 2006). The final decision lay jointly with the USFWS and DLNR, which gave the go-ahead for a translocation to be coordinated by the Maui Forest Bird Recovery Project. From the outset it was acknowledged that the probability was small that a pair bond would be formed between wild Po'o-uli as a consequence of a translocation, but this option was believed to combine the least risk with the highest chances of a successful outcome (Hawai'i Department of Land and Natural Resources and U.S. Fish and Wildlife Service 1999).

Translocation

The translocation strategy was to move one of the two females to the single male (see Fig. 21.1) and to monitor the outcome by radio-tracking both birds throughout the translocation. The safest scenario favored hiking the female between her home range and the male's as opposed to transporting her by helicopter. Consequently, between December 2000 and March 2001, field trials were conducted using the nonendangered Maui 'Alauahio. This work identified the least stressful method of translocation along the route established for the Po'o-uli by measuring the ratios of heterophils to lymphocytes as an index of stress (Groombridge, Massey, Bruch, Malcolm, Brosius, Okada, and Sparklin 2004). These surrogate trials also optimized the radiotelemetry techniques for the extreme topography of Hanawī. Extensive field protocols and veterinary contingencies were developed (Vander-Werf et al. 2003; Groombridge, Massey, Bruch, Malcolm, Brosius, Okada, Sparklin, Fretz, et al. 2004).

The successful translocation of the female Po'o-uli was finally achieved on April 4, 2002, almost five years after the three birds had initially been caught, banded, and their sexes determined (Groombridge, Massey, Bruch, Malcolm, Brosius, Okada, Sparklin, Fretz, et al. 2004). The female was released at dusk into the center of the male's home range, where he had last been seen, and the female's movements were radio-tracked from then onward. She roosted in the male's home range overnight, and her movements were closely monitored throughout the following day. Unfortunately, early on the morning of April 5, it was confirmed that she had left the male's home range and had returned to her own home range by that evening. We cannot be certain that the female failed to encounter the male in his home range, but close monitoring of her movements for nine consecutive days afterward revealed no behavioral differences to suggest that she had.

The reason for the female's return is unknown but perhaps unsurprising in view of the return rates observed for Maui 'Alau-ahio during the surrogate work (Groombridge, Massey, Bruch, Malcolm, Brosius, Okada, and Sparklin 2004) and the low success rates for similar translocations of other passerines (Butler and Merton 1992, Fancy et al. 2001). The female's immediate return was disappointing, but significant progress had already been made while the female was held in field captivity in the male's home range for two hours prior to her release. Remote TV footage of her revealed encouraging signs of acclimation to captivity and consumption of two live succineid snails and 15–20 waxworms that were offered to her, providing a first tentative indication that Po'o-uli might acclimate to longer periods in captivity (Groombridge, Massey, Bruch, Malcolm, Brosius, Okada, Sparklin, Fretz, et al. 2004).

Bringing the Po'o-uli into Captivity

After the translocation attempt, the new information acquired was reviewed alongside the other recovery alternatives to consider the next move. Some have considered it too late to use ecosystem restoration alone without species-level action for critically rare species such as the Po'o-uli (Banko et al. 2001). This option of no manipulation appears unlikely to be sufficient when compared with the options that do involve manipulation, which have repeatedly shown better chances of successful species recovery (Butler and Merton 1992, Jones et al. 1995). The agencies managing Po'o-uli recovery have been realistic about the small chances of success but have constantly recognized the importance of the need to try (Hawai'i Department of Land and Natural Resources 1999).

In June 2002, a thorough evaluation of the probabilities of success of each of the remaining alternatives was conducted by the USFWS and DLNR. All the participating partner agencies were consulted and asked for structured feedback via several formats, including a short written statement, an assignment of ranks to particular attributes of each alternative to reflect their perceived merit, and a contribution to a probability matrix. The probability matrix attempted to compare the relative likelihood of each alternative recovery action to result in the production of eggs, given specific probability values derived from existing knowledge or data sets for the Po'o-uli or, where necessary, from data available for other honeycreeper species. An important part of the decision-making process was to initially arrive at the option with the highest biological probability of success before taking into account additional considerations such as cost, labor, and feasibility.

A consensus was not easy to reach, but, using the results of the structured decision-making process as a guide, the USFWS and DLNR decided that removal of all three birds from the field into captivity was the best hope for recovering the Po'o-uli at that time (VanderWerf et al. 2003). The ZSSD, an important member of the recovery partnership, expressed with reservations how it would attempt to manage and breed the Po'o-uli in captivity. The decision to bring the three birds into captivity stands as perhaps the most difficult managerial decision since the species' discovery. A five-year recovery work plan was drafted in 2003 following this decision (Vander-Werf, Groombridge, et al. 2006).

After many months of effort to capture the remaining Po'o-uli, the first individual (the bird in HR2, Fig. 21.1) was finally caught on September 9, 2004. Following a night in field captivity, the bird was flown by helicopter to the ZSSD's Maui Bird Conservation Center. There the bird was quarantined and acclimated to its captive environment. Closer examination revealed that this bird had only one functioning eye, with the other showing signs of a healed injury previously sustained in the wild. Despite this unusual discovery, the bird showed all the signs of successful acclimation to captivity. Sadly, however, events took a different turn, and the bird died on November 26, 2004, after 78 days in captivity.

Necropsy results revealed several chronic health problems, as well as a subclinical infection of avian malaria (*Plasmodium relictum*). Overall, the necropsy results indicated that the likely cause of death was old age.

Although the bird had avian malaria, the infection did not appear to have been the primary cause of death (B. Rideout, pers. comm.). Fortunately, detailed protocols in place for this eventuality ensured that tissues were successfully taken for cryo-preservation (VanderWerf et al. 2003). The sex of the dead bird was also determined as part of the necropsy, and the bird was found to be a male, not a female as suggested from the DNA testing by UDL. This discrepancy has now brought into question the sexes of the two other wild birds, which from the DNA sexing methods used by UDL had previously been considered to be a male and a female.

The death of the captive Po'o-uli was obviously a serious setback to efforts to recover the species, but the USFWS and the DNLR continued their efforts to capture and remove the two remaining wild birds to captivity as a last attempt to recover the species. Unfortunately, sightings of these birds dwindled and eventually ceased, and attempts to capture them were abandoned in 2005.

WHAT DOES THE FUTURE HOLD FOR THE PO'O-ULI?

The Po'o-uli recovery strategy has demanded that decision makers act in recognition of the existence of only three birds and not hope that additional wild birds might someday be discovered. Now that those three birds appear to be gone, slight hope remains that more birds will be found. Although no other Po'o-uli have been observed since 1996, it remains impossible to declare conclusively that the species is extinct. If additional birds are still alive, they may yet be found in Hanawī NAR, where the Maui Forest Bird Recovery Project has launched population studies of the Maui Parrotbill and maintains a large field presence. Should more Po'o-uli be found, the Rare Bird Discovery Protocol that has been developed would be set in motion by the existing field team (U.S. Fish and Wildlife Service 2006). The protocol describes procedures to locate the remaining population and very likely bring them into captivity if there are only a few birds. The benefits of the long-term fencing and ungulate control that are already in place to restore Po'o-uli habitat, as well as those derived from long-term control of introduced populations of small mammals in Hanawī NAR, will continue in the meantime (Hawai'i Department of Land and Natural Resources and U.S. Fish and Wildlife Service 1999; Groombridge, Massey, Bruch, Malcolm, Brosius, Okada, Sparklin, Fretz, et al. 2004).

SUMMARY

The Po'o-uli, a medium-sized honeycreeper of distinctive appearance, was discovered on Maui Island in 1973. Morphological studies have declared that the Po'o-uli is phenotypically very different from other honeycreepers, but osteological and genetic studies place the Po'o-uli firmly within that clade. Phylogenetic studies have revealed the evolutionary distinctiveness of the Po'o-uli, and on this basis the species has been considered a high priority for conservation.

The cryptic nature of the Po'o-uli, together with its conspicuous absence from museum collections, makes any interpretation of its historical population extremely difficult. Field surveys indicate that the population declined dramatically from 1975 to 1985, and surveys during 1997–2000 located only three birds, the last of which was seen in 2004. These remaining birds occurred in the Hanawī Natural Area Reserve, a 3,035 ha fenced forest reserve where the habitat is now recovering from damage by feral pigs. Each bird resided within a separate, nonoverlapping home range. The Po'o-uli feeds on small insect larvae that it gleans from dead wood and on forest snails, mainly the endemic *Suc-cinea* species, a food source that sets the Po'o-uli apart from other Hawaiian forest birds. The key decisions for recovering the Po'o-uli have also demanded information on its breeding biology, but these data are scarce, and almost all originate from a single breeding pair studied in 1986.

The question of how to save the Po'o-uli has generated a great deal of conservation interest and controversy because of the very small chance of success. A common theme has been a sense of urgency to act without delay, tempered by a measure of caution to get things right. A priority for the Po'o-uli has focused on the creation of a breeding pair, although determination of sex from DNA proved difficult until recently. A translocation was attempted in 2002 to create a wild breeding pair. Although unsuccessful in achieving a breeding pair, the outcome did provide encouraging data on the behavior of captive Po'o-uli. In the same year, the decision was made to attempt to catch all remaining individuals for captive breeding. A male was caught in September 2004 and removed to permanent captivity, but the bird died in November from old age.

The Po'o-uli continues to evoke strong emotion and controversy among those tasked with its recovery. Given the unpredictable field conditions in Hanawī, even simple confirmation that any Po'o-uli are still alive is a considerable task.

ACKNOWLEDGMENTS

The recovery activities for the Po'o-uli were funded by the Hawai'i Natural Areas Reserve System and Hawai'i Division of Forestry and Wildlife (both in the Department of Land and Natural Resources), and the U.S. Fish and Wildlife Service. Funds are administered by the Research Corporation of the University of Hawai'i and the Pacific Cooperative Studies Unit, University of Hawai'i. Other agencies have contributed staff and resources, including the

Bishop Museum, Flycatcher Films, Haleakalā National Park, the Hawai'i Department of Land and Natural Resources, the Honolulu Zoo, the Maui Forest Bird Recovery Project, The Nature Conservancy, Pacific Helicopters, Windward Aviation, the U.S. Fish and Wildlife Service, and the Zoological Society of San Diego, including the Keauhou Bird Conservation Center and Maui Bird Conservation Center. Many individuals have contributed time, expertise, and advice in order to achieve the recovery goals, including P. Baker, P. Banko, A. Banning, C. Brosius, J. Bruch, M. Buck, T. Casey, P. Conry, D. Duffy, A. Engilis, B. Evanson, S. Fancy, R. Fleischer, J. Foster, S. Fretz, B. Gagné, P. Hart, P. Henson, J. Jacobi, J. Jeffrey, C. Kuehler, J. Lease, A. Lieberman, K. Lui, P. Luscomb, T. Malcolm, A. Marmelstein, G. Massey, J. Nelson, M. Okada, G. Olsen, D. Phalen, T. Pratt, M. Reynolds, B. Rideout, K. Rosa, M. Schwartz, J. Scott, J. Simon, R. Smith, T. Snetsinger, B. Sparklin, V. Stein, K. Swinnerton, R. Switzer, E. VanderWerf, and C. Van Riper III.

CHAPTER TWENTY-TWO

Puaiohi

BETHANY L. WOODWORTH,
ALAN A. LIEBERMAN, JAY T. NELSON,
AND J. SCOTT FRETZ

The fabled Alaka'i Swamp on Kaua'i Island is a high-elevation, forested plateau enshrouded in mist and deeply dissected by steep-walled forested stream valleys. The Alaka'i is one of the rainiest places on earth, soaking up 13 m of rain yearly, and long was famous among ornithologists as the last stronghold of some of the most endangered Hawaiian birds. Unlike other islands in the Hawaiian Archipelago, Kaua'i harbored all of its historically known forest birds at least until the late 1960s (Conant et al. 1998). At that time, it seemed reasonable to expect that the Alaka'i, a relatively large and remote wilderness area under no obvious threat, would remain a refuge for rare birds for the foreseeable future. Over the past three decades, however, five bird species have disappeared from the Alaka'i with dizzying speed. Gone are the curve-billed Kaua'i Greater 'Akialoa (*Hemignathus ellisianus stejnegeri*) and Kaua'i Nukupu'u (*H. lucidus hanapepe*), the hook-billed 'Ō'ū (*Psittirostra psittacea*), the charismatic and vocal Kaua'i 'Ō'ō (*Moho braccatus*), and one of the two Hawaiian thrushes endemic to Kaua'i, the Kāma'o (*Myadestes myadestinus*). Of the endangered species (there are also eight forest bird species on Kaua'i not listed as endangered), only the Puaiohi or Small Kaua'i Thrush (*M. palmeri*) (Plate 3) remains (Reynolds and Snetsinger 2001).

How did we lose this remarkable suite of forest birds? Although several of these species were rare in the 1890s–1960s (Perkins 1903, Richardson and Bowles 1964), probably due to introduced predators and disease, their gradual decline was accelerated by two hurricanes that ravaged Kaua'i in the span of 10 years (Hurricane 'Iwa in 1982 and Hurricane 'Iniki in 1992) (Pratt 1994). The persistence of the Puaiohi presents somewhat of a paradox: by all accounts, the Puaiohi was historically one of the rarest birds on Kaua'i, while its larger cousin the Kāma'o was the most abundant forest bird (Munro 1944), outnumbering Puaiohi by at least an order of magnitude (Perkins 1903, Richardson and Bowles 1964, U.S. Fish and Wildlife Service 1983b, Scott et al. 1986). The extinctions that swept over the Alaka'i robbed us of other birds that were historically more common (Scott et al. 1986) and left us the elusive Puaiohi. What is it about the Puaiohi that has allowed it to survive, and what challenges does the Puaiohi face? What have we learned about the biology of this species that will assist us in its recovery? How has its unique conservation setting in a remote, publicly owned wilderness shaped the biological, social, and political aspects of recovery actions? This chapter reviews the history and future of recovery efforts for the Puaiohi in an attempt to gain a larger perspective of challenges, pitfalls, and successes in endangered species recovery.

ECOLOGY AND LIMITING FACTORS

The Puaiohi is one of five *Myadestes* solitaire thrushes that evolved in the Hawaiian Islands; the others are the Kāma'o, 'Āmaui (M. *woahensis*), 'Oloma'o (M. *lanaiensis*), and 'Ōma'o (M. *obscurus*) (Pratt 1982). The Puaiohi is secretive and difficult to follow except during its breeding season (generally March–July), when its distinctive, reedy song gives its location away. The Puaiohi is found in stream valleys and associated ridges above 1,050 m elevation on the southern and central plateau of the Alaka'i Wilderness Preserve, although subfossil evidence from the coast suggests that it was widely distributed prehistorically (Burney et al. 2001). Even in the late 1800s, it was considered exceedingly rare (Perkins 1903). The first population estimates, by J. Sincock (U.S. Fish and Wildlife Service 1983b), yielded the rangewide population estimate of 176 ± 192 birds for the period

1968–1973. Results of the Hawai'i Forest Bird Survey in 1981 suggested that the species was nearly extinct (Scott et al. 1986). Although these estimates probably reflected the difficulties of ascertaining a genuinely small population in a remote habitat, they did not take into account the habitat preference of the Puaiohi for steep-sided ravines. Recent surveys that counted Puaiohi along streams revealed 300–500 birds (Snetsinger et al. 1999). Whether this apparent increase was due to improved survey methodology or rebounding populations is impossible to determine with confidence. However, those recent surveys did show a marked decrease in range, particularly in the western portions of the plateau (Fig. 22.1).

Puaiohi feed on fleshy native fruits, insects, snails, and other invertebrates and nest primarily on cliff faces (Snetsinger et al. 1999). Puaiohi appear quite particular in their nest site requirements, and the amount of suitable cliff-face nesting habitat may be limiting. Invasive vegetation has degraded formerly suitable nesting habitat and appears to be an accelerating threat in much of the current Puaiohi range. In the core of the Puaiohi range, pairs space their nests about 90 m apart along steep-sided stream valleys. Females alone build the nest and incubate the eggs, but both parents feed the chicks. The clutch size is usually two eggs, and successful nests fledge 1–2 chicks. Females quickly renest after success or failure, so pairs raise an average of 3–5 young in favorable years. However, reproductive success varies greatly among years. In 1998, pairs fledged on average only 0.4 young per territory due to a shortened breeding season and increased nest failure (Snetsinger et al. 2005). Overall, about 40% of nests fledge young; the others are lost to rat predation, owl predation, abandonment, severe weather, disturbance by nonnesting Puaiohi, and hatching failure. Breeding and natal dispersal distances may be small, which has important implications for the rate of natural recoloniza-

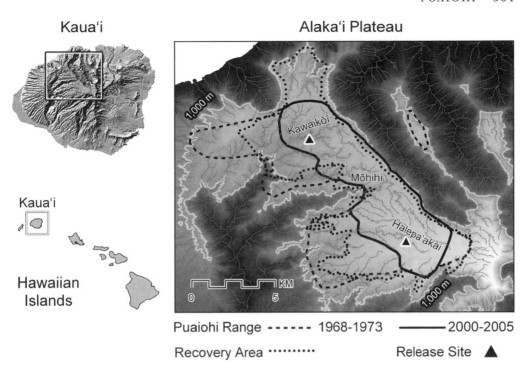

Kaua'i

Alaka'i Plateau

Kaua'i

Hawaiian Islands

Puaiohi Range ----- 1968-1973 ———— 2000-2005

Recovery Area ·········· Release Site ▲

Figure 22.1. Map of the Alaka'i Wilderness Preserve and surrounding areas, Kaua'i. Puaiohi range has decreased over the past 30 years, from approximately 7,713 ha in 1968–1973 to 3,966 ha in 2000. The white border indicates the 1,000 m contour interval, and shading indicates stream drainages. Puaiohi range as determined by 1968–1973 and 2000–2005 surveys is shown along with the recovery area identified in the recovery plan (U.S. Fish and Wildlife Service 2006). The highest density of Puaiohi is concentrated in the Mōhihi stream drainage, and this is where the initial ecological studies of the wild population were conducted. The first reintroductions of captive-bred Puaiohi were made in the Kawaikōī stream drainage, a formerly occupied habitat. The release site in the Halepa'akai contained more pristine habitat and a resident Puaiohi population.

tion of recovering habitat (Snetsinger et al. 2005).

Finally, introduced avian malaria (*Plasmodium relictum*) and avian pox virus (*Avipoxvirus* sp.) and their vector, the introduced mosquito *Culex quinquefasciatus*, are present in the Alaka'i (Chapter 9). The absence of Puaiohi from forested ravines below about 1,050 m suggests that the species may be vulnerable to these diseases (U.S. Fish and Wildlife Service 2006). Of five wild Puaiohi

tested for disease, one had antibodies to malaria, providing hope that at least some Puaiohi may be able to survive infection (Atkinson, Lease, et al. 2001). However, on necropsy three captive-bred birds that were released and died in 2007 were found to have high levels of the malaria parasite, indicating that this was most likely the cause of death. Further research on the response of Puaiohi to malaria infection is in progress.

HISTORY OF RECOVERY ACTIONS

The relatively pristine forests of the high-elevation plateau of Kaua'i have long been recognized as a unique natural area worthy of protection, and in 1964 the Hawai'i Department of Land and Natural Resources established the 4,022 ha Alaka'i Wilderness Preserve. In 1983, a recovery plan was completed for six Kaua'i forest birds in danger of extinction (U.S. Fish and Wildlife Service 1983b). The remote nature of the Alaka'i, along with the apparently intact nature of its avifauna, led ornithologists to regard the Alaka'i as relatively secure; their attentions were focused on recovery sites more obviously threatened and easily restored elsewhere in the island chain. But by the early 1990s, after the passage of Hurricane 'Iniki in 1992, it was clear that several species of Kaua'i forest birds had disappeared entirely. The Hawai'i Forest Bird Recovery Team (HFBRT) called for research and recovery actions that could prevent further extinctions of Kaua'i forest birds. Recognizing that captive propagation and reintroduction capabilities would be needed to save the rarest species, The Peregrine Fund (TPF) was contracted to develop facilities, techniques, and knowledge that had not been previously available (Chapter 19).

It was in this context that a working group was formed for Kaua'i species consisting of representatives of the U.S. Fish and Wildlife Service (USFWS), TPF (which transferred its duties to the Zoological Society of San Diego [ZSSD] in 2000), the National Biological Service (NBS, now the U.S. Geological Survey [USGS]), and the Hawai'i Division of Forestry and Wildlife (DOFAW). This public-private partnership including aviculture, land management, public policy, and research helped to form a broad-based approach to the conservation challenges that lay ahead.

The first meetings of the Kaua'i working group focused on "search and rescue." The general perception at the time was that the Puaiohi numbered in the dozens, that it was going the way of other Kaua'i species thought to be extinct or near extinction, and that there was no time to lose. From a practical standpoint, the Puaiohi was the only endangered forest bird on Kaua'i that could be located, and luckily it was still sufficiently numerous to manage. Because the Puaiohi shared the same habitat as the other endangered species, it was hoped that any management actions aimed at Puaiohi would benefit the rarer

species, if indeed they still survived. Therefore, the working group simultaneously initiated four actions for the Puaiohi in 1995–1996. By initiating surveys and ecological research concurrently with research on captive breeding techniques, the group was taking measures to ensure that the bird was not lost while it was being studied or before recovery methods were available to save the wild population. The four actions were as follows:

1. Mounting surveys to locate remaining populations of Puaiohi.
2. Initiating an ecological study of the breeding biology and limiting factors of Puaiohi.
3. Developing methods for collecting and transporting wild eggs, captive propagation, and release using non-endangered 'Ōma'o as a surrogate.
4. Establishing a captive breeding program for Puaiohi and testing methods for captive breeding and release of the species into formerly occupied habitat.

The first step was to try to get a better idea of the distribution and abundance of Puaiohi in the Alaka'i to determine how dire the situation was. Surveying the Alaka'i is no small task. Helicopters are needed to access almost all corners of the remote and rugged terrain, and progress on foot through the steep-walled valleys and dense vegetation is slow and treacherous. In 1995, a field team led by TPF biologists spent three weeks in the Alaka'i and located several birds along the Mōhihi Trail. In February–May 1996, the NBS Rare Bird Search Team mounted seven survey expeditions to the area, spending a total of 562 search hours over 58 days in eight drainages and facilitating detections by using playback. They estimated a total population size of 145 ± 19 Puaiohi (Reynolds et al. 1997, Reynolds and Snetsinger 2001), with the highest density (33 birds per km^2) located in the Mōhihi. Additional surveys in 1997–1998 would eventually increase the estimated population to 300–500 birds (Snetsinger et al. 1999).

The second concurrent objective was to gain some insight into Puaiohi breeding ecology and limiting factors. To do this, an intensive field study was initiated in the area of highest Puaiohi density. In 1996–1998, a field crew led by USGS biologists located 96 active nests and obtained vital information on the species' breeding ecology, nesting success, productivity, and behavior (Snetsinger et al. 2005). Notably, the first two years of field studies in 1996–1997 indicated that the cliff-face nest sites used by wild Puaiohi were secure from rat predation and that the seasonal fecundity level of females was remarkably high (Snetsinger et al. 1999). In retrospect, these data were from exceptionally good breeding years in the core of the species' range. This conclusion turned out to be unfortunate, for it led the early working group to discount the importance of predators in limiting Puaiohi populations. Only later would research from a third year of field studies in the 1998 El Niño Southern Oscillation drought, along with studies of reintroduced birds in outlying habitat in 1999–2001, reveal the effects of rodent depredation on Puaiohi (Snetsinger et al. 2005, Tweed et al. 2006).

Therefore, the combined results of surveys and field studies supported the idea that Puaiohi were reproducing well in the wild and that their most recent decline had resulted from two major hurricanes striking their habitat in the span of 10 years. Recovery efforts proceeded based on the working hypothesis that as habitat naturally regenerated, birds would reoccupy it. Natural recovery was likely to be slow, and in the meantime the birds would potentially be vulnerable to losses from avian disease or another hurricane. Reintroductions of birds, either translocated or captive-reared, into former parts of the species' range were viewed as a way to assist re-covery by speeding occupation of recovering habitat and by forming additional populations as a hedge against catastrophic events (Fig. 22.2).

In 1995, TPF, USGS, USFWS, and DOFAW cooperated in developing techniques for captive breeding, rearing, and reintroduction with the closely related non-endangered 'Ōma'o as a surrogate for the endangered Puaiohi. The research showed that rearing Hawaiian solitaires in captivity and gradually releasing them to the wild from field aviaries was highly successful (Kuehler et al. 2000). Furthermore, captive-reared yearling 'Ōma'o demonstrated greater site fidelity than translocated adult birds (Fancy et al. 2001). These results influenced the working group's decision to use captive-bred rather than translocated Puaiohi for future reintroductions.

In 1996, five viable wild eggs were collected from the Alaka'i and transported to TPF's temporary incubation facility, where the first egg hatched on April 23, 1996. This marked the beginning of an ambitious captive breeding program at TPF/ZSSD's Keauhou Bird Conservation Center and its sister facility, the Maui Bird Conservation Center. The protocols developed with surrogate 'Ōma'o proved successful with the Puaiohi (Kuehler et al. 2000). From 1996 to 2007, a total of 1,001 Puaiohi eggs were managed (331 eggs were viable), 261 chicks hatched, and 202 chicks were raised using both parent-rearing and hand-rearing techniques, yielding a good success rate of 84% hatchability and 77% survivability. At the time of this review (February 2009), ZSSD's captive propagation program maintained eight breeding pairs and additional birds representing approximately 86% of the original genetic diversity.

In 1999, the working group began a program to reintroduce captive-bred Puaiohi into an area of apparently suitable but unoccupied habitat in the Kawaikōī drainage in the western Alaka'i. In this remote and

Figure 22.2. Scenes of Puaiohi recovery efforts. (A) Adult Puaiohi with two speckle-breasted young. (B) Puaiohi habitat on the Alaka'i Plateau, Kaua'i. (C) Nesting habitat of Puaiohi. (D) Puaiohi chicks bred and raised in captivity at the Keauhou Bird Conservation Center. (E) A hacktower in the Kawaikōī stream drainage in the Alaka'i in which captive-reared Puaiohi are held during acclimatization before being released from the tower into the wild. (The term and concept of hacking is derived from the falconry practice of raising wild-caught nestlings in a state of partial liberty.) *Source:* Photos by the Zoological Society of San Diego, but (A) © Jack Jeffrey.

rugged landscape, much of which can be reached only by helicopter, one of the more important considerations was accessibility. Because of the need to transport birds to the release area with minimum trauma and to monitor them after release, the group chose a release site that was accessible by vehicle and foot trails. The Kawaikōī was selected for the first releases because it was formerly occupied habitat with a high availability of cliff-face nesting sites, native vegetation, and food plants; had a low abundance of disease vectors as measured by mosquito trapping; and was owned and protected by the state of Hawai'i. Unfortunately, no data on rodent abundance were available. Rodents had not been identified as a major source of nesting mortality in the previous field study (Snetsinger et al. 2005). Even so, the release plan included a localized rodent control component (Tweed et al. 2006).

The first year's release of 14 birds in 1999 was a resounding success, with 100% of the birds surviving at least 9.5 weeks postrelease, observations of captive-captive pairings and captive-wild pairings in the wild, and nestings resulting in 6 confirmed young fledged in the first breeding season following release (Kuehler et al. 2000, Tweed et al. 2006). Over the next two years, 20 more birds were released with similar success (Tweed et al. 2006). By radio-tracking released birds, it was discovered that only 20–43% of the released birds remained in the target drainage to breed, with the others dispersing to find mates or breeding territories elsewhere (Tweed et al. 2003). Significantly, nesting studies of released birds in 1999 and 2001 showed for the first time that rodents were causing significant losses of nests and nesting females: 38% of nests and at least two breeding females were depredated, primarily by rats (presumably black rats, *Rattus rattus*), despite rodent control at nests and in the release area (Tweed et al. 2006).

By 2001, after three years of releases in the Kawaikōī, several important pieces of information had come to light. First, although the goal had been to establish a second breeding population away from the core population as a hedge against catastrophic losses, high dispersal rates after release hindered the managers' ability to "target" a particular drainage for repopulation; birds were leaving to find mates or food or for other reasons (Tweed et al. 2003). Second, it had became ap-

parent that rodent predators were likely to be a significant limiting factor for Puaiohi breeding productivity in at least some parts of its habitat (Tweed et al. 2006). As if to underscore these points, the goal of establishing a self-sustaining population of Puaiohi in the Kawaikōī drainage had remained elusive. Field surveys in May and June 2002 (over 90 hours of surveys of 8.9 km of stream valley) were able to locate only three adult birds remaining in the Kawaikōī, including one unmated adult and a nesting pair with four young (USGS, unpubl. data).

The urgency of dealing with the predator issue had come to the forefront. However, experience in rain forests on Hawai'i Island (Nelson et al. 2002) as well as in the Kawaikōī showed that ground-based predator control of limited extent was relatively ineffectual because of influx of animals from untreated areas. At the time, Hawai'i had not yet gained approval to conduct broad-scale (aerial) predator control; moreover, the level of social opposition to any use of toxicants in the Alaka'i was extremely high. The Alaka'i watershed provides drinking water to local communities and is open to public hunting for feral pigs (*Sus scrofa*) that may eat toxic bait (Chapter 18). Therefore, rodent control on the scale needed to assist Puaiohi was deemed infeasible at the time.

Meanwhile, the Keauhou Bird Conservation Center continued to propagate new cohorts of Puaiohi each spring for release in the Alaka'i. In 2002, releases were begun in the Halepa'akai stream drainage, an area of apparently excellent habitat (a high density of nest sites and native vegetation) in a remote part of the Alaka'i (see Fig. 22.1). These marked the first releases of birds into an area already containing a moderate-density population of wild birds. The decision to release birds in occupied habitat marked a fundamental change in the objectives of the releases, from establishing an additional (separate) population to augmenting an extant population. This decision, which was not without controversy at the time, was complicated by a lack of adequate information on the availability of suitable unoccupied habitat and the potential interactions between released and resident wild birds. The working group responded by initiating a new program to develop better census methods for Puaiohi and to improve the spatial coverage of surveys.

In 2002–2006, a total of 79 Puaiohi were released into the Halepa'akai drainage. Because of the remoteness of the release area and the extremely difficult terrain and dense vegetation, estimates of the survival and dispersal of released birds in the Halepa'akai were difficult to determine. Initial observations indicated that nearly all the birds left the area within a few days of the release and that the rate of mortality of the released birds was higher (estimated ~50% within 30 days of release) than that of the first years' releases in the Kawaikōī. (Note that the mortality rates are not directly comparable because cohorts released in the Halepa'akai included older birds. Also, the greater difficulty of tracking birds there may have affected the estimates of higher rates of apparent mortality.) In 2007, the release effort returned to the original Kawaikōī site, where two releases took place in February and October of that year, with 19 and 21 birds released, respectively. The October release was designed as an adaptive management approach to see whether young released in the fall might show a higher degree of site fidelity than those released in the spring. Since 1999, there have been 10 releases over nine consecutive years, with a total of 153 birds released at the two sites: 74 birds in the Kawaikōī and 79 birds in the Halepa'akai.

Significantly, as in the Kawaikōī, birds released in the Halepa'akai began breeding in their first year in the wild. The breeding of released Puaiohi in the wild has generated an important success story for Ha-

waiian avian conservation—"a conservation home run" (*Honolulu Star Bulletin*, July 10, 1999)—and has become a model for other captive propagation and reintroduction efforts in Hawai'i. Moreover, the interagency collaboration that has led us to this point has been highly beneficial and productive. Although much remains to be learned concerning the impact of this strategy on the long-term recovery of Puaiohi, the apparent success of the releases represents a landmark method for rescuing critically rare forest bird species in Hawai'i.

RECOVERY ACTIONS AND OUTCOMES

The emergency plan for Puaiohi recovery initiated in the early 1990s has now matured. A formal revised recovery strategy has been developed in the *Revised Recovery Plan for Hawaiian Forest Birds* (hereafter "recovery plan"), which outlines recovery actions for 21 species of endangered or candidate forest birds, including the Puaiohi (U.S. Fish and Wildlife Service 2006). The recovery strategy for Puaiohi focuses on restoring habitat, especially mitigating alien predators and disease; augmenting the wild population through captive propagation and release; and conducting adaptive management and population modeling that will help us better manage the species in the future.

The Development of Recovery Goals

Throughout the recovery planning process, ornithologists have been challenged by two questions: What does recovery mean for a species that has never been common in historic times and for which native habitat is limited to a small plateau on a single island? And what is the target population's size and its distribution, and how do we know when we get there? The goals are to restore the population to a level that will allow the bird to persist despite demographic and environmental stochasticity, a level that is large enough to allow natural demographic and evolutionary processes to occur (U.S. Fish and Wildlife Service 2006). But how many Puaiohi are enough? After examining known densities of Puaiohi in optimal habitat (where they require 90 m of stream drainage for each pair [Snetsinger et al. 2005]) and the availability of streamside nesting habitat throughout the Alaka'i, the HFBRT settled on a goal of 2,000 adults in at least five subpopulations that constitute a single metapopulation. Furthermore, it was agreed that recovery would require populations to be stable or increasing (a population growth rate greater than or equal to 1.0 over a period of 30 years). Despite the fact that Puaiohi are restricted to a single island, the proven success with captive breeding provides some assurances that a captive population could be used as an "ark" should disaster strike the wild population (for example, another devastating hurricane or the introduction of a new avian disease to Kaua'i). Some biologists have suggested reintroducing Puaiohi to another island in the archipelago that has lost its native thrush (Chapter 25). To date, recovery goals and criteria have been set without the benefit of explicit population models. Such models are needed to refine the goals and confirm that they are adequate for long-term population persistence.

Habitat Protection and Restoration and the Threat of Invasive Species

As part of the recovery planning process, the HFBRT identified nearly 6,700 ha of montane forest habitat considered essential for the recovery of Puaiohi (see Fig. 22.1). This area includes all high-elevation montane wet forest remaining in the Alaka'i-Kōke'e region above approximately 1,000 m, as well as the Alaka'i Wilderness Preserve, portions of Kōke'e State Park, and private lands to the south deemed restorable. (As described in the recovery plan, "recovery area" is habitat that will allow

for the long-term survival and recovery of the species. It is not a legal designation but rather a guide for recovery planning.) Recovery would likely require that portions of Kōke'e State Park be reforested with native forest and protected. Likewise, the southern Alaka'i Plateau (about 1,417 ha), which is privately held, would be suitable for cooperative management agreements or purchase from willing sellers.

Habitat change due to the establishment of nonnative plants is a tremendous threat to the Alaka'i. Some areas have been severely impacted. Kāhili ginger (*Hedychium gardnerianum*) has taken over portions of the Kawaikōī stream drainage where Puaiohi have been released. Even the most remote areas have infestations of strawberry guava (*Psidium cattleianum*) and Australian tree fern (*Cyathea cooperi*), which could transform the landscape in a few decades. A relatively new weed, daisy fleabane (*Erigeron annuus*), grows well on steep surfaces such as the rock walls where Puaiohi make their nests and could pose a special threat to the Puaiohi. The Kaua'i Invasive Species Committee, a partnership of government, private, and nonprofit organizations and individuals, is working to control the most threatening invasive plant and animal species.

Rodent Control

One of the most important lessons of the past decade has been that rodents are an important factor limiting the survival and reproductive success of the Puaiohi. Protecting individual Puaiohi nests through baiting and trapping has proven difficult because of the time and funds needed to locate individual nests or lay effective rodent control grids in rugged, remote habitat and on the vertical cliff faces where the birds nest. A previous attempt to encourage Puaiohi to use predator-proof artificial nest boxes met with minimal success. However, a variety of new nest box designs hold promise and are currently in the early stages of testing in the Alaka'i. Although both these methods may assist recovery on a nest-by-nest basis, neither is likely to provide the landscape-scale protection required for full species recovery.

Large-scale rodent control in the Alaka'i might be achievable only through aerial broadcast. In 2008, aerial broadcast of specific rodenticide products for the protection of native species was approved by the Hawai'i Department of Agriculture and the U.S. Environmental Protection Agency (Chapter 18). However, this application may be feasible only in areas where secondary poisoning of feral pigs, an important game species, can be minimized, for example, in areas that are fenced and already free from pigs. There are currently no such areas of significant size in the Alaka'i (see the later section headed "Control of Feral Ungulates"). Further, community concerns about the impacts of rodenticides in the watershed will also need to be addressed. Although all of these issues will need to be resolved, the Alaka'i is considered a high-priority area for treatment.

Control of Avian Disease

No progress has yet been made to mitigate the potential impact of avian disease in Puaiohi habitat. This is due in part to the lack of quantitative data on its effects on Puaiohi; wild Puaiohi are notoriously difficult to capture, and regulations prohibit conducting challenge experiments with endangered species. Compounding this lack of information is a general sense that avian disease is extremely difficult to mitigate (Chapter 17). Throughout the world, vector control has proven one of the few effective controls for vector-borne diseases, including human malaria (Warrell and Gilles 2002). Thus, a recommended first step is to determine the sources of adult *Culex* mosquito vectors within primary forest. Once sources are identified, they may be eliminated or treated, with an emphasis on larval habitats windward and within 1.9 km of the Puaiohi recovery area

(U.S. Fish and Wildlife Service 2006). Pig control may be necessary to reduce the larval habitat they create within intact forest (U.S. Fish and Wildlife Service 2006; this volume, Chapter 17). Aside from vector control, we have few tools at our disposal to manage avian disease in wild populations (Chapter 17). Lack of funding remains the single greatest obstacle to conducting the disease research and vector control actions recommended in the recovery plan (U.S. Fish and Wildlife Service 2006).

Control of Feral Ungulates

Ungulate control in Hawai'i, as elsewhere, is logistically difficult, expensive to implement and maintain, and socially and politically contentious (U.S. Fish and Wildlife Service 2006; this volume, Chapter 16). In particular, in Hawai'i a highly influential hunting advocacy community is distrustful of government agencies imposing game control policies on public lands. Perhaps nowhere is this conflict more acute than in the high-elevation habitats on publicly owned lands on Kaua'i needed for Puaiohi recovery.

DOFAW recognizes the Alaka'i Wilderness Preserve as an area of the highest-quality native ecosystems and communities and mandates there the implementation of game control policies designed to reduce impacts on native resources. In accordance with its management guidelines, for several decades DOFAW has maintained somewhat liberal public hunting seasons to minimize the forest damage caused by feral pigs and goats (*Capra hircus*) within the Alaka'i Wilderness Preserve. Unfortunately, public hunting accomplishes only a moderate level of control (Chapter 16) and succeeds only in the more accessible areas of the preserve. Ungulates in remote areas remain numerous, and their damage to habitat is quite evident. Fencing and ungulate removal in remote areas are expensive and logistically difficult but also highly effective. Very limited aerial reconnaissance and aerial shooting of feral goats and pigs have been attempted in the most remote regions of the Alaka'i.

There is considerable support in the conservation community for fencing remote portions of the Alaka'i, and several small areas have been fenced to protect rare plants. With the formation of the public-private Kaua'i Watershed Alliance in 2007 and increased funding for this partnership, management actions have increased in the Alaka'i, and plans are now in place to fence and remove ungulates from large areas that may support Puaiohi. These actions represent a positive step in landscape-scale management, and it is hoped that additional funding and management will continue in the near future. Galvanizing support for this work has required a public relations and information effort based on a well-developed, technically feasible, and biologically justifiable plan with financially sufficient means for implementation. Similar support for continued and future work to control ungulates will be needed.

Captive Propagation and Reintroduction Program

In the years since the first Puaiohi egg was collected from the wild, the ZSSD's Puaiohi captive breeding and reintroduction program has matured from pilot research and development to full-scale production, providing birds every year for reintroduction to the wild. Simultaneously, the objectives of the work have shifted from establishing additional breeding populations to augmenting the existing wild metapopulation in light of the birds' high rate of dispersal from target sites and the perceived connectivity of existing subpopulations. Captive breeding and release is a potent tool that is well developed for the recovery of Puaiohi. The job now is to determine whether this tool remains a cost-effective recovery strategy and, if so, how best to apply it for maximum effect on recovery.

All parties agree that the captive release program should ideally be conducted in habitat where released birds will be relatively safe from disease and predators. Although rodent control is routinely applied to areas around the hacktowers before and during releases, the birds quickly disperse to breed in areas that we cannot protect (Tweed et al. 2003), and this fact brings us back to the question of large-scale rodent control, discussed earlier.

Research

Past research has contributed significantly to our understanding and recovery of the species but also points to the need for landscape-level treatments that are going to be expensive and potentially controversial. Population modeling taking into consideration the availability of nesting habitat and the impacts of stochastic events would build confidence in the expected outcomes of predator and ungulate control. Additional research into Puaiohi survival, dispersal, home ranges, and susceptibility to avian disease will further improve our modeling and predictive capability. Finally, experimenting by means of adaptive management with rodent control and habitat improvement by fencing and removing ungulates could yield positive results and demonstrate what is possible for Puaiohi recovery.

FUTURE DIRECTIONS AND CHALLENGES

Lack of sufficient funding is the fundamental obstacle to Puaiohi recovery. The HFBRT and the Puaiohi Working Group are now charged with building on the experience gained to develop and implement a broad-scale plan for the recovery of Puaiohi based on unstable and likely insufficient funding. Although funding for endangered species recovery is severely constrained the world over, Hawai'i suffers a more profound funding shortfall than comparable areas.

Leonard (Chapter 24) discusses this phenomenon and its social and political roots. This severe funding limitation makes it even more crucial that we accurately answer this question: What research and management options, or combination of options, will prove the most effective in recovering the Puaiohi?

One of the most important factors currently being evaluated is the future role of reintroductions in the overall recovery strategy for the species. In light of the size of the wild population, it may be more efficient to tackle limiting factors in the wild (primarily predation, nest site availability, habitat degradation, invasive vegetation, and disease). However, at this time and perhaps into the future, captive releases may be one of the few recovery tools we have until large-scale habitat management can be initiated.

The question of how much longer to continue the release program rests on an assessment of its success and cost-effectiveness relative to other strategies—specifically, how many released birds survive each year, how many of these produce young, and how release dollars might be spent in other ways. Unfortunately, such comparisons are difficult to make with limited demographic data on either wild or released birds. The present plan is to work toward a better assessment of the relative merits of these alternatives by beginning long-term demographic studies in select breeding areas, which can then provide study populations for the implementation of rodent control measures in an adaptive management context. Research is now in progress to document the long-term survival and breeding success of released birds. Our hope is that in the next five years we will obtain the information needed to develop robust population models and use these to compare the cost and efficacy of different types of management.

An analysis of the best use of limited recovery dollars will also require additional research on the relative importance of other methods to assist Puaiohi. That is, how effective could recovery dollars be if directed toward developing predator-proof nest boxes or controlling invasive species or disease vectors? Such research is expensive but necessary to ensure that funds are directed effectively in the long term. In the near future, in addition to gaining important insight into the demography of wild and released populations, DOFAW biologists will be testing varying designs of predator-proof structures that might be used to increase the species' reproductive success. In addition, recent concern about two other Kaua'i forest birds in decline—the 'Akikiki (Oreomystis bairdi) and 'Akeke'e (Loxops caeruleirostris)—has led to the first new disease surveys on Kaua'i in over a decade, currently under way, and the results can be expected to provide valuable information for Puaiohi recovery planning. It is hoped that these data will ultimately translate into increased availability of funding for the vector surveys and control recommended in the recovery plan (U.S. Fish and Wildlife Service 2006).

In the present sociopolitical climate, development of habitat restoration projects using aerial broadcasts of rodenticides faces challenging obstacles. Use of fences to control ungulates in the Alaka'i will continue to be contentious and difficult. The fencing and ungulate eradication work that is now planned by the Kaua'i Watershed Alliance will provide an opportunity to assess the response of habitat to ungulate removal within the Alaka'i. In addition, local and regional support for ungulate control elsewhere in Hawai'i has been rapidly building, and a number of lessons have been learned that may ultimately be applied on Kaua'i. Large-scale habitat management on public lands needs to be a major focus of recovery efforts over the coming decade. To achieve this will require continued research to develop effective broad-scale and cost-effective rodent and ungulate control, public education, and stakeholder involvement on Kaua'i.

SUMMARY

The Puaiohi has survived on the remote and rugged Alaka'i Plateau even as more common forest birds have dwindled and eventually disappeared, contradicting the expectation that rarer species are the most vulnerable to extinction. Its affinity for narrow stream valleys, in combination with a relatively high level of productivity, may have helped it weather two hurricanes. As recently as the mid-1990s, it was thought that only a handful of these birds remained, and an aggressive plan to develop a captive breeding and release methodology, in combination with field studies, was pursued with great success. Annual releases are now conducted to augment the wild population. In the meantime, surveys have shown that the Puaiohi population is larger than previously thought, numbering 300–500 birds. Nonetheless, the Puaiohi is among the rarest bird species in Hawai'i. It is threatened by predation, disease, invasive plant species, and habitat degradation, and its small population size and restricted range leave the population at continued risk from stochastic environmental or catastrophic events and loss of genetic diversity.

The strengths of the Puaiohi recovery effort to date have been the early recognition that the species was dwindling to perilously low numbers, the establishment of a step-by-step process to move the species toward recovery, the participation of a number of diverse agencies with complementary roles, and the dedication of significant (though limited) funds and staff time. All elements of the program have built on one another in a systematic way, each providing key feedback to guide the process. The program benefited because its timing coincided with the initiation of the full-scale captive propagation program in Hawai'i, without which most of its early goals could not have been accomplished, and we were fortunate that the species lent itself so well to captive propagation. Because of limited funding, however, tough decisions have had to be made on when to emphasize one aspect of the program over another and what questions to pursue in detail.

The benefit of hindsight shows that other lessons have been learned as well. The original three-pronged approach—survey, study, and captive breeding—was formed in an atmosphere of crisis. Captive breeding was chosen as the method of choice for many reasons, perhaps foremost among them the fear that "he who hesitates is lost." This approach is understandable, for endangered species have been known to go extinct quickly. The captive breeding program was continued because it was well under way and beginning to show success. Most important, perhaps, landscape-scale predator and ungulate control were logistically difficult, expensive, and politically controversial at the time, and regulatory issues had yet to be resolved.

Thus, the Puaiohi Recovery Project is in the midst of an important transition. At issue are which management options, or combination of options, will prove the most effective in recovering the Puaiohi for the long term. The most important tasks still before us include understanding the demographic impact of releases, ensuring sufficient suitable habitat and safe nest sites, and experimental implementation of broad-scale rodent control. The information gathered to date suggests that reducing rodent predation on Puaiohi over large areas would be highly beneficial to the species. There must be careful analysis of the costs and benefits of the ongoing reintroductions in light of the new population estimates for the wild birds and anticipated population gains from rodent control. And we must remain cognizant of the fact that the further incursion of avian malaria and avian pox virus into the Alaka'i or introduction of new alien species could quickly change the situation for the worse.

Perhaps the biggest challenge to Puaiohi recovery, and to forest bird recovery throughout Hawai'i, is to secure sufficient funding to effectively tackle the limiting factors at a landscape scale. We must solve the problem of how to implement large-scale habitat restoration on public lands in Hawai'i in light of the concerns and issues of a diverse group of stakeholders, including hunters, landowners, watershed residents, and natural resource managers. Getting beyond the present impasse will not be easy, but lessons from watershed partnerships elsewhere in Hawai'i may give us important insight into tools and processes. Without creative action on this front, the future of the Puaiohi will remain in jeopardy, and the rain forest of Kaua'i may lose yet another of its unique songsters.

ACKNOWLEDGMENTS

The Hawai'i Forest Bird Recovery Team and the original Puaiohi Working Group (P. Conry, S. Fancy, J. Jacobi, C. Kuehler, A. Lieberman, K. Rosa, T. Snetsinger, and T. Telfer) were the driving force behind the early recovery actions for Puaiohi and conceived and implemented the original recovery plan. Financial and logistical support for Puaiohi conservation and management was provided by the U.S. Fish and Wildlife Service, the Hawai'i Division of Forestry and Wildlife, The Peregrine Fund, the Zoological Society of San Diego, the U.S. Geological Survey Pacific Island Ecosystems Research Center, and the Pacific Cooperative Studies Unit at the University of Hawai'i at Mānoa. The research on Puaiohi ecology, captive breeding, and reintroduction summarized here was conducted by (in addition to the authors and in alphabetical order) G. Beauprez, T. Casey, S. Fancy, J. Foster, T. Goltz, C. Hermann, J. Kellerman, C. Kuehler, J. Kuhn, M. Kuhn, D. Leonard, W. Monahan, P. Oesterle, R. Pratt, M. Reynolds, P. Roberts, E. Rose, T. Savre, T. Snetsinger, E. Tweed, and many dedicated volunteers. M. Gorresen kindly produced the map in Fig. 22.1. The chapter was improved by comments from S. Fancy, J. Innes, and T. Snetsinger.

CHAPTER TWENTY-THREE

Palila

PAUL C. BANKO, KEVIN W. BRINCK,
CHRIS FARMER, AND STEVEN C. HESS

The Palila (*Loxioides bailleui*) (Plate 4) is a Hawaiian honeycreeper species best known for its specialized seed-eating habit and the ambitious, sometimes contentious, campaign to prolong its existence. The essence of Palila evolution and ecology is obvious to all who see these birds energetically ripping open seed pods of the māmane tree (*Sophora chrysophylla*) high on the slopes of Mauna Kea on the island of Hawai'i. Subalpine Mauna Kea has become the last bastion of the species, but this environment and Palila populations have been battered by waves of invasive species and human activity since Western contact over two centuries ago. However, three decades of ecological research indicate that recovery is possible and, more important, provide a comprehensive, scientific basis for restoring Palila habitats and populations. At the same time, Palila conservation has sparked a local land use controversy and national debate about how to interpret the Endangered Species Act of 1973, as amended (U.S. Congress 1973).

For all of the research, management, and legal focus on the Palila, the long-term fate of this endangered species continues to be tenuous. Reestablishing Palila populations throughout their historical range and restoring their dry woodland habitats seem daunting for biological reasons and because resources are urgently needed for many pressing environmental problems in the Hawaiian Islands. Nevertheless, there is more cause for optimism than for pessimism in considering the Palila's future. The research described in this account has contributed greatly to Palila conservation, but significant management challenges remain.

ECOLOGICAL DIMENSIONS OF PALILA RESTORATION

Like most other endangered Hawaiian forest birds, Palila are feeding specialists that have low reproductive capacity, inhabit a small range that has become restricted historically to higher elevations, and live in small, unstable, or declining populations (Chapters 2, 8). Other than Palila, only two closely related species on the tiny islands of Nihoa and Laysan remain of the once numerous prehistoric seed-eating honeycreeper species (Morin and Conant 2002). The Palila, Laysan Finch (*Telespiza cantans*), and Nihoa Finch (*T. ultima*) are also the last species distributed exclusively in dry forest or shrub habitats. Because the populations of most Hawaiian forest birds are extinct or rare at low elevations, the Palila and other species have been studied only in the uppermost portions of their historical ranges. Moreover, what we know about the ecology of Hawaiian forest birds is distorted by the extinction or decline of many plants and animals that influenced the evolution of bird species, as well as by current interactions (usually negative) with many invasive species. We can only speculate about Palila ecology before humans changed Hawaiian ecosystems; however, Palila occurred prehistorically at low elevations on Oʻahu and Kauaʻi in habitats that were quite different from the subalpine woodlands of Hawaiʻi Island (Olson and James 1982a, Burney et al. 2001).

Food Specialization

At the root of the Palila's vulnerability to extinction is its dietary dependence on the unhardened seeds of the māmane (Banko and Banko 2006). Although the seeds, which are protected inside tough, fibrous pods, are high in protein and lipids, they also are loaded with secondary compounds, particularly some alkaloids that are quite toxic to most vertebrates (Banko, Cipollini, et al. 2002). The seed chemistry may vary sufficiently among trees to influence Palila foraging behavior, because Palila harvest more pods from some trees than from others of similar size and similar standing crops. Nevertheless, Palila seldom remain long in any particular tree before flying off in search of more pods to sample. The habit of endlessly sampling pods, although still not completely understood, indicates more than any other behavior that Palila require more trees in their home range than would seem needed based just on the quantity of the pods available. An extensive forest with many māmane trees seems necessary, therefore, to sustain even a small population of birds.

Māmane offer more than just seeds to Palila, although seeds are the primary attraction. Flower buds, flower parts, whole immature pods, leaf buds, and developing leaves are all eaten. Caterpillars found mainly on māmane are important in the diets of nestlings and fledglings and are common foods of adults (Banko, Johnson, et al. 2002). Palila glean some caterpillars from māmane foliage, but their chief prey, *Cydia plicata* (Tortricidae), are found only inside māmane pods, where they feed and develop on the maturing seeds. One advantage of eating *Cydia* and feeding them to nestlings is that the caterpillars do not sequester the alkaloids that are found abundantly in māmane seeds (Banko, Cipollini, et al. 2002). Additionally, the growth demands of nestlings suggest that caterpillar availability could affect the distribution and demography of Palila and other native forest birds (Perkins 1903; this volume, Chapter 7). The implication for Palila conservation is that habitats must provide an abundance of both caterpillars and māmane pods. Nevertheless, there may be potential value for Palila in alternative foods, such as the fruits, leaves, and flowers of naio (*Myoporum sandwichense*), ʻiliahi or sandalwood (*Santalum paniculatum*), and at least eight other native trees, shrubs, and vines occurring in their habitat. Increasing resources, therefore, should be a key con-

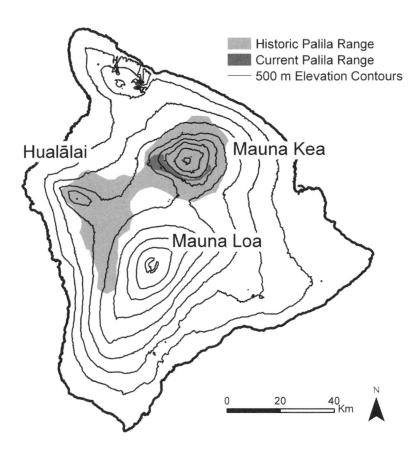

Figure 23.1. Historical distribution of Palila. Palila occurred where māmane dominated large areas and broad
ranges of elevation on Mauna Kea, Hualālai, and Mauna
Loa volcanoes. The present population is concentrated
on western Mauna Kea, where the māmane forest is
most extensive.

servation goal even for this extreme feeding specialist.

Habitat Ecology and Threats

Historically, Palila flourished where
māmane dominated the landscape along
a broad range of elevations (Fig. 23.1).
Most māmane trees tend to flower earlier at
high elevations, and then a slow-moving
wave of flowering rolls downhill. Seed
set follows the wave of flowers, so Palila
track their primary food by moving slowly
from higher to lower slopes over time
(Hess et al. 2001; Banko, Oboyski, et al.

2002). However, seeds became available
to Palila over less of the year as their habitat was truncated, primarily by cattle (*Bos
taurus*) on the lower slopes and by feral
sheep (*Ovis aries*), mouflon sheep (*O. gmelini
musimon*), and feral goats (*Capra hircus*) at
upper elevations. Presumably, a threshold
is reached below which Palila are unable
to find enough seeds year-round and populations cannot be sustained. This relationship between the range of forest elevations
and population viability helps explain why
the Palila population declined everywhere
except on the western slope of Mauna Kea,
where more than 95% of all Palila live and
māmane range from 2,000 to 3,000 m in
elevation (Scott et al. 1984). A key to Palila
restoration, then, is to remove ungulates
and allow habitat recovery above and below the present limits of the forest.

Removing ungulates from Palila habitat
promotes tree recruitment by increasing

seedling and sapling survival (Hess et al. 1999). Twenty years after recent efforts to reduce ungulate populations began in 1980, we observed māmane regeneration on 89% of plots (n = 504; 40 × 40 m) in Mauna Kea Forest Reserve, although only 39% of plots supported seedlings and saplings at relatively high densities (≥100 individuals per ha). In contrast, in adjacent pasturelands, young māmane were found on 42% of plots (n = 84; 20 × 20 m), and only 23% of plots supported high densities (P. Banko, unpubl. data). Small māmane trees tend not to be used by Palila, and it may be several decades before saplings grow sufficiently large to benefit the bird population (Scowcroft and Giffin 1983, Scott et al. 1984). Eventually, increased tree density should benefit Palila because more pods would be available over a longer time span. Although the timing of flowering and seed set is influenced by elevation, reproduction among māmane at the same elevation is not highly synchronized, resulting in extended pod availability (Banko, Oboyski, et al. 2002). Another bonus of reduced browsing is that lower branches of trees soon grow back, promoting tree vigor and reproduction (Scowcroft and Giffin 1983). Palila benefit because they harvest pods from all over the tree, including near ground level (Banko, Johnson, et al. 2002).

Besides the pervasive damage to Palila habitat from ungulates, exotic weeds increase fire risk and suppress the growth and regeneration of māmane and other native plants. About 70% of the 179 species of vascular plants found on recent surveys on Mauna Kea were alien, including 15 highly invasive species (P. Banko, unpubl. data). Several species of smothering vines and a competitive shrub, gorse (Ulex europaeus), were recorded in or near the forest reserve, but the greatest threats to native vegetation are from introduced grasses, which accumulate fine fuels capable of supporting hot, fast-moving fires (Hughes et al. 1991). The recent discovery of an introduced root fungus, Armillaria mellea, which is pathogenic to māmane, indicates yet another potentially serious threat to Palila habitat (Gardner and Trujillo 2001).

Native arthropods are often overlooked as components of forest bird habitats, but Cydia plicata caterpillars are important prey of Palila. Cydia and other caterpillar prey are potentially threatened by introduced predacious ants and wasps but more significantly by alien and native parasitoid wasps, whose larvae slowly eat the caterpillars alive (hence, the wasps are designated parasitoids rather than parasites, which typically do not kill their hosts) (Banko, Oboyski, et al. 2002). Because the impacts of alien parasitoid species tend to diminish as the elevation increases, Cydia caterpillars are more abundant and less frequently parasitized above 2,000 m in elevation (Brenner et al. 2002, Oboyski et al. 2004) (Fig. 23.2). Palila densities are greatest at higher elevations, where both Cydia and māmane pods are more abundant (Banko, Oboyski, et al. 2002). Although the effect of caterpillar parasitism on Palila populations is difficult to isolate, it may nonetheless be important.

Breeding and Demography

Dependent as Palila are on māmane seeds, their nesting behavior, survival, and population dynamics are linked to the availability of pods. In years when māmane trees produce large pod crops, many pairs nest and eggs may be laid over an eight-month period (van Riper 1980b; Pratt, Banko, et al. 1997; Banko, Johnson, et al. 2002) (Fig. 23.3). However, when pod crops are small, as is the case during droughts, very few pairs nest and the breeding season is short (Lindsey, Pratt, et al. 1997). Regardless of pod crop size, each breeding pair typically produces only two eggs and one or two fledglings per nest. Palila sometimes nest again after their first nest attempt fails or even after the first brood fledges. As is the case with other specialist species, their

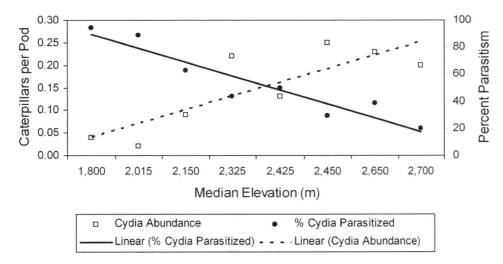

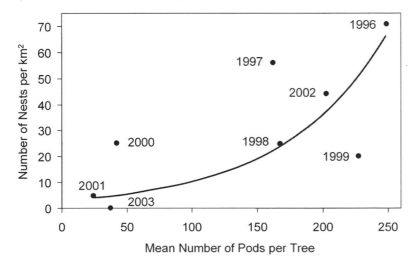

Figure 23.2. Variation in *Cydia plicata* caterpillar abundance and parasitism as a function of elevation. As the elevation increases, the caterpillars inside māmane pods (dashed line) increase in abundance ($R^2 = 0.70$; $t = 3.74$; $P = 0.0096$), while the mortality of caterpillars from parasitoid wasps (solid line) decreases ($R^2 = 0.88$; $t = -6.75$; $P = 0.0005$). Therefore, Palila encounter more caterpillars in the pods they open at higher elevations. Palila tend to be more abundant at higher elevations, perhaps due in part to the greater abundance of these caterpillars. *Source:* Data are from Brenner et al. (2002).

nesting cycle is long: incubation typically lasts for 16–17 days, nestlings leave the nest after about 25 days, and offspring associate with their parents for 3–4 months (Banko, Johnson, et al. 2002). In contrast,

the Hawai'i 'Amakihi (*Hemignathus virens*), a generalist forager ubiquitous in the Palila's habitat, typically produces more eggs and completes its nesting cycle in just 75% of the time required by Palila, usually resulting in two broods per season (van Riper 1987, Lindsey et al. 1998).

Figure 23.3. The relationship between māmane pod abundance and number of Palila nests. The exponential regression curve indicates that the relationship between pod abundance (mean number of pods per tree) and nest density (active nests per km²) is annually variable. This finding suggests that other factors may also influence nesting behavior in some years (e.g., 1997, 1999, and 2000). Data were collected during the breeding season (April–August) on the western slope of Mauna Kea.

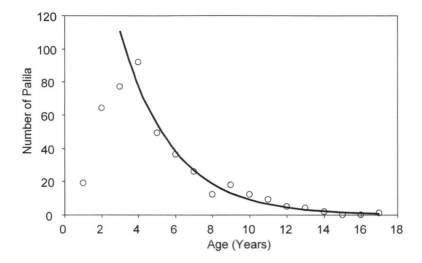

Figure 23.4. Age distribution and number of Palila surviving annually. Data are from birds that were captured more than once and more than a year apart. Most individuals were 2–4 years of age, but a few birds lived very long lives. The total number of birds from ages 2 to 15 was used in a regression to estimate the annual rate of survival of an adult Palila as 0.697 (90% CI = 0.677–0.718). This curve was fitted to the age distribution to estimate the annual rate of survival after the juvenile period, when the mortality rate is assumed to be higher due to the birds' inexperience (see also Lindsey et al. 1995). The probability of a bird being recaptured in a mist net was assumed to be random and independent in terms of avoiding or becoming attracted to nets after being caught initially. Nets were not set at random; instead, they were set where birds seemed common.

Although Palila live as long as 17 years, most are 2–4 years of age when caught in mist nets (Fig. 23.4). There is no significant age-related bias in the sex distribution, although males outnumber females in most age categories (P. Banko, unpubl. data). Females often nest at 1 year of age, whereas males usually delay breeding until their second year. Therefore, most birds apparently have only a few years to nest before they die, although some live and breed for many years. Given their relatively low rate of reproduction, strategies for restoring Palila should strive to maximize survival. Annual survival of adult Palila over

a four-year period declined with the proportion of māmane trees that were barren of green pods at the beginning (April) and end (September) of the breeding season (Lindsey et al. 1995). Although preliminary analyses of additional data do not confirm a significant relationship between survival and pod availability, protecting māmane habitat is central to Palila recovery.

Despite the importance of māmane to Palila ecology, estimates of the Palila population indicate that there is a high degree of annual variability that is not strongly related to estimates of māmane seed availability or other environmental factors (Scott et al. 1984, Jacobi et al. 1996). Annual surveys of the Palila population since 1980 provide the longest monitoring record for any Hawaiian forest bird species, and the results are potentially important for evaluating ecological relationships and conservation actions. The variable circular plot method has been used in all years, but the analysis techniques have been refined and improved over time (Johnson et al. 2006). The long-term population trend, adjusted to provide comparable estimates throughout the monitoring period, indicates that there was a population decline into the mid-1980s, followed by an uneven increase to apparent stability at about 5,000 indi-

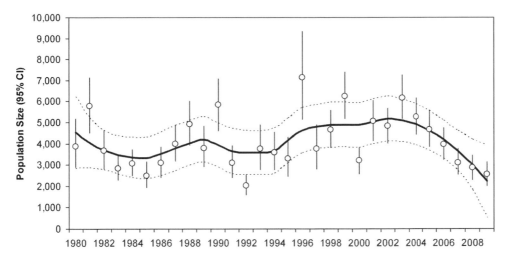

Figure 23.5. Trends in Palila population estimates over 30 years. Annual population estimates (indicated by circles with 95% confidence interval bars) were nearly always obtained during January–February (prior to the nesting season) by the variable circular plot method (Scott et al. 1986). The smooth trend line was obtained by fitting a generalized additive model to the population estimates. The dashed lines represent a confidence envelope of approximately 95% around that curve. The population estimates declined into the mid-1980s, then increased unevenly to a level of apparent stability (at about 5,000 individuals) from the mid-1990s until 2003, when they began a steady decline.

viduals from the mid-1990s to the early 2000s, when the population began a steady decline (Fig. 23.5).

As noted earlier, māmane regeneration increased following the reduction of ungulate populations after 1980. The response of the Palila population to increased māmane has lagged because the birds tend to congregate near tall trees and seldom forage in saplings (P. Banko, unpubl. data). Around the mountain, we found that māmane tree height decreased with increasing elevation (linear regression, $t_{6117} = 16.01$, $P < 0.0001$). During annual surveys (usually in January), Palila were congregated in the most pod-rich trees, which at that time of year were in the upper half of their habitat on the western slope. The survey data indicated that there was greater Palila abundance at sites with greater māmane cover ($n = 285$ stations, $t_{283} = 7.91$, $P < 0.0001$). Nevertheless, as the new cohort of young, short trees continues to grow, more resources will become available to Palila.

Predators and Disease

Reducing the impacts of alien predators would likely extend the lifespan of Palila and increase their productivity, both of which are essential to recover the species. Conceivably, long exposure at nests increases the vulnerability of Palila females and chicks to predators, and relatively low reproductive capacity exacerbates the demographic impacts of predation. Additionally, nestling feces accumulate later in the rearing period, possibly attracting predators (van Riper 1980b). Pletschet and Kelly (1990) reported that only 4% of Palila clutches were destroyed by predators, whereas 21% of broods were depredated. Moreover, predation on adults can be particularly damaging to populations in which breeding is delayed, as it is for Palila (Pratt, Banko, et al. 1997).

Argentine ants (*Linepithema humile*) and other ant species are encroaching on Palila habitat, possibly threatening eggs, nestlings, and nesting adults (Sockman 1997, At-wood and Bontrager 2001, Peterson et al. 2004). However, Palila are more likely at risk from four alien mammal species, whose threats vary with their body size (relative to that of Palila), abundance, and ability to climb trees. In contrast to wet forest communities, where nest depredation by rats is most frequent, feral cats (*Felis catus*) have the greatest impacts in the dry habitat of Palila (Hess et al. 2004; this volume, Chapter 11). Despite their abundance and ability to scale trees, mice (*Mus musculus*) are not considered important predators of eggs or chicks (Amarasekare 1993). Black rats (*Rattus rattus*), although relatively uncommon in the Palila's habitat, are strong climbers and occasionally prey on eggs and chicks (Pletschet and Kelly 1990; Amarasekare 1993, 1994). Small Indian mongooses (*Herpestes auropunctatus*) infrequently climb trees (Hays and Conant 2007) and probably do not represent a significant threat, although Palila occasionally nest low enough for them to reach the nests (Banko, Johnson, et al. 2002).

The threat to Palila from feral cats may have been underestimated by some studies due to the cats' cryptic and primarily nocturnal behavior. Remote video surveillance of 23 nests documented nestling depredation by feral cats on three occasions during the day, when cats are typically less active (Laut et al. 2003, Hess et al. 2004). No other predators were recorded by those cameras. Most cat predation involves nestlings; however, van Riper (1980b) observed a cat killing an adult female on a nest. Of particular relevance to Palila conservation has been research demonstrating the degree to which Palila benefit when feral cats are removed from their habitat. Of 37 active nests monitored in 1999 prior to feral cat removal, forensic evidence of hair, feathers, or scratch marks implicated predation by cats at 4 (11%) nests (Hess et al. 2004). In contrast, no evidence of predation was recorded at 25 nests that were monitored in 2002, when feral cats had been removed from Palila nesting habitat.

Though still vulnerable to introduced predators, Palila are no longer affected by avian malaria (*Plasmodium relictum*) and avian pox virus (*Avipoxvirus* sp.) because their current habitat is above the range of disease-vectoring mosquitoes. However, Palila are sensitive to these diseases (van Riper et al. 1982), which probably played a role in the historical decline of the species in lower elevations of their range. A century ago, many birds may have entered the mosquito zone as they tracked the seasonal availability of māmane seeds from high to low elevations. Today, there is virtually no suitable Palila habitat below about 1,500 m in elevation. However, present or future diseases would likely hinder attempts to reintroduce Palila to restored tracts of mid-elevation forest. Moreover, global warming may allow mosquito-borne diseases to spread upslope, posing serious conservation challenges (Benning et al. 2002).

Movements and Behavior

Palila are not territorial and tend not to nest at the same site from year to year, yet they do not disperse widely or wander far around Mauna Kea (Banko, Johnson, et al. 2002). Palila traverse the landscape on flights that possibly range for hundreds of meters or more. The short-term (two-month) home range of birds fitted with radio transmitters is about 3 km², and Palila are rarely observed more than 2 km from the site where they were first captured and marked with leg bands (Fancy et al. 1993). Although Palila apparently wander occasionally, their relatively sedentary nature suggests that they may be slow to naturally reoccupy former range and that populations must be actively reintroduced to distant areas of improving habitat.

Our knowledge of Palila social behavior is sketchy, but we should not assume that it is irrelevant to restoring populations.

Birds are found in small, loosely associated groups during the nonbreeding season, but prior to nesting, birds congregate where māmane pods are abundant (Banko, Johnson, et al. 2002). Pre-breeding assemblages, although not highly concentrated, are lively as birds sing and chase one another in courtship. After the nesting season, recently independent juveniles associate in small, wandering flocks in which they may develop important social skills, form pair bonds, and learn critical information about their physical environment.

Palila are monogamous during the year and often across multiple years (Banko, Johnson, et al. 2002), and extrapair copulation was not detected in 12 families surveyed using DNA fingerprinting (Fleischer et al. 1994). However, sometimes extra males (usually at least two years of age) and rarely females defend nests and feed females, nestlings, and fledglings (van Riper 1980b; Miller 1998; Banko, Johnson, et al. 2002; Patch-Highfill 2008). Helpers are likely to have been offspring of one or both parents (Patch-Highfill 2008), and Miller (1998) found that nests with helpers were usually successful (eight of nine nests). These and other aspects of social behavior should be considered when planning restoration strategies.

RESTORATION AND MANAGEMENT

The wealth of ecological knowledge available about the Palila provides a foundation for restoration that is unmatched for any other Hawaiian forest bird. Palila have been the target of research largely due to their accessibility, relatively high numbers, and perceived vulnerability to extinction. Moreover, land use conflicts have lent urgency to the need for information to guide management. Following the completion of the Hawai'i Forest Bird Survey by Scott et al. (1986), federal biologists began Palila research to initiate a program of focused

ecological studies of endangered forest birds (the program is now with U.S. Geological Survey [USGS]). By starting with a species that was logistically tractable, it was thought that approaches and methods could be developed with applicability to other, more problematic, species. The research encompassed limiting factors, particularly predation and disease; breeding and demography; food ecology; movements and home range; and experimental population reintroduction (Pratt, Banko, et al. 1997).

USGS studies were not tied to management mandates until 1996, when partner agencies sponsored research to develop mitigation for a roadway that was proposed to pass through Palila critical habitat, a circumstance in which Palila protection supersedes other proposed activities and development. Funding constraints and a wide spectrum of conservation priorities were two reasons that Palila research and management were not integrated before this regulatory mandate. Nevertheless, just as lawsuits had been filed to protect Palila habitat from ungulates (discussed later), regulatory action to mitigate the loss of critical habitat was a stimulus for Palila recovery activities and for long-anticipated studies of vegetation and fire ecology that had been all but overlooked in earlier research.

The U.S. Fish and Wildlife Service Recovery Plan

Endangered species recovery plans prescribe a mix of research, management, and regulatory action intended to reverse population decline and avoid extinction (Clark et al. 2002). Identifying threats and recommending research and management to alleviate them are essential to species recovery, and success is more likely when research and management programs are integrated and historical dimensions of the species' decline are understood (Banko et al. 2001). Two fundamental themes characterize recovery plans for Palila and other Hawaiian forest bird species: (1) rehabilitation or protection of existing populations and habitats that are threatened by a variety of environmental factors and conflicting human interests and (2) reintroduction of populations to restored habitats in species' former range (U.S. Fish and Wildlife Service 1986, 2006). It has long been understood that the Palila is in danger of extinction primarily because loss of habitat has stranded the species in a small portion of its historical range (van Riper et al. 1978, Scott et al. 1984, Banko 1986). The concentrated Palila population is highly vulnerable to both short- and long-term threats. Large annual fluctuations in population estimates suggested that Palila were demographically unstable (Jacobi et al. 1996). Fire has been considered the greatest threat to the species' core habitat; however, extinction could result from a combination of ungulate browsing, weed invasion, predation, and food depletion by alien pests (U.S. Fish and Wildlife Service 2006). Diseases would threaten Palila at lower elevations and could spread to higher elevations as temperatures increase as a result of climate change (Chapter 9).

Threats posed by predation, disease, food depletion, and habitat destruction are challenging to understand and mitigate under ordinary circumstances, but conflicts with human interests, such as sport hunting and grazing, can further complicate species recovery (Juvik and Juvik 1984). Protecting habitat from the effects of introduced game animals is a key element of Palila recovery (U.S. Fish and Wildlife Service 1986, 2006). The Palila is one of only two endangered Hawaiian birds for which critical habitat has been designated, and managing Palila critical habitat has had implications for national endangered species policy and environmental law. The Palila has been the subject of legal contests on six occasions, setting or upholding one of the most important precedents in environ-

mental law in each case (see Houck 2004 for case summaries). Starting in 1979, the central argument in each case has been whether feral goats and sheep (Palila cases 1 and 2) and later mouflon sheep, either as a pure strain or hybridized with feral sheep (Palila cases 3 and 4), harm the Palila by damaging the māmane habitat upon which it depends. Later arguments (Palila cases 5 and 6) were decided in 1999 and upheld the earlier decisions. Through 20 years of contention, the regulatory and legal implications of designating critical habitat under the Endangered Species Act and removing ungulates from Palila critical habitat have been extensively argued. The protracted series of court battles, however, has led to a feeling of disenfranchisement on the part of those land managers who believe that ungulates are necessary to reduce fire fuel loading and are committed to a multiple-use philosophy of land management by which endangered species recovery and sustained-yield sport hunting can be seen as compatible (Juvik and Juvik 1984, Hawai'i Division of Forestry and Wildlife 1997).

Ungulate Removal

Although protecting habitat from ungulates is a high-priority recovery action (U.S. Fish and Wildlife Service 1986, 2006), progress in removing animals has been slow and uneven. Following federal court rulings in 1979 and 1986, public hunting, guided hunting, dog-assisted drives, and aerial shooting were used to remove feral ungulates from Palila critical habitat. Although few or no feral sheep now remain, hybrid sheep persist. During aerial hunts in 1999–2008, marksmen attempted to remove all animals observed from helicopters for two days twice each year. Public hunters were restricted to hunting on foot during nearly continuous open seasons. Data provided by the Hawai'i Division of Forestry and Wildlife (DOFAW) show that although the number of ani-

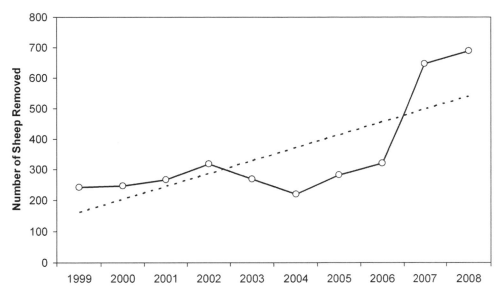

Figure 23.6. Removal of hybrid mouflon sheep from Mauna Kea. Shown is the trend in numbers of hybrid sheep (*Ovis gmelini musimon* × *O. aries*) removed by aerial shooting from Palila critical habitat on Mauna Kea, Hawai'i, beginning in 1999 (the first full year of hunting) through 2008. Data are from the Hawai'i Division of Forestry and Wildlife. Regression analysis (dashed line, $t_8 = 3.15$, $P = 0.014$, $R^2 = 0.55$) indicates that the numbers of animals removed annually by aerial shooting (constant effort) increased, which suggests that the population may also be increasing. An unknown number of hybrids likely jump over or pass through fencing that separates Palila critical habitat from the lands below. Feral goats and pure strains of feral sheep (not mouflon sheep hybrids) apparently have been eliminated from Palila critical habitat.

mals removed annually during 1999–2006 ranged from 220 to 320, the numbers surged to 688 in 2008 (Fig. 23.6). Aerial removals averaged 350.3 ± 53.9 SE animals per year over the ten-year period. The number of animals removed annually by helicopter increased significantly in 1999–2008, with an increment of 41.9 ± 13.3 SE animals per year. Public hunters also seem to have taken increasing numbers of animals in recent years.

Because helicopter hunting has been conducted in a standardized manner, these data indicate that the population of hybrid sheep on Mauna Kea may be increasing rather than declining. To increase the rate of removal, new tactics and improved fencing are warranted. For example, the 70-year-old perimeter fence around Mauna Kea Forest Reserve is permeable to hybrid sheep and mouflon, which are spreading across the interior of the island. Although planning has begun to replace a portion of the old, dilapidated fence, funding to secure the entire perimeter has not yet been identified. Modern attempts at ungulate removal stand in stark contrast to the early forest protection program, when the forest reserve was fenced during 1935–1937 (Bryan 1937a) and nearly 47,000 feral sheep and more than 2,200 other ungulates were removed in the following ten years by territorial foresters and Civilian Conservation Corps workers on foot and horseback (Bryan 1937b, 1947).

Saddle Road Mitigation

Road construction—not recovery plans or court orders—has most recently propelled Palila restoration. In December 1999, a proposal to realign Saddle Road (Highway 200), the interior connection between east and west Hawai'i, through Palila critical habitat was approved on the condition that the loss of habitat would be mitigated

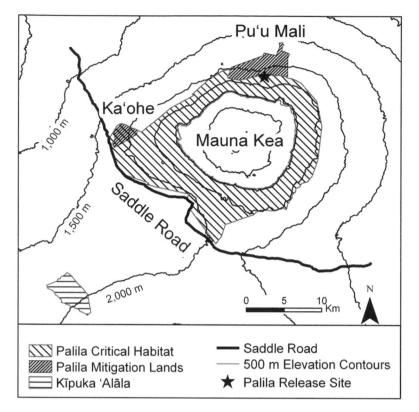

Figure 23.7. Saddle Road mitigation lands and critical habitat. To mitigate the realignment of Saddle Road through Palila critical habitat, cattle grazing was suspended on lands below Mauna Kea Forest Reserve at the Pu'u Mali (northern slope) and Ka'ohe (western slope) leases. Palila were reintroduced to the forest reserve above Pu'u Mali. Forest restoration on the mitigation lands might eventually allow this colony to expand. Palila from the main population on the western slope are expected to naturally reoccupy Ka'ohe as pastures are converted to forest. Following ungulate removal and habitat recovery, it may be possible to reintroduce Palila to Kīpuka 'Alalā in Pōhakuloa Training Area on northwestern Mauna Loa.

(U.S. Fish and Wildlife Service 1998, Federal Highway Administration 1999). A key component of the Saddle Road mitigation was to temporarily suspend cattle grazing leases to allow māmane regeneration in pastures adjacent to Palila range on the western slope (Ka'ohe) and northern slope (Pu'u Mali) of Mauna Kea (Fig. 23.7). An additional area at Kīpuka 'Alalā on the northwestern slope of Mauna Loa was pro-

tected from military training activities and feral ungulates to allow former Palila habitat to recover. The Saddle Road mitigation provided for the reintroduction of Palila at Pu'u Mali, where the species had last been recorded in 1968 (Walker 1968), but it was anticipated that Palila would naturally reoccupy the Ka'ohe parcel from adjacent areas as habitat conditions improved following the removal of cattle. Kīpuka 'Alalā, on the other hand, was isolated from the Mauna Kea Palila population and was severely damaged by ungulates. Many years of habitat recovery seemed necessary before reintroducing Palila to Kīpuka 'Alalā, where they were last seen in 1950 (Banko 1986). Moreover, the species' habitat extended over a relatively narrow range of elevations at Kīpuka 'Alalā, whereas forest recovery at Ka'ohe and Pu'u Mali would extend Palila habitat to lower elevations.

Temporarily withdrawing lands on Mauna Kea from grazing, protecting habitat on Mauna Loa, and experimentally reintroducing Palila to Puʻu Mali have offered conservation agencies new opportunities for testing the feasibility of Palila recovery, as outlined in the *Revised Recovery Plan for Hawaiian Forest Birds* (U.S. Fish and Wildlife Service 2006). Another important feature of the Saddle Road mitigation was to integrate conservation research and management by (1) expanding or continuing earlier studies of Palila ecology and limiting factors, (2) testing predator management techniques to protect Palila populations, and (3) using an adaptive management approach to reintroduce Palila to their former range. Fire ecology and management studies are also supported by the Saddle Road mitigation, and DOFAW is sponsoring research to develop habitat restoration strategies and techniques. Nevertheless, all of these conservation actions would not have been possible without the designation of critical habitat and the unprecedented commitment to research prior to mitigation planning.

To reduce the risk of extinction and develop techniques of potential value to other restoration efforts, Palila have been experimentally reintroduced to the Puʻu Mali area of northern Mauna Kea (see Fig. 23.7) using two methods: translocating wild Palila from the western slope of Mauna Kea and releasing captive-reared birds. Palila were translocated in six trials in the fall, winter, or spring of 1997–1999 and again during 2004–2006, following a five-year hiatus when access across the mitigation lands to the release site was unavailable. In total, 188 Palila were captured in the mornings in the vicinity of Puʻu Lāʻau, transported by vehicle to Puʻu Mali, and released in the afternoons. Early trials included monitoring control birds on the western slope, and we targeted young Palila based on the results of a promising pilot translocation of adult Palila to the eastern slope of Mauna Kea in 1993 (Fancy et al. 1997). However, young Palila persisted at the release site no longer than did older birds, and in later trials we translocated birds of any age. We

moved potential pairs (males and females captured together or otherwise believed to be associated) whenever possible; however, we were unable to incorporate other social considerations into the translocations. After eight weeks (the typical battery life of radio transmitters), 34% of 173 radio-tagged, translocated birds remained in the vicinity of Puʻu Mali, 21% returned to the western slope (some only temporarily), 30% were incompletely monitored due to radio failure, 14% were killed or scavenged by predators, and 1% (2 birds) died from trauma associated with translocation (Fig. 23.8). Fancy et al. (1997), who held Palila for about a day in a common aviary before release, found that 40% (14/35) of translocated birds remained on the eastern slope after eight weeks, a result that was similar to ours ($\chi^2 = 1.36, P = 0.24$). The overall mortality rate of our translocated birds (14.5% [25/173]) was not significantly different from that of control birds (7.7% [3/39]; $\chi^2 = 1.96, P = 0.16$).

Estimating the persistence of birds at Puʻu Mali has been complicated by the fact that at least 10% (6/62) of the Palila translocated in Trial 5 "commuted" between the northern and western slopes on at least one occasion each before their radio transmitters expired. One bird flew back and forth twice before his transmitter expired but seemed to have accepted Puʻu Mali as his primary residence. Additional support for commuting behavior comes from two birds from Trial 4. These birds were not detected in the release area for five to six months during intensive monitoring, but they later reappeared to successfully breed on the northern slope. The distance from the western capture site to the northern release site was about 16 km, which some birds traveled in a day. The movement of individuals between slopes could help to sustain the numbers and increase the genetic diversity of the small colony at Puʻu Mali. Moreover, if a disaster were to strike the western slope, a few returned translocated birds might

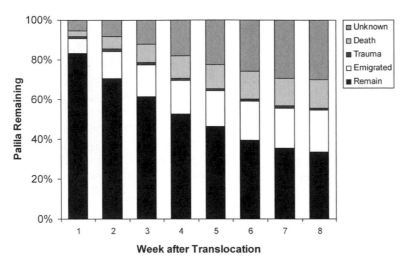

decamp to Puʻu Mali. Even more surprising than the commuters was the occasional appearance on the northern slope of unbanded birds. We do not know whether commuters perhaps influenced other Palila to travel to Puʻu Mali or whether these birds traveled on their own; however, they remained with the small colony of translocated birds for at least several weeks.

Not until 2005 had the population of Palila at Puʻu Mali consisted of birds from more than one translocation trial. This is likely due to the relatively small number of birds translocated during Trials 1–3 and the five-year gap between Trial 3 and Trial 4. During Trials 5 and 6, the Puʻu Mali population contained individuals from Trials 4 and 5. Also contributing to the new population were approximately 6 of 21 Palila that were raised in captivity at the Keauhou Bird Conservation Center and released near Puʻu Mali in December 2003, 2004, and 2005. Breeding attempts between birds from different trials and sources suggest that this small colony could even-

tually become viable. During 2004–2007, 7 of 12 clutches produced by translocated pairs produced 8 fledglings, most of which survived to independence (=3 months). Three additional fledglings were produced by two different mixed pairs (a translocated female and captive-reared male). Captive-reared pairs produced two clutches, but no eggs hatched.

Although the monitoring effort at Puʻu Mali was drastically reduced after mid-2007, at least two successful nesting attempts were detected there in 2008. The most noteworthy nest involved a translocated male and a locally produced female that was the offspring of translocated parents. This pair produced a fledgling, demonstrating that translocation could be a viable method for creating a self-sustaining population of Palila. The other 2008 nest involved a captive-reared male and either a translocated or a locally produced female that together produced a juvenile. In late 2008, the Puʻu Mali population was estimated to consist of 11–21 Palila.

This small breeding nucleus of wild and captive-reared Palila might grow if predator removal and habitat restoration can be continued beyond the limits of the Saddle Road mitigation and perhaps if additional birds are added. Translocating additional wild Palila, preferably adults, may be the most effective option because their level of breeding success is relatively high, the mortality rate associated with translocation is relatively low, individuals readily return to the source population if they do not stay in the new area (meaning that they are not lost from the metapopulation), the breeding of individuals that return to the source population is not deterred by translocation (at least in some cases), and the connection between the new and source populations is strengthened by commuters.

The Future for Recovery Efforts

The population decline that began in 2003 underscores the urgency for additional Palila restoration efforts, despite decades of focused research and management (Leonard et al. 2008). The Saddle Road mitigation has provided critical impetus to implementing and evaluating key elements of the recovery plan for Palila. Nevertheless, mitigation by itself does not fulfill all the requirements designated in the recovery plan, particularly because the mitigation will terminate in 2012. Furthermore, the five-year interruption (2000–2004) in translocation trials and the delay in removing cattle from the mitigation parcels until 2006–2009 reduced the effective implementation of mitigation. The challenge to Hawaii's conservation agencies, therefore, is to strategically plan and develop lines of funding for follow-up recovery action before the conservation easements on grazing lands expire. It can be expected that recovery planning would be sponsored by the Hawai'i Forest Bird Recovery Team.

As called for in the recovery plan (U.S. Fish and Wildlife Service 2006), a major goal of Palila recovery should be to convert degraded landscapes into expansive māmane-dominated habitats ranging broadly in elevation and supporting a diverse community of native plants. Habitat restoration depends upon removing ungulates throughout the Palila's historic range, but removal strategies and tactics could be improved. Ungulate removal also depends on the interests of ranchers and sport hunters and on the value that state and private landowners assign to restoring native ecosystems. Moreover, planting trees and controlling weeds may be needed, because ungulate removal alone may not be sufficient to rapidly transform highly degraded habitats, such as pasturelands, into habitats capable of sustaining Palila and other native birds. Other important goals mentioned in the recovery plan are to prevent grass-fire cycles, suppress highly invasive weeds and pests, and restore native vegetation. To protect populations of Palila, feral cats and other alien predators should be removed, which could also facilitate the restoration of other forest birds and seabirds (U.S. Fish and Wildlife Service 2006). Achieving these goals will require close partnerships between research and management as well as innovative and strategic approaches to program funding and implementation. For example, encouraging public participation in restoration efforts may generate results in the field and popular support for this species.

Although restoration must be focused on Mauna Kea in the short term, a more comprehensive strategy emerges when two sites on northwestern Mauna Loa are included: Kīpuka 'Alalā and the uplands of Kona. Furthermore, the proclivity of many translocated birds to return home suggests a mechanism by which natural movements could link populations on the two volcanoes, especially if habitat can be restored at Kīpuka 'Alalā.

The past may also provide clues for enhancing recovery. Palila that prehistorically inhabited lowland areas on O'ahu and Kaua'i encountered environments very different from their subalpine habitat of today. Restoring native plant and arthropod communities may allow Palila to survive on caterpillars and plant species that are presently uncommon, such as 'iliahi, when māmane seeds are scarce.

SUMMARY

As the last seed specialist in the main Hawaiian islands, the Palila reminds us of the formerly rich and widespread assemblage of finch-billed Hawaiian honeycreepers, which has all but disappeared since ancient Hawaiian times. Palila recovery depends on understanding—from biochemical to landscape levels—the ramifications of the bird's dependency on māmane seeds. Māmane seed availability affects such critical ecological variables as the length of the nesting season, the number of nests per season, bird survival, and the species' movements and home range. Loss of māmane habitat due to ungulates has endangered the Palila, but feral cats and other predators are also important threats. Because Palila occur above the range of mosquitoes, disease is not currently a problem; however, Palila restoration at lower elevations may be hindered by disease.

The recovery plan emphasizes the importance of protecting māmane habitat from ungulates, weeds, and fire, and it recommends reintroducing Palila to rehabilitated areas of their former range to increase the number of populations. Integrated research and management have demonstrated the feasibility of reintroducing Palila to their former range and the value of removing alien predators to protect the birds' populations. Research to rehabilitate habitats and to protect native vegetation from fire has begun, and we hope that these results will spur complementary management. Threats to arthropod prey from alien parasitoids and predators have been identified, but reducing their impacts will require additional research and experimental trials. The Palila population has been surveyed annually since 1980, with the results providing a useful basis for understanding and predicting population responses to changing environmental conditions and management actions.

The Saddle Road mitigation has stimulated Palila restoration on Mauna Kea, although meeting the long-term goals of species recovery requires additional commitment. However, inspiration for achieving the recovery goals can be found in the forest protection program of the 1930s and 1940s, which resulted in the near extermination of the feral ungulates that today continue to degrade Palila habitat. Removing forever the threat of browsing in Palila habitat could powerfully stimulate vital actions for recovering additional native bird species.

ACKNOWLEDGMENTS

This research was made possible thanks to support from the Federal Highway Administration, U.S. Army Garrison, Hawai'i; Hawai'i Division of Forestry and Wildlife; and U.S. Geological Survey wildlife research programs. For permits and access to study sites, we are grateful to the U.S. Fish and Wildlife Service (USFWS), Hawai'i Division of Forestry and Wildlife (DOFAW), Pōhakuloa Training Area (U.S. Army), KK Ranch, SC Ranch, Boteilho Ranch, and Parker Ranch. We especially thank R. David (Rana Productions), J. Nelson (USFWS), and S. Fretz and D. Leonard (DOFAW) for facilitating our work related to the Saddle Road mitigation.

We are deeply indebted and immensely thankful to the dozens of staff and many hundreds of volunteers who have carried out work in the field and laboratory since 1987. Biologists at Kīlauea Field Station who contributed greatly to our understanding of Palila biology prior to 1996 are S. Fancy, J. Jacobi, J. Jeffrey, G. Lindsey, S. Pletschet, T. Pratt, and R. Sugihara. Since 1996, many others have contributed substantially, especially those who worked on the project for two or more years: R. Danner, S. Dougill, D. Goltz, J. Higashino, L. Johnson, M. Laut, J. Leialoha, B. Muffler, C. Murray, P. Oboyski, D. Pollock, K. Rapozo, M. Schwarzfeld, J. Semones, J. Slotterback, and M. Wiley. F. Benevides provided valuable technical assistance with surveillance cameras at Palila nests. Palila were reared in captivity and released to the wild by staff of the Zoological Society of San Diego at the Keauhou Bird Conservation Center. We thank A. Lieberman, M. Morin, and two anonymous reviewers for helpful suggestions to improve the chapter.

PART 5

The Future

CHAPTER TWENTY-FOUR

Social and Political Obstacles to Saving Hawaiian Birds

Realities and Remedies

DAVID L. LEONARD JR.

The conservation status of many species of Hawaiian forest birds is so dire that they deserve immediate conservation attention. Losing more of these birds is unthinkable, yet imminent. The preceding chapters outlined the biology of some of the best-known species and the many threats they face. Unfortunately, in addition to the biological threats to Hawaiian forest birds, their recovery also is hampered by social and political barriers or threats.

A lack of adequate conservation funding is the primary nonbiological threat and is ultimately related to a lack of awareness of and exposure to the extinction crisis that is occurring in Hawai'i (Leonard 2008). The reasons this extinction crisis goes unappreciated are multifaceted. Ultimately, however, lack of awareness and funding limitations are tied to the somewhat unique biological, demographic, and geographic characteristics of Hawai'i, often in interesting and unexpected ways. Perhaps the most poignant circumstance relates to exotic bird diseases (Chapters 9 and 17).

Comparing bird conservation efforts in Hawai'i with those in various island nations reveals the special challenges faced in Hawai'i. In New Zealand, observing native birds and the benefits of conservation efforts requires only a boat ride to a restored offshore island or a trip to a mainland island preserve (Towns 2002, Towns and Broome 2003). There, in a setting of restored native forests, native birds thrive. By contrast, in Hawai'i, disease and historic habitat loss restrict almost all native forest birds to high-elevation forests that are difficult, if not impossible, for the public to access. Hawaii's isolation, its small size in terms of both human population and land area, and the state's disconnect from the U.S. mainland certainly contribute to the lack of needed attention and funds. Inadequate funding is related to political realities tied to the characteristics just mentioned and, although this may seem counterintuitive, to the minimal economic conflict associated with recovering Hawaiian forest birds, as explained later (Leonard 2008).

Increased recovery funding generally improves the likelihood that a listed species or subspecies (hereafter species) will be recovered and removed from the endangered species list (Miller et al. 2002, Male and Bean 2005). However, despite a biologically based system to prioritize recovery expenditures, resulting in priority rankings (Box 24.1), charismatic species or those whose recovery conflicts with economic development often receive more

---

Box 24.1. How the U.S. Fish and Wildlife Service Ranks Endangered Species for Recovery and Funding

The U.S. Fish and Wildlife Service assigns each taxon listed as endangered or threatened a recovery priority number ranging from 1 to 18. Rankings are based on three criteria related to the threats, recovery potential, and genetic distinctiveness of the listed taxa. Taxa with a rank of 1 are those that experience a high degree of threat, have a high recovery potential, and belong to a monotypic genus, and thus should warrant the highest priority for funding allocation. Taxa with a rank of 18 are those that experience a low degree of threat, have a low recovery potential, and belong to a subspecies. In addition to this numerical ranking, taxa whose recovery may result in economic conflict, such as stopping or postponing development, have a "c" associated with their priority rank; these taxa are considered higher priority than those with the same rank but without a potential economic conflict (U.S. Government Accounting Office 2005a). The priority numbering system as applied to Hawaiian forest bird and mainland bird taxa and the numbers of birds of each priority rank as of 2004 are shown below.

Degree of Threat	Recovery Potential	Taxonomic Distinctiveness	Priority Rank	Number of Mainland Bird Taxa	Number of Hawaiian Forest Birds
High	High	Monotypic genus	1	0	2
		Species	2	8	3
		Subspecies	3	16	1
	Low	Monotypic genus	4	1	3
		Species	5	4	6
		Subspecies	6	4	1
Moderate	High	Monotypic genus	7	0	1
		Species	8	2	5
		Subspecies	9	7	0
	Low	Monotypic genus	10	0	0
		Species	11	0	0
		Subspecies	12	0	0
Low	High	Monotypic genus	13	0	0
		Species	14	1	1
		Subspecies	15	0	0
	Low	Monotypic genus	16	0	0
		Species	17	1	0
		Subspecies	18	0	0

Box 24.2. **Watershed Partnerships**

The first Watershed Partnership was established in 1991 on the island of Maui to ensure the conservation of future water supplies. As of October 2006, nine watershed partnerships had been established on all the major Hawaiian islands (Plates 30–32). Forty-five private landowners manage 166,769 ha, and 24 public agencies manage 287,472 ha of public lands. The benefits of managing these lands collaboratively include the ability to address common threats across large landscapes and the pooling of expertise and limited funding. Fortunately, virtually all native Hawaiian ecological community types are found in watershed partnerships. Most management actions are habitat based and focus on removing or controlling ungulates and invasive plants. These actions not only benefit native forests and birds but, by reducing erosion and sedimentation, protect coastal and coral reef areas as well.

---

funds than would be predicted based on their ranking (U.S. Fish and Wildlife Service 1983c; Restani and Marzluff 2001, 2002a). The result is that a few species receive the bulk of recovery funding (Restani and Marzluff 2001, Male and Bean 2005). Those with small geographic ranges, especially island species, are typically greatly underfunded relative to their priority rank (Restani and Marzluff 2002a, Male and Bean 2005).

Island birds comprise a disproportionately high percentage of the birds covered under the Endangered Species Act. Between 1996 and 2004, 52% (50 of 96 species) of the federally listed birds occurred on islands, with 31 of those species endemic to Hawai'i. Despite the threats facing island birds (Johnson and Stattersfield 1990, Steadman 2006) and the high costs of mitigating these threats (Burnett et al. 2006, Martins et al. 2006), far less funding has been provided for Hawaiian birds than for mainland birds (Leonard 2008).

The threats facing Hawaiian birds are monumental and will require many times the amount of funding currently allocated to their recovery. However, there are successes on which greater progress can be built. Relatively new public-private watershed partnerships are protecting large tracts of forest bird habitat (Box 24.2), landscape-scale restoration techniques are successfully transforming former pastures into native forests (Box 24.3), opportunities for public viewing of Hawaiian forest birds are improving, and recently it has been determined that at least one endangered forest bird, the 'Akiapōlā'au (for scientific names of endangered birds, please refer to Tables 24.1 and 24.2), uses young plantations of a valuable native timber tree (Pejchar et al. 2005). Although Hawaii's small human population of 1.3 million limits the amount of funding that can be generated from mechanisms such as tax checkoffs or the sale of wildlife license plates, between 1997 and 2006 an average of 8.1 million tourists visited the islands annually (Hawai'i Department of Business, Economic Development and Tourism 2006), and tourists represent an untapped source of conservation funding.

In this chapter I (1) outline the funding situation of Hawaiian forest birds; (2) review the reasons funding is limited; (3) briefly discuss successful recovery and habitat restoration projects from which greater conservation benefits can be derived, especially those that have engaged the public; and (4) offer solutions for overcoming social and political barriers that are inhibiting funding and review mechanisms likely to be successful in generating

Box 24.3. Hakalau Forest National Wildlife Refuge

The 13,355 ha Hakalau Forest National Wildlife Refuge (NWR) was created in 1985 to protect rain forest that supports endangered forest birds. Located on the windward slope of Mauna Kea Volcano on Hawai'i Island, the refuge contains some of the best remaining examples of native rain forest in the state. However, at the time of purchase, there was approximately 1,620 ha of once forested pasture on these lands. In 1989, refuge staff began restoring habitat. After fencing and removing ungulates from most of the refuge, the restoration efforts expanded to include control of exotic plants and the planting of more than 400,000 native trees, mostly koa trees, and also 5,000 endangered plants. Much of the work has been done by volunteers ranging from school groups to off-island service groups.

The Hakalau Forest NWR is one of the few places in the state where native forest bird populations are stable or increasing. Annual forest bird surveys indicate that the number of the endangered 'Akiapōlā'au is increasing. In 2003, refuge staff observed a juvenile 'Akiapōlā'au within former pasturelands. The chick was heard calling from an area planted with koa trees only eight years prior. Now 'Akiapōlā'au family groups are observed in planted groves and corridors of koa on a regular basis. Annual surveys show that the numbers of Hawai'i 'Amakihi (Hemignathus virens) and 'Apapane (Himatione s. sanguinea) also have increased in koa reforestation areas. In addition to native forest birds, more than 200 Nēnē, endangered Hawaiian Geese, live in the refuge. Twenty-nine rare plant species, a dozen of which are either proposed for or already on the Endangered Species List, also can be found in the refuge. In 2006, the Friends of Hakalau Forest National Wildlife Refuge was established to help the U.S. Fish and Wildlife Service better protect and manage this unique refuge for some of Hawaii's rarest and most beautiful birds, plants, and invertebrates.

---

increased funding for conservation in the fiftieth state.

FUNDING NEEDED AND AVAILABLE

As detailed in previous chapters, almost all of Hawaii's extant endangered passerines (perching birds) are restricted to high-elevation forests where mosquitoes are thermally limited or to small remote islands where these disease vectors are absent. (Some native forest birds are developing a tolerance to avian malaria; therefore, it is also important to protect low-elevation forests for birds, as explained in Chapters 9 and 17.) These habitats present numerous logistical, management, and research challenges. Unlike in the case of listed mainland birds, controlling nonnative predators, plants, and ungulates is requisite for recovery of the species covered in this book. Unfortunately, controlling or eradicating these threats is expensive and requires ongoing efforts and funding (Ikuma et al. 2002; Burnett et al. 2006; Martins et al. 2006; this volume, Chapter 15).

The remoteness of the Northwestern Hawaiian Islands increases the cost of management activities there. For example, nine years and $1.2 million were required to eradicate the southern sandbur (Cenchrus echinatus) from Laysan Atoll (411 ha) (Flint and Rehkemper 2002). On the main Hawaiian Islands, depending on the location and ruggedness of the area, standard ungulate fencing costs $40,000 to $105,000 per km to purchase and install (Hawai'i Division of Forestry and Wildlife, unpubl. data). Installation of fencing is step one; once in place, fences must be periodically checked and repaired, ungulates eradicated, habitats restored, and predators controlled. Ungulate eradication typically costs between $12,000 and $15,000 per km^2 (J. Jeffrey, U.S. Fish and Wildlife Service, pers. comm.). Although aerial broadcast of rodenticide has recently been permitted in Hawai'i, its cost, public perception, and restrictive protocols will limit its widespread application in forests for the near term. Thus rat control is being conducted using snap traps and poison bait stations. Controlling rats in high-elevation wet forests using these methods costs approximately $32,500 per km^2 per year (K. Swinnerton, Island Conservation, pers. comm.). Even if aerial rodenticide drops are eventually possible, these applications will have to be repeated in perpetuity because rodents reinvade treated areas. Predator-proof fences, soon to be tried in Hawai'i (Hawai'i Department of Land and Natural Resources 2007), offer the advantage of excluding all mammals but are many times more expensive than standard ungulate fencing (Day and MacGibbon 2001). Nevertheless, predator-proof fences may be worth the investment in certain situations (Engeman et al. 2003).

The U.S. Fish and Wildlife Service's *Revised Recovery Plan for Hawaiian Forest Birds*—a comprehensive plan that took into account nearly all conceivable recovery actions without limitations for feasibility or funding—estimated the cost of recovering 18 endangered Hawaiian forest passerines at $2.5 billion over 30 years, or $83,333,333 per year (U.S. Fish and Wildlife Service 2006). This estimate does not include the 'Io, 'Alalā, Nihoa Millerbird, Laysan Finch, or Nihoa Finch. The *Revised Recovery Plan for the 'Alalā* estimated the cost of recovering the crow at $2.8 million annually for 2009–2013. With these figures in mind, how much funding is currently dedicated to the recovery of endangered birds in Hawai'i?

Of the total state and federal funds expended toward the recovery of all 96 birds listed in the United States between 1996 and 2004, the 31 Hawaiian species received 4.1% ($30,592,692) (Leonard 2008). The 23 listed Hawaiian forest birds covered in this book received $18,550,591, or approximately 2.5% of all funding available, which amounts to far less than the amount needed annually for 18 endangered Hawaiian forest birds ($2,061,176 vs. $83,333,333). Over the same period, expenditures for the 20 highest-funded birds in the United States totaled $619,458,533 (Table 24.1), or 85% of the funds available for all 96 listed birds. Although no Hawaiian bird was included in this top 20, the Palila ranked number 21. However, for perspective, between 1996 and 2004 the Red-cockaded Woodpecker received $100,854,262—five times the funds provided for all listed Hawaiian forest birds combined. This amount of money was spent for a widely distributed species with relatively easily implemented management needs. Between 1996 and 2004, despite small geographic ranges and populations, endangered Hawaiian forest birds received average annual expenditures ranging from $5,889 (Nihoa Finch and Nihoa Millerbird) to $1,029,735 (Palila). The mean for all forest bird species was $92,935 ± 210,610 (SD) per species (Table 24.2).

Among birds that breed in the United States, only two birds have a recovery priority rank of 1, and both are endemic to Hawai'i—the Maui Parrotbill (Box 24.4) and the Palila. The 'Alalā is the only Hawaiian bird with an economic conflict tied to its recovery; this conflict is related to cattle ranching and koa (*Acacia koa*) harvesting. Compared with mainland birds, more endangered Hawaiian forest birds are full species (more endangered mainland birds are subspecies), all are endangered rather than merely threatened, and on average Hawaiian forest birds have been listed longer

Table 24.1. Summary of the 20 highest-funded birds listed under the Endangered Species Act

Species	T/E	Year Listed	Priority	Population Size	Total Spending (× 1,000)
Red-cockaded Woodpecker (*Picoides borealis*)	E	1970	8c	>15,000	$100,854
Northern Spotted Owl (*Strix occidentalis caurina*)	T	1990	3	>8,000	$75,908
Southwestern Willow Flycatcher (*Empidonax traillii extimus*)	E	1995	3c	1,800–2,200	$63,563
Bald Eagle (*Haliaeetus leucocephalus*)	T	1967	14c	>15,000	$61,547
Marbled Murrelet (*Brachyramphus marmoratus*)	T	1992	3	~17,000[a]	$55,632
Whooping Crane (*Grus americana*)	E	1967	2c	>400	$42,414
Mexican Spotted Owl (*Strix o. lucida*)	T	1993	9c	>2,000	$31,372
Piping Plover (*Charadrius melodus*)	E, T	1985	2c	>6,400	$29,018
Cape Sable Seaside Sparrow (*Ammodramus maritimus mirabilis*)	E	1967	3c	~3,000	$19,991
Least Tern (*Sterna antillarum*), interior populations	E	1985	3c	>4,700	$17,287
Black-capped Vireo (*Vireo atricapillus*)	E	1987	2c	8,000	$16,553
Western Snowy Plover (*Charadrius alexandrinus nivosus*)	T	1993	3c	<4,000	$16,493
Golden-cheeked Warbler (*Dendroica chrysoparia*)	E	1990	2c	10,000–32,000	$15,685
California Gnatcatcher (*Polioptila californica*)	T	1993	3c	6,000	$14,782
San Clemente Loggerhead Shrike (*Lanius ludovicianus mearnsi*)	E	1977	9	135	$14,257
Least Bell's Vireo (*Vireo bellii pusillus*)	E	1986	3c	>2,000	$12,257
California Condor (*Gymnogyps californianus*)	E	1967	4c	289	$12,133
Puerto Rican Parrot (*Amazona vittata*)	E	1967	2	200	$11,742
California Least Tern (*Sterna a. browni*)	E	1970	3c	>14,000	$9,904
Wood Stork (*Mycteria americana*)	E	1984	5	~12,000	$9,806

Source: Excerpted from Leonard (2008: Table 3).

Notes: The table includes the birds' status (T, threatened, or E, endangered), listing year, funding priority (see Box 24.1), and estimated population, along with the total spending (federal and state) for each species for the period 1996–2004 (www.fws.gov/endangered/pubs/index.html). The total spending for these species over nine years was $619,458,533.

[a]California, Oregon, and Washington populations.

than mainland birds (Table 24.3). Despite these differences and the large disparity in funding, the mean priority rankings of the two groups are similar, suggesting that funding is not tied to need.

For all birds listed under the Endangered Species Act, there is no correlation between priority ranks and recovery expenditures (Leonard 2008). However, when only Hawaiian forest birds are considered, this correlation approaches significance (Fig. 24.1). On average, birds with an economic conflict associated with their recovery receive four times more funding than those without a conflict, even though there is no difference in the priority rank of the two groups (Leonard 2008). Although the Palila does not have an economic conflict associated with its recovery, it does exemplify how conflict results in a higher level of funding (discussed later).

In general, there is a positive correlation between the time a bird has been listed and the likelihood that its status will improve (Taylor et al. 2005), and the proportion of taxa that are stable or improving peaks 12–13 years after listing (Male and Bean 2005). Most Hawaiian forest birds

Table 24.2. Summary of Hawaiian forest birds listed under the U.S. Endangered Species Act

Species	T/E	Year Listed	Priority	Population Size or (year last seen)	Annual Mean Spending (SD) ($ × 1,000)	Total Spending ($ × 1,000)
'Io (Hawaiian Hawk, *Buteo solitarius*)	E	1967	14	3,000	60.3 (44.0)	543
Kaua'i 'Ō'ō (*Moho braccatus*)	E	1967	4	(1987)	7.8 (7.7)	70
'Alalā (Hawaiian Crow, *Corvus hawaiiensis*)	E	1967	2C	60	193.0 (82.0)	1,737
O'ahu 'Elepaio (*Chasiempis sandwichensis ibidis*)	E	2000	3	2,000	171.7 (60.4)	859
Millerbird (*Acrocephalus familiaris*)	E	1967	8	<400	6.0 (6.7)	54
Kāma'o (*Myadestes myadestinus*)	E	1970	5	(1985)	8.8 (7.3)	80
Oloma'o (*Myadestes lanaiensis*)	E	1970	5	(1980)	7.9 (7.1)	71
Puaiohi (*Myadestes palmeri*)	E	1967	2	300–500	65.2 (16.9)	587
Laysan Finch (*Telespiza cantans*)	E	1967	8	10,000	9.9 (12.0)	89
Nihoa Finch (*Telespiza ultima*)	E	1967	8	3,000	5.9 (6.5)	53
'Ō'ū (*Psittirostra psittacea*)	E	1967	4	(1989)	10.9 (5.3)	98
Palila (*Loxioides bailleui*)	E	1967	1	3,900	1,029.7 (1,056.3)	9,268
Maui Parrotbill (*Pseudonestor xanthophrys*)	E	1967	1	500	98.4 (40.6)	886
Kaua'i 'Akialoa (*Hemignathus procerus*)	E	1967	5	(1969)	7.4 (7.3)	67
Nukupu'u (*Hemignathus lucidus*)	E	1967	5	(1996)	63.3 (46.4)	570
'Akiapōlā'au (*Hemignathus munroi*)	E	1967	2	1,900	57.5 (54.6)	518
Hawai'i Creeper (*Oreomystis mana*)	E	1975	8	14,000	55.4 (59.7)	499
O'ahu 'Alauahio (*Paroreomyza maculata*)	E	1970	5	(1978)	21.4 (20.7)	193
Kāwāwahie (*Paroreomyza flammea*)	E	1970	5	(1963)	8.2 (6.9)	74
Hawai'i 'Ākepa (*Loxops c. coccineus*)	E	1970	8	12,000	43.5 (30.1)	392
Maui 'Ākepa (*Loxops c. ochraceus*)	E	1970	6	(1980)	47.2 (25.4)	424
'Ākohekohe (*Palmeria dolei*)	E	1967	7	3,800	49.0 (16.8)	441
Po'o-uli (*Melamprosops phaeosoma*)	E	1975	4	(2004)	108.7 (36.0)	978

Source: Excerpted from Leonard (2008: Table 4).

Note: The table includes the birds' status (T, threatened, or E, endangered), listing year, funding priority (see Box 24.1), and estimated population. Annual mean (standard deviation) and total spending refer to federal and state spending for each species for the nine-year period 1996–2004.

have been listed since 1967 (see Table 24.3), yet most are still very vulnerable, and many have declined or their status is unknown (U.S. Fish and Wildlife 2006; this volume, Chapter 5). Miller et al. (2002) reported that increased funding improves the likelihood of recovery, but as threats include more complex biological and political factors, additional funding may not contribute to recovery. Fortunately, there are techniques to combat many alien species in Hawai'i, but given the logistical difficulty of Hawaiian habitats and the need for ongoing efforts, applying them is expensive (Burnett et al. 2006, Martins et al. 2006). Furthermore, the development of new technologies or techniques to control alien species is often expensive (Markin et al. 2002; this volume, Chapter 15). Given the threats facing Hawaii's forest birds and the expense of mitigating these threats, one would think that the recovery funding for Hawaiian forest birds would be larger.

Box 24.4.   The Maui Parrotbill: A High-Priority Species in Need of Funding

The Maui Parrotbill is listed as endangered and is one of two birds given a priority rank of 1 by the U.S. Fish and Wildlife Service. The entire population, an estimated 500 individuals, is restricted to less than 50 km^2 of high-elevation rain forest on northeastern Maui, where it is limited by diseases and historic habitat loss (Simon et al. 1997, U.S. Fish and Wildlife Service 2006). The Maui Parrotbill is characterized by very low fecundity; the clutch size is one egg, and pairs produce a maximum of one fledgling per year (Simon et al. 1997). Unfortunately, the rate of nest success is low, with storms often causing nests to fail (Hawai'i Division of Forestry and Wildlife, unpubl. data). Other than the species' low rates of fecundity and nest success, little is known about factors limiting its population.

Fossil evidence indicates that the species historically occurred in drier forests on the leeward side of the island, where extreme weather is less frequent than in their current range (Simon et al. 1997). Parrotbills obtain most of their insect food by chewing, ripping, and crushing small branches. Historical evidence also suggests that the species utilized extensive mesic koa forests now lost to ungulates. It is anticipated that parrotbills may respond to koa reforestation in the same way that the ecologically similar 'Akiapōlā'au does on Hawai'i Island. Fencing and restoration of degraded mesic koa forest is under way, and a captive flock is being built with the goal of reintroducing the species to leeward Maui. Although the species has the highest possible recovery ranking and despite the high costs of fencing, ungulate removal, and habitat restoration, between 1996 and 2004 the mean annual spending on the Maui Parrotbill was less than $100,000 per year.

Table 24.3. Comparative statistics for mainland and Hawaiian forest birds listed under the Endangered Species Act as of 2004

	All Mainland Birds	Hawaiian Forest Birds
Number of Listed Taxa	44	23
Listed Subspecies[a]	16 (36%)	2 (15%)
Threatened	14 (31.8%)	0 (0.0%)
Endangered	30 (68.2%)	23 (100.0%)
Years Listed[b]	25 ± 11	34 ± 7
Priority Ranking[c]	5.1 ± 3.4 (2–17)	5.2 ± 3.0 (1–14)
Number of "c" Taxa[d]	26 (57.8%)	1 (4.4%)

[a]Number of listed subspecies versus species; $X^2 = 8.18$, df = 1, P = 0.004.

[b]Mean (± SD) years listed as threatened or endangered; t-test = 3.81, df = 65, P < 0.001.

[c]Mean (± SD) USFWS priority rank (1–18; see text for details); t-test = 0.20, df = 65, P = 0.84.

[d]A "c" appended to a taxon's ranking indicates a potential for economic conflict with recovery (see Box 24.1); $X^2 = 9.94$, df = 1, P = 0.002.

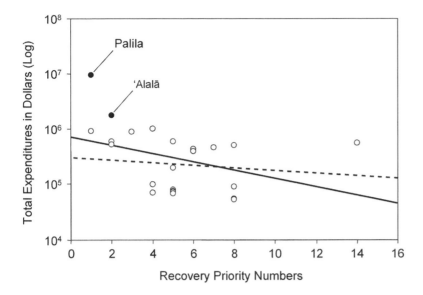

Figure 24.1. Total expenditures for Hawaiian forest birds plotted as a function of each species' recovery priority number (low numbers are meant to receive higher priority). The relationship between expenditure and recovery priority is weak (solid line; Pearson's correlation = −0.38, P = 0.08), particularly when the two highest-funded Hawaiian birds, the Palila and 'Alalā, are removed (dashed line; Pearson's correlation = −0.27, P = 0.24).

WHY THE DISPARITY IN FUNDING?

Hawaiian Birds Are Isolated from People

Why does Hawai'i not receive more funding for the conservation of its endangered endemic forest birds? Most important, there is a lack of awareness of or exposure to Hawaii's avifauna, driven by isolation at two geographic scales. Because of Hawaii's isolation from the rest of the United States, most mainland nongovernmental organizations (NGOs), residents, and birdwatchers generally have little or no acquaintance with Hawaiian birds or their conservation status. However, the very recent State of the Birds report (North American Bird Conservation Initiative 2009) and publications by the American Bird Conservancy (ABC) indicate that awareness of Hawaiian birds and their conservation needs is increasing.

Hawaiian birds are not included in any recent North American field guides or other bird books (Box 24.5), although they were included in the Birds of North America series (Poole and Gill 1992–2002). Nor are they on the American Birding Association (ABA) Checklist. Thus, at the national level, the vast majority of American birdwatchers are not exposed to Hawaii's avifauna and are without the means to develop a sense of engagement with or responsibility for the fate of these birds.

North American bird conservation publications and grant programs often fail to mention Hawaiian forest birds. For example, the 2007 WatchList for United States Birds published in the 107th Christmas Bird Count issue of American Birds did not include Hawaiian birds (National Audubon Society 2007). Given the wide circulation of this publication, this lack of inclusion represented a huge lost opportunity. A recent request for proposals from the National Fish and Wildlife Foundation's (NFWF) Bird Conservation Initiative (www. nfwf.org/AM/Template.cfm?Section=Bird _Conservation2, accessed May 11, 2008) targeted ten species or taxonomic groups; however, no Hawaiian birds or groups were included, although NFWF and ABC are now actively engaged in Hawai'i and

---

Box 24.5.   **Hawaiian Birds Belong in American Field Guides**

A general appreciation of birds is the first step toward public support for and ultimately the success of conservation efforts to save them. Unfortunately, most of Hawaii's endemic birds are unfamiliar to U.S. mainland birdwatchers, NGOs, and government officials. Why is this so? A contributing factor to this lack of recognition and knowledge is the fact that endemic Hawaiian birds are not included in the field guides used by mainland U.S. birdwatchers. One reason for this exclusion is that the ABA Checklist (edition 7.0; 11/08) encompasses only continental North America and offshore islands, does not consider Hawai'i within its geographic scope, and therefore does not list Hawaiian species absent from the U.S. mainland. Therefore, missing from the ABA Checklist are 86 Hawaiian species composed of 41 native species, 34 alien species, 8 Pacific seabirds, and 3 Palearc-

tic migrants (Pyle 2002). Already on the list are 209 other species amounting to 71% of the total 295 species recorded from Hawai'i—mostly north temperate migrants, seabirds, and some alien species.

One would think that it would not matter to the ABA how many birds they have on their list. However, it would concern the authors and publishers of field guides because of all the work necessary to add species accounts, expand and reconfigure plates, and draft new range maps. At present American field guides feature birds from all states but the fiftieth state. The inclusion of Hawaiian birds in national field guides would contribute to the conservation of these unique American species. Greater awareness of Hawaiian birds among all the people of the United States might lead to increased efforts to save them.

---

will likely be providing long-term funding toward Hawaiian forest bird conservation. Hawaiian birds are not included in Partners in Flight's listing of the conservation status of all land birds in the United States and Canada. (At the time of this writing, overtures are being made to change this.) Furthermore, most Hawaiian forest birds are not covered under the Migratory Bird Treaty Act (www.fws.gov/migratorybirds/intrnltr/mbta/mbtintro.html, accessed May 8, 2008). However, as this book was going to press, it was learned that the honeycreepers would be added to the list of those birds covered by the act. The failure to include Hawaiian birds in the previously mentioned lists or cover them under the Migratory Bird Treaty Act has further added to the obscurity of these birds. The information reported here clearly indicates that Hawaiian

forest birds and their plight have until recently received little national exposure.

The second level of isolation is related to the remoteness of the birds within the state itself. Hawaiian forest birds are generally restricted to high mountain forests where access is difficult or impossible, so the opportunities to see native birds are limited. This isolation and the fact that relatively few birds remain in Hawai'i compared to those in the diverse avifaunas of larger islands or continents certainly contribute to Hawaii's not being a top bird watching destination for visitors.

Even more troubling, most Hawai'i residents have little connection to or knowledge of native birds, and without this connection there is little public demand for increased funding to protect them. For most residents, nonnative mynas, bulbuls, cardinals, and finches are Hawaii's familiar birds. Native birds occurring in residential areas are limited to a few forested areas with relatively low human population density. Unfortunately, only one Hawaiian forest bird, the Oʻahu ʻAmakihi (*Hemignathus flavus*), is readily observable on the island of Oʻahu, which supports approximately 90% of the state's human population. Hawaiian children usually grow up without seeing a native forest bird except perhaps on a rare trip to places such as Hawaiʻi Volcanoes National Park or to the Keauhou Bird Conservation Center, both on Hawaiʻi Island. In addition to limiting the exposure of residents and visitors, the isolation of Hawaiian forest birds limits the public's ability to see the benefits of conservation. In New Zealand, the difference in the abundance and diversity of native birds in unmanaged native forests versus managed forests without predators is obvious to anyone and provides tangible evidence of the response of birds to threat abatement (Towns and Broome 2003).

Conflict Is a Double-Edged Sword

In contrast to the situation in the mainland United States, human conversion of habitat for agriculture or development currently threatens few Hawaiian forest birds, although contemporary livestock grazing and logging contributed to the extinction of ʻAlalā in the wild (Chapter 20) and to the range contraction and population instability of the Palila (Leonard et al. 2008; this volume, Chapter 23), and it will complicate the recovery of both species. The ʻAlalā is the only Hawaiian bird that has an economic conflict associated with its recovery. However, the general absence of economic conflict related to other Hawaiian forest birds translates into a lack of political attention to their recovery needs, unlike in the case of mainland birds' conservation. For example, the economic costs resulting from designating critical habitat for the threatened California Gnatcatcher have been estimated at $915 million (U.S.

Fish and Wildlife Service 2004d). The importance of conflict in generating funding is further illustrated by the fact that much of the support for the two Hawaiian forest birds that receive the most funding (the ʻAlalā and Palila) has resulted from lawsuits (U.S. Fish and Wildlife Service 1998, 2006). The designation of critical habitat for the Palila (U.S. Fish and Wildlife Service 1977) has been an important reason that the Palila has received relatively large amounts of funding (U.S. Geological Survey 2006). Thus, litigation requires the U.S. Fish and Wildlife Service (USFWS) to expend disproportionate resources on species named in lawsuits. Mainland species are involved in a majority of such legal actions (Restani and Marzluff 2002a), which can result in the diversion of funds from extinction-prone island species (Restani and Marzluff 2002c).

Hawaii Is a Small State

Finally, there are several interrelated human demographic, geographic, and regulatory factors that result in the underfunding of Hawaiian birds (Leonard 2008). Among the 50 states, Hawaiʻi ranks forty-second in human population (with 1.3 million people in 2006 according to the U.S. Census Bureau), and this small size affects the amount of conservation funds that the state receives or can generate. Hawaii's small population results in a proportionately limited tax base and prevents the generation of substantial conservation funds from initiatives that are successfully used by larger states, such as tax check-offs or the sale of wildlife license plates. For example, New York state generates approximately $1 million and $100,000 annually from tax check-offs and license plate sales, respectively (The Nature Conservancy 2004). New York's population is 15 times that of Hawaiʻi, suggesting that significantly fewer funds would be generated from similar efforts in Hawaiʻi. In addition, Hawaiʻi has only four congressional representatives, which likely reduces the ability of its legislators to garner funding (Restani and Marzluff 2002a).

A major source of state funding for nongame species is the federal State Wildlife Grants (SWG) program. Unfortunately, funds are appropriated among the states not according to need but on the basis of formulas tied to the size of each state's human population and land area. Because of its size, Hawai'i receives the lowest tier of apportionment, about 1% of the grant or approximately $600,000 per year, even though Hawai'i has by far the most imperiled flora and fauna of any state. Over the past eight years, New York state received $23.4 million from the SWG program, while Hawai'i received $4.9 million (www.teaming.com/pdf/SWG_Allocations .pdf., accessed May 10, 2008). New York supports six federally listed plants and two listed birds; Hawai'i is home to 294 listed plants and 31 listed endemic birds (www .fws.gov/pacificislands/wesa/endspindex .html, accessed May 10, 2008).

Compared with most endangered mainland birds, endangered Hawaiian forest birds have small geographic ranges. The smallness of their ranges contributes to reduced funding, because under the Statewide Endangered Wildlife Program (SEWP), recovery funds are allocated by the USFWS partly on the basis of the size of geographic ranges. Those species with large ranges (>1 million acres) receive more funding than those with small ranges (<1,000 acres) (U.S. Government Accountability Office 2005a). Finally, because Hawai'i shares no borders and thus no endangered species' ranges with other states, it cannot share the cost of conservation efforts with other states.

Management of Game Animals Is a Barrier

Many of the previous chapters have identified human-related barriers to the conservation of Hawaiian forest birds. The establishment of nonnative game mammals and birds for hunting is problematic for islands where plants do not have an evolutionary history tied to browsing and grazing pressure (Chapters 6 and 16). Even though all of Hawaii's game animals are alien and many degrade native habitats (Diong 1982, Katahira et al. 1993, Hess et al. 1999), the Hawai'i Division of Forestry and Wildlife has mandates to both conserve endangered species and provide public hunting opportunities, in other words, to support sustainable populations of game animals. This requirement is an important barrier to forest bird conservation, for the presence of alien ungulates is incompatible with the conservation of native forest birds (Mountainspring 1987, Scowcroft and Conrad 1988, Atkinson et al. 1995, Banko et al. 2001). Unfortunately, in addition to being expensive, fencing and ungulate control and/or eradication result in the loss of hunting areas or opportunities, which is unpopular with some in the Division of Forestry and Wildlife and in a small but vocal hunting community. For perspective, between 2001 and 2006, the average number of hunting licenses issued was 8,900 ± 589 SD, and more than 480,000 ha of state, private, and federal lands were available for public hunting (Hawai'i Department of Business, Economic Development and Tourism 2006). Although practiced by a very small proportion of the citizenry of Hawai'i, hunting has strong cultural ties, and subsistence hunting is important in some communities. In contrast, in 2001 it was estimated that 220,000 Hawaiian residents engaged in some form of wildlife watching (U.S. Fish and Wildlife Service and U.S. Census Bureau 2002), yet wildlife sanctuaries and refuges total only 38,400 ha (Hawai'i Department of Business, Economic Development and Tourism 2006).

STARTING POINTS

What Is Working

Despite all the biological and nonbiological impediments to conserving Hawaiian forest birds, the path to saving them is still open. Perhaps most important is the large percentage of land in Hawai'i that is in public ownership. This ownership pattern, strict zoning laws, and the relatively low human population contribute to low development pressure on conservation lands (discussed earlier and in Chapter 16). Although only 19% of forest bird habitat is now designated for active management and only a subset is actually subject to intensive management (Chapter 16), this percentage will likely increase in the near future, in part because of nine watershed partnerships that encompass >330,000 ha across six islands (see Box 24.2). There also are ongoing conservation projects that engage the public and can be used as models for other projects that will ultimately garner public support; the restoration of native forest on former pasture at the Hakalau Forest National Wildlife Refuge is a prime example (see Box 24.3). In addition, it has recently been discovered that young, high-value koa tree plantations can support high densities of the endangered 'Akiapōlā'au (Pejchar et al. 2005). Although more information is needed to determine the long-term conservation value of koa plantations to native forest birds, the tree's fast growth rate and commercial value provides a strong incentive to reforest pastures that are currently not suitable for forest birds (see Box 24.4) (Pejchar and Press 2006).

Engaging the Public

Although there are a number of examples of programs that engage residents in conservation efforts benefiting Hawaiian forest birds, such as Friends of Hakalau, the conservation community could greatly expand these opportunities and educate the public about native habitats and species and their importance (Chapters 16 and 23). New Zealand offers many examples of community conservation efforts (e.g., www.savethekiwi.org.nz/KiwisSavingKiwi/ CommunityEfforts/). Forest birds once

played a unique role in Hawaiian culture (Chapter 3), but unfortunately few Hawaiians are employed by local conservation agencies. New Zealand offers a more progressive program for engaging their indigenous people in conservation (Ramstad et al. 2007). One such Hawaiian program that could be expanded is the Pacific Internship Program for Exploring Science (PIPES; www.UHH.Hawaii.edu/uhintern), which aims to involve local students in conservation research projects.

Though there are limitations because of the logistics of transporting volunteers to remote areas and of living and working in extreme environments, some areas are accessible and more benign. Encouraging public participation in the restoration of dry forests on Mauna Kea to benefit Palila and in leeward East Maui for Maui Parrotbills would certainly contribute to public support of native birds and the forests that support them.

More Hawai'i residents might care more deeply for their native birds if conservation successes were more effectively reported. Although forest bird recovery is daunting due to the many formidable threats that must be overcome, there are opportunities to encourage greater conservation awareness and action. Relatively low-cost projects, such as cat and mongoose control to protect waterbirds and seabirds, would be one way to foster bird conservation. For example, the story of saving the Nēnē (Hawaiian Goose, *Branta sandvicensis*) from extinction is frequently told to Hawaiian schoolchildren (and even many mainland children). Also, not all native forest bird restoration is necessarily difficult, and not all populations are difficult to access. A few species occur in lowland habitats where people can hear and see them, but these are not the glamorous endangered species, so perhaps they are not widely appreciated. If ways can be found to encourage citizens to appreciate and protect these more common species, this increased interest might be the key to opening the door to the public's interest in all native forest birds and their habitats.

Increasing Accessibility

Current opportunities for the public to see communities of rare native forest birds are restricted to the Alaka'i Wilderness Preserve on Kaua'i, The Nature Conservancy's Waikamoi Preserve on Maui, and several kīpuka (forest patches surrounded by lava) off Saddle Road on the Big Island. The O'ahu 'Elepaio can be observed in O'ahu's lowland forests, Palila in the Mauna Kea Forest Reserve on Hawai'i Island, and 'Aki-apōlā'au in the Kapāpala Forest Reserve, also on Hawai'i.

On Hawai'i Island, a kīpuka off Saddle Road, Kīpuka 21 or Kaulana Manu, is being developed for public access, including access for the physically challenged. The 6 ha forest has been fenced to prevent the ingress of ungulates, rare native plants are being planted, and in the fall of 2007, 12 Hawai'i 'Ākepa and 6 Hawai'i Creepers were released at the site. Once opened, this area will provide the public with the opportunity to see most of the Big Island's endangered forest birds in a managed forest. The dry subalpine forest on southwestern Mauna Kea that supports Palila is accessible by vehicle, and a modest investment in trail construction and informational signage would facilitate public access to this unique environment.

In the next several years, the first predator-proof fence in the United States will be constructed on O'ahu at the Ka'ena Point Natural Area Reserve (Hawai'i Department of Land and Natural Resources 2007). Although this fence will protect a coastal area for nesting albatrosses and other seabirds, it will offer the public an opportunity to see the benefits of fencing and predator eradication and will provide a wildlife experience that could otherwise be enjoyed only by visiting one of the Northwestern Hawaiian Islands. In the vicinity of Kaulana Manu, there are numerous unfenced kīpuka, all of which are an easy hike from a highway, and some of these as well as Kaulana Manu could be enclosed with predator-proof fences. There the public could see the stark contrast between fenced and unfenced areas and the benefits to Hawaiian forest birds and plants of ungulate and predator eradication. On O'ahu, a predator-proof fence constructed around the 1,450 ha Honouliuli Preserve would create a similar opportunity.

SOLUTIONS

Setting Cost-Effective Priorities

Insufficient strategic planning by individual agencies, inadequate coordination among agencies, the absence of a scientifically based decision-making process for determining the best use of particular land parcels (particularly hunting versus conservation), and lack of the data necessary to facilitate such determinations complicate the protection of Hawaiian forest birds. For example, manned hunter check stations are uncommon in Hawai'i, so hunters' use of areas is often not adequately quantified.

The Division of Forestry and Wildlife, the state agency responsible for managing approximately 360,000 ha, does have management guidelines that classify state lands based on their quality with respect to native vegetation. These guidelines also outline appropriate recreational, game management, and forest product use levels for the same lands. As currently written, the guidelines allow public hunting on lands that support the highest-quality native vegetation. Although the intent is to reduce ungulate numbers, the stated bag limits and other restrictions are not as liberal as they should be in areas where protection is the stated goal. Further, public hunting is of limited use and generally effective only when animal numbers are high and hunter access is good. The guidelines are currently being revisited with these limitations in mind, along with the fact that ecosystems that developed without mammals cannot be sustained while maintaining populations of game mammals (Chapter 16). By improving the designations of lands under its jurisdiction, particularly areas for conservation versus multiple use, the Hawai'i Department of Land and Natural Resources would clarify and strengthen its primary mission of conservation. Regardless of how the guidelines are revised, without additional funding, protecting landscape-scale areas for forest birds will be problematic.

As outlined in Chapter 16, cost-benefit analyses comparing actions that are expensive but need to be implemented only once (fencing and ungulate eradication) with those that are less expensive but need to be repeated (ungulate control) would inform funding decision making (Engeman et al. 2003). Currently such an analysis is being done for mouflon sheep (*Ovis musimon*) but is lacking for a major culprit in rain forest management—the feral pig (*Sus scrofa*). Analyses examining the cost and success of managing endangered birds in areas supporting ungulates versus the cost of success in ungulate-free areas also could be undertaken.

Because of the differences between the mainland and Hawai'i regarding the feasibility of overlaying conservation and hunting areas, a careful assessment of management to promote game animals is needed. One way the state could reduce the conflict between conservation and hunting is to acquire and develop new hunting areas. Ideally these areas could be in former pasturelands or in nonnative lowland forests that support few, if any, native forest birds. In addition, improving access to other appropriate lands to offset the loss of hunting areas when native habitats are fenced and managed for conservation could lessen the conflicts with some of the public. Finally, recognition that managing wild native birds in their habitat is far less expensive than bringing them into captivity and maintaining them there is a vital consideration when planning the recovery of endangered species and the best use of

particular lands (Chapter 19). Similar analyses have been called for to determine the amount of funding to allocate to the interdiction of alien species (Chapter 15).

Increasing Awareness

The management activities needed to ensure the survival of Hawaiian birds are expensive and can be socially and politically unpopular. If the Hawai'i conservation community were to convince the state legislature of the much larger number of residents who engage in wildlife watching versus hunting and of the number of tourists who also engage in wildlife viewing, perhaps the apportionment of funding and the direction of state conservation agencies could be improved for forest bird recovery (U.S. Fish and Wildlife Service and U.S. Census Bureau 2002). Tourists could be made aware of the fate of Hawaiian forest birds by way of a short video on every flight into Hawai'i, which would be seen by millions of people each year.

There is much to do to improve the awareness of Hawaiian birds nationally. Enstating Hawaiian birds on the ABA Checklist and including them in American bird literature would be an important start. NGOs, especially those involved in bird conservation and birdwatching, could also be lobbied for improved coverage. A recent example has been involvement in grant seeking by the American Bird Conservancy. These organizations have the infrastructure to initiate public information campaigns about the fate of Hawaiian forest birds, as well as the funding needed to protect them.

FUNDING MECHANISMS

Federal Funding Mechanisms

Currently, the Hawai'i Division of Forestry and Wildlife receives approximately $2 million from SEWP and SWG programs. Of this, approximately 60% goes to projects that directly or indirectly benefit Hawaiian forest birds. As mentioned earlier, funds for these programs are allocated to states not according to need but on the basis of criteria that are biased against islands. In addition to these funds, the Ecological Services Division of the Pacific Islands U.S. Fish and Wildlife Service has annually contributed $550,000 to the captive propagation program described in Chapter 19, $50,000–$100,000 to forest bird projects on Maui, and varying amounts from office discretionary funds to other related projects. Recently, funds have been obtained through two agencywide grant programs —Showing Success and Preventing Extinction. In fiscal year 2008, $180,000 was awarded through the latter grant program to increase the capacity for fire response in Palila critical habitat and to increase ungulate control. In the same year, Ecological Services also provided the state with $960,000 to start fencing Palila critical habitat and to increase ungulate control efforts.

NEW FEDERAL FUNDING MECHANISMS

NGOs and their constituents could lobby the U.S. Congress and the U.S. Department of the Interior for consistent and reliable funding or a special funding initiative for Hawaiian forest bird conservation. Many species, especially those for which captive propagation has been an important part of their conservation—such as the Whooping Crane, Puerto Rican Parrot, and California Condor—have consistently received higher levels of annual funding than any Hawaiian forest bird (Leonard 2008). As stipulated in the FY09 federal budget, the Migratory Bird Initiative will provide $9 million to improve the status of bird populations on the mainland through cooperation with private landowners, corporations, nonprofit organizations, and the government of Mexico. In 2000, Congress passed the Comprehensive Everglades Restoration Plan. This 30-year plan has a $10.9 billion price tag, of which the federal government will provide half (Sheikh

and Carter 2006). A Hawaiian initiative similar to the Everglades funding package would provide the level of funding that is needed to carry out the actions called for in the *Revised Recovery Plan for Hawaiian Forest Birds* (U.S. Fish and Wildlife Service 2006). An increase in SEWP and SWG funding and/or changes in their allocation formulas (discussed later) and passage of climate change legislation with funding to mitigate the impact of global warming on wildlife would provide much-needed funds. NGOs also could press for general budget increases for the USFWS (between FY08 and FY09 the agency's budget was cut by 5%).

State of Hawai'i Funding Mechanisms

The largest, and virtually only, dedicated source of state funding for natural resource conservation comes from a real estate transfer or conveyance tax. In recent years the state has spent approximately $2–4 million annually to control or eradicate alien species, and some forest birds benefit from these actions. However, these funds were greatly reduced in 2009 because of the economic downturn. Although only one full-time state biologist is dedicated to forest bird recovery, other biologists and state workers participate in forest bird conservation in varying ways, such as by removing feral cattle (*Bos taurus*) from conservation areas. Other than these contributions, very few state operating funds— approximately $400,000 per year—are invested in conservation efforts that benefit Hawaiian forest birds. In 2008, the Division of Forestry and Wildlife initiated a wildlife license plate program. However, the sale of these plates is unlikely to generate more than $25,000 in the first year, and sales of such plates are expected to decline fairly rapidly thereafter (The Nature Conservancy 2004).

The sale of property in Hawai'i is taxed on a sliding scale, starting at $1 per $1,000 for properties valued at less than $600,000 and rising to $3 per $1,000 for properties valued at more than $1 million. Those purchasing property for investment purposes pay a slightly higher rate. As of 2004, more than 90% of properties sold were valued at less than $600,000. Of the total conveyance tax collected, 25% goes to the Natural Area Reserve Fund, and 10% goes to the Land Conservation Fund. The latter fund provides moneys to purchase land or conservation easements. The former supports four programs: the Watershed Partnership Program, Natural Area Partnership and Forest Stewardship Program, Natural Area Reserve Program, and Youth Conservation Corps. The first three programs all indirectly contribute to forest bird conservation by protecting and managing native habitats. Natural area reserves receive the highest level of management, and currently there are 15 reserves that encompass 44,000 ha. In 2007, the conveyance tax provided $12.4 million to the Natural Area Reserve Fund, although this figure was greatly reduced in 2009.

Hawaii's conveyance tax is one of the lowest in the country, along with those of Alabama, Georgia, Illinois, and Virginia, all of which charge $1 per $1,000 conveyance. Other states charge much more per $1,000 conveyance: Delaware $20, New Hampshire $15, and Florida $7. In 2004, the mean (± SD) revenue per capita generated by the 35 states with real estate transfer taxes was $46.33 ± 85.81 (range $0.05–$485.20); Hawaii's per capita revenue was $14.59 (www.taxadmin.org/fta/rate/Realtytransfer.html).

NEW STATE FUNDING MECHANISMS

The cost of living in Hawai'i is one of the highest in the nation, including the highest housing prices in the country. Thus, although other states charge higher conveyance taxes, increasing the rate in Hawai'i would likely be very unpopular, as would other tax increases. Given Hawaii's small population, other funding mechanisms, such as the nongame tax check-offs that are used by 32 states, would yield relatively little revenue (The Nature Conservancy 2004). Other states use a variety of mechanisms to fund conservation, and these could be evaluated for use in Hawai'i (The Nature Conservancy 2004).

The legislatures of 17 states have established conservation trust funds and provided appropriations to these funds or have provided general appropriations for conservation. For example, in 1983 Indiana passed the Indiana Natural Heritage Protection Act, creating a trust fund for natural lands protection. A one-time $5 million general appropriation established the fund and was matched by a $5 million contribution from The Nature Conservancy. The Hawai'i Legislature could be lobbied to establish a similar fund for Hawaiian forest bird conservation.

In 1986, the Natural Resource Foundation of Wisconsin, a 501(c)(3) charitable organization governed by a board of private citizens, was formed to raise private support for conservation. As of 2006, the foundation had 1,900 members and had been able to contribute $410,000 to worthy conservation projects, which included providing the state's Department of Natural Resources with $230,000 (www.wisconservation.org). A similar organization in Hawai'i would be another way to raise needed conservation funds.

Funds to mitigate against the taking of endangered seabirds are increasing in Hawai'i through USFWS Section 7 consultations under the Endangered Species Act and the development of the habitat conservation plans that are required for the issuance of incidental take licenses and permits by both the federal and state governments. Growing interest in developing wind power will likely result in an increase in such mitigation funds. Although these funds must be directed toward management activities that will benefit seabirds, 'A'o or Newell's Shearwaters (*Puffinus auricularis newelli*) nest in mountain forests mainly on Kaua'i, and actions to assist these birds, such as predator control, also may benefit some forest birds.

Corporate sponsorship and/or support has been successfully used by New Zealand to generate funds for the recovery of kiwis (*Apteryx* spp.). Since 1991, the Bank of New Zealand has contributed more than 4 million NZ dollars toward kiwi recovery programs. In November 2002, the New Zealand Department of Conservation and Bank of New Zealand established the Bank of New Zealand Kiwi Recovery Trust. The formation of an independent charitable trust whose sole focus is kiwi conservation was designed to enhance fundraising opportunities for kiwi recovery as well as assist greater community participation. The development of similar programs in Hawai'i could be investigated. A potential sponsor could be one of the airlines, one of which names its jets after Hawaiian forest birds and another of which has a Hawaiian forest bird logo. The development of an iconic symbol of Hawaiian forest bird conservation would be a worthwhile investment, and the brilliantly colored and widespread 'I'iwi (*Vestiaria coccinea*) would be an excellent candidate.

Hawaii's small human population does limit the utility of many funding mechanisms available. However, the number of tourists visiting Hawai'i annually far exceeds the resident population. In 2006, 8.9 million visitors arrived in Hawai'i (Hawai'i Department of Business, Economic Development and Tourism 2006), although in 2009 the number of visitors was declining because of the economy. The tourist population represents a large untapped source of conservation funds. Although it seems likely that the Hawai'i Tourism Authority would object to taxing tourists for conservation purposes, several countries levy relatively high tourist taxes. For example, Belize charges each person visiting the country by plane or cruise ship a tourist tax of $3.75 that goes directly to conservation. The Seychelles charges a $90 fee to each tourist entering the country (Sweeting et al. 1999). A modest tax of $10 per individual entering Hawai'i would generate ~$75 million per year. This figure approaches the amount estimated to be needed to recover Hawaiian forest birds (U.S. Fish and Wildlife Service 2006). Similar measures, such as a hotel tax, have great potential to generate funds for the conservation of Hawaiian birds.

The sale of carbon offsets is potentially a new way of generating conservation funding that may hold promise for Hawai'i. Carbon offsets are a voluntary mechanism by which individuals can compensate for their CO_2 emissions by donating money to a company that plants trees or invests in green technology. For example, based on the carbon footprint calculator found at CarbonFootPrint.com, a round-trip flight from Atlanta to Honolulu releases 1.67 tons of CO_2; a modest donation of between $25.66 and $72.43 can offset this emission by investing in green technology or by paying for the planting of three native trees in Kenya or the United Kingdom. Given that virtually all tourists arrive in Hawai'i by plane, donating to a Hawai'i-based company or a company focused on reforestation in Hawai'i could benefit Hawaiian forest birds. Given the ongoing restoration projects and the thousands of hectares of former pasture that could be reforested, there are opportunities to cost-share with entities already engaged in conservation activities. Because koa trees grow rapidly and have value as bird habitat and commercial timber, by employing sustainable harvest practices, such a carbon offset enterprise could potentially provide conservation and economic benefits. Given the media attention to global climate change and the number of tourists arriving in Hawai'i annually, an advertising campaign in Hawai'i focused on offsetting the CO_2 produced in traveling to Hawai'i could be very successful.

SUMMARY

Duffy and Kraus (2006) lament the little that Hawai'i has to show for the funds already spent in the state on conservation in general ($63 million per year), on bird research ($37.8 million between 1987 and 1997), and on management and land acquisition ($57.7 million between 1987 and 1997) in particular. However, these amounts pale in comparison to what other states are spending on conservation, especially given the highly unique flora and fauna of Hawai'i, the relatively unusual biological challenges to their protection, and the cost associated with mitigating these challenges. For example, Florida's "Florida Forever" program generates $300 million annually to purchase environmentally sensitive lands. In addition, the Florida Freshwater Fish and Conservation Commission has an annual budget of more than $279 million. In contrast, the annual budget for environmental protection within the Hawai'i Department of Land and Natural Resources is approximately $45 million annually (www.state.hi.us/budget/memos/budget).

Without a profound change in political will, at both the state and federal levels, to dramatically increase funding for the recovery of endangered Hawaiian birds, many taxa that survive today face an unquestionably bleak future. Funding agencies should consider the unique threats faced by Hawaiian birds and the fact that the current criteria for funding allocation provide inadequate resources to address these threats. Funding should be allocated with the goal of minimizing extinction risk. Given the limited funding available, it is essential that conservation agencies work together to develop strategic plans. The state of Hawai'i must recognize that attempting to conserve native birds and their habitats and simultaneously sustaining populations of feral ungulates for hunting cannot co-occur within the same forest (Chapters 16, 17, 20, 22, 23). Attempting both is not ecological or eco-nomically feasible and will ultimately result in the loss of more habitat and endemic forest birds. A scientifically based decision-making process must be developed that will unambiguously determine the best use, whether conservation or recreation, of all lands currently owned and managed by the state. Such a process is now being developed. Because of the potential influx of alien organisms (e.g., West Nile virus), hurricanes, and volcanic activity, this decision-making process needs to incorporate the precautionary principle (Cooney 2004); there is no longer time to postpone protecting Hawaii's endangered forest birds. Once high-priority conservation lands are designated, tapping into the millions of tourists who visit Hawai'i annually could provide the funds needed to overcome the many threats faced by Hawaiian forest birds and the habitats that support them without unduly burdening residents. Although the myriad problems facing Hawaiian forest birds will be expensive and difficult to mitigate and face many nonbiological hurdles, overcoming these obstacles will provide additional benefits, including the protection of endangered plants and invertebrates, ecosystems, and watersheds. In addition, because Hawaiian forest birds will benefit from ongoing efforts to restore forest ecosystems and biota in Hawai'i, there are numerous opportunities to share the responsibilities and costs of habitat protection. Can the barriers described in this chapter be overcome? I believe the answer is yes, but time is running out. Action must be swift and effective.

ACKNOWLEDGMENTS

This chapter was much improved by the careful edits and conceptual suggestions of S. Fretz, C. Roland, and E. VanderWerf.

CHAPTER TWENTY-FIVE

Can Hawaiian Forest Birds Be Saved?

THANE K. PRATT, CARTER T. ATKINSON,
PAUL C. BANKO, JAMES D. JACOBI,
BETHANY L. WOODWORTH,
AND LOYAL A. MEHRHOFF

No nation has lost more species of birds in the past 25 years than the United States, largely as a result of recent extinction events in the Pacific islands. It would take multiple rediscoveries nearly as miraculous as that of the ivorybill to alter this shameful fact.
—Wilcove (2005)

In the previous chapters we have presented the many problems faced by Hawaiian forest birds and the various efforts to halt their extinction. This final chapter takes a look to the future. We discuss the overall fate of Hawaiian forest birds and then explore in detail four major challenges to their conservation: alien species, climate change, the scale of recovery efforts, and social factors. For each of these challenges we review the effects on forest birds and ask, What might we predict for the future? What can be done?

CAN HAWAIIAN FOREST BIRDS BE SAVED?

A Grave Status Quo

Twenty-five years ago, the Hawaiian Forest Bird Survey (1976–1983) brought international attention to the plight of Hawaiian birds (Scott et al. 1986). Since then, 10 species of endemic Hawaiian birds have been lost to extinction—an average of 1 extinction every two years (Table 25.1, Fig. 25.1). Of the 46 historically known forest bird species, only 24 still survive, and of these, 13 are listed as endangered. Given such a track record, is it possible to save the remaining Hawaiian forest birds?

The answer depends on what actions are taken in the next few years. A "business-as-usual" scenario that includes rampant avian diseases, habitat degradation by feral ungulates and invasive plants, minimal control of mammalian predators, and the continued arrival of new plant diseases

Table 25.1. Extinct Hawaiian forest birds: The historically known species

Common Name[a]	Scientific Name[b]	Range[c]	Year of Disappearance[d]
ʻŌʻōʻāʻā, Kauaʻi ʻŌʻō (E)	Moho braccatus	K	1987
Oʻahu ʻŌʻō	Moho apicalis	O	1837
Bishop's ʻŌʻō	Moho bishopi	Mo, Ma	1904
Hawaiʻi ʻŌʻō	Moho nobilis	H	1902
Kioea	Chaetoptila angustipluma	H	1859
Kāmaʻo (E)	Myadestes myadestinus	K	1985
ʻĀmaui	Myadestes woahensis	O	1825
ʻOlomaʻo (E)	Myadestes lanaiensis	Mo, L, Ma	1980
ʻŌʻū (E)	Psittirostra psittacea	Main HI	1989
Lānaʻi Hookbill	Dysmorodrepanis munroi	L	1918
Lesser Koa-Finch	Rhodacanthis flaviceps	H	1891
Greater Koa-Finch	Rhodacanthis palmeri	H	1937
Kona Grosbeak	Chloridops kona	H	1892
Greater ʻAmakihi	Hemignathus sagittirostris	H	1901
Lesser ʻAkialoa	Hemignathus obscurus	H	1940
Kauaʻi Greater ʻAkialoa (E)	Hemignathus ellisianus stejnegeri	K	1969
Oʻahu Greater ʻAkialoa	Hemignathus e. ellisianus	O	1939
Maui-nui Greater ʻAkialoa	Hemignathus e. lanaiensis	Mo, L, Ma	1894
Kauaʻi Nukupuʻu (E)	Hemignathus lucidus hanapepe	K	1990s?
Oʻahu Nukupuʻu	Hemignathus l. lucidus	O	1860?
Maui Nukupuʻu (E)	Hemignathus l. affinis	Ma	1996
Oʻahu ʻAlauahio (E)	Paroreomyza maculata	O	1978
Kākāwahie (E)	Paroreomyza flammea	Mo	1963
ʻUla-ʻai-hāwane	Ciridops anna	H	1892
Hoa, Black Mamo	Drepanis funerea	Mo, Ma	1907
Hawaiʻi Mamo	Drepanis pacifica	H	1899
Laysan Honeycreeper	Himatione sanguinea freethii	Laysan	1923
Poʻo-uli (E)	Melamprosops phaeosoma	Ma	2004

Notes: An additional 20 extinct species have been named from subfossils, and more await description. Many living species once enjoyed a wider geographic range and are now extinct on certain islands. Ten species are likely recently extinct and are still federally listed as endangered (E). There is a small chance that one or more of these birds may be rediscovered. Rediscovery of any species would trigger the Rare Bird Discovery Protocol, which prescribes research and intervention steps (U. S. Fish and Wildlife Service 2006: Section 3, pages 17–21).

[a]Common names mostly from Pyle (2002).
[b]Taxonomy and order generally follow American Ornithologists' Union (1983).
[c]K, Kauaʻi; O, Oʻahu; Mo, Molokaʻi; L, Lānaʻi; Ma, Maui; H, Hawaiʻi; Main HI, main Hawaiian islands.
[d]Years of birds' disappearance from Chapters 2 and 5.

and other habitat-altering alien species will lead to the eventual loss of almost all native forest birds. Mosquito-borne diseases have eliminated the majority of endemic bird species on the islands of Oʻahu, Molokaʻi, and Lānaʻi and are now causing the extinction of birds on Kauaʻi. Bird species on Maui and Hawaiʻi that have high-elevation habitats without mosquitoes have a better chance of survival and are likely to pass through the next quarter century. In all, seven species are at significant risk of extinction in the next 25 years (Fig. 25.2). Ten more species are locally common but are severely restricted in their geographic ranges. Only seven species of

Figure 25.1. They were just here! Ten species or sub-species of Hawaiian forest birds vanished in the two decades after the landmark Hawai'i Forest Bird Survey (1976–1983). More bird species disappeared in Hawai'i during this period than anywhere else in the world. Painted figures © H. Douglas Pratt. Graphics by J. Robinson, Hawai'i Volcanoes National Park.

Hawaiian forest birds have a reasonable chance of surviving without intervention: the 'Io (Hawaiian Hawk), 'Elepaio, 'Ōma'o, and 'Apapane and the three species of 'amakihi.

The Main Threats

If key recovery actions are implemented over sufficiently large and viable areas of habitat, the outlook for Hawaiian forest birds changes from pessimistic to cautiously optimistic. The following important threats must be minimized before such an optimistic view can be realized.

AVIAN DISEASE

First and foremost, introduced mosquito-borne diseases must be effectively controlled. All species of native songbirds are highly threatened by introduced avian diseases to which they have no or, at best, partial resistance. There are two aspects of managing this most immediate threat.

Strategies must be implemented to control mosquito populations in bird habitat, thereby reducing disease transmission. Mosquito control may be achieved through forest management: removing feral pigs, cleaning up artificial breeding sites (tires, water catchments), and converting pasture to forest (Chapter 17). These efforts have now begun across thousands of acres, mainly on Maui and Hawai'i (discussed later), at high elevations where both mosquitoes and the avian malarial parasite are naturally limited by cool temperatures (Figs. 25.3 and 25.4). Significantly, a few native bird species have demonstrated

Critically Endangered

Nihoa

Kaua'i

O'ahu

Maui

Hawai'i

1 Nihoa Finch	3 'Akikiki
2 Nihoa Millerbird	4 Puaiohi

5 O'ahu 'Elepaio	7 Maui Parrotbill
6 'Ākohekohe	8 'Alalā

Figure 25.2. Which species could become extinct next? The eight most vulnerable Hawaiian forest birds are six species designated as critically endangered by the International Union for the Conservation of Nature (IUCN); the O'ahu 'Elepaio, not recognized by the IUCN because of its present subspecies status; and the 'Alalā, which survives only in captivity. Extinction of these species can be prevented if recovery plans are carried out. Recovery plans prescribe a combination of habitat management, active management of wild populations, captive management, and translocations. Painted figures © H. Douglas Pratt. Graphics by J. Robinson, Hawai'i Volcanoes National Park.

some resistance to avian malaria, and others may yet do so. How to hasten the development of disease resistance in native birds remains the Holy Grail for avian conservationists in Hawai'i (Kilpatrick 2006).

The second aspect of managing disease is to prevent the introduction of additional avian diseases, such as West Nile virus (LaDeau et al. 2007), and new disease vectors, such as cold-tolerant mosquitoes.

If the historic effects of avian disease indicate future effects, it is doubtful that the Hawaiian avifauna can survive continual introductions of new diseases and vectors.

THREATS TO HABITAT

The existing habitat of native forest birds must remain viable. There is ample opportunity to improve existing habitat and increase food availability, thereby increasing the capacity to support more birds. There is also plenty of room to expand habitat on Maui and Hawai'i to enlarge species' ranges and populations. Although there are large areas of apparently good habitat for forest birds on Kaua'i, O'ahu, Moloka'i, Maui, and Hawai'i, very little of this land is being managed to ensure its viability; specifically, 3.5% is ungulate-free, and ungulate eradication is planned for 19% (Chapter 16). These unmanaged areas are likely to be degraded further by a variety

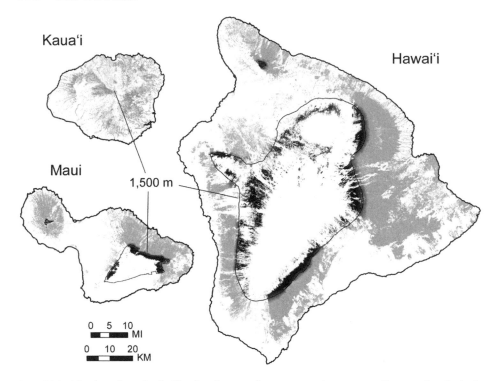

Kaua'i

Hawai'i

Maui

1,500 m

0 5 10
MI
0 10 20
KM

Figure 25.3. Islands within islands: The distribution of disease-safe habitat in Hawai'i. The map of forest above the 1,500 m contour line (dark shading) shows the approximate extent of habitat safe from the mosquito-borne diseases avian malaria (*Plasmodium relictum*) and avian pox virus (*Avipoxvirus* sp.). Light shading indicates forest below this zone (landcover data from the National Oceanic and Atmospheric Administration, Coastal Change Analysis Program 1995). Note how small these areas are and that they are essentially available on two islands only. Native forest birds mainly inhabit rain forest. However, much of the disease-safe vegetation is instead mesic forest, dry forest, and shrubland now depleted of endangered birds. The low stature of this vegetation increases the exposure of birds to cats and rats, particularly during nesting (Chapter 8). Removal of alien mammalian predators from short-statured vegetation could make it more habitable for forest birds, a strategy that may become increasingly necessary with the expected drying conditions of global warming in Hawai'i. Maps drafted by M. Gorresen.

of factors (Chapters 6 and 16). Some of the most important factors that must be addressed follow.

1. *Ungulates—pigs, cattle, sheep, mouflon, deer, and goats.* The long-term presence of feral ungulates has been shown to eliminate forest bird habitat in Hawai'i. Control of ungulates is a precursor to effective habitat conservation.

2. *Invasive plants.* Such plants replace native forests by outcompeting endemic species (Smith et al. 1985). Nonnative grasses and some alien trees, particularly conifers and eucalypts, increase the frequency of forest fires.

3. *Introduced diseases or insect pests of native plants.* Only a few native tree species dominate Hawaiian forest communities. These are highly utilized by native birds, and their loss would be devastating to the avifauna. As in the case of avian diseases, it is imperative to prevent the introduction of plant pathogens and insect pests that could cause the catastrophic loss of dominant forest trees or key matrix species (Ellison et al. 2005).

4. *Climate change.* Predicting how global climate change will be felt in Hawai'i and how native and nonnative species will respond is still a new science.

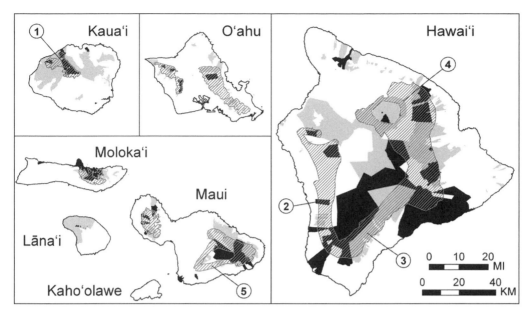

Figure 25.4. How much endangered bird habitat is managed for conservation? The recovery area for endangered forest birds—identified in the *Revised Recovery Plan for Hawaiian Forest Birds*—is indicated with hatching. The recovery area is shown overlapping managed conservation lands (black) and areas designated for management but not managed (gray). Existing concentrations of endangered species could be targeted by prioritizing new areas for habitat management: (1) Alakaʻi Plateau, (2) central Kona, (3) Kaʻū and Kapāpala forest reserves, and (4) subalpine Mauna Kea. Leeward Halcakalā (5) offers an opportunity to expand the ranges of endangered Maui forest birds. Maps drafted by M. Gorresen. GIS polygons provided by J. Price.

Potential changes include increased warming, which may expand the distribution of disease vectors into higher elevations (Benning et al. 2002); droughts and storms, which will deplete food, reduce bird reproduction, and alter plant communities; and increasing the frequencies of fire, to which native forests are intolerant.

INTRODUCED PREDATORS

Rats and feral cats have eliminated native forest birds from shrubland and short-statured forest, and they cause bird mor-

tality in tall rain forests (Chapters 8 and 11). In particular, Oʻahu ʻElepaio and Palila are threatened by mammalian predators. Effective control programs that are supported by local communities are needed (Chapter 18). Predators are absent from the Northwestern Hawaiian Island, and precautions to prevent the spread of predators to these islands are in place to protect the Millerbird, Laysan Finch, and Nihoa Finch. Predators not yet established in the islands, such as the brown treesnake (*Boiga irregularis*), are also a deadly threat that can be prevented by strong biosecurity (Chapter 15).

Changing Course: Reasons for Optimism

If the recovery actions described are undertaken across the archipelago, there is a realistic possibility of saving the remaining Hawaiian forest birds. This optimism is based on seven positive considerations.

SALVAGEABLE POPULATIONS

The populations of the surviving 24 species are still large enough to recover (Table 25.2). All species except the ʻAlalā have wild populations exceeding 300 birds. Populations of this size offer more options

Table 25.2. Conservation status of Hawaiian forest birds

Species	Islands[a]	Population[b]	Trend[b]	Current Range[b] in Hectares (acres)	Recovery Area[c]
'Io, Hawaiian Hawk (*Buteo solitarius*) (E)	H	3,000	Stable	575,500 (1,422,000)	None delineated
Pueo, Short-eared Owl (*Asio flammeus*)	Main HI	?	Decreasing?	?	—
'Alalā, Hawaiian Crow (*Corvus hawaiiensis*) (E)	H	Wild—0 Captivity—60	Extinct in wild	0	None delineated
Kaua'i 'Elepaio (*Chasiempis sandwichensis sclateri*)	K	152,000	Increasing	37,900 (93,700)	—
O'ahu 'Elepaio (*Chasiempis s. ibidis*) (E)	O	<2,000	Decreasing	5,500 (14,000)	26,700 (66,000)
Hawai'i 'Elepaio (*Chasiempis s. sandwicensis*)	H	<200,000	Decreasing locally	394,200 (974,000)	—
Millerbird (*Acrocephalus familiaris*) (E)	N	<400	Stable	40 (100)	700 (1,729)
'Ōma'o (*Myadestes obscurus*)	H	170,000	Decreasing locally	230,300 (569,100)	—
Puaiohi (*Myadestes palmeri*) (E)	K	Wild—300–500 Captivity—~30	Core population stable? Range contracting	4,000 (10,000)	6,700 (16,000)
Laysan Finch (*Telespiza cantans*) (E)	L	10,000	Stable	240 (600)	590 (1,500)
Nihoa Finch (*Telespiza ultima*) (E)	N	3,000	Decreasing?	40 (100)	770 (1,900)
Palila (*Loxioides bailleui*) (E)	H	3,900	Stable? Recently (2003–2009) decreased	13,600 (34,000)	48,900 (121,00)
Maui Parrotbill (*Pseudonestor xanthophrys*) (E)	Ma	Wild—500 Captivity—12	Stable?	5,000 (12,000)	40,600 (100,000)
Hawai'i 'Amakihi (*Hemignathus virens virens*)	H	>800,000	Mostly stable; increasing locally	531,900 (1,315,000)	—
Hawai'i 'Amakihi (*H. v. wilsoni*)	Mo, Ma	50,000	Stable? Increasing locally	84,300 (208,000)	—
O'ahu 'Amakihi (*Hemignathus flavus*)	O	52,000	Stable? Increasing locally	27,300 (67,000)	—
Kaua'i 'Amakihi (*Hemignathus kauaiensis*)	K	51,000	Increasing	37,900 (94,000)	—
'Anianiau (*Hemignathus parvus*)	K	37,500	Increasing	12,700 (31,000)	—

Habitat Protection	Habitat Improvement	Create New Habitat	Disease Management	Predator Management	Introductions	Captive Management
✓	✓	✓			✓	
✓?	✓			✓		
✓	✓	✓	✓	✓	✓	✓
	✓		✓			
	✓		✓	✓	✓	✓
✓	✓	✓	✓	✓		✓
	✓		✓	✓	✓	✓
✓	✓	✓	✓	✓	✓	✓
	✓		✓	✓	✓	✓
	✓	✓	✓	✓	✓	✓
	✓	✓	✓	✓	✓	✓
	✓	✓		✓	✓	✓
✓	✓	✓	✓	✓	✓	✓
✓	✓	✓				
✓	✓	✓			✓	
✓	✓					
	✓					
	✓		✓			

continued

Table 25.2. Continued

Species	Islands[a]	Population[b]	Trend[b]	Current Range[b] in Hectares (acres)	Recovery Area[c]
'Akiapōlā'au (*Hemignathus munroi*) (E)	H	1,900	Range contracting, increasing locally	27,700 (68,400)	238,000 (588,000)
'Akikiki (*Oreomystis bairdi*)	K	3,600	Range contracting	3,900 (9,600)	9,200 (23,000)
Hawai'i Creeper (*Oreomystis mana*) (E)	H	14,000	Range contracting; increasing locally	34,000 (84,000)	199,600 (493,000)
Maui 'Alauahio (*Paroreomyza montana*)	M	>35,000	Increasing?	9,800 (24,000)	—
'Akeke'e (*Loxops caeruleirostris*)	K	7,900	Range contracting	12,700 (31,000)	—
'Ākepa (*Loxops coccineus*) (E)	H	12,000	Decreasing	22,600 (55,800)	
'I'iwi (*Vestiaria coccinea*)	Main HI	360,000	Decreasing on all islands but locally stable	254,000 (628,000)	—
'Ākohekohe (*Palmeria dolei*) (E)	M	3,800	Increasing	6,000 (15,000)	52,200 (129,000)
'Apapane (*Himatione s. sanguinea*)	Main HI	1,300,000	Stable; increasing locally	632,900 (1,564,000)	—

Notes: Federally listed endangered species are indicated with an (E). The checked columns show species management needs and specify actions other than research measures required for all species, viz., population monitoring and research on population ecology and conservation biology. Note that common species as well as rare ones could benefit from a range of conservation measures to improve their status. "Introductions" means there is a need to establish a new population either in or outside the species' historic range by various methods, including translocating wild birds or releasing captive reared ones.
[a]K, Kaua'i; O, O'ahu; Mo, Moloka'i; L, Lāna'i; Ma, Maui; H, Hawai'i; Main HI, main Hawaiian islands.
[b]From Scott et al. (1986); Camp, Gorresen, et al. (2009); and this volume, Chapters 5 and 12. Range areas are derived from the GIS polygons used to generate the range maps for Chapter 5.

for their recovery (in situ management, captive management, and reintroductions) and more time to act than do smaller populations. By contrast, populations of the 10 recently disappeared species were down to the last few individuals 30 years ago (except possibly that of the somewhat more numerous 'Ō'ū).

LARGE AREAS OF HABITAT

Significant blocks of native habitat remain for most of the surviving forest bird species (see Table 25.2, Fig. 25.4). Numerous measures have been taken to protect native biota and ecosystems across landscapes in Hawai'i, with benefit for endemic

Habitat Protection	Habitat Improvement	Create New Habitat	Disease Management	Predator Management	Introductions	Captive Management
✓	✓	✓	✓	✓	✓	✓
	✓		✓	✓	✓	✓
✓	✓	✓	✓	✓	✓	✓
✓	✓	✓	✓		✓	
	✓		✓	✓	✓	✓
✓	✓	✓	✓	✓	✓	✓
✓	✓	✓	✓	✓		✓
✓	✓	✓	✓		✓	✓
✓	✓	✓				

^cRecovery areas, as defined by the *Revised Recovery Plan for Hawaiian Forest Birds*, are "those areas of habitat that will allow for the long-term survival and recovery of endangered Hawaiian forest birds," and they identify lands where recovery is taking place or might take place (U.S. Fish and Wildlife Service 2006: Section 3:7). Only the Palila has designated critical habitat. The Northwestern Hawaiian Island birds lack recovery areas, so for these we include the islands listed for near-term recovery by Morin and Conant (2007), that is, for the Miller-bird, Kure, Lisianski, Laysan, and Nihoa; for the Nihoa Finch, Kure, Midway, and Nihoa; and for the Laysan Finch, Lisianski and Laysan. "—" indicates that the recovery areas do not pertain to nonendangered species.

birds. The amount of protected lands has increased recently, and the extent of forest managed wholly or partly on behalf of forest birds has grown to 147,600 hectares (363,000 acres) (Chapter 16). For the three Northwestern Hawaiian Island songbirds, there are islands to which the birds can be translocated to build additional populations.

THE RESPONSE OF FOREST BIRDS
TO HABITAT IMPROVEMENT

Population gains in forest birds have recently been demonstrated on some managed lands. These gains have included endangered species such as the ʻAkiapōlāʻau and Hawaiʻi Creeper.

Table 25.3. Prioritizing captive management for forest birds

Species	Priority[a]	Captive Strategy	Reared in Captivity?[b]	Current Actions[c]
'Io, Hawaiian Hawk	0	Not needed	No	—
Pueo, Short-eared Owl	0	Not needed	No	—
'Alalā, Hawaiian Crow	1	Build captive flock, then breed for production	Yes	Building of captive flock
Kaua'i 'Elepaio	0	Not needed	No	—
O'ahu 'Elepaio	3	Translocation; rear and release	No	—
Hawai'i 'Elepaio	4	Develop captive technology	Yes	—
Millerbird	1	Research on rear and release; translocation	No	Research on rear and release
'Ōma'o	4	Translocation; rear and release	Yes	—
Puaiohi	1	Captive breeding for release	Yes	Captive breeding for release
Laysan Finch	3	Research on rear and release; translocation	Yes	—
Nihoa Finch	2	Research on rear and release; translocation	No	—
Palila	2	Captive breeding for release	Yes	Captive breeding for release
Maui Parrotbill	2	Captive breeding for release	Yes	Captive breeding for release
Hawai'i 'Amakihi	0	Not needed	Yes	—
Maui 'Amakihi	0	Not needed	Yes	—
O'ahu 'Amakihi	0	Not needed	No	—
Kaua'i 'Amakihi	0	Not needed	No	—
'Anianiau	0	Not needed	No	—
'Akiapōlā'au	2	Develop technology, captive breeding for release	No	—
'Akikiki	2	Develop technology, captive breeding for release	No	—
Hawai'i Creeper	3	Rear and release; captive breeding for release	Yes	—
Maui 'Alauahio	0	Not needed	Yes	—
'Akeke'e	2	Develop technology, captive breeding for release	No	—
'Ākepa	3	Rear and release; captive breeding for release	Yes	—
'I'iwi	4	Develop captive technology	Yes	—
'Ākohekohe	3	Translocation; rear and release	Yes	—
'Apapane	0	Not needed	Yes	—

Sources: Information for this table was derived from the *Revised Recovery Plan for Hawaiian Forest Birds* (U.S. Fish and Wildlife Service 2006: Table 13) and this volume, Chapter 19: Table 19.1. Our table has been updated with recent developments for the Northwestern Hawaiian Island songbirds (Morin and Conant 2007) and Kaua'i forest birds (H. Friefeld, U.S. Fish and Wildlife Service, pers. comm.).

Notes: Captive management of Hawaiian forest birds has come into being in the past 20 years with the establishment of two state-of-the-art facilities on Maui and Hawai'i run by the Hawai'i Endangered Bird Conservation Program, a partnership between the San Diego Zoo, Hawai'i Department of Land and Natural Resources, and U.S. Fish and Wildlife Service (Chapter 19). Fifteen species or subspecies have been raised and maintained in captivity. Note that captive management is under way for all Priority 1 species (2 spp.) and for some Priority 2 species (3 of 7 spp.). There is no active captive management for the Priority 3 and 4 species (5 spp.

TECHNICAL ADVANCES IN RESEARCH AND CONSERVATION

The scientific and conservation achievements of the past three decades have accelerated and laid the groundwork for increased future action. Research has provided methods to control many of the worst threats: various ungulates, certain alien weeds, and mammalian predators. All endangered forest birds have updated and useful recovery plans.

CAPTIVE MANAGEMENT ACHIEVEMENTS

Although too late to save the rarest species, huge advances have recently been made in the development of captive management techniques and programs in Hawaiʻi (Table 25.3). Achievements include research on captive populations for most forest bird species, the establishment of breeding and release programs for several, and saving the ʻAlalā, which now lives only in captivity. There is now considerable capacity to breed endangered species in captivity. Reintroductions with either captive-reared (for example, Puaiohi and Palila) or wild birds (Palila and Northwestern Hawaiian Islands passerines) are still at the development stage but hold much promise (Chapters 19 and 23).

DEVELOPMENT OF DISEASE RESISTANCE

Some species of Hawaiian forest birds show encouraging signs of resistance to avian diseases. Two species of ʻamakihi appear to be adapting to avian malaria and are extending their populations in the lowlands of Oʻahu and Hawaiʻi islands (Chapter 9). These signs may indicate that other species are capable of eventually developing resistance to alien diseases and adapting to nonnative habitats. We must "buy them time" through habitat improvement and other management actions.

AN INCREASE IN PUBLIC INTEREST

Overall public awareness and support for conservation of native Hawaiian species and ecosystems are improving (Dayer et al. 2006).

ALIEN SPECIES

In the remainder of this chapter we explore four major challenges to the conservation of Hawaiian forest birds. Alien species are the first challenge. Our book reveals that alien species pose the largest threat to Hawaiian forest birds today. The number of alien species and their impacts on native biota increase yearly (Chapter 15), and there are few means available to control or eliminate alien species once they become established and widespread (Chapters 15 and 16). The current levels of on-the-ground control and quarantine simply do not keep up with invasions. Despite this gloomy outlook, it is important to remember that the high-elevation habitats upon which forest birds depend are still composed mainly of native species (Chapters 6 and 7).

Protecting native ecosystems from alien species is fundamental to the survival of Hawaiian forest birds. To effectively control alien species already present requires application of abatement procedures at an appropriate scale, about which we say more in the section headed "Implementing

---

and 3 spp., respectively). There is an urgent need to bring to full capacity the captive management of all 9 species of Priority 1 and 2 birds and for research on the Priority 3 and 4 species.
[a]Priority ranks: 1, critical need for captive management; 2, great need for same; 3, need for captive management, but other conservation options are superior; 4, surrogate for other species more in need.
[b]Indicates whether reared in captivity 1987–2009.
[c]Actions ongoing in 2009.

Conservation at an Effective Scale for Recovery." Although the list of established alien species is already long, the number of potential invaders is even longer (Loope et al. 2001). Detecting and eliminating incipient populations of invaders is more feasible and cost-effective than waiting to control them after they are widely established.

The best ways to keep out unwanted species are to further strengthen risk assessment and inspections and resolve jurisdiction issues (Chapter 15). Risk assessments of arrival pathways and taxonomic groups focus on documenting threats by species not yet found in Hawai'i. Protecting the islands from pests coming from the U.S. mainland is a task funded and implemented by the state, which currently has limited resources to apply to the problem, although these have increased in recent years. Rather than allowing entry of all species except those on a restricted entry list, as is currently done, some have advocated the use of a system followed by New Zealand of allowing in only species assessed to be environmentally safe while restricting all other species. Protection from species coming in through international arrivals is generally under federal jurisdiction and is mainly concerned with relatively few pests, rarely those that could threaten Hawaiian ecosystems. Expanding the jurisdiction and resources of the state's Department of Agriculture could further restrict the import of harmful species that could arrive from other countries.

CLIMATE CHANGE

Effects on Hawaiian Ecosystems and Birds

Today, native forest birds face climatic changes unprecedented in both magnitude and speed. Current models predict an overall temperature increase in the tropics of about 2–3° C by the year 2100 (Neelin et al. 2006), and in fact increases in mean temperatures have already occurred in Hawai'i (Giambelluca and Luke 2007). The sea's

level is expected to rise, and that will significantly reduce the size of the Northwestern Hawaiian Islands, especially the lowest islands (Baker et al. 2006). Throughout the archipelago, predicted changes associated with this temperature increase include changes in mean precipitation, with unpredictable effects on local environments; increased occurrence of drought cycles; and increases in the intensity and number of hurricanes (Loope and Giambelluca 1998). On the main islands, vegetation zones are expected to shift upslope. High-elevation habitats of native forest birds may experience dramatic change because they are located in cloud forests just under the tropical inversion layer that controls the altitudinal limits of rainfall. Above the inversion layer, there is a rapid decline in rainfall, which is more important than temperature in determining the upper limit of montane rain forest on Maui and Hawai'i (Giambelluca and Luke 2007). This zone experienced major shifts during past cycles of climate change (Gavenda 1992, Hotchkiss et al. 2000).

Given that most remaining native forest bird habitat is limited to narrow belts around the summits of the higher islands, these changes could lead to a reduction of the remaining habitats and to increasing fragmentation. We can also expect to see the movement upslope of established invasive species that favor warmer temperatures: mosquitoes and bird disease, invasive plants, and parasitoids that deplete the insect prey base of birds (Loope and Giambelluca 1998; Benning et al. 2002; Peck et al. 2008). In short, areas that are currently capable of supporting large populations of forest birds may lose that capability within a matter of decades. Kaua'i is the island of most immediate concern (Fig. 25.5, Appendix 25.1). If climate change happens slowly enough, it is conceivable that native forests on Maui and Hawai'i might be able to stay one step ahead by expanding their ranges into current subalpine habitats. However, some recent analyses indicate

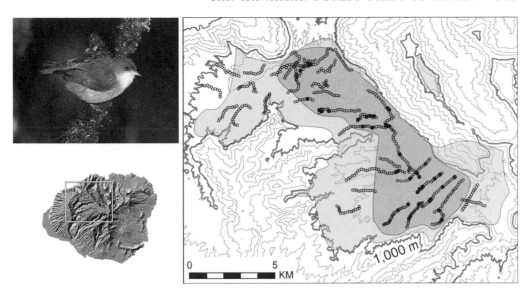

Figure 25.5. The 'Akikiki and the contraction of its range since the 1960s. The 'Akikiki is a rapidly declining species emblematic of the beleaguered Kaua'i avifauna. Kaua'i Island has lost five bird species since the 1960s, and the 'Akikiki could well be next. Main threats to the 'Akikiki are probably mosquito-borne bird diseases and habitat damage from hurricanes. A petition has been submitted to list the 'Akikiki as an endangered species. Restricted to the Alaka'i Plateau (inset, contour intervals at 200 m), the 'Akikiki has contracted its range (dark shading) to about half the area occupied in the late 1960s (light shading). The inset shows 'Akikiki detections in 2000–2005 (points) relative to locations surveyed (circles). Photo © Jack Jeffrey. Maps drafted by M. Gorresen.

that the inversion layer has been stable or increasing in frequency during the past 25 years (Cao et al. 2007, Giambelluca and Luke 2007). If this trend continues in a warming climate, forest habitats will likely remain capped at their current elevations, and forest birds will not be able to escape the upward march of disease and invasive species.

Adapting Bird Conservation to Climate Change

Given the current uncertainty of climate forecasts for the Hawaiian Islands, the best we can do is strive to keep existing forest bird habitat resilient to change. Key to this effort will be work to preserve and foster biodiversity in native ecosystems as a whole and phenotypic or genetic diversity and adaptability within the individual species that comprise these natural systems. Also key to fostering resiliency is the concept of watershed management (Chapter 16), because native plants and animals will need as much range of elevation as possible to be able to adapt to changing conditions. Conservation in Hawai'i is already heading in this direction by consolidating and managing large tracts of native forests through multiagency partnerships, removing ungulates from fenced units, reforesting high-elevation habitats, and identifying and eradicating incipient invasive species.

If climate change approaches worst-case scenarios, we will not be able to maintain native ecosystems as they once were. Three novel approaches may become necessary.

1. We will need to think of ways to save individual bird species in modified habitats with alternative composition of native plant species. In many locations this may mean shifting the tree community toward species better adapted to drier conditions. Similarly, if a stable inversion layer

allows disease to invade the last high-elevation rain forest refugia, we may also need to broaden our restoration efforts beyond the present emphasis on rain forests and shift the focus to montane and subalpine mesic and dry forest habitats, where many native bird species were once more common and where mosquito control is more practical. A key management concern for birds in these habitats is predator control (see Fig. 25.3).

2. Certain bird species may have to be moved outside of their historic ranges, creating with other species artificial associations with no prior historic or evolutionary precedent. Translocation planning for the three Northwestern Hawaiian Islands species must consider that the higher-elevation islands such as Lisianski, Laysan, and Nihoa offer longer-term refuge. One or more Kaua'i forest species may need to be moved to an island with a higher elevation, specifically Maui, to avoid extinction from avian disease. An important consideration for these introductions is the potential for conflict with native species already present, involving issues such as competition with other birds for food, depletion of threatened native food species, and hybridization with closely related species.

3. A last resort is to maintain species in captivity indefinitely (Chapter 19) or to reserve germ plasm if they become extinct, with the ultimate hope that someday we will be able to resuscitate the species.

IMPLEMENTING CONSERVATION AT AN EFFECTIVE SCALE FOR RECOVERY

The Problem of Scale

Bird conservation in Hawai'i tends to be folded into landscape-level protection of native biotic communities (Chapter 16). The explanations for this tendency are numerous and practical. There are so many endangered animal and plant species in Hawai'i (hundreds) that few can be given individual attention. Most endangered species are concentrated in the same geographic hotspots and are threatened by the same alien species, so protection and management have targeted concentrations of biodiversity. It is economically and strategically efficient to coordinate efforts among conservation partners and programs to protect areas of shared concern; the island watershed partnerships are a prime example. The advantage to this approach of saving ecosystems by managing the alien species—the ones that can be managed—is that both habitats and imperiled native species may respond positively.

The drawbacks to relying on ecosystem management alone to recover endangered birds are that some species may continue to decline because specific threats are not controlled, key areas are not protected, or some endangered populations require direct management and therefore focused attention. We must realize that only five forest bird species are being managed directly in the field—including predator control, nest protection, and reintroduction—or in captivity: the 'Alalā, O'ahu Elepaio, Puaiohi, Palila, and Maui Parrotbill (Chapters 20, 22, and 23). In addition, the three Northwestern Hawaiian Island songbird species are being considered for translocation. The remaining five listed endangered species and the two petitioned species must rely on whatever habitat management is done within their geographic range.

Are the recovery efforts for Hawaiian forest birds sufficient? Butchart et al. (2006) reviewed the status of all critically endangered birds worldwide for 1994–2004 and concluded that only 16 species were saved from extinction during this period; not one was a Hawaiian forest bird. When the recovery needs of an individual species are not strategically and specifically implemented, there is increased risk of losing that species.

Recovery Goals for Different Birds

Recovery of endangered Hawaiian forest birds is guided by four recovery plans, one each for the 'Io, the 'Alalā, the three North-western Hawaiian Islands passerines, and all other forest birds (U.S. Fish and Wildlife Service 1984a, 1984b, 2006, 2009). Recovery goals for delisting as stated in the plans are summarized here.

'IO OR HAWAIIAN HAWK

Delisting criteria have not been set. The objective of the plan is a self-sustaining wild population of 1,500–2,500 birds in stable, secure habitat. (At the time this book was being written, the Fish and Wildlife Service was actively attempting to delist the species.)

'ALALĀ OR HAWAIIAN CROW

Delisting criteria have not been set. The objective is to restore multiple self-sustaining populations in the wild in secure habitat within the species' historic range.

NORTHWESTERN HAWAIIAN ISLANDS PASSERINES

Delisting criteria have not been set. The objectives are to maintain the existing populations and protect their habitat. Actions to be taken include introducing the Nihoa Millerbird to Laysan.

ALL OTHER FOREST BIRDS

Downlisting and eventual delisting depend on four criteria: (1) the species occurs in two or more viable populations or a metapopulation, (2) population viability is determined by quantitative population surveys or demographic monitoring, (3) sufficient habitat in recovery areas is protected and managed to achieve criteria 1 and 2, and (4) threats responsible for the species' decline in the first place have been identified and controlled. These recovery criteria are more specifically defined for some species.

GENERAL RECOVERY GOALS

In summary, then, recovery depends upon halting species' declines through habitat protection and improvement and upon achieving multiple populations or a viable metapopulation—a difficult task for species limited to a singe restricted range, such as Kaua'i and Maui island endemics and the Palila on Mauna Kea. Recovery goals also prescribe monitoring to measure progress. In the rest of this section we discuss the scale required for the various recovery tasks, ranging from habitat improvement to population management.

Recovery at the Landscape Scale

HOW MUCH HABITAT?

There is now little forest remaining that is actually habitable for most forest birds. The majority of species require forest above 1,500 m in elevation to escape mosquito-borne disease and other threats. Among the five smaller islands, the extent of such forest ranges from functionally 0 ha on O'ahu, Moloka'i, and Lāna'i, to 9 ha (22 acres) on Kaua'i and 6,500 ha (16,000 acres) on Maui (see Fig. 25.3). On Hawai'i Island, four volcanoes exceed that height and together offer about 56,700 ha (140,000 acres) of "disease-free" forested habitat. If upland pastureland were reforested with native trees, the bird habitat on Maui and Hawai'i could be doubled. The percentages of actively managed high-elevation habitat on each island are currently 0% on Kaua'i, 74% on Maui, and 40% on Hawai'i (Chapter 16).

This approach oversimplifies matters, because nearly all bird populations also occasionally use forests at middle elevations. Furthermore, midelevation forests are required to maintain the ecosystem integrity of the native forest tracts of which high-elevation habitats are a part. By taking in potentially reforested areas and some mid-elevation habitat, the recovery areas drawn in the *Revised Recovery Plan for Hawaiian Forest Birds* (U.S. Fish and Wildlife Service 2006) were expanded to 9,400 ha (23,000 acres) for Kaua'i, 26,700 ha (66,000 acres) for O'ahu, 47,000 ha (116,000 acres) for Maui and Moloka'i, and 282,000 ha (697,000 acres) for Hawai'i (see Fig. 25.4). Two important points can be made from these calculations:

1. There is not much suitable habitat left for Hawaiian forest birds, and only a fraction of this habitat is being actively managed to control threats at an ecologically meaningful scale.
2. Arguably, an advantage of small habitat areas is that many forms of habitat management would be feasible across those areas (given the resources and sociopolitical will available). For example, the recovery area for endangered Kaua'i birds is small enough that ungulate and weed control would be possible across most of that area based on management precedents elsewhere in Hawai'i, such as in the East Maui Watershed Partnership area.

IS THIS ENOUGH HABITAT?

Because of the small amount of safe habitat remaining, the carrying capacity for bird populations is correspondingly limited. Setting sustainable target population sizes for recovery planning for each species is hampered by habitat availability. The *Revised Recovery Plan for Hawaiian Forest Birds* attempts to compensate for this limitation by including midelevation forests,

but as yet there is no way to do this in practice because of mosquito-borne avian disease and other threats at these elevations.

For the foreseeable future, we must expect to work with population sizes for endangered species of a few hundred to a few thousand mainly at high elevations or on a few small predator- and mosquito-free islands. Maintaining small populations in these limited areas will require extensive and stringent control of the most damaging alien species, and locally it will be necessary to reforest with native foundation tree species and restore rare plant species important to the birds.

SCALES OF HABITAT MANAGEMENT

Some extremely important research and management actions are being carried out at a large scale (thousands of hectares). These include forest bird monitoring, reforestation, ungulate removal and fencing, and biocontrol for a few plant species such as miconia (*Miconia calvescens*) and banana poka (*Passiflora tarminiana*). Landscape partnership planning occurs at this scale (Chapter 16). Examples of small-scale actions (tens or hundreds of hectares) are population research, predator control, understory planting, and chemical and mechanical removal of weeds. Reintroduction of forest birds to new sites is still at an early stage of development and has yet to achieve large-scale success. On the whole, there is room to do a great deal more management at meaningful landscape scales. Habitat restoration efforts are accelerating (Chapter 16), so a reversal in habitat and species declines is possible.

MANAGEMENT OF EXISTING HABITAT

Habitat protection is already focused on the large forest blocks, but only small areas within these blocks are managed. There is a need to further improve management across landscapes through partnerships, es-

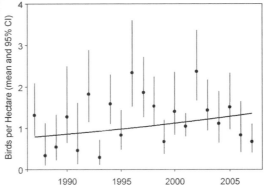

Figure 25.6. The Hawai'i Creeper, its trend line, and Hakalau Forest National Wildlife Refuge, Hawai'i Island.
Habitat improvement has led to stable or increasing
bird populations at Hakalau; note the lines of planted
trees in the photo taken in 2004. The 13,300 ha refuge
was created to protect endangered forest birds. Habitat
management since 1987 has restored damaged native
forest and reforested 2,000 ha that were formerly pasture. Annual monitoring of the eight native songbird
species has shown that their populations are stable or
increasing (Camp, Pratt, et al. 2009). In contrast, native bird populations are decreasing in deteriorating
habitat elsewhere on Hawai'i Island. Habitat management holds great promise for native forest birds across
the state. Photos © Jack Jeffrey.

pecially on Kaua'i and Hawai'i islands,
where proportionately less habitat for
endangered birds is actively managed.
Conservation management is rapidly becoming more widespread and effective in
Hawai'i, where it is aimed both at preserving native biota and ecosystems and at
increasing natural resources such as watersheds and native timber, especially koa
(*Acacia koa*) (Chapter 16).

REFORESTATION TO EXPAND HABITAT

A new and exciting development has
been the reforestation of upland pastures
(manmade grasslands) on Hawai'i Island
and more recently on Maui (Fig. 25.6;

Plate 18) (Chapter 16). Koa are planted or encouraged to recolonize the site, while other native trees and plants fill in the understory, albeit slowly where there is dense grass cover. For the first time, forest bird habitat is beginning to increase at a significant scale, and reconnection of some forest patches seems possible, allowing for expansion and connection of bird populations. Most of this reforestation is happening at high elevations, where the risk of bird diseases is lowest. Although some specialist bird species do not colonize forests at a midseral stage, others, most notably 'Akiapōlā'au, do so readily (Pejchar et al. 2005). If this trend continues, many forest bird populations may stabilize or increase substantially over the short or long term, depending upon the species' response.

ALIEN SPECIES MANAGEMENT

Previously, we underscored the importance of alien species prevention and management. This must be done at an effective scale to protect Hawaiian forest birds and their habitats. Some forms of management, such as ungulate control to improve habitat, are technologically advanced and mainly require expanded application to be effective (Chapters 16). Others, such as predator control, depend on the development of technology and public and regulatory approval for broad-scale application (Chapter 18). Methods are available—but others have yet to be developed—to reduce mosquito populations and disrupt transmission of bird diseases (Chapter 17). However, control of most alien species, particularly plant and insect pests, is not yet feasible at any practical scale, and research on the means for control will first be necessary. Biological control of plant pests, an approach with great potential, is probably most in need of development to protect Hawaiian ecosystems (Denslow and Johnson 2006; Messing and Wright 2006; this volume, Chapter 16). The critical importance of preventing new intro-

ductions has already been emphasized (Chapter 15).

Recovery at the Species Level

SPECIES-LEVEL PROGRAMS

In addition to the current approach of conserving communities of native bird species through ecosystem management, recovery of Hawaiian forest birds needs to focus on species populations (see Appendix 25.1). The actions necessary include ecological studies to understand the nature and effects of limiting factors, strategic planning for species recovery, intervention management to reverse declines of the rarest species in the wild, captive management, reintroductions, and monitoring (see Table 25.3 and Box 25.1).

REINTRODUCTIONS

Reintroducing birds to restored habitat is a potentially productive means of increasing populations that has seldom been attempted in Hawai'i (Chapters 20, 22, and 23). However, for reintroductions to be successful, it is first necessary to eliminate or control the main limiting factors in the habitat, for example, mammalian predators of the Puaiohi and sheep that eat trees upon which Palila depend. New reintroduction techniques could be successfully applied to expand the ranges of endangered birds such as the 'Akiapōlā'au (see Table 25.2 and Appendix 25.1). Some nonendangered native birds would also benefit from range expansion efforts, such as reestablishing the 'Ōma'o in the Kona and Kohala districts and the Maui 'Alauahio on West Maui.

INTRODUCTIONS BEYOND
HISTORIC RANGES

We also recommend assessing the possibility of introducing native forest bird species to Hawaiian islands outside their

> ### Box 25.1. Nine Biological Research Priorities for the Conservation of Hawaiian Forest Birds, Grouped by Topic
>
> Some of the priorities listed do not involve birds directly but nevertheless are fundamental to their long-term survival.
>
> 1. Mitigation of avian diseases of native birds.
>   a. Effective methods to control disease-vectoring mosquitoes over large landscapes
>   b. Development of disease resistance in native birds
> 2. Pathway analyses and techniques for preventing the introduction of new invasive species, especially
>   a. Cold-tolerant mosquitoes
>   b. Avian diseases
>   c. Pathogens of native forest trees
>   d. New predators such as the brown treesnake
>   e. Invasive, cold-tolerant forest trees, vines, and shrubs
> 3. Effective, socially acceptable control methods for alien predators of native birds, specifically rats and feral cats.
> 4. Control techniques for ecosystem-altering invasive plants, especially those that spread into intact high-elevation forests or promote fire.
> 5. Captive propagation techniques for all at-risk bird species that emphasize maintaining the genetic viability of species for extended periods (>100 years).
> 6. Effective methods to introduce or reintroduce forest birds, including postrelease monitoring of sufficient duration.
> 7. Feasibility assessment of translocating forest birds outside of their historic ranges to islands with reduced threats.
> 8. Contingency plans to address the effects of future climate change on forest birds' habitat quality, habitat connectivity, predators, and diseases and vectors.
> 9. Population monitoring to evaluate species' vital rates, response to management, and population trends to aid managers and researchers in assessing management and planning options.

historic range, an option not considered in the four recovery plans (see Appendix 25.1 for specific recommendations). Translocating endangered birds between islands has been an enormously successful strategy for endemics on Indian and Pacific islands, especially in New Zealand (Armstrong and McLean 1995), but also recently in Hawai'i for the Laysan Duck (*Anas laysanensis*) (Reynolds et al. 2008). Such action could serve three purposes for Hawaiian forest birds: (1) increasing the species' populations when there is inadequate recovery habitat in the recent native range (Laysan, Nihoa, and Kaua'i species), (2) creating secondary populations as insurance for the primary population, and (3) restoring ecosystem function vacated by the extinction of a related species, as in the potential case of moving the seed-dispersing Puaiohi or 'Ōma'o to Maui to replace their extirpated counterpart, the 'Oloma'o.

SOCIAL FACTORS IN HAWAIIAN FOREST BIRD CONSERVATION

Forest Birds and People: Worlds Apart

Today, people and forest birds in Hawai'i typically occupy different habitats. Human settlement is mostly along the coast and shared by a few dozen species of alien birds.

In the uplands, settlement is constrained by zoning laws. For the most part, native forest birds live in upland forests on conservation lands.

An unfortunate consequence of the separation of people from birds is that most Hawaiian residents have little, if any, personal experience with the native forest birds (Chapter 24). Nevertheless, whether through cultural awareness (Chapter 3), education, or the news media, most Hawaiian citizens know of some forest birds, extinct and living, and they are certainly aware of their plight. A recent survey, *Wildlife Values of the West*, documented strong statewide support for many conservation activities in Hawai'i, including strong support for protecting endangered species (Dayer et al. 2006). Nevertheless, active public involvement in the conservation of native biota is limited and often compromised by other interests and values. For example, a longstanding, popular competing use of Hawaiian wild lands is for the hunting of feral ungulates, and many hunters want to maintain and promote high numbers of game mammals, especially pigs and mouflon sheep, in spite of the environmental damage they cause (Stone 1985, Stone and Loope 1987). There is substantial opportunity for social scientists to facilitate discussions of conservation values and ethics in Hawai'i and contribute to conservation by understanding and capitalizing upon beliefs and attitudes toward nature held among the varied cultures of the islands.

Much could be done to increase the involvement of Hawaii's people with native birds (Chapter 24). Many rare bird species are inaccessible, even to determined birders. However, there is now an opportunity to provide ready public access to almost all species at key sites and to facilitate personal involvement in recreation (birding), research, and conservation. A good example of birder access is the boardwalk trail system built by the state in the western Alaka'i on Kaua'i Island. Here hikers have a chance to see all the Kaua'i forest birds.

Service trips and volunteer programs are another excellent way to connect people with native birds and their habitat. All the major Hawaiian conservation organizations have such programs. Volunteers not only contribute to conservation but also gain a deeper understanding and appreciation of island ecology and memorable personal experience in research and natural resource management.

A greater problem is the general remoteness of Hawai'i from the rest of the United States (Chapter 24). Visitors rarely have the opportunity to gain little more than a brief understanding and appreciation of rare native birdlife. Hawaiian birds are absent from American field guides and other bird books from which the nation's people learn about their birds (see Box 24.5). The fact that Hawaiian forest birds are foreign to mainland residents does not help to inspire the degree of support needed to save them.

Support for Recovery

It is beyond the scope of this chapter to delve into the complexity of agency conservation programs for Hawaiian forest birds; interested readers are referred to Chapter 24. However, it is important to realize that Hawaii's limited tax base and the comparatively low level of funding for the Hawai'i Department of Land and Natural Resources (HDLNR) translates into limited state funding for native wildlife programs (Tummons 2005a; this volume, Chapter 24). Funding for HDLNR is crucial because all core bird habitats are partly or mostly on state lands managed by this agency: the Alaka'i Plateau on Kaua'i, windward Haleakalā on Maui, and, on Hawai'i, subalpine Mauna Kea, the windward rain forests of Mauna Kea and Mauna Loa, and the vast Ka'ū rain forest. There are many worthy state programs, such as the Hawai'i Natural Area Reserve System, which in 2005 received greatly increased funding from the Natural Area Fund and the state Legacy Act (Dawson 2005). However, because of Hawaii's limited tax revenue, federal funding is also crucial.

Hawaii's geographic separation leads to political isolation at the federal level (Chapter 24). With few exceptions, Hawai'i cannot participate in large conservation programs that spread across state lines, and many funding initiatives need multistate support to win approval from Congress. For example, programs aimed to protect migratory songbirds—the mainland ecological counterparts to Hawaiian forest birds—do not include Hawai'i. Funding for endangered species, for the national parks, and for various state assistance programs (e.g., Pittman-Robertson funds, the U.S. Fish and Wildlife Section 6 program, and the Landowner Incentive Program) is the main source of revenue for forest bird conservation in Hawai'i.

Endangered Species Support: Hawai'i Compared with the Mainland

Restani and Marzluff (2001, 2002a) and Leonard (2008) recently reported that current patterns of spending on endangered species in the United States resulted in proportionately less funding for island species, decreasing these species' long-term viability. Mainland species typically received much more funding, especially when they were widespread, had a high potential for recovery, or had a captive propagation program. Species that stood in the way of economic development or government projects also received higher funding. In Hawai'i, two well-funded endangered forest birds—the 'Alalā and Palila—are the only two species whose recovery involved litigation. How does funding for endangered Hawaiian bird species compare with that for endangered birds on the U.S. mainland?

On a per species basis, Hawaiian forest birds fare very poorly nationwide (Fig.

25.7). We encourage the reader to turn to Chapter 24 for an analysis of Hawaii's funding shortfall. Leonard (2008) analyzed total governmental recovery expenditures on all U.S.-listed bird species for 1996–2004 to compare Hawaiian and mainland birds. Data on recovery priority and expenditure were obtained from state and federal Endangered Species Expenditure reports (1996–2004) and Reports to Congress on the Recovery of Threatened and Endangered Species (1996–2004) (found at www.fws.gov/endangered/pubs/index.html). Although Hawaiian birds amounted to a third of the species listed, they received only 4.1% ($30,592,692) of the total spent on the recovery of all listed birds by federal and state agencies for the nine-year period (Leonard 2008; this volume, Chapter 24). In other words, mainland endangered birds received 15 times more funding on average despite being assigned similar priority rankings. The highest-funded species, the Red-cockaded Woodpecker (*Picoides borealis*), received $11.7 million annually—three times the funding that went to all 31 listed Hawaiian bird species combined. On a per species basis, the annual expenditures on Hawaiian birds were $112,000. This annual per species figure is problematic compared with what is really needed to prevent the extinction of more Hawaiian forest birds.

The Cost of Saving Hawaiian Forest Birds

How much will it cost to save Hawaiian forest birds? Estimating such costs is extremely difficult, and endangered species programs typically lack full funding. The *Revised Recovery Plan for Hawaiian Forest Birds* estimates this cost at $83.3 million per year for 18 listed species, or $4.6 million per year per species (U.S. Fish and Wildlife Service 2006: xiii; this volume, Chapter 24). This total includes all recommended actions in a best-of-all-worlds scenario. If only first-priority actions are funded—

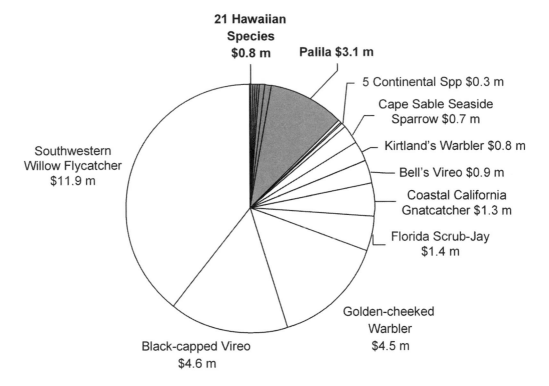

Figure 25.7. Comparative costs of endangered species programs for passerines in Hawai'i versus the continental United States, 2004. These are species-specific, activity-related expenditures in millions of dollars by federal and state governments (not counting land purchase) (U.S. Fish and Wildlife Service 2004c). Species-specific management of Hawaiian forest birds (shaded) lags behind that of their mainland counterparts. Only the Palila is funded at a level on a par with that for the best-funded mainland passerines. The endangered mainland passerines are the Southwestern Willow Flycatcher (*Empydonax triallii extimus*), Black-capped Vireo (*Vireo atricapilla*), Golden-cheeked Warbler (*Dendroica chrysoparia*), Florida Scrub-Jay (*Aphelocoma coerulescens*), Coastal California Gnatcatcher (*Polioptila californica californica*), Least Bell's Vireo (*Vireo bellii pusillus*), Kirtland's Warbler (*Dendroica kirtlandii*), and Cape Sable Seaside Sparrow (*Ammodramus maritimus mirabilis*).

"those actions that must be taken to prevent extinction or prevent the species from declining irreversibly in the foreseeable future"—the cost is $39.4 million per year for all species, or $1.9 million per species per year. Consider, however, that many of these costs involve forest management—costs that are already borne by or could be

shared with land management agencies for the conservation of watersheds and native ecosystems. Returning to Leonard's (2008) comparison, the projected costs of recovering Hawaiian birds fall within ballpark range of the costs of recovering the 20 top-funded U.S. birds (none of which are Hawaiian forest birds), which average $3.5 million per species per year.

Despite the absence of this level of funding to date, some progress has already been made to locally stabilize and, in some cases, increase forest bird populations in Hawai'i. Both the Fish and Wildlife Service and the National Park Service have acquired important blocks of habitat to create refuges for forest birds, notably at Hakalau and more recently at Kahuku on Hawai'i Island. On the upper windward slopes of Haleakalā, Maui, HDLNR has fenced and removed pigs from key areas of recovery habitat. Collaboration with private landowners has been especially important in strategizing and implementing forest protection and management, particularly through The Nature Conservancy of Hawai'i; Kamehameha Schools; Alexander and Baldwin, Inc.; Haleakalā Ranch Company; Maui Land and Pineapple Company; and Moloka'i Ranch. Additionally, The Peregrine Fund and the San Diego Zoo, in collaboration with HDLNR, have developed a sophisticated program for the captive management of forest birds. Conservation interests within Hawai'i have accomplished a great deal, but will it be enough?

The past two decades of research and conservation have shown us how to save the Hawaiian forest birds. Nonetheless, without a greater conservation effort most of these birds will continue down the road to extinction. Upward trends are achievable, as we can see from local results in managed habitat such as Hakalau Forest National Wildlife Refuge (Camp, Pratt, et al. 2009) (see Fig. 25.6). Increasing the value of native birds both locally in Hawai'i and on the mainland would lead to an increase in conservation efforts and improve the status of at least some bird species. In general, data on national endangered species programs statistically show an expected pattern: "Species that have higher proportional spending have an improved chance of achieving a status of improving or stable" (Miller et al. 2002: 167; see also Male and Bean 2005). The status of Hawaiian forest birds, too, could well improve with a greater commitment to their conservation.

FINAL THOUGHTS

Hawaii's birds have always been considered one of the most spectacular and unique avifaunas in the world—the ultimate showcase of bird evolution on islands. Before humans arrived in Hawai'i, evolution had crafted more than 110 species of birds found nowhere else on earth. Sadly, only about half of the historically known forest birds survive today, 24 species in all. Those that remain, however, are special indeed,

and we have the opportunity to help these species continue into the next century. Much of the information and planning needed to guide recovery is available. Investing in the recovery of at least several species has demonstrated the gains that can be realized by not waiting until populations are reduced to tiny remnants. But Hawaiian bird conservation is now at a critical juncture. To prevent further extinctions, accelerated conservation efforts are urgently needed over much larger areas in the archipelago. Success in this endeavor will require dedicated funding and strong leadership to both build recovery at landscape scales and coordinate strategic action to restore all endangered bird species. These measures will also keep the common bird species common. To accomplish these goals, we, as a nation and a state, face a crucial challenge —summoning the will to save the last representatives of the world's most unique island avifauna while there is still time.

ACKNOWLEDGMENTS

The editors especially thank J. M. (Mike) Scott for his inspiration, encouragement, and lively discussion on the future of Hawaiian forest birds. We thank our colleagues for their insights, ideas, and commitment to saving native forest birds and their habitats. We also thank the many authors of the preceding book chapters for their generous and supportive involvement with this book and for providing information pertinent to the conservation of Hawaiian forest birds. We also thank D. Buffington and other high-level biologists within the U.S. Geological Survey for initiating and supporting the science behind this book project. G. Butcher, S. Conant, D. Duffy, D. Helweg, F. Kraus, D. La-Pointe, A. Lieberman, T. Male, J. Nelson, M. Reynolds, J. M. Scott, C. Smith, and D. Wilcove reviewed early versions of this chapter and offered improvements for the final draft.

APPENDIX 25.1. PREVENTING FURTHER EXTINCTIONS OF HAWAIIAN BIRDS

Recovery plans prescribe a comprehensive range of measures to recover endangered Hawaiian forest birds (U.S. Fish and Wildlife Service 1984a, 1984b, 2003, 2006). However, recovery has been difficult to strategically implement across so many species. Here we itemize the most-needed species-specific actions. Species are ordered by island. 'Akikiki and 'Akeke'e are not yet listed as endangered.

The Northwestern Hawaiian Islands

The following three songbirds are threatened because they live on a few tiny and remote islands still free from mammalian predators, bird disease, and other threats found on the main Hawaiian islands. Any one of these populations will vanish quickly if rats, disease, fire, prolonged drought, or some other ecological calamity befalls its island.

At the time of this writing, a translocation plan for these species is being drafted by the U.S. Fish and Wildlife Service whereby each species would be introduced to one or more unoccupied islands (Morin and Conant 2007). These new island populations would greatly increase the overall population size and range of each species and reduce the risk of extinction from events on any one island. But most of these islands are low—no higher than a few meters above sea level. If we consider the likely shrinking or loss of some Northwestern Islands due to rising sea levels resulting from global warming (Baker et al. 2006), one conclusion we might reach is that it may be prudent in the long run to attempt translocating these species to higher islands in the main Hawaiian chain, such as Lehua or Kaho'olawe, or elsewhere in the Pacific (Wake Island). However, first it will be necessary to remove mosquitoes and mammalian predators from these islands.

MILLERBIRD (NIHOA)

The Millerbird, rarest of the Northwestern Islands songbirds, is being proposed for introduction to Laysan—an island the species once occupied—and to other islands in the Northwestern chain (Kure and Lisianski). Research on the Millerbird in captivity is needed to enhance the likelihood of successful introductions of this insectivorous species.

LAYSAN FINCH (LAYSAN ISLAND AND PEARL AND HERMES REEF)

There is a good chance that the adaptable Laysan Finch could be successfully introduced to other Northwestern islands, particularly Lisanski Island, which offers habitat similar to that on Laysan, although first it will be necessary to create freshwater sources on the island.

NIHOA FINCH (NIHOA)

Translocations of Nihoa Finches might be complicated by the absence of the finch's usual nest sites—rock cavities—on the sandy islands of the Northwestern chain (Laysan Finches nest in vegetation). Experimentation with artificial nest structures would show whether these are acceptable to Nihoa Finches. Kure and Midway islands have been proposed as translocation sites, but mosquitoes must first be eliminated from Midway. A further consideration is the concurrent need to enlarge the range of the Laysan Finch. Because the two closely related finches would compete for food and might hybridize, they would best be kept separated. For the Nihoa Finch, rocky islets off the main Hawaiian islands, such as Lehua Island, offer potential habitat but are within reach of disease-carrying mosquitoes. De-

CAN HAWAIIAN FOREST BIRDS BE SAVED? 577

spite such difficulties, experimental translocations seem advisable.

Kaua'i

Survival of endemic Kaua'i birds is of grave concern because the island peaks at 1,597 m, affording only a narrow band of high-elevation forest where transmission of bird disease is rare. As atmospheric temperatures rise, that safety zone may disappear entirely (Benning et al. 2002). Within the past 40 years, five Kaua'i bird species have vanished and another three have declined (Chapter 5). Disease appears to be the main threat. However, habitat for these species continues to deteriorate. The numbers of pigs, deer, and goats are not controlled except by hunting by the public near roads and trails, and ungulate damage to vegetation is severe in the heart of Kaua'i bird habitat. Several aggressive weeds are spreading into the interior of the mountains.

PUAIOHI

Habitat improvement through removal of ungulates and small mammalian predators is urgently needed for the Puaiohi. Reintroduction of captive-bred birds has demonstrated some success, but the wild population will not likely expand until these two limiting factors are brought under control at an effective scale (Chapter 22). Should transmission of mosquito-borne avian disease increase in the species' range and the wild population finally dwindle toward extinction, translocations of Puaiohi to disease-free elevations on Maui could be considered. Captive propagation and reintroduction will continue in recovery attempts.

'AKIKIKI

With a small and rapidly declining population, the 'Akikiki is the subject of a petition to the U.S. Fish and Wildlife Ser-

vice for listing as an endangered species (see Fig. 25.5). Much of the species' forest habitat was toppled by two hurricanes in 1982 and 1992. The 'Akikiki and the following species, the 'Akeke'e, are among the most poorly known Hawaiian birds. Field research on both species is urgently needed to formulate recommendations for their recovery, as is research on these birds in captivity.

'AKEKE'E

Together with the 'Akikiki, this declining species is the subject of a petition for listing as an endangered species. Ecological studies are needed to guide recovery, and captive management research on this species could assist its recovery.

O'ahu

O'ahu lost most of its native forest birds over a century ago. Today, no habitat on this island is safe from mosquitoes and avian disease because the mountains do not reach high enough elevations. Native forest still covers the mountains, but it is extensively modified by numerous nonnative plants, pigs, goats, and other alien species.

O'AHU ELEPAIO

Somewhat tolerant of disease, this declining species lives in a much fragmented metapopulation. Local populations have responded well to the rat control currently employed at a number of sites (Chapter 18; E. VanderWerf, pers. comm.). If expanded to a larger scale and strategically implemented, predator control could reverse regional declines. Reintroductions by means of translocating wild birds, coupled with predator control, could reestablish extirpated subpopulations. The proximity of the bird to the Honolulu metropolis affords much opportunity for community involvement on public and private lands.

Its occupation of military lands has led to recovery efforts by the Department of Defense.

Maui

The high windward slopes of Haleakalā Volcano offer some of the most intact forest bird habitat in the state and also some of the best possibilities for restoring native forest on pasturelands. Windward rain forests are well protected by a national park, state forest reserves, a Nature Conservancy reserve, private lands, and the all-encompassing Windward Maui Watershed Partnership. These partners are increasingly managing the forest, particularly by controlling pigs, goats, deer, and alien plants. Certain deforested lands on the drier, leeward slopes are slated for restoration. West Maui mountain forests are also well protected and managed but are small in area and do not reach elevations safe for endangered forest birds.

MAUI PARROTBILL

Habitat improvement is the main tool for recovering the parrotbill. Ongoing habitat restoration in rain forests and new reforestation initiatives in former pastures on the leeward slopes of Haleakalā should enable the population to grow. Habitat improvement may require predator control to increase adult survivorship and nesting success for this low-fecundity species. The parrotbill's ability to utilize fast-growing young koa forests is yet unknown but will set the pace for expanding its range. Long-term planning includes reintroduction to West Maui and Moloka'i if disease transmission there can be controlled. Captive birds are being bred for introduction to newly restored sites.

ʻĀKOHEKOHE

Recovery of the ʻĀkohekohe also depends upon habitat improvement and reforesta-tion. Removal of feral pigs across large reaches of the species' range is a positive development for the species. Understory vegetation provides a seasonal source of nectar for this territorial honeycreeper, allowing the bird to remain at higher elevations and avoid traveling at times of flower scarcity to lower elevations where the incidence of infection from mosquito-borne disease increases. Like the parrotbill, the ʻĀkohekohe may be encouraged to colonize restored habitat on leeward Haleakalā, assisted by translocations if necessary. (The aggressive birds have not been bred in captivity and have proven difficult to maintain under captive conditions.) Translocations to the species' former range on West Maui and Molokaʻi would be feasible only if disease can be managed at those sites.

Island of Hawaiʻi

Forest bird conservation on the "Big Island" of Hawaiʻi has much greater potential than on the smaller islands. The high-elevation habitat there is ten times the area of that on Maui, the only other island with similar mountainous geography. Also, habitat blocks on its three massive volcanoes allow for large multiple populations of endangered birds on Hawaiʻi, which is not possible on other islands. A national park, a national wildlife refuge, two Nature Conservancy reserves, some private conservation lands, and small areas of state lands actively manage some of the recovery habitat for endangered forest birds. In addition, other large areas, mostly on state lands, are not actively managed or require additional management in order to sustain endangered forest birds for the long term.

ʻIO

This adaptable raptor occupies a wide range of wooded habitats where it preys on rodents, birds, and insects (Chapter 12). It is neither affected by the diseases that have decimated native songbirds nor at present is it threatened to the same degree. The 'Io is considered lower risk or near threatened by the International Union for Conservation of Nature. As this book is going to press, the U.S. Fish and Wildlife Service is moving ahead to delist the 'Io as an endangered species. Threats include conversion of habitat to pasture and forests of alien species, urbanization, and the risk of introduction of West Nile virus.

'ALALĀ

A large reintroduction program is needed for the 'Alalā, which has been a most difficult species to recover and now lives only in captivity. Restoring former habitat and reestablishing the species in the wild are crucial goals for recovery (Chapter 20). Restoring this species where its native predator, the endangered 'Io, or Hawaiian Hawk, is abundant will be challenging, but increasing the forest cover may reduce hawk attacks.

PALILA

A long-term commitment for Palila management has yet to be agreed upon. The bird is confined to subalpine forest on Mauna Kea Volcano, where it is dependent on one species of tree, the māmane (*Sophora chrysophylla*), for food and nest sites. Habitat improvement depends on fire suppression, cat control, and eliminating mouflon and feral sheep from Mauna Kea. Stopping the spread of ants and other potential food competitors may also be important (Chapter 23).

'AKIAPŌLĀ'AU

The 'Akiapōlā'au currently has only two populations, one on windward Mauna Kea and Mauna Loa volcanoes and the other in Ka'ū on southeastern Mauna Loa. The species feeds on insects found in tree bark, and it readily colonizes young forests of koa, a fast-growing foundation tree of restoration projects and also a valuable timber tree. Because young koa can attain a size useful to 'Akiapōlā'au in less than ten years, the potential for increasing 'Akiapōlā'au populations through reforestation is very encouraging. However, the species is limited by a low rate of fecundity, so populations cannot easily recover from mortality due to disease, predation, and other factors. Hence, recovery must be targeted for forests above the zone of disease transmission. Protecting old forests and new restoration sites from ungulate damage is key to recovery of this species. Captive management or translocations of wild birds could assist reintroduction of 'Akiapōlā'au to restored, unoccupied habitat.

HAWAI'I CREEPER

The Hawai'i Creeper also takes insects from bark. It occupies the same range as the 'Akiapōlā'au and additionally occurs in remnant populations on western Mauna Loa and northern Hualālai. It also responds to forest restoration but is less dependent on koa. Forest protection and restoration throughout its high-elevation range are the species' main recovery needs.

'ĀKEPA

This species occupies the same four core areas as the Hawai'i Creeper, but 'Ākepa populations are more localized. Unlike the previous two species, the 'Ākepa has not yet been shown to increase its numbers in managed areas. Proposed reasons for its decline include avian disease, loss of nest trees, and competition for food from Japanese White-eyes (*Zosterops japonicus*) and nonnative wasps. ('Ākepa take small insects mainly from new foliage of 'ōhi'a trees, a foraging site where insects are also available as a food resource to these competitors.) 'Ākepa nest in tree cavities and require large trees to provide these cavities; as older trees with nest sites fall, they are slow to be replaced in forests long grazed by livestock and feral ungulates. Therefore, nest site management is key for the near-term recovery of 'Ākepa. This management must include restoring tree recruitment in old-growth forest and controlling nest predators. Research on artificial nest sites has so far yielded limited positive results but may still hold promise with further experimentation. 'Ākepa are likely to expand into reforestation sites only when trees reach the large size required to supply nest sites.

Hawaiian Pronunciation Guide and Glossary

VERNA L. U. AMANTE-HELWEG, LINDA W. PRATT, AND THANE K. PRATT

Although the Hawaiian language is written with seven consonants (h, k, l, m, n, p, w), it is pronounced with eight, as the consonant v is pronounced interchangeably with the consonant w (see the following). Hawaiian has five vowels (a, e, i, o, u) that are pronounced much as in Spanish. As in all languages, pronunciation is variable, and there can be dialectal differences. For example, Hawai'i can be "Hah-wai ee" or "Hah-vai ee."

Consonants

p and k pronounced about as in English but with less aspiration

h, l, m, and n about as in English

w like "w" or a soft "v," dependent on the accompanying vowel:

1. after i and e, usually like "v"
2. after o and u, usually like "w"
3. initially and after a, like either "w" or "v"

Vowels (Unstressed)

a as in "above" (close to "uh," as in "up")

e as in "bet"

i = "ee," as in "street"

o = "oh," as in "sole"

u = "oo," as in "moon"

Vowels (Stressed)

The Hawaiian kahakō, or macron, is a straight line over a vowel and indicates that the vowel sound is held longer (similar to the long vowel sound used in the English language) but is not as strong as in English; rather the sound is elongated or

581

lengthened. For example, "ah" becomes "ahh."

> ā as in "far"
> ē like ay in "play"
> ī as in "see"
> ō as in "bold"
> ū as in "bloom"

Glottal Stop

The Hawaiian 'okina ('), like the glottal stop, means our breath stops briefly as between the two parts of the English term "oh-oh."

Diphthongs

The diphthongs *ei, eu, oi, ou, ai, ae, ao,* and *au* are always stressed on the first letter but are not as closely joined as in English. The diphthong *ai* is pronounced as in the English word "eye."

Syllables

Every syllable ends with a vowel, and every syllable is pronounced—there are no silent letters. For example, the name of the Hawaiian god Kāne is pronounced "Kah-nay," that is, it does not sound like "cane," as in "sugar cane." *'O Ka 'Ōlelo Ke Ka'ā; O Ka Mauli.* (Oh kah oh-LEH-loh keh kah ahh; oh kah mah-oo-lee. "Language is the fiber that binds us to our cultural identity.")

Syllable Stress

Stress is a system that marks one syllable as more prominent than those around it. Knowledge of patterns of syllable stress requires familiarity with the language.

Pronunciation of Hawaiian Bird Names

We have attempted in the plate captions to help readers with the pronunciation of Hawaiian bird names. Our pronunciations are derived from the *Hawaiian Dictionary* (Pukui

and Elbert 1986) with interpretation by K. Awong and B. Camara at Hawai'i Volcanoes National Park. Stressed vowels are shown with an extra "h." A glottal stop is shown with a break. Where to stress syllables is shown by means of capitalization. For instance, po'o-uli = poh oh-OO-lee, and 'Ākohekohe = ahh-KOH-heh-KOH-heh.

GLOSSARY

This glossary is intended mainly to be used with the chapter headed "Hawaiian Culture and Forest Birds" but also includes the Hawaiian words appearing in the book except for place names. Most of the definitions in this glossary are adapted from the *Hawaiian Dictionary* (Pukui and Elbert 1986). For some words, we have taken the meanings as given in original sources or as inferred from the context in these sources.

'a'ā   one of the two types of lava; this type has a rough, slaglike surface.

'a'ali'i (*Dodonaea viscosa*)   a shrub or small tree in the soapberry family (Sapindaceae) widespread in the Hawaiian Islands and a community co-dominant of many lowland, montane, and subalpine shrublands.

a'e (*Sapindus saponaria*)   a large tree of the soapberry family (Sapindaceae) indigenous to the island of Hawai'i, where it is a component of rare mesic forests.

Ae'o (Black-necked Stilt, *Himantopus mexicanus knudseni*)   a native shorebird found mainly in brackish and freshwater ponds near the coast.

'aha'aina ho'omana'o   a great feast of commemoration; associated with Makahiki.

āhinahina (silverswords, *Argyroxiphium* spp.)   a group of rare endemic shrubs in the sunflower family (Asteraceae). These are iconic plants of subalpine communities and wet montane bogs.

ahupua'a   a land unit extending from upland to the sea; a traditional unit of resource management.

'ahu'ula   feathered capes and cloaks.

'Akeke'e (*Loxops caeruleirostris*)   a Hawaiian honeycreeper on Kaua'i that is the counterpart of the 'Ākepa.

'Ākepa (*Loxops coccineus*)   the smallest of the Hawaiian honeycreepers. In Hawaiian the name means "sprightly."

'akialoa (Greater 'Akialoa, *Hemignathus ellisianus*; Lesser 'Akialoa, *H. obscurus*; also 'akihi a loa)   a group of extinct honeycreeper species characterized by the extraordinarily long bill deployed for probing the hiding places of insects and for sipping nectar. The name might translate as "little bird of the long" bill.

'Akiapōlā'au (*Hemignathus munroi*)   a honeycreeper found only on Hawai'i Island. The Hawaiian name, 'akihi po'o lā'au, might translate to "little bird that pecks wood," alluding to its bizarre bill and woodpecker feeding habits.

'Akikiki (*Oreomystis bairdi*)   a small honeycreeper formerly known as Kaua'i Creeper that jerkily moves about on branches and trunks while searching for insects. It is endemic to Kaua'i.

'Ākohekohe (Crested Honeycreeper, *Palmeria dolei*)   the only honeycreeper with a crest; now restricted to Maui Island. The name is onomatopoeic for the bird's song.

'akoko (*Chamaesyce* spp.)   a group of fleshy-stemmed shrubs or small trees in the spurge family (Euphorbiaceae) that are components of lowland dry shrublands and montane dry forests.

akua   the highest-ranking spiritual force or power, i.e., god, goddess, spirit.

akua 'aumakua   extensions of the highest-ranking spiritual power; supernatural with dual roles as akua to unrelated persons and 'aumakua to relatives.

akua maoli   a god in human form.

akua 'ohana   family spirits or personal gods.

akua pā'ani   an image presenting the god of sports.

akua pili kino   high-ranking ancestral gods of chiefs and maka'āinana (commoners) who can appear in any chosen form.

akua poko   the god Lono displayed on a staff in the squat form of a man wearing a feathered helmet.

'Alae ke'oke'o or 'Alae kea (Hawaiian Coot, *Fulica alai*)   an endemic member of the rail family that is black and sports a prominent knob on its forehead that is white, red, or yellow. It lives mainly in wetlands near the coast.

'Alae 'ula (Common Moorhen, *Gallinula chloropus sandvicensis*)   a bird similar and related to the 'Alae ke'oke'o but with a red knob and more terrestrial habits.

alahe'e (*Psydrax odorata*)   a small tree or shrub in the coffee family (Rubiaceae) that is an important component of Hawaii's remaining lowland forests and shrublands.

'alaiaha   a gray forest bird about the size of the 'Elepaio; the meaning of the name is now obscure.

'Alalā (Hawaiian Crow, *Corvus hawaiiensis*)   a species endemic to Hawai'i Island that now survives only in captivity. Another onomatopoeic name.

alaneo   a cape or cloak of a single color, e.g., only of mamo feathers.

'alauahio (O'ahu 'Alauahio *P. maculata* and Maui 'Alauahio *Paroreomyza montana*)   two species of small green and yellow warblerlike honeycreepers.

alawī   a name said to have been used for a young 'Anianiau or Kaua'i 'Amakihi.

ali'i   chief, chiefess, ruling class.

ali'i ai moku   ruler of a whole district.

ali'i nui   king.

'amakihi (Kaua'i 'Amakihi, *H. kauaiensis*; O'ahu 'Amakihi, *H. flavus*; Hawai'i 'Amakihi, *H. virens*)   several species of honeycreeper that once inhabited all the main Hawaiian islands and are still relatively common (but extinct on Lāna'i).

'ama'u (*Sadleria cyatheoides*)   a large fern with a thick fibrous trunk that is widespread in both lowland and montane shrublands of all the main Hawaiian islands. This genus of ferns is endemic to the Hawaiian Islands and has been placed in the blechnum family (Blechnaceae). The fronds of this fern were used in the thatching and decoration of Hawaiian houses and temples, and the cooked stems were fed to domestic pigs.

'Āmaui (*Myadestes woahensis*) the native thrush on O'ahu.

'ānai a severe curse that could cripple, sicken, kill, or punish the accursed.

'Anianiau (*Hemignathus parvus*) a small green honeycreeper on Kaua'i.

'A'o (Newell Shearwater, *Puffinus auricularis newelli*) an endemic seabird that nests in burrows dug into the mountainside, especially under the cover of uluhe ferns. It now resides mainly on Kaua'i.

'Apapane (*Himatione s. sanguinea*) an abundant crimson honeycreeper. Its plumage was seldom used in Hawaiian featherwork. Its bill is slightly decurved and black. Variants 'akakane and 'akakani. The name may be onomatopoeic.

au as used in Chapter 3, a branch of a kāhili.

'Auku'u (Black-crowned Night-Heron, *Nycticorax nycticorax*) the only indigenous species of heron.

'aumakua (pl., 'aumākua) a personal god or family spirit that could appear in many physical forms, e.g., bird, shark, plant, rock, rainbow, and so on.

'aumākua hulu manu feathered icons of royal battle gods.

'āweoweo (*Chenopodium oahuense*) a shrub in the goosefoot family (Chenopodiaceae) that is common in montane dry forests and is less often seen in the dry lowlands.

'ē'ē yellow feathers taken from the underwing of the 'ō'ō bird.

'e'ea a forest bird name of antiquity, said to have been used for an adult 'alaiaha.

'Elepaio (*Chasiempis sandwichensis*) an endemic monarch flycatcher with three subspecies, one each on Kaua'i, O'ahu, and Hawai'i islands. The bird was believed to be the goddess of canoe makers.

Its name is onomatopoeic for the bird's song.

haku-lei a braided or plaited lei (garland).

hakupapa a feather band worn on the head.

hala (*Pandanus tectorius*) the pandanus tree or screw pine (Pandanaceae), a tree of many uses, especially the leaves, used for plaiting.

hala pepe (*Pleomele* spp.) a group of fleshy-stemmed trees or shrubs in the agave family (Agavaceae) native to dry and mesic forests on the main Hawaiian islands. Several members of the genus are listed as endangered species.

halulu a legendary man-eating bird.

hānai an adopted child; to care for or feed.

hāpapa two sticks made out of 'ōlapa wood and joined together to form a cross.

hāpu'u (*Cibotium* spp.) a group of tree ferns endemic to the Hawaiian Islands. Members of the dicksonia family (Dicksoniaceae), these large ferns are common in the understory of most montane wet forests in Hawai'i. The fiddleheads of the fern are sometimes used as food, and both the coarse fibers and the starchy core of the trunk have been used as commercial forest products.

hau (*Hibiscus tiliaceus*) a lowland tree in the mallow or hibiscus family (Malvaceae). The inner bark of the hau is used by Hawaiians to make cordage.

hāwane another name for loulu palms (*Pritchardia* spp.). One species of hāwane or loulu native to Kohala mountain, *P. lanigera*, is now rare, but was reputedly an important food plant for the extinct 'Ula 'ai hāwane

heiau ki'i sacred temple; altar.

hō'awa (*Pittosporum* spp.) a genus of understory trees in the pittosporum family (Pittosporaceae) found in wet, mesic, and dry forests, as well as in subalpine woodlands. The woody capsules of the tree open to reveal an orange interior and black, shiny seeds, which were formerly a favored food of the 'Alalā or Hawaiian Crow.

hulu   feather(s).

hulu-kua   feathered combs.

'ie'ie (*Freycinetia arborea*)   an endemic woody, branching climber related to the hala or pandanus, growing in forests mainly at altitudes of about 300–600 meters.

'I'iwi (*Vestiaria coccinea*)   a brilliant scarlet honeycreeper that was the main source of red feathers for Hawaiian feather work. Still fairly common at high elevations on the larger islands. Several color forms are recognized in the Hawaiian language (e.g., 'i'iwi polena is yellow), and other names are applied ('i'iwi alokele, 'i'iwi popolo, 'iwi, kikiwi). The name 'i'iwi is likely an onomatopoeia for the bird's creaking call.

'iliahi (sandalwood, *Santalum* spp.)   trees with several species (Santalaceae) that are native to the Hawaiian Islands, where they occur in both lowland wet forests and montane dry forests and woodlands. Formerly more abundant in Hawai'i, 'iliahi trees were intensively harvested in the 1800s for their fragrant heartwood.

'ilima (*Sida fallax*)   an indigenous shrub or prostrate woody plant found in coastal strand and dry lowland vegetation. This member of the mallow or hibiscus family (Malvaceae) bears bright yellow-orange flowers that are favored by Hawaiians for lei-making.

imu   an underground oven.

'Io (Hawaiian Hawk, *Buteo solitarius*)   the islands' endemic hawk, now restricted to Hawai'i Island; it once signified royalty. The name is probably onomatopoeic for the hawk's scream.

'Iwa (Great Frigatebird, *Fregata minor*)   a large, handsome, black seabird that steals food from smaller birds and also captures flying fish and similar prey on its own. Its feathers were used in the construction of kāhili.

kahakai   seashore, seacoast; near-shore and marine resources.

kāhili   feather standards used to signify and tend to ali'i.

kahu   honored attendant of the ali'i.

kahuna (pl., kāhuna)   a priest, spiritual adviser, minister, specialist, sorcerer.

Kākāwahie (*Paroreomyza flammea*)   a small, extinct, insect-eating honeycreeper with flame-colored plumage in the male (females were brown) that inhabited only Moloka'i Island. The name literally means "wood-chopping," which this bird's call notes may have suggested.

kākū'ai   a special ceremony that was performed for a deceased relative in which the family would choose the form of 'aumakua their deceased relative would embody.

kalo (taro, *Calocasia esculenta*)   a large herb with a starchy corm in the aroid family (Araceae). This plant, the most important crop plant of the Hawaiians, was often grown in flooded fields but was also cultivated in partly irrigated or non-irrigated dryland field systems.

kama'āina   native-born; one born in a place.

Kāma'o (or Large Kaua'i Thrush, *Myadestes myadestinus*)   the larger of the two thrush species on Kaua'i.

Kanaloa   one of the primary Hawaiian gods, typically associated with ocean phenomena. More generally, it means "man" or "male."

Kāne   one of the primary Hawaiian gods, typically associated with forest phenomena.

kapa (tapa)   cloth made from the bark of various trees and shrubs.

kapa wai-kaua   a battle cloak.

kapu   taboo, prohibition; forbidden; sacred privilege.

kauila (also kauwila, *Alphitonia ponderosa*)   a native tree in the buckthorn family (Rhamnaceae) with especially hard wood.

kāwa'u (*Ilex anomala*)   a common understory tree in wet forests of the Hawaiian

Islands. Also known as Hawaiian holly, this is the only native member of the holly family (Aquifoliaceae) in the islands. The fleshy fruits are eaten by the ʻŌmaʻo.

kēpau   a sticky substance (gum) used by poʻe kia manu to catch birds; analogous to the "birdlime" used by European bird catchers.

kia   a wooden pole or rod used in snaring birds with kēpau (gum).

kia manu (pl., poʻe kia manu)   bird catcher(s).

kino lau   the ability to take many forms or bodies.

Kioea (*Chaetoptila angustipluma*)   a large, extinct nectar-feeding native bird. See the glossary entry for ʻŌʻō.

kīpuka   a patch of vegetation surrounded by younger lava substrate.

kīwaʻa   the name of a mythical bird.

kō (sugar cane, *Saccharum officinarum*)   a tall grass (Poaceae) introduced to the islands by Polynesians as a minor food plant. Until recently, it was the most important commercial crop in modern times.

koa (*Acacia koa*)   a tall, leguminous tree (Fabaceae) that is co-dominant is some montane wet and mesic forests. It is currently the most valuable timber tree in Hawaiʻi.

Koaʻeʻula (Red-tailed Tropicbird, *Phaethon rubricauda*)   a white seabird with two long, red central tail feathers. Koaʻe kea (White-tailed Tropicbird, *Phaethon lepturus*) is smaller than the previous species, with white tail plumes.

Koloa (Hawaiian Duck, *Anas wyvilliana*)   the native duck living on the main Hawaiʻi Islands.

konohiki   a land manager in ancient Hawaiʻi.

koʻomamo   the yellow uppertail coverts of the Hawaiʻi Mamo.

Kū   one of the primary Hawaiian gods, often associated with human activities, machinations, and war.

kua   an anvil.

Kū-huluhulu-manu   (lit. "Ku-doubly-feathered")   the god of bird catchers and people who did feather work.

Kūkāʻilimoku   "Kū the land-grabber," a war god whose feathered image was displayed during expeditions.

Kū-ka-ʻōhiʻa-Laka   a forest god of ʻōhiʻa trees whose image was adorned with feathers.

kukui (candlenut tree, *Aleurites moluccana*)   a member of the spurge family (Euphorbiaceae) brought to Hawaiʻi by Polynesian settlers and now a common tree of windward lowland forests. Gum can be extracted from its bark and oil from the kernel of its seed.

kumu   a tree stem; pole.

Kumulipo   Hawaiian cosmogonic genealogy; creation chant.

lama (*Diospyros sandwicensis*)   a dominant tree of some lowland dry and mesic forests. A member of the ebony family (Ebenaceae), this tree has hard wood that was used by Hawaiians for construction and was a source of medicine and offerings for the altars of some temples.

lehua   the flower of the ʻōhiʻa tree (*Metrosideros polymorpha*).

lei   necklace or garland.

Lono   one of the primary Hawaiian gods, associated with the Makahiki festival and agriculture.

loulu (*Pritchardia* spp.)   a group of palms (Arecaceae) native to coastal communities and both lowland and montane wet forests in Hawaiʻi. Prior to human colonization of the islands, species of this genus were dominant trees in some lowland forests. The genus contains 19 species, each endemic to just one island; several species are listed as endangered.

mahiole   a feathered helmet.

makaʻāinana   commoners.

Makahiki Festival   a time sanctified for peace, rest, and worship outside the ordinary religious observances.

malihini   a stranger, foreigner, newcomer.

malo   a loincloth.

māmaki (*Pipturus albidus*)   a large shrub or small tree in the nettle family (Urticaceae) that is a common understory species in both wet and mesic forests. It is used in the production of kapa cloth.

māmane (*Sophora chrysophylla*)   a small to medium-sized tree widespread in the Hawaiian Islands and distributed from the lowlands to the treeline of high mountains. A member of the pea family (Fabaceae), this is the forest dominant in dry subalpine forests, and its flowers are an important source of nectar for Hawaiian honeycreepers. Its pods are the staple food of the endangered Palila.

mamo (Hawai'i Mamo, *Drepanis pacifica*)   a relatively large, black honeycreeper with a long, sickle-shaped bill that is endemic to the island of Hawai'i. The bird is renowned for is highly valued yellow rump and under-tail feathers.

mana   divine power or authority; spiritual essence.

manu   bird; any winged creature.

maunu   bait used by po'e kia manu to bring birds into snares (e.g., rats, 'i'iwi, 'Apapane, 'amakihi, 'ō'ō).

moa (domestic chicken, *Gallus gallus*)   the only bird brought to the islands by the Polynesians.

moa-nalo   a recently coined name for five species of large flightless Hawaiian birds known from subfossils. Although shaped like geese, moa-nalo share anatomical features with ducks and therefore were likely descended from them.

Moho (Hawaiian Rail, *Porzana sandwichensis*)   one of numerous species of closely related small, flightless birds that once inhabited the Hawaiian Islands. This species lived on Hawai'i Island and survived until the latter half of the nineteenth century.

na'ena'e (*Dubautia* spp.)   a group of shrubs and small trees in the sunflower family (Asteraceae). This genus of 21 species is endemic to the Hawaiian Islands and is considered part of the "silversword alliance." Several endangered species belong to this genus, which also includes a number of common forest understory plants and important components of subalpine shrublands.

naio (*Myoporum sandwicense*)   a tree widely distributed in the Hawaiian Islands that occurs in several vegetation types from the coast to the treeline of high mountains. This is the only member of the myoporum family (Myoporaceae) in Hawai'i. The seeds of this tree were important in the diet of the extinct Kona Grosbeak.

naupaka kuahiwi   a group of endemic species of shrubs in the goodenia family (Goodeniaceae) that occur in upland wet and dry forests of the main Hawaiian islands. The tubular "half-flowers" and fleshy fruits of these shrubs are a resource for native birds.

Nēnē (Hawaiian Goose, *Branta sandvicensis*)   a medium-sized, native goose. It was saved from extinction at the eleventh hour and is now honored as the state bird.

noni (*Morinda citrifolia*)   a small native tree or shrub in the coffee family (Rubiaceae) used for making dyes and medicines.

Nukupu'u (*Hemignathus lucidus*)   a species of honeycreeper with a deeply curved bill. Like its close relative the 'Akiapōlā'au, this species possesses a lower mandible much shorter than the upper mandible. Three forms, one each on Kaua'i, O'ahu, and Maui, are all likely extinct now.

'oha (*Clermontia hawaiiensis* and related species)   a medium-sized rain forest shrub in the bellflower family (Campanulaceae) that bears long, curved, tubular flowers much sought by nectar-feeding birds.

'ohana   family, kin group.

'ōhelo (*Vaccinium reticulatum*)   a common shrub species in the heath family (Ericaceae) found in open shrublands of several vegetation zones and a co-dominant species in subalpine woodlands. The fleshy berries are an important food item for the ʻŌmaʻo and the Nēnē.

ʻohe makai (*Reynoldsia sandwicensis*) an uncommon tree in the ginseng family (Araliaceae) that is a component of dry lowland forests.

ʻōhiʻa or ʻōhiʻa lehua (*Metrosideros polymorpha*) a dominant forest tree in the myrtle family (Myrtaceae) found from coast to timberline. Its flowers (lehua) are the most important nectar source for native birds.

ʻōlapa (*Cheirodendron* spp.) a medium-sized tree frequent in wet forests. This member of the ginseng family (Araliaceae) produces copious fleshy fruit important to frugivorous native birds.

ʻōlelo noʻeau a proverb, folk saying.

ʻOlomaʻo (or Molokaʻi Thrush, *Myadestes lanaiensis*) a thrush historically endemic to Molokaʻi and Lānaʻi islands.

olonā (*Touchardia latifolia*) an endemic shrub in the nettle family (Urticaceae) whose bark provides a strong, durable fiber used for making nets to snare birds and as a base for ʻahuʻula.

olopua (*Nestegis sandwicensis*) a tree relatively uncommon in dry and mesic forests of the Hawaiian Islands. A member of the olive family (Oleaceae), this tree was formerly the dominant species in some forest communities.

ʻŌmaʻo (or Hawaiian Thrush, *Myadestes obscurus*) a drab thrush endemic to Hawaiʻi Island; related species with similar names inhabited other Hawaiian islands.

ʻōʻō (three species, each named for its home island, except Bishop's, which lived on Molokaʻi: Oʻahu ʻŌʻō, *Moho apicalis*; Bishop's ʻŌʻō, *M. bishopi*; and Hawaiʻi ʻŌʻō, *M. nobilis*) a glossy black bird with a long, graduated tail. Although these birds are at present classified as honeyeaters (Meliphagidae, a South Pacific bird family), recent DNA evidence places them together with the ʻŌʻōʻāʻā and Kioea in their own family

unique to Hawaiʻi. The bright yellow tufts of axillary plumes emerging from under each wing were extensively used in Hawaiian feather work. The Hawaiian name is onomatopoeic for the birds' loud call, but it also means "to pierce." All species are extinct.

ʻŌʻōʻāʻā (Kauaʻi ʻŌʻō, *Moho braccatus*) a small ʻōʻō endemic to Kauaʻi Island, black with short yellow thigh feathers; its name means "dwarf ʻōʻō." Now extinct.

ʻōpeʻapeʻa (Hawaiian Hoary Bat, *Lasiurus cinereus semotus*) the sole species of Hawaiian bat, although a second, undescribed species is known from the fossil record.

ōpuhe (*Urera sandwicensis*) a small tree or shrub found in the understory of many wet and mesic forests of the islands. Like other members of the nettle family (Urticaceae), this tree was a source of material for cordage and inferior kapa.

ʻŌʻū (*Psittirostra psittacea*) a bright green parrotlike honeyeater; males had a bright yellow head and canarylike song. Once widespread in the islands, the species is now extinct. Its green feathers were occasionally used in feather work.

Palila (*Loxioides bailleui*) a finch-billed honeycreeper that lives on Mauna Kea, Hawaiʻi Island. The name is onomatopoeic for the bird's call.

pāpala, pāpala kēpau (*Pisonia brunoniana*) a small tree whose sap was used to make kēpau (birdlime). This tree, in the four-o'clock family (Nyctaginaceae), is an understory component of montane mesic forests.

pāpale a hat or cap.

pāʻū a skirt.

pīlali the black and white outer tail feathers of the ʻōʻō.

pili (*Heteropogon contortus*) a grass (Poaceae) indigenous to the Hawaiian Islands and also found in many tropical lands. Formerly it was a dominant ground cover in the dry lowlands, where the grass was used as thatching material by Hawaiians. Fire was used to stimulate the growth of pili grass.

pilo (*Coprosma* spp.)   a member of a group of understory trees in the coffee family (Rubiaceae). Pilo trees are often common in wet and mesic forests, where their fruits are a food source for native frugivorous birds such as the ʻŌmaʻo.

pīpī   a female ʻōʻō.

Pō   a deep metaphysical concept in several Polynesian cultures, including the Ha-waiian and Maori cosmogonies and mythologies. Pō is formative darkness; night; invisibility; a realm of ancient and primal gods; a realm of lightless eternity or limbo; a realm of afterlife, either good or bad.

poʻe kia manu   see kia manu.

Poʻo-uli   (*Melamprosops phaeosoma*)   a black-masked, brown honeyeater endemic to Maui Island and recently extinct. The name means "black head."

pōpolo (*Solanum* spp.)   a member of a group of native shrubs in the night-shade family (Solanaceae) that bear black or orange berries. Most native members of the group are rare; only *S. americanum* is a common species.

Puaiohi (or Small Kauaʻi Thrush, *Myadestes palmeri*)   the smaller of the two thrush species found on Kauaʻi.

pue   feathers from the rump of a bird. In the context of feather work, refers to the yellow feathers taken from Hawaiʻi Mamo.

Pueo (Hawaiian Short-eared Owl, *Asio flammeus sandwichensis*)   a subspecies thought to be endemic that is active both day and night, usually at dawn and dusk or when overcast. There were several mythical forms: Pueo kani kaua, Pueo nui akea, and Pueo nui o kona.

pūkiawe (*Leptecophylla tameiameiae*)   a shrub found in many natural vegetation types from the dry lowlands to montane forests and subalpine shrublands. It is one of the most common and widespread native

species of the Hawaiian Islands. Its white to pink fruits are important to native frugivorous birds. Formerly considered a member of the epacris family (Epacridaceae), the species is now placed in the heath family (Ericaceae).

puleʻaha   a prayer or direct appeal to the gods.

ti-leaf (*Cordyline terminalis*, ki)   a leaf of an introduced shrub in the agave family (Agavaceae) cultivated in both ancient and modern times for many purposes, especially as a food wrapper.

ʻuala (sweet potato, *Ipomoea batatas*)   a vine in the morning glory family (Convolvulaceae) that was introduced to Hawaiʻi by Polynesians and was the most important food crop of dry, unirrigated fields.

ʻUaʻu (Hawaiian Petrel, *Pterodroma sandwichensis*)   a species of gadfly petrel endemic to Hawaiʻi. Formerly abundant, the species is locally common at a few remote locations on the main islands where it can nest safe from mammalian predators.

ʻuhane   a spirit or soul.

ʻUla-ʻai-hāwane (*Ciridops anna*)   an extinct honeyeater. The name means "red bird eating palm fruit."

ʻūlei (*Osteomeles anthyllidifolia*)   a shrub indigenous to the Hawaiian islands that has strong, tough woody stems used in ancient times for digging sticks and spears. A member of the rose family (Rosaceae), ʻūlei is found in dry lowland vegetation and montane shrublands.

ʻulu (breadfruit tree, *Artocarpus altilis*)   a member of the mulberry family (Moraceae) that was brought to Hawaiʻi by Polynesian settlers who ate its large, edible fruit; the sap is used to make kēpau (birdlime).

uluhe (*Dicranopteris linearis*)   a large, mat-forming fern, known elsewhere as false staghorn fern, that is a co-dominant of many lowland wet forests and cliff communities. A member of the gleichenia family (Gleicheniaceae), this fern is indigenous to the Hawaiian Islands and also found throughout the tropics.

waiawi (strawberry guava, *Psidium cattleianum*) a tree in the myrtle family (Myrtaceae) introduced to Hawai'i in the early 1800s for its edible fruits and now one of the most invasive alien species in lowland and montane wet and mesic forests.

wao akua upland and often uninhabited areas; the land of the gods.

wao kanaka lower-elevation inland lands for habitation and cultivation.

wao ma'u kele or wao kele the rain forest belt.

wao nahele an inland forest region.

wiliwili (coral tree, *Erythrina sandwicensis*) a relatively rare endemic tree in the pea family (Fabaceae) with beautiful white, orange, or red curved flowers whose nectar is useful to native birds. Formerly this species was a co-dominant of some dry lowland forests.

References

Anonymous. 1937. Exotic birds of Hawaii. Paradise of the Pacific 49: 2.

————. 1953. Hui Manu. 'Elepaio 14: 11–14.

————. 1970. Hui Manu. 'Elepaio 31: 24–28.

Aaron, J. L., and R. M. May. 1982. The population dynamics of malaria. Pp. 139–179 in M. Anderson (editor), Population dynamics of infectious diseases: Theory and applications. Chapman and Hall, London.

Abbott, I. A. 1992. La'au Hawai'i: Traditional Hawaiian uses of plants. Bishop Museum Press, Honolulu.

Abrams, P. A. 2000. Character shifts of prey species that share predators. American Naturalist 156 (Supplement): S45–S61.

Adkisson, C. S. 1996. Red Crossbill (*Loxia curvirostra*). In A. Poole and F. Gill (editors), The birds of North America, No. 256. Birds of North America Inc., Philadelphia.

————. 1999. Pine Grosbeak (*Pinicola enucleator*). In A. Poole and F. Gill (editors), The birds of North America, No. 456. Birds of North America Inc., Philadelphia.

Aguon, C. F., E. W. Campbell III, and J. M. Morton. 2002. Efficacy of electrical barriers used to protect Mariana crow nests. Wildlife Society Bulletin 30: 703–708.

Ahumada, J. A., D. LaPointe, and M. D. Samuel. 2004. Modeling the population dynamics of *Culex quinquefasciatus* (Diptera: Culicidae), along an elevational gradient in Hawaii. Journal of Medical Entomology 41: 1157–1170.

Ainley, D. G., T. C. Telfer, and M. H. Reynolds. 1997. Newell's and Townsend's Shearwater (*Puffinus auricularis*). In A. Poole and F. Gill (editors), The birds of North America, No. 297. Birds of North America Inc., Philadelphia.

Alexander, W. D. 1891. Brief history of the Hawaiian people. American Book Company, New York.

Algar, D. A., A. A. Burbidge, and G. J. Angus. 2002. Cat eradication on Hermite Island, Montebello Islands, Western Australia. Pp. 14–18 in C. R. Veitch and M. N. Clout (editors), Turning the tide: The eradication of invasive species. Proceedings of the International Conference on Eradication of Island Invasives. Auckland, New Zealand.

Alicata, J. E. 1964. Parasitic infections of man and animals in Hawaii. Hawaii Agricultural Experiment Station, Technical Bulletin 61. Honolulu.

Allen, J. 1997. Pre-contact landscape transformation and cultural change in windward O'ahu. Pp. 230–247 in P. V. Kirch and T. L. Hunt (editors),

Historical ecology in the Pacific Islands: Prehistoric environmental and landscape change. Yale University Press, New Haven, CT.

Allen, M. L., D. A. O'Brochta, P. W. Atkinson, and C. S. Levesque. 2001. Stable, germ-line transformation of *Culex quinquefasciatus* (Diptera: Culicidae). Journal of Medical Entomology 38: 701–710.

Allen, R. H., M. Gottlieb, E. Clute, M. J. Pongsiri, J. Sherman, and G. I. Obrams. 1997. Breast cancer and pesticides in Hawaii: The need for further study. Environmental Health Perspectives 105: 679–683.

Altman, I., and M. M. Chemers. 1980. Culture and environment. Brooks/Cole, Pacific Grove, CA.

Alto, B. W., and S. A. Juliano. 2001. Precipitation and temperature effects on populations of *Aedes albopictus* (Diptera: Culicidae): Implications for range expansion. Journal of Medical Entomology 38: 646–656.

Aly, C., and M. S. Mulla. 1987. Effects of two microbial insecticides on aquatic predators of mosquitoes. Journal of Applied Entomology 103: 113–118.

Amadon, D. 1950. The Hawaiian honeycreepers (Aves, Drepaniidae). Bulletin of the American Museum of Natural History 95: 151–262.

Amarasekare, P. 1993. Potential impact of mammalian nest predators of endemic forest birds of western Mauna Kea, Hawaii. Conservation Biology 7: 316–324.

———. 1994. Ecology of introduced small mammals on western Mauna Kea, Hawaii. Journal of Mammalogy 75: 24–38.

American Ornithologists' Union. 1983. Check-list of North American birds. 7th edition. American Ornithologists' Union, Washington, DC.

American Society of Mammalogists. 1998. Guidelines for the capture, handling, and care of mammals as approved by the American Society of Mammalogists. Journal of Mammalogy 79: 1416–1431.

American Veterinary Medical Association. 2001. 2000 report of the AVMA panel on euthanasia. Journal of the American Veterinary Medical Association 218: 669–696.

American Zoo Association Small Population Management Advisory Group. 2000. Artificial selection in captive wildlife populations: A statement from the AZA Small Population Management Advisory Group, July 5–9, Cornell University, NY. American Zoo Association Communiqué, September 2000, page 31.

Anders, A. D., D. C. Dearborn, J. Faaborg, and F. R. Thompson III. 1997. Juvenile survival in a population of Neotropical migrant birds. Conservation Biology 11: 698–707.

Anderson, R. C., D. E. Gardner, C. C. Daehler, and F. C. Meinzer. 2002. Dieback of *Acacia koa* in Hawaii: Ecological and pathological characteristics of affected stands. Forest Ecology and Management 162: 273–286.

Anderson, R. M., and R. M. May. 1978. Regulation and stability of host–parasite population interactions, I: Regulatory processes. Journal of Animal Ecology 47: 219–247.

———. 1982. Coevolution of hosts and parasites. Parasitology 85: 411–426.

Anderson, S. B. 1999a. Axis deer research–Maui, Hawaii. www.maui.net/~brooke/.

———. 1999b. Axis deer overview and profile, December 20. www.hear.org/hnis/reports/HNIS-AxiAxiVO1.pdf.

Anderson, S. J., and C. P. Stone. 1994. Indexing sizes of feral pig populations in a variety of Hawaiian natural areas. Transactions of the Western Section of the Wildlife Society 30: 26–39.

Andreadis, T. G., J. J. Becnel, and S. E. White. 2003. Infectivity and pathogenicity of a novel baculovirus, CuniNPV from *Culex nigripalpus* (Diptera: Culicidae) for thirteen species and four genera of mosquitoes. Journal of Medical Entomology 40: 512–517.

Apo, S. U. 1946. Ka'i'iwi, the sacred owl. Paaheo Press, Honolulu.

Apps, P. J. 1983. Aspects of the ecology of feral cats on Dassen island, South Africa. South African Journal of Zoology 18: 393–399.

Armitage, G. T., and H. P. Judd. 1944. Ghost dog and other Hawaiian legends. Advertiser Publishing, Honolulu.

Armstrong, D. 2000. Re-introductions of the New Zealand robins: A key component of ecological restoration. Re-introduction News 19: 44–47.

Armstrong, D., R. Jakob-Hoff, and U. S. Seal (editors). 2003. Animal movements and disease risks: A workbook. 5th edition. Conservation Breeding Specialist Group, Apple Valley, MN.

Armstrong, D. P., and I. G. McLean. 1995. New Zealand translocations: Theory and practice. Pacific Conservation Biology 2: 39–54.

Arnold, K. E., and I. P. F. Owens. 1998. Cooperative breeding in birds: A comparative test of the life history hypothesis. Proceedings of the Royal Society of London, Series B 265: 739–745.

Aruch, S., C. T. Atkinson, A. F. Savage, and D. A. LaPointe. 2007. Prevalence and distribution of pox-like lesions, avian malaria and mosquito vectors in Kīpahulu Valley, Haleakalā National Park, Hawai'i. Journal of Wildlife Diseases 43: 567–575.

Ash, S. J., and C. E. Adams. 2003. Public preferences for free-ranging domestic cat (*Felis catus*) management options. Wildlife Society Bulletin 31: 334–339.

Askins, R. A. 2000. Restoring North America's birds: Lessons from landscape ecology. Yale University Press, New Haven, CT.

Asman, S. M., P. T. McDonald, and T. Prout. 1981. Field studies of genetic control systems for mosquitoes. Annual Review of Entomology 26: 289–318.

Athens, J. S. 1997. Hawaiian native lowland vegetation in prehistory. Pp. 248–270 in P. V. Kirch and T. L. Hunt (editors), Historical ecology in the Pacific islands. Yale University Press, New Haven, CT.

Athens, J. S., M. W. Kaschko, and H. F. James. 1991. Prehistoric bird hunters: High altitude resource exploitation on Hawai'i Island. Bishop Museum Occasional Papers 31: 63–84.

Athens, J. S., J. V. Ward, and S. Wickler. 1992. Late Holocene lowland vegetation, O'ahu, Hawai'i. New Zealand Journal of Archaeology 14: 9–34.

Atkinson, C. T. 2008. Avian malaria. Pp. 35–53 in C. T. Atkinson, N. J. Thomas, and D. B. Hunter (editors), Parasitic diseases of wild birds. Wiley-Blackwell, Ames, IA.

Atkinson, C. T., R. J. Dusek, and J. K. Lease. 2001. Serological responses and immunity to superinfection with avian malaria in experimentally-infected Hawaii Amakihi. Journal of Wildlife Disease 37: 20–27.

Atkinson, C. T., R. J. Dusek, K. L. Woods, and W. M. Iko. 2000. Pathogenicity of avian malaria in experimentally infected Hawaii Amakihi. Journal of Wildlife Diseases 36: 197–204.

Atkinson, C. T., J. K. Lease, B. M. Drake, and N. P. Shema. 2001. Pathogenicity, serological responses, and diagnosis of experimental and natural malarial infections in native Hawaiian thrushes. Condor 103: 209–218.

Atkinson, C. T., J. K. Lease, R. J. Dusek, and M. D. Samuel. 2005. Prevalence of pox-like lesions and malaria in forest bird communities on leeward Mauna Loa Volcano, Hawaii. Condor 107: 537–546.

Atkinson, C. T., and C. van Riper III. 1991. Pathogenicity and epizootiology of avian haemoatozoa: Plasmodium, Leucocytozoon, and Haemoproteus. Pp. 19–48 in J. E. Loye and M. Zuk (editors), Bird–parasite interactions, ecology, evolution and behavior. Oxford University Press, New York.

Atkinson, C. T., K. L. Woods, R. J. Dusek, L. S. Sileo, and W. M. Iko. 1995. Wildlife disease and conservation in Hawaii: Pathogenicity of avian malaria (Plasmodium relictum) in experimentally infected Iiwi (Vestiaria coccinea). Parasitology 111: S59–S69.

Atkinson, I. A. E. 1970. Successional trends in the coastal and lowland forest of Mauna Loa and Kilauea Volcanoes, Hawaii. Pacific Science 24: 387–400.

———. 1973. Spread of the ship rat (Rattus r. rattus L.) in New Zealand. Journal of the Royal Society of New Zealand 3: 457–472.

———. 1977. A reassessment of factors, particularly Rattus rattus L., that influenced the decline of endemic forest birds in the Hawaiian Islands. Pacific Science 31: 109–133.

———. 1985. The spread of commensal species of Rattus to oceanic islands and their effects on island avifaunas. Pp. 35–81 in P. J. Moors (editor), Conservation of island birds. Technical Publication 3. International Council for Bird Preservation. Cambridge, England.

———. 1989. Introduced animals and extinctions. Pp. 54–75 in D. Western and M. C. Pearl (editors), Conservation for the twenty-first century. Oxford University Press, New York.

Atwood, J. L., and D. R. Bontrager. 2001. California Gnatcatcher (Polioptila californica). In A. Poole and F. Gill (editors), The birds of North America, No. 574. Birds of North America Inc., Philadelphia.

Au, S., and G. Swedburg. 1966. A progress report on the introduction of the Barn Owl (Tyto alba pratincola) to the island of Kaua'i. 'Elepaio 26: 58–60.

Aye, P. P. 1994. Study of biology, ecology, and mortality of owls in Hawai'i. M.S. thesis. University of Hawai'i at Mānoa, Honolulu.

Bacon, P. J. (editor). 1985. Population dynamics of rabies in wildlife. Academic Press, New York.

Badyaev, A. V. 1997. Avian life history variation along altitudinal gradients: An example with cardueline finches. Oecologia 111: 365–374.

Badyaev, A. V., and C. K. Ghalambor. 2001. Evolution of life histories along elevational gradients: Trade-off between parental care and fecundity. Ecology 82: 2948–2960.

Baker, A. J. 1980. Morphometric differentiation in New Zealand populations of the house sparrow (Passer domesticus). Evolution 34: 638–653.

Baker, A. J., and A. Moeed. 1987. Rapid genetic differentiation and founder effect in colonizing populations of Common Mynas (Acridotheres tristis). Evolution 41: 525–538.

Baker, H., and P. E. Baker. 2000. Maui 'Alauahio (Paroreomyza montana). In A. Poole and F. Gill (editors), The birds of North America, No. 504. Birds of North America Inc., Philadelphia.

Baker, J. D., C. L. Littnan, and D. W. Johnston. 2006. Potential effects of sea level rise on the terrestrial habitats of endangered and endemic megafauna of the Northwestern Hawaiian Islands. Endangered Species Research 2: 21–30.

Baker, J. K., and M. S. Allen. 1979. Roof rat depredations on Hibiscadelphus (Malvaceae) trees. Pp. 2–5 in C. W. Smith (editor), Proceedings Second Conference in Natural Sciences, Hawaii Volcanoes National Park. Cooperative National Park Resources Studies Unit, University of Hawaii at Mānoa, Honolulu.

Baker, J. K., and D. W. Reeser. 1972. Goat management problems in Hawaii Volcanoes National Park: A history, analysis, and management plan. Natural Resources Report 2. U.S. Department of the Interior, National Park Service (NPS-112), Washington, DC.

Baker, P. E. 1998. A description of the first live Poʻo-uli captured. Wilson Bulletin 110: 307–310.

Baker, P. E. 2001. Status and distribution of the Poʻo-uli in the Hanawi Natural Area Reserve between December 1995 and June 1997. Studies in Avian Biology 22: 144–150.

Baker, P. E., and H. Baker. 2000. Kākāwahie (Paroreomyza flammea) and Oʻahu ʻAlauahio (Paroreomyza maculata). In A. Poole and F. Gill (editors), The birds of North America, No. 503. Birds of North America Inc., Philadelphia.

Baldwin, H. S. 1972. Letters of Winston Banko. Located at the archives of the U.S. Geological Survey, Pacific Island Ecosystems Research Center, Hawaiʻi Volcanoes National Park, Volcano, HI.

Baldwin, J. A. 1980. The domestic cat, Felis catus L., in the Pacific islands. Carnivore Genetics Newsletter 4: 57–66.

Baldwin, P. H. 1941. Checklist of the birds of the Hawaii National Park, Kilauea-Mauna Loa Section, with remarks on their present status and a field key for their identification. Internal report. Hawaii National Park, Volcano, HI.

———. 1945. The Hawaiian goose, its distribution and reduction in numbers. Condor 47: 27–37.

———. 1953. Annual cycle, environment and evolution in the Hawaiian honeycreepers (Aves: Drepaniidae). University of California Publications in Zoology 52: 285–398.

———. 1969. The ʻAlala (Corvus tropicus) of western Hawaii Island. ʻElepaio 30: 41–45.

Baldwin, P. H., and T. L. C. Casey. 1983. A preliminary list of foods of the Poʻo-uli. ʻElepaio 43: 53–56.

Baldwin, P. H., and G. O. Fagerlund. 1943. The effect of cattle grazing on koa reproduction in Hawaii National Park. Ecology 24: 118–122.

Baldwin, P. H., C. W. Schwartz, and E. R. Schwartz. 1952. Life history and economic status of the mongoose in Hawaii. Journal of Mammalogy 33: 335–356.

Banko, P. C. 1992. Constraints on productivity of wild Nene or Hawaiian Geese Branta sandvicensis. Wildfowl 43: 99–106.

Banko, P. C., D. L. Ball, and W. E. Banko. 2002. Hawaiian Crow (Corvus hawaiiensis). In A. Poole and F. Gill (editors), The birds of North America, No. 648. Birds of North America Inc., Philadelphia.

Banko, P. C., and W. E. Banko. 1980. Historical trends of passerine populations in Hawaii Volcanoes National Park and vicinity. Pp. 108–125 in Proceedings of the Second Conference on Scientific Research in the National Parks (San Francisco, CA, November 26–30, 1979), Vol. 8: Endangered and Threatened Species; Exotic Species. U.S. Department of the Interior, National Park Service, Washington, DC.

———. 2006. Food specialization and radiation of Hawaiian honeycreepers. Acta Zoologica Sinica 52 (Supplement): 253–256.

Banko, P. C., M. L. Cipollini, G. W. Breton, E. Paulk, M. Wink, and I. Izhaki. 2002. Seed chemistry of Sophora chrysophylla (mamane) in relation to the diet of specialist avian seed predator Loxioides bailleui (palila) in Hawaii. Journal of Chemical Ecology 78: 1393–1410.

Banko, P. C., R. E. David, J. D. Jacobi, and W. E. Banko. 2001. Conservation status and recovery strategies for endemic Hawaiian birds. Studies in Avian Biology 22: 359–376.

Banko, P. C., S. C. Hess, L. Johnson, and S. J. Dougill. 1998. Palila population estimate for 1997. ʻElepaio 58: 11–15.

Banko, P. C., and J. D. Jacobi. 1992. Recent surveys of the Hawaiian crow. Research Information Bulletin 77. U.S. Fish and Wildlife Service, Washington, DC.

Banko, P. C., L. Johnson, G. D. Lindsey, S. G. Fancy, T. K. Pratt, J. D. Jacobi, and W. E. Banko. 2002. Palila (Loxioides bailleui). In A. Poole and F. Gill (editors), The birds of North America, No. 679. Birds of North America Inc., Philadelphia.

Banko, P. C., P. T. Oboyski, J. W. Slotterback, S. J. Dougill, D. M. Goltz, L. Johnson, M. E. Laut, and T. C. Murray. 2002. Availability of food resources, distribution of invasive species, and conservation of a Hawaiian bird along a gradient of elevation. Journal of Biogeography 29: 789–808.

Banko, W. E. 1968. Rediscovery of Maui Nukupuʻu, Hemignathus lucidus affinis, and sighting of Maui Parrotbill, Pseudonestor xanthophrys, Kipahulu Valley, Maui, Hawaii. Condor 70: 265–266.

———. 1979. History of endemic Hawaiian birds: Specimens in museum collections. Avian History Report 2. Cooperative National Park Resources Studies Unit, University of Hawaiʻi at Mānoa, Honolulu.

———. 1980. History of endemic Hawaiian birds, Part 1: Population histories—species accounts, forest birds: Hawaiian thrushes. Avian History Report 6C-6D. Cooperative National Park Resources Studies Unit, University of Hawaii, Honolulu.

———. 1980–1987. History of endemic Hawaiian birds, Part 1: Population histories—species accounts, forest birds. Avian History Reports 6–11. Cooperative National Park Resources Studies Unit, University of Hawaii, Honolulu.

———. 1986. History of endemic Hawaiian birds, Part 1: Population histories–species accounts, forest birds: Maui Parrotbill, ʻOʻu, Palila, Greater Koa Finch, Lesser Koa Finch, and Grosbeak Finch. Avian History Report 10. Cooperative National Park Resources Studies Unit, University of Hawaii, Honolulu.

Banko, W. E., and P. Banko. 1976. Role of food depletion by foreign organisms in historical decline of Hawaiian forest birds. Pp. 29–34 in C. W. Smith (editor), Proceedings of the First Conference in Natural Science in Hawaii. Cooperative National Park Resources Study Unit, University of Hawaii, Honolulu.

Banko, W. E., and P. C. Banko. 1980. History of endemic Hawaiian birds, Part 1: Population histories —species accounts, forest birds: Hawaiian Raven/ Crow ('Alalā). Avian History Report 6B. Cooperative National Park Resources Studies Unit, University of Hawaii, Honolulu.

Barbosa, P., and J. C. Schultz (editors). 1987. Insect outbreaks. Academic Press, San Diego, CA.

Barbour, A. G., and D. Fish. 1993. The biological and social phenomenon of Lyme disease. Science 260: 1610–1616.

Barclay, J. H. 1988. Peregrine restoration in the eastern United States: Peregrine Falcon populations; Their management and recovery. Peregrine Fund Inc., Boise, ID.

Barrett, G., A. Silcocks, S. Barry, R. Cunningham, and R. Poulter. 2003. The new atlas of Australian birds. CSIRO Publishing, Collingwood, Victoria, Australia.

Barrett, R. H. 1983. Hunting as a control method for wild pigs in Hawaii Volcanoes National Park. Unpublished report to the Superintendent of Hawaii Volcanoes National Park, Volcano, HI.

Barry, S. C., and A. H. Welsh. 2001. Distance sampling methodology. Journal of the Royal Statistical Society, Series B, 63: 31–54.

Bart, J. 2005. Monitoring the abundance of bird populations. Auk 122: 15–25.

Bart, J., K. P. Burnham, E. H. Dunn, C. M. Francis, and C. J. Ralph. 2004. Goals and strategies for estimating trends in landbird abundance. Journal of Wildlife Management 68: 611–626.

Bart, J., S. Droege, P. Geissler, B. Peterjohn, and C. J. Ralph. 2004. Density estimation in wildlife surveys. Wildlife Society Bulletin 32: 1242–1247.

Baskin, Y. 2002. A plague of rats and rubbervines: The growing threat of species invasions. Island Press, Washington, DC.

Bastian, O. 2001. Landscape ecology—towards a unified discipline? Landscape Ecology 16: 757–766.

Bataillon, T., and M. Kirkpatrick. 2000. Inbreeding depression due to mildly deleterious mutations in finite populations: Size does matter. Genetical Research 75: 75–81.

Bay, E. C. 1985. Other larvivorous fish. Pp. 18–24 in H. C. Chapman (editor), Biological control of mosquitoes. Bulletin 6. American Mosquito Control Association, Fresno, CA.

Beadell, J. S., and R. C. Fleischer. 2005. A restriction enzyme-based assay to distinguish between avian hemosporidians. Journal of Parasitology 91: 683–685.

Beadell, J. S., E. Gering, J. Austin, J. P. Dumbacher, M. A. Peirce, T. K. Pratt, C. T. Atkinson, and R. C. Fleischer. 2004. Prevalence and differential host-specificity of two avian blood parasite genera in the Australo-Papuan region. Molecular Ecology 13: 3829–3844.

Beadell, J. S., F. Ishtiaq, R. Covas, M. Melo, B. H. Warren, C. T. Atkinson, S. Bensch, G. R. Graves, Y. V. Jhala, M. A. Peirce, A. R. Rahmani, D. M. Fonseca, and R. C. Fleischer. 2006. Global phylogeographic limits of Hawaii's avian malaria. Proceedings of the Royal Society of London, Series B: Biological Sciences 273: 2935–2944.

Beaglehole, J. C. (editor). 1967. The journals of Captain James Cook on his voyages of discovery, Vol. 3, Parts 1 and II: The voyage of the Resolution and Discovery, 1776–1780. Hakluyt Society Extra Series 36. Cambridge University Press, Cambridge, England.

Beamish, R. 2004. Sleepless in Hawaii. Smithsonian 34 (January): 21–22.

Beardsley, J. W. Jr. 1979. New immigrant insects in Hawaii: 1962 through 1976. Proceedings, Hawaiian Entomological Society 13: 35–44.

Beasley, V. R., and W. B. Buck. 1982. Warfarin and other anticoagulant poisonings. Pp. 101–106 in R. E. Kirk (editor), Current veterinary therapy, vol. 8. W. B. Saunders, Philadelphia.

Beccari, O., and J. F. Rock. 1921. A monographic study of the genus Pritchardia. Memoirs of the B. P. Bishop Museum, Vol. 8, No. 1. Bishop Museum Press, Honolulu.

Bechard, M. J., and J. K. Schmutz. 1995. Ferruginous Hawk (Buteo regalis). In A. Poole and F. Gill (editors), The birds of North America, No. 172. Birds of North America Inc., Philadelphia.

Bechard, M. J., and T. R. Swem. 2002. Rough-legged Hawk (Buteo lagopus). In A. Poole and F. Gill (editors), The birds of North America, No. 641. Birds of North America Inc., Philadelphia.

Beck, B. B. 1992. Report from the AAZPA Advisory Group: Guidelines for reintroduction of animals held in captivity. American Association of Zoological Parks and Aquariums. Published in Captive Breeding Specialist Group, IUCN/SSC/CBSG Briefing Book, August 1.

Beck, B. B., L. G. Rapaport, M. R. Stanley Price, and A. C. Wilson. 1994. Reintroduction of captive-born animals. Pp. 265–286 in P. J. S. Olney, G. M. Mace, and A. T. C. Feistner (editors), Creative conservation: Interactive management of wild and captive animals. Chapman and Hall, London.

Becker, R. E. 1976. The phytosociological position of tree ferns (Cibotium spp.) in the montane rain forest of the island of Hawaii. Ph.D. dissertation. University of Hawai'i at Mānoa, Honolulu.

Beckwith, M. W. 1951. The Kumulipo. University of Hawai'i Press, Honolulu, HI.

———. 1970. Hawaiian mythology. University of Hawai'i Press, Honolulu, HI.

Becnel, J. J., S. E. White, B. A. Moser, T. Fukuda, M. J. Rotstein, A. H. Undeen, and A. Cockburn. 2001. Epizootiology and transmission of a newly discovered baculovirus from the mosquito *Culex nigripalpus* and *C. quinquefasciatus*. Journal of General Virology 82: 275–282.

Bednarz, J. C., and R. J. Raitt. 2002. Chihuahuan Raven (*Corvus cryptoleucus*). In A. Poole and F. Gill (editors), The birds of North America, No. 606. Birds of North America Inc., Philadelphia.

Beidler, P. 1979. Animals and theme in ceremony. American Indian 5: 13–18.

Belkin, J. N. 1962. The mosquitoes of the South Pacific (Diptera: Culicidae), Vol. 1. University of California Press, Berkeley, CA.

Bell, B. D., and D. V. Merton. 2002. Critically endangered bird populations and their management. Conservation Biology 7: 105–138.

Bell, D., S. Robertson, and P. R. Hunter. 2004. Animal origins of the SARS coronavirus: Possible links with the international trade in small carnivores. Philosophical Transactions of the Royal Society of London, Series B 359: 1107–1114.

Benedict, M. Q., and A. S. Robinson. 2003. The first releases of transgenic mosquitoes: An argument for the sterile insect technique. Trends in Parasitology 19: 349–355.

Benevides, F. L. Jr., H. Hansen, and S. C. Hess. 2008. Design and evaluation of a simple signaling device for live-traps. Journal of Wildlife Management 72: 1434–1436.

Benkman, C. W. 1992. White-winged Crossbill (*Loxia leucoptera*). In A. Poole and F. Gill (editors), The birds of North America, No. 27. Birds of North America Inc., Philadelphia.

———. 1993. Adaptation to single resources and the evolution of crossbill (*Loxia*) diversity. Ecological Monographs 63: 305–325.

Bennett, G. F., M. Whiteway, and C. Woodworth-Lynas. 1982. A host-parasite catalogue of the avian haematozoa. Occasional Papers in Biology 5. Memorial University of Newfoundland, St. John's.

Bennett, P. M., and I. P. F. Owens. 1997. Variation in extinction risk among birds: Chance or evolutionary predisposition? Proceedings of the Royal Society of London, Series B 264: 401–408.

———. 2002. Evolutionary ecology of birds. Oxford University Press, Oxford.

Benning, T. L., D. LaPointe, C. T. Atkinson, and P. M. Vitousek. 2002. Interactions of climate change with biological invasions and land use in the Hawaiian Islands: Modeling the fate of endemic birds using a geographic information system. Proceedings of the National Academy of Sciences (USA) 99: 14246–14249.

Bensch, S., and M. Akesson. 2005. Ten years of AFLP in ecology and evolution: Why so few animals? Molecular Ecology 14: 2899–2914.

Berger, A. J. 1972. Hawaiian birdlife. University of Hawaii Press, Honolulu.

———. 1974. History of exotic birds in Hawaii. 'Elepaio 35: 60–65.

———. 1975. History of exotic birds in Hawaii, Part 2. 'Elepaio 35: 72–80.

———. 1980. Longevity of Hawaiian honeycreepers in captivity. Wilson Bulletin 92: 263–264.

———. 1981. Hawaiian Birdlife. 2nd edition. University Press of Hawaii, Honolulu.

Berlin, K. E., T. K. Pratt, J. C. Simon, J. R. Kowalsky, and J. S. Hatfield. 2000. Plant phenology in a cloud forest on the island of Maui, Hawaii. Biotropica 32: 90–99.

Berlin, K. E., J. C. Simon, T. K. Pratt, J. R. Kowalsky, and J. S. Hatfield. 2001. 'Ākohekohe response to flower availability: Seasonal abundance, foraging, breeding, and molt. Studies in Avian Biology 22: 202–212.

Berlin, K. E., and E. M. VanGelder. 1999. 'Ākohekohe (*Palmeria dolei*). In A. Poole and F. Gill (editors), The birds of North America, No. 400. Birds of North America Inc., Philadelphia.

Bester, M. N., J. P. Bloomer, R. J. Van Aarde, B. H. Erasmus, P. J. J. Van Rensburg, J. D. Skinner, P. G. Howell, and T. W. Naude. 2002. A review of the successful eradication of feral cats from sub-Antarctic Marion Island, Southern Indian Ocean. South African Journal of Wildlife Research 32: 65–73.

Bianchi, F. A. 1959. Entomological changes in the sugarcane fields of Hawaii, 1930–1959. Proceedings of the 10th Congress of the International Society of Sugar Cane Technology: 989–994.

Bibby, C. J., N. D. Burgess, D. A. Hill, and S. Mustoe. 2000. Bird census techniques. 2nd edition. Academic Press, London.

Big Island Invasive Species Committee. 2005. BIISC invasive species target lists. www.hear.org/biisc/.

Bildstein, K. L. 2004. Raptor migration in the Neotropics: Patterns, processes, and consequences. Ornitologia Neotropical 15: 83–99.

Bird, I. L. (Mrs. Bishop). 1906. The Hawaiian archipelago: Six months amongst the palm groves, coral reefs, and volcanoes of the Sandwich Islands. John Murray, London.

BirdLife International. 2000. Threatened birds of the world. Lynx Edicions and BirdLife International, Barcelona, Spain, and Cambridge, England.

———. 2004. Threatened birds of the world 2004. CD-ROM. BirdLife International, Cambridge, England.

Black, J. H. 1906. Feather hunting at Christmas. Paradise of the Pacific 19: 20–22.

Black, J. M., and P. C. Banko. 1994. Is the Hawaiian goose *Branta sandvicensis* saved from extinction? Pp. 394–410 in P. J. Olney, G. Mace, and A. Feistner (editors), Creative conservation: Interactive management of wild and captive animals. Chapman and Hall, London.

Blackburn, T. M., P. Cassey, R. P. Duncan, K. L. Evans, and K. J. Gaston. 2004. Avian extinction and mammalian introductions on oceanic islands. Science 305: 1955–1958.

Blackburn, T. M., and R. P. Duncan. 2001. Determinants of establishment success in introduced birds. Nature 414: 195–197.

Blackmore, M., and P. M. Vitousek. 2000. Cattle grazing, forest loss, and fuel loading in a dry forest ecosystem at Pu'u Wa'aWa'a Ranch, Hawai'i. Biotropica 32: 625–632.

Blanvillain, C., J. M. Salducci, G. Tutururai, and M. Maeura. 2003. Impact of introduced birds on the recovery of the Tahiti Flycatcher (*Pomarea nigra*), a critically endangered forest bird of Tahiti. Biological Conservation 109: 197–205.

Blondel, J., C. Ferry, and B. Frochot. 1970. La méthode des indices ponctuels d'abondance (I.P.A.) ou des relevés d'avifaune par "stations d'écoute." Alauda 38: 55–71.

Boag, P. T. 1987. Effects of nestling diet on growth and adult size of Zebra Finches (*Poephila guttata*). Auk 104: 155–166.

Boag, P. T., and P. R. Grant. 1981. Intense natural selection in a population of Darwin's finches (Geospizinae) in the Galapagos. Science 214: 82–85.

Boarman, W. I., and B. Heinrich. 1999. Common Raven (*Corvus corax*). In A. Poole and F. Gill (editors), The birds of North America, No. 476. Birds of North America Inc., Philadelphia.

Bock, W. J. 1978. Tongue morphology and affinities of the Hawaiian honeycreeper *Melamprosops phaeosoma*. Ibis 120: 467–479.

Boersma, P. D., and J. K. Parrish. 1998. Threats to seabirds: Research, education, and societal approaches to conservation. Pp. 237–259 in J. M. Marzluff and R. Sallabanks (editors), Avian conservation: Research and management. Island Press, Washington, DC.

Bonneaud, C., J. Mazuc, G. Gonzalez, C. Haussy, O. Chastel, B. Faivre, and G. Sorci. 2003. Assessing the cost of mounting an immune response. American Naturalist 161: 367–379.

Bonnet, D. D., and B. Gross. 1951. Susceptibility of *Rattus hawaiiensis* Stone to warfarin. Science 118: 44–45.

Bonnet, D. D., and S. M. K. Hu. 1951. The introduction of *Toxorhynchites brevipalpis* Theobald into the Territory of Hawaii. Proceedings of the Hawaiian Entomological Society 14: 237–242.

Borth, W. B., D. E. Gardner, and T. L. German. 1990. Association of double-stranded RNA and filamentous virus like particles with *Dodonaea* yellows disease. Plant Disease 74: 434–437.

Bourtzis, K., and S. O'Neill. 1998. *Wolbachia* infections and arthropod reproduction. Bioscience 48: 287–293.

Bowen, R. V. 1997. Townsend's Solitaire (*Myadestes townsendi*). In A. Poole and F. Gill (editors), The birds of North America, No. 269. Birds of North America Inc., Philadelphia.

Boyd, S., and I. Castro. 2000. Translocation history of Hihi (stitchbird), an endemic New Zealand honeyeater. Re-introduction News 19: 28–30.

Boyer, A. G. 2008. Extinction patterns in the avifauna of the Hawaiian Islands. Diversity and Distributions 14: 509–517.

Brackenridge, W. D. 1841. Journal kept while on the U.S. exploring expedition, 1838–1841. Unpublished manuscript at the Maryland Historical Society, Baltimore.

Bradley, J. E., and J. M. Marzluff. 2003. Rodents as nest predators: Influences on predatory behavior and consequences to nesting birds. Auk 120: 1180–1187.

Brazil, M. A. 1991. The birds of Japan. Smithsonian Institution Press, Washington DC.

Breese, P. L. 1959. Information on Cattle Egret, a bird new to Hawaii. 'Elepaio 20: 33–34.

Breitwisch, R. 1989. Mortality patterns, sex ratios, and parental investment in monogamous birds. Current Ornithology 6: 1–50.

Brenner, G. J., P. T. Oboyski, and P. C. Banko. 2002. Parasitism of *Cydia* spp. (Lepidoptera; Tortricidae) on *Sophora chrysophylla* (Fabaceae) along an elevation gradient of dry subalpine forest on Mauna Kea, Hawaii. Pan-Pacific Entomologist 78: 101–109.

Brigham, W. T. 1899–1903. Hawaiian feather work. Memoirs of the B. P. Bishop Museum 1: 1–81.

Briker, O. P., and M. A. Ruggiero. 1998. Toward a national program for monitoring environmental resources. Ecological Applications 8: 326–329.

Bristol, R. 2005. Conservation introductions of Seychelles fody and warbler to Denis Island, Seychelles. Re-introduction News 24: 35–36.

Brock, V. E. 1960. The introduction of aquatic animals into Hawaiian waters. Internationale Revue der gesamten Hydrobiologie 45: 463–464.

Brogdon, W. G., and J. C. McAllister. 1998. Insecticide resistance and vector control. Emerging Infectious Diseases 4: 605–613.

Brooks, T. M., R. A. Mittermeier, C. G. Mittermeier, G. A. B. da Fonseca, A. B. Rylands, W. R. Konstant, P. Flick, J. Pilgrim, S. Oldfield, G. Mace, and C. Hilton-Taylor. 2002. Habitat loss and extinction in the hotspots of biodiversity. Conservation Biology 16: 1523–1739.

Brown, J. H. 1971. Mammals on mountaintops: Nonequilibrum insular biogeography. American Naturalist 105: 467–478.

Brown, J. H., and A. Kodric-Brown. 1977. Turnover rates in insular biogeography: Effect of immigration on extinction. Ecology 58: 445–449.

Brown, J. H., and M. V. Lomolino. 1995. Biogeography. 2nd edition. Sinauer Associates, Sunderland, MA.

Brown, J. H., D. W. Mehlman, and G. C. Stevens. 1995. Spatial variation in abundance. Ecology 76: 2028–2043.

Brown Tree Snake Control Committee. 1996. The brown tree snake control plan. Prepared by the Brown Tree Snake Control Committee, Aquatic Nuisance Species Task Force, Honolulu, June.

Brumfield, R. T., P. Beerli, D. A. Nickerson, and S. V. Edwards. 2003. The utility of single nucleotide polymorphisms in inferences of population history. Trends in Ecology and Evolution 18: 249–256.

Bryan, E. H. Jr. 1933. Hawaiian nature notes. Honolulu Star-Bulletin, Honolulu.

———. 1958. Checklist and summary of Hawaiian Birds. Books about Hawaii, Honolulu.

Bryan, L. W. 1937a. The big fence on the Big Island. Paradise of the Pacific 49: 15, 30.

———. 1937b. Wild sheep in Hawaii. Paradise of the Pacific 49: 19, 31.

———. 1937c. Letter to Gordon Scruton, Hilo Chamber of Commerce. In the letters of Winston Banko. Located at the archives of the U.S. Geological Survey, Pacific Island Ecosystems Research Center, Hawai'i Volcanoes National Park, Volcano, HI.

———. 1947. Twenty-five years of forestry work on the Island of Hawaii. Hawaii Planters' Record 51: 1–80.

———. 1961. History of Hawaiian forestry up to and including 1920. Unpublished manuscript in the L. W. Bryan Forestry Collection, B. P. Bishop Museum, Honolulu.

Bryan, W. A. 1903. The Short-eared Owl taken far out at sea. Auk 20: 212–213.

———. 1908. Some birds of Molokai. Bishop Museum Occasional Papers 4: 133–176.

———. 1912. The introduction of birds into Hawaii. Proceedings of the Hawaiian Entomological Society 2: 168–175.

Buckland, S. T., D. R Anderson, K. P. Burnham, and J. L. Laake. 1993. Distance sampling: Estimating abundance of biological populations. Chapman and Hall, London.

Buckland, S. T., D. R. Anderson, K. P. Burnham, J. L. Laake, D. L. Borchers, and L. Thomas. 2001. Introduction to distance sampling: Estimating abundance of biological populations. Oxford University Press, Oxford.

——— (editors). 2004. Advanced distance sampling. Oxford University Press, Oxford.

Buckle, A. P., and M. G. P. Fenn. 1992. Rodent control in the conservation of endangered species. Proceedings of the Vertebrate Pest Conference 15: 36–41.

Burger, J., and M. Gochfeld. 1994. Predation and effects of humans on island nesting seabirds. Pp. 39–67 in D. N. Nettleship, J. Burger, and M. Gochfeld (editors), Seabirds on islands: Threats, case studies and action plans. Bird Conservation Series 1. BirdLife International, Cambridge, England.

Burgess, S. L. 2005. Phylogeography and conservation genetics of the Elepaio. Ph.D. dissertation. University of Hawai'i at Mānoa, Honolulu.

Burgess, S. L., and R. C. Fleischer. 2006. Isolation and characterization of polymorphic microsatellite loci in the Hawaiian flycatcher, the elepaio (Chasiempis sandwichensis). Molecular Ecology Notes 6: 14.

Burnett, K., B. Kaiser, B. A. Pitafi, and J. Roumasset. 2006. Prevention, eradication, and containment of invasive species: Illustrations from Hawaii. Agricultural and Resource Economics Review 35: 63–77.

Burney, D. A., R. V. DeCandido, L. P. Burney, F. N. Kostel-Hughes, T. W. Stafford Jr., and H. F. James. 1995. A Holocene record of climate change, fire ecology, and human activity from montane Flat Top Bog, Maui. Journal of Paleolimnology 13: 209–217.

Burney, D. A., H. F. James, L. P. Burney, S. L. Olson, W. Kikuchi, W. L. Wagner, M. Burney, D. McCloskey, D. Kikuchi, F. V. Grady, R. Gage II, and R. Nishek. 2001. Fossil evidence for a diverse biota from Kaua'i and its transformation since human arrival. Ecological Monographs 71: 615–641.

Burney, L. P., and D. A. Burney. 2003. Charcoal stratigraphies for Kaua'i and the timing of human arrival. Pacific Science 57: 211–226.

Burnham, K. P., and D. R. Anderson. 2002. Model selection and multimodel inference: A practical information-theoretic approach. Springer, New York.

Burns, K. J., S. J. Hackett, and N. K. Klein. 2002. Phylogenetic relationships and morphological diversity in Darwin's finches and their relatives. Evolution 56: 1240–1252.

Burton, P. 1980. Plant invasion into an ohia-treefern rain forest following experimental canopy opening. Pp. 21–39 in C. W. Smith (editor), Proceedings, Third Conference in Natural Sciences, Hawaii Volcanoes National Park. Cooperative National Park Resources Studies Unit, University of Hawai'i, Honolulu.

Burwash, M. D., M. E. Tobin, A. D. Woolhouse, and T. P. Sullivan. 1998. Laboratory evaluation of predator odors for eliciting an avoidance response in roof rats (Rattus rattus). Journal of Chemical Ecology 24: 49–66.

Butchart, S. H. M., A. J. Stattersfield, L. A. Bennun, S. M. Shutes, H. R. Akçakaya, J. E. M. Baillie, S. N. Stuart, C. Hilton-Taylor, and G. M. Mace. 2004. Measuring global trends in the status of biodiversity: Red List Indices for birds. Public Library of Science Biology 2: e383. www.plosbiology.org.

Butchart, S. H. M., A. J. Stattersfield, and N. J. Collar. 2006. How many bird extinctions have we prevented? Oryx 40: 266–278.

Butcher, G. S. 1990. Audubon Christmas bird counts. Pp. 5–13 in J. R. Sauer and S. Droege (editors), Survey designs and statistical methods for the estimation of avian population trends. Biological Report 90 (1). U.S. Department of the Interior, Fish and Wildlife Service, Washington, DC.

Butler, D., and D. Merton. 1992. The Black Robin: Saving the world's most endangered bird. Oxford University Press, Auckland, New Zealand.

Butler, G. D. Jr., and R. L. Usinger. 1963. Insects and other invertebrates from Laysan Island. Atoll Research Bulletin 98: 1–30.

Bystrak, D. 1981. The North American Breeding Bird Survey. Studies in Avian Biology 6: 34–41.

Cabin, R. J., S. G. Weller, D. H. Lorence, T. W. Flynn, A. K. Sakai, D. Sandquist, and L. J. Hadway. 2000. Effects of long-term ungulate exclusion and recent alien species control on the preservation and restoration of a Hawaiian tropical dry forest. Conservation Biology 14: 439–453.

Cade, T. J. 1983. Restoration of Bald Eagles in New York and elsewhere. Bird Conservation 1: 109–112.

Cade, T. J., and B. Burnham (editors). 2003. Return of the Peregrine. Peregrine Fund Publications, Boise, ID.

Camp, R. J., P. M. Gorresen, T. K. Pratt, and B. L. Woodworth. 2009. Population trends of native Hawaiian forest birds: 1976–2008. Technical Report HCSU-012. Hawai'i Cooperative Studies Unit, University of Hawai'i at Hilo, Hilo.

Camp, R. J., T. K. Pratt, C. Bailey, and D. Hu. In press. Landbirds vital sign monitoring protocol—Pacific Island Network. Natural Resources Report NPS/PWR/PACN/NRR–2008/000. National Park Service, Oakland, CA.

Camp, R. J., T. K. Pratt, P. M. Gorresen, J. J. Jeffrey, and B. L. Woodworth. 2009. Passerine bird trends at Hakalau Forest National Wildlife Refuge. Technical Report HCSU-011. Hawai'i Cooperative Studies Unit, University of Hawai'i at Hilo, Hilo.

Campbell, E. W. III. 1991. The effect of introduced roof rats on bird diversity of Antillean Cays. Journal of Field Ornithology 62: 343–348.

Campbell, E. W. III, A. E. Koehler, R. T. Sugihara, and M. E. Tobin. 1998. The development of an integrated pest management plan for roof rats in Hawaiian macadamia nut orchards. Proceedings of the Vertebrate Pest Conference 18: 171–175.

Campbell, F. C., and S. E. Schlarbaum. 2002. Fading forests 2: Trading away North America's natural heritage. Healing Stones Foundation in cooperation with the American Lands Alliance and the University of Tennessee, Knoxville, TN.

Campbell, J. 1973. The hero with a thousand faces. Princeton University Press, Princeton, NJ.

Cannarella, R. 1998. Forest industry. P. 249 in S. P. Juvik and J. O. Juvik (editors), Atlas of Hawaii, 3rd edition. University of Hawaii Press, Honolulu.

Cao, G., T. W. Giambelluca, D. E. Stevens, and T. A. Schroeder. 2007. Inversion variability in the Hawaiian trade wind regime. Journal of Climate 20: 1145–1160.

Captive Breeding Specialist Group. 1990. Bali mynah starling recovery plan: Population viability analysis. Captive Breeding Specialist Group, Apple Valley, MN.

Carlquist, S. 1966. The biota of long-distance dispersal: Principles of dispersal and evolution. Quarterly Review of Biology 41: 247–270.

———. 1974. Island biology. Columbia University Press, New York.

———. 1980. Hawaii: A natural history. 2nd edition. Pacific Tropical Botanical Garden, Lawai, HI.

Carlstead, K., and K. Hunt. 2002. Stress responses to environmental disturbances in captive Hawaiian honeycreepers (*Aves, Passeriformes*). Annual Conference Proceedings. Association of Zoos and Aquariums, Silver Spring, MD.

Carothers, J. H. 1986a. Behavioral and ecological correlates of interference competition among some Hawaiian drepanidanae. Auk 103: 564–574.

———. 1986b. The effect of retreat site quality on interference-related behavior among Hawaiian honeycreepers. Condor 88: 421–426.

Carothers, J. H., and R. B. Hansen. 1982. Occurrence of the Japanese Bush-Warbler on Maui. 'Elepaio 43: 17–18.

Carpenter, F. L. 1976. Plant-pollinator interactions in Hawaii: Pollination energetics of *Metrosideros collina* (Myrtaceae). Ecology 57: 1125–1144.

Carpenter, F. L., and R. E. MacMillen. 1973. Interactions between Hawaiian honeycreepers and *Metrosideros collina* on the island of Hawaii. Technical Report 33. Island Ecosystems Integrated Research Program, U.S. International Biological Program, Department of Botany, University of Hawaii, Honolulu.

———. 1976a. Threshold model of feeding territoriality and test with a Hawaiian honeycreeper. Science 194: 639–642.

———. 1976b. Energetic costs of feeding territories in a Hawaiian honeycreeper. Oecologia 26: 213–223.

———. 1980. Resource limitation, foraging strategies, and community structure in Hawaiian honeycreepers. Proceedings of the 17th International Ornithological Congress: 1100–1104.

Carr, G. D. 1985. Monograph of the Hawaiin Madiinae (Asteraceae): *Argyroxiphium, Dubautia,* and *Wilkesia.* Allertonia 4: 1–123.

Carson, H. L., and D. A. Clague. 1995. Geology and biogeography of the Hawaiian Islands. Pp. 14–29 in W. L. Wagner and V. A. Funk (editors), Hawaiian biogeography: Evolution on a hot spot archipelago. Smithsonian Institution Press, Washington, DC.

Case, T. J. 1996. Global patterns in the establishment and distribution of exotic birds. Biological Conservation 78: 69–96.

Case, T. J., and D. T. Bolger. 1991. The role of introduced species in shaping the distribution and abundance of island reptiles. Evolutionary Ecology 5: 272–290.

Casey, T. L. C., and J. D. Jacobi. 1974. A new genus and species of bird from the Island of Maui, Hawaii (Passeriformes: Drepanididae). Bishop Museum Occasional Papers 24: 216–226.

Cassey, P., and B. H. McArdle. 1999. An assessment of distance sampling techniques for estimating animal abundance. Environmetrics 10: 261–278.

Castelletta, M., N. S. Sodhi, and R. Subaraj. 2000. Heavy extinctions of forest avifauna in Singapore: Lessons for biodiversity conservation in Southeast Asia. Conservation Biology 14: 1870–1880.

Castle, A. L. 1935. The bird blessing of "Hui Manu." Paradise Holiday Number 1935: 14–16.

Catteruccia, F., T., Nolan, T. G. Loukeris, C. Blass, C. Savakis, F. C. Kafatos, and A. Cristanti. 2000. Stable germline transformation of the malaria mosquito *Anopheles stephensi*. Nature 405: 959–962.

Caum, E. L. 1933. The exotic birds of Hawaii. Bishop Museum Occasional Papers 10: 1–55.

Chalfoun, A. D, F. R. Thompson III, and M. J. Ratsnaswamy. 2002. Nest predators and fragmentation: A review and meta-analysis. Conservation Biology 16: 306–318.

Channell, R., and M. V. Lomolino. 2000. Dynamic biogeography and conservation of endangered species. Nature 403: 84–86.

Chimera, C. G., A. C. Medeiros, L. L. Loope, and R. H. Hobdy. 2000. Status of management and control efforts for the invasive alien tree *Miconia calvescens* DC. (Melastomataceae) in Hana, East Maui. Technical Report 128. Pacific Cooperative Studies Unit, University of Hawai'i, Honolulu.

Christensen, C. C., and P. V. Kirch. 1986. Nonmarine mollusks and ecological change at Barbers Point, O'ahu, Hawai'i. Bishop Museum Occasional Papers 26: 52–80.

Cid del Prado Vera, I., A. R. Maggenti, and C. van Riper III. 1985. New species of Spiruridae (Nemata: Spirurida) from endemic Hawaiian honeycreepers (Passeriformes: Drepanididae), the Japanese White-eye (Passeriformes: Zosteropidae) and a new species of Acuariidae (Nemata: Spirurida) from the Japanese White-eye collected on the island of Hawaii. Proceedings of the Helminthological Society of Washington 52: 247–259.

Clague, D. 1998. Geology. Pp. 37–46 in S. P. Juvik and J. O. Juvik (editors), Atlas of Hawai'i. 3rd edition. University of Hawai'i Press, Honolulu.

Clark, J., and P. V. Kirch (editors). 1983. Archaeological investigations of the Mudlane–Waimea–Kawaihae Road Corridor, Island of Hawai'i: An interdisciplinary study of the environmental transect. Report Series 83-1. Department of Anthropology, Bishop Museum, Honolulu.

Clark, J. A., J. M. Hoekstra, P. D. Boersma, and P. Kareiva. 2002. Improving U.S. Endangered Species Act recovery plans: Key findings and recommendations of the SCB Recovery Plan Project. Conservation Biology 16: 1523–1739.

Clarke, T. 2002. Mosquitoes minus malaria. Nature 419: 429–430.

Clarkson, K. E., and L. P. Laniawe. 2000. Hawaiian Hawk (*Buteo solitarius*). In A. Poole and F. Gill (editors), The birds of North America, No. 523. Birds of North America Inc., Philadelphia.

Cleaveland, S., M. K. Laurenson, and L. H. Taylor. 2001. Diseases of humans and their domestic mammals: Pathogen characteristics, host range and the risk of emergence. Philosophical Transactions of the Royal Society of London, Series B 356: 991–999.

Clegg, S. M., and I. P. F. Owens. 2002. The "island rule" in birds: Medium body size and its ecological explanation. Proceedings of the Royal Society of London, Series B 269: 1359–1365.

Clout, M. N., K. Denyer, R. E. James, and I. G. McFadden. 1995. Breeding success of New Zealand pigeons (*Hemiphaga novaeseelandiae*) in relation to control of introduced mammals. New Zealand Journal of Ecology 19: 209–212.

Clutton-Brock, J. 1989. A natural history of domesticated animals. University of Texas Press, Austin.

Clyde, D. F., V. C. McCarthy, R. M. Miller, and W. E. Woodward. 1975. Immunization of man against falciparum and vivax malaria by use of attenuated sporozoites. American Journal of Tropical Medicine and Hygiene 24: 397–401.

Cochran, W. G. 1977. Sampling techniques. 3rd edition. John Wiley and Sons, New York.

Cody, M. L. 1974. Competition and the structure of bird communities. Monographs in Population Biology 7. Princeton University Press, Princeton, NJ.

Cohen, J. E. 1988. Statistical power analysis for the behavioral sciences. Erlbaum, Hillsdale, NJ.

Cole, F. R., L. L. Loope, A. C. Medeiros, J. A. Raikes, and C. S. Wood. 1994. Conservation implications of introduced game birds in high-elevation Hawaiian shrubland. Conservation Biology 9: 306–313.

Cole, F. R., A. C. Medeiros, L. L. Loope, and W. W. Zuehlke. 1992. Effects of Argentine ant on arthropod fauna of Hawaiian high-elevation shrubland. Ecology 73: 1313–1322.

Colum, P. 1924. At the gateways of the day. Yale University Press, New Haven, CT.

———. 1937. Legends of Hawai'i. Yale University Press, New Haven, CT.

Colvin, B. A., M. W. Fall, L. A. Fitzgerald, and L. L. Loope. 2005. Review of brown treesnake problems and control programs: Report of observations and recommendations. Prepared at the request of the U.S. Department of the Interior, Office of Insular Affairs, for the Brown Treesnake Control Committee, March. www.invasivespeciesinfo.gov/animals/controlplans.shtml (accessed March 23, 2008).

Conant, P., A. C. Medeiros, and L. L. Loope. 1997. A multi-agency containment program for miconia (*Miconia calvescens*), an invasive tree in Hawaiian rain forests. Pp. 249–254 in J. Luken and J. Thieret (editors), Assessment and management of invasive plants. Springer, New York.

Conant, S. 1975. Spatial distribution of bird species on the east flank of Mauna Loa. Technical Report 74, Island Ecosystems International Research Program, U.S. International Biology Program, University of Hawaii, Honolulu.

———. 1977. The breeding biology of the Oahu 'Elepaio. Wilson Bulletin 89: 193–210.

———. 1980. Recent records of the 'Ua'u (Dark-rumped Petrel) and the 'A'o (Newell's Shearwater) in Hawai'i. 'Elepaio 41: 11–13.

———. 2005. Honeycreepers in Hawaiian material culture. Pp. 278–284 in H. D. Pratt. The Hawaiian honeycreepers: Drepanidinae. Oxford University Press, Oxford.

Conant, S., C. C. Christensen, P. Conant, W. C. Gagné, and M. L. Goff. 1983. The unique terrestrial biota of the Northwestern Hawaiian Islands. Pp. 77–94 in Proceedings of the Second Symposium on Resource Investigations in the Northwestern Hawaiian Islands, 25–27 May. University of Hawaii Sea Grant College Report UNIHI-SEAGRANT-MR-84-01, Honolulu.

Conant, S., M. S. Collins, and C. J. Ralph. 1981. Effects of observers using different methods upon the total population estimates of two resident island birds. Studies in Avian Biology 6: 377–381.

Conant, S., and M. Morin. 2001. Why isn't the Nihoa Millerbird extinct? Studies in Avian Biology 22: 338–346.

Conant, S., H. D. Pratt, and R. J. Shallenberger. 1998. Reflections on a 1975 expedition to the lost world of the Alaka'i and other notes on the natural history, systematics, and conservation of Kaua'i birds. Wilson Bulletin 110: 1–22.

Congressional Office of Technology Assessment. 1993. Harmful non-indigenous species in the United States. Congressional Office of Technology Assessment OTA-F-565. U.S. Government Printing Office, Washington, DC.

Conry, P. 1991. Oahu forest bird survey indicates decline in native species. Hawaii's Forests and Wildlife 6: 2.

Conry, P., P. Schuyler, and M. Collins. 2000. Po'o-uli (Black-faced honeycreeper). Pp. 235–240 in R. P. Reading and B. Miller (editors), Endangered animals: A reference guide to conflicting issues. Greenwood, Westport, CT.

Conry, P. J. 1994. Diphacinone approved for rodent control in forests and other natural areas. Hawaii's Forest and Wildlife 9: 2–10.

Conway, C. J., and T. E. Martin. 2000. Evolution of passerine incubation behavior: Influence of food, temperature, and nest predation. Evolution 54: 670–685.

Conway, W. 1995. Wild and zoo animal interactive management and habitat conservation. Biodiversity and Conservation 4: 573–594.

———. 1997. The changing role of zoos in international conservation and the WCS. Society for Conservation Biology Newsletter 4: 1–3.

Conway, W. G. 1986. The practical difficulties and financial implications of endangered species breeding programmes. International Zoo Yearbook 24–25: 210–219.

Cook, J., and J. King. 1784. A voyage to the Pacific Ocean . . . performed in His Majesty's ships *Resolution* and *Discovery* in the years 1776, 1777, 1778, 1779, and 1780. 3 vols. G. Nicol and T. Cadell, London.

Cooney, R. 2004. The precautionary principle in biodiversity conservation and natural resource management: An issues paper for policy-makers, researchers and practitioners. International Union for Conservation of Nature, Gland, Switzerland, and Cambridge, England.

Cooper, A. C., J. Rhymer, H. F. James, S. L. Olson, M. Sorenson, R. C. Fleischer, and C. McIntosh. 1996. Ancient DNA and island endemics. Nature 381: 484.

Cooray, R. G., and D. Mueller-Dombois. 1981. Feral pig activity. Pp. 309–317 in D. Mueller-Dombois, K. W. Bridges, and H. L. Carson (editors), Island ecosystems: Biological organization in selected Hawaiian communities. Hutchinson Ross, Stroudsburg, PA, and Woods Hole, MA.

Coordinating Group on Alien Pest Species. 1996. The silent invasion. Developed collaboratively with the Coordinating Group on Alien Pest Species by InfoGrafik Inc., Honolulu. Part available online at www.hear.org/intro/contents.htm (accessed March 23, 2008).

Cordell, S., R. J. Cabin, S. G. Weller, and D. H. Lorence. 2002. Simple and cost-effective methods control fountain grass in dry forests (Hawaii). Ecological Restoration 20: 139–140.

Cordell, S., G. Goldstein, D. Mueller-Dombois, D. Webb, and P. M. Vitousek. 1998. Physiological and morphological variation in *Metrosideros polymorpha*, a dominant Hawaiian tree species, along an altitudinal gradient: The role of phenotypic plasticity. Oecologia 113: 188–196.

Cortinas, M. N., N. Koch, J. Thain, B. D. Wingfield, and M. J. Wingfield. 2004. First record of the

Eucalyptus stem canker pathogen, *Coniothyrium zuluense,* from Hawaii. Australasian Plant Pathology 33: 309–312.

Côte, I. M., and W. J. Sutherland. 1997. The effectiveness of removing predators to protect bird populations. Conservation Biology 11: 395–405.

Courchamp, F., and S. J. Cornell. 2000. Virus-vectored immunocontraception to control feral cats on islands: A mathematical model. Journal of Applied Ecology 37: 903–913.

Courchamp, F., M. Langlais, and G. Sugihara. 1999. Cats protecting birds: Modelling the mesopredator release effect. Journal of Animal Ecology 68: 282–292.

———. 2000. Rabbits killing birds: Modelling the hyperpredation process. Journal of Animal Ecology 69: 154–164.

Coutinho, T. A., M. J. Wingfield, A. C. Alfenas, and P. W. Crous. 1998. Eucalyptus rust: A disease with the potential for serious international implications. Plant Disease 82: 819–825.

Cowie, R. H. 1995. Variation in species diversity and shell shape in Hawaiian land snails: In situ speciation and ecological relationships. Evolution 49: 1191–1202.

———. 1998. Patterns of introduction of nonindigenous non-marine snails and slugs in the Hawaiian Islands. Biodiversity and Conservation 7: 349–368.

Cowie, R. H., N. J. Evenhuis, and C. C. Christensen. 1995. Catalog of the native land and freshwater mollusks of the Hawaiian Islands. Backhuys, Leiden, The Netherlands.

Crampton, J., A. Morris, G. Lycett, A. Warren, and P. Eggleston. 1990. Transgenic mosquitoes: A future vector control strategy? Parasitology Today 6: 31–36.

Crandall, K. A., O. R. P. Bininda-Edmonds, G. M. Mace, and R. K. Wayne. 2000. Considering evolutionary process in conservation biology. Trends in Ecology and Evolution 15: 290–295.

Crick, H. Q. P., S. R. Baillie, D. E. Balmer, R. I. Bashford, C. Dudley, D. E. Glue, R. D. Gregory, J. H. Marchant, W. J. Peach, and A. M. Wilson. 1997. Breeding birds in the wider countryside: Their conservation status (1971–1995). Research Report 187. British Trust for Ornithology, Thetford, UK.

Crocoll, S. T. 1994. Red-shouldered Hawk (*Buteo lineatus*). In A. Poole and F. Gill (editors), The birds of North America, No. 107. Birds of North America Inc., Philadelphia.

Crosby, A. W. Jr. 1974. Virgin soil epidemics as a factor in the aboriginal depopulation in America. William and Mary Quarterly 33: 291–292.

Cuddihy, L. W., and C. P. Stone. 1990. Alteration of native Hawaiian vegetation: Effects of humans, their activities and introductions. University of Hawai'i Press, Honolulu.

Culliney, J. L. 1988. Islands in a far sea: Nature and man in Hawaii. Sierra Club Books, San Francisco.

Cunningham, A. A. 1996. Disease risks of wildlife translocations. Conservation Biology 10: 349–353.

Daehler, C. C. 1998. The taxonomic distribution of invasive Angiosperm plants: Ecological insights and comparison to agricultural weeds. Biological Conservation 84: 167–180.

Daehler, C. C., J. S. Denslow, S. Ansari, and H. Kuo. 2004. A risk assessment system for screening out invasive pest plants from Hawaii and other Pacific islands. Conservation Biology 18: 360–368.

Danner, R. M, D. M. Goltz, S. C. Hess, and P. C. Banko. 2007. Evidence of feline immunodeficiency virus, feline leukemia virus, and *Toxoplasma gondii* in feral cats on Mauna Kea, Hawaii. Journal of Wildlife Diseases 43: 315–318.

D'Antonio, C. M., and S. E. Hobbie. 2005. Plant species effects on ecosystem processes: Insights from invasive species. Pp. 65–84 in D. V. Sax, J. J. Stachowicz, and S. D. Gaines (editors), Species invasions: Insights into ecology, evolution, and biogeography. Sinauer Associates, Sunderland, MA.

D'Antonio, C. M., and P. M. Vitousek. 1992. Biological invasions by exotic grasses, the grass/fire cycle, and global change. Annual Review of Ecology and Systematics 23: 63–87.

Darwin, C. R. 1859. On the origin of species by means of natural selection, or the preservation of favoured races in the struggle for life. John Murray, London.

Daszak, P., A. A. Cunningham, and A. D. Hyatt. 2000. Emerging infectious diseases of wildlife—threats to biodiversity and human health. Science 287: 443–449.

Davies, K. F., C. R. Margules, and J. F. Lawrence. 2000. Which traits of species predict population declines in experimental forest fragments? Ecology 81: 1450–1461.

Davies, N. B. 2000. Cuckoos, cowbirds and other cheats. T. and A. D. Poyser, London.

Davis, B. D. 1978. Human settlement and environmental change at Barbers Point, O'ahu. Pp. 87–97 in C. W. Smith (editor), Proceedings of the 2nd Conference in Natural Science. Cooperative National Park Resources Study Unit, University of Hawai'i, Honolulu.

Davis, J. N. 1999. Lawrence's Goldfinch (*Carduelis lawrencei*). In A. Poole and F. Gill (editors), The birds of North America, No. 480. Birds of North America Inc., Philadelphia.

Dawson, J. W., and L. Stemmermann. 1999. *Metrosideros* Banks ex Gaertn. Pp. 964–970 in W. L. Wagner, D. R. Herbst, and S. H. Sohmer (editors), Manual of the flowering plants of Hawai'i. Bishop Museum Special Publication 97. University of Hawaii Press and Bishop Museum Press, Honolulu.

Dawson, T. 2005. NARs, precious lands poised for windfall with Legacy Act's conveyance tax increase. Environment Hawai'i 16 (2): 1, 6–7.

Dawson, W. R. 1997. Pine Siskin (*Carduelis pinus*). In A. Poole and F. Gill (editors), The birds of North America, No. 280. Birds of North America Inc., Philadelphia.

Day, T., and R. MacGibbon. 2001. Cost estimates for Xcluder™ pest proof fences. Unpublished. Xcluder Pest Proof Fencing Company, Cambridge, New Zealand.

Dayer, A. A., M. J. Manfredo, T. L. Teel, and A. D. Bright. 2006. State report for Hawai'i from the research project entitled "Wildlife Values in the West." Project Report 68 for the Hawai'i Department of Land and Natural Resources. Colorado State University, Human Dimensions in Natural Resources Unit, Fort Collins.

Degener, O. 1930. Plants of Hawai'i National Park: Illustrative of plants and customs of the south seas. Edwards Brothers, Ann Arbor, MI.

Denslow, J. S., and M. T. Johnson. 2006. Biological control of tropical weeds: Research opportunities in plant-herbivore interactions. Biotropica 38: 139–142.

Derrickson, S. R. 1998. Appendix C: Summary of 05 June 98. In Environmental assessment for the proposed management actions to save the Po'o-uli. Cooperative Hawaiian forest bird propagation project summary. State of Hawaii, Department of Land and Natural Resources and U.S. Department of the Interior, Fish and Wildlife Service, Honolulu.

Derrickson, S. R., and N. F. R. Snyder. 1992. Potentials and limits of captive breeding in parrot conservation. Pp. 133–163 in S. R. Beissinger and N. F. R. Snyder (editors), New World parrots in crisis: Solutions from conservation biology. Smithsonian Institution Press, Washington, DC.

Desowitz, R. S. 1991. The malaria capers. More tales of parasites and people, research and reality. W. W. Norton and Company, New York.

———. 2000. The malaria vaccine: Seventy years of the great immune hope. Parasitologia 42: 173–182.

Devick, W. S. 1991. Patterns of introductions of aquatic organisms to Hawaiian freshwater habitats. Pp. 189–213 in W. S. Devick (editor), New directions in research, management and conservation of Hawaiian freshwater stream ecosystems: Proceedings of the 1990 symposium on freshwater stream biology and fisheries management. State of Hawaii, Department of Land and Natural Resources, Division of Aquatic Resources, Honolulu.

Diamond, J. M. 1984. Historical extinctions: A rosetta stone for understanding prehistoric extinctions. Pp. 824–862 in P. S. Martin and R. G. Klein (editors), Quaternary extinctions: A prehistoric revolution. University of Arizona Press, Tucson, AZ.

———. 1985. Population processes in island birds: Immigration, extinction and fluctuations. Pp. 17–21 in P. J. Moors (editor), Conservation of island birds: Case studies for the management of threatened island species. Technical Publication 3. International Council of Bird Preservation, Cambridge, England.

Diamond, J. M., and C. R. Veitch. 1981. Extinctions and introductions in the New Zealand avifauna: Cause and effect? Science 211: 499–501.

Dieckmann, O., J. A. P. Heesterbeek, and J. A. J. Metz. 1990. On the definition and the computation of the basic reproduction ratio R_0 in models for infectious-diseases in heterogeneous populations. Journal of Mathematical Biology 28: 365–382.

Dilks, P., M. Willans, M. Pryde, and I. Fraser. 2003. Large scale stoat control to protect mohua (*Mohoua ochrocephala*) and kaka (*Nestor meridionalis*) in the Eglinton Valley, Fiordland, New Zealand. New Zealand Journal of Ecology 27: 1–9.

Dillingham, W. F. 1938. Hui Manu—friend of birds. Pacific Holiday Number 1938: 20–21.

Diong, C. H. 1982. Population biology and management of the feral pig (*Sus scrofa* L.) in Kipahulu Valley, Maui. Ph.D. dissertation. University of Hawai'i at Mānoa, Honolulu.

Di Stefano, J. 2003. How much power is enough? Against the development of an arbitrary convention for statistical power calculations. Functional Ecology 17: 707–709.

Dobson, A. 2004. Population dynamics of pathogens with multiple host species. American Naturalist 164: S64–S78.

Dobson, A. P., and J. Foufopoulos. 2001. Emerging infectious pathogens of wildlife. Philosophical Transactions of the Royal Society of London, Series B 356: 1001–1012.

Dobson, A. P., and R. M. May. 1986. Patterns of invasions by pathogens and parasites. Pp. 58–76 in H. A. Mooney and J. A. Drake (editors), Ecology of biological invasions of North American and Hawaii. Springer, New York.

Dohm, D. J., M. L. O'Guinn, and M. J. Turell. 2002. Effect of environmental temperature on the ability of *Culex pipiens* (Diptera: Culicidae) to transmit West Nile virus. Journal of Medical Entomology 39: 221–225.

Domm, S., and J. Messersmith. 1990. Feral cat eradication on a barrier reef island, Australia. Atoll Research Bulletin 338.

Donagho, W. 1970. Observations of the Edible Nest Swiftlet on Oahu. 'Elepaio 30: 64–65.

Doolan, D. L., M. Sedegah, R. C. Hedstrom, J. C. Aguiar, and S. L. Hoffman. 1996. DNA vaccination against malaria. Advanced Drug Delivery Reviews 21: 49–61.

Doty, R. E. 1938. The prebaited feeding-station method of rat control. Hawaiian Planters' Record 42: 39–76.

————. 1945. Rat control on Hawaiian sugarcane plantations. Hawaiian Planters' Record 49: 71–239.

Dow, R. P., and G. M. Gerrish. 1970. Day-to-day change in relative humidity and activity of *Culex nigripalpus* (Diptera—Culicidae). Annals of the Entomological Society of America 63: 995–999.

Drake, D. R. 1998. Relationships among the seed rain, seed bank and vegetation of a Hawaiian forest. Journal of Vegetation Science 9: 103–112.

Drake, J. A., H. A. Mooney, F. di Castri, R. H. Groves, F. J. Kruger, M. Rejmanek, and M. Williamson (editors). 1989. Ecology of biological invasions: A global perspective. John Wiley and Sons, Chichester, England.

Dubey, J. P., and C. P. Beattie. 1998. Toxoplasmosis of animals and man. CRC, Boca Raton, FL.

Dubey, J. P., and K. Odening. 2001. Toxoplasmosis and related infections. Pp. 478–519 in W. M. Samuel, M. J. Pybus, and A. A. Kocan (editors), Parasitic diseases of wild mammals. Iowa State University Press, Ames.

Duckworth, W. D., T. J. Cade, H. L. Carson, S. Derrickson, J. Fitzpatrick, and F. C. James. 1992. The scientific bases for the preservation of the Hawaiian Crow. National Academy Press, Washington, DC.

Duffy, D. C., and F. Kraus. 2006. Science and the art of the solvable in Hawaii's terrestrial extinction crisis. Environment Hawai'i 16: 3–6.

Duncan, R. P. 1997. The role of competition and introduction effort in the success of passeriform birds introduced to New Zealand. American Naturalist 149: 903–915.

Duncan, R. P., and T. M. Blackburn. 2002. Morphological over-dispersion in game birds (Aves: Galliformes) successfully introduced to New Zealand was not caused by interspecific competition. Evolutionary Ecology Research 4: 551–561.

Duncan, R. P., T. M. Blackburn, and D. Sol. 2003. The ecology of bird introductions. Annual Review of Ecology Evolution and Systematics 34: 71–98.

Duncan, R. P., M. Bomford, D. M. Forsyth, and L. Conibear. 2001. High predictability in introduction outcomes and the geographical range size of introduced Australian birds: A role for climate. Journal of Animal Ecology 70: 621–632.

Dunlevy, P. A., and E. W. Campbell III. 2002. Assessment of hazards to non-native mongooses (*Herpestes auropunctatus*) and feral cats (*Felis catus*) from the broadcast application of rodenticide bait in native Hawaiian forests. Proceedings of the Vertebrate Pest Conference 20: 277–281.

Dunlevy, P. A., E. W. Campbell III, and G. D. Lindsey. 2000. Broadcast application of a placebo rodenticide bait in a native Hawaiian forest. International Biodeterioration and Biodegradation 45: 199–208.

Dunn, E. H. 2002. Using decline in bird populations to identify needs for conservation action. Conservation Biology 16: 1632–1637.

Duvall, F. P., and S. Conant. 1986. Current status of the Hawaiian Crow. Endangered Species Technical Bulletin 3: 1–2.

Ebenhard, J. 1988. Introduced birds and mammals and their ecological effects. Swedish Wildlife Research 13: 5–107.

Eddinger, C. R. 1969. Experiences with hand-raising passerine birds in Hawaii. Aviculture Magazine 75: 12–14.

————. 1970. A study of the breeding biology of four species of Hawaiian honeycreepers (Drepanididae). Ph.D. dissertation. University of Hawai'i at Mānoa, Honolulu.

————. 1971. Hand-raising Hawaii's endemic honeycreepers. Avicultural Magazine 77: 112–114.

————. 1972a. Discovery of the nest of the Kauai Akepa. Wilson Bulletin 84: 95–97.

————. 1972b. Discovery of the nest of the Kauai Creeper. Auk 89: 673–674.

Edman, J. D., and J. S. Haeger. 1977. Feeding patterns of Florida mosquitoes V. *Wyeomyia*. Journal of Medical Entomology 14: 477–479.

Eggert, L. S., J. Beadell, A. McClung, C. E. McIntosh, and R. C. Fleischer. 2009. Evolution of microsatellite loci in the adaptive radiation of Hawaiian honeycreepers. Journal of Heredity 100: 137–147.

Eggert, L. S., and R. C. Fleischer. 2004. Isolation of polymorphic microsatellite loci in the Hawaii 'amakihi (*Hemignathus virens*) and their use in other honeycreeper species. Molecular Ecology Notes 4: 725–727.

Eggert, L. S., L. A. Terwilliger, B. A. Woodworth, P. J. Hart, D. Palmer, and R. C. Fleischer. 2008. Genetic structure along an elevational gradient in Hawaiian honeycreepers reveals contrasting evolutionary responses to avian malaria. BMC Evolutionary Biology 8: 315.

Eidson, M., L. Kramer, Y. Hagiwara, K. Schmit, and W. Stone. 2001. Dead bird surveillance as an early warning system for West Nile. Emerging Infectious Diseases 7: 631–636.

Eisemann, J. D., and C. E. Swift. 2006. Ecological and human health hazards from broadcast application of 0.005% diphacinone rodenticide baits in native Hawaiian ecosystems. Proceedings of the Vertebrate Pest Conference 22: 413–433.

Elder, W. H., and D. H. Woodside. 1958. Biology and management of the Hawaiian goose. Transactions of the North American Wildlife Conference 23: 198–215.

Eldredge, L. G., and N. L. Evenhuis. 2003. Hawaii's biodiversity: A detailed assessment of the numbers of species in the Hawaiian Islands. Records of the Hawaii Biological Survey for 2001–2002. Bishop Museum Occasional Papers 76: 1–28.

Eldredge, L. G., and S. E. Miller. 1995. How many species are there in Hawaii? Records of the Hawaii Biological Survey for 1994. Bishop Museum Occasional Papers 41: 1–18.

Eliasson, U. 1995. Patterns of diversity in island plants. Pp. 35–50 in P. M. Vitousek, L. L. Loope, and H. Adersen (editors), Islands biological diversity and ecosystem function. Springer, Berlin.

Ellegren, H. 1996. First gene on the avian W chromosome (CHD) provides a tag for universal sexing of non-ratite birds. Proceedings of the Royal Society of London, Series B 263: 1635–1644.

Ellis, S., C. Kuehler, R. Lacy, K. Hughes, and U. S. Seal. 1992. Hawaiian forest birds: Conservation assessment and management plan. Captive Breeding Specialist Group, International Union for Conservation of Nature–The World Conservation Union / Species Survival Commission. U.S. Department of the Interior, Fish and Wildlife Service, Honolulu.

Ellis, W. 1827 [reprinted 1963]. Journal of William Ellis: Narrative of a tour of Hawai'i or Owhyhee, with remarks on the history traditions, manners, customs, and language of the inhabitants of the Sandwich Islands. Advertiser Publishing, Honolulu.

Ellison, A. M., M. S. Bank, B. D. Clinton, E. A. Colburn, K. Elliott, C. R. Ford, D. R. Foster, B. D. Kloeppel, J. D. Knoepp, G. M. Lovett, J. Mohan, D. A. Orwig, N. L. Rodenhouse, W. V. Sobczak, K. A. Stinson, J. K. Stone, C. M. Swan, J. Thompson, B. Von Holle, and J. R. Webster. 2005. Loss of foundation species: Consequences for the structure and dynamics of forested ecosystems. Frontiers in Ecology and the Environment 3: 479–486.

Ely, C. A., and R. B. Clapp. 1973. The natural history of Laysan Island, Northwestern Hawaiian Islands. Atoll Research Bulletin 171: 1–361.

Emerson, B. C. 2002. Evolution on oceanic islands: Molecular phylogenetic approaches to understanding pattern and process. Molecular Ecology 11: 951–966.

Emerson, N. B. 1895. The bird-hunters of ancient Hawai'i. Pp. 101–111 in T. G. Thrum (editor), Hawaiian almanac and annual. Press Publishing Co. Steam Print, Honolulu.

Emlen, J. T. 1971. Population densities of birds derived from transect counts. Auk 88: 332–342.

Emlen, S. T., and L. W. Oring. 1977. Ecology, sexual selection, and the evolution of mating systems. Science 197: 215–223.

Engeman, R. M., S. A. Shwiff, F. Cano, and B. Constantin. 2003. An economic assessment of the potential for predator management to benefit Puerto Rican parrots. Ecological Economics 46: 283–292.

Engilis, A. Jr. 1990. Field notes on native forest birds in the Hanawi Natural Area Reserve, Maui. 'Elepaio 50: 67–72.

Engilis, A. Jr., T. K. Pratt, C. B. Kepler, A. Ecton, and M. Fluetsch. 1996. Description of adults, eggshells, nestling, fledgling, and the nest of the Poouli. Wilson Bulletin 108: 607–619.

England, A. S., M. J. Bechard, and C. S. Houston. 1997. Swainson's Hawk (Buteo swainsoni). In A. Poole and F. Gill (editors), The birds of North America, No. 265. Birds of North America Inc., Philadelphia.

Englund, R. A. 1999. The impacts of introduced poeciliid fish and Odonata on the endemic Megalagrion (Odonata) damselflies of Oahu Island, Hawaii. Journal of Insect Conservation 3: 225–243.

Evans Mack, D., and W. Yong. 2000. Swainson's Thrush (Catharus ustulatus). In A. Poole and F. Gill (editors), The birds of North America, No. 540. Birds of North America Inc., Philadelphia.

Faaborg, J. 1986. Reproductive success and survivorship of the Galapagos Hawk (Buteo galapagoensis): Potential costs and benefits of cooperative polyandry. Ibis 128: 337–347.

Faaborg, J., and W. J. Arendt. 1995. Survival rates of Puerto Rican birds: Are islands really that different? Auk 112: 503–507.

Fagerstone, K. A., R. W. Bullard, and C. A. Ramey. 1990. Politics and economics of maintaining pesticide registrations. Proceedings of the Vertebrate Pest Conference 14: 8–11.

Faith, D. P. 1992. Conservation evaluation and phylogenetic diversity. Biological Conservation 61: 1–10.

Fancy, S. G. 1997. A new approach for analyzing bird densities from variable circular-plot counts. Pacific Science 51: 107–114.

Fancy, S. G., J. T. Nelson, P. Harrity, J. Kuhn, M. Kuhn, C. Kuehler, and J. G. Giffin. 2001. Reintroduction and translocation of 'Oma'o: A comparison of methods. Studies in Avian Biology 22: 347–353.

Fancy, S. G., T. K. Pratt, G. D. Lindsey, C. K. Harada, A. H. Parent, and J. D. Jacobi. 1992. Identifying sex and age of apapane and iiwi on Hawaii. Journal of Field Ornithology 64: 262–269.

Fancy, S. G., and C. J. Ralph. 1997. 'Apapane (Himatione sanguinea). In A. Poole and F. Gill (editors), The birds of North America, No. 296. Birds of North America Inc., Philadelphia.

———. 1998. 'I'wi (Vestiaria coccinea). In A. Poole and F. Gill (editors), The birds of North America, No. 327. Birds of North America Inc., Philadelphia.

Fancy, S. G., S. A. Sandin, M. H. Reynolds, and J. D. Jacobi. 1996. Distribution and population status of the endangered 'Akiapola'au. Pacific Science 50: 355–362.

Fancy, S. G., T. J. Snetsinger, and J. D. Jacobi. 1997. Translocation of the Palila, an endangered Hawaiian honeycreeper. Pacific Conservation Biology 3: 39–46.

Fancy, S. G., R. T. Sugihara, J. J. Jeffrey, and J. D. Jacobi. 1993. Site tenacity of the endangered Palila. Wilson Bulletin 105: 587–596.

Farias, M. E. M. 2008. Variation in the trap gene of Plasmodium relictum: Prevalence of single nucleotide polymorphisms in infected Hawai'i 'Amakihi (*Hemignathus virens*) on the east side of Hawai'i Island. M.S. thesis. University of Hawai'i at Hilo, Hilo.

Farias, M. E. M., and S. I. Jarvi. 2009. A nucleotide-constrained single base extension method for improved detection of minority alleles in *Plasmodium*. Molecular and Biochemical Parasitology 163: 114–118.

Farquhar, C. C. 1992. White-tailed Hawk (*Buteo albicaudatus*). In A. Poole and F. Gill (editors), The birds of North America, No. 30. Birds of North America Inc., Philadelphia.

Federal Highway Administration. 1999. Record of decision: Saddle Road (State Route 200), Mamalahoa Highway (State Route 190) to Milepost 6. County of Hawai'i, State of Hawai'i. U.S. Department of Transportation, Federal Highway Administration, Central Federal Lands Highway Division, Lakewood, CO, October 30.

Feldman, R. A., L. A. Freed, and R. L. Cann. 1995. A PCR test for avian malaria in Hawaiian birds. Molecular Ecology 4: 663–673.

Ferguson, N. M., C. Fraser, C. A. Donnelly, A. C. Ghani, and R. M. Anderson. 2004. Public health risk from the avian H5N1 influenza epidemic. Science 304: 968–969.

Fessel, S., and S. Tebbich. 2002. *Philornis downsi*—a threat for Darwin's finches? Ibis 144: 445–451.

Fewster, R. M., and S. T. Buckland. 2004. Assessment of distance sampling estimators. Pp. 281–306 in S. T. Buckland, D. R. Anderson, K. P. Burnham, J. L. Laake, D. L. Borchers, and L. Thomas (editors), Advanced distance sampling. Oxford University Press, Oxford.

Fewster, R. M., J. L. Laake, and S. T. Buckland. 2005. Line transect sampling in small and large regions. Biometrics 61: 856–861.

Figuerola, J., and A. J. Green. 2000. Haematozoan parasites and migratory behaviour in waterfowl. Evolutionary Ecology 14: 143–153.

Filardi, C. E., and R. G. Moyle. 2005. Single origin of a pan-Pacific bird group and upstream colonization of Australasia. Nature 438: 216–219.

Fisher, H. I. 1948. The question of avian introductions in Hawaii. Pacific Science 2: 59–64.

Fisher, H. I., and P. H. Baldwin. 1946. War and the birds of Midway Atoll. Condor 48: 3–15.

———. 1947. Notes on the Red-billed Leiothrix in Hawaii. Pacific Science 1: 45–51.

Fisher, R. A. 1958. The genetical theory of natural selection. Dover, New York.

Fitzgerald, M. 1992. Ecology of feral cats in New Zealand. Pp. 11–121 in D. Veitch, M. Fitzgerald, J. Innes, and E. Murphy (editors), Proceedings of the National Predator Management Workshop, Craigieburn, Canterbury. Threatened Species Occasional Publications 3. Department of Conservation, Wellington, New Zealand.

Fitzpatrick, J. W., M. Lammertink, M. D. Luneau Jr., T. W. Gallagher, B. R. Harrison, G. M. Sparling, K. V. Rosenberg, R. W. Rohrbaugh, E. C. H. Swarthout, P. H. Wrege, S. Barker Swarthout, M. S. Dantzker, R. A. Charif, T. R. Barksdale, J. V. Remsen Jr., S. D. Simon, and D. Zollner. 2005. Ivory-billed Woodpecker (*Campephilus principalis*) persists in continental North America. Science 308: 1460–1462.

Flack, J. A. D. 1978. Interisland transfers of New Zealand black robins. Pp. 365–372 in S. A. Temple (editor), Endangered birds: Management techniques for preserving threatened species. University of Wisconsin Press, Madison.

Fleischer, R. C. 1998. Genetics and avian conservation. Pp. 29–47 in J. Marzluff and J. Sallabanks (editors), Avian conservation: Research and management. Island Press, Washington, DC.

———. 2003. Genetic analysis of captive Alala (*Corvus hawaiiensis*). Unpublished report to U.S. Fish and Wildlife Service, Honolulu.

Fleischer, R. C., S. C. Conant, and M. Morin. 1991. Population bottlenecks and genetic variation in native and introduced populations of the Laysan Finch (*Telespiza cantans*). Heredity 66: 125–130.

Fleischer, R. C., G. Fuller, and D. B. Ledig. 1995. Genetic structure of endangered clapper rail (*Rallus longirostris*) populations in southern California. Conservation Biology 9: 1234–1243.

Fleischer, R. C., H. F. James, and J. Kirchman. 2003. Identification of *Corvus* subfossil bones on Maui with mitochondrial DNA sequences. Report to the U.S. Fish and Wildlife Service—Hawaii for Order 1448-12200-1-M057. U.S. Fish and Wildlife Service, Honolulu.

Fleischer, R. C., H. F. James, and S. L. Olson. 2008. Convergent evolution of Hawaiian and Australo-Pacific honeyeaters from distant songbird ancestors. Current Biology 18: 1927–1931.

Fleischer, R. C., and C. E. McIntosh. 2001. Molecular systematics and biogeography of the Hawaiian avifauna. Studies in Avian Biology 22: 51–60.

Fleischer, R. C., C. E. McIntosh, and C. L. Tarr. 1998. Evolution on a volcanic conveyor belt: Using phylogeographic reconstructions and K-Ar based ages of the Hawaiian Islands to estimate molecular evolutionary rates. Molecular Ecology 7: 533–545.

Fleischer, R. C., S. L. Olson, H. F. James, and A. C. Cooper. 2000. Identification of the extinct Hawaiian Eagle (*Haliaeetus*) by mtDNA sequence analysis. Auk 117: 1051–1056.

Fleischer, R. C., B. Slikas, J. Beadell, C. Atkins, C. E. McIntosh, and S. Conant. 2007. Genetic variability and taxonomic status of the Nihoa and Laysan Millerbirds. Condor 109: 954–962.

Fleischer, R. C., C. L. Tarr, H. F. James, B. Slikas, and C. E. McIntosh. 2001. Phylogenetic placement of the Po'ouli, *Melamprosops phaeosoma*, based on mitochondrial DNA sequence and osteological characters. Studies in Avian Biology 22: 98–103.

Fleischer, R. C., C. L. Tarr, and T. K. Pratt. 1994. Genetic structure and mating system in the palila, an endangered Hawaiian honeycreeper, as assessed by DNA fingerprinting. Molecular Ecology 3: 383–392.

Flint, E., and C. Rehkemper. 2002. Control and eradication of the introduced grass, *Cenchrus echintus*, at Laysan Island, central Pacific Ocean. Pp. 110–115 in C. R. Veitch and M. N. Clout (editors), Turning the tide: The eradication of invasive species. IUCN SSC Invasive Species Specialist Group. International Union for Conservation of Nature, Gland, Switzerland, and Cambridge, England.

Follett, P. A., P. Anderson-Wong, M. T. Johnson, and V. P. Jones. 2003. Revegetation in dead *Dicranopteris* (Gleicheniaceae) fern patches associated with Hawaiian rain forests. Pacific Science 57: 347–357.

Fonseca, D. M., C. T. Atkinson, and R. C. Fleischer. 1998. Microsatellite primers for *Culex pipiens quinquefasciatus*, the vector of avian malaria in Hawaii. Molecular Ecology 7: 1617–1618.

Fonseca, D. M., D. A. LaPointe, and R. C. Fleischer. 2000. Bottlenecks and multiple introductions: Population genetics of the vector of avian malaria in Hawaii. Molecular Ecology 9: 1803–1814.

Fonseca, D. M., J. L. Smith, R. C. Wilkerson, and R. C. Fleischer. 2006. Pathways of expansion and multiple introductions illustrated by large genetic differentiation among worldwide populations of the southern house mosquito. American Journal of Tropical Medicine and Hygiene 74: 284–289.

Food and Agriculture Organization. 2004. FAO recommendations on the prevention, control and eradication of highly pathogenic avian influenza (HPAI) in Asia. FAO, Rome.

Foose, T. J., R. Lande, N. R. Flesness, G. Rabb, and B. Read. 1986. Propagation plans. Zoo Biology 5: 139–146.

Forman, R. T. T., and M. Godron. 1986. Landscape ecology. John Wiley and Sons, New York.

Fornander, A. 1916–1920. Fornander collection of Hawaiian antiquities and folk-lore. 6 vols. Bishop Museum Press, Honolulu.

Fosberg, F. R. 1948. Derivation of the flora of the Hawaiian Islands. Pp. 107–119 in E. C. Zimmerman (editor), Insects of Hawaii, Vol. 1. University Press of Hawaii, Honolulu.

———. 1972. Guide to excursion III, Tenth Pacific Science Conference. Department of Botany, University of Hawaii, with assistance from the Hawaiian Botanical Gardens Foundation Inc., Honolulu.

Foster, J. T. 2005. Exotic bird invasions into forests of Hawaii: Demography, competition, and seed dispersal. Ph.D. dissertation. University of Illinois, Urbana-Champaign.

Foster, J. T., and S. K. Robinson. 2007. Introduced birds and the fate of Hawaiian rainforests. Conservation Biology 21: 1248–1257.

Foster, J. T., J. M. Scott, and P. W. Sykes Jr. 2000. 'Akikiki (*Oreomystis bairdi*). In A. Poole and F. Gill (editors), The birds of North America, No. 552. Birds of North America Inc., Philadelphia.

Foster, J. T., E. J. Tweed, R. J. Camp, B. L. Woodworth, C. D. Adler, and T. Telfer. 2004. Long-term population changes of native and introduced birds in the Alaka'i Swamp, Kaua'i. Conservation Biology 18: 716–725.

Foster, J. T., B. L. Woodworth, L. E. Eggert, P. J. Hart, D. Palmer, D. C. Duffy, and R. C. Fleischer. 2007. Genetic structure and evolved malaria resistance in Hawaiian honeycreepers. Molecular Ecology 16: 4738–4746.

Foster, M. S. 1975. The overlap of molting and breeding in some tropical birds. Condor 77: 304–314.

Foufopoulos, J., and A. R. Ives. 1999. Reptile extinctions on land-bridge islands: Life-history attributes and vulnerability to extinction. American Naturalist 153: 1–25.

Fox, A. M., and L. L. Loope. 2007. Globalization and invasive species issues in Hawai'i: Role-playing some local perspectives. Journal of Natural Resources and Life Sciences Education 36: 147–157.

Fox, M. 2003. Testimony of Mark R. Fox, Director of External Affairs, The Nature Conservancy, Hawai'i Program, Field Hearing on Invasive Species. Subcommittee on National Parks, U.S. Senate Committee on Energy and Natural Resources, August 9, Hawai'i Volcanoes National Park, Volcano, HI.

Frame, W. V., and R. H. Horwitz. 1965. Public land policy in Hawaii: The multiple-use approach. Report 1. Legislative Reference Bureau. University of Hawaii, Honolulu.

Frankel, O. H., and M. E. Soulé. 1981. Conservation and evolution. Cambridge University Press, Cambridge, England.

Frankham, R. 1997. Do island populations have lower genetic variation than mainland populations? Heredity 78: 311–327.

Frankham, R., J. D. Ballou, and D. A. Briscoe. 2002. Introduction to conservation genetics. Cambridge University Press, New York.

Fraser, J. G. 1922. The golden bough: A study in magic and religion. MacMillan, New York.

Freed, L. A. 1999. Extinction and endangerment of Hawaiian honeycreepers: A comparative approach. Pp. 137–162 in L. F. Landweber and A. P. Dobson (editors), Genetics and the extinction of species: DNA and the conservation of biodiversity. Princeton University Press, Princeton, NJ.

———. 2001. Significance of old-growth forest to the Hawai'i 'Ākepa. Studies in Avian Biology 22: 173–184.

Freed, L. A., R. L. Cann, and G. R. Bodner. 2008. Incipient extinction of a major population of the Hawaii akepa owing to introduced species. Evolutionary Ecology Research 10: 931–965.

Freed, L. A., S. Conant, and R. C. Fleischer. 1987. Evolutionary ecology and radiation of Hawaiian passerine birds. Trends in Ecology and Evolution 2: 196–203.

Freedman, M. D. 1992. Oral anticoagulants: Pharmacodynamics, clinical indications and adverse effects. Journal of Clinical Pharmacology 32: 196–209.

Fretz, J. S. 2000. Relationship of canopy arthropod prey to distribution and life history of the Hawai'i 'Ākepa. Ph.D. dissertation. University of Hawai'i at Mānoa, Honolulu.

Friend, M., R. L. McLean, and F. J. Dean. 2004. Disease emergence in birds: Challenges for the twenty-first century. Auk 118: 290–303.

Friesen, V. L, B. C. Congdon, H. E. Walsh, and T. P. Birt. 1997. Intron variation in marbled murrelets detected using analyses of single-stranded conformational polymorphisms. Molecular Ecology 6: 1047–1058.

Fritts, T. H., and G. H. Rodda. 1998. The role of introduced species in the degradation of island ecosystems: A case history of Guam. Annual Review of Ecology and Systematics 29: 113–140.

Fry, H. (editor). 2002. Foundation for the conservation of the Bearded Vulture. Annual Report. Frankfurt Zoological Society, Frankfurt, Germany.

Fujimori, L. 2003. Isle bird businesses are hit by second avian disease. Honolulu Star-Bulletin, February 1.

Fullaway, D. T. 1947. Notes and exhibitions. Proceedings of the Hawaii Entomological Society 13: 3–4.

Futuyma, D. J., and D. Moreno. 1988. The evolution of ecological specialization. Annual Review of Ecology and Systematics 19: 207–233.

Gagné, B. H., L. L. Loope, A. C. Medeiros, and S. J. Anderson. 1992. Miconia calvescens: A threat to native forests in the Hawaiian Islands (Abstract). Pacific Science 46: 390–391.

Gagné, W. C. 1979. Canopy-associated arthropods in Acacia koa and Metrosideros tree communities along an altitudinal transect on Hawaii Island. Pacific Insects 21: 56–82.

———. 1988. Conservation priorities in Hawaiian natural systems. Bioscience 38: 264–271.

Gagné, W. C., and C. C. Christensen. 1985. Conservation status of native terrestrial invertebrates in Hawaii. Pp. 105–126 in C. P. Stone and J. M. Scott (editors), Hawaii's terrestrial ecosystems: Preservation and management. Cooperative National Park Resources Studies Unit, University of Hawaii, Honolulu.

Gagné, W. C., and L. W. Cuddihy. 1999. Vegetation. Pp. 45–114 in W. L. Wagner, D. R. Herbst, and S. H. Sohmer (editors), Manual of the flowering plants of Hawai'i. Bishop Museum Special Publication 97. University of Hawai'i Press and Bishop Museum Press, Honolulu.

Gagné, W. C., and F. G. Howarth. 1981. Arthropods associated with foliar crowns of structural dominants. Pp. 275–288 in D. Mueller-Dombois, K. W. Bridges, and H. L. Carson (editors), Island ecosystems: Biological organization in selected Hawaiian communities. U.S. International Biological Program Synthesis Series 15. Hutchinson Ross, Stroudsburg, PA, and Woods Hole, MA.

Galván, J. P., G. Howald, A. Samaniego, B. Keitt, J. Russell, M. Pascal, M. Browne, K. Broome, J. Parkes, and B. Tershy. 2006. A review of commensal rodent eradication on islands. Pp. 158–159 in Proceedings of the 13th Australasian Vertebrate Pest Conference, Wellington, New Zealand.

Gambino, P., A. C. Medeiros, and L. L. Loope. 1987. Introduced vespids Paravespula pensylvanica prey on Maui's endemic arthropod fauna. Journal of Tropical Ecology 3: 169–170.

Gandon, S. 2004. Evolution of multi-host parasites. Evolution 58: 455–469.

Garcia, M., N. Narang, W. M. Reed, and A. M. Fadly. 2003. Molecular characterization of reticuloendotheliosis virus insertions in the genome of field and vaccine strains of fowl poxvirus. Avian Diseases 47: 343–354.

Gardner, D. E. 1997a. Additions to the rust fungi of Hawai'i. Pacific Science 51: 174–182.

———. 1997b. Botryosphaeria mamane sp. nov. associated with witches'-brooms on the endemic forest tree Sophora chrysophylla in Hawaii. Mycologia 89: 298–303.

Gardner, D. E., and E. E. Trujillo. 2001. Association of Armillaria mellea with mamane decline at Pu'u La'au. Newsletter of the Hawaiian Botanical Society 40: 33–34.

Garmendia, A. E., H. J. VanKruiningen, and R. A. French. 2001. The West Nile virus: Its recent emergence in North America. Microbes and Infection 3: 223–229.

Garnham, P. C. C. 1948. The incidence of malaria at high elevations. Journal of the National Malaria Society 7: 275–284.

———. 1966. Malaria parasites and other haemosporidia. Blackwell Scientific, New York.

Gassmann-Duvall, R. 1988. Update on owl die-off. 'Elepaio 48: 94.

Gaston, K. J. 1990. Patterns in the geographical ranges of species. Biological Reviews 65: 105–129.

Gavenda, R. T. 1992. Hawaiian Quaternary paleoenvironments: A review of geological, pedological, and botanical evidence. Pacific Science 46: 295–307.

Geertz, C. 1973. The interpretation of cultures. Basic Books, New York.

George, T. L. 2000. Varied Thrush (*Ixoreus naevius*). In A. Poole and F. Gill (editors), The birds of North America, No. 541. Birds of North America Inc., Philadelphia.

Gerrish, G., D. Mueller-Dombois, and K. W. Bridges. 1988. Nutrient limitation and *Metrosideros* forest dieback in Hawai'i. Ecology 69: 723–727.

Gerrodette, T. 1987. A power analysis for detecting trends. Ecology 68: 1364–1372.

Giambelluca, T. W., and M. S. A. Luke. 2007. Climate change in Hawaii's mountains. Mountain Views 1: 13–18.

Giambelluca, T. W., M. A. Nullet, and T. A. Schroeder. 1986. Rainfall atlas of Hawaii. Report R76. Water Resources Research Center, University of Hawai'i at Mānoa, Hawaii Department of Land and Natural Resources, Honolulu.

Giambelluca, T. W., and T. A. Schroeder. 1998. Climate. Pp. 49–59 in S. P. Juvik and J. O. Juvik (editors), Atlas of Hawai'i. 3rd edition. University of Hawai'i Press, Honolulu.

Gibbs, J. P. 2000. Monitoring populations. Pp. 213–252 in L. Boitani and T. K. Fuller (editors), Research techniques in animal ecology: Controversies and consequences. Columbia University Press, New York.

Gibbs, J. P., S. Droege, and P. C. Eagle. 1998. Monitoring local populations of plants and animals. Bioscience 48: 935–940.

Giffard, W. M. 1918. Some observations on Hawaiian forests and forest cover in their relation to water supply. Hawaiian Planters' Record 18: 513–538.

Giffin, J. G. 1973. Ecology of the feral pig, final report. Unpublished report. Hawai'i Department of Land and Natural Resources, Honolulu.

———. 1976. Ecology of the feral sheep on Mauna Kea. Unpublished report. Hawai'i Department of Land and Natural Resources, Honolulu.

———. 1980. Ecology of the mouflon sheep on Mauna Kea. Unpublished report. Hawai'i Department of Land and Natural Resources, Honolulu.

———. 1983. Alala investigation. Unpublished report, Pittman-Robertson Project W-18-R, Study R-II-B. Hawai'i Department of Land and Natural Resources, Honolulu.

———. 1989a. Final job progress report, Project EW-1-6 covering the period July 1, 1989, to June 30, 1990. Hawai'i Department of Land and Natural Resources, Honolulu.

———. 1989b. Captive propagation of birds. Pp. 103–106 in C. P. Stone and D. S. Stone (editors), Conservation biology in Hawai'i. University of Hawai'i Press, Honolulu.

———. 2007. A comparison of moth diversity at Kilauea (1911–1912) and Upper Waiakea Forest Reserve (1998–2000), Island of Hawaii. Proceedings of the Hawaiian Entomological Society 39: 15–26.

Giffin, J. G, J. M. Scott, and S. Mountainspring. 1987. Habitat selection and management of the Hawaiian Crow. Journal of Wildlife Management 51: 485–494.

Gillespie, R. G. 1999. Naiveté and novel perturbations: Conservation of native spiders on an oceanic island. Journal of Insect Conservation 3: 263–272.

Gillihan, S. W., and B. Byers. 2001. Evening Grosbeak (*Coccothraustes vespertinus*). In A. Poole and F. Gill (editors), The birds of North America, No. 599. Birds of North America Inc., Philadelphia.

Gilman, A. G., L. S. Goodman, T. W. Rall, and F. Murad (editors). 1985. The pharmacological basis of therapeutics. 7th ed. Macmillan, New York.

Givnish, T. J. 1998. Adaptive plant evolution on islands: Classical patterns, molecular data, new insights. Pp. 281–304 in P. R. Grant (editor), Evolution on islands. Oxford University Press, Oxford.

Givnish, T. J., K. J. Systma, J. F. Smith, and W. J. Hahn. 1994. Thorn-like prickles and heterophylly in *Cyanea*: Adaptations to extinct avian browsers on Hawaii? Proceedings of the National Academy of Sciences (USA) 91: 2810–2814.

Glen, M., A. C. Alfenas, E. A. V. Zauza, M. J. Wingfield, and C. Mohammed. 2007. *Puccinia psidii*: A threat to the Australian environment and economy—a review. Australasian Plant Pathology 36: 1–16.

Glenn, T. C., W. Stephan, and M. J. Braun. 1999. Effects of a population bottleneck on Whooping Crane mitochondrial DNA variation. Conservation Biology 13: 1097–1107.

Goergen, E., and C. C. Daehler. 2001. Inflorescence damage by insects and fungi in native pili grass (*Heteropogon contortus*) versus alien fountain grass (*Pennisetum setaceum*) in Hawai'i. Pacific Science 55: 129–136.

Goff, M. L., and C. van Riper III. 1980. Distribution of mosquitoes (Diptera: Culicidae) on the east flank of Mauna Loa Volcano, Hawaii. Pacific Insects 22: 178–188.

Goldstein, D. B., and D. D. Pollock. 1997. Launching microsatellites: A review of mutation processes and methods of phylogenetic inference. Journal of Heredity 88: 335–342.

Goltz, D. M., S. C. Hess, K. W. Brinck, P. C. Banko, and R. M. Danner. 2008. Home range and movements of feral cats on Mauna Kea, Hawai'i. Pacific Conservation Biology 14: 177–184.

Gon, S. M. III, A. Allison, R. J. Cannarella, J. D. Jacobi, K. Y. Kaneshiro, M. H. Kido, M. Lane-Kamahele, and S. E. Miller. 2006. Hawaii Gap Analysis Project, final report. U.S. Geological Survey, National Gap Analysis Program, Moscow, ID.

Goodrich, L. J., S. C. Crocoll, and S. E. Senner. 1996. Broad-winged Hawk (*Buteo platypterus*). In A. Poole and F. Gill (editors), The birds of North America, No. 218. Birds of North America Inc., Philadelphia.

Goodwin, D. 1986. Crows of the world. 2nd edition. University of Washington Press, Seattle.

Gorresen, P. M., R. J. Camp, J. L. Klavitter, and T. K. Pratt. 2008. Abundance, distribution and population trend of the Hawaiian hawk: 1998–2007. Hawai'i Cooperative Studies Unit Technical Report HCSU-009. University of Hawai'i at Hilo, Hilo.

Gorresen, P. M., R. J. Camp, and T. K. Pratt. 2007. Forest bird distribution, density and trends in the Ka'ū region of Hawai'i Island. Open-File Report 2007-1076. U.S. Geological Survey, Biological Resources Division, Reston, VA.

Gorresen, P. M., R. J. Camp, T. K. Pratt, and B. L. Woodworth. 2005. Status of forest birds in the central windward region of Hawai'i Island: Population trend and power analyses. Open-File Report 2005-1441. U.S. Geological Survey, Biological Resources Division, Reston, VA.

Gorresen, P. M., G. P. McMillan, R. J. Camp, and T. K. Pratt. 2009. A spatial model of bird abundance as adjusted for detection probability. Ecography 32: 291–298.

Gorrochotegui-Escalante, N., I. Fernandez-Salas, and H. Gomez-Dantes. 1998. Field evaluation of Mesocyclops longisetus (Copepoda: Cyclopoidea) control of larval Aedes aegypti (Diptera: Culicidae) in Northeastern Mexico. Journal of Medical Entomology 35: 699–703.

Grant, B. R., and P. R. Grant. 1989. Evolutionary dynamics of a natural population: The Large Cactus Finch of the Galápagos. University of Chicago Press, Chicago.

Grant, P. R. 1994. Population variation and hybridization: Comparison of finches from two archipelagos. Evolutionary Ecology 8: 598–617.

———. 1995. Commemorating extinctions. American Scientist 83: 420–422.

——— (editor). 1998a. Evolution on islands. Oxford University Press, Oxford.

———. 1998b. Patterns on islands and microevolution. Pp. 1–17 in P. R. Grant (editor), Evolution on islands. Oxford University Press, Oxford.

———. 1998c. Radiations, communities, and biogeography. Pp. 198–209 in P. R. Grant (editor), Evolution on islands. Oxford University Press, Oxford.

Grant, P. R., and B. R. Grant. 2002. Adaptive radiation of Darwin's Finches. American Scientist 90: 130–139.

Grant, P. R., B. R. Grant, K. Petren, and L. F. Keller. 2005. Extinction behind our backs: The possible fate of one of the Darwin's finch species on Isla Floreana, Galápagos. Biological Conservation 122: 499–503.

Grant, T. D., and C. Lynch. 2004. Analysis and breeding recommendations of the San Clemente Island loggerhead shrike (Lanius ludovicianus mearnsi). Population biology management plan: Analysis and breeding recommendations. Applied Conservation, Zoological Society of San Diego, San Diego, CA.

Greaves, J. H. 1994. Resistance to anticoagulant rodenticides. Pp. 197–217 in A. P. Buckle and R. H. Smith (editors), Rodent pests and their control. CAB International, Wallingford, England.

Green, A. J., J. Figuerola, and M. I. Sánchez. 2002. Implications of waterbird ecology for the dispersal of aquatic organisms. Acta Oecologica 23: 177–189.

Green, L. S. 1928. Folk tales from Hawai'i. Hawaiian Board Book Room, Honolulu.

Green, R. E. 1997. The influence of numbers released on the outcome of attempts to introduce exotic bird species to New Zealand. Journal of Animal Ecology 66: 25–35.

Greenwood, J. J. D., S. R. Baillie, and H. Q. P. Crick. 1994. Long-term studies and monitoring of bird populations. Pp. 343–364 in R. A. Leigh and A. E. Johnston (editors), Long-term experiments in agricultural and ecological sciences. CAB International, Wallingford, England.

Greenwood, P. J., and P. H. Harvey. 1982. The natal and breeding dispersal of birds. Annual Review of Ecology and Systematics 13: 1–21.

Grenfell, B. T., O. N. Bjornstad, and J. Kappey. 2001. Traveling waves and spatial hierarchies in measles epidemics. Nature 414: 716–723.

Grenfell, B. T., and B. M. Bolker. 1998. Cities and villages: Infection hierarchies in a measles metapopulation. Ecology Letters 1: 63–70.

Grenfell, B. T., K. Dietz, and M. G. Roberts. 1995. Modelling the immuno-epidemiology of macroparasites in naturally-fluctuating host populations. Pp. 362–383 in B. T. Grenfell and D. P. Dobson (editors), Ecology of infectious diseases in natural populations. Cambridge University Press, Cambridge, England.

Griffin, C. R. 1985. Biology of the Hawaiian Hawk (Buteo solitarius). Ph.D. dissertation. University of Missouri, Columbia.

———. 1989. Raptors in the Hawaiian Islands. Pp. 155–160 in B. Pendleton (editor), Western raptor management symposium and workshop. Science Technical Series 12. National Wildlife Federation, Washington, DC.

Griffin, C. R., C. M. King, J. A. Savidge, F. Cruz, and J. B. Cruz. 1989. Effects of introduced predators on island birds: Contemporary case histories from the Pacific. Pp. 687–698 in H. Ouelle (editor), Proceedings of the XIX Ornithological Congress, Vol. 1.

Griffin, C. R., P. W. Paton, and T. S. Baskett. 1998. Breeding ecology and behavior of the Hawaiian Hawk. Condor 100: 654–662.

Griffith, B., J. M. Scott, J. W. Carpenter, and C. Reed. 1989. Translocation as a species conservation tool: Status and strategy. Science 245: 277–480.

Griffiths, R., S. Daan, and C. Dijkstra. 1996. Sex identification in birds using two CHD genes. Proceedings of the Royal Society of London, Series B 263: 1249–1254.

Griffiths, R., M. C. Double, K. Orr, and R. J. G. Dawson. 1998. A DNA test to sex most birds. Molecular Ecology 7: 1071–1075.

Grim, K. C., T. McCutchan, J. Li, M. Sullivan, T. K. Graczyk, G. McConkey, and M. Cranfield. 2004. Preliminary results of an anticircumsporozoite DNA vaccine trial for protection against avian malaria in captive African black-footed penguins (*Spheniscus demersus*). Journal of Zoo and Wildlife Medicine 35: 154–161.

Grinnell, J. 1912. A name for the Hawaiian Linnet. Auk 29: 24–25.

Groombridge, J. J., M. W. Bruford, C. G. Jones, and R. A. Nichols. 2001. Estimating the severity of the population bottleneck in the Mauritius Kestrel *Falco punctatus* from ringing records using MCMC estimation. Journal of Animal Ecology 70: 401–409.

Groombridge, J. J., C. G. Jones, M. W. Bruford, and R. A. Nichols. 2000. "Ghost" alleles of the Mauritius Kestrel. Nature 403: 616.

Groombridge, J. J., J. G. Massey, J. C. Bruch, T. Malcolm, C. N. Brosius, M. M. Okada, and B. Sparklin. 2004. Evaluating stress-levels in a Hawaiian honeycreeper following translocation using different container designs. Journal of Field Ornithology 75: 183–187.

Groombridge, J. J., J. G. Massey, J. C. Bruch, T. Malcolm, C. N. Brosius, M. M. Okada, B. Sparklin, J. S. Fretz, and E. A. WanderWerf. 2004. An attempt to recover the Po'ouli by translocation and an appraisal of recovery strategy for bird species of extreme rarity. Biological Conservation 118: 365–375.

Groombridge, J. J., B. Sparklin, T. Malcolm, C. N. Brosius, M. M. Okada, and J. C. Bruch. 2006. Patterns of spatial use and movement of the Po'o-uli—a critically endangered Hawaiian honeycreeper. Biodiversity and Conservation 15: 3357–3368.

Gruner, D. S. 2004. Attenuation of top-down and bottom-up forces in a complex terrestrial community. Ecology 85: 3010–3022.

———. 2005. Biotic resistance to an invasive spider conferred by generalist insectivorous birds on Hawaii Island. Biological Invasions 7: 541–546.

Gubler, D. J. 2001. Resurgent vector-borne diseases as a global health problem. Emerging Infectious Diseases 4: 442–450.

Gulland, F. M. D. 1995. The impact of infectious diseases on wild animal populations—a review. Pp. 20–51 in B. T. Grenfell and A. P. Dobson (editors), Ecology of infectious diseases in natural populations. Cambridge University Press, Cambridge, England.

Gunderson, L., and C. Folke. 2003. Toward a "science of the long view." Conservation Ecology 7: 15. www.consecol.org/vol7/iss1/art15.

Hadfield, M. G., S. E. Miller, and A. H. Carwile. 1993. The decimation of endemic Hawaiian tree snails by alien predators. American Zoologist 33: 610–622.

Hahn, T. P. 1996. Cassin's Finch (*Carpodacus cassinii*). In A. Poole and F. Gill (editors), The birds of North America, No. 240. Birds of North America Inc., Philadelphia.

Haldane, J. B. S. 1948. The theory of a cline. Journal of Genetics 48: 277–284.

———. 1949. Suggestions as to the quantitative measurement of rates of evolution. Evolution 3: 51–56.

Halford, F. J. 1954. Nine doctors and God. University of Hawaii Press, Honolulu.

Hall, L. S., M. L. Morrison, and P. H. Bloom. 1997. Population status of the endangered Hawaiian Hawk. Journal of Raptor Research 31: 11–15.

Hall, W. L. 1904. The forests of the Hawaiian Islands. U.S. Department of Agriculture, Bureau of Forestry Bulletin 48: 1–29.

Halupka, L. 1996. Breeding ecology of the Sedge Warbler *Acrocephalus schoenobaenus* in the Biebrza Marshes (NE Poland). Ornis Hungarica 6: 9–14.

Halupka, L., and J. Wróblewski. 1998. Ekologia rozrodu trzcinniczka (*Acrocephalus scirpaceus*) na stawach milickich w roku 1994. Ptaki Śląska 12: 5–15.

Hamilton, W. D., and M. Zuk. 1982. Heritable true fitness and bright birds: A role for parasites? Science 218. 384–387.

Handy, E. S. C., and E. G. Handy. 1972. Native planters in old Hawai'i: Their life, lore, and environment. Bishop Museum Bulletin 233: 1–287.

Hansell, M. 2000. Bird nests and construction behaviour. Cambridge University Press, Cambridge, England.

Hansen, H., S. C. Hess, D. Cole, and P. C. Banko. 2007. Using population genetic tools to develop a control strategy for feral cats (*Felis catus*) in Hawaii. Wildlife Research 34: 587–596.

Hardy, D. E. 1960. Diptera: Nematocera–Brachycera (except Dolichopodidae). P. 368 in E. C. Zimmerman (editor), Insects of Hawaii, Vol. 10. University of Hawaii Press, Honolulu.

Hardy, J. W. 1973. Feral exotic birds in southern California. Wilson Bulletin 85: 506–512.

Harris, M. 1985. Culture, people, nature. 4th edition. Harper and Row, New York.

Harrison, G. 1978. Mosquitoes, malaria and man: A history of hostilities since 1880. E. P. Dutton, New York.

Hart, P. J. 2001. Demographic comparisons between high and low density populations of Hawai'i 'Ākepa. Studies in Avian Biology 22: 185–193.

———. 2003. Structure and dynamics of mixed-species flocks in a Hawaiian rainforest. Auk 120: 82–95.

Harvell, C. D., C. E. Mitchell, J. R. Ward, S. Altizer, A. P. Dobson, R. S. Ostfeld, and M. D. Samuel. 2002. Climate warming and disease risks for terrestrial and marine biota. Science 296: 2158–2162.

Harvey, N. C., S. M. Farabaugh, and B. B. Druker. 2002. Effects of early rearing experience on adult behavior and nesting in captive Hawaiian Crows (Corvus hawaiiensis). Zoo Biology 21: 59–75.

Hasibeder, G., and C. Dye. 1988. Population dynamics of mosquito-borne disease: Persistence in a completely heterogeneous environment. Theoretical Population Biology 33: 31–53.

Hastings, B. E., D. Kenny, L. J. Lowenstine, and J. W. Foster. 1991. Mountain gorillas and measles: Ontogeny of a wildlife vaccination program. Pp. 198–205 in R. E. Junge (editor), Proceedings of the Annual Meeting of the American Association of Zoo Veterinarians. American Association of Zoo Veterinarians, Media, Philadelphia.

Hatfield, J. S., W. R. Gould IV, B. A. Hoover, M. R. Fuller, and E. L. Lindquist. 1996. Detecting trends in raptor counts: Power and Type I error rates of various statistical tests. Wildlife Society Bulletin 24: 505–515.

Hauff, R. 2005. New forest pests invade Hawaii. Na Leo o ka Aina: Newsletter of the Division of Forestry and Wildlife, Hawaii State Department of Land and Natural Resources 1: 1.

Hawaii Audubon Society. 1941. A resolution passed by the Hawaiian Audubon Society at a meeting held June 9th. 'Elepaio 2: 8.

Hawai'i Department of Agriculture. 2002a. Kahului Airport pest risk assessment. Plant Quarantine Division, Department of Agriculture, State of Hawaii, Honolulu, HI. www.hawaiiag.org/PQ/KARA%20Report%20Final.pdf (accessed March 23, 2008).

———. 2002b. News release: Hawaii Department of Agriculture imposes embargo on birds shipped through U.S. Mail. Hawai'i Department of Agriculture, Honolulu, September 19. www.hawaiiag.org/hdoa/newsrelease/02-17.htm (accessed March 23, 2008).

Hawai'i Department of Business, Economic Development and Tourism. 1998. Private residential construction and demolition authorized by permits, by counties: 1993 to 1998. www.hawaii.gov/dbedt/db98/index.html (accessed May 1, 2000).

———. 2006. The State of Hawai'i data book: A statistical abstract. Research and Economic Analysis Division, Hawai'i Department of Business, Economic Development and Tourism, Honolulu.

Hawai'i Department of Land and Natural Resources. 1974. 1972 report of Nene restoration program. 'Elepaio 34: 136–142.

———. 1988. Hanawī Natural Area Reserve management plan. Hawai'i Department of Land and Natural Resources, Honolulu.

———. 1996. Final environmental assessment for a fence project to protect the east Maui watershed. Hawai'i Department of Land and Natural Resources, Honolulu.

———. 1999. The Department of Land and Natural Resources announces its recommendations to prevent the extinction of the Po'o uli. 'Elepaio 59: 3–5.

———. 2007. Draft environmental assessment: Kaena Point ecosystem restoration project. Hawai'i Division of Forestry and Wildlife, Honolulu.

Hawai'i Department of Land and Natural Resources and Department of Planning and Economic Development. 1976. Forestry potentials for Hawaii. U.S. Forest Service Region 5, Honolulu.

Hawai'i Department of Land and Natural Resources and U.S. Fish and Wildlife Service. 1999. Environmental assessment for proposed management actions to save the Po'ouli. U.S. Department of the Interior, Fish and Wildlife Service, Honolulu.

Hawai'i Division of Forestry and Wildlife. 1997. Draft Mauna Kea ecosystem wildland fire management plan. Unpublished document, 37 pp. plus appendixes. Hawai'i Department of Land and Natural Resources, Honolulu.

———. 2001. Maui Forest Bird Recovery Project, quarterly report, April 1–June 30, 2001. Hawai'i Division of Forestry and Wildlife, Kahului.

Hawaiian Ethnographic Notes. 1922. Kahilis, Lahilahi Webb Collection of Notes. Data from Lucy K. Peabody, January 22, translated by Mary Kawena Pukui. Hawaiian Ethnological Notes, Vol. 1. Typescript in the Bishop Museum Archives, Honolulu.

Hayes, J., and T. D. Downs. 1980. Seasonal-changes in an isolated population of Culex pipiens quinquefasciatus (Diptera, Culicidae)—time-series analysis. Journal of Medical Entomology 17: 63–69.

Hays, W. S. T., and S. Conant. 2007. Biology and impacts of Pacific island invasive species, 1: A worldwide review of effects of the small Indian mongoose, Herpestes javanicus (Carnivora: Herpestidae). Pacific Science 61: 3–16.

Heddle, A. L. 2003. Systematics and phylogenetics of the endemic genus Scotorythra (Lepidoptera: Geometridae) in the Hawaiian Islands. Ph.D. dissertation. University of California, Berkeley.

Hedrick, P. W. 1999. Balancing selection and MHC. Genetica 104: 207–214.

Hedrick, P. W., and T. Kim. 2000. Genetics of complex polymorphisms: Parasites and maintenance of MHC variation. Pp. 204–234 in R. Singh and C. Krimbas (editors), Evolutionary genetics: From molecules to morphology. Cambridge University Press, New York.

Heelas, P., and A. Lock. 1981. Indigenous psychologies: The anthropology of the self. Academic Press, London.

Henke, L. A. 1929. A survey of livestock in Hawaii. Research Publication 5. University of Hawaii, Honolulu.

Henneman, M. L., and J. Memmott. 2001. Infiltration of a Hawaiian community by introduced biological control agents. Science 293: 1314–1316.

Henshaw, H. W. 1901. Introduction of foreign birds into the Hawaiian Islands with notes on some of the introduced species. Thrum, Honolulu.

————. 1902. Birds of the Hawaiian Islands, being a complete list of the birds of the Hawaiian possessions with notes on their habits. Thrum, Honolulu.

Hershey, A. E., A. R. Lima, G. J. Niemi, and R. R. Regan. 1998. Effects of Bacillus thuringiensis israelensis (Bti) and methoprene on nontarget macroinvertebrates in Minnesota wetlands. Ecological Applications 8: 41–60.

Hess, S. C., P. C. Banko, G. J. Brenner, and J. D. Jacobi. 1999. Factors related to the recovery of subalpine woodland on Mauna Kea, Hawaii. Biotropica 31: 212–219.

Hess, S. C., P. C. Banko, D. M. Goltz, R. M. Danner, and K. W. Brinck. 2004. Strategies for reducing feral cat threats to endangered Hawaiian birds. Proceedings of the Vertebrate Pest Conference 21: 21–26.

Hess, S. C., P. C. Banko, M. H. Reynolds, G. J. Brenner, L. P. Laniawe, and J. D. Jacobi. 2001. Drepanidine movements in relation to food availability in subalpine woodlands on Mauna Kea, Hawai'i. Studies in Avian Biology 22: 154–163.

Hess, S. C., H. Hansen, D. Nelson, R. Swift, and P. C. Banko. 2007. Diet of feral cats in Hawai'i Volcanoes National Park. Pacific Conservation Biology 13: 244–249.

Hess, S. C., B. Kawakami Jr., D. Okita, and K. Medeiros. 2006. A preliminary assessment of mouflon abundance at the Kahuku unit of Hawaii Volcanoes National Park. Open-File Report 2006-1193. U.S. Geological Survey, Biological Resources Division, Reston, VA.

Heu, R. A., D. M. Tsuda, W. T. Nagamine, and T. H. Suh. 2006. Erythrina gall wasp. Quadrastichus erythrinae Kim (Hymenoptera: Eulophidae). New Pest Advisory 05-03. Department of Agriculture, State of Hawaii, Honolulu. Updated February. www.hawaiiag.org/hdoa/npa/npa05-03-EGW.pdf (accessed March 23, 2008).

Hewitt, R. 1940. Bird malaria. American Journal of Hygiene Monograph, Serial 15. John Hopkins University Press, Baltimore, MD.

Hill, A. V. S., and D. J. Weatherall. 1998. Host genetic factors in resistance to malaria. Pp. 445–455 in I. W. Sherman (editor), Malaria: Parasite biology, pathogenesis and protections. American Society for Microbiology Press, Washington, DC.

Hill, G. E. 1993. House Finch (Carpodacus mexicanus). In A. Poole and F. Gill (editors), The birds of North America, No. 46. Academy of Natural Sciences, Philadelphia, and the American Ornithologists' Union, Washington, DC.

Hillebrand, W. 1981. Flora of the Hawaiian Island: A description of their phanerogams and vascular cryptogams. Facsimile of the 1888 edition. Lubrecht and Cramer, Monticello, NY.

Hillis-Starr, Z., and G. W. Witmer. 2004. The eradication of introduced roof rats on the U.S. Buck Island Reef National Monument, St. Croix, U.S. Virgin Islands (Abstract). Second National Invasive Rodent Summit, Fort Collins, CO.

Hilton, H. W., W. H. Robison, A. H. Teshima, and R. D. Nass. 1972. Zinc phosphide as a rodenticide for rats in Hawaiian sugarcane. Congress of the International Society of Sugar Cane Technologists 14: 561–570.

Hodges, C. S., K. T. Adee, J. D. Stein, H. B. Wood, and R. D. Doty. 1986. Decline of ohia (Metrosideros polymorpha) in Hawaii: A review. General Technical Report PSW-86. Pacific Southwest Forest Range Experiment Station, U.S. Department of Agriculture, Forest Service, Berkeley, CA.

Hodges, C. S., and D. E. Gardner. 1984. Hawaiian forest fungi, IV: Rusts on endemic Acacia species. Mycologia 76: 332–349.

Hoffman, S. L., L. M. L. Goh, T. C. Luke, I. Schneider, T. P. Le, D. L. Doolan, J. Sacci, P. de la Vega, M. Dowler, C. Paul, D. M. Gordan, J. A. Stoute, L. W. Preston Church, M. Sedegah, D. G. Heppner, W. P. Ballou, and T. L. Richie. 2002. Protection of humans against malaria by immunization with radiation-attenuated Plasmodium falciparum sporozoites. Journal of Infectious Diseases 185: 1155–1164.

Hogue, C. E. 1957. Hui Manu Society helped save island's feathered friends. Advertiser Centennial, Honolulu, p. 8.

Hoi-Leitner, M., M. Romero-Pujante, H. Hoi, and A. Pavlova. 2001. Food availability and immune capacity in serin (Serinus serinus) nestlings. Behavioral Ecology and Sociobiology 49: 333–339.

Holcomb, R. T. 1987. Eruptive history and long-term behavior of Kilauea Volcano. Pp. 261–350 in R. W. Decker, T. L. Wright, and P. H. Stauffer (editors), Volcanism in Hawaii. U.S. Geological Survey Professional Paper 1350. U.S. Government Printing Office, Washington, DC.

Holdaway, R. N. 1999. Introduced predators and avifaunal extinction in New Zealand. Pp. 189–238 in R. D. MacPhee (editor), Extinctions in near time: Causes, contexts, and consequences. Kluwer Academic/Plenum Publishers, New York.

Hölldobler, B., and E. O. Wilson. 1990. The ants. Belknap Press of Harvard University Press, Cambridge, MA.

Hollingsworth, R. D., and L. L. Loope. 2007. Learning from quarantine successes. Proceedings of the Hawaiian Entomological Society 39: 57–61.

Holshuh, H. J., A. E. Sherrod, C. R. Taylor, B. F. Andrews, and E. B. Howard. 1985. Toxoplasmosis in a northern fur seal. Journal of the American Veterinary Medical Association 187: 1229–1230.

Holt, A. 1996. An alliance of biodiversity, health, agriculture, and business interests for improved alien species management in Hawaii. Pp. 155–160 in O. T. Sandlund, P. J. Schei, and A. Viken (editors), Proceedings of the Norway/UN Conference on alien species. Directorate for Nature Management and Norwegian Institute for Nature Research, Trondheim, Norway.

Holt, D. W., and S. M. Leasure. 1993. Short-eared Owl (Asio flammeus). In A. Poole and F. Gill (editors), The birds of North America, No. 62. Birds of North America Inc., Philadelphia.

Holt, R. A., et al. (123 authors). 2002. The genome sequence of the malaria mosquito Anopheles gambiae. Science 298: 129–149.

Hone, J., and C. P. Stone. 1989. A comparison and evaluation of feral pig management in two national parks. Wildlife Society Bulletin 17: 419–425.

Honnold, S. P., R. Braun, D. P. Scott, C. Sreekumar, and J. P. Dubey. 2005. Toxoplasmosis in a Hawaiian monk seal (Monachus schauinslandi). Journal of Parasitology 91: 695–697.

Hoover, W. 2003. Empty nest? New federal law could bleed local gamecock breeding businesses. Honolulu Advertiser, Honolulu, April 28.

Hosmer, R. S. 1959. The beginning five decades of forestry in Hawaii. Journal of Forestry 57: 83–89.

Hotchkiss, S., P. M. Vitousek, O. A. Chadwick, and J. Price. 2000. Climate cycles, geomorphological change, and the interpretation of soil and ecosystem development. Ecosystems 3: 522–533.

Hotchkiss, S. C. 1998. Quaternary vegetation and climate of Hawai'i. Ph.D. dissertation. University of Minnesota, Saint Paul.

Hotchkiss, S. C., and J. O. Juvik. 1999. A late quaternary pollen record from Ka'au Crater, O'ahu, Hawai'i. Quaternary Research 52: 115–128.

Houck, O. A. 2004. More unfinished stories: Lucas, Atlanta Coalition, and Palila/Sweet Home. University of Colorado Law Review 75: 331–432.

Howald, G. H., K. R. Faulkner, B. Tershey. B. Keitt, H. Gellermani, E. M. Creel, M. Grinnell, S. T. Ortega, and D. A. Croll. 2005. Eradication of black rats from Anapapa Island: Biological and social considerations. Proceedings of the Sixth California Islands Symposium, Ventura, California, December 1–3, 2003. Technical Publication CHIS-05-01. National Park Service Institute for Wildlife Studies, Arcata, CA.

Howarth, F. G. 1985. Impacts of alien land arthropods and mollusks on native plants and animals in Hawaii. Pp. 149–179 in C. P. Stone and J. M. Scott (editors), Hawai'i's terrestrial ecosystems: Preservation and management. University of Hawaii Cooperative National Park Resources Studies Unit, University of Hawaii Press, Honolulu.

———. 1990. Hawaiian terrestrial arthropods: An overview. Bishop Museum Occasional Papers 30: 4–26.

———. 1991. Environmental impacts of classical biological control. Annual Review of Entomology 36: 485–509.

Howarth, F. G., and W. P. Mull. 1992. Hawaiian insects and their kin. University of Hawaii Press, Honolulu.

Hoyt, D. F. 1979. Practical methods of estimating volume and fresh weight of bird eggs. Auk 96: 73–77.

Hsu, V. P., M. J. Hossain, U. D. Parashar, M. M. Ali, T. G. Ksiazek, I. Kuzmin, M. Niezgoda, C. Rupprecht, J. Bresee, and R. F. Breiman. 2004. Nipah virus encephalitis reemergence, Bangladesh. Emerging Infectious Diseases 10: 2082–2087.

Hu, D., C. Gildden, J. S. Lippert, L. Schnell, J. S. MacIvor, and J. Meisler. 2001. Habitat use and limiting factors in a population of Hawaiian Dark-rumped Petrels on Mauna Loa, Hawai'i. Studies in Avian Biology 22: 234–242.

Hudson, P. J., A. P. Dobson, and D. Newborn. 1998. Prevention of population cycles by parasite removal. Science 282: 2256–2258.

Hugh, W. J., T. Tanaka, J. C. Nolan Jr., and L. K. Fox. 1986. The livestock industry in Hawaii. Information Text Series 025. Hawaii Institute of Tropical Agriculture and Human Resources, College of Tropical Agriculture and Human Resources, University of Hawaii, Honolulu.

Hughes, A. L. 2004. A statistical analysis of factors associated with historical extinction and current endangerment of non-passerine birds. Wilson Bulletin 116: 330–336.

Hughes, A. L., and M. Yeager. 1998. Natural selection at major histocompatibility loci of vertebrates. Annual Review of Genetics 32: 415–435.

Hughes, F., P. M. Vitousek, and T. Tunison. 1991. Alien grass invasion and fire in the seasonal submontane zone of Hawai'i. Ecology 72: 743–746.

Hughes, R. F., and J. S. Denslow. 2005. Invasion by a N_2-fixing tree alters function and structure in wet lowland forests of Hawaii. Ecological Applications 15: 1615–1628.

Hurley, T. 1991. Miconia: Fast-growing weed tree in the sights of scientists. Maui News (Wailuku, HI), May 17, pp. A1, A3.

Husby, M. 2003. Point count census using volunteers of terrestrial breeding birds in Norway, and its status after six years. Ornis Hungarica 12–13: 63–72.

Hutchins, M., and C. Wemmer. 1991. In defense of captive breeding. Endangered Species Update 8: 5–6.

Hutchins, M., R. J. Wiese, and K. Willis. 1997. Captive breeding and conservation. Conservation Biology 11: 1.

Hutchinson, M. F. 1990. Robust-calibration of seasonally varying stochastic weather models using periodic smoothing splines. Mathematics and Computers in Simulation 32: 125–130.

Hutto, R. L., S. M. Pletschet, and P. Hendricks. 1986. A fixed-radius point count method for nonbreeding and breeding-season use. Auk 103: 593–602.

Ikuma, E. K., D. Sugano, and J. K. Mardfin. 2002. Filling the gaps in the fight against invasive species. Hawai'i Legislative Reference Bureau, Honolulu.

Innes, J., and G. Barker. 1999. Ecological consequences of toxin use for mammalian pest control in New Zealand—an overview. New Zealand Journal of Ecology 23: 111–127.

Innes, J., B. Warburton, D. Williams, H. Speed, and P. Bradfield. 1995. Large-scale poisoning of ship rats (Rattus rattus) in indigenous forests of the North Island, New Zealand. New Zealand Journal of Ecology 19: 5–17.

Innes, J. G., and J. R. Hay. 1991. The interactions of New Zealand forest birds with introduced fauna. ACTA XX Congressus Intaernationalis Ornithologici 20: 2523–2533.

International Species Information System—ISIS. www.isis.org/CMSHOME/.

International Union for Conservation of Nature. 1995. IUCN/SSC guidelines for re-introductions. 41st meeting of the IUCN Council. Species Survival Commission Re-introduction Specialist Group, Gland, Switzerland.

———. 2001. IUCN red list categories and criteria version 3.1. IUCN Species Survival Commission, Gland, Switzerland, and Cambridge, England.

———. 2003. Guidelines on the implementation of the "IUCN Policy Statement on the Research Involving Species at Risk of Extinction," with special reference to scientific collecting of threatened species. IUCN Species Survival Commission, Gland, Switzerland, and Cambridge, England.

'Io Recovery Working Group. 2001. Products from the 26–27 February 2001 IRWG meeting. U.S. Department of the Interior, Fish and Wildlife Service, Honolulu.

Irwin, B. P. 1936. In Menehune land. Printshop, Honolulu.

Ishtiaq, F., J. S. Beadell, A. J. Baker, A. R. Rahmani, Y. V. Jhala, and R. C. Fleischer. 2006. Prevalence and evolutionary relationships of haematozoan parasites in native versus introduced populations of common myna Acridotheres tristis. Proceedings of the Royal Society of London, Series B 273: 587–594.

Jacobi, J. D. 1983. Metrosideros dieback in Hawai'i: A comparison of adjacent dieback and non-dieback rain forest stands. New Zealand Journal of Ecology 6: 79–97.

———. 1989. Vegetation maps of the upland plant communities on the islands of Hawai'i, Maui, Moloka'i, and Lāna'i. Technical Report 68. Cooperative National Park Resources Studies Unit, Department of Botany, University of Hawai'i, Honolulu.

———. 1990. Distribution maps, ecological relationships, and status of native plant communities on the island of Hawai'i. Ph.D. dissertation. University of Hawai'i at Mānoa, Honolulu.

———. 1993. Distribution and dynamics of Metrosideros dieback on the island of Hawai'i: Implications for management programs. Pp. 236–242 in R. F. Huettl and D. Mueller-Dombois (editors), Forest decline in the Atlantic and Pacific region. Springer, Heidelberg, Germany.

Jacobi, J. D., and C. T. Atkinson. 1995. Hawaii's endemic birds. Pp. 376–381 in E. T. LaRoe, G. S. Farris, C. E. Puckett, P. D. Doran, and M. J. Mac (editors), Our living resources: A report to the nation on the distribution, abundance, and health of U.S. plants, animals, and ecosystems. U.S. Department of the Interior, National Biological Service, Washington, DC.

Jacobi, J. D., S. G. Fancy, J. G. Giffin, and J. M. Scott. 1996. Long-term population variability in the Palila, an endangered Hawaiian honeycreeper. Pacific Science 50: 363–370.

Jacobi, J. D., G. Gerrish, and D. Mueller-Dombois. 1983. 'Ohi'a dieback in Hawai'i: Vegetation changes in permanent plots. Pacific Science 37: 327–337.

Jacobi, J. D., G. Gerrish, D. Mueller-Dombois, and L. Whiteaker. 1988. Stand-level dieback and Metrosideros regeneration in the montane rain forest of Hawaii. GeoJournal 17: 193–200.

Jacobi, J. D., and J. M. Scott. 1985. An assessment of the current status of native upland habitats and associated endangered species on the island of Hawai'i. Pp. 3–22 in C. P. Stone and J. M Scott (editors), Hawai'i's terrestrial ecosystems: Preservation and management. Cooperative National Park Resources Studies Unit, University of Hawaii, Honolulu.

Jacobi, J. D., and F. R. Warshauer. 1992. Distribution of six alien plant species in upland habitats on the island of Hawai'i. Pp. 155–188 in C. P. Stone, C. W. Smith, and J. T. Tunison (editors), Alien plant invasions in native ecosystems of Hawai'i: Management and research. University of Hawaii Cooperative National Park Resources Studies Unit. University of Hawaii Press, Honolulu.

Jacobs, W. W. 1992. Vertebrate pesticides no longer registered and factors contributing to loss of registration. Proceedings of the Vertebrate Pest Conference 15: 142–148.

James, F. C., C. E. McCulloch, and D. A. Wiedenfeld. 1996. New approaches to the analysis of population trends in land birds. Ecology 77: 13–27.

James, H. F. 1995. Prehistoric extinctions and ecological changes on oceanic islands. Ecological Studies 115: 87–102.

———. 2004. The osteology and phylogeny of the Hawaiian finch radiation (Fringillidae: Drepanidini), including extinct taxa. Zoological Journal of the Linnean Society 141: 207–255.

James, H. F., and S. L. Olson. 1991. Descriptions of thirty-two new species of birds from the Hawaiian Islands, Part 2: Passeriformes. Ornithological Monographs 46: 1–88.

———. 2003. A giant new species of Nukupuu (Fringillidae: Drepanidini: Hemignathus) from the island of Hawaii. Auk 120: 970–981.

———. 2005. The diversity and biogeography of koa-finches (Drepanidini: Rhodacanthis), with descriptions of two new species. Zoological Journal of the Linncan Society 144: 527–541.

———. 2006. A new species of Hawaiian finch (Drepanidini: Loxioides) from Makauwahi Cave, Kaua'i. Auk 123: 335–344.

James, H. F., T. W. Stafford Jr., D. W. Steadman, S. L. Olson, P. S. Martin, A. J. T. Jull, and P. C. McCoy. 1987. Radiocarbon dates on bones of extinct birds from Hawai'i. Proceedings of the National Academy of Sciences (USA) 84: 2350–2354.

Jamieson, I., W. Lee, and J. Maxwell. 2000. Fifty years of conservation management and re-introductions of the takahe in New Zealand. Re-introduction News 19: 30–32.

Janzen, D. H. 1973. Sweep samples of tropical foliage insects: Effects of seasons, vegetation types, elevation, time of day, and insularity. Ecology 54: 687–708.

———. 1988. Tropical dry forests, the most endangered major tropical ecosystem. Pp. 130–137 in E. O. Wilson (editor), Biodiversity. National Academy Press, Washington, DC.

Jarvi, S. I., C. T. Atkinson, and R. C. Fleischer. 2001. Immunogenetics and resistance to avian malaria in Hawaiian honeycreepers (Drepanidinae). Studies in Avian Biology 22: 254–263.

Jarvi, S. I., and P. C. Banko. 2000. Application of a PCR-based approach to identify sex in Hawaiian honeycreepers (Drepanidinae). Pacific Conservation Biology 6: 14–17.

Jarvi, S. I., and K. R. Bianchi. 2006. Genetic analysis of captive Alala (Corvus hawaiiensis) using AFLP analyses. Open-File Report 2006-1349. U.S. Geological Survey, Biological Resources Division, Reston, VA.

Jarvi, S. I., and M. E. M. Farias. 2006. Molecular sexing and sources of CHD1-Z/W sequence variation in Hawaiian birds. Molecular Ecology Notes 6: 1003–1005.

Jarvi, S. I., M. E. M. Farias, and C. T. Atkinson. 2008. Genetic characterization of Hawaiian isolates of Plasmodium relictum reveals mixed-genotype infections. Biology Direct 3: 25.

Jarvi, S. I., M. E. M. Farias, H. Baker, H. B. Freifeld, P. E. Baker, E. Van Gelder, J. G. Massey, and C. T. Atkinson. 2003. Detection of avian malaria (Plasmodium spp.) in native land birds of American Samoa. Conservation Genetics 4: 629–637.

Jarvi, S. I., M. M. Miller, R. M. Goto, G. F. Gee, and W. E. Briles. 2001. Evaluation of the major histocompatibility complex (Mhc) in cranes: Applications to conservation efforts. Proceedings of the North American Crane Workshop 8: 223.

Jarvi, S. I., J. J. Schultz, and C. T. Atkinson. 2002. PCR diagnostics underestimate the prevalence of avian malaria (Plasmodium relictum) in experimentally-infected passerines. Journal of Parasitology 88: 153–158.

Jarvi, S. I., C. L. Tarr, C. E. McIntosh, C. T. Atkinson, and R. C. Fleischer. 2004. Natural selection of the major histocompatibility complex (Mhc) in Hawaiian honeycreepers (Drepanidinae). Molecular Ecology 13: 2157–2168.

Jarvi, S. I., D. Triglia, A. Giannoulis, M. Farias, K. Bianchi, and C. T. Atkinson. 2008. Diversity, origins and virulence of avipoxviruses in Hawaiian forest birds. Conservation Genetics 9: 339–348.

Jeffrey, J. J., S. G. Fancy, G. D. Lindsey, P. C. Banko, T. K. Pratt, and J. D. Jacobi. 1993. Sex and identification of Palila. Journal of Field Ornithology 64: 490–499.

Jenkins, C. D., S. A. Temple, C. van Riper III, and W. R. Hansen. 1989. Disease-related aspects of conserving the endangered Hawaiian Crow. International Council for Bird Preservation Technical Publication 10: 77–87.

Jenkins, I. 1983. Hawaiian furniture and Hawaii's cabinet makers, 1820–1940. Daughters of Hawaii, Editions Ltd., Honolulu.

Jensen, T., S. P. Lawler, and D. A. Dritz. 1999. Effects of ultra-low volume pyrethrin, malathion, and permethrin on nontarget invertebrates, sentinel mosquitoes, and mosquitofish in seasonally impounded wetlands. Journal of the American Mosquito Control Association 15: 330–338.

Johnson, K. P., F. R. Adler, and J. L. Cherry. 2000. Genetic and phylogenetic consequences of island biogeography. Evolution 54: 387–396.

Johnson, K. P., and J. Seger. 2001. Elevated rates of nonsynonymous substitution in island birds. Molecular Biology and Evolution 18: 874–881.

Johnson, L., R. J. Camp, K. W. Brinck, and P. C. Banko. 2006. Long-term population monitoring: Lessons learned from an endangered passerine in Hawai'i. Wildlife Society Bulletin 34: 1055–1063.

Johnson, M. S. 1945. Rodent control on Midway Islands. U.S. Naval Medical Bulletin 45: 384–398.

Johnson, N. K., J. A. Martin, and C. J. Ralph. 1989. Genetic evidence for the origin and relationships of Hawaiian honeycreepers (Aves: Fringillidae). Condor 91: 379–396.

Johnson, R. E. 2002. Black Rosy-Finch (*Leucosticte atrata*). In A. Poole and F. Gill (editors), The birds of North America, No. 678. Birds of North America Inc., Philadelphia.

Johnson, R. E., P. Hendricks, D. L. Pattie, and K. B. Hunter. 2000. Brown-capped Rosy-Finch (*Leucosticte australis*). In A. Poole and F. Gill (editors), The birds of North America, No. 536. Birds of North America Inc., Philadelphia.

Johnson, T. H., and A. J. Stattersfield. 1990. A global review of island endemic birds. Ibis 132: 167–180.

Johnston, J. J., W. C. Pitt, R. T. Sugihara, J. D. Eisemann, T. M. Primus, M. J. Holmes, J. Crocker, and A. Hart. 2005. Probabilistic risk assessment for snails, slugs, and endangered honeycreepers in diphacinone rodenticide baited areas on Hawaii, USA. Environmental Toxicology and Chemistry 24: 1557–1567.

Johnston, J. P., W. J. Peach, R. D. Gregory, and S. A. White. 1997. Survival rates of tropical and temperate passerines: A Trinidadian perspective. American Naturalist 150: 771–789.

Johnston, R. F., and R. K. Selander. 1964. House sparrows: Rapid evolution of races in North America. Science 144: 548–550.

———. 1971. Evolution of the house sparrow, 2: Adaptive differentiation in North American populations. Evolution 25: 1–28.

Johnston, S. 2000. Building a species recovery program on trust: The case of the Hawaiian crow ('alala). Conservation in Practice 1: 35–37.

Johnstone, G. W. 1985. Threats to birds on subantarctic islands. Pp. 101–121 in P. J. Moors (editor), Conservation of island birds. Technical Publication 3. International Council for Bird Preservation, Cambridge, England.

Jones, C. G., W. Heck, R. E. Lewis, Y. Mungroo, G. Slade, and T. Cade. 1995. The restoration of the Mauritius Kestrel *Falco punctatus* population. Ibis 137 (Supplement 1): 173–180.

Jones, P. W., and T. M. Donovan. 1996. Hermit Thrush (*Catharus guttatus*). In A. Poole and F. Gill (editors), The birds of North America, No. 261. Birds of North America Inc., Philadelphia.

Jones, S. M. (editor). 1925. Diary of Andrew Bloxham. Bernice P. Bishop Museum Special Publication 10: 1–96.

Jones, T. R., and S. L. Hoffman. 1994. Malaria vaccine development. Clinical Microbiology Reviews 7: 303–310.

Jones, V. P., P. Anderson-Wong, P. A. Follett, P. Yang, D. Westcot, D. E. Ullman, J. Hu, and D. Foote. 2000. Feeding damage of the introduced leafhopper *Sophonia rufofascia* (Homoptera: Cicadellidae) to plants in forests and watersheds of the Hawaiian Islands. Environmental Entomology 29: 171–180.

Joyce, C. R., and P. Y. Nakagawa. 1963. *Aedes vexans nocturnes* (Theobald) in Hawai'i. Proceedings of the Hawaiian Entomological Society 18: 273–280.

Judd, C. S. 1918. Forestry as applied in Hawaii. Hawaiian Forester and Agriculturist 15: 117–133.

———. 1927a. The natural resources of the Hawaiian forest regions and their conservation. Hawaiian Forester and Agriculturalist 24: 40–47.

———. 1927b. Factors deleterious to the Hawaiian forest. Hawaiian Forester and Agriculturalist 24: 47–53.

———. 1931. Forestry in Hawaii for water conservation. Journal of Forestry 29: 363–367.

Julien, M. H., and M. W. Griffiths (editors). 1998. Biological control of weeds: A world catalogue of agents and their target weeds. 4th edition. C.A.B. International, Wallingford, England.

Juvik, J. O., and A. P. Austring. 1979. The Hawaiian avifauna: Biogeographic theory in evolutionary time. Journal of Biogeography 6: 205–224.

Juvik, J. O., and S. P. Juvik. 1984. Mauna Kea and the myth of multiple use: Endangered species and mountain management in Hawaii. Mountain Research and Development 4: 191–202.

Juvik, S. P., and J. O. Juvik. 1998. Atlas of Hawai'i. 3rd edition. University of Hawai'i Press, Honolulu.

Kaeppler, A. 1970. Feather cloaks, ship captains, and lords. Bishop Museum Occasional Papers 24: 91–114.

———. 1978. Artificial curiosities. Bishop Museum Press, Honolulu.

———. 1985. Hawaiian art and society: Traditions and transformations. Memoirs of the Polynesian Society 45: 105–131.

Kaiser, B. A., and K. M. Burnett. 2006. Economic impacts of E. coqui frogs in Hawaii. Interdisciplinary Environmental Review 8: 1–11.

Kalakaua, D. 1888. The legends and myths of Hawaii. C. L. Webster, New York.

———. 1972. The legends and myths of Hawai'i: Fables and folk-lore of a strange people. Charles Tuttle, Rutland, VT.

Kamakau, S. M. 1964. Ka po'e kahiko: The people of old. Bishop Museum Special Publication 51. Bishop Museum Press, Honolulu.

———. 1991. Tales and traditions of the people of old. Bishop Museum Press, Honolulu.

———. 1992. Ruling chiefs of Hawai'i. Revised edition. Kamehameha School Press, Honolulu.

Kami, H. T. 1964. Food of the mongoose in the Hamakua District, Hawaii. Zoonoses Research 3: 165–170.

———. 1966. Foods of rodents in the Hamakua District, Hawaii. Pacific Science 20: 367–373.

Kāne, H. K. 1997. Ancient Hawai'i. Kawainui Press, Captain Cook.

Karr, J. R. 1982a. Avian extinction on Barro Colorado Island, Panama: A reassessment. American Naturalist 119: 220–239.

———. 1982b. Population variability and extinction in the avifauna of a tropical land bridge island. Ecology 63: 1975–1978.

Karr, J. R., J. D. Nichols, M. K. Klimkiewicz, and J. D. Brawn. 1990. Survival rates of birds of tropical and temperate forests: Will the dogma survive? American Naturalist 136: 277–291.

Katahira, L. K. 1980. The effects of feral pigs on a montane rain forest in Hawaii Volcanoes National Park. Pp. 173–178 in C. W. Smith (editor), Proceedings Third Conference in Natural Sciences, Hawaii Volcanoes National Park. Cooperative National Park Resources Studies Unit, University of Hawai'i, Honolulu.

Katahira, L. K., P. Finnegan, and C. P. Stone. 1993. Eradicating feral pigs in montane mesic habitat at Hawai'i Volcanoes National Park. Wildlife Society Bulletin 21: 269–274.

Katz, A. R., V. E. Ansdell, P. V. Effler, C. R. Middleton, and D. R. Sasaki. 2002. Leptospirosis in Hawaii, 1974–1998: Epidemiologic analysis of 353 laboratory-confirmed cases. American Journal of Tropical Medicine and Hygiene 66: 61–70.

Kaufman, J., and J. Salomonsen. 1997. The "minimal essential Mhc" revisited: Both peptide-binding and cell surface expression level of MHC molecules are polymorphisms selected by pathogens in chickens. Hereditas 127: 67–73.

Kaukeinen, D. E. 1982. A review of the secondary poisoning hazard potential to wildlife from the use of anticoagulant rodenticides. Proceedings of the Vertebrate Pest Conference 10: 151–158.

Kawakami, K., and H. Higuchi. 2003. Interspecific interactions between the native and introduced White-eyes in the Bonin Islands. Ibis 145: 583–592.

Kay, B. H., S. A. Lyons, J. S. Holt, M. Holynska, and B. M. Russell. 2002. Point source inoculation of Mesocyclops (Copepoda: Cyclopidae) gives widespread control of Ochlerotatus and Aedes (Diptera: Culicidae) immatures in service manholes and pits in North Queensland, Australia. Journal of Medical Entomology 39: 469–474.

Kear, J., and A. J. Berger. 1980. The Hawaiian Goose: An experiment in conservation. Buteo Books, Vermillion, SD.

Keast, A. 1996. Wing shape in insectivorous passerines inhabiting New Guinea and Australian rain forest and Eucalypt forest / Eucalypt woodland. Auk 113: 94–104.

Keauokalani, K. 1865. Short notes pertaining to Hawaiian life: Descriptions of Hawaiian birds. Hawaiian Ethnological Notes, Vol. 1: 1127–1155. Courtesy of the Bishop Museum Archives, Honolulu.

Keeling, M. J., M. E. Woolhouse, R. M. May, G. Davies, and B. T. Grenfell. 2003. Modeling vaccination strategies against foot-and-mouth disease. Nature 421: 136–142.

Keith, J. O., D. N. Hirata, D. L. Espy, S. Greiner, and D. Griffin. 1989. Field evaluation of 0.00025% diphacinone for mongoose control in Hawaii. Unpublished report. U.S. Department of Agriculture, Animal and Plant Health Inspection Service, Denver Wildlife Research Center, Denver.

Keller, L. F., P. Arcese, J. N. M. Smith, W. M. Hochachka, and S. C. Stearns. 1994. Selection against inbred song sparrows during a natural population bottleneck. Nature 372: 356–357.

Kelly, J. A. 1998. Agriculture. Pp. 246–251 in S. P. Juvik and J. O. Juvik (editors), Atlas of Hawai'i. 3rd edition. University of Hawai'i Press, Honolulu.

Kelly, M. 1983. Na mala o Kona: Gardens of Kona, a history of land use in Kona Hawai'i. Pacific Anthropological Records 6. Anthropology Department, B. P. Bishop Museum, Honolulu.

Kepler, C. B., T. K. Pratt, A. M. Ecton, A. Engilis Jr., and K. M. Fluetsch. 1996. Nesting behavior of the Poo-uli. Wilson Bulletin 108: 620–638.

Kerlinger, P. 1989. Flight strategies of migrating hawks. University of Chicago Press, Chicago.

Kerr, P. 2002. Immune responses to myxoma virus. Viral Immunology 15: 229–246.

Keyghobadi, N., D. LaPointe, R. C. Fleischer, and D. M. Fonseca. 2006. Fine-scale population genetic structure of a wildlife disease vector: The southern house mosquito on the island of Hawaii. Molecular Ecology 15: 3919–3930.

Killgore, E. M., and R. A. Heu. 2007. A rust disease on 'ohi'a. New Pest Advisory 05-04. Department of Agriculture, State of Hawaii. Updated December. www.hawaiiag.org/hdoa/npa/npa05-04-ohiarust.pdf (accessed March 23, 2008).

Kilpatrick, A. M. 2003. The evolution of resistance to avian malaria (Plasmodium relictum) in Hawaiian birds. Ph.D. dissertation. University of Wisconsin, Madison.

———. 2006. Facilitating the evolution of resistance to avian malaria in Hawaiian birds. Biological Conservation 128: 475–485.

Kilpatrick, A. M., Y. Gluzberg, J. Burgett, and P. Daszak. 2004. Quantitative risk assessment of the pathways by which West Nile virus could reach Hawaii. EcoHealth 1: 205–209.

Kilpatrick, A. M., S. L. LaDeau, and P. P. Marra. 2007. Ecology of West Nile virus transmission and its impact on birds in the Western Hemisphere. Auk 124: 1121–1136.

Kilpatrick, A. M., D. A. LaPointe, C. T. Atkinson, B. L. Woodworth, J. K. Lease, M. E. Reiter, and K. Gross. 2006. Effects of chronic avian malaria (Plasmodium relictum) infection on reproductive success of Hawaii Amakihi (Hemignathus virens). Auk 123: 764–774.

Kim, I.-K., G. Delvare, and J. Lasalle. 2004. A new species of Quadrastichus (Hymenoptera: Eulophidae): A gall-inducing pest on Erythrina spp. (Fabaceae). Journal of Hymenopteran Research 13: 243–249.

Kim, T., and D. N. Tripathy. 2006. Evaluation of pathogenicity of avian poxvirus isolates from endangered Hawaiian wild birds in chickens. Avian Diseases 50: 288–291.

King, C. 1984. Immigrant killers: Introduced predators and the conservation of birds in New Zealand. Oxford University Press, Auckland, New Zealand.

King, W. B. 1985. Island birds: Will the future repeat the past? Pp. 3–15 in P. J. Moors (editor), Conservation of island birds: Case studies for the management of threatened island species. Technical Publication 3. International Council of Bird Preservation, Cambridge, England.

Kingsland, S. E. 1995. Modeling nature. University of Chicago Press, Chicago.

Kirch, P. V. 1982. The impact of the prehistoric Polynesians on the Hawaiian ecosystem. Pacific Science 36: 1–14.

———. 1983. Man's role in modifying tropical and subtropical Polynesian ecosystems. Archaeology of Oceania 18: 26–31.

———. 1985. Feathered gods and fishhooks: An introduction to Hawaiian archaeology and prehistory. University of Hawai'i Press, Honolulu.

———. 2007. Hawaii as a model system for human ecodynamics. American Anthropologist 109: 8–26.

Kirch, P. V., and M. Kelly (editors). 1975. Prehistory and ecology in a windward Hawaiian valley: Halawa Valley, Molokai. Pacific Anthropological Records 24. Department of Anthropology, Bishop Museum, Honolulu.

Kitayama, K., D. Mueller-Dombois, and P. M. Vitousek. 1995. Primary succession of Hawaiian montane rain forest on a chronosequence of eight lava flows. Journal of Vegetation Science 6: 211–222.

Kitron, U. 1998. Landscape ecology and epidemiology of vector-borne diseases: Tools for spatial analysis. Journal of Medical Entomology 35: 435–445.

Kjargaard, M. S. 1994. Alien plant-disperser interactions in Hawaiian forest ecosystems. Ph.D. dissertation. University of Hawaii at Mānoa, Honolulu.

Klasing, K. C. 1998a. Comparative avian nutrition. CABI Publishing, Oxon, England, and New York.

———. 1998b. Nutritional modulation of resistance to infectious diseases. Poultry Science 77: 1119–1125.

Klassen, W. 2003. Edward F. Knipling: Titan and driving force in ecologically selective area-wide pest management. Journal of the American Mosquito Control Association 19: 94–103.

Klavitter, J. L. 2000. Survey methodology, abundance, and demography of the endangered Hawaiian Hawk: Is delisting warranted? M.S. thesis. University of Washington, Seattle.

Klavitter, J. L., and J. M. Marzluff. 2007. Methods to correct for density inflation biases in surveys using attractant calls: A case study of Hawaiian Hawks. Journal of Raptor Research 41: 81–89.

Klavitter, J. L., J. M. Marzluff, and M. S. Vekasy. 2003. Abundance and demography of the Hawaiian hawk: Is delisting warranted? Journal of Wildlife Management 67: 165–176.

Kleiman, D. G., M. R. Stanley Price, and B. B. Beck. 1994. Criteria for reintroductions. Pp. 287–303 in P. J. S. Olney, G. M. Mace, and A. T. C. Feistner (editors), Creative conservation: Interactive management of wild and captive animals. Chapman and Hall, London.

Klein, J. 1986. Natural history of the histocompatibility complex. Wiley, New York.

Klein, J., Y. Satta, N. Takahata, and C. O'hUigin. 1993. Trans-species MHC polymorphisms and the origin of species in primates. Journal of Medical Primatology 22: 57–64.

Kluckhohn, F., and F. Stodtbeck. 1961. Variations in value orientations. Row and Peterson, Evanston, IL.

Knox, A. G., and P. E. Lowther. 2000a. Common Redpoll (Carduelis flammea). In A. Poole and F. Gill (editors), The birds of North America, No. 543. Birds of North America Inc., Philadelphia.

———. 2000b. Hoary Redpoll (Carduelis hornemanni). In A. Poole and F. Gill (editors), The birds of North America, No. 544. Birds of North America Inc., Philadelphia.

Knudsen, E. A. 1946. Teller of Hawaiian tales. Mutual Publishing, Honolulu.

Kocan, R. M., and W. E. Banko. 1974. Trichomoniasis in the Hawaiian Barred Dove. Journal of Wildlife Diseases 10: 359–360.

Koebele, A. 1901. Report of Professor Koebele on destruction of forest trees, Hawaii. Pp. 50–66 in Report of the Commission on Agriculture and Forestry for 1900. Commission on Agriculture and Forestry, Honolulu.

Koenig, W. D. 1982. Ecological and social factors affecting hatchability of eggs. Auk 99: 526–536.

Komdeur, J. 1991. Inter-island transfers and population dynamics of Seychelles warblers. Bird Conservation International 7: 7–26.

Konuma, J., and S. Chiba. 2007. Trade-offs between force and fit: Extreme morphologies associated with feeding behavior in carabid beetles. American Naturalist 170: 90–100.

Kowalsky, J. R., T. K. Pratt, and J. C. Simon. 2002. Prey taken by feral cats (Felis catus) and Barn Owls (Tyto alba) in Hanawi Natural Area Reserve, Maui, Hawai'i. 'Elepaio 62: 127–130.

Koyama, J., H. Kahinohana, and T. Miyatake. 2004. Eradication of the melon fly, Bactrocera cucurbitae, in Japan: Importance of behavior, ecology, genetics and evolution. Annual Review of Entomology 49: 331–349.

Krajewski, C. 1994. Phylogenetic measures of bio-diversity: A comparison and critique. Biological Conservation 69: 33–39.

Kraus, F. 2002. New records of alien reptiles in Hawai'i. Bishop Museum Occasional Papers 69: 48–52.

———. 2003. Invasion pathways for terrestrial vertebrates. Pp. 68–92 in G. Ruiz and J. Carlton (editors), Invasive species: Vectors and management strategies. Island Press, Washington, DC.

———. 2004. New records of alien reptiles and amphibians in Hawai'i. Bishop Museum Occasional Papers 79: 62–64.

———. 2007. Using pathway analysis to inform prevention strategies for alien reptiles and amphibians. Pp. 94–103 in G. W. Witmer, W. C. Pitt, and K. A. Fagerstone (editors), Managing vertebrate invasive species: Proceedings of an international symposium. USDA/APHIS/WS National Wildlife Research Center, Fort Collins, CO.

———. 2008. Alien reptiles and amphibians: A scientific compendium and analysis. Springer, New York.

Kraus, F., and E. W. Campbell III. 2002. Human-mediated escalation of a formerly eradicable problem: The invasion of Caribbean frogs in the Hawaiian Islands. Biological Invasions 4: 327–332.

Kraus, F., E. W. Campbell, A. Allison, and T. K. Pratt. 1999. Eleutherodactylus frog introductions to Hawaii. Herpetological Review 30: 21–25.

Kraus, F., and D. Cravalho. 2001. The risk to Hawaii from snakes. Pacific Science 55: 409–417.

Kreuder, C., A. R. Irizarry-Rovira, E. B. Janovitz, P. J. Deitschel, and D. B. DeNicola. 1999. Avian pox in sanderlings from Florida. Journal of Wildlife Diseases 35: 582–585.

Krone, O., R. Alten Kamp, and N. Kenntner. 2005. Prevalence of Trichomonas gallinae in Northern Goshawks from the Berlin area of northeastern Germany. Journal of Wildlife Diseases 41: 304–309.

Krushelnycky, P. D., L. L. Loope, and N. J. Reimer. 2005. The ecology, policy and management of ants in Hawaii. Proceedings of the Hawaiian Entomological Society 37: 1–25.

Kuehler, C., and J. Good. 1990. Artificial incubation of bird eggs at the Zoological Society of San Diego. International Zoo Yearbook 29: 118–136.

Kuehler, C., P. Harrity, A. Lieberman, and M. Kuhn. 1995. Reintroduction of hand-reared Alala Corvus hawaiiensis in Hawaii. Oryx 29: 261–266.

Kuehler, C., M. Kuhn, J. E. Kuhn, A. Lieberman, N. Harvey, and B. Rideout. 1996. Artificial incubation, hand-rearing, behavior, and release of Common 'Amakihi (Hemignathus virens virens): Surrogate research for restoration of endangered Hawaiian forest birds. Zoo Biology 15: 541–553.

Kuehler, C., M. Kuhn, B. McIlraith, and G. Campbell. 1994. Artificial incubation and hand-rearing of 'Alala (Corvus hawaiiensis) eggs removed from the wild. Zoo Biology 13: 257–266.

Kuehler, C., and A. Lieberman. 2005. Alala Studbook. SPARKS (Small Population Animal Record Keeping System). Hawai'i Endangered Bird Conservation Program, Volcano, HI.

Kuehler, C., A. Lieberman, P. Harrity, M. Kuhn, J. Kuhn, B. McIlraith, and J. Turner. 2001. Restoration techniques for Hawaiian forest birds: Collection of eggs, artificial incubation and hand-rearing of chicks, and release to the wild. Studies in Avian Biology 22: 354–358.

Kuehler, C., A. Lieberman, B. McIlraith, W. Everett, T. A. Scott, M. L. Morrison, and C. Winchell. 1993. Artificial incubation and hand-rearing of loggerhead shrikes. Wildlife Society Bulletin 21: 165–171.

Kuehler, C., A. Lieberman, P. Oesterle, T. Powers, M. Kuhn, J. Kuhn, J. Nelson, T. Snetsinger, C. Herrmann, P. Harrity, E. Tweed, S. Fancy, B. Woodworth, and T. Telfer. 2000. Development of restoration techniques for Hawaiian thrushes: Collection of wild eggs, artificial incubation, hand-rearing, captive-breeding, and re-introduction to the wild. Zoo Biology 19: 263–277.

Kuehler, C., A. Lieberman, A. Varney, P. Unitt, R. M. Sulpice, J. Azua, and B. Tehevini. 1997. Translocation of Ultramarine Lories Vini ultramarina in the Marquesas Islands: Ua Huka to Fatu Hiva. Bird Conservation International 7: 69–79.

Kuiken, T., G. Rimmelzwaan, D. van Riel, G. van Amerongen, M. Baars, R. Fouchier, and A. Osterhaus. 2004. Avian H5N1 influenza in cats. Science 305: 1385.

Kuris, A. M., and S. F. Norton. 1985. Evolutionary importance of overspecialization: Insect parasitoids as an example. American Naturalist 126: 387–391.

Kuykendall, R. S. 1938 [reprinted 1965]. The Hawaiian kingdom, Vol. 1: 1778–1854, foundation and transformation. University of Hawaii Press, Honolulu.

Kuykendall, R. S. 1967. The Hawaiian kingdom, Vol. 3: 1874–1893, the Kalakaua Dynasty. University of Hawaii Press, Honolulu.

Lacey, L. A., J. Day, and C. M. Heitzman. 1987. Long term effects of Bacillus sphaericus on Culex quinquefasciatus. Journal of Invertebrate Pathology 49: 116–123.

Lacey, L. A., and M. S. Mulla. 1990. Safety of Bacillus thuringiensis var. israelensis and Bacillus sphaericus to nontarget organisms in the aquatic environment. Pp. 169–188 in M. Laird, L. A. Lacey, and E. W. Davidson (editors), Safety of microbial insecticides. CRC Press, Boca Raton, FL.

Lacey, L. A., and B. K. Orr. 1994. The role of bio-logical control of mosquitoes in integrated vector control. American Journal of Tropical Medicine and Hygiene 50 (Supplement): 97–115.

Lack, D. 1947. The significance of clutch-size. Ibis 89: 302–352.

———. 1976. Island biology illustrated by the land birds of Jamaica. Studies in Ecology, Vol. 3. Black-well Scientific, Oxford.

Lacy, R. C. 1994. Managing genetic diversity in cap-tive populations of animals. Pp. 63–89 in M. L. Bowles and C. J. Whelan (editors), Restoration and recovery of endangered plants and animals. Cambridge University Press, Cambridge, England.

LaDeau, S. L., A. M. Kilpatrick, and P. P. Mara. 2007. West Nile virus emergence and large-scale declines of North American bird populations. Nature 447: 710–713.

Laird, M., and C. van Riper III. 1981. Questionable reports of Plasmodium from birds in Hawaii with the recognition of P. relictum ssp. capistranoae (Rus-sell, 1932) as the avian malaria parasite there. Pp. 159–165 in E. U. Canning (editor), Parasito-logical topics: A presentation volume to P.C.C. Garnham, F.R.S. on the occasion of this 80th birthday. Special Publication 1. Society of Proto-zoologists, Allen Press, Lawrence, KS.

Lande, R. 1987. Extinction thresholds in demo-graphic models of territorial populations. Amer-ican Naturalist 130: 624–635.

———. 1988. Genetics and demography in bio-logical conservation. Science 241: 1455–1460.

LaPointe, D. A. 2000. Avian malaria in Hawai'i: The distribution, ecology and vector potential of forest-dwelling mosquitoes. Ph.D. dissertation. University of Hawai'i at Mānoa, Honolulu.

———. 2007. Current and potential impacts of mosquitoes and the pathogens they vector in the Pacific region. Proceedings of the Hawaiian Ento-mological Society 39: 1–7.

———. 2008. Dispersal of Culex quinquefasciatus (Diptera: Culicidae) in a Hawaiian rain forest. Journal of Medical Entomology 45: 600–609.

LaPointe, D. A., M. L. Goff, and C. T. Atkinson. 2005. Comparative susceptibility of introduced forest-dwelling mosquitoes in Hawai'i to avian malaria, Plasmodium relictum. Journal of Parasitology 91: 843–849.

LaPointe, D. A., E. Hofmeister, C. T. Atkinson, R. E. Porter, and R. J. Dusek. 2009. Experimental infec-tion of Hawai'i Amakihi (Hemignathus virens) with West Nile virus and competence of a co-occurring vector, Culex quinquefasciatus: Potential impacts on endemic Hawaiian avifauna. Journal of Wildlife Diseases. 45: 257–271.

Larish, L. B., and H. M. Savage. 2005. Introduction and establishment of Aedes (Finlaya) japonicus japoni-cus (Theobald) on the island of Hawaii: Implica-tions for arbovirus transmission. Journal of the American Mosquito Control Association 21: 318–321.

Larkin, R. P., T. R. Van Deelen, R. M. Sabick, T. E. Gos-selink, and R. E. Warner. 2003. Electronic signal-ing for prompt removal of an animal from a trap. Wildlife Society Bulletin 31: 392–398.

La Rosa, A. M. 1984. The biology and ecology of Passiflora mollissima in Hawaii. Technical Report 50. Cooperative National Park Resources Studies Unit, University of Hawai'i, Honolulu.

Latchininsky, A. V. 2008. Grasshopper outbreak challenges conservation status of a small Hawai-ian Island. Journal of Insect Conservation 12: 343–357.

Lauer, N. C. 2008. Lingle wants to nix invasive species fee: Tax on incoming freight too burdensome, given poor economy. Hawaii Tribune-Herald, Oc-tober 11, 2008. www.hawaiitribune-herald.com/ articles/2008/10/11/front/ (accessed February 12, 2009).

Laut, M. E., P. C. Banko, and E. M. Gray. 2003. Nest-ing behavior of Palila, as assessed from video recordings. Pacific Science 57: 385–392.

Laven, H. 1959. Speciation by cytoplasmic isolation in the Culex pipiens complex. Cold Spring Harbor Symposium of Quantitative Biology 24: 166–173.

———. 1967. Eradication of Culex pipiens fatigans through cytoplasmic incompatibility. Nature 216: 383–384.

Lee, C. E. 2002. Evolutionary genetics of invasive species. Trends in Ecology and Evolution 17: 386–391.

Lennox, C. G. 1948. Are forests essential to Hawaii's water economy? Biennial Report. Board of Com-missioners of Agriculture and Forestry, Honolulu, June 30.

Lenton, G. M. 1984. The feeding and breeding ecology of Barn Owls (Tyto alba) in peninsular Malaysia. Ibis 126: 551–575.

Lenz, L. S. 2000. The dieback of an invasive tree in Hawai'i: Interactions between the two-spotted leafhopper (Sophonia rufofascia) and faya tree (Myrica faya). M.S. thesis. University of Hawai'i at Mānoa, Honolulu.

Leonard, D. L. Jr. 2008. Recovery expenditures for birds listed under the US Endangered Species Act: The disparity between mainland and Hawaiian taxa. Biological Conservation 141: 2054–2061.

Leonard, D. L. Jr., P. C. Banko, K. W. Brinck, C. Farmer, and R. J. Camp. 2008. Recent surveys indicate rapid decline of Palila population. 'Elepaio 68: 27–30.

Lepson, J. K. 1997. 'Anianiau (Hemignathus parvus). In A. Poole and F. Gill (editors), The birds of North America, No. 312. Birds of North America Inc., Philadelphia.

Lepson, J. K., and L. A. Freed. 1995. Variation in male plumage and behavior of the Hawaii Akepa. Auk 112: 402–414.

———. 1997. 'Ākepa (*Loxops coccineus*). In A. Poole and F. Gill (editors), The birds of North America, No. 294. Birds of North America Inc., Philadelphia.

Lepson, J. K., and S. M. Johnston. 2000. Greater 'Akialoa (*Hemignathus ellisianus*) and Lesser 'Akialoa (*Hemignathus obscurus*). In A. Poole and F. Gill (editors), The birds of North America, No. 512. Birds of North America Inc., Philadelphia.

Lepson, J. K., and H. D. Pratt. 1997. 'Akeke'e (*Loxops caeruleirostris*). In A. Poole and F. Gill (editors), The birds of North America, No. 295. Birds of North America Inc., Philadelphia.

Lepson, J. K., and B. L. Woodworth. 2002. Hawaii Creeper (*Oreomystis mana*). In A. Poole and F. Gill (editors), The birds of North America, No. 680. Birds of North America Inc., Philadelphia.

Leukering, T., M. F. Carter, A. Panjabi, D. Faulkner, and R. Levad. 2000. Monitoring Colorado's birds: The plan for count-based monitoring. Colorado Bird Observatory, Brighton.

Lever, C. 1987. Naturalized birds of the world. Longman, New York.

Levey, D. J., and W. H. Karasov. 1989. Digestive responses of temperate birds switched to fruit or insect diets. Auk 106: 675–686.

Levins, R. 1969. Some demographic and genetic consequences of environmental heterogeneity for biological control. Bulletin of the Entomological Society of America 15: 237–240.

———. 1970. Extinction. Pp. 77–107 in M. Gerstenhaber (editor), Some mathematical questions in biology, Vol. 2. American Mathematical Society, Providence, RI.

Lewin, V. 1971. Exotic game birds of the Puu Waawaa Ranch, Hawaii. Journal of Wildlife Management 35: 141–155.

Lewin, V., and G. Lewin. 1984. The Kalij Pheasant, a newly established game bird on the island of Hawaii. Wilson Bulletin 96: 634–646.

Lewis, E. H. 1937. Birds introduced to Hawaii by the "Hui Manu." Paradise of the Pacific 49: 3.

Lieberman, A., and R. Hayes. 2004. Palila: Honeycreeper of the Hawaiian highlands. Zoonooz 77: 16–19.

Lieberman, A., C. Kuehler, A. Varney, P. Unitt, R. M. Sulpice, J. Azua, and B. Tehevini. 1997. A note on the 1997 survey of the translocated Ultramarine Lory *Vini ultramarina* population on Fatu Hiva, Marquesas Islands, French Polynesia. Bird Conservation International 7: 291–292.

Lieberman, A., J. V. Rodriquez, J. M. Paez, and J. Wiley. 1993. The reintroduction of the Andean condor into Colombia, South America: 1989–1991. Oryx 27: 83–90.

Lieberman, A., W. Toone, and C. Kuehler. 1989. Breeding record for the Zoological Society of San Diego, California, in 1988. Avicultural Magazine 95: 103–110.

———. 1990. Bird species bred at the Zoological Society of San Diego, 1989. Avicultural Magazine 96: 189–200.

Lincoln, R., G. Boxshall, and P. Clark. 1998. A dictionary of ecology, evolution, and systematics. 2nd edition. Cambridge University Press, Cambridge, England.

Lindsay, D. S., M. V. Collins, S. M. Mitchell, R. A. Cole, G. J. Flick. C. N. Wetch, A. Lindquist, and J. P. Dubey. 2003. Sporulation and survival of *Toxoplasma gondii* ooysts in seawater. Journal of Eukaryotic Microbiology 50: 687–688.

Lindsay, D. S., J. P. Dubey, and B. L. Blagburn. 1991. *Toxoplasma gondii* infections in Red-tailed Hawks inoculated orally with tissue cysts. Journal of Parasitology 77: 322–325.

Lindsey, G. D., C. T. Atkinson, P. C. Banko, G. J. Brenner, E. W. Campbell III, R. E. David, D. Foote, C. M. Forbes, M. P Morin, T. K. Pratt, M. H. Reynolds, W. M. Steiner, R. T. Sugihara, and F. R. Warshauer. 1997. Technical options and recommendations for faunal restoration of Kaho'olawe. U.S. Geological Survey, Pacific Island Ecosystems Research Center, Hawai'i Volcanoes National Park, Volcano, HI.

Lindsey, G. D., S. G. Fancy, M. H. Reynolds, T. K. Pratt, K. A. Wilson, P. C. Banko, and J. D. Jacobi. 1995. Population structure and survival of Palila. Condor 97: 528–535.

Lindsey, G. D., and S. M. Mosher. 1994. Tests indicate minimal hazard to 'Io from diphacinone baiting. Hawaii's Forest and Wildlife 9: 2–3.

Lindsey, G. D., S. M. Mosher, S. G. Fancy, and T. D. Smucker. 1999. Population structure and movements of introduced rats in an Hawaiian rainforest. Pacific Conservation Biology 5: 94–102.

Lindsey, G. D., R. D. Nass, and G. A. Hood. 1971. An evaluation of bait stations for controlling rats in sugarcane. Journal of Wildlife Management 35: 440–444.

Lindsey, G. D., T. K. Pratt, M. H. Reynolds, and J. D. Jacobi. 1997. Response of six species of Hawaiian forest birds to a 1991–1992 El Nino drought. Wilson Bulletin 109: 339–343.

Lindsey, G. D., E. A. VanderWerf, H. Baker, and P. E. Baker. 1998. Hawai'i (*Hemignathus virens*), Kaua'i (*Hemignathus kauaiensis*), O'ahu (*Hemignathus chloris*), and Greater 'Amakihi (*Hemignathus sagittirostris*). In A. Poole and F. Gill (editors), The birds of North America, No. 360. Birds of North America Inc., Philadelphia.

Liss, C. A. 1997. The public is attracted by the use of repellents. Pp. 429–433 in J. R. Mason (editor), Proceedings of the Second DWRC Special Symposium (August 8–10, 1995, Denver, Colorado). National Wildlife Research Center, Fort Collins, CO.

Little, E. L. Jr., and R. G. Skolmen. 1989. Common forest trees of Hawaii (native and introduced). Forest Service, United States Department of Agriculture, Washington, DC.

Locey, F. H. 1937. Introduced game birds of Hawaii. Paradise of the Pacific 49: 5–29.

Lockwood, J. A., F. G. Howarth, and M. F. Purcell (editors). 2001. Balancing nature: Assessing the impact of importing non-native biological control agents (an international perspective). Entomological Society of America, Lanham, MD.

Lockwood, J. L., M. P. Moulton, and S. K. Anderson. 1993. Morphological assortment and the assembly of communities of introduced passeriforms on oceanic islands—Tahiti versus Oahu. American Naturalist 141: 398–408.

Lockwood, J. L., M. P. Moulton, and K. L. Balent. 1999. Introduced avifaunas as natural experiments in community assembly. Pp. 108–129 in E. Weiher and P. Keddy (editors), Ecological assembly rules: Perspectives, advance, retreats. Cambridge University Press, New York.

Lockwood, J. P., and P. W. Lipman. 1987. Holocene eruptive history of Mauna Loa Volcano. Pp. 509–535 in R. W. Decker, T. L. Wright, and P. H. Stauffer (editors), Volcanism in Hawaii. U.S. Geological Survey Professional Paper 1350. U.S. Government Printing Office, Washington, DC.

Loh, R. K., and J. T. Tunison. 1999. Vegetation recovery following pig removal in 'Ola'a-Koa rainforest unit, Hawaii Volcanoes National Park. Technical Report 123. Pacific Cooperative Studies Unit, University of Hawai'i, Honolulu.

Loi P., G. Ptak, B. Barboni, J. Fulka Jr., P. Cappai, and M. Clinton. 2001. Genetic rescue of an endangered mammal by cross-species nuclear transfer using post-mortem somatic cells. Nature Biotechnology 19: 962–964.

Lomolino, M. V. 1985. Body size of mammals on islands: The island rule reexamined. American Naturalist 125: 310–316.

Long, J. L. 1981. Introduced birds of the world. David and Charles, London.

Loope, L. L. 1992. An overview of problems with introduced plant species in National Parks and Biosphere Reserves of the United States. Pp. 3–29 in C. P. Stone, C. W. Smith, and J. T. Tunison (editors), Alien plant invasions in native ecosystems of Hawai'i: Management and research. University of Hawaii Cooperative National Park Resources Studies Unit. University of Hawaii Press, Honolulu.

———. 1998. Hawaii and Pacific islands. Pp. 747–774 in M. J. Mac, P. A. Opler, C. E. Puckett Haecker, and P. D. Doran (editors), Status and trends of the nation's biological resources, Vol. 2. U.S. Geological Survey, Reston, VA.

———. 2004. New Zealand's border protection quarantine and surveillance: A potential model for Hawai'i. Ecological Restoration 22: 69–70.

Loope, L. L., and T. W. Giambelluca. 1998. Vulnerability of island tropical montane cloud forests to climate change, with special reference to East Maui, Hawaii. Climatic Change 39: 503–517.

Loope, L. L., F. G. Howarth, F. Kraus, and T. K. Pratt. 2001. Newly emergent and future threats of alien species to Pacific birds and ecosystems. Studies in Avian Biology 22: 291–304.

Loope, L. L, and S. P. Juvik. 1998. Protected areas. Pp. 154–157 in S. P. Juvik and J. O. Juvik (editors), Atlas of Hawai'i. 3rd edition. University of Hawai'i Press, Honolulu.

Loope, L., and A. M. La Rosa. 2008. An analysis of the risk of introduction of additional strains of the rust *Puccinia psidii* Winter ('Ohi'a rust) to Hawai'i. Open-File Report 2008-1008. U.S. Geological Survey, Biological Resources Division, Reston, VA.

Loope, L. L., A. C. Medeiros, W. Minyard, S. Jessel, and W. Evanson. 1992. Strategies to prevent establishment of feral rabbits on Maui, Hawaii. Pacific Science 46: 402–403.

Loope, L. L., and D. Mueller-Dombois. 1989. Characteristics of invaded islands. Pp. 257–280 in J. Drake, F. Decastri, R. Goves, F. Kruger, H. Mooney, M. Rejmanek, and M. Williamson (editors), Biological invasions: A global perspective. John Wiley and Sons, Chichester, England.

Loope, L. L., and P. G. Scowcroft. 1985. Vegetation response within exclosures in Hawaii: A review. Pp. 377–402 in C. P. Stone and J. M Scott (editors), Hawai'i's terrestrial ecosystems: Preservation and management. Cooperative National Park Resources Studies Unit, University of Hawaii, Honolulu.

Loope, L. L., F. Starr, and K. M. Starr. 2004. Management and research for protecting endangered plant species from displacement by invasive plants on Maui, Hawaii. Weed Technology 18: 1472–1474.

Lorence, D. H., and K. R. Wood. 1994. *Kanaloa*, a new genus of Fabaceae (Mimosoideae) from Hawaii. Novon 4: 137–145.

Losos, J. B. 1998. Ecological and evolutionary determinants of the species-area relationship in Caribbean anoline lizards. Pp. 210–224 in P. R. Grant (editor), Evolution on islands. Oxford University Press, Oxford.

Lovegrove, T. G. 1996a. A comparison of the effects of predation by Norway (*Rattus norvegius*) and Polynesian rats (*R. exulans*) on the Saddleback (*Philesturnus carunculatus*). Notornis 93: 91–112.

———. 1996b. Island releases of saddlebacks *Philesturnus caruculatus* in New Zealand. Biological Conservation 77: 151–157.

Lovette, I. J, E. Bermingham, and R. E. Ricklefs. 2002. Clade-specific morphological diversification and adaptive radiation in Hawaiian songbirds. Proceedings of the Royal Society of London, Series B 269: 32–42.

Lowther, P. E., C. C. Rimmer, B. Kessel, S. L. Johnson, and W. G. Ellison. 2001. Gray-cheeked Thrush (*Catharus minimus*). In A. Poole and F. Gill (editors), The birds of North America, No. 591. Birds of North America Inc., Philadelphia.

Lund, M. 1988. Anticoagulant rodenticides. Pp. 341–351 in I. Prakash (editor), Rodent pest management. CRC Press, Boca Raton, FL.

Lyon, H. L. 1909. The forest disease on Maui. Hawaiian Planters' Record 1: 151–159.

———. 1918. The forests of Hawaii. Hawaiian Planters' Record 20: 276–280.

MacArthur, R. H., J. M. Diamond, and J. R. Karr. 1972. Density compensation in island faunas. Ecology 53: 330–342.

MacArthur, R. H., and E. O. Wilson. 1967. The theory of island biogeography. Princeton University Press, Princeton, NJ.

MacCaughey, V. 1918. History of botanical exploration in Hawaii. Hawaiian Forester and Agriculturalist 15: 388–396 and 417–429.

MacDougall-Shackleton, S. A., R. E. Johnson, and T. P. Hahn. 2000. Gray-crowned Rosy-Finch (*Leucosticte tephrocotis*). In A. Poole and F. Gill (editors), The birds of North America, No. 559. Birds of North America Inc., Philadelphia.

Mack, M., and C. D'Antonio. 2003. Exotic grasses alter controls over soil nitrogen dynamics in a Hawaiian woodland. Ecological Applications 13: 154–166.

Mack, R. N., D. Simberloff, W. M. Lonsdale, H. Evans, M. Clout, and F. A. Bazzaz. 2000. Biotic invasions: Causes, epidemiology, global consequences, and control. Ecological Applications 10: 689–710.

MacKenzie, D. I., J. D. Nichols, J. E. Hines, M. G. Knutson, and A. B. Franklin. 2003. Estimating site occupancy, colonization, and local extinction when a species is detected imperfectly. Ecology 84: 2200–2207.

MacKenzie, D. I., J. D. Nichols, G. B. Lachman, S. Droege, J. A. Royle, and C. A. Langtimm. 2002. Estimating site occupancy rates when detection probabilities are less than one. Ecology 83: 2248–2255.

MacMillen, R. E. 1981. Nonconformance of standard metabolic rate with body mass in Hawaiian honeycreepers. Oecologia 49: 340–343.

Madge, S., and H. Burn. 1994. Crows and jays: A guide to the crows, jays, and magpies of the world. Houghton Mifflin, Boston and New York.

Magin, C. D., T. H. Johnson, B. Groombridge, M. Jenkins, and H. Smith. 1994. Species extinctions, endangerment and captive breeding. Pp. 3–30 in P. J. S. Olney, G. M. Mace, and A. T. C. Feistner (editors), Creative conservation: Interactive management of wild and captive animals. Chapman and Hall, London.

Malakoff, D. 2002. Bird advocates fear that West Nile virus could silence the spring. Science 297: 1989–1990.

Malcolm, T. R., K. J. Swinnerton, J. J. Groombridge, B. D. Sparklin, C. N. Brosius, J. P. Vetter, and J. T. Foster. 2008. Ground-based rodent control in a remote Hawaiian rainforest on Maui. Pacific Conservation Biology 14: 206–214.

Male, T., and W. Loeffler. 1997. Patterns of distribution and seed predation in a population of *Pritchardia hillebrandii*. Newsletter of the Hawaiian Botanical Society 36: 1–11.

Male, T. D., and M. J. Bean. 2005. Measuring and understanding progress in U.S. endangered species conservation. Ecological Letters 8: 986–992.

Male, T. D., and T. J. Snetsinger. 1998. Has the Red-billed Leiothrix disappeared from Kaua'i? 'Elepaio 58: 39–43.

Malo, D. 1951. Hawai'i antiquities. Bishop Museum Press, Honolulu.

Maloney, R., and D. Murray. 2000. Summary of Kaki (black stilt) releases in New Zealand. Reintroduction News 19: 34–36.

Mandon-Dalger, I., P. Clergeau, J. Tassin, J. N. Riviere, and S. Gatti. 2004. Relationships between alien plants and an alien bird species on Reunion Island. Journal of Tropical Ecology 20: 635–642.

Manea, S. J., D. M. Sasaki, J. K. Ikeda, and P. P. Bruno. 2001. Clinical and epidemiological observations regarding the 1998 Kaua'i murine typhus outbreak. Hawaii Medical Journal 60: 7–11.

Manne, L. L., T. M. Brooks, and S. L. Pimm. 1999. Relative risk of extinction of passerine birds on continents and islands. Nature 399: 258–262.

Marchant, J. H. 1994. The new Breeding Bird Survey. British Birds 87: 26–28.

Marchant, J. H., A. M. Wilson, D. E. Chamberlain, R. D. Gregory, and S. R. Baillie. 1997. Opportunistic bird species: Enhancements for the monitoring of populations. Research report 176. British Trust for Ornithology, Thetford, England.

Margalit, J., and D. Dean. 1985. The story of *Bacillus thuringiensis* var. *israelensis* (B.t.i.). Journal of the American Mosquito Control Association 1: 1–7.

Markin, G. P., P. Conant, E. Killgore, and E. Yoshioka. 2002. Biological control of gorse in Hawai'i: A program review. Pp. 53–61 in Workshop on biological control of invasive plants in native Hawaiian ecosystems. Tech. Rep. 129. Cooperative National Park Resources Studies Unit, Botany Department, University of Hawaii at Mānoa, Honolulu.

Marques, F. F. C., and S. T. Buckland. 2003. Incorporating covariates into standard line transect analyses. Biometrics 59: 924–935.

Marquis, R. J., and C. J. Whelan. 1994. Insectivorous birds increase growth of white oak through consumption of leaf-chewing insects. Ecology 75: 2007–2014.

Marra, P. P., S. Griffing, C. Caffrey, A. M. Kilpatrick, R. McLean, C. Brand, E. Saito, A. P. Dupuis, L. Kramer, and R. Novak. 2004. West Nile virus and wildlife. Bioscience 54: 393–402.

Marsh, R. E. 1986. Role of anticoagulant rodenticides in agriculture. Paper presented at the Pacific Northwest Forest and Orchard Animal Damage Control Conference (October 27–30), Washington State University, Wenatchee.

Marten, G. G. 1984. Impact of the copepod *Mesocyclops leuckarti pilosa* and the green alga *Kirchneriella irregularis* upon larval *Aedes albopictus* (Diptera: Culicidae). Bulletin of the Society of Vector Ecologists 9: 1–5.

Marti, C. D. 1992. Barn Owl (*Tyto alba*). In A. Poole and F. Gill (editors), The birds of North America, No. 1. Birds of North America Inc., Philadelphia.

Martin, K., and S. Ogle. 2000. The use of alpine habitats by migratory birds in B.C. parks: 1998 summary. Unpublished report, Department of Forest Sciences, University of British Columbia and Canadian Wildlife Service, Pacific and Yukon region. www.forestry.ubc.ca/alpine/docs/alpmig-2.pdf.

Martin, T. E. 1988. On the advantage of being different: Nest predation and the coexistence of bird species. Proceedings of the National Academy of Sciences (USA) 85: 2196–2199.

———. 1992a. Interaction of nest predation and food limitation in reproductive strategies. Current Ornithology 9: 163–197.

———. 1992b. Breeding productivity considerations: What are the appropriate habitat features for management? Pp. 455–473 in J. M. Hagen III and D. W. Johnston (editors), Ecology and conservation of Neotropical migrant landbirds. Smithsonian Institution Press, Washington, DC.

———. 1993. Nest predation and nest sites: New perspectives on old patterns. BioScience 43: 523–532.

———. 1995. Avian life history evolution in relation to nest sites, nest predation, and food. Ecological Monographs 65: 101–127.

———. 2004. Avian life-history evolution has an eminent past: Does it have a bright future? Auk 121: 289–301.

Martin, T. E., and P. Li. 1992. Life history traits of open- vs. cavity-nesting birds. Ecology 73: 579–592.

Martin, T. E., P. R. Martin, C. R. Olson, B. J. Heidinger, and J. J. Fontaine. 2000. Parental care and clutch sizes in North and South American birds. Science 287: 1482–1485.

Martin, T. E., J. Scott, and C. Menge. 2000. Nest predation increases with parental activity: Separating nest site and parental activity effects. Proceedings of the Royal Society of London, Series B 267: 2287–2293.

Martins, T. L. F., M. D. L. Brooke, G. M. Hilton, S. Farnsworth, J. Gould, and D. J. Pain. 2006. Costing eradications of alien mammals from islands. Animal Conservation 9: 439–444.

Marzluff, J. M., and T. Angell. 2005. In the company of crows and ravens. Yale University Press, New Haven, CT, and London.

Marzluff, J. M., L. Valutis, and K. D. Whitmore. 1995. Captive propagation and reintroduction of social birds. 1995 Annual Report. Sustainable Ecosystems Institute, Meridian, ID.

Massey, G., L. Valutis, and J. Marzluff. 1997. Secondary poisoning effects of diphacinone on Hawaiian crows: A study using American crows as surrogates. Unpublished report. U.S. Fish and Wildlife Service, Honolulu.

Massey, J. G., T. K. Graczyk, and M. R. Cranfield. 1996. Characteristics of naturally acquired *Plasmodium relictum capistranoae* infections in naïve Hawaiian Crows (*Corvus hawaiiensis*) in Hawaii. Journal of Parasitology 82: 182–185.

Matisoo-Smith, E., and J. H. Robins. 2004. Origins and dispersals of Pacific peoples: Evidence from mtDNA phylogenies of the Pacific rat. Proceedings of the National Academy of Sciences (USA) 24: 9167–9172.

May, R. M., and R. M. Anderson. 1978. Regulation and stability of host–parasite population interactions, 2: Destabilising processes. Journal of Animal Ecology 47: 249–267.

———. 1983. Epidemiology and genetics in the coevolution of parasites and hosts. Proceedings of the Royal Society of London, Series B 219: 281–313.

Mayfield, H. F. 1961. Nesting success calculated from exposure. Wilson Bulletin 73: 255–261.

———. 1975. Suggestions for calculating nest success. Wilson Bulletin 87: 456–466.

Mayr, E. 1942. Systematics and the origin of species. Columbia University Press, New York.

———. 1943. The zoogeographic position of the Hawaiian Islands. Condor 45: 45–48.

Mayr, E., and J. M. Diamond. 2001. The birds of Northern Melanesia. Oxford University Press, New York.

McCallum, H. 2000. Population parameters: Estimation for ecological models. Blackwell Science, Oxford.

McClung, A. 2005. A population viability analysis of the Laysan Finch (*Telespiza cantans*). Ph.D. dissertation. University of Hawai'i, Honolulu.

McConkey, K. R., D. R. Drake, H. J. Meehan, and N. Parsons. 2003. Husking stations provide evidence of seed predation by introduced rodents in Tongan rain forests. Biological Conservation 109: 221–225.

McCutchan, T. F., K. C. Grim, J. Li, W. Weiss, D. Rathore, M. Sullivan, T. K. Graczyk, S. Kumar, and M. R. Cranfield. 2004. Measuring the effects of an ever-changing environment on malaria control. Infection and Immunology 72: 2248–2253.

McDonald, T. L. 2003. Review of environmental monitoring methods: Survey designs. Environmental Monitoring and Assessment 85: 277–292.

McDowell, S. B. 1984. Results of the Archbold Expeditions 112: The snakes of the Huon Peninsula, Papua New Guinea. American Museum Novitates 2775: 1–28.

McEldowney, H. 1983. A description of major vegetation patterns in the Waimea–Kawaihae area during the early historic period. Report 16. Pp. 407–448 in J. T. Clark and P. V. Kirch (editors), Archaeological investigations of the Mudlane–Waimea–Kawaihae road corridor, island of Hawai‘i: An interdisciplinary study of an environmental transect. Report 83-1. Anthropology Department, B. P. Bishop Museum, Honolulu.

McGowan, K. J. 2001. Fish Crow (Corvus ossifragus). In A. Poole and F. Gill (editors), The birds of North America, No. 589. Birds of North America Inc., Philadelphia.

McGregor, R. C. 1973. The emigrant pests: A report to Dr. Francis Mulhern, administrator, Animal and Plant Health Inspection Service, Berkeley, California. Unpublished report on file at Hawaii Department of Agriculture, Honolulu, HI. www.hear.org/articles/mcgregor1973 (accessed March 23, 2008).

McKillop, I. G., and R. M. Sibley. 1988. Animal behavior at electric fences and implications for management. Mammal Review 18: 91–102.

McNab, B. K. 1994a. Energy conservation and the evolution of flightlessness in birds. American Naturalist 144: 628–642.

———. 1994b. Resource use and the survival of land and freshwater vertebrates on oceanic islands. American Naturalist 144: 643–660.

———. 2002. Minimizing energy expenditure facilitates vertebrate persistence on oceanic islands. Ecology Letters 5: 693–704.

Medeiros, A. C. 2004. Phenology, reproductive potential, seed dispersal and predation, and seedling establishment of three invasive plant species in a Hawaiian rainforest. Ph.D. dissertation. University of Hawaii at Mānoa, Honolulu.

Medeiros, A. C., L. L. Loope, P. Conant, and S. McElvaney. 1997. Status, ecology, and management of the invasive plant, Miconia calvescens DC (Melastomataceae) in the Hawaiian Islands. Bishop Museum Occasional Papers 48: 23–35.

Medeiros, A. C., L. L. Loope, and R. Hobdy. 1995. Conservation of cloud forests in Maui County (Maui, Molokai, and Lanai) Hawaiian Islands. Pp. 223–233 in L. S. Hamilton, J. O. Juvik, and F. N. Scatena (editors), Tropical montane cloud forests. Springer, New York.

Medeiros, A. C., L. L. Loope, and R. A. Holt. 1986. Status of native flowering plant species on the south slope of Haleakala, East Maui, Hawaii. Technical Report 59. Cooperative National Park Studies Unit, University of Hawaii at Mānoa, Honolulu.

Medway, D. G. 1981. The contribution of Cook's third voyage to the ornithology of the Hawaiian Islands. Pacific Science 35: 105–175.

Meehan, H. J., K. R. McConkey, and D. R. Drake. 2002. Potential disruptions to seed dispersal mutualisms in Tonga, Western Polynesia. Journal of Biogeography 29: 695–712.

Mehrhoff, L. A. 1998. Endangered and threatened species. Pp. 150–153 in S. P. Juvik and J. O. Juvik (editors), Atlas of Hawai‘i. 3rd edition. University of Hawai‘i Press, Honolulu.

Meine, C., and G. W. Archibald. 1996. Ecology, status, and conservation. Pp. 263–292 in D. H. Ellis, G. F. Gee, and C. M. Mirande (editors), Cranes: Their biology, husbandry and conservation. Hancock House, Blaine, WA.

Meisch, M. V. 1985. Gambusia affinis affinis. Pp. 3–17 in H. C. Chapman (editor), Biological control of mosquitoes. Bulletin 6. American Mosquito Control Association, Fresno, CA.

Menzies, A. 1920. Hawai‘i Nei 128 years ago. The New Freedom, Honolulu.

Merton, D., C. Reed, and D. Crouchley. 1999. Recovery strategies and techniques for three free-living, critically-endangered New Zealand birds: Kakapo Strigops habroptilus, Black Stilt Himantopus novaezelandiae and Takahe Porphyrio mantelli. Pp. 151–162 in T. L. Roth, W. F. Swanson, and L. K. Blattman (editors), Proceedings 7th world conference on breeding endangered species. Cincinnati Zoo and Botanical Garden, Cincinnati, OH.

Merton, D. V. 1977. Controlling introduced predators and competitors on islands. Pp. 121–128 in S. A. Temple (editor), Endangered birds: Management techniques for preserving threatened species. University of Wisconsin Press, Madison.

Messing, R. H., and M. G. Wright. 2006. Biological control of invasive species: Solution or pollution? Frontiers of Ecology and the Environment 4: 132–140.

Metzger, B. 1929. Tales told in Hawai‘i. Frederick A. Stokes, New York.

Meyer, J.-Y. 1996. Status of Miconia calvescens (Melastomataceae), a dominant invasive tree in the Society Islands (French Polynesia). Pacific Science 50: 66–76.

———. 1998. Observations on the reproductive biology of Miconia calvescens DC (Melastomataceae), an alien invasive tree on the island of Tahiti (South Pacific Ocean). Biotropica 30: 609–624.

Meyer, J.-Y., and J. Florence. 1996. Tahiti's native flora endangered by the invasion of *Miconia calvescens* DC (Melastomataceae). Journal of Biogeography 23: 775–781.

Meyer, J.-Y., and P.-P. Malet. 1997. Study and management of the alien invasive tree *Miconia calvescens* DC (Melastomataceae) in the islands of Raiatea and Tahaa (Society Islands, French Polynesia): 1992–1996. Technical Report 111. Cooperative National Park Resources Studies Unit, University of Hawaii at Mānoa, Honolulu.

Meyer, J.-Y., and C. W. Smith (editors). 1998. Proceedings of the First Regional Conference on Miconia Control, August 26–29, 1997, Papeete, Tahiti, French Polynesia. Gouvernement de Polynésie française / University of Hawai'i at Mānoa / Centre ORSTOM de Tahiti.

Mian, L. S., M. S. Mulla, H. Axelrod, J. D. Chaney, and M. S. Dhillon. 1990. Studies on the bioecological aspects of adult mosquitoes in the Prado Basin of southern California. Journal of the American Mosquito Control Association 6: 64–71.

Middleton, A. L. A. 1993. American Goldfinch (*Carduelis tristis*). In A. Poole and F. Gill (editors), The birds of North America, No. 80. Birds of North America Inc., Philadelphia.

Middleton, C. R., V. E. Ansdell, and D. M. Sasaki. 2001. Of mice and mongooses . . . A history of leptospirosis research in Hawaii. Hawaii Medical Journal 60: 179–186.

Migaki, G. T., R. Sawa, and J. P. Dubey. 1990. Fatal disseminated toxoplasmosis in a spinner dolphin (*Stenella longirostris*). Veterinary Pathology 27: 463–464.

Mikkola, H. 1983. Owls of Europe. Buteo Books, Vermillion, SD.

Miller, J. K., J. M. Scott, C. R. Miller, and L. P. Waits. 2002. The Endangered Species Act: Dollars and sense? BioScience 52: 163–168.

Miller, K. E., and K. D. Meyer. 2002. Short-tailed Hawk (*Buteo brachurus*). In A. Poole and F. Gill (editors), The birds of North America, No. 674. Birds of North America Inc., Philadelphia.

Miller, L. J. 1998. Behavioral ecology of juvenile Palila (*Loxioides bailleui*): Foraging development, social dynamics, and helping behavior. M.S. thesis. University of Maryland, College Park.

Miller, M. J., E. Bermingham, and R. E. Ricklefs. 2007. Historical biogeography of the New World solitaires (*Myadestes* sp.). Auk 124: 868–885.

Miller, S. E., and L. G. Eldredge. 1996. Numbers of Hawaiian species: Supplement 1. Bishop Museum Occasional Papers 45: 8–17.

Mineau, P., M. R. Fletcher, L. C. Glaser, N. J. Thomas, C. Brassard, L. K. Wilson, J. E. Elliot, L. A. Lyon, C. H. Henny, T. Bollinger, and S. L. Porter. 1999. Poisoning of raptors with organophosphorus and carbamate pesticides with emphasis on Canada, U.S. and U.K. Journal of Raptor Research 33: 1–37.

Minerbi, L. 1999. Indigenous management models and protection of the ahupua'a. Social Process in Hawai'i 39: 208–225.

Mitchell, C., C. Ogura, D. W. Meadows, A. Kane, L. Strommer, S. Fretz, D. Leonard, and A. McClung. 2005. Hawaii's comprehensive wildlife conservation strategy. Department of Land and Natural Resources, Honolulu.

Mitsch, W. J., and J. G. Gosselink. 1993. Wetlands. 2nd edition. Van Nostrand Reinhold, New York.

Møller, A. P. 1994. Sexual selection and the Barn Swallow. Oxford University Press, Oxford.

Mooney, H. A., and J. A. Drake (editors). 1986. Ecology of biological invasions of North America and Hawaii. Springer, New York.

Moorhouse, R., T. Greene, P. Dilks, R. Powlesland, L. Moran, G. Taylor, A. Jones, J. Knegtmans, D. Wills, M. Pryde, I. Fraser, A. August, and C. August. 2003. Control of introduced mammalian predators improves kaka *Nestor meridionalis* breeding success: Reversing the decline of a threatened New Zealand parrot. Biological Conservation 110: 33–44.

Moors, P. J. 1985. Conservation of island birds: Case studies for the management of threatened island species. Technical Publication 3. International Council of Bird Preservation, Cambridge, England.

Moors, P. J., I. A. E. Atkinson, and G. H. Sherley. 1992. Reducing the rat threat to island birds. Bird Conservation International. 2: 93–114.

Moreau, R. E. 1944. Clutch size: A comparative study, with reference to African birds. Ibis 86: 286–347.

Morel, C. M., Y. T. Toure, B. Dobrokhotov, and A. M. J. Oduola. 2002. The mosquito genome—a breakthrough for public health. Science 298: 79.

Morgan, G. S., and C. A. Woods. 1986. Extinctions and the zoogeography of West Indian land mammals. Biological Journal of the Linnean Society 28: 167–203.

Morgan, J. 1983. Hawai'i: A geography. Westview, Boulder, CO.

Morin, M., and S. Conant. 1998. Laysan Island ecosystem restoration plan. Unpublished U.S. Fish and Wildlife Service report on file at Hamilton Library, University of Hawaii at Mānoa, Honolulu.

Morin, M. P. 1992a. The breeding biology of an endangered Hawaiian honeycreeper, the Laysan Finch. Condor 94: 646–667.

———. 1992b. Laysan Finch nest characteristics, nest spacing and reproductive success in two vegetation types. Condor 94: 344–357.

Morin, M. P., and S. Conant. 1994. Variables influencing population estimates of an endangered passerine. Biological Conservation 67: 73–84.

————. 2002. Laysan Finch (*Telespiza cantans*) and Nihoa Finch (*Telespiza ultima*). In A. Poole and F. Gill (editors), The birds of North America, No. 639. Birds of North America Inc., Philadelphia.

————. 2007. Summary of scoping, evaluation, and recommendations for Northwestern Hawaiian Islands passerines' translocation sites. Unpublished report to the U.S. Fish and Wildlife Service, Honolulu.

Morin, M. P., S. Conant, and P. Conant. 1997. Laysan and Nihoa Millerbird (*Acrocephalus familiaris*). In A. Poole and F. Gill (editors), The birds of North America, No. 302. Academy of Natural Sciences, Philadelphia, and American Ornithologists' Union, Washington, DC.

Moritz, C. 1994. Defining "evolutionarily significant units" for conservation. Trends in Ecology and Evolution 9: 373–375.

Morrell, T. E., B. Ponwith, P. Craig, T. Ohashi, J. Murphy, and E. Flint. 1991. Eradication of Polynesian rats (*Rattus exulans*) from Rose Atoll National Wildlife Refuge, American Samoa. Biological Report Series 20. Department of Marine and Wildlife Resources, Pago Pago, American Samoa.

Morris, P. 2002. Minutes from the meeting of the Hawaii Veterinary Consortium (C. Atkinson, A. Lieberman, G. Massey, P. Morris, J. Nelson, B. Rideout, Thierry Work). Hawaii Veterinary Consortium. Captive Propagation Working Group, U.S. Fish and Wildlife Service, Honolulu, October 22.

Moskoff, W. 1995. Veery (*Catharus fuscescens*). In A. Poole and F. Gill (editors), The birds of North America, No. 142. Birds of North America Inc., Philadelphia.

Mostello, C. S. 1996. Diets of the pueo, the barn owl, the cat, and the mongoose in Hawai'i: Evidence for competition. M.S. thesis. University of Hawai'i at Mānoa, Honolulu.

Moulton, M. P. 1985. Morphological similarity and coexistence of congeners—an experimental test with introduced Hawaiian birds. Oikos 44: 301–305.

————. 1993. The all-or-none pattern in introduced Hawaiian passeriforms—the role of competition sustained. American Naturalist 141: 105–119.

Moulton, M. P., and J. L. Lockwood. 1992. Morphological dispersion of introduced Hawaiian finches—evidence for competition and a Narcissus effect. Evolutionary Ecology 6: 45–55.

Moulton, M. P., K. E. Miller, and E. A. Tillman. 2001. Patterns of success among introduced birds in the Hawaiian Islands. Studies in Avian Biology 22: 31–46.

Moulton, M. P., and S. L. Pimm. 1983. The introduced Hawaiian avifauna: Biogeographic evidence for competition. American Naturalist 121: 669–690.

Moulton, M. P., and M. E. T. Scioli. 1986. Range sizes and abundances of passerines introduced to Oahu, Hawaii. Journal of Biogeography 13: 339–344.

Mount, G. A. 1998. A critical review of ultralow-volume aerosols of insecticide applied with vehicle-mounted generators for adult mosquito control. Journal of the American Mosquito Control Association 14: 305–334.

Mountainspring, S. 1987. Ecology, behavior, and conservation of the Maui Parrotbill. Condor 89: 24–39.

Mountainspring, S., T. L. C. Casey, C. B. Kepler, and J. M. Scott. 1990. Ecology, behavior, and conservation of the Poo-uli (*Melamprosops phaeosoma*). Wilson Bulletin 102: 109–122.

Mountainspring, S., and J. M. Scott. 1985. Interspecific competition among Hawaiian forest birds. Ecological Monographs 55: 219–239.

Mueller-Dombois, D. 1980. The 'ohi'a dieback phenomenon in the Hawaiian rain forest. Pp. 153–161 in J. J. Cairns Jr. (editor), The recovery process in damaged ecosystems. Ann Arbor Science Publishers, Ann Arbor, MI.

————. 1981. Fire in tropical ecosystems. Pp. 137–176 in H. A. Mooney, T. M. Bonnicksen, N. L. Christensen, J. E. Lotan, and W. A. Reiners (editors), Fire regimes and ecosystem properties. Proceedings Conference December 11–15, 1978, Honolulu. Forest Service Technical Report WO-26. U.S. Department of Agriculture, Washington, DC.

————. 1983. Canopy dieback and successional processes in Pacific forests. Pacific Science 37: 317–325.

————. 1985. Ohia dieback in Hawaii: 1984 synthesis and evaluation. Pacific Science 39: 150–170.

————. 1986. Perspectives for an etiology of stand-level dieback. Annual Review of Ecology and Systematics 17: 221–243.

————. 2005. A silvicultural approach to restoration of native Hawaiian rainforests. Lyonia 8: 59–63.

Mueller-Dombois, D., R. G. Cooray, J. E. Maka, G. Spatz, W. C. Gagné, F. G. Howarth, J. L. Gressitt, G. A. Samuelson, S. Conant, and P. Q. Tomich. 1981. Structural variation of organism groups studied in the Kilauea Forest. Pp. 231–317 in D. Mueller-Dombois, K. W. Bridges, and H. L. Carson (editors), Island ecosystems: Biological organization in selected Hawaiian communities. US/IBP Synthesis Series 15. Hutchinson Ross Publishing, Stroudsburg, PA, and Woods Hole, MA.

Mueller-Dombois, D., and F. R. Fosberg. 1998. Vegetation of the tropical Pacific Islands. Ecological Studies 132. Springer, New York.

Mueller-Dombois, D., and G. Spatz. 1975. The influence of feral goats on the lowland vegetation in Hawaii Volcanoes National Park. Phytocoenologia 3: 1–29.

Mueller-Dombois, D., and N. Wirawan. 2005. The Kahana Valley ahupuaʻa, a PABITRA study site on Oʻahu, Hawaiian Islands. Pacific Science 59: 293–314.

Mulla, M. S. 1994. Mosquito control then, now and in the future. Journal of the American Mosquito Control Association 10: 574–584.

Munkwitz, N. M., J. M. Turner, E. L. Kershner, S. M. Farabaugh, and S. R. Heath. 2005. Predicting release success of captive-reared loggerhead shrikes (*Lanius ludovicianus*) using pre-release behavior. Zoo Biology 24: 447–458.

Munro, G. C. 1895. Hawaiian birds. Paradise of the Pacific 8: 49–50.

———. 1944. Birds of Hawaiʻi. Tongg Publishing, Honolulu.

———. 1946. The Hawaiian bird survey of 1935–1937: Bird protection. ʻElepaio 7: 22–24.

———. 1960. Birds of Hawaii. Revised edition. Charles E. Tuttle, Rutland, VT, and Tokyo.

Murakami, G. M. 1983. Analysis of charcoal from archaeological contexts. Report 20. Pp. 514–526 in J. T. Clark and P. V. Kirch (editors), Archaeological investigations of the Mudlane–Waimea–Kawaihae road corridor, island of Hawaiʻi; an interdisciplinary study of an environmental transect. Report 83-1. Anthropology Department, B. P. Bishop Museum, Honolulu.

Murphy, E., and P. Bradfield. 1992. Change in diet of stoats following poisoning of rats in a New Zealand forest. New Zealand Journal of Ecology 16: 137–140.

Murphy, E. C., B. K. Clapperton, P. M. F. Bradfield, and H. J. Speed. 1998. Effects of rat-poisoning operations on abundance and diet of mustelids in New Zealand podocarp forests. New Zealand Journal of Zoology 25: 315–328.

Murphy, J. G. 1997a. Rat eradication on Eastern and Spit Island, Midway Atoll National Wildlife Refuge. Unpublished report. U.S. Department of Agriculture, Animal and Plant Health Inspection Service, Animal Damage Control, Honolulu.

———. 1997b. Rat eradication on Sand Island, Midway Atoll National Wildlife Refuge. Unpublished report. U.S. Department of Agriculture, Animal and Plant Health Inspection Service, Animal Damage Control, Honolulu.

Murphy, J. G., and T. J. Ohashi. 1991. Report of rat eradication operations conducted under specific emergency exemption to use Talon-G containing brodifacoum in a field situation on Rose Atoll National Wildlife Refuge, American Samoa. U.S. Department of Agriculture, Animal and Plant Health Inspection Service, Animal Damage Control, Honolulu.

Murray, B. G. Jr. 1985. Evolution of clutch size in tropical species of birds. Ornithological Monographs 36: 505–519.

———. 2001. The evolution of passerine life histories on oceanic islands, and its implications for the dynamics of population decline and recovery. Studies in Avian Biology 22: 281–290.

Murray, J. D., E. A. Stanley, and D. L. Brown. 1986. On the spatial spread of rabies among foxes. Proceedings of the Royal Society of London, Series B 229: 111–150.

Nagata, K. M. 1985. Early plant introductions in Hawaiʻi. Hawaiian Journal of History 19: 35–61.

Nagendran, M., R. P. Urbanek, and D. H. Ellis. 1996. Reintroduction techniques. Pp. 231–240 in D. H. Ellis, G. F. Gee, and C. M. Mirande (editors), Cranes: Their biology, husbandry and conservation. Hancock House, Blaine, WA.

Nagy, K. A. 2001. Food requirements of wild animals: Predictive equations for free-living mammals, reptiles, and birds. Nutrition Abstracts and Reviews, Series B: Livestock Feeds and Feeding 71: 1–12.

Nakagawa, P. Y. 1963. Status of *Toxorhynchites* in Hawaiʻi. Proceedings of the Hawaiian Entomological Society 18: 291–293.

———. 1964. Mosquito control in Hawaii. Pest Control 32: 24, 27, 28, 30.

Nakagawa, P. Y., and J. Ikeda. 1969. Biological control of mosquitoes with larvivorous fishes in Hawaii. WHO/VBC/69.173. World Health Organization, Geneva.

Nakamura, J. J. M. 1999. The archaeology of human foraging and bird resources on the island of Hawaiʻi: The evolutionary ecology of avian predation, resource intensification. Ph.D. dissertation. University of Hawaiʻi at Mānoa, Honolulu.

Nakuina, E. M. 1907. Ahuula: A legend of Kanikaniaula and the first feather cloak. Pp. 147–155 in T. G. Thrum (editor), Hawaiian folk tales. A. C. McClurg, Chicago.

Nalimu, H. B. No date. Bird trapping. Hawaiian Ethnological Notes, Vol. 1: 824–825. T. Kelsey Collection. Courtesy of the Bishop Museum Archives, Honolulu.

Nam, V. S., N. T. Yen, T. V. Phong, T. U. Ninh, L. Q. Mai, L. V. Lo, L. T. Nghia, A. Bektas, A. Briscombe, J. G. Aaskov, P. A. Ryan, and B. H. Kay. 2005. Elimination of dengue by community programs using *Mesocyclops* (Copepoda) against *Aedes aegypti* in central Vietnam. American Journal of Tropical Medicine and Hygiene 72: 67–73.

Nass, R. D, G. A. Hood, and G. D. Lindsey. 1971. Influence of gulch-baiting on rats in adjacent sugarcane fields. Journal of Wildlife Management 35: 357–360.

National Audubon Society. 2007. The 107th Christmas Bird Count. American Birds 61: 18–25.

National Oceanic and Atmospheric Administration. 2004. Coastal change analysis program: Hawaiʻi land cover, Main Islands. NOAA Coastal Services Center, Charleston, SC.

National Oceanic and Atmospheric Administration, Coastal Change Analysis Program. 1995. Guidance for Regional Implementation. NOAA Technical Report NMFS123. Department of Commerce, Washington, DC.

National Oceanic and Atmospheric Administration, Coastal Change Analysis Program. 2000. www.csc.noaa.gov/crs/lca/hawaii.html.

National Plant Board. 1999. Safeguarding American plant resources: A stakeholder review of the APHIS–PPQ safeguarding system. Report prepared for the U.S. Department of Agriculture, Washington, DC.

National Research Council. 2002. Predicting invasions of nonindigenous plants and plant pests in the United States. National Academy Press, Washington, DC.

Natividad-Hodges, C. S., and R. J. Nagata, Sr. 2001. Effects of predator control on the survival and breeding success of the endangered Hawaiian Dark-rumped Petrel. Studies in Avian Biology 22: 308–318.

Nayar, J. K., and A. Ali. 2003. A review of monomolecular surface films as larvicides and pupicides of mosquitoes. Journal of Vector Ecology 28: 190–199.

Neelin, J. D., M. Münnich, H. Su, J. E. Meyerson, and C. E. Holloway. 2006. Tropical drying trends in global warming models and observations. Proceedings of the National Academy of Sciences (USA) 103: 6110–6115.

Neihardt, J. G. 1961. Black Elk speaks. University of Nebraska Press, Lincoln.

Nelson, J. T., and S. G. Fancy. 1999. A test of the variable circular-plot method where exact density of a bird population was known. Pacific Conservation Biology 5: 139–143.

Nelson, J. T., and A. Vitts. 1998. First reported sighting of Japanese Bush-warbler (Cettia diphone) on the island of Hawaii. 'Elepaio 58: 1.

Nelson, J. T., B. L. Woodworth, S. G. Fancy, G. D. Lindsey, and E. J. Tweed. 2002. Effectiveness of rodent control and monitoring techniques for a montane rainforest. Wildlife Society Bulletin 30: 82–92.

Nelson, R. E. 1965. A record of forest plantings in Hawaii. Forest Service Resource Bulletin PSW-1: 1–18. U.S. Department of Agriculture, Honolulu.

Newcomb, W. 1852. Report of the committee on worms and other injurious vermin. Transactions of the Royal Hawaiian Agricultural Society 1: 97.

Newman, T. S. 1972. Man in the prehistoric Hawaiian ecosystem. Pp. 559–603 in E. A. Kay (editor), A natural history of the Hawaiian Islands: Selected readings. University Press of Hawaii, Honolulu.

Newmark, W. D. 1987. A land-bridge island perspective on mammalian extinctions in western North American parks. Nature 325: 430–432.

Newton, I. 1972. Finches. William Collins Sons, London.

———. 1979. Population ecology of raptors. Buteo Books, Vermillion, SD.

———. 1998. Population limitation in birds. Academic Press, San Diego, CA.

———. 1999. Lifetime reproduction in birds. Academic Press, London.

———. 2003. The speciation and biogeography of birds. Academic Press, San Diego, CA.

Niare, O., K. Markianos, J. Volz, F. Odoul, A. Toure, M. Bagayoko, D. Sanare, S. F. Traore, R. Wang, C. Blass, G. Dolo, M. Bouare, F. C. Kafatos, and K. D. Vernick. 2002. Genetic loci affecting resistance to human malaria parasites in a West African mosquito vector population. Science 298: 213–216.

Nichol, S. T., C. F. Spiroupoulou, S. Morzunov, P. W. Rollin, T. G. Ksiazek, H. Feldmann, A. Sanchez, J. Childs, S. Zaki, and C. J. Peters. 1993. Genetic identification of a hantavirus associated with an outbreak of acute respiratory illness. Science 262: 914–917.

Nielsen, B. M. B. 2000. Nesting ecology of Apapane (Himatione sanguinea). Ph.D. dissertation. University of Idaho, Moscow, ID.

Nishibayashi, E. 1995. Fish meal bait blocks with 0.005% diphacinone—palatability to mongoose. Unpublished report. Haleakala National Park, Maui, Hawaii.

Nishida, G. M. (editor). 2002. Hawaiian terrestrial arthropod checklist. 4th edition. Bishop Museum Technical Report 22. Hawaii Biological Survey, Bishop Museum, Honolulu.

Nogales, M., A. Martín, B. R. Tershy, C. J. Donlan, D. Veitch, N. Puerta, B. Wood, and J. Alonso. 2004. A review of feral cat eradication on islands. Conservation Biology 18: 310–319.

Nordling, D., M. Andersson, S. Zohari, and L. Gustafsson. 1998. Reproductive effort reduces specific immune response and parasite resistance. Proceedings of the Royal Society of London, Series B 265: 1291–1298.

North American Bird Conservation Initiative, U.S. Committee. 2009. The state of the birds, United States of America, 2009. U.S. Department of the Interior, Washington, DC.

Noske, B. 1989. Humans and other animals: Beyond the boundaries of anthropology. Pluto Press, London.

Nussenzweig, R. S., J. Vanderberg, H. Most, and C. Orten. 1967. Protective immunity produced by injection of x-irradiated sporozoites of Plasmodium berghei. Nature 216: 160–162.

Oboyski, P. T., J. W. Slotterback, and P. C. Banko. 2004. Differential parasitism of seed-feeding Cydia (Lepidoptera: Tortricidae) by native and alien wasp species relative to elevation in subalpine Sophora (Fabaceae) forests on Mauna Kea, Hawaii. Journal of Insect Conservation 8: 229–240.

Oda, T., K. Uchida, A. Mori, M. Mine, Y. Eshita, K. Kurokawa, K. Kato, and H. Tahara. 1999. Effects of high temperature on the emergence and survival of adult *Culex pipiens molestus* and *Culex quinquefasciatus* in Japan. Journal of the American Mosquito Control Association 15: 153–156.

O'Donnell, C. F. J. 1996. Predators and the decline of New Zealand forest birds: An introduction to the hole-nesting bird and predator programme. New Zealand Journal of Zoology 23: 213–219.

Ohashi, T. J. 2001. Environmental assessment for the rat eradication on Palmyra Atoll. Prepared by the U.S. Department of Agriculture, Animal and Plant Health Inspection Service, Wildlife Services, Honolulu.

Ohashi, T. J., and J. G. Oldenburg. 1992. Endangered species in the Pacific islands: The role of animal damage control. Proceedings of the Vertebrate Pest Conference 15: 32–35.

Oliver, J. A., and C. E. Shaw. 1953. The amphibians and reptiles of the Hawaiian Islands. Zoologica (New York) 38: 64–95.

Olson, S. L. 1999a. Kona Grosbeak (*Chloridops kona*), Greater Koa-Finch (*Rhodacanthis palmeri*), and Lesser Koa-Finch (*Rhodacanthis flaviceps*). In A. Poole and F. Gill (editors), The birds of North America, No. 424. Birds of North America Inc., Philadelphia.

———. 1999b. Laysan Rail (*Porzana palmeri*) and Hawaiian Rail (*Porzana sandwichensis*). In A. Poole and F. Gill (editors), The birds of North America, No. 426. Birds of North America Inc., Philadelphia.

Olson, S. L., and H. F. James. 1982a. Prodromus of the fossil avifauna of the Hawaiian Islands. Smithsonian Contributions to Zoology 365: 1–59.

———. 1982b. Fossil birds from the Hawaiian Islands: Evidence for wholesale extinction by man before Western contact. Science 217: 633–635.

———. 1991. Descriptions of thirty-two new species of birds from the Hawaiian Islands, Part 1: Non-passeriformes. Ornithological Monographs 45: 1–88.

———. 1994. A chronology of ornithological exploration in the Hawaiian Islands, from Cook to Perkins. Studies in Avian Biology 15: 91–102.

———. 1997. Prehistoric status and distribution of the Hawaiian Hawk (*Buteo solitarius*), with the first fossil record from Kaua'i. Bishop Museum Occasional Papers 49: 65–69.

Olson, S. L., and A. C. Ziegler. 1995. Remains of land birds from Lisianski Island, with observations on the terrestrial avifauna of the Northwestern Hawaiian Islands. Pacific Science 49: 111–125.

Omland, K. E., J. Marzluff, W. Boarman, C. L. Tarr, and R. C. Fleischer. 2000. Cryptic genetic variation and paraphyly in ravens. Proceedings of the Royal Society of London, Series B 267: 2475–2482.

Oniki, Y. 1985. Why robin eggs are blue and birds build nests: Statistical tests for Amazonian birds. Ornithological Monographs 36: 536–545.

Oreinstein, R. 1968. Birds of the Crowder "Birds of Hawai'i" tour, December 23, 1967 to January 5, 1968. 'Elepaio 29: 21–27.

Ortiz, A. 1972. Ritual drama and the Pueblo world view. Pp. 135–161 in A. Ortiz (editor), New perspectives on the Pueblo. University of New Mexico Press, Albuquerque, NM.

Otis, D. L., K. P. Burnham, G. C. White, and D. R. Anderson. 1978. Statistical inference from capture data on closed animal populations. Wildlife Monographs 62: 1–135.

Owens, I. P. F., and P. M. Bennett. 2000. Ecological basis of extinction risk in birds: Habitat loss versus human persecution and introduced predators. Proceedings of the National Academy of Sciences (USA) 97: 12144–12148.

Owre, O. T. 1973. A consideration of the exotic avifauna of southeastern Florida. Wilson Bulletin 85: 491–500.

Painter, M. K., K. J. Tennessen, and T. D. Richardson. 1996. Effects of repeated applications of *Bacillus thuringiensis israelensis* on the mosquito predator *Erythemis simplicicollis* (Odonata: Libellulidae) from hatching to final instar. Environmental Entomology 25: 184–191.

Pank, L. F., D. P. Fellows, D. N. Hirata, and R. T. Sugihara. 1978. The efficacy and hazards of zinc phosphide rat control adjacent to macadamia orchards. Annual Proceedings of the Hawaii Macadamia Producers Association 18: 27–41.

Paoletti, E. 1996. Applications of pox virus vectors to vaccination: An update. Proceedings of the National Academy of Sciences (USA) 21: 11349–11353.

Papp, R. P., J. T. Kliejunas, R. S. Smith Jr., and R. F. Scharpf. 1979. Association of *Plagithmysus bilineatus* (Coleoptera: Cerambycidae) and *Phytophthora cinnamomi* with the decline of ohia lehua forests on the island of Hawaii. Forest Science 25: 187–196.

Patch-Highfill, L. 2008. Estimating genetic diversity of Palila (*Loxioides bailleui*) and familial relationships of helper males. M.S. thesis. Tropical Conservation Biology and Environmental Science, University of Hawai'i at Hilo, Hilo.

Patterson, R. S., R. E. Lowe, B. J. Smittle, D. A. Dame, M. D. Boston, and A. L. Cameron. 1977. Release of radio-sterilized males to control *Culex pipiens quinquefasciatus*. Journal of Medical Entomology 14: 299–304.

Patterson, R. S., D. E. Weidhaas, H. R. Ford, and C. S. Lofgren. 1970. Suppression and elimination of an island population of *Culex pipiens quinquefasciatus* with sterile males. Science 168: 1368–1370.

Paxinos, E. E., H. F. James, S. L. Olson, J. D. Ballou, J. A. Leonard, and R. C. Fleischer. 2002. Prehistoric decline of genetic diversity in the Nene. Science 296: 1827.

Paxinos, E. E., H. F. James, S. L. Olson, M. D. Sorenson, J. Jackson, and R. C. Fleischer. 2002. MtDNA from fossils reveals a radiation of Hawaiian geese recently derived from the Canada goose (*Branta canadensis*). Proceedings of the National Academy of Sciences (USA) 99: 1399–1404.

Pearson, H. L., and P. M. Vitousek. 2001. Stand dynamics, nitrogen accumulation, and symbiotic nitrogen fixation in regenerating stands of *Acacia koa*. Ecological Applications 11: 1381–1394.

Pearson, R. J., P. V. Kirch, and M. Peitrusewsky. 1971. An early prehistoric site at Bellows Beach, Waimanalo, Oahu, Hawaiian Islands. Archaeology and Physical Anthropology in Oceania 6: 204–234.

Pease, C. M., and J. A. Grzybowski. 1995. Assessing the consequences of brood parasitism and nest predation on seasonal fecundity in passerine birds. Auk 112: 343–363.

Peck, R. W. 1993. The influence of arthropods, forest structure, and rainfall on insectivous Hawaiian forest birds. M.S. thesis. University of Hawai'i at Mānoa, Honolulu.

Peck, R. W., P. C. Banko, M. Schwarzfeld, M. Euaparadorn, and K. W. Brinck. 2008. Alien dominance of the parasitoid wasp community along an elevation gradient on Hawai'i Island. Biological Invasions 10: 1441–1455.

Peirce, M. A., A. G. Greenwood, and K. Swinnerton. 1997. Pathogenicity of *Leucocytozoon marchouxi* in the pink pigeon (*Columba mayer*) in Mauritius. Veterinary Record 140: 155–156.

Pejchar, L. 2004. Ecology of an endangered Hawaiian honeycreeper and implications for conservation on private lands. Ph.D. dissertation. University of California, Santa Cruz.

Pejchar, L., K. D. Holl, and J. L. Lockwood. 2005. Hawaiian honeycreeper home range size varies with habitat: Implications for native *Acacia koa* forestry. Ecological Applications 15: 1053–1061.

Pejchar, L., and D. M. Press. 2006. Achieving conservation objectives through production forestry: The case of *Acacia koa* on Hawaii Island. Environmental Science and Policy 9: 439–447.

Pekelo, N. Jr. 1963. Nature notes from Molokai. 'Elepaio 24: 17–18.

Pemberton, C. E. 1936. Letter to A. T. Spalding, Honomu Sugar Company. In letters of Winston Banko. Located at the archives of the U.S. Geological Survey, Pacific Island Ecosystems Research Center, Hawai'i Volcanoes National Park, Volcano, HI.

Pemberton, R. W. 2000. Predictable risk to native plants in weed biological control. Oecologia 125: 489–494.

———. 2002. Predictable risk to native plants in biological control of weeds in Hawaii. Pp. 91–102 in C. W. Smith, J. Denslow, and S. Hight (editors),

Biological control of invasive plants in native Hawaiian ecosystems. Technical Report 129. Pacific Cooperative Studies Unit, University of Hawaii at Mānoa, Honolulu.

Pennock, D. S., and W. W. Dimmick. 1997. Critique of the evolutionarily significant unit as a definition for "distinct population segments" under the U.S. Endangered Species Act. Conservation Biology 11: 611–619.

Penukahi, T. N. 1871. He mau wahi kamaaina o ka ua tuahine, ua nalowale (The natives of the land of the tuahine rains are lost). Short notes pertaining to Hawaiian life: Descriptions of Hawaiian birds. Ke Au Okoa, June 29. Courtesy of the Bishop Museum Archives, Honolulu.

Perkins, R. C. L. 1893. Notes on collecting in Kona, Hawaii. Ibis 1893: 101–113.

———. 1903. Vertebrata. Pp. 365–466 in D. Sharp (editor), Fauna Hawaiiensis or the zoology of the Sandwich (Hawaiian) Isles, Vol. 1. University Press, Cambridge, England.

———. 1913. Introduction: Being a review of the land-fauna of Hawaii. Pp. xv–ccxxviii in D. Sharpe (editor), Fauna Hawaiiensis or the zoology of the Sandwich (Hawaiian) Isles, Vol. 1. Cambridge University Press, Cambridge, England.

Perkins, R. C. L., and O. H. Swezey. 1926. The introduction into Hawaii of insects that attack lantana. Bulletin of the Experiment Station of the Hawaiian Sugar Planters' Association 16: 1–83.

Perkins, S. L., S. M. Osgood, and J. J. Schall. 1998. Use of PCR for detection of subpatent infections of lizard malaria: Implications for epizootiology. Molecular Ecology 7: 1589–1590.

Peterjohn, B. G., J. R. Sauer, and C. S. Robbins. 1995. Population trends from the North American Breeding Bird Survey. Pp. 3–39 in T. E. Martin and D. M. Finch (editors), Ecology and management of neotropical migratory birds. Oxford University Press, New York.

Peterson, B. L., B. E. Kus, and D. H. Deutschman. 2004. Determining nest predators of the Least Bell's Vireo through point counts, tracking stations, and video photography. Journal of Field Ornithology 75: 89–95.

Philipp, P. F. 1953. Diversified agriculture of Hawaii. University of Hawaii Press, Honolulu.

Pimm, S. L. 1991. The balance of nature? Ecological issues in the conservation of species and communities. University of Chicago Press, Chicago.

Pimm, S. L., and R. A. Askins. 1995. Forest losses predict bird extinctions in Eastern North America. Proceedings of the National Academy of Sciences (USA) 92: 9343–9347.

Pimm, S. L., M. P. Moulton, and L. J. Justice. 1995. Bird extinctions in the central Pacific. Pp. 75–87 in J. H. Lawton and R. M. May (editors), Extinction rates. Oxford University Press, Oxford.

Pimm, S. L., and J. W. Pimm. 1982. Resource use, competition, and resource availability in Hawaiian honeycreepers. Ecology 63: 1468–1480.

Pinkney, A. E., P. C. McGowan, D. R. Murphy, T. P. Lowe, D. W. Sparling, and L. C. Ferrington. 2005. Effects of the mosquito larvicides temephos and methoprene on insect population in experimental ponds. Environmental Toxicology and Chemistry 19: 678–684.

Pitt, W. C., J. Eisemann, K. Swift, R. Sugihara, B. Dengler-Germain, and L. Driscoll. 2005. Diphacinone residues in free-ranging wild pigs following aerial broadcast of rodenticide bait in a Hawaiian forest. Good Laboratory Practices Report to the Environmental Protection Agency. U.S. Department of Agriculture, Animal and Plant Health Inspection Service, National Wildlife Research Center, Hilo, HI.

Pletschet, S. M., and J. F. Kelly. 1990. Breeding biology and nesting success of Palila. Condor 92: 1012–1021.

Poché, R. M. 1992. How GLP provisions influence costs of rodenticide field evaluations. Proceedings of the Vertebrate Pest Conference 15: 245–248.

Pokras, M. A., E. B. Wheeldon, and C. J. Sedgwich. 1993. Trichomoniasis in owls: Report on a number of clinical cases and a survey of the literature. Pp. 239–245 in P.T. Redig, J. E. Cooper, C. J. Remple, and D. B. Hunter (editors), Raptor biomedicine. University of Minnesota Press, Minneapolis.

Pollock, C. G. 2001. Silent Spring revisted: A 21st-century look at the effect of pesticides on wildlife. Journal of Avian Medicine and Surgery 15: 50–53.

Pollock, K. H., J. D. Nichols, C. Brownie, and J. E. Hines. 1990. Statistical inference for capture-recapture experiments. Wildlife Monographs 107: 1–97.

Poole, A., and F. Gill (editors). 1992–2002. The birds of North America. Birds of North America Inc., Philadelphia.

Porter, W. P., N. Vakharia, W. D. Klousie, and D. Duffy. 2006. Po'ouli landscape bioinformatics models predict energetics, behavior, diets, and distribution on Maui. Integrative and Comparative Biology 46: 1143–1158.

Powell, A. 2008. The race to save the world's rarest bird: The discovery and death of the po'ouli. Stackpole Books, Mechanicsburg, PA.

Pradel, R. 1996. Utilization of capture-mark-recapture for the study of recruitment and population growth rate. Biometics 52: 703–709.

Prakash, I. 1988. Bait shyness and poison aversion. Pp. 321–329 in I. Prakash (editor), Rodent pest management. CRC Press, Boca Raton, FL.

Pranty, B., and K. L. Garrett. 2003. The parrot fauna of the ABA area: A current look. Birding 35: 248–261.

Pratt, H. D. 1979. A systematic analysis of the endemic avifauna of the Hawaiian islands. Ph.D. dissertation. Louisiana State University, Baton Rouge, LA.

———. 1982. Relationships and speciation of the Hawaiian thrushes. Living Bird 19: 73–90.

———. 1992. Is the Poo-uli a Hawaiian honeycreeper (Drepanidinae)? Condor 94: 172–180.

———. 1994. Avifaunal change in the Hawaiian Islands, 1893–1993. Studies in Avian Biology 15: 103–118.

———. 2001. Why the Hawai'i Creeper is an Oreomystis: What phenotypic characters reveal about the phylogeny of Hawaiian Honeycreepers. Studies in Avian Biology 22: 81–97.

———. 2002. Enjoying birds and other wildlife in Hawaii. 3rd edition. Mutual Publishing, Honolulu.

———. 2005. The Hawaiian honeycreepers: Drepanidinae. Oxford University Press, Oxford.

Pratt, H. D., P. L. Bruner, and D. G. Berrett. 1987. A field guide to the birds of Hawaii and the tropical Pacific. Princeton University Press, Princeton, NJ.

Pratt, H. D., and T. K. Pratt. 2001. The interplay of species concepts, taxonomy, and conservation: Lessons from the Hawaiian avifauna. Studies in Avian Biology 22: 68–80.

Pratt, L. W., L. L. Abbott, and D. K. Palumbo. 1999. Vegetation above a feral pig barrier fence in rainforests of Kilauea's East Rift, Hawaii Volcanoes National Park. Technical Report 124. Cooperative National Park Resources Studies Unit, University of Hawaii at Mānoa, Honolulu.

Pratt, L. W., and S. M. Gon III. 1998. Terrestrial ecosystems. Pp. 121–129 in S. P. Juvik and J. O. Juvik (editors), Atlas of Hawai'i. 3rd edition. University of Hawai'i Press, Honolulu.

Pratt, T. K. 1988. Recent observations June and July. 'Elepaio 48: 113–114.

———. 2004. The 104th Christmas Bird Count: Hawaii / Pacific Islands, Commonwealth of Northern Mariana Islands and Guam. American Birds 58: 102–104.

Pratt, T. K., P. C. Banko, S. G. Fancy, G. D. Lindsey, and J. D. Jacobi. 1997. Status and management of the Palila, an endangered Hawaiian honeycreeper, 1987–1996. Pacific Conservation Biology 3: 330–340.

Pratt, T. K., S. G. Fancy, and C. J. Ralph. 2001. 'Akiapōlā'au (Hemignathus munroi) and Nukupu'u (Hemignathus lucidus). In A. Poole and F. Gill (editors), The birds of North America, No. 600. Birds of North America Inc., Philadelphia.

Pratt, T. K., J. G. Giffin, and F. P. Duvall III. 1989. Resent observations of 'Akepa and other endangered forest birds in central Kona, Hawai'i Island. 'Elepaio 49: 62–64.

Pratt, T. K., C. B. Kepler, and T. L. C. Casey. 1997. Po'ouli (Melamprosops phaeosoma). In A. Poole and F. Gill (editors), The birds of North America, No. 272. Academy of Natural Sciences, Philadelphia, and American Ornithologists' Union, Washington, DC.

Pratt, T. K., and R. L. Pyle. 2000. Nukupu'u in the twentieth century: Endangered species or phantom presence? 'Elepaio 60: 35–41.

Pratt, T. K., J. C. Simon, B. P. Farm, K. E. Berlin, and J. R. Kowalsky. 2001. Home range and territoriality of two Hawaiian honeycreepers, the 'Ākohekohe and Maui Parrotbill. Condor 103: 746–755.

Preston, C. R., and R. D. Beane. 1993. Red-tailed Hawk (Buteo jamaicensis). In A. Poole and F. Gill (editors), The birds of North America, No. 52. Birds of North America Inc., Philadelphia.

Price, J., and J. D. Jacobi. 2007. Rapid assessment of vegetation at six potential 'Alalā release sites on the island of Hawai'i. Unpublished report issued by the U.S. Geological Survey, Pacific Island Ecosystems Research Center at Hawai'i Volcanoes National Park to the U.S. Fish and Wildlife Service, Honolulu.

Price, J. P. 2004. Floristic biogeography of the Hawaiian Islands: Influences of area, environment and paleogeography. Journal of Biogeography 31: 487–500.

Price, J. P., and D. A. Clague. 2002. How old is the Hawaiian biota? Geology and phylogeny suggest recent divergence. Proceedings of the Royal Society of London, Series B 269: 2429–2435.

Price, J. P., and D. Elliott-Fisk. 2004. Topographic history of the Maui Nui complex, Hawaii, and its implications for Biogeography. Pacific Science 58: 27–45.

Price, J. P., and W. L. Wagner. 2004. Speciation in Hawaiian angiosperm lineages: Cause, consequence, and mode. Evolution 58: 2185–2200.

Pukui, M. K. 1951. The water of Kane. Kamehameha Schools Press, Honolulu.

———. 1960. Tales of the Menehune and other short legends of the Hawaiian Islands. Kamehameha Schools Press, Honolulu.

Pukui, M. K., and S. H. Elbert. 1986. Hawaiian dictionary. Revised and enlarged edition. University of Hawaii Press, Honolulu.

Pukui, M. K., E. W. Haertig, and C. A. Lee. 1972. Nana i ke kumu (Look to the source), Vols. 1 and 2. Hui Hanai, Honolulu.

Pyle, R. L. 1981. Hawaiian bird observations August 1979 through July 1980. 'Elepaio 41: 72–79.

———. 1985a. Hawaiian Islands region. American Birds 37: 351–353.

———. 1985b. Hawaiian Islands region. American Birds 37: 964–965.

———. 1987. Hawaiian Islands region. American Birds 41: 491–493.

———. 1989. Hawaiian Islands region. American Birds 43: 369–371.

———. 1992. Checklist of the birds of Hawaii. 'Elepaio 52: 53–62.

———. 1993. Hawaiian Islands region. American Birds 47: 302–304.

———. 2002. Checklist of the birds of Hawaii—2002. 'Elepaio 62: 137–148.

———. 2003. SIGHTINGS database. At the B. P. Bishop Museum, Honolulu.

Ragg, J. R., H. Moller, and K. A. Waldrup. 1995. The prevalence of bovine tuberculosis (Mycobacterium bovis) infections in feral populations of cats (Felis catus), ferrets (Mustela furo) and stoats (Mustela erminea) in Otago and Southland, New Zealand. New Zealand Veterinary Journal 43: 333–337.

Rahbeck, C. 1993. Captive breeding—a useful tool in the preservation of biodiversity? Biodiversity and Conservation 2: 426–437.

Ralph, C. J., and S. G. Fancy. 1994a. Demography and movements of the endangered Akepa and Hawaii Creeper. Wilson Bulletin 106: 615–628.

———. 1994b. Timing of breeding and molting in six species of Hawaiian honeycreepers. Condor 96: 151–161.

———. 1994c. Demography and movements of the Omao (Myadestes obscurus). Condor 96: 503–511.

———. 1995. Demography and movements of Apapane and Iiwi in Hawaii. Condor 97: 729–742.

———. 1996. Aspects of the life history and foraging ecology of the endangered Akiapolaau. Condor 98: 312–321.

Ralph, C. J., G. R. Geupel, P. Pyle, T. E. Martin, and D. F. DeSante. 1993. Handbook of field methods for monitoring landbirds. General Technical Report PSW-GTR-144. U.S. Department of Agriculture, Forest Service, Pacific Southwest Research Station, Albany, CA.

Ralph, C. J., and B. R. Noon. 1988. Foraging interactions of small Hawaiian forest birds. Pp. 1992–2006 in ACTA XIX Congressus Internationalis Ornithologici, 1986, Vol. 2. University of Ottawa Press, Ottawa.

Ralph, C. J., J. R. Sauer, and S. Droege. 1995. Monitoring bird populations by point counts. General Technical Report PSW-GTR-149. U.S. Department of Agriculture, Forest Service, Pacific Southwest Research Station, Albany, CA.

Ralph, C. J., and C. van Riper III. 1985. Historical and current factors affecting Hawaiian native birds. Bird Conservation 2: 7–42.

Ramsey, F. L., and J. M. Scott. 1978. Use of circular plot surveys in estimating the density of a population with Poisson scattering. Technical Report 60. Department of Statistics, Oregon State University, Corvallis.

———. 1979. Estimating population densities from variable circular plot surveys. Pp. 155–181 in R. Cormack, G. Patil, and D. Robson (editors), Sampling biological populations. International Co-operative Publishing House, Fairland, MD.

Ramstad, K. M., N. J. Nelson, G. Paine, D. Beech, A. Paul, P. Paul, F. W. Allendorf, and C. H. Daugherty. 2007. Species and cultural conservation in New Zealand: Maori traditional ecological knowledge of tuatara. Conservation Biology 21: 455–464.

Randall, R. P. 2002. A global compendium of weeds. R. G. and F. J. Richardson, Melbourne, Australia.

Rappole, J. H., S. R. Derrickson, and Z. Hubalek. 2000. Migratory birds and spread of West Nile virus in the Western Hemisphere. Emerging Infectious Diseases 6: 319–328.

Rasgon, J. L., and F. Gould. 2005. Transposable element insertion location bias and the dynamics of gene drive in mosquito populations. Insect Molecular Biology 14: 493–500.

Rauzon, M. J. 1985. Feral cats on Jarvis Island: Their effects and their eradication. Atoll Research Bulletin 282: 1–32.

———. 2001. Isles of refuge. University of Hawai'i Press, Honolulu.

Rauzon, M. J., W. T. Everett, D. Boyle, L. Bell, and J. Gilardi. 2008. Eradication of feral cats at Wake Atoll. Atoll Research Bulletin 560: 1–21.

Rauzon, M. J., and E. N. Flint. 2002. Seabird recolonisation after cat eradication on equatorial Jarvis, Howland, and Baker Islands, USA, Central Pacific. P. 411 in C. R. Veitch and M. N. Clout (editors), Turning the tide: The eradication of invasive species. Proceedings of the International Conference on Eradication of Island Invasives, Auckland, New Zealand.

Rave, E. H. 1995. Genetic analyses of wild populations of Hawaiian geese using DNA fingerprinting. Condor 97: 82–90.

Rave, E. H., R. C. Fleischer, F. Duvall, and J. M. Black. 1994. Genetic analyses through DNA fingerprinting of captive populations of nene. Conservation Biology 8: 744–751.

———. 1998. Factors influencing reproductive success in captive populations of Hawaiian geese Branta sandvicensis. Wildfowl 49: 36–44.

Reding, D. M., J. T. Foster, H. F. James, H. D. Pratt, and R. C. Fleischer. 2008. Convergent evolution of "creepers" in the Hawaiian honeycreeper radiation. Biology Letters (doi: 10.1098/rsbl.2008. 0589).

Redman, C. L., and A. P. Kinzig. 2003. Resilience of past landscapes: Resilience theory, society, and the Longue Durée. Conservation Ecology 7 (1): 14. www .consecol.org/vol1/iss1/art14.

Reeser, D. W. 2002. Crossing boundaries at Haleakala: The struggle to get improved quarantine protection prior to expansion of Maui's airport. Pp. 107–111 in D. Harmon (editor), Crossing boundaries in park management. Proceedings of the 11th Conference on Research and Resource Management in Parks and on Public Lands, George Wright Society, Denver, CO, April 2001 .www.georgewright.org/19reeser.pdf (accessed March 23, 2008).

Reichard, S. H., and C. W. Hamilton. 1997. Predicting invasions of woody plants introduced into North America. Conservation Biology 11: 193–203.

Reichard, S. H., and P. S. White. 2001. Horticulture as a pathway of invasive plant introductions in the United States. BioScience 51: 103–113.

Reisen, W. K., R. P. Meyer, C. H. Tempelis, and J. J. Spoehel. 1990. Mosquito abundance and bionomics in residential communities in Orange and Los Angeles Counties, California. Journal of Medical Entomology 27: 356–367.

Reisen, W. K., M. M. Milby, and R. P. Meyer. 1992. Population dynamics of adult Culex mosquitoes (Diptera: Culicidae) along the Kern River, Kern County, California, in 1990. Journal of Medical Entomology 29: 531–543.

Reiter, M. E., and D. A. LaPointe. 2007. Landscape factors influencing the spatial distribution and abundance of mosquito vector Culex quinquefasciatus (Diptera: Culicidae) in a mixed residential-agricultural community in Hawai'i. Journal of Medical Entomology 44: 861–868.

Rejmanek, M., and D. M. Richardson. 1996. What attributes make some plant species more invasive? Ecology 77: 1655–1661.

Ren, P., P. Y. Stark, R. L. Johnson, and R. G. Bell. 1977. Mechanism of action of anticoagulants: Correlation between the inhibition of prothrombin synthesis and the regeneration of vitamin K_1 from vitamin K_1 epoxide. Journal of Pharmacology and Experimental Therapeutics 201: 541–546.

Rensberger, B. 1977. The cult of the wild. Anchor Press / Doubleday, Garden City / New York.

Restani, M., and J. M. Marzluff. 2001. Avian conservation under the Endangered Species Act: Expenditures versus recovery priorities. Conservation Biology 15: 1292–1299.

———. 2002a. Funding extinction? Biological needs and political realities in the allocation of resources to endangered species recovery. BioScience 52: 169–177.

———. 2002b. Response from Restani and Marzluff. BioScience 52: 548–551.

———. 2002c. Response from Restani and Marzluff. BioScience 52: 868–870.

Reynolds, M. H., R. J. Camp, B. M. B. Nielsen, and J. D. Jacobi. 2003. Evidence of change in a low-elevation forest bird community of Hawai'i since 1979. Bird Conservation International 13: 175–187.

Reynolds, M. H., and K. Kozar. 2000. History and current status of the Laysan Duck (Anas laysanensis) in captivity. 'Elepaio 60: 59–65.

Reynolds, M. H., N. E. Seavy, M. S. Vekasy, J. L. Klavitter, and L. P. Laniawe. 2008. Translocation and early post-release demography of endangered Laysan Teal. Animal Conservation 11: 160–168.

Reynolds, M. H., and T. J. Snetsinger. 2001. The Hawai'i rare bird search 1994–1996. Studies in Avian Biology 22: 133–143.

Reynolds, M. H., T. J. Snetsinger, and C. M. Herrmann. 1997. Kauai's endangered solitaires: Update on population status and distribution 1996. Transactions of the Western Section of the Wildlife Society 33: 49–55.

Reynolds, R. T., J. M. Scott, and R. A. Nussbaum. 1980. A variable circular-plot method for estimating bird numbers. Condor 82: 309–313.

Rhoades, D. F. 1979. Evolution of plant chemical defense against herbivores. Pp. 3–54 in G. A. Rosenthal and D. H. Janzen (editors), Herbivores: Their interaction with secondary plant metabolites. Academic Press, New York.

Rhymer, J. M. 2001. Evolutionary relationships and conservation of the Hawaiian anatids. Studies in Avian Biology 22: 61–67.

Richardson, F., and J. Bowles. 1964. A survey of the birds of Kauai, Hawaii. B. P. Bishop Museum Bulletin 227: 1–51.

Ricklefs, R. E. 1966. The temporal component of diversity among species of birds. Evolution 20: 235–242.

———. 1968. Patterns of growth in birds. Ibis 110: 419–451.

———. 1969. The nesting cycle of songbirds in tropical and temperate regions. Living Bird 8: 165–175.

———. 1992. Embryonic development period and the prevalence of avian blood parasites. Proceedings of the National Academy of Sciences (USA) 89: 4722–4725.

———. 2000. Lack, Skutch, and Moreau: The early development of life-history thinking. Condor 102: 3–8.

Ricklefs, R. E., and E. Bermingham. 2002. The concept of the taxon cycle in biogeography. Global Ecology and Biogeography 11: 353–361.

Ricklefs, R. E., and S. S. Renner. 1994. Species richness within families of flowering plants. Evolution 49: 1619–1636.

Ries, L., R. J. Fletcher Jr., J. Battin, and T. D. Sisk. 2004. Ecological responses to habitat edges: Mechanisms, models, and variability explained. Annual Review of Ecology, Evolution, and Systematics 35: 491–522.

Riesing, M. J., L. Kruckenhauser, A. Gamauf, and E. Haring. 2003. Molecular phylogeny of the genus Buteo (Aves: Accipitridae) based on mitochondrial marker sequences. Molecular Phylogenetics and Evolution 27: 328–342.

Rimmer, C. C., K. P. McFarland, W. G. Ellison, and J. E. Goetz. 2001. Bicknell's Thrush (Catharus bicknelli). In A. Poole and F. Gill (editors), The birds of North America, No. 592. Birds of North America Inc., Philadelphia.

Ritchie, B. W. 1995. Avian viruses function and control. Winger Publishing, Lake Worth, FL.

Rivers, T. M. 1937. Viruses and Koch's postulates. Journal of Bacteriology 33: 1–12.

Riviere, F., and R. Thirel. 1981. La prédation du copepoda Mesocyclops leuckarti pilosa (Crustacea) sur les larvae de Aedes (Stegomyia) aegypti et de Ae. (St.) polynesiensis (Diptera: Culicidae): Essais preliminaires d'utilisation comme agent de lutte biologique. Entomophaga 26: 427–439.

Robertson, H. A., J. R. Hay, D. K. Saul, and G. V. McCormack. 1994. Recovery of the Kakerori: An endangered forest bird of the Cook Islands. Conservation Biology 8: 1078–1086.

Robinson, B. W., and D. S. Wilson. 1998. Optimal foraging, specialization, and a solution to Liem's paradox. American Naturalist 151: 223–235.

Robinson, J. M. 2001. The dynamics of avicultural markets. Environmental Conservation 28: 76–85.

Rocamora, G. 2002. Successful inter-island transfer of Seychelles white-eyes. World Birdwatch 24: 6.

Rock, J. F. 1913. The indigenous trees of the Hawaiian Islands. Published privately, Honolulu, HI. Reproduced in 1974, with an introduction by S. Carlquist and an addendum by D. R. Herbst, by Charles E. Tuttle, Rutland, VT.

Rodda, G. H., and K. Dean-Bradley. 2002. Excess density compensation of island herpetofaunal assemblages. Journal of Biogeography 29: 1–10.

Rodda, G. H., T. H. Fritts, and D. Chiszar. 1997. The disappearance of Guam's wildlife. Bioscience 47: 565–574.

Rodda, G. H., T. H. Fritts, and P. J. Conry. 1992. Origin and population growth of the brown tree snake, Boiga irregularis, on Guam. Pacific Science 46: 46–57.

Rodda, G. H., T. H. Fritts, M. J. McCoid, and E. W. Campbell III. 1999. An overview of the biology of the brown treesnake (Boiga irregularis), a costly introduced pest on Pacific islands. Pp. 44–80 in G. H. Rodda, Y. Sawai, D. Chiszar, and H. Tanaka (editors), Problem snake management: The habu and the brown treesnake. Comstock, Ithaca, NY, and London.

Rodda, G. H., G. Perry, R. J. Rondeau, and J. Lazell. 2001. The densest terrestrial vertebrate. Journal of Tropical Ecology 17: 331–338.

Rodriguez, J. P. 2002. Range contraction in declining North American bird populations. Ecological Applications 12: 238–248.

Rodwell, J. 2001. The history of forest conservation in Hawai'i. Newsletter of the Hawaiian Botanical Society 40: 7–14.

Roff, D. A. 2002. Life history evolution. Sinauer, Sunderland, MA.

Rohner, R. P. 1984. Toward a conception of culture for cross-cultural psychology. Journal of Cross-Cultural Psychology 5: 111–138.

Rojek, N. A. 1994. Notes on the behavior of released Nene: Interactions with 'Io and a lava tube accident. 'Elepaio 54: 41–43.

Rolstad, J. 1991. Consequences of forest fragmentation for the dynamics of bird populations: Conceptual issues and the evidence. Biological Journal of the Linnean Society 42: 149–163.

Roots, C. 1970. Softbilled birds. Arco, New York.

Rose, R. G., S. Conant, and E. P. Kjellgren. 1993. Hawaiian standing kāhili in Bishop Museum: An ethnological and biological analysis. Journal of Polynesian Society 102: 273–304.

Rose, R. I. 2001. Pesticides and public health: Integrated methods of mosquito control. Emerging Infectious Diseases 7: 17–23.

Rosendahl, P. 1972. Aboriginal agriculture and residence patterns in upland Lapakahi, Island of Hawai'i. Ph.D. dissertation. University of Hawai'i at Mānoa, Honolulu.

Rosenstock, S. S., D. A. Anderson, K. M. Giesen, T. Leukering, and M. F. Carter. 2002. Landbird counting techniques: Current practices and an alternative. Auk 119: 46–53.

Ross, D. 1964. Aristotle. University Paperbacks, London.

Roth, R. R., M. S. Johnson, and T. J. Underwood. 1996. Wood Thrush (Hylocichla mustelina). In A. Poole and F. Gill (editors), The birds of North America, No. 246. Birds of North America Inc., Philadelphia.

Rothschild, W. 1892. Descriptions of seven new species of birds for the Sandwich Islands. Annual Magazine of Natural History 60: 108–112.

———. 1893–1900. The avifauna of Laysan and the neighboring islands: With a complete history to date of birds of the Hawaiian possessions. 3 vols. R. H. Porter, London.

Rueda, L. M., K. J. Patel, R. C. Axtell, and R. E. Stinner. 1990. Temperature-dependent development and survival rates of Culex quinquefasciatus and Aedes aegypti (Diptera: Culiciadae). Journal of Medical Entomology 27: 892–898.

Ruiz, G., and J. Carlton (editors). 2003. Invasive species: Vectors and management strategies. Island Press, Washington, DC.

Russell, C. A. 1980. Food habits of the roof rat (Rattus rattus) in two areas of Hawaii Volcanoes National Park. Pp. 269–272 in C. W. Smith (editor), Proceedings Third Conference in Natural Sciences, Hawaii Volcanoes National Park. Cooperative National Park Resources Studies Unit, University of Hawaii at Mānoa, Honolulu.

Ryder, O. A. 2002. Cloning advances and challenges for conservation. Trends in Biotechnology 2: 231–232.

Ryttman, H. 1994. Estimates of survival and population development of the Osprey (Pandion haliaetus), Common Buzzard (Buteo buteo), and Sparrowhawk (Accipiter nisus) in Sweden. Ornis Svec 4: 159–172.

Sabo, S. R. 1982. The rediscovery of Bishop's 'O'o on Maui. 'Elepaio 42: 69–70.

Sæther, B.-E., and Ø. Bakke. 2000. Avian life history variation and contribution of demographic traits to the population growth rate. Ecology 81: 642–653.

Sæther, B.-E., S. Engen, and E. Matthysen. 2002. Demographic characteristics and population dynamical patterns of solitary birds. Science 295: 2070–2073.

Sakai, A. K., W. L. Wagner, and L. A. Mehrhoff. 2002. Patterns of endangerment in the Hawaiian flora. Systematic Biology 51: 276–302.

Sakai, H. F. 1988. Avian response to mechanical clearing of a native rainforest in Hawaii. Condor 90: 339–348.

Sakai, H. F., and J. R. Carpenter. 1990. The variety and nutritional value of foods consumed by Hawaiian Crow nestlings, an endangered species. Condor 92: 220–228.

Sakai, H. F., and T. C. Johanos. 1983. The nest, egg, young, and aspects of the life history of the endangered Hawaii Creeper. Western Birds 14: 73–84.

Sakai, H. F., C. J. Ralph, and C. D. Jenkins. 1986. Foraging ecology of the Hawaiian Crow, an endangered generalist. Condor 88: 211–219.

Sallabanks, R., and F. C. James. 1999. American Robin (Turdus migratorius). In A. Poole and F. Gill (editors), The birds of North America, No. 462. Birds of North America Inc., Philadelphia.

Sandercock, B. K., S. R. Beissinger, S. H. Stoleson, R. R. Melland, and C. R. Hughes. 2000. Survival rates of a neotropical parrot: Implications for latitudinal comparisons of avian demography. Ecology 81: 1351–1370.

Sanderson, E. W., K. H. Redford, A. Vedder, P. B. Coppolillo, and S. E. Ward. 2002. A conceptual model for conservation planning based on landscape species requirements. Landscape and Urban Planning 58: 41–56.

Santana, C. E., and S. A. Temple. 1988. Breeding biology and diet of Red-tailed Hawks in Puerto Rico. Biotropica 20: 151–160.

Sardelis, M. R., M. J. Turell, and R. G. Andre. 2003. Experimental transmission of St. Louis Encephalitis virus by Ochlerotatus j. japonicus. Journal of the American Mosquito Control Association 19: 159–162.

Sarr, Z., N. P. Shema, and C. P. Stone. 1998. Nesting success and population status of the 'Elepaio (Chasiempis sandwichensis) in the Mauna Loa strip section of Hawai'i Volcanoes National Park. Technical Report 118. Cooperative National Park Studies Unit, University of Hawai'i, Honolulu.

Sasaki, D. M., and J. K. Ikeda. 2000. Beware Hawaiian critters! Disease reservoirs and vectors in Hawai'i. Communicable Disease Report May–June. Hawai'i Department of Health, Honolulu.

Sasaki, D. M., P. Kitsutani, and M. R. Bomgaars. 2003. Murine typhus in Hawai'i. Communicable Disease Report March–April. Hawai'i Department of Health, Honolulu.

Sasaki, D. M., L. Pang, H. P. Minette, C. K. Wakida, W. J. Fujimoto, S. J. Manea, R. Kunioka, and C. R. Middleton. 1993. Active surveillance and risk factors for leptospirosis in Hawai'i. American Journal of Tropical Medicine and Hygeine 48: 35–43.

Saunders G. R., D. Choquenot, J. C. McIlroy, and R. Packwood. 1999. Initial effects of rabbit haemorrhagic disease on free-living rabbit (*Oryctolagus cuniculus*) populations in central–western New South Wales. Wildlife Research 26: 69–74.

Savarie, P. J. 1991. The nature, modes of action, and toxicity of rodenticides. Pp. 589–598 *in* D. Pimental (editor), CRC handbook of pest management in agriculture. 2nd edition, Vol. 2. CRC Press, Boca Raton, FL.

Savidge, J. A. 1987. Extinction of an island forest avifauna by an introduced snake. Ecology 68: 660–668.

Sax, D. F., and J. H. Brown. 2000. The paradox of invasion. Global Ecology and Biogeography 9: 363–371.

Schluter, D. 2000a. The ecology of adaptive radiation. Oxford University Press, Oxford.

———. 2000b. Ecological character displacement in adaptive radiation. American Naturalist 156 (Supplement): 4–16.

Schluter, D., and P. R. Grant. 1984. Determinants of morphological patterns in communities of Darwin's Finches. American Naturalist 123: 175–196.

Schluter, D., and R. R. Repasky. 1991. Worldwide limitation of finch densities by food and other factors. Ecology 72: 1763–1774.

Schmitt, R. C. 1977. Historical statistics of Hawaii. University Press of Hawaii, Honolulu.

Schmutz, J. K., K. A. Rose, and R. G. Johnson. 1989. Hazards to raptors from strychnine poisoned ground squirrels. Journal of Raptor Research 23: 147–151.

Schwartz, C. W., and E. R. Schwartz. 1949. A reconnaissance of the game birds in Hawaii. Board of Commissioners of Agriculture and Forestry. Hawaii News Printshop, Hilo, HI.

Scott, J. M., J. D. Jacobi, and F. L. Ramsey. 1981. Avian surveys of large geographical areas: A systematic approach. Wildlife Society Bulletin 9: 190–200.

Scott, J. M., C. B. Kepler, P. Stine, H. Little, and K. Taketa. 1987. Protecting endangered forest birds in Hawaii: The development of a conservation strategy. Pp. 348–363 *in* Transactions of the 52nd North American Wildlife and Natural Resources Conference.

Scott, J. M., S. Mountainspring, F. L. Ramsey, and C. B. Kepler. 1986. Forest bird communities of the Hawaiian Islands: Their dynamics, ecology, and conservation. Studies in Avian Biology 9. Cooper Ornithological Society. Allen Press, Lawrence, KS.

Scott, J. M., S. Mountainspring, C. van Riper III, C. B. Kepler, J. D. Jacobi, T. A. Burr, and J. G. Giffin. 1984. Annual variation in the distribution, abundance, and habitat response of the Palila (*Loxioides bailleui*). Auk 101: 647–664.

Scott, J. M., F. L. Ramsey, and C. B. Kepler. 1981. Distance estimation as a variable in estimating bird numbers from vocalizations. Studies in Avian Biology 6: 334–340.

Scott, J. M., F. L. Ramsey, M. Lammertink, K. V. Rosenberg, R. Rohrbaugh, J. A. Wiens, and J. M. Reed. 2008. When is an "extinct" species really extinct? Gauging the search efforts for Hawaiian forest birds and the Ivory-billed Woodpecker. Avian Conservation and Ecology 3:3. www.ace-eco.org/vol3/iss2/art3/.

Scott, J. M., D. H. Woodside, and T. L. C. Casey. 1977. Observations of birds in the Molokai Forest Reserve, July 1975. 'Elepaio 38: 25–27.

Scott, T. W., W. Takken, B. G. J. Knols, and C. Boete. 2002. The ecology of genetically modified mosquitoes. Science 298: 117–119.

Scowcroft, P. G. 1971. Koa—monarch of Hawaiian forests. Newsletter of the Hawaiian Botanical Society 10: 23 26.

———. 1983. Tree cover changes in mamane (*Sophora chrysophylla*) forests grazed by sheep and cattle. Pacific Science 37: 109–119.

Scowcroft, P. G., and C. E. Conrad. 1988. Restoring critical habitat for Hawaii's endangered Palila by reducing ungulate populations. Transactions of the Western Section of the Wildlife Society 24: 72–79.

Scowcroft, P. G., and J. G. Giffin. 1983. Feral herbivores suppress the regeneration of mamane and other browse species on Mauna Kea, Hawaii. Journal of Range Management 36: 638–645.

Scowcroft, P. G., and R. Hobdy. 1987. Recovery of goat-damaged vegetation in an insular tropical montane forest. Biotropica 19: 208–215.

Scowcroft, P. G., and H. F. Sakai. 1984. Stripping of *Acacia koa* bark by rats on Hawaii and Maui. Pacific Science 38: 80–86.

Seddon, P. J. 1998. Improving the rigour of reintroduction project assessment, planning and execution: Report on the round table discussions on bird re-introductions. Re-introduction News 16: 3–5.

Serena, M. (editor). 1995. Reintroduction biology of Australian and New Zealand fauna. Surry Beatty and Sons, Chipping Norton, New South Wales, Australia.

Seto, N. W., and S. Conant. 1996. The effects of rat (*Rattus rattus*) predation on the reproductive success of the Bonin petrel (*Pterodroma hypoleuca*) on Midway Atoll. Colonial Waterbirds 19: 171–185.

Shallenberger, R. J., and H. D. Pratt. 1978. Recent observations and field identification of the Oahu creeper (*Loxops maculata maculata*). 'Elepaio 38: 135–140.

Shallenberger, R. J., and G. K. Vaughn. 1978. Avifaunal survey in the central Ko'olau Range, Oahu. Ahuimanu Productions, Honolulu.

Shapin, L. 2000. Black-hooded Oriole spotted in Kailua and Ahuimanu. 'Elepaio 60: 22.

Shea, K., H. P. Possingham, W. W. Murdoch, and R. Roush. 2002. Active adaptive management in insect pest and weed control: Intervention with a plan for learning. Ecological Applications 12: 927–936.

Shehata, C., L. Freed, and R. L. Cann. 2001. Changes in native and introduced bird populations on Oʻahu: Infectious diseases and species replacement. Studies in Avian Biology 22: 264–273.

Sheikh, P. A., and N. T. Carter. 2006. Everglades restoration: The federal role in funding. CRS Report for Congress. Library of Congress, Washington, DC.

Sheldon, B. C., and S. Verhulst. 1996. Ecological immunology: Costly parasite defenses and trade-offs in evolutionary ecology. Trends in Ecology and Evolution 11: 317–321.

Shepherdson, D. J., K. C. Carlstead, and N. Wielebnowski. 2004. Cross-institutional assessment of stress responses in zoo animals using longitudinal monitoring of faecal corticoids and behaviour. Animal Welfare 13: S105–S113.

Shroyer, D. A. 1981. Establishment of Wyeomyia mitchellii on the island of Oahu, Hawaiʻi. Mosquito News 41: 805.

Shweder, R. A. 1991. Thinking through cultures: Expeditions in cultural psychology. Harvard University Press, Cambridge, MA.

Sibley, C. G., and J. E. Ahlquist. 1982. The relationships of the Hawaiian Honeycreepers (Drepaninini) as indicated by DNA–DNA hybridization. Auk 99: 130–140.

Siepielski, A. M. 2006. A possible role for red squirrels in structuring breeding bird communities in lodgepole pine forests. Condor 108: 232–238.

Sillett, T. S., and R. T. Holmes. 2002. Variation in survivorship of a migratory songbird throughout its annual cycle. Journal of Animal Ecology 71: 296–308.

Simberloff, D. 1995a. Why do introduced species appear to devastate islands more than mainland areas? Pacific Science 49: 87–97.

———. 1995b. Habitat fragmentation and population extinction of birds. Ibis 137: S105–S111.

———. 1998. Flagships, umbrellas, and keystones: Is single-species management passé in the landscape era? Biological Conservation 83: 247–257.

Simberloff, D., and W. Boecklen. 1991. Patterns of extinction in the introduced Hawaiian avifauna: A reexamination of the role of competition. American Naturalist 138: 300–327.

Simberloff, D., and L. Gibbons. 2003. Now you see them, now you don't!—population crashes of established introduced species. Biological Invasions 6: 161–172.

Simberloff, D., and P. Stiling. 1996. How risky is biological control? Ecology 77: 1965–1974.

Simon, J. C., P. E. Baker, and H. Baker. 1997. Maui Parrotbill (Pseudonestor xanthophrys). In A. Poole and F. Gill (editors), The birds of North America, No. 311. Birds of North America Inc. and Academy of Natural Sciences, Philadelphia, and American Ornithologists' Union, Washington, DC.

Simon, J. C., T. K. Pratt, K. E. Berlin, and J. R. Kowalsky. 2000. Reproductive ecology of the Maui Parrotbill. Wilson Bulletin 112: 482–490.

———. 2001. Reproductive ecology and demography of the ʻĀkohekohe. Condor 103: 736–745.

Simon, J. C., T. K. Pratt, K. E. Berlin, J. R. Kowalsky, S. G. Fancy, and J. S. Hatfield. 2002. Temporal variation in bird counts within a Hawaiian rainforest. Condor 104: 469–481.

Simons, T. R. 1983. Biology and conservation of the endangered Hawaiian Dark-rumped Petrel (Pterodroma phaeopygia sandwichensis). CPSU/UW 83-2. National Park Service, Cooperative Studies Unit, University of Washington, Seattle.

Sincock, J. L., and E. Kridler. 1977. The extinct and endangered endemic birds of the Northwestern Hawaiian Islands. Unpublished report. U.S. Fish and Wildlife Service, Honolulu.

Singer, S. 1985. Bacillus sphaericus (Bacteria). Pp. 123–131 in H. C. Chapman (editor), Biological control of mosquitoes. American Mosquito Control Association Bulletin 6, Fresno, CA.

Sinoto, A. 1978. Archaeological and paleontological salvage at Barbers Point, Oʻahu. Typescript report at the Department of Anthropology, Bishop Museum, Honolulu.

Skinner, C. M. 1971. Myths and legends of our new possessions and protectorate. Gryphon, Ann Arbor, MI.

Skolmen, R. G. 1979. Plantings on the forest reserves of Hawaii 1910–1960. U.S. Forest Service, Pacific Southwest Forest and Range Experiment Station, Institute of Pacific Islands Forestry, Honolulu.

———. 1986. Where koa can be grown. Paper presented at the Koa Forest Conference, Hilo, HI, December 17–19.

Skolmen, R. G., and D. M. Fujii. 1980. Growth and development of a pure stand of koa (Acacia koa) at Keauhou–Kilauea. Pp. 301–310 in C. W. Smith (editor), Proceedings Third Conference in Natural Sciences, Hawaii Volcanoes National Park. Cooperative National Park Resources Studies Unit, University of Hawaii at Mānoa, Honolulu.

Skutch, A. F. 1949. Do tropical birds rear as many young as they can nourish? Ibis 91: 430–455.

———. 1985. Clutch size, nesting success, and predation on nests of Neotropical birds, reviewed. Ornithological Monographs 36: 575–594.

———. 1989. Birds asleep. University of Texas Press, Austin, TX.

Slate, D., C. E. Rupprecht, J. A. Donovan, D. H. Lein, and R. B. Chipman. 2005. Status of oral rabies vaccination in wild carnivores in the United States. Virus Research 111: 68–76.

Slikas, B., S. L. Olson, H. F. James, and R. C. Fleischer. 2002. Rapid, independent evolution of flightlessness in four species of Pacific island rails (Rallidae): An analysis based on mitochondrial sequence data. Journal of Avian Biology 33: 5–14.

Slippers, B., J. Stenlid, and M. J. Wingfield. 2005. Emerging pathogens: Fungal host jumps following anthropogenic introduction. Trends in Ecology and Evolution 20: 420–421.

Smales, I., B. Quin, D. Krake, D. Dobrozczyk, and P. Menkorst. 2000. Re-introduction of helmeted honeyeaters, Australia. Re-introduction News 19: 34–36.

Smart, N. 1983. Worldviews: Crosscultural explorations of human beliefs. Charles Scribner's Sons, New York.

Smathers, G. A., and D. Mueller-Dombois. 1974. Invasion and recovery of vegetation after a volcanic eruption in Hawaii. National Park Service Scientific Monograph Series 5. Island Ecosystems IRP/IBP Hawaii. NPS 118. Government Printing Office, Washington, DC.

Smith, C. W. 1985. Impact of alien plants on Hawai'i's native biota. Pp. 180–250 in C. P. Stone and J. M Scott (editors), Hawai'i's terrestrial ecosystems: Preservation and management. Cooperative National Park Resources Studies Unit, University of Hawaii, Honolulu.

———. 1991. The alien plant problem in Hawaii. Pp. 327–338 in T. D. Center, R. F. Doren, R. L. Hofstetter, R. L. Myers, and L. D. Whiteaker (editors), Proceedings of the Symposium on Exotic Plant Pests. U.S. Department of the Interior, National Park Service, Washington, DC.

Smith, C. W., and J. T. Tunison. 1992. Fire and alien plants in Hawai'i: Research and management implications for native ecosystems. Pp. 394–408 in C. P. Stone, C. W. Smith, and J. T. Tunison (editors), Alien plant invasions in native ecosystems of Hawai'i: Management and research. University of Hawaii Cooperative National Park Resources Studies Unit. University of Hawaii Press, Honolulu.

Smith, D. G., J. T. Polhemus, and E. A. VanderWerf. 2000. Efficacy of fish-flavored diphacinone blocks for controlling small Indian mongoose (Herpestes auropunctatus) populations in Hawai'i. 'Elepaio 60: 47–51.

———. 2002. Comparison of managed and unmanaged Wedge-tailed Shearwater colonies on O'ahu: Effects of predation. Pacific Science 56: 451–457.

Smith, D. L., B. Lucey, L. A. Waller, J. E. Childs, and L. A. Real. 2002. Predicting the spatial dynamics of rabies epidemics on heterogeneous landscapes. Proceeding of the National Academy of Sciences (USA) 99: 3668–3672.

Smith, K. F., A. P. Dobson, F. E. McKenzie, L. A. Real, D. L. Smith, and M. A. Wilson. 2005. Ecological theory to enhance infectious disease control and public health policy. Frontiers in Ecology and the Environment 3: 29–37.

Smith, T. B. 1987. Bill size polymorphism and interspecific niche utilization in an African finch. Nature 329: 717–719.

Smith, T. B., L. A. Freed, J. K. Lepson, and J. H. Carothers. 1995. Evolutionary consequences of extinctions in populations of a Hawaiian honeycreeper. Conservation Biology 9: 107–113.

Smucker, T. D., G. D. Lindsey, and S. M. Mosher. 2000. Home range and diet of feral cats in Hawaii forests. Pacific Conservation Biology 6: 229–237.

Snetsinger, T. J., S. G. Fancy, J. C. Simon, and J. D. Jacobi. 1994. Diets of owls and feral cats in Hawai'i. 'Elepaio 54: 47–50.

Snetsinger, T. J., C. M. Herrmann, D. E. Holmes, C. D. Hayward, and S. G. Fancy. 2005. Breeding ecology of the Puaiohi (Myadestes palmeri). Wilson Bulletin 117: 72–84.

Snetsinger, T. J., M. H. Reynolds, and C. M. Herrmann. 1998. 'Ō'ū (Psittirostra psittacea) and Lānai Hookbill (Dysmorodrepanis munroi). In A. Poole and F. Gill (editors), The birds of North America, No. 335–336. Birds of North America Inc., Philadelphia.

Snetsinger, T. J., K. M. Wakelee, and S. G. Fancy. 1999. Puaiohi (Myadestes palmeri). In A. Poole and F. Gill (editors), The birds of North America, No. 461. Birds of North America Inc., Philadelphia.

Snyder, N. F. R., S. R. Derrickson, S. R. Beissinger, J. W. Wiley, T. B. Smith, W. D. Toone, and B. Miller. 1997a. Letter to the editor. Conservation Biology 11: 3–4.

———. 1997b. Limitations of captive breeding: Reply to Gippoliti and Carpento. Conservation Biology 11: 808–810.

Snyder, N. F. R., and H. Snyder. 2000. The California Condor: A saga of natural history and conservation. Academic Press, San Diego, CA.

Sockman, K. W. 1997. Variation in life-history traits and nest-site selection affects risk of nest predation in the California Gnatcatcher. Auk 114: 324–332.

Sodhi, N. S., L. W. Liow, and F. A. Bazzaz. 2004. Avian extinctions from tropical and subtropical forests. Annual Review of Ecology, Evolution, and Systematics 35: 323–345.

Sokal, R. R., and F. J. Rohlf. 1995. Biometry: The principles and practice of statistics in biological research. 3rd edition. W. H. Freeman, New York.

Sol, D. 2000. Are islands more susceptible to be invaded than continents? Birds say no. Ecography 23: 687–692.

Sol, D., S. Timmermans, and L. Lefebvre. 2002. Behavioural flexibility and invasion success in birds. Animal Behaviour 63: 495–502.

Soorae, P. S., and P. J. Seddon (editors). 1998. Re-introduction practitioners directory. Published jointly by the International Union for Conservation of Nature Species Survival Commission's Reintroduction Specialist Group, Nairobi, Kenya, and the National Commission for Wildlife Conservation and Development, Riyadh, Saudi Arabia.

——— (editors). 2000. Special bird issue. Reintroduction News 20.

Sorenson, M. D., A. Cooper, E. E. Paxinos, T. W. Quinn, H. F. James, S. L. Olson, and R. C. Fleischer. 1999. Relationships of the extinct moa-nalos, flightless Hawaiian waterfowl, based on ancient DNA. Proceedings of the Royal Society of London, Series B 266: 2187–2193.

Soulé, M., M. Gilpin, W. Conway, and T. Foose. 1986. The millenium ark: How long a voyage, how many staterooms, how many passengers? Zoo Biology 5: 101–113.

Soulé, M. E. 1980. Thresholds for survival: Maintaining fitness and evolutionary potential. Pp. 151–159 in M. E. Soulé and B. A. Wilcox (editors), Conservation biology: An evolutionary-ecological perspective. Sinauer, Sunderland, MA.

Spalding, A. T. 1936. Letter to Hilo Chamber of Commerce. In the letters of Winston Banko, located at the archives of the U.S. Geological Survey, Pacific Island Ecosystems Research Center, Hawai'i Volcanoes National Park, Volcano, HI.

Spatz, G., and D. Mueller-Dombois. 1973. The influence of feral goats on koa tree reproduction in Hawaii Volcanoes National Park. Ecology 54: 870–876.

Speakman, J. R. 1991. The impact of predation by birds on bat populations in the British Isles. Mammal Review 21: 123–142.

Spiegel, C. S., P. J. Hart, B. L. Woodworth, E. J. Tweed, and J. J. LeBrun. 2006. Distribution and abundance of native forest birds in low-elevation areas on Hawai'i Island: Evidence of range expansion. Bird Conservation International 16: 175–185.

Spielman, A. 1994. Why entomological antimalaria research should not focus on transgenic mosquitoes. Parasitology Today 10: 374–376.

Spielman, A., U. Kitron, and R. J. Pollack. 1993. Time limitation and the role of research in the worldwide attempt to eradicate malaria. Journal of Medical Entomology 30: 6–19.

Spurr, E. B. 2002. Bait buckets for helicopters. Contract Report LC0102/062. Manaaki Whenua Landcare Research, Lincoln, New Zealand.

Spurr, E. B., G. D. Lindsey, C. Forbes Perry, and D. Foote. 2003a. Effectiveness of hand broadcast application of pelletised bait containing 0.005% diphacinone in reducing rat populations in Hawaiian forests. Unpublished report QA-02a. Pacific Island Ecosystems Research Center, Hawai'i Volcanoes National Park, Volcano, HI.

———. 2003b. Efficacy of aerial broadcast of pelletised bait containing 0.005% diphacinone in reducing rat populations in Hawaiian forests. Un-published report QA-02b. Pacific Island Ecosystems Research Center, Hawai'i Volcanoes National Park, Volcano, HI.

Stannard, D. E. 1989. Before the horror: The population of Hawaii on the eve of Western contact. Social Science Research Institute, University of Hawai'i, Honolulu.

Staples, G. W., and D. R. Herbst. 2005. A tropical garden flora: Plants cultivated in the Hawaiian Islands and other tropical places. Bishop Museum Press, Honolulu.

Staples, G. W., D. R. Herbst, and C. T. Imada. 2000. Survey of invasive or potentially invasive plants in Hawaii. Bishop Museum Occasional Papers 65: 1–35.

Stattersfield, A. J., M. J. Crosby, A. J. Long, and D. C. Wege. 1998. Endemic bird areas of the world: Priorities for biodiversity conservation. BirdLife Conservation Series 7. BirdLife International, Cambridge, England.

Steadman, D. W. 1995. Prehistoric extinctions of Pacific island birds: Biodiversity meets zoo-archaeology. Science 267: 1123–1131.

———. 2006. Extinction and biogeography of tropical Pacific birds. University of Chicago Press, Chicago.

Stearns, S. C. 1992. The evolution of life histories. Oxford University Press, New York.

Steffan, W. A. 1968. Hawaiian Toxorhynchites (Diptera: Culicidae). Proceedings of the Hawaiian Entomology Society 20: 141–155.

Steidl, R. J. 2001. Practical and statistical considerations for designing population monitoring programs. Pp. 284–288 in R. Field, R. J. Warren, H. Okarma, and P. R. Sievert (editors), Wildlife, land and people: Priorities for the 21st century. Proceedings of the Second International Wildlife Management Congress, Wildlife Society, Bethesda, MD.

Stemmermann, L., and T. Ihsle. 1993. Replacement of Metrosideros polymorpha, 'ōhi'a, in Hawaiian dry forest succession. Biotropica 25: 36–45.

St. John, H. 1947. The history, present distribution, and abundance of sandalwood on Oahu, Hawaiian Islands. Hawaiian Plant Studies 14. Pacific Science 1: 5–20.

St. Louis, V. L., and J. C. Barlow. 1991. Morphometric analyses of introduced and ancestral populations of the Eurasian Tree Sparrow. Wilson Bulletin 103: 1–12.

Stokstad, E. 2004. Hawaii girds itself for arrival of West Nile virus. Science 306: 603.

Stone, C. P. 1985. Alien animals in Hawai'i's native ecosystems: Toward controlling the adverse effects of introduced vertebrates. Pp. 251–297 in C. P. Stone and J. M Scott (editors), Hawai'i's terrestrial ecosystems: Preservation and management. Cooperative National Park Resources Studies Unit, University of Hawai'i, Honolulu.

———. 1995. Toward ethical treatment of animals in Hawai'i's natural areas. Pacific Science 49: 98–108.

Stone, C. P., L. W. Cuddihy, and J. T. Tunison. 1992. Responses of Hawaiian ecosystems to removal of feral pigs and goats. Pp. 666–704 in C. P. Stone, C. W. Smith, and J. T. Tunison (editors), Alien plant invasions in native ecosystems of Hawai'i: Management and research. University of Hawaii Cooperative National Park Resources Studies Unit. University of Hawai'i Press, Honolulu.

Stone, C. P., M. Dusek, and M. Aeder. 1995. Use of an anticoagulant to control mongooses in Nene breeding habitat. 'Elepaio 54: 73–78.

Stone, C. P., and L. L. Loope. 1987. Reducing negative effects of introduced animals on native biotas in Hawaii: What is being done, what needs doing, and the role of national parks. Environmental Conservation 14: 245–258.

Stone, C. P., and L. W. Pratt. 2002. Hawai'i's plants and animals: Biological sketches of Hawaii Volcanoes National Park. Hawaii Natural History Association, National Park Service, and University of Hawaii Cooperative National Park Resources Studies Unit. University of Hawai'i Press, Honolulu.

Stone, W. B., J. C. Okoniewski, and J. R. Stedelin. 1999. Poisoning of wildlife with anticoagulant rodenticides in New York. Journal of Wildlife Diseases 35: 187–193.

Stoskopf, M. K., and J. Beier. 1979. Avian malaria in African black-footed penguins. Journal of the American Veterinary Medicine Association 175: 944–947.

Stratford, J. A., and W. D. Robinson. 2005. Gulliver travels to the fragmented tropics: Geographic variation in mechanisms of avian extinction. Frontiers in Ecology and Environment 3: 91–98.

Stutchbury, B. J. M., and E. S. Morton. 2001. Behavioral ecology of tropical birds. Academic Press, San Diego, CA.

Sugihara, R. T. 1997. Abundance and diets of rats in two native Hawaiian forests. Pacific Science 51: 189–198.

———. 2002. Rodent damage research in Hawaii: Changing times and priorities. Proceedings of the Vertebrate Pest Conference 20: 40–45.

Sugihara, R. T., M. E. Tobin, and A. E. Koehler. 1995. Zinc phosphide baits and prebaiting for controlling rats in Hawaiian sugarcane. Journal of Wildlife Management 59: 882–889.

Sweeting, J. E. N., A. G. Bruner, and A. B. Rosenfeld. 1999. The green host effect: An integrated approach to sustainable tourism and resort development. Conservation International Policy Paper. Conservation International, Washington, DC.

Swenson, J. 1986. Is Manana Island now "Rabbitless Island?" 'Elepaio 46: 125–126.

Swezey, O. H. 1926. Reports on defoliation of Maui koa forests by caterpillars. Hawaiian Planter's Record 30: 354–362.

———.1927. Hyposoter exigue (Viereck) in Hawaii. Proceedings of the Hawaii Entomological Society 6: 404–407.

———. 1954. Forest entomology in Hawaii: An annotated check-list of the insect faunas of the various components of the Hawaiian forests. B. P. Bishop Museum Special Publication 44: 1–266.

Swift, C. E. 1998. Laboratory assays with wild-caught roof (Rattus rattus) and Polynesian (R. exulans) rats to determine minimum amounts of Ramik® Green (0.005% diphacinone) and exposure times for field broadcast applications in Hawaii. M.S. thesis. University of Hawai'i at Mānoa, Honolulu.

Swinnerton, K. J., C. G. Jones, R. Lam, S. Paul, R. Chapman, K. A. Murray, and K. Freeman. 2000. Conservation of the pink pigeon in Mauritius. Re-introduction News 19: 10–12.

Swinnerton, K. J., M. A. Peirce, A. Greenwood, R. E. Chapman, and C. G. Jones. 2005. Prevalence of Leucocytozoon marchouxi in the endangered Pink Pigeon Columba mayer. Ibis 147: 725–737.

Sykes, P. W. Jr., A. K. Kepler, C. B. Kepler, and J. M. Scott. 2000. Kaua'i 'Ō'ō (Moho braccatus), O'ahu 'Ō'ō (Moho apicalis), Bishop's 'Ō'ō (Moho bishopi), Hawai'i 'Ō'ō (Moho nobilis), and Kioea (Chaetoptila angustipluma). In A. Poole and F. Gill (editors), The birds of North America, No. 535. Birds of North America Inc., Philadelphia.

Tabachnick, W. J. 2003. Reflections on the Anopheles gambiae genome sequence, transgenic mosquitoes and the prospect for controlling malaria and other vector borne diseases. Journal of Medical Entomology 40: 597–606.

Tanaka, K., K. Mizusawa, and E. S. Saugstad. 1979. A revision of the adult and larval mosquitoes of Japan (including the Ryukyu Archipelago and the Ogasawara Islands) and Korea (Diptera: Culicidae). Contributions of the American Entomological Institute (Ann Arbor, MI) 16: 1–987.

Tanner, C., and M. Nunn. 1998. Australian quarantine post the Nairn Review. Australian Journal of Agricultural and Resource Economics 42: 445–458.

Tarr, C. L., S. Conant, and R. C. Fleischer. 1998. Founder events and variation at microsatellite loci in an insular passerine bird, the Laysan Finch (Telespiza cantans). Molecular Ecology 7: 719–731.

Tarr, C. L., and R. C. Fleischer. 1993. Mitochondrial DNA variation and evolutionary relationships in the 'amakihi complex. Auk 110: 825–831.

———. 1995. Evolutionary relationships of the Hawaiian honeycreepers (Aves: Drepanidinae). Pp. 147–159 in W. Wagner and V. Funk (editors), Hawaiian biogeography: Evolution in a hot-spot archipelago. Smithsonian Institution Press, Washington, DC.

———. 1999. Population boundaries and genetic diversity in the endangered Mariana crow (*Corvus kubaryi*). Molecular Ecology 8: 941–949.

Taylor, B. L., and T. Gerrodette. 1993. The use of statistical power in conservation biology: The vaquita and northern spotted owl. Conservation Biology 7: 489–500.

Taylor, D. 1994. Restoring endangered species in Hawaii Volcanoes National Park. Endangered Species Technical Bulletin 19: 18–19.

Taylor, M. F. J., K. F. Suckling, and J. J. Rachlinski. 2005. The effectiveness of the endangered species act: A quantitative analysis. BioScience 55: 360–367.

Taylor, T. H. 1979. How the Macquarie Island parakeet became extinct. New Zealand Journal of Ecology 2: 42–45.

Teauotalani, Z. 1865. Short notes pertaining to Hawaiian life: Descriptions of Hawaiian birds. Hawaiian Ethnological Notes, Vol. 1: 192–196. Courtesy of the Bishop Museum Archives, Honolulu, HI.

Telfer, T. C. 1977. Report on the apparent establishment of Small Indian Mongoose (*Herpestes auropuctatus*) on Kauai. Unpublished report. Hawaii Division of Fish and Game, Honolulu.

———. 1988. Status of black-tailed deer on Kauai. Transactions of the Western Section of the Wildlife Society 24: 53–60.

Tempelis, C. H., R. O. Hayes, A. D. Hess, and W. C. Reeves. 1970. Blood-feeding habitats of four species of mosquito found in Hawaii. American Journal of Tropical Medicine and Hygiene 19: 335–341.

Temple, S. A. 1984. Final report on an evaluation of Hawaii's endangered species captive rearing project (Contract UC1983-1(W)042). Unpublished report to the State of Hawaii, Department of Land and Natural Resources, Honolulu, dated September 21.

———. 1986. The problem of avian extinctions. Current Ornithology 3: 453–485.

Temple, S. A., and D. A. Jenkins. 1981. Unpublished final progress report: 1979 and 1980 Alala Research. Submitted to the U.S. Forest Service, Pacific Southwest Forest and Range Experiment Station, Institute of Pacific Islands Forestry, Honolulu.

TenBruggencate, J. 1992. Mystery of the starving owls still unsolved. Honolulu Star-Bulletin, March 15.

Te Rangi Hiroa. 1957. Arts and crafts of Hawai'i, Section 11: Religion. Bernice P. Bishop Museum Special Publication 45: 502–532.

Terborgh, J., L. Lopez, P. Nuñez V., M. Rao, G. Shahabuddin, G. Orihuela, M. Riveros, R. Ascanio, G. H. Adler, T. D. Lambert, and L. Balbas. 2001. Ecological meltdown in predator-free forest fragments. Science 294: 1923–1926.

Terborgh, J., and B. Winter. 1980. Some causes of extinction. Pp. 119–134 in M. E. Soulé and B. A. Wilcox (editors), Conservation biology: An evolutionary-ecological perspective. Sinauer, Sunderland, MA.

Teshima, A. H. 1976. Anticoagulants—a problem of distribution for the Hawaiian sugar industry. Proceedings of the Vertebrate Pest Conference 7: 121–124.

The Nature Conservancy. 2004. State level conservation funding mechanisms. Government Relations Department, The Nature Conservancy, Arlington, VA.

The Nature Conservancy of Hawaii and Natural Resources Defense Council. 1992. The alien pest invasion in Hawaii: Background study and recommendations for interagency planning. The Nature Conservancy of Hawai'i and Natural Resources Defense Council. Joint agency report. www.hear.org/articles/pdfs/nrdctnch1992.pdf (accessed March 23, 2008).

Thibault, J.-C., J.-L. Martin, A. Penloup, and J.-Y. Meyer. 2002. Understanding the decline and extinction of monarchs (Aves) in Polynesian Islands. Biological Conservation 108: 161–174.

Thistle, A. 1959. Report on Barn Owls. 'Elepaio 19: 81.

Thomas, L. 1997. Retrospective power analysis. Conservation Biology 11: 276–280.

Thomas, L., J. L. Laake, S. Strindberg, F. F. C. Marques, S. T. Buckland, D. L. Borchers, D. R. Anderson, K. P. Burnham, S. L. Hedley, J. H. Pollard, J. R. B. Bishop, and T. Marques. 2006. Distance 5.0. Release 2. Research Unit for Wildlife Population Assessment, University of St. Andrews, Scotland. www.ruwpa.st-and.ac.uk/distance/.

Thompson, R. 2003. Aid increased for injured birds on Big Isle. Honolulu Star-Bulletin, May 2.

Thrum, T. G. 1907. Hawaiian folk tales. A. C. McClurg and Co., Chicago.

Timm, R. M. 1994. Anticoagulants, description of active ingredients. Pp. G26–G29 in S. E. Hygnstrom, R. M. Timm, and G. E. Larson (editors), Prevention and control of wildlife damage, Vol. 2. University of Nebraska Cooperative Extension, Lincoln.

Tobin, M. E. 1992. Rodent damage in Hawaiian macadamia orchards. Proceedings of the Vertebrate Pest Conference 15: 272–276.

Tobin, M. E., A. E. Koehler, R. T. Sugihara, and M. E. Burwash. 1997. Repellency of mongoose feces and urine to rats (*Rattus* spp.). Pp. 285–300 in J. R. Mason (editor), Proceedings of the Second DWRC Special Symposium. National Wildlife Research Center, Fort Collins, CO.

Tobin, M. E., and R. T. Sugihara. 1992. Abundance and habitat relationships of rats in Hawaiian sugarcane fields. Journal of Wildlife Management 56: 816–822.

Tobin, M. E., R. T. Sugihara, and A. E. Koehler. 1997. Bait placement and acceptance by rats in macadamia orchards. Crop Protection 16: 507–510.

Tobin, M. E., R. T. Sugihara, A. E. Koehler, and G. R. Uenten. 1996. Seasonal activity and movements of *Rattus rattus* (Rodentia, Muridae) in a Hawaiian macadamia orchard. Mammalia 60: 3–13.

Tomich, P. Q. 1962. Notes on the Barn Owl in Hawaii. 'Elepaio 23: 16–17.

———. 1971. Notes on foods and feeding behavior of raptorial birds in Hawaii. 'Elepaio 31: 111–113.

———. 1981a. Rodents. Pp. 105–110 in D. Mueller-Dombois, K. Bridges, and H. L. Carson (editors), Island ecosystems: Biological organization in selected Hawaiian communities. U.S. International Biological Program Synthesis Series 15. Hutchinson Ross, Stroudsburg, PA.

———. 1981b. Community structure of introduced rodents and carnivores. Pp. 301–309 in D. Mueller-Dombois, K. Bridges, and H. L. Carson (editors), Island ecosystems: Biological organization in selected Hawaiian communities. U.S. International Biological Program Synthesis Series 15. Hutchinson Ross, Stroudsburg, PA.

———. 1986. Mammals in Hawai'i: A synopsis and notational bibliography. 2nd edition. Bishop Museum Special Publication 76. Bishop Museum Press, Honolulu.

Tomich, P. Q., A. M. Barnes, W. S. Devick, H. H. Higa, and G. E. Haas. 1984. Evidence for the extinction of plague in Hawaii. American Journal of Epidemiology 119: 261–273.

Torchin, M. E., K. D. Lafferty, A. P. Dobson, V. J. McKenzie, and A. M. Kuris. 2003. Introduced species and their missing parasites. Nature 421: 628–630.

Towns, D. R. 2002. Korapuki Island as a case study for restoration of insular ecosystems in New Zealand. Journal of Biogeography 29: 593–607.

Towns, D. R., and K. G. Broome. 2003. From small Maria to massive Campbell: Forty years of rat eradications from New Zealand islands. New Zealand Journal of Zoology 30: 377–398.

Towns, D. R., D. Simberloff, and I. A. E. Atkinson. 1997. Restoration of New Zealand islands: Redressing the effects of introduced species. Pacific Conservation Biology 3: 99–124.

Townsend, D. S., and M. M. Stewart. 1994. Reproductive ecology of the Puerto Rican frog *Eleutherodactylus coqui*. Journal of Herpetology 28: 34–40.

Townson, H. 2002. *Wolbachia* as a potential tool for suppressing filarial transmission. Annals of Tropical Medicine and Parasitology 96 (Supplement 2): S117–S127.

Tripathy, D. 1993. Poxviridae. Pp. 1–15 in J. B. McFerran and M. S. McNulty (editors), Virus infections of birds. Elsevier Science, New York.

Tripathy, D. N., W. M. Schnitzlein, P. J. Morris, D. L. Janssen, J. K. Zuba, G. Massey, and C. T. Atkinson. 2000. Characterization of poxviruses from forest birds in Hawaii. Journal of Wildlife Diseases 36: 225–230.

Trujillo, E. E. 1985. Biological control of Hamakua pamakani with *Cercosporella* sp. in Hawaii. Pp. 661–671 in E. S. Delfosse (editor), Proceedings of the VI symposium on the biological control of weeds. Agriculture Canada, Vancouver.

Trujillo, E. E., C. Kadooka, V. Tanimoto, S. Bergfeld, G. Shishido, and G. Kawakami. 2001. Effective biomass reduction of the invasive weed species banana poka by Septoria leaf spot. Plant Disease 85: 357–361.

Trujillo, E. E., F. M. Latterell, and A. E. Rossi. 1986. *Colletotrichum gloeosporioides*, a possible biological control agent for *Clidemia hirta* in Hawaiian forests. Plant Disease 70: 974.

Tummons, P. 1991a. Pu'uwa'awa'a burns, and the DLNR fiddles. Environment Hawai'i 1 (9): 1–7.

———. 1991b. State must live down the past if the 'Alala is to have a future. Environment Hawai'i 1 (10): 1–7.

———. 1991c. Private rights, public interests at center of 'Alala lawsuit. Environment Hawai'i 1 (10): 3.

———. 1991d. A decade-long standoff over access to wild birds. Environment Hawai'i 1 (10): 4.

———. 1996. Changing of the guard at Olinda: State disowns 'Alala breeding facility. Environment Hawai'i 6 (9): 1–5.

———. 1997. At Hakalau Refuge, hunter pressure overrides conservationists' concerns. Environment Hawai'i 8: 1–3.

———. 2000a. The Peregrine Fund transfers State contract to San Diego Zoo. Environment Hawai'i 10 (7): 10–11.

———. 2000b. Cattle, sheep, goats trample Pu'uwa'awa'a treasures. Environment Hawai'i 11 (3): 6–8.

———. 2003a. Government pays $8 million for refuge in South Kona, but has no legal access. Environment Hawai'i 14 (4): 1, 8–11.

———. 2003b. Ranchers press federal government to pay to move junk, wild cattle off refuge land. Environment Hawai'i 14 (5): 1–9.

———. 2005a. Land and Natural Resources Department suffers punishing blows in 2005 budget. Environment Hawai'i 15 (5): 1, 4–9.

———. 2005b. Eight years after purchasing Kona land, federal government finally gains access. Environment Hawai'i 15 (9): 3.

Tunison, J. T., C. M. D'Antonio, and R. K. Loh. 2001. Fire and invasive plants in Hawaii Volcanoes

National Park. Pp. 122–131 in K. E. M. Galley and T. P. Wilson (editors), Proceedings of the invasive species workshop: The role of fire in the control and spread of invasive species. Fire conference 2000: The first national congress on fire ecology, prevention, and management. Miscellaneous Publication 11. Tall Timbers Research Station, Tallahassee, FL.

Tunison, J. T., and J. Leialoha. 1988. The spread of fire in alien grasses after lightning strikes in Hawaii Volcanoes National Park. Newsletter of the Hawaiian Botanical Society 27: 102–109.

Tunison, J. T., R. K. Loh, and J. A. K. Leialoha. 1995. Fire effects in the submontane seasonal zone, Hawai'i Volcanoes National Park. Technical Report 97. Cooperative National Parks Studies Unit. University of Hawai'i at Mānoa, Honolulu.

Tunison, J. T., A. A. McKinney, and W. L. Markiewicz. 1995. The expansion of koa forest after cattle and goat removal, Hawai'i Volcanoes National Park. Technical Report 99. Cooperative National Parks Studies Unit. University of Hawai'i at Mānoa, Honolulu.

Tunison, J. T., N. G. Zimmer, M. R. Gates, and R. M. Mattos. 1994. Fountain grass control in Hawai'i Volcanoes National Park 1985–1992. Technical Report 91. Cooperative National Parks Studies Unit. University of Hawai'i, Honolulu.

Turell, M. J., M. L. O'Guinn, D. J. Dohm, and J. W. Jones. 2001. Vector competence of North American mosquitoes (Diptera: Culicidae) for West Nile virus. Journal of Medical Entomology 38: 130–134.

Tweed, E. J., J. T. Foster, B. L. Woodworth, W. B. Monahan, J. L. Kellerman, and A. Lieberman. 2006. Breeding biology and success of a reintroduced population of the critically endangered Puaiohi (Myadestes palmeri). Auk 123: 753–763.

Tweed, E. J., J. T. Foster, B. L. Woodworth, P. Oesterle, C. Kuehler, A. A. Lieberman, A. T. Powers, K. Whitaker, W. B. Monahan, J. Kellerman, and T. Telfer. 2003. Survival, dispersal, and home-range establishment of reintroduced captive-bred puaiohi, Myadestes palmeri. Biological Conservation 111: 1–9.

Tweed, E., P. M. Gorresen, R. J. Camp, P. J. Hart, and T. K. Pratt. 2007. Forest bird inventory of the Kahuku Unit of Hawai'i Volcanoes National Park. Technical Report 143. Pacific Cooperative Studies Unit, University of Hawai'i, Honolulu.

Urquhart, N. S., and T. M. Kincaid. 1999. Trend detection in repeated surveys of ecological responses. Journal of Agricultural, Biological and Environmental Statistics 4: 404–414.

U.S. Bureau of the Census. 1998. County population estimates and demographic components of population change: Annual time series, July 1, 1990 to July 1, 1998. www.hawaii.gov/abedt/db98/01/014798.pdf (accessed May 1, 2000).

U.S. Congress. 1973. Endangered Species Act of 1973. Public Law 93-205, 93rd Congress, December 28.

U.S. Department of Agriculture, Animal and Plant Health Inspection Service. 2004. Nursery stock regulations: Advance notice of proposed rulemaking and request for comments. U. S. Department of Agriculture, Animal and Plant Health Inspection Service. Federal Register 69 (237): 71736–71744.

U.S. Department of the Interior. 1994. Environmental assessment for the proposed captive propagation facility for endangered forest birds. U.S. Fish and Wildlife Service, Honolulu, December 2.

———. 2000. Policy regarding controlled propagation of species listed under the Endangered Species Act. Federal Register 65 (183): 56916–56922.

U.S. Environmental Protection Agency. 2000. Guidance for the data quality objectives process. EPA QA/G-4 EPA/600/R-96-055. U.S. Environmental Protection Agency, Washington, DC.

U.S. Fish and Wildlife Service. 1975. Endangered and threatened wildlife, listings of endangered and threatened fauna. Federal Register 40: 44151.

———. 1977. Determination of critical habitat for six endangered species. Federal Register 42: 40685–40690.

———. 1982a. Alala recovery plan. U.S. Fish and Wildlife Service, Portland, OR.

———. 1982b. The Hawaii forest bird recovery plan. U.S. Fish and Wildlife Service, Portland, OR.

———. 1983a. Nene recovery plan. U.S. Fish and Wildlife Service, Portland, OR.

———. 1983b. Kauai forest birds recovery plan. U.S. Fish and Wildlife Service, Portland, OR.

———. 1983c. Endangered and threatened species listing and recovery priority guidelines. Federal Register 48: 43098, 51985.

———. 1984a. Hawaiian Hawk recovery plan. U.S. Fish and Wildlife Service, Portland, OR.

———. 1984b. Recovery plan for the Northwestern Hawaiian Islands passerines. U.S. Fish and Wildlife Service, Portland, OR.

———. 1984c. Maui-Molokai forest birds recovery plan. U.S. Fish and Wildlife Service, Portland, OR.

———. 1986. Recovery plan for the Palila. U.S. Fish and Wildlife Service, Portland, OR.

———. 1994. Endangered and threatened wildlife and plants: Withdrawal of proposed rule to reclassify the Hawaiian Hawk (Buteo solitarius) from endangered to threatened. U.S. Fish and Wildlife Service, Honolulu.

———. 1996a. Policy regarding the recognition of distinct vertebrate population segments under the endangered species act. Federal Register 61: 4722.

———. 1996b. Feral ungulate management plan. U.S. Fish and Wildlife Service, Portland, OR.

———. 1998. Biological opinion of the U.S. Fish and Wildlife Service for the Saddle Road realignment and improvement project. U.S. Fish and Wildlife Service, Honolulu, July 27.

———. 2004a. Draft revised recovery plan for the Nēnē. Region 1, U.S. Fish and Wildlife Service, Portland, OR.

———. 2004b. Draft revived recovery plan for Laysan Duck (*Anas laysanensis*). U.S. Fish and Wildlife Service, Region 1, Portland, OR.

———. 2004c. Federal and state endangered and threatened species expenditures, Fiscal Year 2004. U.S. Fish and Wildlife Service, Washington, DC.

———. 2004d. Endangered and threatened wildlife and plants: Notice of the availability of draft economic analyses, and of a public hearing for the proposed designations of critical habitat for the Coastal California Gnatcatcher (*Polioptila californica californica*) and the San Diego Fairy Shrimp (*Branchinecta sandiegonensis*). Federal Register 69: 18516.

———. 2005. Draft revised recovery plan for the Aga or Mariana Crow (*Corvus kubaryi*). U.S. Fish and Wildlife Service, Region 1, Portland, OR.

———. 2006. Revised recovery plan for Hawaiian forest birds. U.S. Fish and Wildlife Service, Region 1, Portland, OR.

———. 2008a. Endangered and threatened wildlife and plants: Withdrawal of proposed reclassification of the Hawaiian Hawk or ʻIo (*Buteo solitarius*) from endangered to threatened; proposed rule to remove the Hawaiian Hawk from the federal list of endangered and threatened wildlife. Federal Register 73: 45680–45689.

———. 2008b. Report on U.S. Fish and Wildlife Service's Implementing Recovery for Endangered Forest Bird Species in Hawaiʻi Workshop, October 8–10, Hilo, HI. Unpublished report. U.S. Fish and Wildlife Service, Honolulu, HI.

———. 2009. Revised recovery plan for the ʻAlalā (*Corvus hawaiiensis*). U.S. Fish and Wildlife Service, Region 1, Portland, OR.

U.S. Fish and Wildlife Service and U.S. Census Bureau. 2002. 2001 National survey of fishing, hunting, and wildlife-associated recreation. U.S. Fish and Wildlife Service and U.S. Census Bureau, Washington, DC.

U.S. Geological Survey. 2006. Palila restoration: Lessons from long-term research. USGS Fact Sheet 2006-3104. U.S. Geological Survey, Reston, VA.

U.S. Government Accountability Office. 2005a. Fish and Wildlife Service generally focuses recovery funding on high priority species, but needs to periodically assess its funding decisions. USGAO Report GAO-05-211. U.S. Government Accountability Office, Washington, DC.

———. 2005b. Homeland security: Much is being done to protect agriculture from a terrorist attack, but important challenges remain. Report to Congressional Requesters, GAO-05-214. U.S. Government Accountability Office, Washington, DC.

Valkiūnas, G. 2005. Avian malaria parasites and other haemosporidia. CRC Press, Boca Raton, FL.

Valkiūnas, G., P. Zehtindjiev, O. Hellgren, M. Ilieva, T. A. Iezhova, and S. Bensch. 2007. Linkage between mitochondrial cytochrome b lineages and morphospecies of two avian malaria parasites, with a description of *Plasmodium* (*Novyella*) *ashfordi* sp. n. Parasitology Research 100: 1311–1322.

Valutis, L. L., and J. M. Marzluff. 1999. The appropriateness of puppet-rearing birds for reintroduction. Conservation Biology 13: 584–591.

Van Bael, S. A., J. D. Brawn, and S. K. Robinson. 2003. Birds defend trees from herbivores in a neotropical forest canopy. Proceedings of the National Academy of Sciences (USA) 100: 8304–8307.

VanderWerf, E. A. 1994. Intraspecific variation in ʻElepaio foraging behavior in Hawaiian forests of different structure. Auk 111: 915–930.

———. 1997. Oʻahu ʻAmakihi nest in Mānoa Valley. ʻElepaio 57: 125–126.

———. 1998a. ʻElepaio (*Chasiempis sandwichensis*). In A. Poole and F. Gill (editors), The birds of North America, No. 344. Birds of North America Inc., Philadelphia.

———. 1998b. Breeding biology and territoriality of the Hawaii Creeper. Condor 100: 541–545.

———. 2001a. Distribution and potential impacts of avian poxlike lesions in ʻElepaio at Hakalau Forest National Wildlife Refuge. Studies in Avian Biology 22: 247–253.

———. 2001b. Rodent control decreases predation on artificial nests in Oʻahu ʻElepaio habitat. Journal of Field Ornithology 72: 448–457.

———. 2001c. Two-year delay in plumage maturation of male and female ʻElepaio. Condor 103: 756–766.

———. 2004. Demography of Hawaiʻi ʻElepaio: Variation with habitat disturbance and population density. Ecology 85: 770–783.

———. 2007. Biogeography of ʻElepaio: Evidence from inter-island song playbacks. Wilson Journal of Ornithology 119: 325–333.

VanderWerf, E. A., M. D. Burt, J. L. Rohrer, and S. M. Mosher. 2006. Distribution and prevalence of mosquito-borne diseases in Oʻahu ʻElepaio. Condor 108: 770–777.

VanderWerf, E. A., A. Cowell, and J. L. Rohrer. 1997. Distribution, abundance, and conservation of Oʻahu ʻElepaio in the southern leeward Koʻolau Range. ʻElepaio 57: 99–106.

VanderWerf, E. A., J. J. Groombridge, J. S. Fretz, and K. J. Swinnerton. 2006. Decision-analysis to guide recovery of the Poʻouli, a critically endangered Hawaiian honeycreeper. Biological Conservation 129: 383–392.

VanderWerf, E. A., T. R. Malcolm, J. S. Fretz, J. G. Massey, A. Lieberman, J. J. Groombridge, B. D. Sparklin, M. M. Okada, and C. N. Brosius. 2003. Update on recovery efforts for the Poʻo-uli. ʻElepaio 63: 25–30.

VanderWerf, E. A., and J. L. Rohrer. 1996. Discovery of an ʻIʻiwi population in the Koʻolau Mountains of Oʻahu. ʻElepaio 56: 25–28.

VanderWerf, E. A., J. L. Rohrer, D. G. Smith, and M. D. Burt. 2001. Current distribution and abundance of the Oʻahu ʻElepaio. Wilson Bulletin 113: 10–16.

VanderWerf, E. A., and D. G. Smith. 2002. Effects of alien rodent control on demography of the Oʻahu ʻElepaio, an endangered Hawaiian forest bird. Pacific Conservation Biology 8: 73–81.

Van Dine, D. L. 1904. Mosquitoes in Hawaii. Hawaii Agricultural Experimental Station Bulletin 6: 7–30.

———. 1907. The introduction of top-minnows (natural enemies of the mosquitoes) into the Hawaiian Islands. Hawaiian Forester and Agriculturist 20: 1–10.

Van Driesche, J. 2002. Using the internet to build a conservation network. Conservation in Practice 3: 36–40.

Van Driesche, J., and R. Van Driesche. 2000. Nature out of place: Biological invasions in the global age. Island Press, Washington, DC.

VanGelder, E. M., and T. B. Smith. 2001. Breeding characteristics of the ʻĀkohekohe on east Maui. Studies in Avian Biology 22: 194–201.

van Rensburg, P. J., J. D. Skinner, and R. J. van Aarde. 1987. Effects of feline panleucopaenia on the population characteristics of feral cats on Marion Island. Journal of Applied Ecology 24: 36–73.

van Riper, C. III. 1973. Island of Hawaii land bird distribution and abundance. ʻElepaio 34: 1–3.

———. 1978. Breeding ecology of the Amakihi (Loxops virens) and Palila (Psittirostra bailleui) on Mauna Kea, Hawaii. Ph.D. dissertation. University of Hawaii at Mānoa, Honolulu.

———. 1980a. The phenology of the dryland forest of Mauna Kea, Hawaii, and the impact of recent environmental perturbations. Biotropica 12: 282–291.

———. 1980b. Observations on the breeding of the Palila Psittirostra bailleui of Hawaii. Ibis 122: 462–475.

———. 1984. The influence of nectar resources on nesting success and movement patterns of the Common Amakihi. Auk 101: 38–46.

———. 1987. Breeding ecology of the Hawaii Common Amakihi. Condor 89: 85–102.

———. 1995. Ecology and breeding biology of the Hawaii Elepaio (Chasiempis sandwichensis bryani). Condor 97: 512–527.

van Riper, C. III, C. T. Atkinson, and T. M. Seed. 1994. Plasmodia of birds. Pp. 73–140 in J. P. Krier (editor), Parasitic protozoa, Vol. 7. Academic Press, New York.

van Riper C. III, and D. J. Forrester. 2007. Avian pox. Pp. 131–176 in N. J. Thomas, D. B. Hunter, and C. T. Atkinson (editors), Infectious diseases of wild birds. Blackwell, Ames, IA.

van Riper, C. III, M. D. Kern, and M. K. Sogge. 1993. Changing nest placement of Hawaiian Common Amakihi during the breeding cycle. Wilson Bulletin 105: 436–447.

van Riper, C. III, and J. M. Scott. 1979. Observations on distribution, diet, and breeding of the Hawaiian Thrush. Condor 81: 65–71.

———. 2001. Limiting factors affecting Hawaiian native birds. Studies in Avian Biology 22: 221–233.

van Riper, C. III, J. M. Scott, and D. M. Woodside. 1978. Distribution and abundance patterns of the Palila on Mauna Kea, Hawaii. Auk 95: 518–527.

van Riper, C. III, S. G. van Riper, M. L. Goff, and M. Laird. 1982. The impact of malaria on birds in Hawaii Volcanoes National Park. Technical Report 47. Cooperative National Park Resources Studies Unit, University of Hawaiʻi at Mānoa, Honolulu.

———. 1986. The epizootiology and ecological significance of malaria in Hawaiian land birds. Ecological Monographs 56: 327–344.

van Riper, C. III, S. G. van Riper, and W. R. Hansen. 2002. Epizootiology and effect of avian pox on Hawaiian forest birds. Auk 119: 929–942.

van Riper, S. G., and C. van Riper III. 1985. A summary of known parasites and diseases recorded from the avifauna of the Hawaiian Islands. Pp. 298–371 in C. P. Stone and J. M. Scott (editors), Hawaiʻi's terrestrial ecosystems: Preservation and management. Cooperative National Park Resources Studies Unit, University of Hawaiʻi, Honolulu.

van Wilgen, B. W., M. P. de Wit, and H. J. Anderson. 2004. Costs and benefits of biological control of invasive alien plants: Case studies from South Africa. South African Journal of Science 100: 113–122.

Veillet A., R. Shrestha, and D. K. Price. 2008. Polymorphic microsatellites in nēnē, the endangered Hawaiian Goose (Branta sandvicensis). Molecular Ecology Resources 8: 1158–1160.

Veitch, C. R. 1985. Methods of eradicating feral cats from offshore islands in New Zealand. Pp. 125–155 in P. J. Moors (editor), Conservation of island birds. International Council Bird Preservation, Cambridge, England.

Veitch, C. R., and M. N. Clout (editors). 2002. Turning the tide: The eradication of invasive species. IUCN SSC Invasive Species Specialist Group. International Union for Conservation of Nature, Gland, Switzerland, and Cambridge, England.

Veltman, C. J., S. Nee, and M. J. Crawley. 1996. Correlates of introduction success in exotic

New Zealand birds. American Naturalist 147: 542–557.

Verbeek, N. A. M., and R. W. Butler. 1999. Northwestern Crow (*Corvus caurinus*). In A. Poole and F. Gill (editors), The birds of North America, No. 407. Birds of North America Inc., Philadelphia.

Verbeek, N. A. M., and C. Caffrey. 2002. American Crow (*Corvus brachyrhynchos*). In A. Poole and F. Gill (editors), The birds of North America, No. 647. Birds of North America Inc., Philadelphia.

Vince, M. 1996. Softbills: Care, breeding and conservation. Hancock House, Blaine, WA.

Vincek, V., C. O'Huigin, Y. Satta, Y. Takahata, P.T. Boag, P. R. Grant, B. R. Grant, and J. Klein. 1997. How large was the founding population of Darwin's finches? Proceedings of the Royal Society of London, Series B 264: 111–118.

Vitousek, P. M. 1990. Biological invasions and ecosystem process—towards an integration of population biology and ecosystem studies. Oikos 57: 7–13.

Vitousek, P. M., T. N. Ladefoged, P. V. Kirch, A. S. Hartshorn, M. W. Craves, S. C. Hotchkiss, S. Tuljapurkar, and O. A. Chadwick. 2004. Soils, agriculture, and society in precontact Hawai'i. Science 304: 1665–1669.

Vitousek, P. M., and L. R. Walker. 1989. Biological invasion by *Myrica faya* in Hawaii: Plant demography, nitrogen fixation and ecosystem effects. Ecological Monographs 59: 247–265.

Vitousek, P. M., L. R. Walker, L. D. Whiteaker, D. Mueller-Dombois, and P. A. Matson. 1987. Biological invasion by *Myrica faya* alters ecosystem development in Hawaii. Science 238: 802–804.

Viverette, C. B., S. Struve, L. J. Goodrich, and K. L. Bildstein. 1996. Decreases in migrating Sharpshinned Hawks (*Accipiter striatus*) at traditional raptor-migration watch-sites in eastern North America. Auk 113: 32–40.

Voge, M., and B. S. Davis. 1953. Studies on the cestode genus *Anonchotaenia* (Dilepididae, Paruterininae) and related forms. University of California Publications in Zoology 59: 1–30.

Vogel, G. 2002. An elegant but imperfect tool. Science 298: 94–95.

Vogl, R. J. 1977. Fire: A destructive menace or a natural process? Pp. 261–289 in J. Cairns Jr., K. L. Dickson, and E. E. Herricks (editors), Recovery and restoration of damaged ecosystems: Proceedings of the international symposium, March 23–25, 1975, Blacksburg, VA. University Press of Virginia, Charlottesville.

Vos, P., R. Hogers, M. Bleeker, M. Reijans, T. van de Lee, M. Hornes, A. Friters, J. Pot, J. Paleman, M. Kuiper, and M. Zabeau. 1995. AFLP: A new technique for DNA fingerprinting. Nucleic Acids Research 21: 4407–4414.

Vreysen, M. J. B., K. M. Saleh, M. Y. Ali, A. M. Abdulla, Z. R. Zhu, K. G. Juma, V. A. Msangi, P. A. Mkonyi, and H. U. Feldmann. 2000. *Glossina austeni* (Diptera: Glossinidae) eradicated on the island of Unguja, Zanzibar, using the sterile insect technique. Journal of Economic Entomology 93: 123–135.

Wagner, W. H. Jr. 1995. Evolution of Hawaiian ferns and fern allies in relation to their conservation status. Pacific Science 49: 31–41.

Wagner, W. L., M. M. Bruegmann, D. R. Herbst, and J. Q. C. Lau. 1999. Hawaiian vascular plants at risk: 1999. Bishop Museum Occasional Papers 60: 1–58.

Wagner, W. L., and V. A. Funk (editors). 1995. Hawaiian biogeography: Evolution on a hot spot archipelago. Smithsonian Institution, Washington, DC, and London.

Wagner, W. L., D. R. Herbst, and S. H. Sohmer. 1999. Revised edition. Manual of the flowering plants of Hawai'i. Bishop Museum Special Publication 97. University of Hawai'i Press and Bishop Museum Press, Honolulu.

Wakelee, K. M. 1996. Life history of the Omao (*Myadestes obscurus*). M.S. thesis. University of Hawai'i at Mānoa, Honolulu.

Wakelee, K. M., and S. G. Fancy. 1999. 'Ōma'o (*Myadestes obscurus*), Kāma'o (*Myadestes myadestinus*), Oloma'o (*Myadestes lanaiensis*), and 'Āmaui (*Myadestes woahensis*). In A. Poole and F. Gill (editors), The birds of North America, No. 460. Birds of North America Inc., Philadelphia.

Walker, R. L. 1967. A brief history of exotic game bird and mammal introductions into Hawaii with a look to the future. 'Elepaio 28: 39–43.

———. 1968. Field note from Ronald L. Walker, February 12: Mauna Kea. 'Elepaio 28: 98–99.

Wallace, G. D. 1973. The role of the cat in the natural history of *Toxoplasma gondii*. American Journal of Tropical Medicine and Hygiene 22: 313–322.

Wallace, G. D., J. W. Bass, J. G. Bennet, and C. J. Straehley. 1969. Toxoplasmosis in Hawaii. Hawaii Medical Journal 29: 101–105.

Walsh, H. E., and V. L. Friesen. 2003. A comparison of intraspecific patterns of DNA sequence variation in mitochondrial DNA, α-enolase, and MHC class II B loci in auklets (Charadriiformes: Alcidae). Journal of Molecular Evolution 57: 681–693.

Walters, M. J. 2006. Seeking the sacred raven: Politics and extinction on a Hawaiian Island. Island Press, Washington, DC.

Walther, M. 2006. History and status of the 'Apapane on Lana'i. 'Elepaio 66: 31–33.

Walton, C., N. Ellis, and P. Pheloung. 1999. A manual for using the Weed Risk Assessment system (WRA) to assess new plants. Australian Quarantine and Inspection Service, Canberra, Australia.

Warner, R. E. 1960. A forest dies on Mauna Kea. Pacific Discovery 13: 6–14.

———. 1968. The role of introduced diseases in the extinction of the endemic Hawaiian avifauna. Condor 70: 101–120.

Warrell, D. A., and H. M. Gilles (editors). 2002. Essential malariology. 4th edition. Arnold, London.

Warren, P. 2003. Elements of an ideal biosecurity system for biodiversity protection purposes. Unpublished article. www.hear.org/articles/warren 200305/.

Warshauer, F. R. 1998. Alien species and threats to native ecology. Pp. 146–149 in S. P. Juvik, J. O. Juvik, and T. R. Paradise (editors), Atlas of Hawai'i. 3rd edition. University of Hawai'i Press, Honolulu.

Warshauer, F. R., and J. D. Jacobi. 1982. Distribution and status of *Vicia menziesii* Spreng. (Leguminosae): Hawai'i's first officially listed endangered plant species. Biological Conservation 23: 111–126.

Warshauer, F. R., J. D. Jacobi, A. M. La Rosa, J. M. Scott, and C. W. Smith. 1983. The distribution, impact and potential management of the introduced vine *Passiflora mollissima* (Passifloraceae) in Hawaii. Technical Report 48. Cooperative National Park Resources Studies Unit, University of Hawai'i, Honolulu.

Watt, D. J., and E. J. Willoughby. 1999. Lesser Goldfinch (*Carduelis psaltria*). In A. Poole and F. Gill (editors), The birds of North America, No. 392. Birds of North America Inc., Philadelphia.

Weatherhead, P. J., and G. Blouin-Demers. 2004. Understanding avian nest predation: Why ornithologists should study snakes. Journal of Avian Biology 35: 185–190.

Weathers, W. W., and K. A. Sullivan. 1989. Juvenile foraging proficiency, parental effort, and avian reproductive success. Ecological Monographs 59: 223–246.

Weathers, W. W., and C. van Riper III. 1982. Temperature regulation in two endangered Hawaiian honeycreepers: The Palila (*Psittirostra bailleui*) and the Laysan Finch (*Psittirostra cantans*). Auk 99: 667–674.

Webster's Revised Unabridged Dictionary. 1913. Merriam, Springfield, MA.

Weissenbock, H., J. Kolodziejek, A. Url, H. Lussy, B. Rebel-Bauder, and N. Nowotny. 2002. Emergence of Usutu virus, an Africian mosquito-borne Flavivirus of the Japanese encephalitis virus group, central Europe. Emerging Infectious Diseases 8: 652–656.

Weldon, P. J., and J. H. Rappole. 1997. A survey of birds odorous or unpalatable to humans: Possible indications of chemical defense. Journal of Chemical Ecology 23: 2609–2633.

Westemeier, R. L., J. D. Brawn, S. A. Simpson, T. L. Esker, R. W. Jansen, J. W. Walk, E. L. Kershner, J. L. Bouzat, and K. N. Paige. 1998. Tracking the long-term decline and recovery of an isolated population. Science 282: 1695–1698.

Wester, L. 1992. Origin and distribution of adventive alien flowering plants in Hawai'i. Pp. 99–154 in C. P. Stone, C. W. Smith, and J. T. Tunison (editors), Alien plant invasions in native ecosystems of Hawaii: Management and research. Cooperative National Park Resources Studies Unit, University of Hawai'i at Mānoa, Honolulu.

Westervelt, W. D. 1963. Hawaiian legends of old Honolulu. Tuttle, Rutland, VT.

Westneat, D. F., and T. R. Birkhead. 1998. Alternative hypotheses linking the immune system with mate choice for good genes. Proceedings of the Royal Society of London, Series B 265: 1065–1073.

White, C. M., and L. F. Kiff. 2000. Biodiversity, island raptors and species concepts. Pp. 633–652 in R. D. Chancellor and B.-U. Meyburg (editors), Raptors at risk. Hancock House and World Working Group on Birds of Prey and Owls, Blaine, WA, and Berlin, Germany.

White, L. A. 1964. The world view of the Keresan Pueblo Indians. Pp. 83–94 in S. Diamond (editor), Primitive view of the world. Columbia University Press, New York.

White, N. J. 2004. Antimalarial drug resistance. Journal of Clinical Investigations 113: 1084–1092.

White, T. H. Jr., J. A. Collazo, and F. J. Vilella. 2005. Survival of captive-reared Puerto Rican Parrots released in the Caribbean National Forest. Condor 107: 424–432.

Whitesell, C. D. 1964. Silvical characteristics of koa (*Acacia koa* Gray). Resource Paper PSW-16: 1–12. U.S. Forest Service, Pacific Southwest Forest and Range Experimental Station, Berkeley, CA.

Whitney, L. D., E. Y. Hosaka, and J. C. Ripperton. 1939. Reprinted 1964. Grasses of the Hawaiian ranges. Bulletin 82. Hawaii Agricultural Experiment Station, University of Hawaii, Honolulu.

Whittaker, R. J. 1998. Island biogeography: Ecology, evolution, and conservation. Oxford University Press, New York.

Wichman, J. R., and H. St. John. 1990. A chronicle and flora of Niihau. National Tropical Botanical Garden, Lawai, HI.

Wiens, J. A. 1989. The ecology of bird communities. Cambridge University Press, New York.

———. 1995. Habitat fragmentation: Island v landscape perspectives on bird conservation. Ibis 137: 97–104.

Wiersma, P., A. Munoz-Garcia, A. Walker, and J. B. Williams. 2007. Tropical birds have a slow pace of life. Proceedings of the National Academy of Sciences (USA) 104: 9340–9345.

Wikelski, M., L. Spinney, W. Schelsky, A. Scheuerlein, and E. Gwinner. 2003. Slow pace of life in tropical sedentary birds: A common-garden experiment on four stonechat populations from different latitudes. Proceedings of the Royal Society of London, Series B 270: 2382–2388.

Wilcove, D. S. 2005. Rediscovery of the Ivory-billed Woodpecker. Science 308: 1422–1423.

Wilcove, D. S., and J. Terborgh. 1984. Patterns of population decline in birds. American Birds 38: 10–13.

Wilkinson, D., G. C. Smith, R. J. Delahay, and C. L. Cheeseman. 2004. A model of bovine tuberculosis in the badger *Meles meles*: An evaluation of different vaccination strategies. Journal of Applied Ecology 41: 492–501.

Williams, D. G., and R. A. Black. 1994. Drought response of a native and introduced Hawaiian grass. Oecologia 97: 512–519.

Williams, D. G., R. N. Mack, and R. A. Black. 1995. Ecophysiology of introduced *Pennisetum setaceum* on Hawaii: The role of phenotypic plasticity. Ecology 76: 1569–1580.

Williams, J. M. 2000. New Zealand under siege: A review of the management of biosecurity risks to the environment. Office of the Parliamentary Commissioner for the Environment, Wellington, New Zealand. www.pce.govt.nz/reports/all reports/0_908804_93_8.shtml (accessed March 23, 2008).

Williams, R. N. 1987. Alien birds on Oahu: 1944–1985. 'Elepaio 47: 87–92.

Wilmut, I., A. E. Schnieke, J. McWhir, A. J. Kind, and K. H. Campbell. 1997. Viable offspring derived from fetal and adult mammalian cells. Nature 385: 810–813.

Wilson, A. C., and M. R. Stanley Price. 1994. Reintroduction as a reason for captive breeding. Pp. 243–264 in P. J. S. Olney, G. M. Mace, and A. T. C. Geistner (editors), Creative conservation: Interactive management of wild and captive animals. Chapman and Hall, London.

Wilson, E. O. 1996. Hawaii: A world without social insects. Bishop Museum Occasional Papers 45: 3–7.

Wilson, E. O., and R. W. Taylor. 1967. The ants of Polynesia (Hymenoptera: Formicidae). Pacific Insects Monographs 14: 1–109.

Wilson, J. T. 1963. A possible origin of the Hawaiian Islands. Canadian Journal of Physics 41: 135–138.

Wilson, S. B., and A. H. Evans. 1890–1899. Aves Hawaiiensis: The birds of the Sandwich Islands. R. H. Porter, London.

Wingfield, M. J., B. Slippers, J. Roux, and B. D. Wingfield. 2001. Worldwide movement of exotic forest fungi, especially in the tropics and the Southern Hemisphere. BioScience 51: 134–140.

Winter, L. 2003. Popoki and Hawai'i's native birds. 'Elepaio 63: 43–46.

Witmer, G. W. 2002. Captive Canada geese acceptability and toxicity trials with two formulations of 0.005% diphacinone rodenticide baits. Unpublished report QA-770. National Wildlife Research Center, Fort Collins, CO.

Wobeser, G. 2002. Disease management strategies for wildlife. Revue Scientifique et Technique (International Office of Epizootics) 21: 159–178.

Wobeser, G. A. 2006. Essentials of disease in wild animals. Blackwell, Ames, IA.

Wodarz, D., and M. A. Nowak. 2002. Mathematical models of HIV pathogenesis and treatment. BioEssays 24: 1178–1187.

Wolf, C. M., B. Griffith, C. Reed, and S. A. Temple. 1996. Avian and mammalian translocation: Update and reanalysis of 1987 survey data. Conservation Biology 10: 1142–1154.

Wolfe, E. W., and J. Morris. 1996. Geologic map of the island of Hawaii. Miscellaneous Investigation Series, Map 1-2524-A. U.S. Geological Survey, Reston, VA.

Wood, B., B. R. Tershy, M. A. Hermosillo, C. J. Donlan, J. A. Sanchez, B. S. Keitt, D. A. Croll, G. R. Howald, and N. Biavaschi. 2002. Removing cats from islands in northwest Mexico. Pp. 374–380 in C. R. Veitch and M. N. Clout (editors), Turning the tide: The eradication of invasive species. International Union for Conservation of Nature, Gland, Switzerland.

Woodward, S. A., P. M. Vitousek, K. Matson, F. Hughes, K. Benvenuto, and P. A. Matson. 1990. Use of the exotic tree *Myrica faya* by native and exotic birds in Hawai'i Volcanoes National Park. Pacific Science 44: 88–93.

Woodworth, B. L., C. T. Atkinson, D. A. LaPointe, P. J. Hart, C. S. Spiegel, E. J. Tweed, C. Henneman, J. LeBrun, T. Denette, R. DeMots, K. L. Kozar, D. Triglia, D. Lease, A. Gregor, T. Smith, and D. Duffy. 2005. Host population persistence in the face of introduced vector-borne diseases: Hawaii amakihi and avian malaria. Proceedings of the National Academy of Sciences (USA) 102: 1531–1536.

Woodworth, B. L., J. T. Nelson, E. J. Tweed, S. G. Fancy, M. P. Moore, E. B. Cohen, and M. S. Collins. 2001. Breeding productivity and survival of the endangered Hawai'i Creeper in a wet forest refuge on Mauna Kea, Hawai'i. Studies in Avian Biology 22: 164–172.

Woolaver, L., R. Nichols, and E. Williams. 2005. Captive field propagation and experimental release of the Eastern loggerhead shrike in Ontario, Canada. Re-introduction News 24: 29–32.

Woolbright, L. L., A. H. Hara, C. M. Jacobsen, W. J. Mautz, and F. L. Benevides Jr. 2006. Population densities of the coqui, *Eleutherodactylus coqui* (Anura:

Leptodactylidae) in newly invaded Hawaii and in native Puerto Rico. Journal of Herpetology 40: 122–126.

Woolhouse, M. E., M. Chase-Topping, D. Haydon, J. Friar, L. Matthews, G. Hughes, D. Shaw, J. Wilesmith, A. Donaldson, S. Cornell, M. Keeling, and B. Grenfell. 2001. Epidemiology: Foot-and-mouth disease under control in the UK. Nature 411: 258–259.

Wootton, J. T. 1996. Purple Finch (Carpodacus purpureus). In A. Poole and F. Gill (editors), The birds of North America, No. 208. Birds of North America Inc., Philadelphia.

Work, T. M., D. Ball, and M. Wolcott. 1999. Erysipelas in a free-ranging Hawaiian crow (Corvus hawaiiensis). Avian Diseases 43: 338–341.

Work, T. M., and J. Hale. 1996. Causes of owl mortality in Hawaii, 1992 to 1994. Journal of Wildlife Diseases 32: 266–273.

Work, T. M., J. G. Massey, D. S. Lindsay, and J. P. Dubey. 2002. Toxoplasmosis in three species of native and introduced Hawaiian birds. Journal of Parasitology 88: 1040–1042.

Work, T. M., J. G. Massey, B. A. Rideout, C. H. Gardiner, D. B. Ledig, O. C. Kwok, and J. P. Dubey. 2000. Fatal toxoplasmosis in free-ranging endangered 'Alala from Hawaii. Journal of Wildlife Diseases 36: 205–212.

Work, T. M., C. U. Meteyer, and R. A. Cole. 2004. Mortality in Laysan Ducks (Anas laysanensis) by emaciation complicated by Echinuria uncinata on Laysan Island, Hawaii, 1993. Journal of Wildlife Diseases 40: 110–114.

Work, T. M., and R. A. Rameyer. 1996. Haemoproteus iwa n. sp. in great frigatebirds (Fregata minor [Gmelin]) from Hawaii: Parasite morphology and prevalence. Journal of Parasitology 82: 489–491.

Worthy, T. H., and R. N. Holdaway. 2002. The lost world of the moa: Prehistoric life of New Zealand. Canterbury University Press, Christchurch, New Zealand.

Wright, S. 1931. Evolution in Mendelian populations. Genetics 16: 97–159.

Yen, J. H., and A. R. Barr. 1971. New hypothesis of the cause of cytoplasmic incompatibility in Culex pipiens. Nature 232: 657–658.

Yocom, C. F. 1967. Ecology of feral goats in Haleakala National Park, Maui, Hawaii. American Midland Naturalist 77: 418–451.

Yorinks, N., and C. T. Atkinson. 2000. Effects of malaria on activity budgets of experimentally infected juvenile Apapane (Himatione sanguinea). Auk 117: 731–738.

Yu, C. C., Y. H. Atallah, and D. M. Whitacre. 1982. Metabolism and disposition of diphacinone in rats and mice. Drug Metabolism and Disposition: The Biological Fate of Chemicals 10: 645–648.

Zahiri, N. S., and M. S. Mulla. 2003. Susceptibility profile of Culex quinquefasciatus (Diptera: Culicidae) to Bacillus sphaericus on selection with rotation and mixture of B. sphaericus and B. thuringiensis israelensis. Journal of Medical Entomology 40: 672–677.

Zanette, L., P. Doyle, and S. M. Trémont. 2000. Food shortage in small fragments: Evidence from an area-sensitive passerine. Ecology 81: 1654–1666.

Zar, J. H. 1996. Biostatistical analysis. 3rd edition. Prentice-Hall, Upper Saddle River, NJ.

Zelenak, J. R., and J. J. Rotella. 1997. Nest success and productivity of Ferruginous Hawks in northern Montana. Canadian Journal of Zoology 75: 1035–1041.

Zhong, H., J. Dukes, M. Greer, P. Hester, M. Shirley, and B. Anderson. 2003. Ground deposition impact of aerially applied fenthion on the fiddler crab Uca pugliator. Journal of the American Mosquito Control Association 19: 47–52.

Zhong, S., B. Yang, and A. C. Alfenas. 2008. Development of microsatellite markers for the guava rust fungus, Puccinia psidii. Molecular Ecology Resources 18: 348–350.

Ziegler, A. C. 2002. Hawaiian natural history, ecology, and evolution. University of Hawai'i Press, Honolulu.

Zimmerman, E. C. 1948. Insects of Hawaii, Vol. 1: Introduction. University of Hawaii Press, Honolulu.

———. 1958a. Insects of Hawaii, Vol. 7: Macrolepidoptera. University of Hawaii Press, Honolulu.

———. 1958b. Insects of Hawaii, Vol. 8: Lepidoptera; Pyraloidea. University of Hawaii Press, Honolulu.

———. 1965. Nature of the land biota. Pp. 57–63 in F. R. Fosberg (editor), Man's place in the island ecosystem. Bishop Museum Press, Honolulu.

Zoological Society of San Diego. 2001. Five-year workplan: 2001–2004. Report submitted to the Hawaii Endangered Bird Conservation Partnership (U.S. Fish and Wildlife Service, Hawai'i Division of Forestry and Wildlife, and Zoological Society of San Diego), Honolulu.

———. 2007. Annual report submitted to the U.S. Fish and Wildlife Service and the Hawai'i Division of Forestry and Wildlife, Honolulu.

Contributors

Jorge A. Ahumada
Technical Director
Tropical Ecology Assessment and
   Monitoring Network
Center for Applied Biodiversity Science
Conservation International
Arlington, VA

Verna L. U. Amante-Helweg
Cultural Psychologist and Founder
EthnoSphere
Hawai'i National Park, HI

Carter T. Atkinson
Research Microbiologist
Pacific Island Ecosystems Research Center
U.S. Geological Survey
Hawai'i Volcanoes National Park, HI

Paul C. Banko
Research Wildlife Biologist
Pacific Island Ecosystems Research Center
U.S. Geological Survey
Hawai'i Volcanoes National Park, HI

Winston E. Banko
Research Wildlife Biologist (retired)
Pacific Island Ecosystems Research Center
U.S. Geological Survey
Hawai'i Volcanoes National Park, HI

Kevin W. Brinck
Quantitative Ecologist
Hawai'i Cooperative Studies Unit
Pacific Aquaculture and Coastal Resources
   Center
University of Hawai'i at Hilo
Hilo, HI

Richard J. Camp
Project Coordinator
Hawai'i Forest Bird Interagency Database
    Project
Hawai'i Cooperative Studies Unit
Pacific Aquaculture and Coastal Resources
    Center
University of Hawai'i at Hilo
Hilo, HI

Earl W. Campbell III
Coordinator
Invasive Species Division
Pacific Islands Fish and Wildlife Office
U.S. Fish and Wildlife Service
Honolulu, HI

Sheila Conant
Professor
Department of Zoology
University of Hawai'i at Mānoa
Honolulu, HI

Andrew P. Dobson
Professor
Department of Ecology and Evolutionary
    Biology
Princeton University
Princeton, NJ

David C. Duffy
Professor
Pacific Cooperative Studies Unit
Department of Botany
University of Hawai'i at Mānoa
Honolulu, HI

Lori S. Eggert
Assistant Professor
Division of Biological Sciences
University of Missouri
Columbia, MO

Chris A. Farmer
Palila Restoration Project Manager
Hawai'i Cooperative Studies Unit
Pacific Aquaculture and Coastal Resources
    Center
University of Hawai'i at Hilo
Hilo, HI

Robert C. Fleischer
Head
Center for Conservation and Evolutionary
    Genetics
National Zoological Park
National Museum of Natural History
Smithsonian Institution
Washington, DC

Jeffrey T. Foster
Research Associate
Center for Microbial Genetics and
    Genomics
Northern Arizona University
Flagstaff, AZ

J. Scott Fretz
Wildlife Program Manager
Division of Forestry and Wildlife
Hawai'i Department of Land and Natural
    Resources
Honolulu, HI

P. Marcos Gorresen
Project Support Specialist
Hawai'i Forest Bird Interagency Database
    Project
Hawai'i Cooperative Studies Unit
Pacific Aquaculture and Coastal Resources
    Center
University of Hawai'i at Hilo
Hilo, HI

Jim J. Groombridge
Senior Lecturer in Biodiversity
    Conservation
Durrell Institute for Conservation and
    Ecology
University of Kent
Canterbury, UK

Steven C. Hess
Research Wildlife Biologist
Pacific Island Ecosystems Research Center
U.S. Geological Survey
Hawai'i Volcanoes National Park, HI

Peter H. F. Hobbelen
Postdoctoral Researcher
Department of Forest and Wildlife
    Ecology
University of Wisconsin
Madison, WI

James D. Jacobi
Biologist
Pacific Island Ecosystems Research
    Center
U.S. Geological Survey
Honolulu, HI

Susan I. Jarvi
Associate Professor
Biology Department
University of Hawai'i at Hilo
Hilo, HI

John L. Klavitter
Wildlife Biologist
Midway Atoll National Wildlife Refuge
U.S. Fish and Wildlife Service
Honolulu, HI

Fred Kraus
Vertebrate Zoologist
Department of Natural Sciences
Bernice P. Bishop Museum
Honolulu, HI

Cynthia M. Kuehler
Avian Conservation Coordinator
Zoological Society of San Diego
San Diego, CA

Dennis A. LaPointe
Research Ecologist
Pacific Island Ecosystems Research Center
U.S. Geological Survey
Hawai'i Volcanoes National Park, HI

David L. Leonard Jr.
Wildlife Biologist
Division of Forestry and Wildlife
Hawai'i Department of Land and Natural
    Resources
Honolulu, HI

Alan A. Lieberman
Conservation Program Manager
Zoological Society of San Diego
San Diego, CA

Gerald D. Lindsey (1938–2001)
Research Wildlife Biologist
Pacific Island Ecosystems Research Center
U.S. Geological Survey
Hawai'i Volcanoes National Park, HI

Lloyd L. Loope
Research Scientist
Pacific Island Ecosystems Research Center
U.S. Geological Survey
Makawao, Maui, HI

Loyal A. Mehrhoff
Center Director
Pacific Island Ecosystems Research Center
U.S. Geological Survey
Honolulu, HI

Jay T. Nelson
Fish and Wildlife Biologist
Pacific Islands Fish and Wildlife Office
U.S. Fish and Wildlife Service
Honolulu, HI

Linda W. Pratt
Botanist
Pacific Island Ecosystems Research Center
U.S. Geological Survey
Hawai'i Volcanoes National Park, HI

Thane K. Pratt
Research Wildlife Biologist
Pacific Island Ecosystems Research Center
U.S. Geological Survey
Hawai'i Volcanoes National Park, HI

Jonathan P. Price
Assistant Professor
Department of Geography and
    Environmental Studies
University of Hawai'i at Hilo
Hilo, HI

Michelle H. Reynolds
Research Wildlife Biologist
Pacific Island Ecosystems Research Center
U.S. Geological Survey
Hawai'i Volcanoes National Park, HI

Michael D. Samuel
Assistant Unit Leader
Wisconsin Cooperative Wildlife Research
    Unit
U.S. Geological Survey
University of Wisconsin
Madison, WI

J. Michael Scott
Unit Leader
Idaho Cooperative Fish and Wildlife
    Research Unit
U.S. Geological Survey
University of Idaho
Moscow, ID

Clifford W. Smith
Professor Emeritus and Research
    Associate (retired)
Department of Botany
University of Hawai'i at Mānoa
Honolulu, HI

Robert T. Sugihara
Wildlife Technician
National Wildlife Research
    Center–Hawai'i
U.S. Department of Agriculture / Animal
    and Plant Health Inspection Service
    Wildlife Services
Hilo, HI

Catherine E. Swift
Predator Control / Toxicant Registration
    Specialist
Invasive Species Division
Pacific Islands Fish and Wildlife Office
U.S. Fish and Wildlife Service
Honolulu, HI

Fredrick R. Warshauer
Botanist (retired)
Hawai'i Cooperative Studies Unit
Pacific Aquaculture and Coastal Resources
    Center
University of Hawai'i at Hilo
Hilo, HI

Bethany L. Woodworth
Instructor
Department of Environmental Studies
University of New England
Biddeford, ME

Index

Page numbers followed by the letters b, f, and t indicate boxes, figures, and tables, respectively.